SOC[7]

Introduction to Sociology

Nijole V. Benokraitis

Cengage

Australia • Brazil • Canada • Mexico • Singapore • United Kingdom • United States

Cengage

SOC[7]: Introduction to Sociology

SOC[7]: Introduction to Sociology
Nijole V. Benokraitis

WCN: 01-100-371

ISBN: 979-8-214-57026-6

Cengage
5191 Natorp Boulevard
Mason, OH 45040
USA

Cengage is a leading provider of customized learning solutions with employees residing in nearly 40 different countries and sales in more than 125 countries around the world. Find your local representative at: **www.cengage.com.**

To learn more about Cengage platforms and services, register or access your online learning solution, or purchase materials for your course, visit **www.cengage.ca.**

SOC[7] Brief Contents

Contents

About the Author

Nijole V. Benokraitis, professor of sociology at the University of Baltimore, received a B.A. in sociology and English from Emmanuel College, an M.A. in sociology from the University of Illinois at Urbana, and a doctorate in sociology from the University of Texas at Austin. She has served as both chair and graduate program director of the University of Baltimore's Department of Sociology and has chaired numerous university committees. She has authored, coauthored, edited, or coedited eight books, including *Contemporary Ethnic Families in the United States: Characteristics, Variations, and Dynamics*; *Feuds About Families: Conservative, Centrist, Liberal, and Feminist Perspectives; Modern Sexism: Blatant, Subtle, and Covert Discrimination*; and *Seeing Ourselves: Classic, Contemporary, and Cross-Cultural Readings in Sociology*. In addition, she has published numerous articles and book chapters on such topics as institutional racism and discrimination against women in government and higher education.

Laura C. Atkins is associate dean of arts and sciences and professor of sociology at Jacksonville University. She received a B.A. in sociology and women's studies from the University of Wisconsin at Milwaukee and an M.A. and Ph.D. both in sociology from the University of Illinois at Urbana-Champaign. At Jacksonville University, she has served as director for three areas: service-learning, experiential learning, and the university's quality enhancement plan. In addition, she has published articles on service-learning and leadership and is Co-PI on the NSF S-STEM grant "Project-Based Service Learning and Career-Training to Support Graduation and Career Readiness of Engineering and Computing Science Majors." Dr. Atkins has also served on committees for Sociologists for Women in Society and the Southern Sociological Society as well as been a reviewer for the Southern Association of Colleges and Schools Commission on Colleges.

1 Thinking Like a Sociologist

John M Lund Photography Inc/DigitalVision/Getty Images

Learning Objectives

After studying this chapter, you will be able to…

1-1 Explain what sociology is and how sociology differs from common sense.

1-2 Explain how and why a sociological imagination helps people understand society.

1-3 Identify why sociology is worthwhile for people to study.

1-4 Explain the origins of sociology including why sociology developed and its most influential early theorists.

1-5 Compare the four contemporary sociological perspectives.

True or False?

How Much Do You Know about Thinking Like a Sociologist?

1. Both divorce and marriage rates are increasing in the United States (refer to Chapter 7).
2. On average, women still earn less than men who are in comparable jobs (refer to Chapters 9 and 11).
3. The poverty rate in the United States has been declining (refer to Chapter 8).
4. Latina/o/xs make up the largest racial or ethnic minority group in the United States (refer to Chapter 10).
5. Since sociology is a social science, it does not use the scientific method (refer to Chapter 2).
6. The quality of health care in the United States is better compared with other high income nations (refer to Chapter 14).

The answers are at the end of the chapter.

Texting and emailing are associated with the highest risk of car crashes, and hands-free devices aren't much safer than handheld versions. Almost 90 percent of U.S. drivers say that distracted driving is a serious safety threat. (AAA Foundation for Traffic Safety, 2015; AT&T Newsroom, 2015; National Center for Statistics and Analysis, 2016). However, in 2020, distracted driving claimed the lives of 3,142 people in the United States (Stewart, 2022).

Why is there such a disconnection between many people's attitudes and their behavior? This chapter examines these and other questions. Let's begin by considering what sociology is (and isn't) and how a "sociological imagination" can give us more control over our lives. We'll then examine how sociologists grapple with complex theoretical issues in explaining social life.

Be sure to take the True or False quiz in the chapter opener to better understand how much you know about these topics.

What do you think?

Sociology is basically common sense.

1	2	3	4	5	6	7
	strongly agree				strongly disagree	

1-1 What Is Sociology?

Stated simply, **sociology** is the scientific study of human behavior in society. Sociologists study behavior patterns that occur between individuals, among small groups (e.g., families), large organizations (e.g., Apple), and entire societies (e.g., the United States). But, you might protest, "I'm unique."

1-1a Are You Unique?

Yes and no. Each of us is unique in the sense that you and I are like no one else on earth. Even identical twins, who have the same physical characteristics and genetic matter, often differ in personality and interests. One of my colleagues likes to tell a story about his 3-year-old twin girls who received the same doll. One twin chattered that the doll's name was Lori, that she loved Lori, and that she would take good care of her. The second twin muttered, "Her name is Stupid," and she flung the doll into a corner.

Despite some individual differences, identical twins, you, and I are all like other people in many ways. Around the world, we experience grief when a loved one dies, participate in rituals that celebrate marriage or the birth of a child, and want to have healthy and happy lives. Some actions, such as terrorist attacks, are less

Marriage Before Love? No Way!

When I ask my students, "Would you marry someone you're not in love with?" most laugh, raise an eyebrow, or stare at me in disbelief. "Of course not!" they exclaim. In fact, the "open" courtship and dating systems common in Western nations, including the United States, are less common in other areas of the world. In many African, Asian, Mediterranean, and Middle Eastern countries, it is more common and accepted for marriages to be arranged: They forge bonds between families rather than individuals and preserve family continuity along religious and socioeconomic lines. Love isn't a prerequisite for marriage in societies that value kin groups more than individual choices (refer to Chapters 9 and 12).

sociology the scientific study of human behavior in society.

predictable. For the most part, however, people conform to expected and acceptable behavior. From the time we get up until we go to bed, we follow a variety of different rules and customs about what we eat, how we drive, how we act in different social situations, and how we dress for work, classes, and leisure activities.

"So what?" you might shrug. Isn't it "obvious" that we dress differently for classes than for job interviews? Isn't all of this just plain old common sense?

1-1b Isn't Sociology Just Common Sense?

No. Sociology goes well beyond conventional wisdom, what we call common sense, in several ways:

- **Common sense is subjective.** If a woman crashes into my car, I might conclude, according to conventional wisdom and statements we've heard over the years, that "women are terrible drivers." In fact, most drivers involved in crashes are men—especially teenagers (ages 16 to 19) and those age 70 and older (Insurance Institute for Highway Safety, 2022). Thus, *objective* data show that, overall, men are worse drivers than women.
- **Common sense ignores facts.** Because common sense is subjective, it ignores facts that challenge cherished beliefs. For example, many Americans are most concerned about street crimes, such as robbery or homicide. FBI and sociological data show, however, that we're much more likely to be assaulted or murdered by someone we know or live with (refer to Chapters 7 and 12).
- **Common sense varies across groups and cultures.** Many Americans believe that working harder decreases poverty. In contrast, Europeans tend to think that poverty is due to forces outside people's control (refer to Chapters 3 and 8). Thus, common sense notions about economic success vary considerably across countries.
- **Much of our common sense is based on myths and misconceptions.** A common myth is that living together is a good way to find out whether partners will get along after marriage. Generally, however, couples who live together before marriage have higher divorce rates than those who don't (refer to Chapter 12).

Sociology, in contrast to conventional wisdom, examines claims and beliefs critically, considers many points of view, and enables us to move beyond established ways of thinking. The *sociological perspective* analyzes how social context influences people's lives. The "sociological imagination" is at the center of the sociological perspective.

1-2 What Is a Sociological Imagination?

According to sociologist C. Wright Mills (1916–1962), social factors such as religion, ethnicity, and politics affect our behavior. Mills (1959) called this ability to find the relationship between individual experiences and larger social influences the **sociological imagination**. The sociological imagination emphasizes the connection between personal troubles (biography) and structural (public and historical) issues.

Consider unemployment. If only a small group of people can't find a job, it's a *personal trouble* that may be due, in part, to an individual's low educational attainment, lack of specific skills that employers want, not searching for work, and so on. If unemployment is widespread, it's a *public issue* because economic problems are also the result of structural factors such as mass layoffs, jobs moving overseas, technological changes, and restrictive hiring policies (refer to Chapter 11). Thus, people may be unemployed regardless of skills, a college degree, and job searches.

A sociological imagination helps us understand how larger social forces impact our everyday lives. It identifies why personal troubles are often due to larger public issues and policies over which we have little, if any, control. A sociological imagination relies on both micro- and macro-level approaches to understand our social world.

1-2a Microsociology: How People Affect Our Everyday Lives

Microsociology examines the patterns of individuals' social interaction in specific settings. In most of our relationships, we interact with others on a micro, or "small," level (e.g., members of a work group discussing who will perform which tasks). These everyday interactions involve what people think, say, or do on a daily basis.

1-2b Macrosociology: How Social Structure Affects Our Everyday Lives

Macrosociology focuses on large-scale patterns and processes that characterize society as a whole. Macro, or "large," approaches are especially useful in

sociological imagination finding the relationship between individual experiences and larger social influences.

microsociology examines the patterns of individuals' social interaction in specific settings.

macrosociology examines the large-scale patterns and processes that characterize society as a whole.

PictureLux/Alamy Stock Photo

HBO's teen drama Euphoria revolves around a group of teenagers facing a variety of challenges, including Rue, a 17 year old who recently left drug rehabilitation treatment. For sociologists, the series illustrates the connection between micro-level individual behavior (such as Rue opting to use drugs) and macro-level factors (family structures, societal expectations related to sexuality, and violence affecting youth).

understanding some of the constraints—such as economic forces and public policies.

Microsociology and macrosociology differ conceptually but are interrelated. Consider the reasons for divorce. On a micro level, sociologists study factors like extramarital affairs, substance misuse, arguments about money, and other everyday interactions that fuel marital tension and unhappiness, leading to divorce. On a macro level, sociologists look at how the economy, laws, cultural values, and technology affect divorce rates (refer to Chapter 12). Examining micro, macro, and micro–macro forces is one of the reasons why sociology is a powerful tool in understanding (and changing) individual behavior and society at large (Ritzer, 1992).

1-3 Why Study Sociology?

Sociology offers explanations that can greatly improve the quality of your everyday life. These explanations can influence choices that range from your personal decisions to expanding your career opportunities.

1-3a Making Informed Decisions

Sociology can help us make more informed decisions. We often hear that grief counseling is essential after the death of a loved one. Only about 10 percent of those going through bereavement show ongoing, elevated grief symptoms (Mehta, 2013). Grief is normal, and most people work through their losses on their own, while counseling sometimes prolongs depression and anxiety (Stroebe et al., 2000).

1-3b Understanding Diversity

The racial and ethnic composition of the United States is changing. By 2030, only 56 percent of the U.S. population is projected to identify as non-Hispanic White, down from 61 percent in 2016 and 86 percent in 1950 (Passel et al., 2011; Vespa et al., 2020). As you'll find out in later chapters, this racial/ethnic shift has already affected interpersonal relationships as well as education, politics, religion, and other spheres of social life.

Recognizing and understanding diversity is one of sociology's central themes. A person's gender, social class, marital status, ethnicity, sexual orientation, and age—among other factors—shape their beliefs, behavior, and experiences. If, for example, you're a White middle-class female who attends a private college, your experiences may be very different from those of a male Vietnamese immigrant who is struggling to pay expenses at a community college.

Increasingly, nations around the world are intertwined through political and economic ties. What happens in other societies often has a direct or indirect impact on contemporary U.S. life. Decisions in oil-producing countries, for example, affect gas prices, spur the development of hybrid cars that are less dependent on oil, and stimulate research on alternative sources of energy.

1-3c Shaping Social and Public Policies and Practices

Sociology is valuable in applied, clinical, and policy settings because many jobs require an understanding of society and research to create social change. According to a director of a research institute, sociology increased her professional contributions: "I can look at problems of concern to the National Institutes of Health and say 'here's a different way to solve this problem'" (Nyseth et al., 2011, p. 48).

1-3d Thinking Critically

We develop a sociological imagination not only when we understand and can apply the concepts but also when we can think, speak, and write critically. Much of our thinking and decision making is impulsive and emotional. In contrast, critical thinking involves knowledge and problem solving (Paul and Elder, 2007).

Critical sociological thinking goes even further because we begin to understand how our individual lives, choices, and troubles are shaped by race, gender, social class, and social institutions like the economy,

Snapshots

Jason Love/Cartoonstock.com

"I love our lunches out here, but I always get the feeling that we're being watched."

politics, and education (Eckstein et al., 1995; Grauerholz and Bouma-Holtrop, 2003). **Table 1.1** summarizes some of the basic elements of critical sociological thinking.

1-3e Expanding Your Career Opportunities

A degree in sociology is a springboard for many jobs and professions. A national survey of undergraduate sociology majors found that 44 percent were in administrative or management positions, 22 percent were employed in social service and counseling, 18 percent were in sales and marketing, and 12 percent were teachers (Senter et al., 2014; refer also to Senter and Spalter-Roth, 2016).

What specific skills do sociology majors learn that are useful in their jobs? Some of the most important are

Table 1.1 What Is Critical Sociological Thinking?

Critical sociological thinking requires a combination of skills. Some of the basic elements include the ability to:
• rely on reason rather than emotion
• ask questions, avoid snap judgments, and examine popular and unpopular beliefs
• recognize one's own and others' assumptions, prejudices, and points of view
• remain open to alternative explanations and theories
• examine competing evidence (refer to Chapter 2)
• understand how public policies affect private troubles

being better able to work with diverse teams (68 percent), writing reports that non-sociologists understand (64 percent), developing evidence-based arguments (69 percent), and awareness of professionalism and work ethic learned through identifying ethical issues in sociological research (69 percent) (Ciabattari et al., 2018).

In other cases, students major in sociology because it provides a broad liberal arts foundation for professions such as law, education, and social work. The Medical College Admission Test (MCAT) now includes material from sociology because "Being a good physician is about more than scientific knowledge. It's about understanding people—how they think, interact, and make decisions" (Olsen, 2016, p. 72).

Even if you don't major in sociology, developing your sociological imagination can enrich your job skills. Sociology courses help you learn to think abstractly and critically, formulate problems, ask incisive questions, search for data in the most reliable and up-to-date sources, organize material, and improve your oral presentations (ASA Research Department, 2013; Spalter-Roth et al., 2013).

Some well-known people who were sociology majors: Rev. Martin Luther King, Ronald Reagan, Michelle Obama, Robin Williams, and Joe Theismann. Library of Congress Prints and Photographs Division[Leffler, Warren K/ LC-DIG-ds-00836]; U.S. National Archives and Records Administration (NARA); Official White House Photo; Everett Collection/Shutterstock.com; Nate Fine/Getty Images Sport/Getty Images; iStock.com/belterz

1-4 Some Origins of Sociological Theory

During college, most of my classmates and I postponed taking theory courses (regardless of our major) as long as possible. "This stuff is boring, boring, boring," we'd grump, "and has nothing to do with the real world." In fact, theorizing is part of our everyday lives. Every time you try to explain why your family and friends behave as they do, for example, you're theorizing.

As people struggle to understand human behavior, they develop theories. A **theory** is a set of statements that explains why a phenomenon occurs. Theories produce knowledge, guide our research, help us analyze our findings, and, ideally, offer solutions for social problems.

Sociologist James White (2005, pp. 170–171) describes theories as "tools" that don't profess to know "the truth" but "may need replacing" over time as our understanding of society becomes more sophisticated. In effect, theories evolve over time because of cultural and technological changes. For example, we'll soon explore how sociological theories changed considerably after the rise of feminist scholarship during the late 1960s.

Sociological theories didn't emerge overnight. Nineteenth-century thinkers grappled with some of the same questions that sociologists try to answer today: Why do people behave as they do? What holds society together? What pulls it apart? Of the many early sociological theorists, some of the most influential were Auguste Comte, Harriet Martineau, Émile Durkheim, Karl Marx, Max Weber, Jane Addams, and W. E. B. Du Bois.

1-4a Auguste Comte

Hulton Archive/Getty Images

Auguste Comte

Auguste Comte (1798–1857) coined the term *sociology* and is often described as the "father of sociology." Comte maintained that the study of society must be **empirical**. That is, information should be based on observations, experiments, or other data collection rather than on ideology, religion, intuition, or conventional wisdom.

He saw sociology as the scientific study of two aspects of society: social statics and social dynamics. *Social statics* investigates how principles of social order explain a particular society and the interconnections between institutions. *Social dynamics* explores how individuals and societies change over time. Comte's emphasis on social order and change within and across societies is still useful today because many sociologists examine the relationships between education and politics (social statics) and how such interconnections change over time (social dynamics).

theory a set of statements that explains why a phenomenon occurs.

empirical information that is based on observations, experiments, or other data collection rather than on ideology, religion, intuition, or conventional wisdom.

1-4b Harriet Martineau

Spencer Arnold/Hulton Archive/Getty Images

Harriet Martineau

Harriet Martineau (1802–1876), an English author, published several dozen books on a wide range of topics in social science, politics, literature, and history. Her translation and condensation of Auguste Comte's difficult material for popular consumption was largely responsible for the dissemination of Comte's work. Her work was so important that we might say there were two founders of sociology: Comte and Martineau. She emphasized the importance of systematic data collection through observation and interviews and an objective analysis of data to explain events and behavior. She also published the first sociology research methods textbook.

Martineau, a feminist and strong opponent of slavery, denounced many aspects of capitalism as alienating and degrading and criticized dangerous workplaces that often led to injury and death. Martineau promoted improving women's positions in the workforce through education, nondiscriminatory employment, and training programs. She advocated for women's admission into medical schools and emphasized issues such as infant care, the rights of the aged, suicide prevention, and other social problems (Hoecker-Drysdale, 1992).

After a long tour of the United States, Martineau described American women as being socialized to be subservient and dependent rather than equal marriage partners. She also criticized American and European religious institutions for expecting women to be pious and passive rather than educating them in philosophy and politics. Most scholars, including sociologists, ridiculed and dismissed such ideas as too radical.

1-4c Émile Durkheim

Pictorial Press Ltd/Alamy Stock Photo

Émile Durkheim

Émile Durkheim (1858–1917), a French sociologist and writer, agreed with Comte that societies are characterized by unity and cohesion because their members are bound together by common interests and attitudes.

While Comte acknowledged the importance of using scientific methods to study society, Durkheim actually did so by poring over official statistics to test a theory about suicide (Adams and Sydie, 2001).

Social Facts

To be scientific, Durkheim maintained, sociology must study **social facts**—aspects of social life, external to the individual, that can be measured. Sociologists can determine *material facts* by examining demographic characteristics such as age, place of residence, and population size. They can gauge *nonmaterial facts*, like communication processes, by observing everyday behavior and how people relate to each other (refer to Chapters 3 to 6). For contemporary sociologists, social facts also include collecting and analyzing data on *social currents* such as collective behavior and social movements (refer to Chapter 16).

Division of Labor

One of Durkheim's central questions was how people can be autonomous and individualistic while being integrated in society. **Social solidarity**, or social cohesiveness and harmony, according to Durkheim, is maintained by a **division of labor**—an interdependence of different tasks and occupations, characteristic of industrialized societies, that produces social unity and facilitates change.

As the division of labor becomes more specialized, people become increasingly dependent on others for specific goods and services. Today, for example, many couples who marry often contract with other professionals (e.g., photographers, florists, musicians, caterers, bartenders, travel agents, and even "wedding planners").

Social Integration

Durkheim was one of the first sociologists to test a theory using data. In his classic study, *Suicide*, Durkheim (1897/1951) relied on extensive data collection to test his theory that suicide is associated with social integration. He concluded that people who experience meaningful social relationships in families, social groups, and communities are less likely to die by suicide than those who feel alone, helpless, or hopeless. Thus, many seemingly isolated individual acts, including suicide, are often the result of structural arrangements, such as weak social ties.

We typically hear about high teenage suicide rates, but they're much higher at later ages. Similar to Durkheim's day, men currently have higher suicide rates than women across all age groups (**Figure 1.1**), and White males aged 75 and older are the most likely to take their own lives (National Institute of Mental Health, 2020; Centers for Disease Control and Prevention, 2023).

Durkheim's connection of social integration to the suicide rate is still relevant today. The high suicide rates of older White men are due to a complex interplay of depression, substance misuse, access to guns,

social facts aspects of social life, external to the individual, that can be measured.

social solidarity social cohesiveness and harmony.

division of labor an interdependence of different tasks and occupations, characteristic of industrialized societies, that produces social unity and facilitates change.

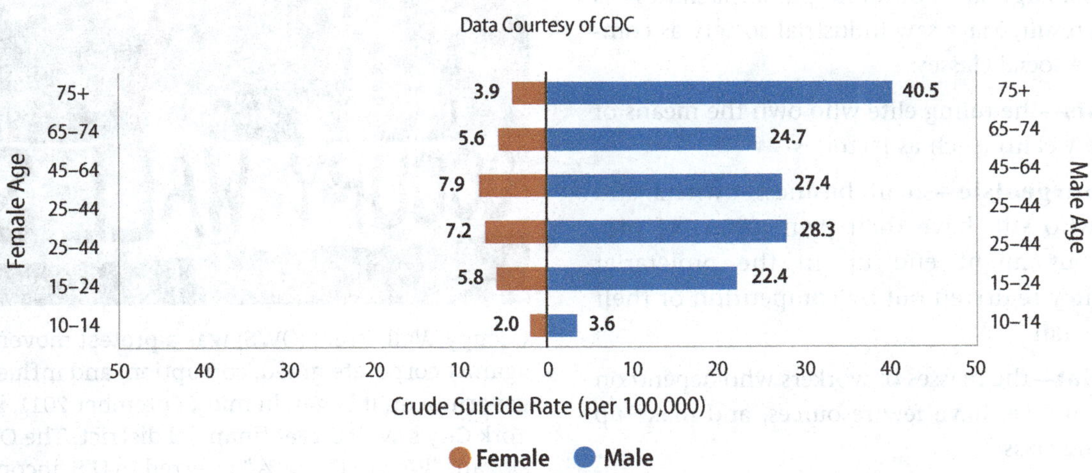

Figure 1.1 Suicide Rates by Age Group (2020)
Suicide deaths per 100,000 population.
Source: National Institute of Mental Health, 2020.

hopelessness because of terminal illnesses, and not being "connected" to family, friends, community groups, and support systems that women tend to develop throughout their lives (American Association of Suicidology, 2009; refer also to Chapters 9 and 12).

1-4d Karl Marx

Karl Marx

Karl Marx (1818–1883), a German social philosopher, is often described as the most influential social scientist who ever lived. Marx, like Comte and Durkheim, tried to explain the societal changes that were taking place during the Industrial Revolution.

The Industrial Revolution began in England around 1780 and spread throughout Western Europe and the United States during the nineteenth century. A number of technological inventions—like the spinning wheel, the steam engine, and large weaving looms—enabled the development of large-scale manufacturing and mining industries over a relatively short period of time. The extensive mechanization shifted agricultural and home-based work to factories in cities. As masses of people migrated from small farms to factories to find jobs, urbanization and capitalism grew rapidly.

Capitalism

Unlike his predecessors and contemporaries, Marx (1867/1967, 1964) maintained that economic issues produce divisiveness rather than social solidarity. For Marx, the most important social change was the development of **capitalism**, an economic system in which the ownership of the means of production—such as land, factories, large sums of money, and machines—is private. As a result, Marx saw industrial society as composed of three social classes:

- **capitalists**—the ruling elite who own the means of producing wealth (such as factories)
- **petit bourgeoisie**—small business owners and workers who still have their own means of production but might end up in the proletariat because they're driven out by competition or their businesses fail
- **proletariat**—the masses of workers who depend on wages to survive, have few resources, and make up the working class

---

capitalism an economic system based on the private ownership of property and the means of production.

alienation feeling separated from one's group or society.

Class Conflict

Marx believed that society is divided into the haves (capitalists) and the have-nots (proletariat). For Marx, capitalism is a class system in which conflict between the classes is common and society is anything but cohesive. Instead, class antagonisms revolve around struggles between the capitalists, who increase their profits by exploiting workers, and workers, who resist but give in because they depend on capitalists for jobs.

Marx argued that there's a close relationship between inequality, social conflict, and social class. History, he maintained, is a series of class struggles between capitalists and workers. As wealth becomes more concentrated in the hands of a few capitalists, Marx predicted, the ranks of an increasingly dissatisfied proletariat would swell, leading to bloody revolution and eventually a classless society. The Occupy Wall Street movement showed that thousands of Americans were very unhappy about the growing inequality between the haves and the have-nots. Yet, in recent times in the United States and other nations, various protests, riots, and uprisings have been more common than complete revolutions.

Alienation

In industrial capitalist systems, Marx (1844/1964) contended, **alienation**—feeling separated from one's group or society—is common across all social classes. Workers feel alienated because they don't own or control either

Occupy Wall Street (OWS) was a protest movement against corporate greed, corruption, and influence on government. It began in mid-September 2011, in New York City's Wall Street financial district. The OWS slogan, "We are the 99%," referred to U.S. income and wealth inequality between the wealthiest 1 percent and the rest of the population. OWS received global attention and spawned similar movements worldwide but was short-lived (refer to Chapter 16).

the means of production or the product. Because meaningful labor is what makes us human, Marx maintained, the workplace has alienated workers "from the essence of our humanness." Instead of collaboration, a capitalistic society encourages competition, backstabbing, and "looking out for number one."

According to Marx, capitalists are also alienated. They regard goods and services as important simply because they are sources of profit. Capitalists may not care who buys or sells their products, how the workers feel about the products they make, or whether buyers value the products. The major focus, for capitalists, is on increasing profits as much as possible rather than feeling "connected" to the products or services they sell. Every year, for example, companies must recall cars, pharmaceutical items, toys, and food that cause injuries, illness, or death.

1-4e Max Weber

Hulton Archive/Getty Images

Max Weber

Max Weber (pronounced VAY-ber; 1864–1920) was a German sociologist, economist, legal scholar, historian, and politician. Unlike Marx's emphasis on economics as a major factor in explaining society, Weber focused on social organization, a subjective understanding of behavior, and a value-free sociology.

Social Organization

For Weber, economic factors were important, but ideas, religious values, ideologies, and charismatic leaders were just as crucial in shaping and changing societies. He maintained that a complete understanding of society requires an analysis of the social organization and interrelationships among economic, political, and cultural institutions. In his *Protestant Ethic and the Spirit of Capitalism*, for example, Weber (1920/1958) argued that the self-denial fostered by Calvinism supported the rise of capitalism and shaped many of our current values about working hard (refer to Chapters 3 and 6).

Subjective Understanding

Weber posited that an understanding of society requires a "subjective" understanding of behavior. Such understanding, or *verstehen* (pronounced fer-SHTAY-en), involves knowing how people perceive the world in which they live. Weber described two types of *verstehen*. In *direct observational understanding*, the social scientist observes a person's facial expressions and gestures and listens to their words. In *explanatory understanding*, the social scientist tries to grasp the intention and context of the behavior.

If a person bursts into tears (direct observational understanding), the observer typically knows the set of emotions the person may be feeling (anger, sorrow, and so on). An explanatory understanding goes a step further by spelling out the reason for the behavior (rejection by a loved one, frustration if you lose your smartphone, humiliation if a boss yells at you in public).

Value-Free Sociology

One of Weber's most lasting and controversial views was the notion that sociologists must be as objective, or "value free," as possible in analyzing society. A researcher who is **value free** is one who separates their personal values, opinions, ideology, and beliefs from scientific research.

During Weber's time, the government and other organizations demanded that university faculty teach the "right" ideas. Weber encouraged everyone to be involved as citizens, but he maintained that educators and scholars should be as dispassionate as possible politically and ideologically. The task of the teacher, Weber argued, was to provide students with knowledge and scientific experience, not to "imprint" the teacher's personal political views and value judgments (Gerth and Mills, 1946). "Is It Possible to Be a Value-Free Sociologist?" examines this issue further.

1-4f Jane Addams

Bettmann/Getty Images

Jane Addams with a child at Hull House

Jane Addams (1860–1935) was a social worker who cofounded Hull House, one of the first settlement houses in Chicago that served neighborhood community members that would in today's terms be described as living below the poverty line. An active reformer throughout her life, Jane Addams was a leader in the women's suffrage movement and, in 1931, was the first American woman to be awarded the Nobel Peace Prize for her advocacy of negotiating, rather than waging war, to settle disputes.

Sociologist Mary Jo Deegan (1986) describes Jane Addams as "the greatest woman sociologist of her day." However, she was ignored by her colleagues at the University of Chicago (the first sociology

---

value free separating one's personal values, opinions, ideology, and beliefs from scientific research.

Is It Possible to Be a Value-Free Sociologist?

Max Weber was concerned about popular professors who took political positions that pleased many of their students. He felt that these professors were behaving improperly because science, including sociology, must be "value free." Faculty must set their personal values aside to make a contribution to society. According to a sociologist who agrees with Weber, sociology's weakness is its tendency toward moralism and ideology:

> Many people become sociologists out of an impulse to reform society, fight injustice, and help people. Those sentiments are noble, but unless they are tempered by skepticism, discipline, and scientific detachment, they can be destructive. Especially when you are morally outraged and burning with a desire for action, you need to be cautious (Massey, 2007, p. B12).

Some argue that being value free is a myth because it's impossible for a scholar's attitudes and opinions to be totally divorced from their scholarship (Gouldner, 1962). Many sociologists, after all, do research on topics that they consider significant and about which they have strong views.

Others maintain that one's values should be passionately partisan, should frame research issues, and should improve society (Feagin, 2001). Sociologists shouldn't apologize for being subjective in their teaching and research. By staying silent, social scientists "cede the conversation to those with the loudest voices or deepest pockets . . . people with megaphones who spread sensational misinformation" that deprives the public of the best available data (Wang, 2015, p. A48).

> Can sociologists be value free— especially when they have strong feelings about many societal issues? *Should they be?*

Can sociologists *really* be value free—especially when they have strong feelings about many societal issues? Should they be?

department established in the United States in 1892) because discrimination against women sociologists was "rampant" (p. 8).

Despite such discrimination, Addams published articles in many popular and scholarly journals as well as many books on the everyday life of urban neighborhoods, especially the effects of social disorganization and immigration. Much of her work contributed to symbolic interaction, an emerging school of thought you'll read about shortly. One of Addams' greatest intellectual legacies was her emphasis on applying knowledge to everyday problems. Her pioneering work in criminology included ecological maps of Chicago that were later credited to men (Moyer, 2003).

1-4g W. E. B. Du Bois

Hulton Archive/Hulton Archive/Getty Images

W. E. B. Du Bois

W. E. B. Du Bois (pronounced Do-BOICE; 1868–1963) was a prominent Black sociologist, writer, editor, social reformer, and orator. The author of almost two dozen books on Africans and Black Americans, Du Bois spent most of his life responding to the critics and detractors of Black life. He was the first Black individual to receive a Ph.D. from Harvard University but once remarked, "I was in Harvard but not of it."

Du Bois helped found the National Association for the Advancement of Colored People (NAACP) and edited its journal, *Crisis*. The problem of the twentieth century, he wrote, is the problem of the color line. Du Bois believed that the race problem was one of ignorance and advocated a "cure" for prejudice and discrimination. Such cures included promoting Black political power and civil rights and providing Black individuals with a higher education rather than funneling them into technical schools.

These and other writings were unpopular at a time when Booker T. Washington, a well-known Black educator, encouraged Black people to be patient instead of demanding equal rights. As a result, Du Bois was dismissed as a radical by his contemporaries but was rediscovered by a new generation of Black scholars during the 1970s and 1980s. Among his many contributions, Du Bois examined the oppressive effects of race and social class, advocated women's rights, and played a key role in reshaping Black–White relations in America (Du Bois, 1986; Lewis, 1993).

All of these and other early thinkers agreed that people are transformed by each other's actions, social patterns, and historical changes. They and other scholars shaped contemporary sociological theories.

1-5 Contemporary Sociological Theories

How one defines "contemporary sociological theory" is somewhat arbitrary. The mid-twentieth century is a good starting point because "the late 1950s and 1960s have, in historical hindsight, been regarded as significant years of momentous changes in the social and cultural life of most Western societies" (Adams and Sydie, 2001, p. 479). Some of the sociological perspectives had earlier origins, but all matured during this period.

Sociologists typically use more than one theory to explain behavior. The theories view our social world somewhat differently, but all of them analyze why society is organized the way it is and why we behave as we do. Four of the most influential theoretical perspectives are functionalism, conflict theory, feminist theories, and symbolic interaction.

1-5a Functionalism

Functionalism (also known as *structural functionalism*) maintains that society is a complex system of interdependent parts that work together to ensure a society's survival. Much of contemporary functionalism grew out of the work of Auguste Comte and Émile Durkheim, both of whom believed that human behavior is a result of social structures that promote order and integration in society.

One of their contemporaries, English philosopher Herbert Spencer (1820–1903), used an organic analogy to explain the evolution of societies. To survive, Spencer (1862/1901) wrote, our vital organs—like the heart, lungs, kidneys, liver, and so on—must function together. Similarly, the parts of a society, like the parts of a body, work together to maintain the whole structure.

Society Is a Social System

Prominent American sociologists, especially Talcott Parsons (1902–1979) and Robert K. Merton (1910–2003), developed the earlier ideas of structure and function. For these and other functionalists, a society is a system that is composed of major institutions such as government, religion, the economy, education, medicine, and the family.

Each institution or other social group has *structures*, or organized units, that are connected to each other and within which behavior occurs. Education structures like colleges, for instance, aren't only organized internally in terms of who does what and when but depend on other structures like government (to provide funding), business (to produce textbooks and construct buildings), and medicine (to ensure that students, staff, and faculty are healthy).

Functions and Dysfunctions

Each structure fulfills certain *functions*, or purposes and activities, to meet different needs that contribute to a society's stability and survival (Merton, 1938). The purpose of education, for instance, is to transmit knowledge to students, to teach them to be good citizens, and to prepare them for jobs (refer to Chapter 13).

Dysfunctions are social patterns that have a negative impact on a group or society. When one part of society isn't working, it affects other parts, generating conflict, divisiveness, and social problems. Consider religion. In the United States, the Catholic Church's stance on issues such as not ordaining women to be priests and denouncing abortion and non-heterosexuality has produced a rift between those who embrace and those who question papal edicts. In other countries, religious intolerance has led to wars and terrorism (refer to Chapter 13).

Manifest and Latent Functions

There are two kinds of functions. **Manifest functions** are intended and recognized; they're present and clearly evident. **Latent functions** are unintended and unrecognized; they're present but not immediately obvious.

Consider the marriage ceremony. Its primary manifest function is to publicize the formation of a new family unit and to legitimize sexual intercourse and childbirth (even though both might occur outside of marriage, and both might not occur within marriage). The latent functions of a marriage ceremony include communicating a "hands-off" message to suitors, providing the new couple with household goods and products through wedding showers and gifts, and redefining family boundaries to include in-laws or stepfamily members.

Critical Evaluation

You'll explore in later chapters how functionalism is useful for laying out the "big picture" of interrelated structures and functions. Its influence waned during the

functionalism (*structural functionalism*) maintains that society is a complex system of interdependent parts that work together to ensure a society's survival.

dysfunctions social patterns that have a negative impact on a group or society.

manifest functions purposes and activities that are intended and recognized; they are present and clearly evident.

latent functions purposes and activities that are unintended and unrecognized; they are present but not immediately obvious.

Sociology and Other Social Sciences: What's the Difference?

How would different social scientists study the same phenomenon, such as homelessness? Criminologists might examine whether crime rates are higher among homeless people than in the general population. Economists might measure the financial impact of programs for the homeless. Political scientists might study whether and how government officials respond to homelessness. Psychologists might be interested in how homelessness affects individuals' emotional and mental health. Social workers are most likely to try to provide needed services (e.g., food, shelter, medical care, and jobs). Sociologists have been most interested in examining homelessness across gender, age, and social class and explaining how this social problem devastates families and communities.

According to sociologist Herbert Gans (2005), sociologists "study everything." The American Sociological Association currently recognizes over 50 different sections, or subfields, in sociology, and the number continues to increase because sociologists' interests range across many areas.

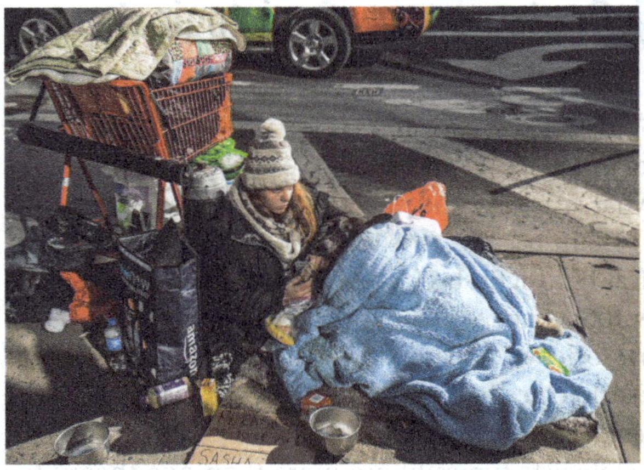
Keep Smiling Photography/Shutterstock.com

1960s and 1970s, however, because functionalism was so focused on order and stability that it often ignored social change. For example, functionalism couldn't explain the many rapid changes sparked by the civil rights, women's, and gay movements.

A second and related criticism is that functionalism often glosses over the widespread inequality that a handful of powerful people create and maintain. Conflict theorists, especially, have pointed out that what is functional for some privileged groups is dysfunctional for many others.

1-5b Conflict Theory

In contrast to functionalism—which emphasizes order, stability, cohesion, and consensus—**conflict theory** examines how and why groups disagree, struggle over power, and compete for scarce resources (like property, wealth, and prestige). Conflict theorists think disagreement between groups is natural and inevitable. They also think that change in society is desirable.

Sources of Conflict

The conflict perspective has a long history. As you saw earlier, Karl Marx predicted that conflict would result from widespread economic inequality, and W. E. B. Du Bois criticized U.S. society for its ongoing and divisive racial discrimination. Since the 1960s, many sociologists—especially feminist and Black, indigenous, and people of color (BIPOC) scholars—have emphasized that the key sources of economic inequity in any society include race, ethnicity, gender, age, and sexual orientation.

> Sociologists typically use more than one theory to explain behavior and why society is organized the way it is.

Conflict theorists agree with functionalists that many societal arrangements are functional. But, conflict theorists ask, who benefits? And who loses? When corporations merge, workers in lower-end jobs are often laid off while the salaries and benefits of corporate executives soar and the value of stocks (usually held by higher social classes) rise. Thus, mergers might be functional for people at the upper end of the socioeconomic ladder but dysfunctional for those on the lower rungs.

conflict theory examines how and why groups disagree, struggle over power, and compete for scarce resources.

skynesher/E+/Getty Images

Among other manifest functions, schools transmit knowledge and prepare children for adult economic roles. Among their latent functions, schools provide matchmaking opportunities. What are some other examples of education's manifest and latent functions?

Social Inequality

Unlike functionalists, conflict theorists consider society not as cooperative and harmonious but as a system of widespread inequality. For conflict theorists, there's a continuous tension between the haves and the have-nots, most of whom are children, women, racial minorities, and people with low incomes.

Many conflict theorists focus on how those in power—typically wealthy White Anglo-Saxon Protestant males (WASPs)—dominate political and economic decision making in U.S. society. This group controls a variety of institutions—like education, criminal justice, and the media—and passes laws that primarily benefit people like themselves (refer to Chapters 8 and 11).

> "Sometimes the best man for the job isn't."
>
> Author Unknown

Critical Evaluation

Conflict theory explains how societies create, struggle with, and navigate disagreements. However, some have criticized conflict theorists for overemphasizing competition and coercion at the expense of order and stability. Inequality exists, and struggles over scarce resources occur, critics agree, but conflict theorists often ignore cooperation and harmony. Voters, for example, can vote dominant White men out of office and replace them with women and BIPOC group members. Critics also point out that the have-nots can increase their power through negotiation, bargaining, lawsuits, and strikes.

1-5c Feminist Theories

As noted earlier, influential male theorists generally overlooked or marginalized early female sociologists' contributions. Until the feminist activism of the 1960s and 1970s, men—who dominated universities and scholarship—were largely "blind to the importance of gender" (Kramer and Beutel, 2015, p. 17).

Feminist scholars agree with contemporary conflict theorists that much of society is characterized by tension and struggle, but **feminist theories** go a step further by focusing on women's social, economic, and political inequality. The theories maintain that women often suffer injustice primarily because of their gender rather than personal inadequacies like low educational levels or not caring about success. Feminist scholars assert that people should be treated fairly and equally regardless not only of their sex but also of other characteristics such as their race, ethnicity, national origin, age, religion, class, sexual orientation, or disability. They emphasize that women should be freed from traditionally oppressive expectations, constraints, roles, and behavior (refer to Reger, 2012).

Focusing on Gender

Feminist scholars have documented women's historical exclusion from most sociological analyses (refer to, for example, Smith, 1987, and Adams and Sydie, 2001). Before the 1960s women's movement in the United States, very few sociologists published anything about gender roles, women's sexuality, fathers, or intimate partner violence. According to sociologist Myra Ferree (2005, p. B10), during the 1970s, "the Harvard social-science library could fit all its books on gender inequalities onto a single half-shelf." Because of feminist scholars, many researchers now routinely include gender as an important research variable on both micro and macro levels.

In most countries, solid majorities of both women and men support gender equality and agree that women should be able to work outside the home. When jobs are scarce, however, many women and men believe that men should be given preferential treatment ("Gender Equality...," 2010). Thus, even equal rights proponents place a higher priority on men's economic rights.

---

feminist theories examine women's social, economic, and political inequality.

> "I myself have never been able to find out precisely what feminism is; I only know that people call me a feminist whenever I express sentiments that *differentiate me from a doormat*."
>
> Rebecca West, British journalist

Listening to Many Voices

Feminist scholars contend that gender inequality is central to *all* behavior, ranging from everyday interactions to political and economic institutions, but feminist theories encompass many perspectives. For example, *liberal feminism* endorses social and legal reform to create equal opportunities for women. *Radical feminism* indicates that male dominance in social institutions (e.g., the economy and politics) is the major cause of women's inequality. *Global feminism* focuses on how the intersection of gender with race, social class, and colonization has exploited women in the developing world (refer to Lengermann and Niebrugge-Brantley, 1992). Many people align with feminism because they endorse equal opportunities for all in the economy, politics, education, and other institutions.

Critical Evaluation

Feminist scholars have challenged employment discrimination, particularly practices that routinely exclude women who aren't part of the "old boy network" (Wenneras and Wold, 1997). One criticism, however, is that many feminists are part of an "old girl network" that hasn't always welcomed different points of view from Black, Asian American, Native American, Muslim, Latina, lesbian, working-class, and women with disabilities (Lynn and Todoroff, 1995; Jackson, 1998; Sánchez, 2013).

A second criticism is that feminist perspectives often overlook the differences among women, including sexual orientation, social class, and generational gaps. Examples of this are evident in the 2016 and 2020 presidential elections. Donald Trump's 2016 presidential victory prompted women to organize the biggest one-day protest in U.S. history. Forecasters for the 2020 presidential election predicted that Trump would lose with a notable double-digit gender gap (Becker, 2020). Yet in 2020, a slightly larger share of women cast votes for Trump compared to 2016; 44 versus 39 percent (Igielnik et al., 2021).

In addition, at least 90 percent of people vote for their party's candidate, but candidates can appeal to voters for a variety of reasons. Trump earned 2 percent more votes from White women compared to Clinton in 2016 (47 vs. 45 percent), and in 2020, Trump brought in an even greater share of votes from that group (53 percent) (Kurtzleben, 2021; Igielnik et al., 2021) (**Table 1.2**).

There are many reasons for a presidential candidate's victory, but some observers have attributed support of Trump, particularly from working-class White women, to frustrations about diminished possibilities for their husbands and sons to provide for their families, fears about downward mobility and poverty, concerns about a growing number of immigrants, and a scarcity of jobs in small towns and rural areas (Featherstone, 2016; Morin, 2016; Roberts and Ely, 2016). In contrast, Trump's approach resonated with millions of voters, especially those without college degrees, who feel economically disenfranchised.

Bernie Sanders, a senator from Vermont, who is part of the Silent Generation (people born between 1928 to 1945) had notable support from Millennials (people

Table 1.2 How Women Voted in the 2016 and 2020 Elections, by Selected Characteristics

	2016 Election		2020 Election	
	Clinton (Democrat)	**Trump (Republican)**	**Biden (Democrat)**	**Trump (Republican)**
Women	54%	39%	55%	44%
White women, non-Latina	39%	54%	43%	55%
Black women, non-Latina	91%	6%	92%	8%
Latina women	66%	28%	59%	38%

Source: Based on Igielnik, Keeter, and Hartig, 2021

born between 1981 and 1996) in both the 2016 and 2020 presidential elections. Sanders' platform called for the most progressive and drastic changes to the U.S. political and economic structures (e.g., free tuition, changes in energy policies, and greater equality of wealth). Some analysts believe that millennial enthusiasm for Sanders is an example of a "feminist generation gap" that has increased because younger and older feminists have different values, convictions, and goals (Norman, 2016; Rosen, 2016).

Some critics, including feminists, also question whether feminist scholars have lost their bearings by concentrating on personal issues like greater sexual freedom rather than broader social issues, particularly wage inequality (Chesler, 2006; Shteir, 2013; Rosen, 2016).

1-5d Symbolic Interaction

Symbolic interaction theory (sometimes called *interactionism*) is a micro-level perspective that examines people's everyday behavior through the communication of knowledge, ideas, beliefs, and attitudes. While functionalist, conflict, and some feminist theories emphasize structures and large (macro) systems, symbolic interactionist theory focuses on *process* and keeps the *person* at the center of the analysis.

There have been many influential symbolic interactionists whom we'll cover in later chapters. In brief, George Herbert Mead's (1863–1931) assertion that the human mind and self arise in the process of social communication became the foundation of the symbolic interaction schools of thought in sociology and social psychology. Herbert Blumer (1900–1987), who coined the term *symbolic interactionism* in 1937, developed Mead's ideas and proposed that people interpret or "define" each other's actions, especially through symbols, instead of merely reacting to them.

Erving Goffman (1922–1982) enriched these earlier theories by examining human interaction in everyday situations ranging from jobs to funerals. Among his other contributions, Goffman used "dramaturgical analysis" to compare everyday social interaction to a theatrical presentation (refer to Chapter 5).

Constructing Meaning

Our actions are based on **social interaction** in the sense that when people calculate their own behaviors, they consider others as part of the assessment. Thus, we act differently in different social settings and continuously adjust our behavior, including our body language, as we interact (Goffman, 1959; Blumer, 1969). A woman's interactions with her husband differ from those with her children. And she will interact still differently when she is teaching, talking to a colleague in the hall, or addressing an audience at a professional conference.

For symbolic interactionists, society is *socially constructed* through human interpretation (O'Brien and Kollock, 2001). That is, meanings aren't inherent but are created and modified through interaction with others. For example, a daughter who has batting practice with her dad will probably interpret her father's behavior as loving and involved. In contrast, she'll views batting practice with her baseball coach as less personal and more goal-oriented. In this sense, our interpretations of the same behavior, such as batting practice, vary across situations and depend on the people with whom we interact.

Symbols and Shared Meanings

Symbolic interaction looks at subjective, interpersonal meanings and how we interact with and influence each other by communicating through *symbols*—words, gestures, or pictures that stand for something and that can have different meanings for different individuals.

After the 9/11 terrorist attacks, many Americans displayed the flag on buildings, bridges, homes, and cars to show their solidarity and pride in the United States. In contrast, some groups in the Middle East burned the U.S. flag to show their contempt for U.S. culture and policies. Thus, symbols are powerful forms of communication that show how people feel and interpret a situation.

To interact effectively, our symbols must have *shared meanings*, or agreed-on definitions. One of the most important of these shared meanings is the *definition of the situation,* or the way we perceive reality and react to it. Relationships often end, for example, because people view emotional closeness differently ("We broke up because my partner wanted more sex. I wanted more communication."). We typically learn our definitions of the situation through interaction with *significant others*—especially parents, friends, relatives, and teachers—who play an important role in our socialization (refer to Chapters 4 and 5).

---

symbolic interaction theory (*interactionism*) examines people's everyday behavior through the communication of knowledge, ideas, beliefs, and attitudes.

social interaction a process in which people take each other into account in their own behavior.

Critical Evaluation

Unlike other theorists, symbolic interactionists show how people play an active role in shaping their lives on a micro level. One of the most common criticisms is that symbolic interaction overlooks the widespread impact of macro-level factors (e.g., economic forces, social movements, and public policies) on our everyday behavior and relationships. During economic downturns, for example, unemployment and ensuing financial problems create considerable interpersonal conflict among couples and families (refer to Chapters 11 and 12). Symbolic interaction rarely considers such macro-level changes in explaining everyday behavior.

A related criticism is that interactionists sometimes have an optimistic and unrealistic view of people's everyday choices. Most of us enjoy little flexibility in our daily lives because deeply embedded social arrangements and practices benefit those in power. For instance, people are usually powerless when corporations transfer jobs overseas or cut the pension funds of retired employees.

Some also believe that interaction theory is flawed because it ignores the irrational and unconscious aspects of human behavior (LaRossa and Reitzes, 1993). People don't always consider the meaning of their actions or behave as reflectively as interactionists assume. Instead, we often act impulsively or say hurtful things without weighing the consequences of our actions or words.

1-5e Other Theoretical Approaches

Table 1.3 summarizes the major sociological perspectives that you've just read about. However, new theoretical perspectives arise because society is always changing.

Table 1.3 Leading Contemporary Perspectives in Sociology

Theoretical Perspective	Functionalism	Conflict	Feminist	Symbolic Interaction
Level of Analysis	**Macro**	**Macro**	**Macro and Micro**	**Micro**
Key Points	• Society is composed of interrelated, mutually dependent parts. • Structures and functions maintain a society's or group's stability, cohesion, and continuity. • Dysfunctional activities that threaten a society's or group's survival are controlled or eliminated.	• Life is a continuous struggle between the haves and the have-nots. • People compete for limited resources that are controlled by a small number of powerful groups. • Society is based on inequality in terms of ethnicity, race, social class, and gender.	• Women experience widespread inequality in society because, as a group, they have little power. • Gender, ethnicity, race, age, sexual orientation, and social class—rather than a person's intelligence and ability—explain many of our social interactions and lack of access to resources. • Social change is possible only if we change our institutional structures and our day-to-day interactions.	• People act on the basis of the meaning they attribute to others. Meaning grows out of the social interactions that we have with others. • People continuously reinterpret and reevaluate their knowledge and information in their everyday encounters.
Key Questions	• What holds society together? How does it work? • What is the structure of society? • What functions does society perform? • How do structures and functions contribute to social stability?	• How are resources distributed in a society? • Who benefits when resources are limited? Who loses? • How do those in power protect their privileges? • When does conflict lead to social change?	• Do men and women experience social situations in the same way? • How does our everyday behavior reflect our gender, social class, age, race, ethnicity, sexual orientation, and other factors? • How do macro structures (such as the economy and the political system) shape our opportunities? • How can we change current structures through social activism?	• How does social interaction influence our behavior? • How do social interactions change across situations and between people? • Why does our behavior change because of our beliefs, attitudes, values, and roles? • How is "right" and "wrong" behavior defined, interpreted, reinforced, or discouraged?
Example	• A college education increases one's job opportunities and income.	• Most low-income families can't afford to pay for a college education.	• Gender affects decisions about a major and which college to attend.	• College students succeed or fail based on their degree of academic engagement.

Chelsea Victoria/Stocksy/Adobe Stock Photos

José Nicolas/Sygma/Getty Images

For many people, a diamond, especially in an engagement ring, signifies love and commitment. For others, diamonds represent Western exploitation of people in Africa who are paid next to nothing for their backbreaking labor in mining these stones.

For example, *postmodern theory* analyzes contemporary societies that are characterized by postindustrialization, consumerism, and global communications.

Sociology, like other social sciences, has subfields. The subfields—such as socialization, deviance, and social stratification—offer specific theories that reinforce and illustrate functionalist, conflict, feminist, and interactionist approaches. No single theory explains social life completely. Each theory, however, provides different insights that guide sociological research, the topic of Chapter 2.

Chapter Review

Key Takeaways

1-1 **What is sociology and how does it differ from common sense?** *Sociology* is the systematic study of human behavior in society, behavior that takes place among individuals, small groups (such as families), large organizations (such as Apple), and entire societies (such as the United States). Sociology goes well beyond common sense and conventional wisdom because it examines claims and beliefs critically, considers many points of view, and enables us to move beyond established ways of thinking.

Example: Critical Thinking versus Common Sense and Conventional Wisdom

When thinking critically, it's important to differentiate between common sense/conventional wisdom and data-based facts. Here are a few examples:

Myth: Older people make up the largest group of people with the lowest incomes.

Fact: Children younger than 6, not older people, make up the largest group of people with the lowest incomes.

Myth: Divorce rates are higher today than ever before.

Fact: Divorce rates are lower today than they were between 1980 and 1990.

Now, based on the material you read in this chapter, construct your own "myth" and "fact."

1-2 **What is a sociological imagination?** The *sociological imagination* emphasizes the connection between personal troubles (biography) and structural (public and historical) issues in understanding the social world. *Microsociology* examines the relationships between individuals, and *macrosociology* focuses on large-scale patterns and processes that characterize society as a whole. Macro-level systems shape society, often limiting our personal options on the micro level.

1-3 **Why study sociology?** Regardless of your major, this course will help you (1) make more informed decisions, (2) understand diversity, (3) shape social policies and practices, (4) think critically, and (5) expand your career opportunities.

1-4 **What were the origins of sociological theory?** Sociologists use *theories* to produce knowledge and to offer solutions to everyday social problems. Some of the most influential theorists have included Auguste Comte, Harriet Martineau, Émile Durkheim, Karl Marx, Max Weber, Jane Addams, and W. E. B. Du Bois. All of these early thinkers shaped contemporary sociological theories.

1-5 **What are the similarities and differences among the contemporary sociological theories?** Sociologists typically use more than one theory to explain human behavior. Four major perspectives help us understand society:

- *Functionalism maintains that society is a complex system of interdependent parts that work together to ensure a society's survival.* Critics contend that functionalism is so focused on order and stability that it often ignores social change, social inequality, and social conflict.
- *Conflict theory* views *disagreement and the resulting changes in society as natural, inevitable, and even desirable.* Critics argue that conflict theory ignores the importance of harmony and cooperation.
- *Feminist theories*, which build on conflict theory, *emphasize women's social, economic, and political inequality*. Critics claim that these theories are too narrowly focused.
- *Symbolic interaction focuses on individuals' everyday behavior through the communication of knowledge, ideas, beliefs, and attitudes*. Critics maintain that this theoretical perspective overlooks the impact of macro-level factors on our everyday behavior.

Key Terms

Sociology (p. 2)
Sociological imagination (p. 3)
Microsociology (p. 3)
Macrosociology (p. 3)
Theory (p. 6)
Empirical (p. 6)
Social facts (p. 7)
Social solidarity (p. 7)
Division of labor (p. 7)
Capitalism (p. 8)
Alienation (p. 8)
Value free (p. 10)
Functionalism (*structural functionalism*) (p. 11)
Dysfunctions (p. 11)
Manifest functions (p. 11)
Latent functions (p. 11)
Conflict theory (p. 12)
Feminist theories (p. 13)
Symbolic interaction theory (*interactionism*) (p. 15)
Social interaction (p. 15)

Discussion and Critical Thinking Questions

1. From a functionalist perspective, what have been some of the manifest functions, latent functions, and dysfunctions of the Affordable Care Act (also called "Obamacare")?
2. If they were alive today, do you think that Harriet Martineau and Jane Addams would think sex discrimination is a problem? And would Du Bois consider race a major problem? Why or why not?
3. Using the symbolic interaction perspective, give two examples of how your recent everyday interactions have affected other people. Did they react as you expected? Or did you spark unintended misunderstandings?
4. Do you think that your professors, especially those in the social sciences (including sociology), should be value free? Or do you think that they should encourage you to participate in activities to change society?

5. Using your sociological imagination, explain whether dropping out of college is due to personal troubles, public issues, or both. Why?

6. It's "common sense" for a college student who pays thousands of dollars to attend every class and read all assigned materials. Why do so many students cut classes and spend little time studying?

True or False?

How Much Do You Know about Thinking Like a Sociologist?

The answers to 2, 3, and 4 are true. The answers to 1, 5, and 6 are false.

1. U.S. marriage and divorce rates rose steadily during the twentieth century but have both declined since 1990 (refer to Chapter 7).
2. Although the gender pay gap is narrowing, women still earn only 82 cents per every dollar men earn. This gap exists across all occupations, even those that require advanced degrees (refer to Chapters 9 and 11).
3. From 2011 until 2022, the poverty rate in the United States declined by approximately 8 percent (refer to Chapter 8).
4. Latina/o/xs constitute the largest racial or ethnic minority group in the United States with nearly 19 percent of the nation's population (refer to Chapter 10).
5. Sociologists rely on the scientific method, a body of objective and systematic techniques used to investigate phenomena, acquire knowledge, and test hypotheses and theories (refer to Chapter 2).
6. In an overall assessment of health care quality, the United States ranked last compared with other high-income nations (refer to Chapter 14).

2 Examining Our Social World

Franck Boston/Shutterstock.com

Learning Objectives

After studying this chapter, you will be able to...

2-1 Compare knowledge based on tradition, authority, and research.

2-2 Explain why sociological research is important in our everyday lives.

2-3 Describe the scientific method.

2-4 Describe the basic steps of the research process.

2-5 Compare the five most common sociological data collection methods, including their strengths and limitations.

2-6 Explain why ethics are important in scientific research.

True or False?

How Much Do You Know about Examining Our Social World?

1. Knowledge based on tradition, authority, and sociological research all serve the same purposes.
2. Sociologists use the scientific method in their work.
3. Sociologists only conduct qualitative (nonnumerical) research.
4. Surveys have too many complications for sociologists to use them in their research.
5. Some sociologists do field research involving observation of people in their natural surroundings.
6. Researchers today operate under stricter guidelines than they did in the past.

The answers are at the end of the chapter.

In an experiment on handwashing, 383 people were recorded by cameras in test kitchens as they prepared and cooked turkey burgers and made salads. Beforehand, half of the group had watched a food safety video, while the other half did not. Regardless, in both groups, only around one-third of the people even attempted to wash their hands (Yee, 2018). If you suspect that this number is too high or too low and wonder how this research project was done, you're thinking like a researcher, which is the focus of this chapter.

Be sure to take the True or False quiz in the chapter opener to better understand how much you know about these topics.

What do you think?

People can find data to support any opinion they have.

1	2	3	4	5	6	7
	strongly agree				strongly disagree	

2-1 How Do We Know What We Know?

Much of our knowledge is based on *tradition*, a handing down of statements, beliefs, and customs from generation to generation ("The groom's parents should pay for the wedding rehearsal dinner"). Another common source of knowledge is *authority*, a socially accepted source of information that includes "experts," parents, government officials, the police, judges, and religious leaders ("My mom says that . . ." or "According to the American Heart Association . . .").

Knowledge based on tradition and authority simplifies our lives because it provides us with basic rules about socially and legally acceptable behavior. The information can be misleading or wrong, however. Suppose a 2-year-old throws a temper tantrum at a family barbecue. One adult comments, "What that kid needs is a smack on the behind." Someone else immediately disagrees: "All kids go through this stage. Just ignore it."

Who's right? To answer this and other questions, sociologists rely on **research methods**, organized and systematic procedures to gain knowledge about a particular topic. Much research shows, for example, that neither ignoring a problem nor inflicting physical punishment (like spanking) stops a toddler's misbehavior. Instead, most young children's misbehavior can be curbed by having simple rules, being consistent in discipline, praising good behavior, and setting a good example (refer to Benokraitis, 2015).

2-2 Why Is Sociological Research Important in Our Everyday Lives?

In contrast to knowledge based on tradition and authority, sociological research is important in our everyday lives for several reasons:

1. **It counteracts misinformation.** Blatant dishonesty and misinformation spread rapidly through digital communication—creating frameworks to mischaracterize, amplify messages, and discredit authoritative voices (Feingold, 2022). For example, public messaging often suggests that crime rates have increased. Yet, the 2021 FBI crime statistics indicate that levels of certain types of crimes have indeed increased, while other forms of crime have gone down. Misperceptions about high levels of crime can be fostered because existing gaps in how the information is presented allows opportunities for misinformation to take hold (Dirks, 2022).
2. **It exposes myths.** According to many newspapers and shows, suicide rates are highest during the winter holidays. In fact, suicide rates are lowest in December and highest in the spring and fall (but the

research methods organized and systematic procedures to gain knowledge about a particular topic.

"Facebook Causes 20 Percent of Today's Divorces." What?!

The founder and self-described leader of the United Kingdom's online divorce site (Divorce-Online) sent out a press release titled "Facebook Is Bad for Your Marriage—Research Finds," and claimed that Facebook causes 20 percent of today's divorces. News media around the world ran stories about this press release with headlines such as "Facebook to Blame for Divorce Boom." You'll learn in Chapter 12 that there's no divorce boom, so where did the 20 percent number come from?

In 2009, the managing director of Divorce-Online scanned its online divorce petition database for the use of the word *Facebook* and found 989 instances in about 5,000 petitions. Divorce-Online never said that the petitions were only those filed by members of the American Academy of Matrimonial Lawyers, who comprise a very small percentage of all divorce attorneys. Two years later, many Internet sites and blogs were still spreading the fiction that "Facebook Causes Divorce" (refer to Bialik, 2011). In reality, as Chapter 12 will explain, there are a number of interrelated macro- and micro-level reasons for divorce; there's no single "cause," much less Facebook.

reasons for these peaks are unclear). Another myth is that more women are victims of domestic violence on Super Bowl Sunday than on any other day of the year, presumably because men become intoxicated and abusive. In fact, intimate partner violence rates are high, and consistent, throughout the year (Romer, 2011; "Super Bull Sunday," 2015).

3. **It helps explain *why* people behave as they do.** A 2011 study predicted that older drivers, particularly those age 70 and older, would be more likely than younger drivers to have fatal crashes but found the opposite (Braitman et al., 2011). The researchers couldn't explain why their prediction turned out to be false. Answering the why question is important for understanding people's behavior. Possible explanations may include that fewer people 70 and older are licensed to drive, and those who are drive fewer miles (Insurance Institute for Highway Safety, 2022).
4. **It affects social policies.** According to the captain of a large North Carolina Police Department, "Research dictates everything that officers do, whether we realize it or not," in reducing crime. Examples include tracking criminal activity in high-risk locations, implementing research-based policies in training patrol officers and detectives, and managing limited resources more efficiently (Nolette, 2015).
5. **It sharpens critical thinking skills.** Many Americans rely on talk shows and podcasts for information on a number of topics. Joe Rogan, host of the most listened to podcast program, has featured and applauded guests who promote misinformation about topics including climate science and COVID-19 (Dewan, 2022; Sauer, 2022). As will be highlighted in Chapters 14 and 15, these issues can endanger human health, negatively impact the environment, and harm communities.

In his first 100 days in office, President Trump gave 17 false statements, while Obama told 1 and Biden told 2 (Markowitz, 2021). *Fake news*—misinformation that deliberately misleads people for financial, political, or other gain—has been around for a long time. About 84 percent of Americans are confident that they can identify false news (Barthel et al., 2016).

Many of us, however, are susceptible to *confirmation bias*, a tendency to embrace and recall information that confirms our beliefs and ignores or downplays contrary evidence. The scientific method, which requires the critical thinking skills you read about in Chapter 1, strengthens our ability to separate fact from fiction, but do people always believe scientific findings?

2-3 The Scientific Method

Sociologists rely on the **scientific method**, a body of objective and systematic techniques used to investigate phenomena, acquire knowledge, and test hypotheses and theories. The techniques include careful data collection, exact measurement, accurate recording and analysis of the findings, thoughtful interpretation of the results, and, when appropriate, generalization of the findings to a larger group.

scientific method a body of objective and systematic techniques used to investigate phenomena, acquire knowledge, and test hypotheses and theories.

Before collecting any data, however, social scientists must grapple with a number of research-related issues. Let's begin with concepts, variables, and hypotheses.

2-3a Concepts, Variables, and Hypotheses

A basic element of the scientific method is a **concept**—an abstract idea, mental image, or general notion that represents some aspect of the world. Some examples of concepts are "blood pressure," "religion," and "marriage."

Because concepts are abstract, scientists use variables to measure (*operationalize*) concepts. A **variable** is a characteristic that can change in value or magnitude under different conditions. Variables can be attitudes, behaviors, or traits (e.g., ethnicity, age, and social class).

An **independent variable** is a characteristic that has an effect on the **dependent variable**, the outcome. A **control variable** is a characteristic that is constant and unchanged during the research process.

Scientists can simply ask a research question ("Why are people economically disadvantaged?"), but they usually begin with a **hypothesis**—a statement of the expected relationship between two or more variables—such as "Unemployment increases poverty." In this example, "unemployment" is the independent variable and "poverty" is the dependent variable.

Researchers might also use control variables, like education, to explain the relationship between unemployment and poverty. For example, people with at least a college degree generally have lower poverty rates than those with lower educational levels because the former are less likely to experience long periods of unemployment.

2-3b Deductive and Inductive Reasoning

Deduction and induction are two different but equally valuable approaches to examining the relationship between variables. Generally, **deductive reasoning**

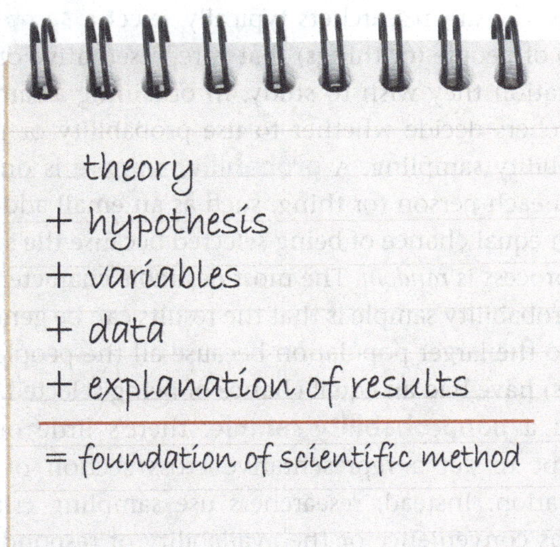

Image Source/Getty Images

SOPA Images Limited / Alamy Stock Photo; Erik Pendzich / Alamy Stock Photo; AFF / Alamy Stock Photo; AFF / Alamy Stock Photo

Are *The Voice* voters an example of a probability or nonprobability sample of the show's fans?

begins with a theory, prediction, or general principle that is then tested through data collection. An alternative mode of inquiry, **inductive reasoning**, begins with specific observations, followed by data collection, a conclusion about patterns or regularities, and the formulation of hypotheses that can lead to theory construction (**Figure 2.1**).

Taking a deductive approach, you might decide to test a theory of academic success using the following hypothesis: "Students who study in groups perform better on exams than those who study alone." You would collect the data, ultimately confirming or rejecting your hypothesis (or theory).

Alternatively, you might notice that your classmates who participate in study groups seem to get higher grades on exams than those who study alone. Using an inductive approach, you would collect the data systematically

concept an abstract idea, mental image, or general notion that represents some aspect of the world.

variable a characteristic that can change in value or magnitude under different conditions.

independent variable a characteristic that has an effect on the dependent variable.

dependent variable the outcome that may be affected by the independent variable.

control variable a characteristic that is constant and unchanged during the research process.

hypothesis a statement of the expected relationship between two or more variables.

deductive reasoning begins with a theory, prediction, or general principle that is then tested through data collection.

inductive reasoning begins with a specific observation, followed by data collection, a conclusion about patterns or regularities, and the formulation of hypotheses that can lead to theory construction.

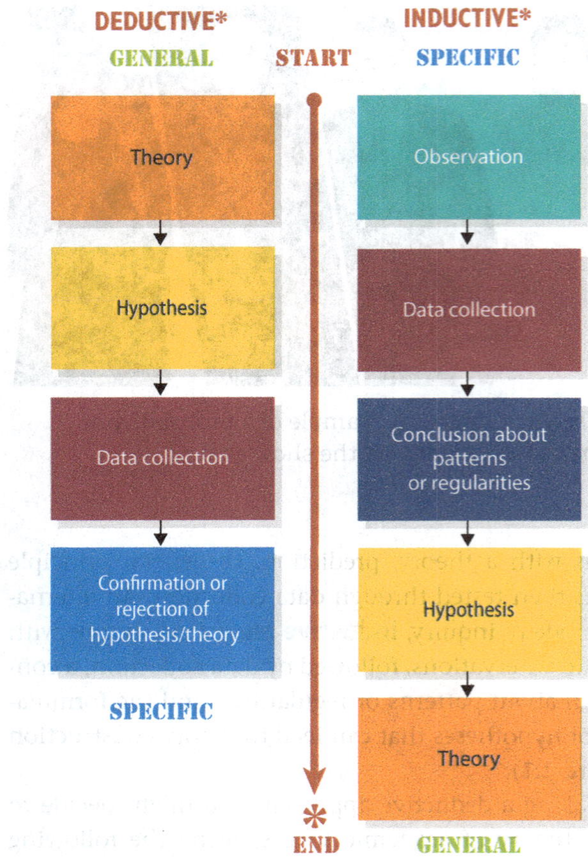

Figure 2.1 Deductive and Inductive Approaches

and formulate hypotheses (or suggest a theory) that could then be tested deductively. Most social science research involves both inductive and deductive reasoning.

2-3c Reliability and Validity

Sociologists are always concerned about reliability and validity. **Reliability** is the *consistency* with which the same measure produces similar results time after time. If, for example, you ask "How old are you?" on two subsequent days, and a respondent gives two different answers, such as 25 and 30, there's either something wrong with the question or the respondent is lying. Respondents might lie, but scientists must make sure that their measures are as reliable as possible.

Validity is the degree to which a measure is accurate and *really* measures what it claims to measure. Consider student course evaluations. The measures of a "good" professor often include items like whether they were "interesting," "fair," or "knowledgeable about the course content."

Because we don't know what students mean by "interesting" and "fair," and students don't know how "knowledgeable" an instructor is, how accurate are such measures in differentiating between "good" and "bad" professors? A study at two large public universities found that a third of the students admitted being dishonest in end-of-semester course evaluations. Some fibbed to make their instructors look good, but most lied to "punish" professors they didn't like, especially when they received lower grades than they thought they deserved (Clayson and Haley, 2011). Such research findings raise questions about the accuracy and usefulness of student course evaluations in measuring an instructor's actual performance.

2-3d Sampling

Early in the research process, sociologists decide what sampling procedures to use. Ideally, researchers would like to study all the units of the population in which they're interested—say, all adolescents who use drugs. A **population** is any well-defined group of people (or things) that researchers want to know something about. Obtaining information about and from populations is problematic, however. The population may be so large that it would be too expensive and time consuming to conduct the research. In other cases—such as all adolescents who use drugs—it's impossible even to identify the population.

As a result, researchers typically select a **sample**, a group of people (or things) that is representative of the population they wish to study. In obtaining a sample, researchers decide whether to use probability or nonprobability sampling. A **probability sample** is one in which each person (or thing, such as an email address) has an equal chance of being selected because the selection process is *random*. The most desirable characteristic of a probability sample is that the results can be generalized to the larger population because all the people (or things) have had an equal chance of being selected.

In a **nonprobability sample**, there's little or no attempt to get a representative cross section of the population. Instead, researchers use sampling criteria such as convenience or the availability of respondents

reliability the consistency with which the same measure produces similar results time after time.

validity the degree to which a measure is accurate and really measures what it claims to measure.

population any well-defined group of people (or things) that researchers want to know something about.

sample a group of people (or things) that is representative of the population researchers wish to study.

probability sample each person (or thing) has an equal chance of being selected because the selection is random.

nonprobability sample there is little or no attempt to get a representative cross section of a population.

or information. Nonprobability samples are especially useful when sociologists are exploring a new topic or want to get people's insights on a particular topic before launching a larger study (Babbie, 2021).

News programs and entertainment media outlets often provide a toll-free number, a texting number, or a website and encourage viewers to vote on an issue (such as whether marijuana should be legal in all states). How representative are these voters of the general population? And how many enthusiasts skew the results by voting more than once?

But, you might think, if as many as 100,000 people respond, doesn't such a large number indicate what most people think? No. Because the respondents are self-selected and don't comprise a random sample, they're not representative of a population.

A study that looked at data from almost 209,000 people found that drinking one to five cups of coffee a day was associated with lower mortality and lower risks of death from heart and neurological diseases. Although the sample size was huge, it was composed primarily of White medical and health professionals. Thus, the results aren't generalizable to other populations (Ding et al., 2015).

2-3e The Time Dimension

Researchers compare variables in two ways: cross-sectional studies and longitudinal studies. The data can be *longitudinal* (collected at two or more points in time from the same or different samples of respondents) or *cross-sectional* (collected at one point in time). **Figure 2.2** shows the change over time in Americans' attitudes toward same-sex marriage; this is an example of a longitudinal study. If the researchers had collected data at only one point in time (2017, 2000, or 1996), this would have been a cross-sectional study. Cross-sectional studies provide valuable information, but longitudinal studies are especially useful in examining trends in behavior or attitudes; a researcher can compare similar populations across different years or follow a particular group of people over time.

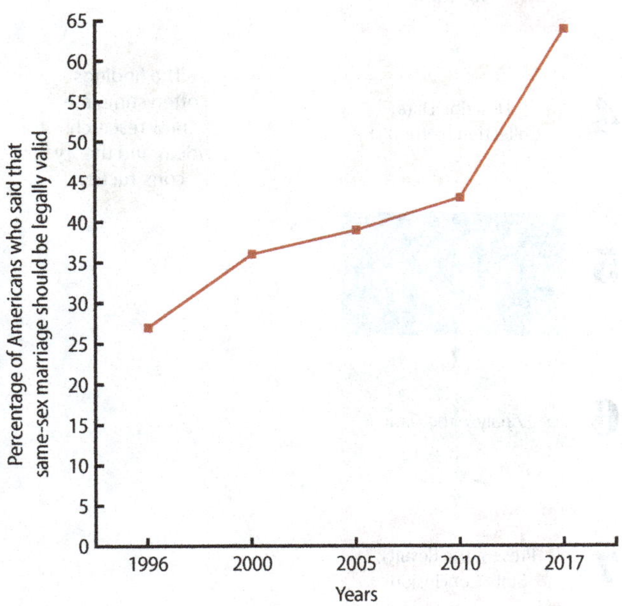

Figure 2.2 Acceptance of Same-Sex Marriage Has Increased

Source: Based on McCarthy, 2017.

2-3f Qualitative and Quantitative Approaches

In **qualitative research**, sociologists examine and interpret nonnumerical material. In a study of mental health supports for veterans, researchers interviewed 30 patients at a Veterans Health Administration medical center who had depression symptoms. The researchers developed a codebook and sorted data into themes. Results showed that veterans felt an important quality of their social supports was a sense of them "being there," which included perceived availability, physical nearness, or frequency or responsiveness of contact (Teo et al., 2020).

In **quantitative research**, sociologists focus on a numerical analysis of people's responses or specific characteristics, studying a wide range of attitudes, behaviors, and traits (e.g., homeowners versus renters). In one survey study of 666 veterans and Canadian Armed Forces (CAF) members who were seeking treatment, information was collected about the relationship between mental health and both perceived and actual levels of social support. Results included that one-third of participants indicated having high levels of social support, participants with less social support had higher symptoms of post-traumatic stress disorder (PTSD), and levels of social support varied by factors such as age and gender (Ketcheson et al., 2018).

Which approach should a researcher use? It depends on their purpose. Consider college attrition. Quantitative data provide information on characteristics such as national college graduation rates. Qualitative data, in contrast, yield in-depth descriptions of why some college students leave college before completing their degrees, while others graduate. In many studies, sociologists use both approaches.

---

qualitative research examines and interprets nonnumerical material.

quantitative research focuses on a numerical analysis of people's responses or specific characteristics.

2-3g Correlation Is Not Causation

Ideally, researchers would like to determine **causation**, a relationship in which one variable is the direct consequence of another. For instance, media headlines proclaimed that, according to a World Health Organization report, eating red meat (e.g., beef, lamb, and pork) and processed meat (e.g., bacon, hot dogs, and ham) causes cancer. The headlines triggered considerable public anxiety. In fact, the study concluded that "high consumption" of red and processed meat "may be associated with a small risk" of several types of cancer (Bouvard et al., 2015).

Because it's difficult to determine causation, sociologists and other scientists calculate a **correlation**, the relationship between two or more variables. For example, much research shows that there's an association between alcohol misuse and domestic violence, and the more frequent the alcohol misuse, the greater the likelihood of domestic violence (refer to Chapter 12).

Alcohol misuse and domestic violence often occur together, but this doesn't mean that one *causes* the other. Domestic violence also occurs when people don't drink alcohol, and not all people who misuse alcohol become violent or aggressive. Instead, there may be other factors that affect alcohol misuse, domestic violence, or both variables (**Figure 2.3**). Sociologists rarely use the term *cause* because they can't *prove* that there's a cause-and-effect relationship. Instead, a researcher might conclude that alcohol misuse is "associated (or correlated) with," "contributes to," or "increases the likelihood of" rather than "causes" domestic violence.

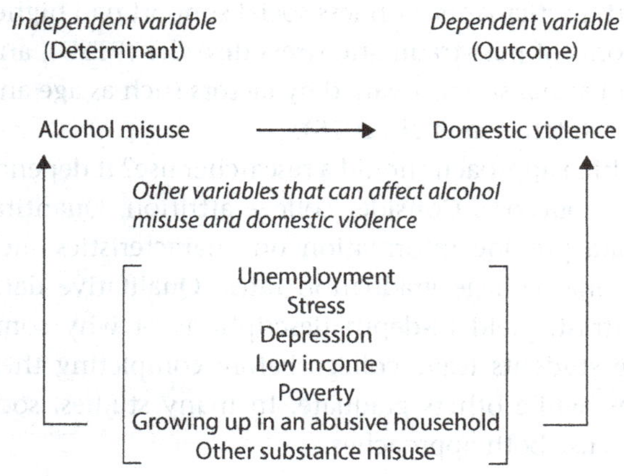

Figure 2.3 An Example of Correlation

causation a relationship in which one variable is the direct consequence of another.

correlation the relationship between two or more variables.

2-4 Basic Steps in the Research Process

Hypothesis construction, deductive and inductive reasoning, establishing reliability and validity, and sampling are some of the preliminary and often most challenging steps in the research process. **Figure 2.4** outlines the scientific method, using a deductive approach that begins with an idea and ends with writing up (and sometimes publishing) the results. Later in this chapter, we'll examine some studies that use an inductive approach.

1. **Choose a topic to study.** The topic can be general or very specific. Some sociologists begin with a new question or idea; others extend or refine previous research findings. A topic can generate new information, replicate a previous study, or propose an intervention (e.g., a new substance treatment program).

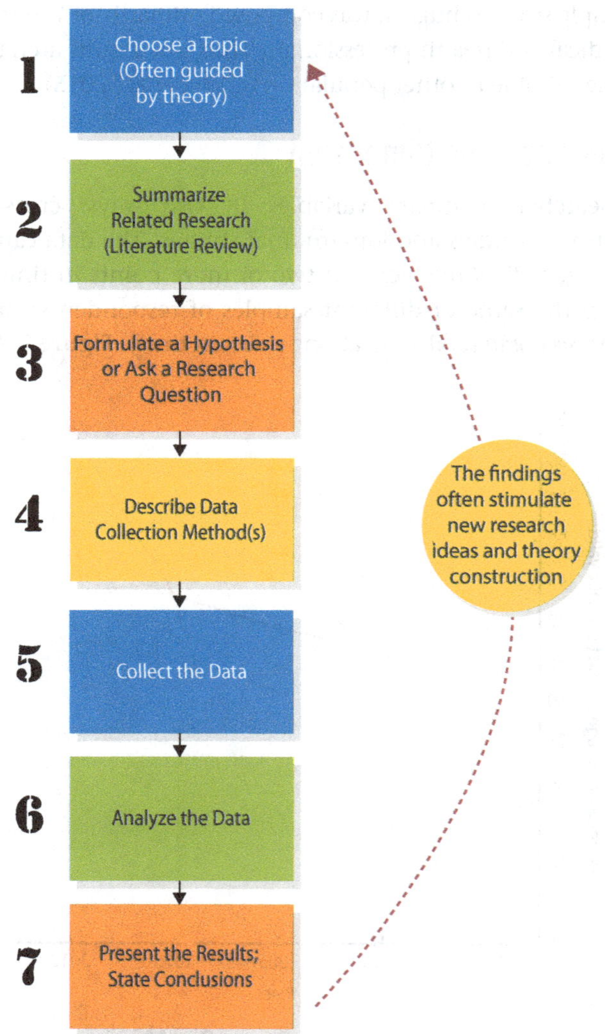

Figure 2.4 Steps in the Scientific Method

Using the Deductive Approach

2. **Summarize the related research.** In what is often called a *literature review*, a sociologist summarizes the pertinent research, shows how their topic is related to previous and current research, and indicates how the study will extend the body of knowledge. If the research is applied, a sociologist also explains how the proposed service or program will improve people's lives.
3. **Formulate a hypothesis or ask a research question.** A sociologist next states a hypothesis or asks a research question. In either case, they have to be sure that the measures of the variables are as reliable and valid as possible.
4. **Describe the data collection method(s).** A sociologist describes which method or combination of methods (sometimes called *methodology, procedure,* or *research design*) is best for testing a hypothesis or answering a research question. This step also describes sampling, the sample size, and the respondents' characteristics.
5. **Collect the data.** The actual data collection might rely on fieldwork, surveys, experiments, or existing sources of information like Census Bureau statistics.
6. **Analyze the data.** After coding (tabulating the results) and running statistical tests, a sociologist presents the findings as clearly as possible.
7. **Present and explain the results.** After analyzing the data, a sociologist explains why the findings are important. This can be done in many ways. They might show how the results provide new information, enrich our understanding of behavior or attitudes that researchers have examined previously, or refine existing theories or research approaches.

In drawing conclusions about the study, sociologists typically discuss its implications. For instance, does a study of juvenile arrests suggest that new policies should be implemented, that existing ones should be changed, or that current police practices may be affecting the arrest rates? That is, the researcher answers the question "So what?" by showing the importance and usefulness of the study.

Farris, Joseph/Cartoonstock.com

"How long have you been dead? Do you have any complaints about your treatment here? Do you have any suggestions?"

2-5 Some Major Data Collection Methods

Sociologists typically use one or more of the following major data collection methods: surveys, field research, content analysis, experiments, or secondary analysis of existing data. Because each method has strengths and weaknesses, researchers must decide which will provide the most accurate information, given time and budget constraints.

2-5a Surveys

Many sociologists use **surveys** that include questionnaires, face-to-face or phone interviews, or a combination of these techniques. Two important elements in survey research are sampling and constructing a series of questions for *respondents*, the people who answer the questions.

Selecting a Sample

Random sample surveys are preferred because the results can be generalized to a larger population. Researchers can obtain representative samples through *random digit dialing*, which involves selecting area codes and exchanges (the next three numbers) followed by four random digits. In the procedure called *computer-assisted telephone interviewing* (CATI), the interviewer uses a computer to select random phone numbers, reads the questions to the respondent from a computer screen, and then enters the answers in precoded spaces, saving time and expense by not having to reenter the data after the interview.

Questionnaires and Interviews

A survey must have a specific plan for asking questions and recording answers. The most common way to do this is to give respondents a **questionnaire**, a series of

survey a data collection method that includes questionnaires, face-to-face or phone interviews, or a combination.

questionnaire a series of written questions that ask for information.

written questions that ask for information. The questions can be *closed-ended* (the researcher provides a list of answers from which a respondent chooses), *open-ended* (the researcher asks respondents to answer questions in their own words) (**Table 2.1**), or a combination. Questionnaires can be mailed, sent electronically (by email), used during an interview, or self-administered (e.g., student course evaluations, Web surveys).

The **interview**, in which a researcher directly asks respondents a series of questions, is another way to collect survey data. Interviews can be conducted face-to-face, by phone, or remotely (e.g., through Zoom). *Structured interviews* use closed-ended questions, while *unstructured interviews* use open-ended questions that allow respondents to answer as they wish.

Online Surveys

Two types of online surveys are becoming increasingly popular. One is sent via email, either as text in the body of the message or as an attachment. The second type is the more familiar survey that's posted on a website. Like the more traditional surveys that use self-administered questionnaires, phones, and face-to-face interviews, online surveys have strengths and weaknesses.

Strengths

Surveys are usually inexpensive, simple to administer, and have a fast turnaround. Self-administered questionnaires can be mailed, or emailed, to thousands of respondents. Having a large number of cases is important in describing, analyzing, and explaining many variables simultaneously.

interview a researcher directly asks respondents a series of questions.

Phone interviews are popular because they're a relatively cost-effective way to collect data. Face-to-face interviews have high response rates (often up to 85 percent) because they involve more personal contact. People are more likely to discuss sensitive issues (e.g., drug use, sexual behavior) in an interview than via a mailed questionnaire, a phone survey, or an electronic survey. If respondents don't understand a question, are reluctant to answer, or give incomplete answers, interviewers can clarify, keep respondents from digressing, or *probe* by asking respondents to elaborate on an answer (Babbie, 2021).

With the innovation of "robo-polls," the entire interview is conducted by a programmed recording that interprets the respondent's spoken answers, records them, and determines how to continue the interview. This method cuts out the cost of hiring people, but it's unclear whether respondents are more reluctant or more willing to express socially unacceptable views to a machine than to a person (Barro, 2015).

Online surveys, like mail surveys, are relatively inexpensive to administer, and the responses can be tabulated quickly. Online surveys are particularly appropriate for certain targeted groups (e.g., people who access health sites), and can increase the pool of people who are increasingly difficult to reach by phone. Online surveys can also provide respondents with visual material, including videos, to look at and respond to (Keeter, 2010; DeSilver, 2014).

Limitations

Unlike questionnaires and phone surveys, face-to-face interviews can be very expensive. A major limitation of surveys that use mailed questionnaires is a low response rate, often only about 10 percent (Higgins, 2015). If the questions are unclear, complicated, or considered to be offensive, respondents may simply throw the

Table 2.1 One Question—Two Formats

Open-Ended Version		
How would you describe your current financial situation?		
Closed-Ended Version		
How would you describe your current financial situation?		
[ ] Excellent	[ ] Fair	[ ] Terrible
[ ] Very good	[ ] Poor	[ ] Not sure
[ ] Good		

Suppose you're the researcher on an income study. Do you think you'd use the open- or closed-ended version? Why?
Now, suppose you're the respondent. Do you think you'd be more likely to answer the open- or closed-ended version? Why?

questionnaire away or give answers on subjects they know nothing about (Babbie, 2021).

A survey's wording can also substantially affect people's answers. According to a Gallup poll, for instance, 51 percent of Americans agreed that doctors should be allowed to "assist the patient to commit suicide," but a whopping 70 percent said that doctors should be allowed to "end the patient's life by some painless means." Thus, not using the word *suicide* was much more acceptable, even though the patient's outcome was the same (Saad, 2013).

Another concern is *social desirability bias*—the tendency of respondents to give the answer they think they "should" give or that will cast them in a favorable light. For example, the proportion of Americans who say they voted in a given election is always much higher than the actual number of votes cast (Taylor and Lopez, 2013).

People also underreport (or lie about) socially undesirable behaviors and traits (e.g., using illicit drugs or experiencing mental health problems), but exaggerate attendance at religious services, donating to charity, and washing their hands after using the restroom. In one survey, 28 percent of the respondents admitted that they had lied to their doctors about smoking, drinking alcohol, using illicit drugs, not exercising, having unsafe sex with multiple partners, and other risky behavior (Reddy, 2013; Bharadwaj et al., 2015; Keeter et al., 2015).

Because most online surveys are based on nonprobability samples, they underrepresent people who fit into one or more of the groups that are less likely to use the Internet. These groups include Black, Latina/o/x, older, lower-income earning individuals and those with lower educational attainment. During the 2016 election, polls underestimated Donald Trump's level of support because, among other reasons, less-educated voters—who comprised a strong pro-Trump segment of the population—refused to respond to online surveys (Kennedy et al., 2016; Mercer et al., 2016).

Because survey research is often used in sociology and many other disciplines, it's important to be an informed consumer. You can't simply assume that a survey, including a public opinion poll, is accurate or representative of a larger population. Asking a few basic questions about a survey, such as those in **Table 2.2**, will help you assess its credibility.

2-5b Field Research

In **field research**, sociologists collect data by systematically observing people in their natural surroundings. In *participant observation*, researchers interact with the people they're studying; they may or may not reveal their identities as researchers. If you recorded interaction patterns between students and professors during your classes, you would be engaging in participant observation. In *nonparticipant observation*, researchers study phenomena without being part of the situation (observing young children in classrooms through one-way mirrors). Researchers sometimes combine participant and nonparticipant observation.

Some field research studies are short term (e.g., observing, over a few weeks or months, whether and how parents discipline their unruly children in grocery stores). Others, called *ethnographies*, require a considerable amount of time in the field. Sudhir Venkatesh (2008), while a graduate student at the University of Chicago, spent more than 6 years studying the culture and members of the Black Kings, a gang in Chicago's inner city that sold crack cocaine and engaged in other illicit activities.

Observational studies are usually highly structured and carefully designed projects in which data are recorded and then converted to quantitative summaries. These studies may examine complex communication patterns, measure the frequency of acts (e.g., the number of head nods or angry statements), or note the duration of a particular behavior (e.g., the length of eye contact) (Stillars, 1991). Thus, observational studies are much more complex and sophisticated than they appear to the general public.

Strengths

Unlike other data collection methods, field research provides rich detail in describing and understanding attitudes and behavior in the "real world." The researcher

Table 2.2 Can I Trust These Numbers?

To determine a survey's credibility, ask:
• Who sponsored the survey? A government agency, a political organization, a business, or a group that's lobbying for change?
• What's the purpose of the survey? To provide objective information, to promote an idea or a political candidate, or to get attention through sensationalism?
• How was the sample drawn? Randomly? Or was it a "self-selected opinion poll" (a SLOP, according to Tanur, 1994)?
• How were the questions worded? Were they clear and objective or biased? If the survey questions aren't provided, why not?
• How did the researchers report their findings? Were they objective, or did they make value judgments?

field research data collected by observing people in their natural surroundings.

bbernard/Shutterstock

Field researchers study a variety of topics, including immersing themselves in high-end dinner and nightclub activities with the global leisure class to understand their consumption practices (refer to Mears, 2020, for a description of this study).

can explore new topics from the respondents' viewpoint using observation or in-depth interviews. Because observation usually doesn't disrupt the natural surroundings, the researcher doesn't directly influence the subjects. Field research is also more flexible than some other methods. The researcher can modify the research design, for example, by deciding to interview (rather than just observe) key people after the research has started.

Validity is usually a strength of field research because researchers can determine how participants define and interpret concepts such as "success," "inequality," or "marital infidelity." Compared with other data collection methods, field research is also more flexible if a situation changes. Studying crime in lower-income neighborhoods, for example, may require shifting the focus of the analysis if the key participants are killed or imprisoned (refer to Goffman, 2014). Finally, field research is a powerful data collection method because it describes a phenomenon using the subjects' own words.

Limitations

Observation can be expensive if a researcher needs elaborate recording equipment, must travel far or often, or has to live in a different society or community for an extended period. Researchers who study other cultures must often learn a new language, which is a time-consuming task.

A second limitation is the problem of *reactivity*: If people know they're being observed, they might behave differently. A study of 392 restaurants from five different chains (e.g., the Olive Garden) found that employee theft decreased by 22 percent after management installed surveillance cameras (Pierce et al., 2013). In the case of participant observation, anything the researcher does or doesn't do can affect what's being observed because the researcher's presence changes a group's dynamics, behavior, and outcomes (refer to Chapter 6 on the Hawthorne effect).

A field researcher may encounter other data collection challenges. Shelters for people who are leaving situations of domestic violence or who are without homes, for example, are usually—and understandably—wary of researchers intruding on their residents' privacy. Even if the researcher has access to such groups, it's often difficult to be objective while collecting and interpreting the data because poverty and domestic violence can evoke strong emotional reactions (e.g., anxiety, anger against perpetrators, sympathy for victims). Finally, the findings can't be generalized because the data come from nonprobability samples.

2-5c Content Analysis

Content analysis systematically examines some form of communication. Researchers can apply this unobtrusive data collection method to almost any form of written or oral communication: speeches, programs, blogs,

AP Images/Mel Evans

Counting people who are unhoused is an ongoing research problem. Communities must provide accurate numbers to qualify for federal funds, but census takers have difficulty getting such counts because safety rules prohibit them from entering private property (such as warehouses) or dark alleys. In one approach, the YWCA in Trenton, New Jersey, gave free haircuts at a fair for unhoused people in the community to attract them to a place where they could be counted (Jonsson, 2007). Is such field research unethical because it violates people's privacy? Or is it innovative in helping a community get more federal funding for those who may benefit from access to housing?

content analysis a data collection method that systematically examines some form of communication.

advertisements, office emails, songs, tweets, advice columns, poems, or social media chatter, to name just a few.

The researcher develops categories for coding the material, sorts and analyzes the data in terms of frequency, intensity, or other characteristics, and draws conclusions about the results. Consider the three baby cards in **Figure 2.5**. What messages do they send about gender roles? These are only three cards, but we could do a content analysis of all birth congratulations at a particular store or from a number of stores or online sites to determine whether such cards reinforce or challenge stereotypical gender role expectations.

Sociologists have used content analysis to examine a number of topics. A few examples are images of women and men in video games and music videos, gender and ethnic differences in yearbook photographs, and gender disparities in media coverage (Downs and Smith, 2010; Zhang et al., 2010; Wallis, 2011; Wondergem and Friedlmeier, 2012; Shor et al., 2019).

Strengths

Content analysis is usually inexpensive and often less time consuming than other data collection methods, particularly field research. If, for example, you wanted to examine the content of commercials during football games, you wouldn't need fancy equipment, a travel budget, or a research staff.

A second advantage is that researchers can recode errors fairly easily. This isn't the case with surveys. If you mail a questionnaire with poorly constructed items, it's too late to change anything.

Third, content analysis is unobtrusive. Because researchers aren't interacting with human subjects, they generally don't need permission to do this type of research, and they don't have to worry about influencing the respondents' attitudes or behavior.

Finally, researchers can use content analysis in both cross-sectional and longitudinal studies. In one study, 400 Instagram posts from female fitness influencers and fitness brands were analyzed from 2019 through 2021. Content was explored for ideal body depiction, fitness focus, objectification, and sexualization (Ahrens et al., 2022).

Limitations

Content analysis can be very labor intensive, especially if a project is ambitious. In one study, the researchers examined the amount and intensity of violence in children's animated movies that were released between 1938 and 1999 (Yokota and Thompson, 2000). It took several years to code one or more of the major characters' words, expressions, and actions.

A second disadvantage is that the coding may be subjective. Having several researchers on a project can increase coding objectivity, but only one researcher often codes the content.

A third limitation is that content analysis often reflects social class bias. Because most books, articles, speeches, films, and so forth are produced by people in upper socioeconomic levels, content analysis rarely captures the behavior or attitudes of working-class people and those with lower socioeconomic status. Even when documents created by lower-income individuals or groups are available, it's difficult to determine whether the coding reflects a researcher's social class prejudices.

Finally, content analyses can't always tell us *why* people behave as they do. We'd have to turn to studies that use different data collection methods—like surveys

Figure 2.5 How Are Views about Gender Roles Represented in Everyday Items?

As these cards illustrate, traditional male and female gender roles and expectations can be reinforced from the moment a baby is born (or even before). Yet, it is becoming more common for these gender binaries to be challenged through items such as gender-neutral baby cards.

An Experiment That Went Awry

In a well-known 1971 experiment, psychologist Philip Zimbardo created a mock prison, using 24 undergraduate volunteers as prisoners and guards. He stopped the experiment after 6 days because some of the participants experienced intense negative reactions. For more information, refer to the Stanford Prison Experiment, prisonexp.org. We'll examine this study, including the ethical issues it raised, in more detail in Chapter 6.

Gary Friedman/Getty Images

and field research—to understand why, for example, people produce video games that portray gender biases in characters' behaviors (refer to Lindner, 2012).

2-5d Experiments

An **experiment** is a controlled artificial situation that allows researchers to manipulate variables and measure the effects. The classic experimental design includes two equal-sized groups that are similar in characteristics such as gender, age, ethnicity, race, and education.

Suppose our hypothesis is "Watching a film on race discrimination reduces prejudice." In the **experimental group**, the participants are exposed to the independent variable (they watch the film on race discrimination). In the **control group**, participants don't watch the film because they're not exposed to the independent variable. Before the experiment, we'll measure the dependent variable (prejudice) in both groups using a *pretest* (a prejudice scale). After the experimental group watches the film, we'll measure both groups again using a *posttest* (the same prejudice scale). If we find a difference in the scores of the dependent variable (prejudice), we might conclude that the independent variable affects the dependent variable (e.g., watching the film on race discrimination reduced or increased prejudice). **Figure 2.6** illustrates this basic experimental design.

Strengths

Experiments come closer than other data collection methods in suggesting a possible cause-and-effect relationship. They're also usually less expensive and time consuming than other data collection techniques (especially large surveys and multiyear field research), and there's often no need to purchase special equipment. A second and related strength is that because researchers recruit students or other volunteers (as in medical studies), participants are usually readily available and don't expect much, if any, monetary compensation.

A third advantage is that experiments can be *replicated* (repeated) many times with different participants. Replication strengthens the researchers' confidence

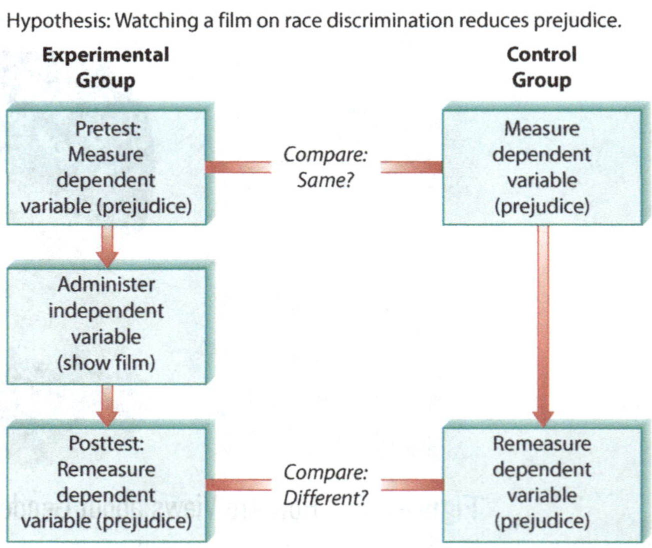

Figure 2.6 Diagram of Basic Experimental Design

Source: Based on Babbie, 2021, Figure 8.1.

experiment a controlled artificial situation that allows researchers to manipulate variables and measure the effects.

experimental group the participants who are exposed to the independent variable.

control group the participants who aren't exposed to the independent variable.

in the validity and reliability of the measures and the study's results. For example, better-designed experiments that replicated earlier studies of the effects of hormone pills on menopausal women found that the pills increased the risk for breast cancer and strokes (Ioannidis, 2005). Now, patients and healthcare providers more carefully weigh the risks and benefits of this form of treatment (Mayo Clinic Staff, 2022).

Limitations

One disadvantage of laboratory experiments is their reliance on student volunteers or paid respondents. Students often feel obligated to participate as part of their grade, or they may fear antagonizing an instructor who's conducting a study. Participants might also give the answers they think the researcher expects (the social desirability bias described earlier). In the case of paid subjects, those who are the busiest, don't need the cash, move, or get sick may not participate fully or stop participating in the study.

Another limitation is that experiments, especially those conducted in laboratories, are artificial. People *know* they're being observed and may behave very differently than they would in a natural setting (Gray et al., 2007; also refer to Chapter 6 on the Hawthorne effect).

Third, the results of experimental studies can't be generalized to a larger population because they come from small, nonrepresentative, or self-selected samples. For example, college students who participate in experiments aren't necessarily representative of other college students, much less of people who aren't in college.

Fourth, experiments aren't suitable for studying large groups of people, a major focus in sociology. As was highlighted in Chapter 1, many sociologists examine macro-level issues (like employment trends) that can't be observed in a laboratory or other experimental setting.

Finally, even if a researcher finds an association between variables (e.g., watching a film reduces prejudice), it doesn't mean that the former "causes" the latter. The control group's responses in the posttest may have been influenced by other factors, including communicating with the members of the experimental group (Cook and Campbell, 1979) or living in a racially diverse neighborhood, which tends to diminish racial and ethnic prejudice (refer to Chapter 10). That is, experiments have few control variables that might explain the relationship between the independent and dependent variables.

Because of such limitations, consumers are often confused and frustrated by conflicting advice. Different experiments have concluded that caffeine can relieve *or* lead to chronic headaches, inhibit *or* improve memory, increase *or* reduce depression, and that some high-fat foods are healthier than low-fat foods (Bartlett, 2011; Mozzafarian and Ludwig, 2015).

2-5e Secondary Analysis

Sociologists rely heavily on **secondary analysis**, an unobtrusive data collection method that examines information collected by someone else. The data may be historical materials (e.g., court proceedings), personal documents (e.g., letters and diaries), public records (e.g., state archives on births, marriages, and deaths), or official statistics (e.g., U.S. Bureau of Labor reports). Researchers with slightly different interests could analyze or reanalyze the data or examine only specific variables. For instance, the federal government compiles information on a large number of variables—including population, age, and education—that social scientists, marketing analysts, health practitioners, and other researchers analyze for their own purposes.

Strengths

Secondary analysis is usually accessible, convenient, and inexpensive. A staggering amount of data on topics like income and employment—for the United States and worldwide—are readily available at libraries, colleges, and online.

Second, there's a huge savings in time. Because many of the data sets are stored electronically, researchers can spend most of their time analyzing, rather than collecting, the data.

A third advantage of secondary analysis is the data's high quality. The federal government and nationally known survey organizations have budgets and well-trained staff equipped to address any data collection problems. Because the samples are representative of national populations, researchers can be more confident about generalizing the findings.

Fourth, data sets often contain hundreds of variables. Thus, researchers can access national, international, and historical information that they themselves can't collect (Boslaugh, 2007). Finally, because many of the data sets are longitudinal, researchers can look at trends and changes over time.

Limitations

Secondary analysis has several drawbacks. First, it may be challenging to access historical materials because the documents are deteriorating, housed in only a few libraries,

secondary analysis examination of data that have been collected by someone else.

or part of private collections. It may also be difficult to determine the accuracy and authenticity of historical data.

Second, existing sources may not have the exact data a researcher needs. In 2000, the Census Bureau changed how it counted race and ethnicity to include the growing numbers of recent immigrants and multiracial Americans. Doing so increased the accuracy of gauging racial and ethnic diversity but created 63 possible racial-ethnic combinations. As a result, today's numbers are different from those before 2000, making comparisons on race and ethnicity across time problematic (Saulny, 2011; also refer to Pratt et al., 2015). Changes to the U.S. census continue to occur; most recently, major changes were proposed for the 2030 census that would transform how people identifying as Middle Eastern or North African as well as those of Latina/o/x descent may be counted (Wang, 2023).

A third and related limitation is that even the same source may modify its measures of variables because of legal and social changes. Regarding marital and family status, for example, the Census Bureau added "separated" only in 1950, "unmarried partner" in 1990, and started categorizing same-sex married couples as families only in 2013 (Bates and DeMaio, 2013; Morello, 2014). Such changes are necessary. They can be frustrating, however, if, for instance, you wanted to compare the number of separated Americans between 1940 and 2015 or determine the number of same-sex married couples since 2004, when Massachusetts became the first state to legalize same-sex marriage.

Finally, because the researcher didn't collect the data, they have no control over the specific variables used or how they were measured, and they may not know how seriously problems such as low response rates, validity, and reliability affect the data. Most federal and other established data sets provide such information, but researchers can't simply assume that all data sets are problem-free.

Table 2.3 summarizes the data collection methods described in this chapter. Often, researchers use a combination of methods because "the social world is a multifaceted and multi-layered reality that reveals itself only in part with any single method" (Jacobs, 2005, p. 4). You should also be familiar with evaluation research, an increasingly important research tool.

2-5f Evaluation Research: Pulling It All Together

Evaluation research examines whether a social intervention has produced the intended result. *Social interventions* are typically programs and strategies that seek to prevent or change negative outcomes (e.g., teenage pregnancy or substance misuse).

Evaluation research "refers to a research purpose rather than a specific research method" (Babbie, 2021, p. 356). Thus, sociologists use a variety of data collection methods, such as secondary analysis and content analysis to examine an organization's records, surveys to gauge employee and client satisfaction, and interviews with the staff and program recipients for in-depth information on an organization's processes.

Evaluation research is *applied* because it's intended to have some real-world effect (Weiss, 1998). Examples include *needs assessment studies* (e.g., to improve or initiate a new service like an after-school program), *cost-benefit studies* (e.g., to determine whether funding teen abstinence programs reduces pregnancies), and *monitoring studies* (e.g., to examine the effect of new policing policies on crime rates). The findings are generally used to improve the efficiency and effectiveness of a particular policy or group.

Strengths

Evaluation research is important for several reasons. First, because the use of available funds is a topic for careful assessment, and evaluation research can help policymakers make decisions about how funds are allocated. Results from evaluation research can help identify which programs show promise and could be expanded and which initiatives aren't working as well and should be scaled back or replaced (Pew-MacArthur Research First Initiative, 2018)

Second, evaluation research is versatile because it includes qualitative and quantitative approaches. It can address almost any topic, such as the effectiveness of driver education programs, foreign aid policies, job training programs, and premarital counseling, to name just a few. The research costs can also be low if, for example, researchers can use secondary analysis rather than collect new data.

Third, evaluation research addresses real-life problems that confront many families and communities. Because state and local resources are shrinking and there are hundreds of intervention strategies, evaluation research can be valuable to program directors or agency heads in deciding which programs to keep, improve, or cut (Peterson et al., 1994; Hattry, 2020).

Limitations

Evaluations can usually address only one or a few of the many factors that affect behavior. For instance,

evaluation research examines whether a social intervention has produced the intended result.

Table 2.3 Some Data Collection Methods in Sociological Research

Method	Example	Advantages	Disadvantages
Surveys	Sending questionnaires and/or interviewing students on why they succeeded in college or dropped out	Questionnaires are fairly inexpensive and simple to administer; interviews have high response rates; findings are often generalizable	Mailed questionnaires may have low response rates; respondents tend to be self-selected; interviews are usually expensive
Secondary analysis	Using data from the National Center for Education Statistics (or similar organizations) to examine why students leave college before completing their degrees	Usually accessible, convenient, and inexpensive; often longitudinal and historical	Information may be incomplete; some documents may be inaccessible; some data can't be collected over time
Field research	Observing classroom participation and other activities of first-year college students with high and low grade-point averages (GPAs)	Flexible; offers deeper understanding of social behavior; usually inexpensive	Difficult to quantify and to maintain observer/subject boundaries; the observer may be biased or judgmental; findings are not generalizable
Content analysis	Comparing the transcripts of college graduates and those leaving before completing their degree on variables such as gender, race/ethnicity, and social class	Usually inexpensive; can recode errors easily; unobtrusive; permits comparisons over time	Can be labor intensive; coding is often subjective (and may be distorted); may reflect social class bias
Experiments	Providing tutors to some students with low GPAs to find out if such resources increase college graduation rates	Usually inexpensive; plentiful supply of subjects; can be replicated	Subjects aren't representative of a larger population; the laboratory setting is artificial; findings can't be generalized

adolescent substance misuse is not only due to individual high-risk decisions but also to multiple interrelated variables that include peers, school policies, and cultural values (Bandy, 2012).

Another limitation is that the social context affects evaluation research because of politics, vested interests, and conflicts of interest. Groups that fund the evaluation may pressure researchers to present only positive findings. Also, the results may not be well received if they contradict deeply held beliefs, challenge politicians' favored projects, or conflict with official points of view (Olson, 2010; Babbie, 2021).

Consider the popular DARE (Drug Abuse Resistance Education) program, introduced in 1983 to teach elementary school children about the dangers of drug use. Social scientists evaluated DARE in 1994, 1998, and 1999 and found no significant difference in the incidence of drug use among students who had and hadn't completed the DARE curriculum. When DARE funding was threatened, its promoters dismissed the research, referring to it in pejorative language, suggesting that it was false science (Miller, 2001).

PJF Military Collection / Alamy Stock Photo

DARE programs have been popular with schools, police departments, parents, and politicians across the country, even though evaluation research has found that the curriculum doesn't reduce students' drug use.

2-6 Ethics and Social Research

Researchers today operate under much stricter guidelines than they did in the past. In conducting their research, what ethical and political dilemmas do sociologists (and other social scientists) encounter?

2-6a What Is Ethical Research?

Any agency (such as a university or hospital) wishing to receive federal research support must establish an institutional review board (IRB)—a panel of individuals with professional experience (such as faculty) who can oversee necessary scientific and ethical review. The charge of the IRB is to review all proposals involving human subjects. The chief responsibility of an IRB is to ensure that the subjects' rights will be protected (Office for Human Research Protections, 2016).

Because so much research relies on human participants, the federal government, IRBs, and many professional organizations have formulated codes of ethics to protect participants. Regardless of the discipline or the research methods used, all ethical standards have at least three golden rules:

- First, *do no harm* by causing participants physical, psychological, or emotional pain.
- Second, the researcher must get the participant's *informed consent* to be in a study. This includes the participant knowing what the study is about and how the results will be used.
- Third, researchers must always protect a participant's *confidentiality,* even if the participant has broken a law that they disclose to the researcher. **Table 2.4** lists some of the basic ethical principles in sociological research.

2-6b Scientific Misconduct

Ethics violations are most common when the research is funded by for-profit companies and organizations. Medical researchers, particularly, have been accused of considerable scientific dishonesty. Some of the alleged violations have included the following: changing research results to please the corporation (such as tobacco or pharmaceutical companies) that sponsored the research; being paid by companies to deliver speeches to health practitioners that endorse specific drugs, even if the medications don't reduce health problems; allowing drug manufacturers to ghostwrite articles that are published in prestigious medical journals (and even draft textbooks); and falsifying data. A study of more than 2,000 articles that were published in some of the most prestigious medical journals found that almost 68 percent were withdrawn because of misconduct and fraud rather than errors (Project on Government Oversight, 2010; Fang et al., 2012; McNeill, 2012). Yet, paper retractions are fairly rare (only about four out of every 10,000) and retractions may be the result of a positive shift—greater oversight at many journals (Brainard and You, 2018).

Table 2.4 Some Basic Principles of Ethical Sociological Research

In its most recent update, the American Sociological Association (1999) has reinforced its previous ethical codes and guidelines for researchers. Researchers must:
• Obtain all subjects' consent to participate in the research and their permission to quote from their responses and comments.
• Not exploit subjects or research assistants involved in the research for personal gain.
• Never harm, humiliate, abuse, or coerce the participants in their studies, either physically or psychologically.
• Honor all guarantees to participants of privacy, anonymity, and confidentiality.
• Use the highest methodological standards and be as accurate as possible.
• Describe the limitations and shortcomings of the research in their published reports.
• Identify the sponsors who funded the research.
• Acknowledge the contributions of research assistants (usually underpaid and overworked graduate students) who participate in the research project.

In the social sciences, some data collection methods are more prone to ethical breaches than others. Surveys, secondary analysis, and content analysis are less vulnerable to these issues because the researchers typically don't interact directly with or affect participants. In contrast, experiments and field research can raise ethical questions because of deception or influencing the participants' feelings, attitudes, or behavior.

2-6c External Pressures on Researchers

Social science research is likely to be challenged when it focuses on sensitive or controversial issues. Federal agencies have avoided funding nearly all research on gun violence partly in response to powerful lobbyists, like the National Rifle Association, that pressure Congress to prevent any studies that deal with firearms (Kelderman, 2016; Raphelson, 2018).

Some religious groups and school administrators have successfully derailed research on adolescent sexual behavior because it undermines "traditional family values" and might make a school district look bad (e.g., if a study reports a high incidence of drug use or sexual activity) (Kempner et al., 2005).

Reviewers of scientific publications can also play a role in limiting whether research is shared. For instance, research on Asian Americans and Pacific Islanders (AAPIs) has been stifled by journal reviewers who suggest such things as, "the struggles of Asian Americans were 'not as bad as for Blacks and Latinos.'" Unfortunately, without research on these groups, important social problems, such as health disparities, poverty, and crime, go unacknowledged, unstudied, and have less chances of being addressed (Yee, 2021).

In the first days of the Trump administration in 2017, the White House "stifled" federal scientists who were working on projects (e.g., pollution, food safety, climate) that some of the most powerful corporations opposed (Jackson, 2017; Mann, 2017). The CEO of a powerful energy company wanted the University of Oklahoma to dismiss scientists who found a possible link between fracking for oil and gas and earthquakes and offered to sit on a search committee to fill the vacancies. The CEO is like many other politicians and corporate leaders "who embrace science when it's convenient but work to undermine, discredit, and inhibit science when it conflicts with their narrow self-interests" (Ganem, 2016, p. 25).

2-6d Do People Believe Scientific Findings?

Not always. For example, 55 percent of Gallup poll respondents indicated that humans evolved over time while 40 percent indicated humans were created in their present form, and 5 percent had no opinion (Brenan, 2019). And although there is evidence that the Earth is round, 2 percent of Americans indicate that they have always believed that the Earth is flat (Nace, 2018).

Why do many people reject scientific findings? One reason may be that when social media publicizes unscientific polls and falsehoods that people already believe, they're unlikely to change their minds about an issue, even after being presented with scientific data (Taub and Nyhan, 2017).

Second, research findings often challenge personal attitudes and beliefs that people cherish. For example, "About one-in-ten Americans say climate change is not serious because 'God is in control of the climate'" (Alper, 2022).

Third, an explosion of information can create confusion, especially when online searches oversimplify complex phenomena or present misleading or unreliable content. Average users "don't know how to work through all the information [and misinformation] that's thrown at them on a daily basis" (Dreid, 2016, p. A24).

Finally, many scientists, including social scientists, spend little or no time engaging in policy debates or explaining their research to the public and journalists (Basken, 2016). As a result, "lobbyists and activists can promote their ideological agendas . . . over those of good science and public policy" (Lynas, 2015).

Rzucamokiem/Shutterstock.com

Wong Sze Yuen/Shutterstock.com

Fisher-Price and other toy companies claim that, based on their research, their iPhone and iPad apps can teach infants as young as 9 months old to read and count (Singer, 2013). In fact, studies have found that such products do more harm than good because infants who watch the apps learn fewer words than those whose parents talk or read to them (refer to Chapter 12).

Chapter Review

Key Takeaways

2-1 **How do we know what we know?** Much of our knowledge is based on tradition and authority. In contrast, sociologists rely on *research methods* to get information about a particular topic.

2-2 **Why is sociological research important in our everyday lives?** In contrast to tradition and authority, sociological research challenges overgeneralizations, exposes myths, helps explain why people behave as they do, influences social policies, and sharpens critical thinking skills about issues that affect our everyday lives.

2-3 **What are the components of the scientific method?** The *scientific method* incorporates careful data collection, exact measurement, accurate recording and analysis of findings, thoughtful interpretation of results, and, when appropriate, a generalization of the findings to a larger group. A research question or a *hypothesis* examines the association between an *independent variable* and a *dependent variable*. Sociologists use both *qualitative* and *quantitative* approaches and are always concerned about the *reliability* and *validity* of their measures.

2-5 **What are the major data collection methods used in sociology?** Five data collection methods are especially useful in sociology (refer to **Table 2.3** on page 35). In designing their studies, sociologists weigh the advantages and limitations of each data collection method. *Evaluation research*, which uses one or more data collection methods, examines whether a social intervention has produced the intended result. Because sociologists don't conduct research in a cultural vacuum, many groups use the findings to change current policies and practices.

Example: The Uneasy Relationship between Research and Practice Many supporters of the DARE (Drug Abuse Resistance Education) program were unhappy when more than 30 research studies showed that DARE had negligible long-term effects in reducing teen drug use. About 75 percent of U.S. school districts revised the DARE curriculum but continued to use it because they believed that the program built a positive relationship between police, students, parents, and educators (refer to the text).

2-6 **What are ethical considerations of social research?** Sociological research demands a strict code of ethics to avoid mistreating participants. For example, participants must give informed consent and must not be harmed, humiliated, mistreated, or coerced; researchers must honor their guarantees of privacy, confidentiality, and/or anonymity. Still, sociologists often encounter pressure from policymakers and others to limit their research to non-controversial topics.

Key Terms

Research methods (p. 21)
Scientific method (p. 22)
Concept (p. 23)
Variable (p. 23)
Independent variable (p. 23)
Dependent variable (p. 23)
Control variable (p. 23)
Hypothesis (p. 23)
Deductive reasoning (p. 23)
Inductive reasoning (p. 23)
Reliability (p. 24)
Validity (p. 24)
Population (p. 24)
Sample (p. 24)
Probability sample (p. 24)
Nonprobability sample (p. 24)
Qualitative research (p. 25)
Quantitative research (p. 25)
Causation (p. 26)
Correlation (p. 26)
Survey (p. 27)
Questionnaire (p. 27)
Interview (p. 28)
Field research (p. 29)

Content analysis (p. 30)
Experiment (p. 32)
Experimental group (p. 32)
Control group (p. 32)
Secondary analysis (p. 33)
Evaluation research (p. 34)

Discussion and Critical Thinking Questions

1. In a national study of 25,000 first-year college students, many reported their ACT and SAT scores incorrectly—most often by inflating them—because of faulty recall, to give a good impression, or to enhance their self-esteem. In another study of 3,000 students at 19 colleges, many of the students' glowing self-reports didn't correspond with objective findings (Schmidt, 2008).
   a. What are the implications of such findings for higher education in assessing student performance?
   b. How do these findings call into question the results of many national studies, such as the National Survey of Student Engagement, that rely on self-reported data?
2. On a number of websites (e.g., pickaprof.com, ratemyprofessors.com), students can say anything they want, anonymously, about faculty. Are the comments representative of all the students in a course? Would it be ethical for faculty to set up similar public sites and evaluate students by name?
3. Americans generally trust scientific data (e.g., on climate change, nuclear power, gun control) when the findings are consistent with their own values and beliefs and not when the findings threaten them (Kahan et al., 2011). Would an understanding of the scientific method decrease people's distrust of research results with which they disagree? How does believing or not believing scientific findings affect social policy?
4. A humanities faculty member I knew was fed up with students texting during his classes. He considered changing their behavior by falsely telling students that his research from the previous semester showed that the students who texted earned the lowest grades. Should he have done so to discourage texting during class?

True or False?

How Much Do You Know about Examining Our Social World?

The answers to 2, 5, and 6 are true. The answers to 1, 3, and 4 are false.

1. Knowledge based on sociological research serves different purposes than knowledge based on tradition or authority. Knowledge generated from sociological research counteracts misinformation, exposes myths, helps explain *why* people behave as they do, affects social policies, and sharpens critical thinking skills.
2. In their work, sociologists use the scientific method for research which includes careful data collection, exact measurement, accurate recording and analysis of the findings, thoughtful interpretation of the results, and, sometimes, generalization of the findings to a larger group. The scientific method involves objective and systematic techniques to investigate phenomena, acquire knowledge, and test hypotheses and theories.
3. Sociologists can use either qualitative (nonnumerical) or quantitative (numerical) research approaches. Some sociologists use both.
4. Sociologists may use surveys which are usually inexpensive, simple to administer, and have a fast turnaround.
5. Sociologists may do field research observing people in their natural surroundings. This can include participant observation involving researchers interacting with the people they're studying or nonparticipant observation involving researchers studying phenomena without being part of the situation.
6. To better address concerns of research, researchers today must follow stricter guidelines for their work. For instance, Institutional Review Boards (IRBs) review all research proposals to ensure that human subjects are protected.

3 Culture

Learning Objectives

After studying this chapter, you will be able to...

3-1 Describe the characteristics of culture.

3-2 Explain the significance of symbols, language, values, and norms.

3-3 Discuss cultural similarities.

3-4 Discuss cultural variations.

3-5 Differentiate between high culture and popular culture.

3-6 Explain how and why technology affects cultural change.

3-7 Compare and evaluate the theoretical explanations of culture.

True or False?

How Much Do You Know about Culture?

1. Culture influences very limited areas of a person's life.
2. Rules of behavior, such as raising one's hand in the classroom or sitting in one's assigned seat in an airplane, are examples of nonmaterial culture.
3. Ethnocentrism is the belief that no one culture is better than another and each should be judged by their own standards.
4. There is cultural variation *across* societies but not *within* the same society.
5. Mass media is a form of culture that has enormous power in shaping public attitudes and behavior.
6. When new forms of material culture (such as new technology) emerge, they always have the necessary support from nonmaterial culture (such as norms and laws).

The answers are at the end of the chapter.

Once, when I returned a set of exams, a student who was unhappy with his grade blurted out an obscenity. A voice from the back of the classroom snapped, "You ain't got no culture, man!" Is it true that people who use vulgar language "ain't got no culture"?

Be sure to take the True or False quiz in the chapter opener to better understand how much you know about these topics.

What do you think?

I shape my culture; my culture doesn't shape me.

1	2	3	4	5	6	7
	strongly agree				strongly disagree	

3-1 Culture and Society

As popularly used, *culture* often suggests an appreciation of the finer things in life, such as Shakespeare's sonnets, the opera, and using civil language. In contrast, sociologists use the term in a much broader sense. **Culture** refers to the learned and shared behaviors, beliefs, attitudes, values, and material objects that characterize a particular group or society. Thus, culture shapes people's total way of life.

Most human behavior isn't random or haphazard. Among other things, culture influences what you eat; how you were raised and may raise children if you have them; if, when, and whom you will marry; how you make and spend money; and what you read. Even people who pride themselves on their individualism conform to most cultural rules. The next time you're in an American classroom, for example, count how many students are *not* wearing jeans, T-shirts, sweatshirts, or sneakers—clothes that are the prevalent choices of adolescents, college students, and many adults in Western cultures. How might this be different in other cultures?

A **society** is a group of people who share a culture and a defined territory. Society and culture go hand in hand; neither can exist without the other. Because of this interdependence, social scientists sometimes use the terms *culture* and *society* interchangeably.

3-1a Some Characteristics of Culture

All societies, despite their diversity, share some cultural characteristics and functions (Murdock, 1940; refer to Aguilar and Stokes, 1995). We don't observe culture directly, but it shapes our attitudes and behaviors.

1. **Culture is learned.** Culture isn't innate but learned, and it affects how people think, feel, and behave. If a child is born in one region of the world but raised in another, they will learn the customs, attitudes, and beliefs of the host culture.
2. **Culture is transmitted from one generation to the next.** People learn many customs, habits, and attitudes informally through parents, relatives, friends, and the media. People also learn culture formally in settings such as schools, workplaces, and community organizations. Whether learning is formal or informal, culture is cumulative because each generation transmits cultural information to the next one.
3. **Culture is shared.** Culture brings members of a society together. People have a sense of belonging because they share similar beliefs, values, and attitudes about what's right and wrong. Imagine the chaos if each person did what they wanted (e.g., physically assaulting an annoying neighbor) or if people couldn't make numerous daily assumptions about other people's behavior (e.g., stopping at red traffic lights).
4. **Culture is adaptive and always changing.** Culture changes over time. New generations discard technological aspects of culture that are no longer practical, such as replacing typewriters with personal computers and landlines with cell phones. Attitudes can also change over time. Compared with several generations ago, for example, many Americans now say that nonmarital sex is acceptable (**Figure 3.1**).

culture the learned and shared behaviors, beliefs, attitudes, values, and material objects that characterize a particular group or society.

society a group of people who share a culture and defined territory.

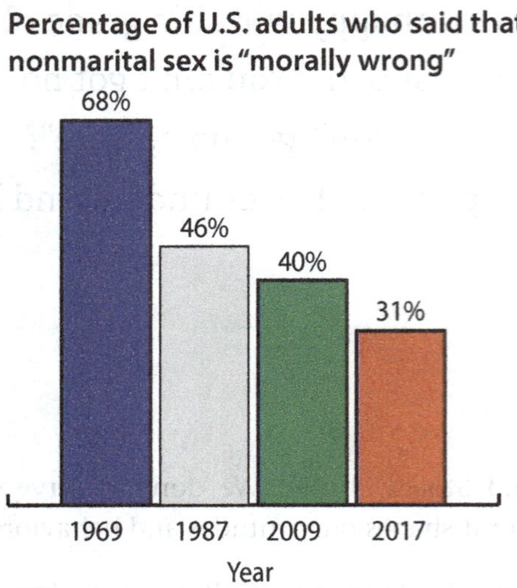

Figure 3.1 Is Nonmarital Sex Wrong?

Sources: Based on Saad, 2001; Jones, 2017.

Culture shapes who we are, but remember that it's people who create culture. As a result, culture changes as people adapt to their surroundings. Since the attacks on September 11, 2001, most people worldwide, including Americans, don't complain at airports when required to pass through a metal detector or full body scanner or have baggage X-rayed or searched. Regardless of the extra steps, 79 percent of Americans feel screening for security threats at airports should be prioritized over convenience for travelers (Orth, 2022). Thus, as the dangers in society have grown, people have passed laws or implemented rules that have increased their security but decreased their privacy.

material culture the physical objects that people make, use, and share.

nonmaterial culture the ideas that people create to interpret and understand the world.

3-1b Material and Nonmaterial Culture

Cultures that people construct are both material and nonmaterial (Ogburn, 1922). **Material culture** consists of the physical objects that people make, use, and share. These objects include buildings, furniture, music, weapons, jewelry, hairstyles, and the Internet. **Nonmaterial culture** consists of the ideas that people create to interpret and understand the world. Beliefs about the supernatural, customs, and rules of behavior are examples of nonmaterial culture.

Because material and nonmaterial culture are created by people, physical objects and ideas can vary considerably across societies (e.g., the farming tools people create, how they form governments, or how they raise their children). Frequently, as you'll read shortly, there's a "cultural lag" because material culture (invention of the automobile and the cell phone) changes faster than nonmaterial culture (laws about pollution and texting while driving).

3-2 The Building Blocks of Culture

Seventy-four-year-old James Davis granted his dying wife's wish to be buried in their front yard. The Alabama Supreme Court agreed with a lower court judge that

Communication symbols vary across societies and change over time. The earliest known map of the world (700–500 B.C.E.) used cuneiform script; modern tablets are loaded with icons that provide a variety of information, including maps.

family members can't bury people anywhere they like. Davis refused to budge because the town doesn't have an ordinance governing burials, and his family members and neighbors supported his decision. After four years of court hearings and appeals, a city crew exhumed the body (Reeves, 2013).

Who's right—Davis, his family and neighbors, or the judges? To answer such questions, we must understand the building blocks of culture, particularly symbols, language, values, norms, and rituals.

3-2a Symbols

A **symbol** is anything that stands for something else and has a particular meaning for people who share a culture. In most societies, for example, a handshake communicates friendship or courtesy, a wedding ring signals that a person isn't a potential dating partner, and a siren denotes an emergency. People influence each other through the use of symbols. A smile and a frown communicate different information and elicit different responses. Through symbols, we engage in *symbolic interaction* (refer to Chapter 1).

Symbols Take Many Forms

Written words are the most common symbols, but we also communicate by tattooing our bodies or not, celebrating or disregarding wedding anniversaries, and purchasing or avoiding goods and services that indicate our social status. Gestures (e.g., raised fists, hugs, and stares) convey important messages about people's feelings and attitudes.

Tim Macpherson/Stone/Getty Images

The U.S. Flag Code describes the flag as a "living thing" and forbids its display on clothing, bedding, drapery, or in advertising (Luckey, 2008). Is this man disrespecting a cherished symbol? Or showing his patriotism?

Symbols Distinguish One Culture from Another

Every nation's flag, including its colors and emblems, is an important symbol of national identity. A third of the world's flags have religious symbols (Theodorou, 2014). Countries usually create or change their flags after becoming independent to create and reinforce national consciousness and pride.

In specific societies, many women wear a head scarf (*hijab* or *hejab*) that covers their hair and neck, a veil that hides part of the face from just below the eyes (*nijab*), or a garment that covers the entire body (*chador* or *burqa*). In Iran and Afghanistan, a headcovering is required for women, while in other countries with large numbers of Islamic followers, such as Morocco, it is an optional cultural practice. Whatever form it takes, the veil is more than a piece of clothing. It means different things to people across and within cultures.

Symbols Can Unify or Divide a Society

Symbols unify people. The American flag is an important symbol within American society. About 50 percent of Americans say that they have a flag in their home (Jaimungal, 2020). Every Fourth of July, many Americans celebrate independence from Britain with parades, firecrackers, barbecues, and speeches by local and national politicians. Because such behavior signifies freedom and democracy, some immigrants purposely choose July 4th to be naturalized.

Naturalization ceremonies themselves teem with symbolic objects and behavior—flags, national anthems, oaths and pledges of allegiance, uniformed military members, and photographs of past immigrants. Such images and rituals "work together to elicit an emotional sense of belonging, pride, and patriotism" that reinforce a sense of community (Aptekar, 2016, p. 49).

Symbols can also be divisive. In 2020, the Mississippi state legislature voted to replace its state flag; it was the last state in the United States to display the Confederate symbol (Smith, 2020). The Confederate flag represents a time in U.S. history when White people oppressed and enslaved Black people. Protesters who think the Confederate flag is a proud emblem of Southern heritage have denounced decisions to remove this flag (Stewart, 2021).

---

symbol anything that stands for something else and has a particular meaning for people who share a culture.

Symbols Can Change Over Time

Societies create new symbols all the time. There are now recycling symbols on many everyday items, restroom symbols that include transgender people, red ribbons that communicate support for people with HIV and AIDS, ever-changing emoticons that express a person's feelings or mood, and hundreds of text and online chat abbreviations (e.g., afaik, "as far as I know") that vary across groups depending on the user's age, occupation, and other characteristics.

Symbols communicate information that varies across societies and may change over time. In 1986, the International Red Cross changed its name to the International Red Cross and the Red Crescent Movement to encompass a number of Arab branches and adopted a crescent emblem in addition to the well-known cross. Israel wanted to use a red Star of David, rejecting the cross as a Christian symbol and the crescent as Islamic. Red Cross officials offered a red diamond as the new shape, but some countries rejected the diamond because it represents bloody conflicts in many African countries that mine diamonds (Whitelaw, 2000). As of now, the Red Cross symbols include a red cross, red crescent, and red crystal (diamond) (The American National Red Cross, 2020).

The meaning of even simple symbols varies across groups and can change over time. The # is commonly called the pound and number sign (#2 pencil). For musicians, # denotes a note that's sharp. For proofreaders, # means "insert a space." The # is also common on phone pads, and, since the ascent of social media, # is a hashtag used to mark keywords or topics in a tweet (#culture).

3-2b Language

Perhaps the most powerful of all human symbols is **language**, a system of shared symbols that enables people to communicate with one another. Language is a human invention that communities of people have endowed with meaning. In every society, children begin to grasp the essential structure of their language at a very early age and without any instruction. Babbling rapidly leads to uttering words and combinations of words that represent an idea, feeling, or physical object.

Anthropologists Edward Sapir (1929) and Benjamin Whorf (1956) theorized that thoughts and behavior are determined by language. Their theory, known as the *Sapir-Whorf hypothesis,* posited that language provides people with a framework for interpreting social reality and the world around them. Thus, as languages vary, so do interpretations of social reality.

AP Images/Keystone/Salvatore Di Nolfi

FABRICE COFFRINI/Getty Images

The cross, crescent, and now crystal are the three symbols of the International Movement of the Red Cross and the Red Crescent.

Hallmark Cards, Inc.

Every year, Hallmark produces "Keepsake" Christmas ornaments and apparel, many of which are collectibles. In 2013, one of the new pieces paid tribute to "Deck the Halls," a traditional Christmas carol. Hallmark changed the lyrics from "Don we now our gay apparel" to "Don we now our fun apparel" because the meaning of gay has changed since the song's publication in the 1800s (Wong, 2013). Because of some negative reactions, Hallmark apologized for the change (Murphy, 2013). Why do you think one word generated so much controversy?

language a system of shared symbols that enables people to communicate with one another.

Subsequent research found little support for the Sapir-Whorf hypothesis. Some critics maintained that language affects but does not "determine" culture. Others argued that language influences some thinking processes, but not all (e.g., Berlin and Kay, 1969; Heider and Olivier, 1972; Schlesinger, 1991). Still, the Sapir-Whorf hypothesis has generated considerable research across several disciplines, including sociology, in exploring the relationship between language, thought, and culture.

Why Language Is Important

Language makes us human: It conveys ideas, transmits information, and influences people's attitudes and behavior. Language also directs our thinking, controls our actions, and shapes our expression of emotions. The meanings of words change over time, but some changes are more controversial than others. Both birds and humans now tweet. Spam is a canned meat product and also an unsolicited electronic message. Hallmark generated considerable controversy, however, when it changed "gay" to "fun" on a line of Christmas decorations and accessories, including a sweater (refer to photo on page 44).

In a survey of 14 countries, including the United States, large majorities of people said that speaking the country's language is very important to being "one of us." Language was far more important in being "part of the nation" than birthplace, religious affiliation, education, clothing preferences, and celebrating the host country's holidays (Stokes, 2017). For many nations, then, language is the cornerstone of acceptance into a group.

Language and Gender

Many professors routinely use phrases such as "Okay, guys, in class today..." and no one objects. One of my colleagues illustrates the linkage between language and gender and how it affects our thinking by saying "Okay, gals, in class today...." "I always get a reaction of gaping mouths, laughs, and bewildered looks," he says. The students react differently to "guys" and "gals" because we've internalized male terms (e.g., *guys*, *policeman*, and *maintenance man*) as "normal" and acceptable.

Language has a profound influence on how people think about and act toward people. Those who adhere to traditional language contend that nouns such as *fireman*, *chairman*, *mailman*, and *mankind* and pronouns like *he* refer to both women and men and that women who object to such usage are too sensitive. Suppose, however, that all of your professors used only *she*, *her*, and *women* when referring to all people. Would the men in the class feel excluded? One solution is to use sentences

Wanted: Someone Who Can Translate Federal Gibberish into Plain English

It's almost impossible to find a U.S. government document that people can understand. In 2010, President Obama signed the Plain Writing Act, which requires federal documents to be rewritten in plain English. Here's a "before" and "after" example from the Department of Health and Human Services Handbook that assures health care for low-income HIV/AIDS patients:

Before: Title I of the CARE Act creates a program of formula and supplemental competitive grants to help metropolitan areas with 2,000 or more reported AIDS cases meet emergency care needs of low-income HIV patients. Title II of the Ryan White Act provides formula grants to States and territories for operation of HIV service consortia in the localities most affected by the epidemic, provision of home and community-based care, continuation of insurance coverage for persons with HIV infection, and treatments that prolong life and prevent serious deterioration of health. Up to 10 percent of the funds for this program can be used to support Special Projects of National Significance.

After: Low-income people living with HIV/AIDS gain, literally, years through the advanced drug treatments and ongoing care supported by HRSA's Ryan White Comprehensive AIDS Resources Emergency (CARE) Act.

Language is supposed to help people communicate. Does a government have more control if people do or do not understand its rules and regulations?

Sources: Based on Health Resources and Services Administration (2010); "Government Resolves...," (2011).

with plural, gender-neutral pronouns (e.g., *they*, *them*) and gender-neutral nouns (e.g., *firefighter*, *police officer*, and *mail carrier*).

Language, Race, and Ethnicity

Words—written and spoken—create and reinforce both positive and negative images about race and ethnicity. Someone might receive a *black mark*, and a *white lie* isn't "really" a lie. A person can *blackball* someone, *blacken* someone's reputation, denounce *blackguards* (villains), and prosecute people in the *black market*. In contrast, a *white knight* rescues people in distress, the *white hope* brings glory to a group, and the "good guys" wear *white hats*.

Language is structured by power relations between and among speakers. Seemingly harmless words, given certain circumstances and power relations, can become derogatory and harmful. For example, in the Jim Crow South (1865–1877), using the term "boy" to refer to a Black man was derogatory because it undermined the man's status. When minoritized groups reclaim and repurpose the derogatory terms used against them, however, they can build solidarity within groups, and the terms can become tools of social change (Due, 1995). For example, the hashtag #BlackBoyJoy is a way of sharing positive images of Black boys and men. It started because many Black boys have been denied the joys of, and sometimes their own lives during, childhood. As a result, Black men have fewer chances to recall childhood memories that are joyous and free of violence or fear. #BlackBoyJoy is also reshaping the meaning of the term "boy" (Young, 2016).

Through language, children learn about their cultural heritage and develop a sense of group identity. Spanish is the fastest-growing and most spoken non-English language in the United States today. That is, in part, because Latina/o/xs now constitute 19% of the total U.S. population. Yet in the United States, 91 percent of U.S. born Latina/o/xs are proficient in the English language, and the use of Spanish within the home has decreased to 57 percent, which is down from 67 percent in 1980 (Funk and Lopez, 2022). In many Latinx and other immigrant families, the native language begins to disappear after just one generation. Assimilation has positive effects, but it can also lead to communication problems between Americanized children and their non-English-speaking neighbors and kin in the homeland (Tobar, 2009; refer also to Chapter 10).

Language and Social Change

Language is dynamic and changes over time. The most recent edition of the New American Bible, an annual best seller, has changed the word "booty," which now has a sexual connotation, to "spoils of war." "Cereal," which many think of as breakfast food, is now "grain" to reference loads of wheat (*Huffington Post*, 2011).

U.S. English is composed of hundreds of thousands of words borrowed from many countries and groups that were in the Americas before the colonists arrived (Carney, 1997). **Table 3.1** provides a few examples of English words borrowed from other languages.

In response to cultural and technological changes, languages continually add new words. Recently, Merriam-Webster added 370 new words to its dictionary (Merriam-Webster, 2022). These include *laggy*, *baller*, *Galentine's Day*, *LARP*, *subvariant*, and *pumpkin spice*, among other new words. And, to some people's

Table 3.1 U.S. English Is a Mixed Salad

Here are a few examples of English words that originated in other languages that are rooted in various areas of the world. What words would you add to the list?	
• Africa: apartheid, Kwanzaa, safari	• Spain: anchovy, bizarre
• Alaska and Siberia: husky, igloo, kayak	• Turkey: baklava, caviar, kebob
• Hungary: coach, goulash, paprika	• France: bacon, police, ballet
• India: bandana, cheetah, shampoo	• Japan: geisha, judo, sushi
• Iran and Afghanistan: bazaar, caravan, tiger	• Norway: iceberg, rig, walrus
• Israel: kosher, rabbi, Sabbath	• Mexico: avocado, chocolate, coyote
• Italy: fresco, spaghetti, piano	• Germany: strudel, vitamin, sauerkraut, kindergarten

dismay and others' delight, some writers are now substituting *partner* for the traditional *spouse*, *wife*, or *husband*.

3-2c Values

Values are the standards by which people define what is good or bad, moral or immoral, proper or improper, desirable or undesirable, beautiful or ugly. These widely shared standards provide *general guidelines* for everyday behavior rather than concrete rules for specific situations. Honesty, for example, is an important but abstract U.S. value. Thus, faculty typically define in course syllabi what they mean by cheating, such as using material from the Internet or other sources without proper acknowledgment, signing an attendance sheet for another student, or having someone else complete an assignment.

Major U.S. Values

Sociologist Robin Williams (1970, p. 452–500) has identified a number of core U.S. values. All are central to the American way of life. They're widespread, have endured over time, and indicate what we believe is important in life.

1. **Achievement and success.** U.S. culture stresses personal achievement, especially occupational success. Many Americans believe that if they work hard, apply themselves, and save their money, they'll be successful in the future.
2. **Activity and work.** Americans want to "make things happen." They respect people who are focused and disciplined and assume that hard work will be rewarded. Journalists and others often praise those who work past their retirement age.
3. **Humanitarianism.** U.S. society emphasizes concern for others, helpfulness, kindness, and support. During natural disasters—like earthquakes, floods, fires, and tornadoes—at home and abroad, many Americans are enormously generous (refer to Chapter 5).
4. **Efficiency and practicality.** Americans emphasize technological innovation, up-to-dateness, practicality, and getting things done. Many American colleges now tout their programs or courses as being practical instead of emphasizing only knowledge attainment and intellectual growth.
5. **Progress.** Americans focus on the future rather than the present or the past. The next time you walk down the aisle of a grocery store, note how many products are "new," "improved," and "better than ever."
6. **Material possessions.** Americans "have more stuff now than any society in history" (even if they can't afford the products). When the stuff accumulates, people buy stuff to stuff it into. Consequently, an entire industry, self-storage, has emerged to house extra belongings (Sanburn, 2015, p. 46).
7. **Freedom and equality.** Countless U.S. documents affirm freedom of speech, freedom of the press, and freedom of worship. U.S. laws also tell Americans that they're "created equal" and have the same legal rights regardless of race, ethnicity, gender, religion, disability, age, or social class.
8. **Conformity.** Most people don't want to be labeled as "strange," "peculiar," or "different." They conform because they want to be accepted, liked, hired, and promoted. As a result, people "toe the line," "go along to get along," and try to "fit in."
9. **Democracy.** Democracy promises the average citizen equal political rights. It emphasizes equality, freedom, and faith in the people rather than giving power to a monarch, dictator, or emperor.
10. **Individualism.** U.S. culture values each person's development. Thus, we often encourage children to be independent, creative, and self-motivated.

Values Sometimes Conflict

Williams acknowledged that these core American values are sometimes contradictory and even hypocritical. For example,

- Americans believe in religious freedom, yet 82 percent of Americans say that it's important that Christians be able to worship freely, while only 61 percent say the same for Muslims (Zoll and Swanson, 2015).
- Americans cherish freedom of speech, but protesters physically attacked political candidates' supporters during the 2016 presidential campaign. At many colleges, campus speakers have been shouted down or disinvited because a small number of vocal students "demand the right to be free from speech that they find offensive or upsetting" (Stone, 2016, p. B10; refer also to Kueppers, 2016).
- Americans prize individualism but reward people who conform to a group's expectations (refer to Chapter 6).

values the standards by which people define what is good or bad, moral or immoral, proper or improper, desirable or undesirable, beautiful or ugly.

- Americans encourage responsibility but often blame popular culture rather than parenting for children's bad behavior (refer to Chapter 12).
- Americans rail against waste, rising food prices, and the growing number of landfills but waste 119 billion pounds of food in the United States each year (Feeding America, 2023).

Values Vary Across Cultures and Change Over Time

Cultural values can change for many reasons, including technological advances, immigration, public health shifts, and contact with outsiders. For example, during the COVID-19 pandemic, there was some resistance to wearing face coverings. From a survey during summer 2020, the number one reported reason was, "it's my right as an American to not wear a mask" which is rooted in an "individual's underlying sense of personal autonomy" (Vargas and Sanchez, 2020). Yet, during that same time period, more than half of U.S. states had mask mandates to slow the spread of the virus (Harring, 2020). Concern over public health took priority over valuing people's individual choice.

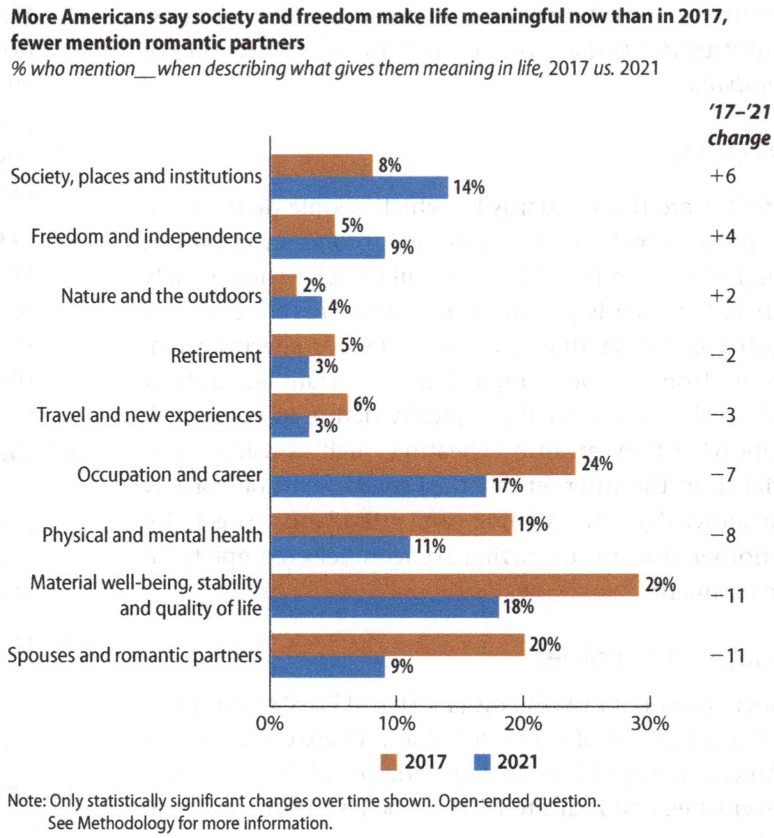

Figure 3.2 Why do you think these shifts in value have occurred?

Source: Van Kessel and Silver, 2021.

In the United States, surveys show a shift in values, which may be attributed to the COVID-19 pandemic. In 2021, more Americans (9%) listed freedom and independence as important, which included free time, work-life balance, or political freedoms such as freedom of speech; this was an increase from 2017 (5%). Yet, there was a decrease in those who found meaning in their romantic partners (13% down from 31% in 2017) as well as a decline in those placing value on physical or mental health (11% down from 19%) (Van Kessel and Silver, 2021) (**Figure 3.2**).

An analysis of census data shows that a greater share of U.S. adults ages 25 through 54 were unpartnered (not married nor living with a partner) in 2019 (38%) compared to 1990 (29%). Of this group, only 53% were married in 2019 (Fry and Parker, 2021). As we'll discuss in Chapter 12, marriage rates vary by social class, among other factors, but values also affect marital status. For example, in a recent survey on people's attitudes about marriage and childbearing, 64 percent of respondents indicated that "society is just as well off if people have priorities other than marriage and children" (Kramer, 2021).

iStock.com/David Henderson

Every year, Americans throw out more food than plastic, paper, metal, and glass. In France, supermarkets are fined up to approximately 80,000 U.S. dollars (75,000 euros) for throwing away food that is almost expired, too unattractive to sell, or could be used for animal feed (Komyati, 2018).

3-2d Norms

Values are general standards, while **norms** are a society's specific rules of right and wrong behavior. Norms tell people what they *should, ought,* and *must* do as well as what they *should not, ought not,* and *must not* do: don't talk in church, wait their turn in line, and so on. People may not like many of the rules, but they make everyday

norms specific rules of right and wrong behavior.

life more orderly and predictable. Here are some general characteristics of norms:

- Most are *unwritten,* passed down orally from generation to generation (use the good dishes and tablecloth on special occasions).
- They're *instrumental* because they serve a specific purpose (get rid of garbage because it attracts roaches and rats).
- Some are *explicit* (save your money "for a rainy day"), while others are *implicit* (be respectful at a wake or funeral).
- They *change* over time (it's now more acceptable to have a child as a single person but much less acceptable to smoke or drink alcohol while pregnant).
- Most are *conditional* because they apply in specific situations (slipping out of your smelly shoes may be fine at home, but not on an airplane).
- Because they're situational, norms can be *rigid* ("You *must* turn in your term paper at the beginning of class on May 5") or *flexible* ("Turn in your term papers by the end of the semester").

Norms regulate our behavior but can differ because of characteristics such as age, social class, race, and gender. For example, a female professor's students sometimes comment on her clothes in their course evaluations and want her to "dress like a professor." In contrast, her husband, also a professor, "has yet to hear a single student comment about his wardrobe." So, she changes her businesslike outfits every day while her husband usually wears khaki pants (or jeans) and a polo shirt day in and day out (Johnston, 2005). Thus, these norms differ based on expectations associated with gender.

Sociologists differentiate three types of norms—folkways, mores, and laws—that vary because some rules are more important than others. As a result, a society punishes some wrongdoers more severely than others.

Folkways

Folkways are norms that involve everyday customs, practices, and interaction. Etiquette rules are good examples of folkways: Cover your mouth when you cough or sneeze, say "please" and "thank you," knock before entering someone's office, and don't pick your ear wax or toenails in public.

People automatically practice numerous folkways because they've internalized them since birth. People often don't realize that they conform to folkways until someone violates one, such as picking one's nose or popping pimples in public or talking loudly on a phone when around others.

Yonhap/AFP/Getty Images/Newscom

South Korea's national media criticized Microsoft founder Bill Gates for his "rude" handshake when meeting President Park Geun-Hye in 2013. As one person tweeted, "How can he put his hand in his pocket when meeting a leader of a state?!" A one-hand shake is disrespectful in South Korea and other parts of Asia and is usually reserved for someone younger or a good friend (Agence France-Presse, 2013; Evans, 2013). Which types of norms did Gates violate—folkways, mores, or laws?

Folkways vary from one country to another. Punctuality is more important in Western countries than in parts of Latin America and the Mediterranean. Tipping is practically nonexistent in Australia, China, and Japan but expected in the United States. Austrians, Germans, and the Swiss consider chewing gum in public to be vulgar. There are also many differences in table manners: Europeans keep their hands above the table at all times, not in their laps; "cleaning your plate" may be insulting in the Philippines, North Africa, and some regions of China because an empty plate suggests that the host didn't serve enough; and in some countries, such as India, it's customary to eat with one's hands instead of using knives and forks (Becker, 2015; Louie, 2015; eDiplomat, 2016).

Folkways can change in response to macro-level changes. The growth of technology and changes in wider society, such as the stressors associated with the COVID-19 pandemic, have altered many campus folkways. In the past, a syllabus was typically a few pages long, outlining basic course requirements and deadlines. Now, many syllabi—some as long as 10 pages—look more like legal documents. They describe, often in great detail, a

folkways norms that involve everyday customs, practices, and interactions.

variety of rules ranging from attendance to laptop and phone use during class and campus resources for mental health (Wasley, 2008; Flaherty, 2020).

Mores

Mores (pronounced "MORE-ayz") are norms that people consider very important because they maintain moral and ethical behavior. According to U.S. cultural mores, people must be sexually faithful to their spouse or partner, loyal to their country, and must not kill another person (except during war or in self-defense).

Folkways emphasize *ought to* behavior, while mores define *must* behavior. The Ten Commandments are a good example of both civil and religious mores. Other mores include ethical guidelines (don't cheat, don't lie), expectations about behavior (don't use illicit drugs), and rules about sexual partners (don't have sexual intercourse with children or family members).

Mores, like folkways, can change. In the 1980s, the median age for women's first birth was 22.6 years (Eickmeyer et al., 2017). As a result of many factors, including more social acceptance of women pursuing college educations and careers, the median age has increased and most recently reached 30 years (Associated Press, 2022).

AFP/Getty Images

China's "Elderly Rights Law" allows aging parents to sue children who don't visit them "often." The judge ordered the daughter of this 77-year-old woman to visit her mother every two months and on at least two of the country's major holidays (Makinen, 2013).

mores norms that people consider very important because they maintain moral and ethical behavior.

laws formally defined norms about what is legal or illegal.

sanctions rewards for good or appropriate behavior and/or penalties for bad or inappropriate behavior.

The strictest mores are *taboos*, which forbid acts that violate social customs, religious or moral beliefs, or laws. Cannibalism, incest, and infanticide (killing infants) are examples of tabooed behavior.

Taboos define and reinforce mores, but there are many violations. For instance, 89 percent of Americans say that having an extramarital affair is morally wrong (Brenan, 2022). However, 23 percent of men and 12 percent of women admit to cheating (Psych Central, 2021).

Laws

Laws are formally defined norms about what is legal or illegal. Unlike folkways and most mores, laws are deliberate, formal, "precisely specified in written texts," and "enforced by a specialized bureaucracy," usually the police and courts, which have the power to punish violators (Hechter and Opp, 2001, p. xi).

Laws change over time. Every year, states pass thousands of bills, ranging from drones to school sports. So far, 41 states and the District of Columbia have legalized medical marijuana and/or the compound CBD, and 21 states plus the District of Columbia and Guam have legalized marijuana for recreational use (DISA, 2023; Hansen et al., 2023). Every state now permits gambling (e.g., lotteries, casinos, racetrack betting). Some people believe that legalizing marijuana and gambling violates mores, but taxing both of these activities has greatly augmented state coffers.

Laws also vary across societies. You'll learn shortly and throughout this textbook that countries' laws differ considerably regarding, for example, alcohol, adultery, elections, health care, religious worship, freedom of speech, and adoption.

Sanctions

Most people conform to norms because of **sanctions**, rewards for good or appropriate behavior and/or penalties for bad or inappropriate behavior. Children learn norms through both *positive* sanctions (praise, hugs, smiles, new toys) and *negative* sanctions (frowning, scolding, spanking, withdrawing love) (refer to Chapter 4).

Negative sanctions vary in the degree of punishment. When we violate folkways, the sanctions are relatively mild (e.g., gossip, raised eyebrows, ridicule) because folkways aren't critical for a society's survival. If a person doesn't bathe or brush their teeth, they may not be invited to parties because others may think they are crude, but not sinful or evil.

The sanctions for violating mores and some laws can be severe: loss of employment, expulsion from college, whipping, torture, banishment, imprisonment, and even execution. The punishment is usually harsh because the behavior threatens a society's moral foundations. For Hindus, who make up more than 80 percent of India's 1.2 billion population, the cow is a revered religious symbol. Much of India bans its slaughter, sale, and consumption. Hindu mobs have beaten to death Muslims that they suspected of killing, stealing, or smuggling cows. In one case, and as the family had maintained, a forensic test found that the murdered victim had eaten goat meat (Raj, 2015).

Sanctions aren't always consistent, despite universally held norms. For instance, someone who can hire a good attorney often receives no penalty or a light one for a serious crime. In other cases, laws may have loopholes that can unevenly impact specific groups of people. For example, there are only six countries that have built gender equality into their laws regarding work to ensure equal rights for men and women; Belgium, Denmark, France, Latvia, Luxembourg, and Sweden (Lamble, 2019). Also, laws are sometimes enforced selectively or not at all. In the United States and elsewhere, only a fraction of offenders who violate rape, sexual harassment, and domestic violence laws are prosecuted (Jewkes et al., 2013; refer also to Chapters 7 and 12).

3-2e Rituals

Rituals are formal and repeated behaviors that unite people. They include giving gifts during holidays, oaths of allegiance, wedding and funeral ceremonies, worship rites, wearing Halloween costumes, graduations, veterans' parades, and even shaking hands. What all rituals have in common is the transmission and reinforcement of norms that unite us and strengthen relationships (Rossano, 2012).

Noah Seelam/Getty Images

In some Indian states, violators who possess or sell beef can be punished by up to five years in prison (Bagri and Najar, 2015).

Rituals are also outward symbols of values. During the playing of our national anthem, for example, everyone stands and faces the flag. Many people place their hands over their hearts, and men remove their caps and hats. Such ritualistic behavior demonstrates respect and reinforces shared values about patriotism and democracy.

3-3 Some Cultural Similarities

You've learned that there's considerable diversity across societies in symbols, language, values, and norms. There are also some striking similarities because of cultural universals, ideal versus real culture, ethnocentrism, and cultural relativism.

3-3a Cultural Universals

Cultural universals are customs and practices that are common to all societies. Anthropologist George Murdock and his associates studied hundreds of societies and compiled a list of 88 activities they found among all cultures. (**Table 3.2** provides some examples.)

There are many cultural universals, but specific behaviors vary across cultures, from one group to another in the same society, and over time. For example, all societies have food taboos, but they differ across societies. In approximately 80 percent of nations, bugs are a normal component of people's food consumption (Troitino, 2017). Vietnamese restaurants offer cat on the menu, and many people with lower incomes in rural China eat dog meat because they can't afford poultry, pork, and beef. Increasingly, however, urban, middle-class Chinese households "have learned to love dogs at the end of a leash rather than on a skewer" ("Dog-lovers . . . ," 2015, p. 46; refer also to Wan, 2011).

3-3b Ideal versus Real Culture

The **ideal culture** comprises the beliefs, values, and norms that people say they hold or follow. In every

rituals formal and repeated behaviors that unite people.

cultural universals customs and practices that are common to all societies.

ideal culture the beliefs, values, and norms that people say they hold or follow.

Table 3.2 Some Cultural Universals

Athletics	Food taboos	Housing	Medicine
Cooking	Funeral rites	Inheritance rules	Music
Courtship	Games	Kin terminology	Property rights
Dancing	Gift giving	Language	Religious rituals
Division of labor	Greetings	Magic	Sexual restrictions
Etiquette	Hairstyles	Marriage	Status differentiation

Source: Based on Murdock, 1940.

society, however, ideal culture differs from **real culture**, or people's actual everyday behavior.

Consider parent–child relationships. Americans say they love and cherish their children, but every year, hundreds of thousands of children experience abuse and neglect on a daily basis. Indeed, 92 percent of the offenders are parents (U.S. Department of Health and Human Services, 2017; refer also to Chapter 12).

In China, the 2,500-year-old Confucian ideal of *filial piety* (respecting and taking care of one's parents) still runs deep. "It is estimated that there are more than 100 million people aged 60 and above who are living alone or with their spouses only," and in response, the Chinese government proposed a new plan to encourage children to live with or near their elderly parents (Zuo, 2021). Thus, ideal culture and actual behavior are often inconsistent.

Kevin Foy/Alamy Stock Photo

In approximately 80 percent of countries, people consume insects, which are high in protein, vitamins, and fiber, and usually low in fat. More than two out of every three American adults are overweight or obese, the highest percentage on the planet (refer to Chapter 14). Are people who eat insects healthier than those who consume burgers and french fries?

3-3c Ethnocentrism and Cultural Relativism

Ethnocentrism is the belief that one's culture, society, or group is inherently superior to others. Thus, "*our* customs, *our* laws, *our* food, *our* traditions, *our* music, *our* religion, *our* beliefs and values, and so forth, are somehow better than those of other societies" (Smedley, 2007, p. 32). Because people internalize their culture and tend to consider their way of life to be the best and the most natural, they often disparage those with differing attitudes and behavior as inferior, wrong, or backward. How do you feel about eating insects, cats, or dogs? How does it fit with your culture?

Some of my Black students argue that it's impossible for Blacks in the United States to be ethnocentric because they experience so much prejudice and discrimination. *Any* group can be ethnocentric, however (Rose, 1997). An immigrant from Nigeria who believes that native-born Black Americans are lazy and criminal is just as ethnocentric as a native-born Black American who describes Nigerians as arrogant and "uppity."

Ethnocentrism can be functional. Appreciating one's own country and heritage promotes loyalty and cultural unity, reinforces conformity, and maintains stability. Members of a society become committed to their particular values and customs and transmit them to the next generation. As a result, life is (generally) orderly and predictable.

Ethnocentrism is also dysfunctional because viewing others as inferior generates hatred, discrimination, and conflict. Many current wars and battles—such as the civil war in Ethiopia and the conflict between Israelis and Palestinians—are due to religious, ethnic, or political

real culture people's actual everyday behavior.

ethnocentrism the belief that one's own culture, society, or group is inherently superior to others.

intolerance (refer to Chapter 13). Thus, ethnocentrism thwarts intergroup understanding and cooperation.

The opposite of ethnocentrism is **cultural relativism**, the belief that no culture is better than another and should be judged by its own standards. In the United States, families are often child-focused and aim to support children's individualism. Parents are more likely to structure schedules and activities around children, serve children meals to suit their own tastes (think chicken nuggets and macaroni), and even center holiday celebrations around the kids. Yet, in France, parents tend to have a parenting approach focused on setting boundaries. Children are expected to eat the same foods as parents, function on parents' schedules, and enjoy holidays that are less focused on catering to them (Hendricks, 2018; Williams, 2022). Thus, the French and American approaches to parenting may differ, but one culture's practice isn't better or worse than the other.

Vatican Pool-Corbis/Corbis News/Getty Images

Dozens of rules and customs govern what an American first lady should and shouldn't wear while visiting other countries. In keeping with Vatican protocol, Melania Trump wore long sleeves, formal black clothing, and a black veil in a meeting with Pope Francis.

3-4 Some Cultural Variations

There's considerable cultural variation not only *across* societies but also *within* the same society. *Subcultures* and *countercultures* account for some of the complexity within a society.

3-4a Subcultures

A **subculture** is a group within society that has distinctive norms, values, beliefs, lifestyle, or language. A subculture is part of the larger, dominant culture but has its own particular ways of thinking, feeling, and behaving. U.S. society contains thousands of subcultures based on a variety of characteristics, interests, or activities:

- **Ethnicity** (Irish, Mexican, Vietnamese)
- **Religion** (evangelical Christians, atheists, Mormons)
- **Politics** (Maine Republicans, Southern Democrats, independents)
- **Sexuality** (heterosexual, lesbian, transgender)
- **Age** (older widows, middle schoolers, college students)
- **Occupation** (surgeons, sex workers, truck drivers)
- **Music and art** (jazz aficionados, country music buffs, art lovers)
- **Social class** (billionaires, working class, middle class)
- **Recreation** (mountain bikers, poker players, dancers)

As you read this list, you'll realize that most of us are members of numerous subcultures. We usually participate in subcultures without having much commitment to any of them (e.g., basketball fans, guitar players, collectors of ceramic frogs). On the other hand, members of racial or ethnic subcultures may intentionally live in the same neighborhoods, associate with each other, have close personal relationships, and marry others who are similar to themselves.

To fit in, members of most subcultures adapt to the larger society, but they may also maintain some of their traditional customs. Among Ghanaian immigrants, for instance, funerals are lavish celebrations rather than sad gatherings. The funerals are "all-night affairs with open bars and window-rattling music." As in Ghana, the guests need not know the deceased or even the family. Many people attend simply to meet other Ghanaians and to "cut loose on the dance floor" (Dolnick, 2011, p. A1; refer also to Bax, 2013).

In many instances, subcultures arise because of technological or other societal changes. After the Internet emerged, subcultures appeared that identified themselves as hackers, techies, and computer geeks.

3-4b Countercultures

A **counterculture** is a group within society that openly opposes and/or rejects some of the dominant culture's

---

cultural relativism the belief that no culture is better than another and should be judged by its own standards.

subculture a group within society that has distinctive norms, values, beliefs, lifestyle, or language.

counterculture a group within society that openly opposes and/or rejects some of the dominant culture's norms, values, or laws.

norms, values, or laws. Countercultures usually emerge when people believe they can't achieve their goals within the existing society. As a result, the groups develop values and practices that run counter to those of the dominant society. Some countercultures are small and informal, but others have millions of members and are highly organized.

Most countercultures are law-abiding. Some, however, are violent and extremist. There are active hate groups in every state (**Figure 3.3**), but their number has increased from 599 in 2000 to 733 in 2021 (Southern Poverty Law Center, 2023). Hate groups include antigovernment militias, white supremacists, abortion foes, domestic Islamist radicals, neo-Nazis, and a small but growing number of Black separatists. These groups target people because of their race, ethnicity, religion, or sexual orientation (Potok, 2017).

There was a 55 percent growth in White nationalist organizations during the Trump administration (Wilson, 2020). In recent years, the number of White nationalist, neo-Nazi, and anti-government extremist groups have declined (Morrison, 2022).

3-4c Multiculturalism

Multiculturalism (sometimes called *cultural pluralism*) refers to the coexistence of several cultures in the same geographic area without one culture dominating another. Many applaud multiculturalism because it encourages intercultural dialogue. This includes schools offering programs in African American, Latinx, Arabic, and Asian studies as well a focus on diversity, equity, and inclusion (DEI). Supporters hope that emphasizing multiculturalism—especially in schools and the workplace—will decrease ethnocentrism, racism, sexism, and other forms of discrimination.

Despite its benefits, not everyone is enthusiastic about multiculturalism. Not learning the language or following

multiculturalism (sometimes called *cultural pluralism*) the coexistence of several cultures in the same geographic area without one culture dominating another.

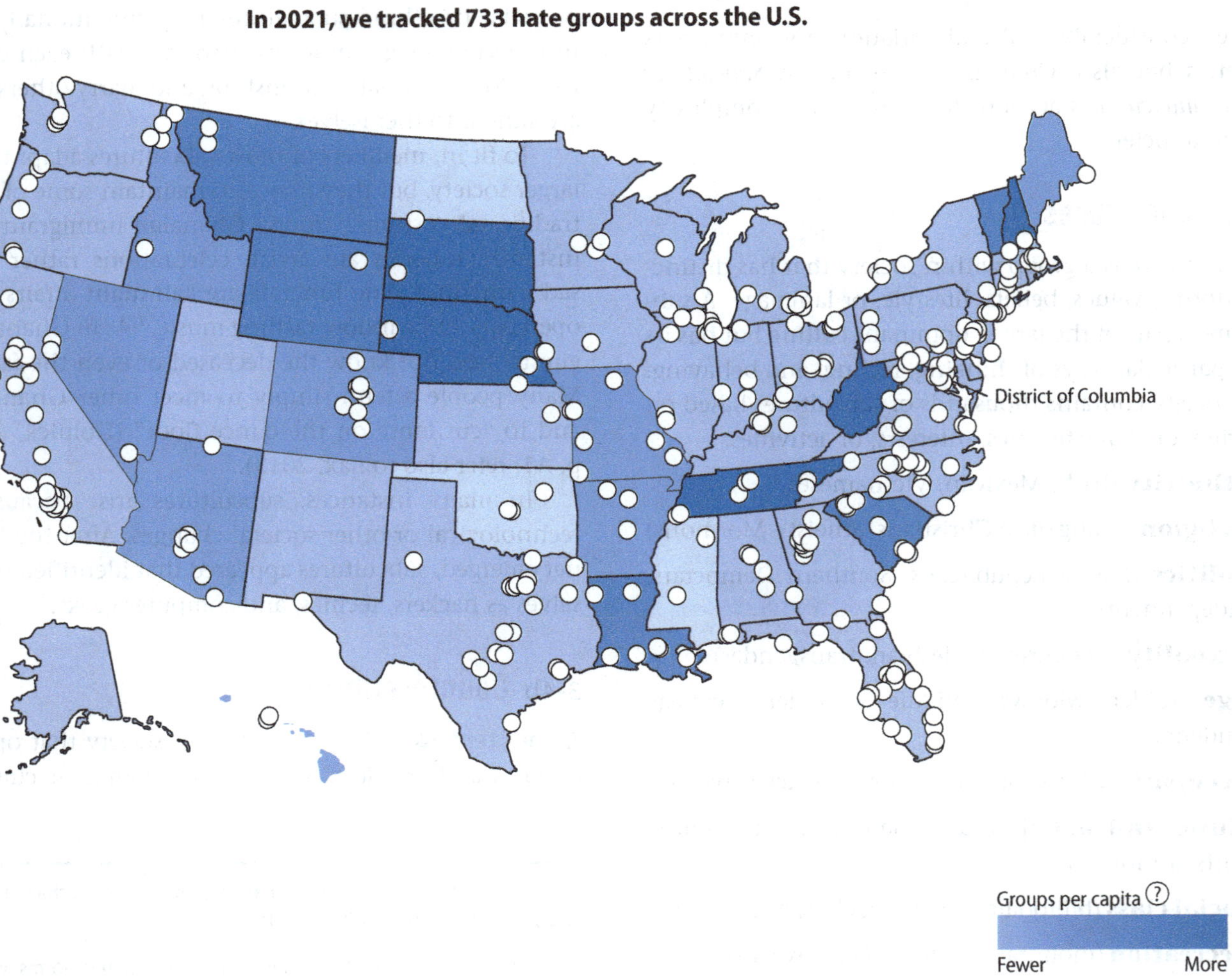

Figure 3.3 Active U.S. Hate Groups, 2021

Source: Southern Poverty Law Center.

the customs of the country where one lives and works can spur tension and conflict. European political leaders in Germany, Great Britain, and France have described multiculturalism, particularly integrating Muslims, as a "failed policy" and a "nightmare" that's had "disastrous results." Instead of adopting the dominant culture's values and beliefs, according to critics, some ethnic and religious subcultures "build tiny nations" within the host nation, thereby jeopardizing the host country's national identity (Marquand, 2011; Kymlicka, 2012; Tharoor, 2015; refer also to Chapter 13). And host nations themselves play a role in the challenges of attaining true multiculturalism. For instance, France banning clothing that may be worn by Muslim women including burkas and burkini (swimwear) (Associated Press, 2022; Peck, 2017).

3-4d Culture Shock

People who travel to other countries often experience **culture shock**—confusion, disorientation, or anxiety that accompanies exposure to an unfamiliar way of life. Conventional cues about how to behave are missing or have a different meaning. Culture shock affects some people more than others, but the most stressful changes involve differences in food, clothes, etiquette, values, language, the general pace of life, and a lack of privacy (Spradley and Phillips, 1972; Pedersen, 1995).

To some degree, everyone is *culture bound* because we've internalized cultural norms and values. During a tour to North Korea, an American college student stole a political poster as a souvenir. North Korea's Supreme Court sentenced him to 15 years of prison and hard labor. Many Americans were outraged: They dismissed the theft as a small and harmless college-style prank and attributed the harsh punishment to the escalating tensions with the United States over nuclear weapons. But in North Korea, where the leader is treated as a deity and "political slogans are sacrosanct," harming slogans bearing leaders' names is "one of the most serious crimes" and is considered an "act of hostility against the state" (Sang-Hun, 2016, p. A4; Dawson, 2017).

3-5 High Culture and Popular Culture

High culture refers to the cultural expression of a society's highest social classes. Examples include opera, ballet, theatre, and classical music. In contrast, **popular culture** refers to the beliefs, practices, activities,

culture shock confusion, disorientation, or anxiety that accompanies exposure to an unfamiliar way of life.

high culture the cultural expression of a society's highest social classes.

popular culture the beliefs, practices, activities, and products that are widespread among a population.

Does Pop Culture Contribute to Kids Being Overweight?

The prevalence of obesity among American children ages 2 to 19 years has more than tripled, increasing from 5 percent in the 1960s to 17 percent in 2014 (Fryar et al., 2016). There are many reasons for the increase, but physicians and researchers lay much of the blame on popular culture, especially the advertising industry. Advertising uses cartoon characters like SpongeBob SquarePants and Scooby-Doo to sell sugary cereals, cookies, candy, and other high-calorie snacks. In contrast, some countries, such as the United Kingdom, Nigeria, and Thailand, have guidelines, limitations, or oversight bodies focused on food advertising directed towards children (Food Marketing Workgroup, 2007). After considerable pressure from the federal government, the nation's largest food companies agreed to curb advertising, but children are watching more fast-food ads on TV than ever before. Only six companies are responsible for more than 70 percent of all TV fast-food ads viewed by children ages 18 and younger ("Food Companies Propose...," 2011; Harris et al., 2013).

iStock.com/Juanmonino

and products that are widespread among a population. Examples include shows, podcasts, music, social media, advertising, sports, hobbies, fads, fashions, movies, the food people eat, the gossip people share, and the jokes people tell (Levine, 1998; Gans, 1999).

Social class affects our participation in high and popular culture. Members of the upper class have greater ability to pursue the fine arts and similar activities because they have **cultural capital**—resources such as knowledge, verbal and social skills, education, and other assets that give a group advantages (Bourdieu, 1986). Resources alone, however, don't determine a person's cultural tastes and interests. For example, in Baltimore (as in many other cities), tickets for the symphony, opera, or ballet are usually much less expensive than those for a Ravens football game or rock concert.

Cultural capital sets up boundaries between social classes, but our cultural expressions and products often contain elements of both popular and high culture. For instance, Black jazz musicians have inspired European classical music composers, and they perform classical music themselves (Salamone, 2005; Lawn, 2013).

Queen Elizabeth II, during her reign, conferred knighthood—one of the highest honors given to an individual in the United Kingdom—on pop singers like Elton John and Paul McCartney. Since 1998, the John F. Kennedy Center for the Performing Arts in Washington, D.C., has presented an annual Mark Twain Prize for American Humor to popular culture comedians like Richard Pryor, Tina Fey, Bill Murray, and Adam Sandler—the upcoming awardee in 2023.

Millions of Americans who think of themselves as "cultured" watch "trashy TV" like *Hoarders, The Bachelor/The Bachelorette, The Real Wives of . . .*, and *Keeping Up with the Kardashians*. Doing so allows them to feel superior to the shows' conventional "low-brow" viewers ("Thank goodness I'm not like that!") (McCoy and Scarborough, 2015). Thus, the divide between popular and high culture isn't as rigid as one might think.

3-5a The Influence of Mass Media

People don't always believe everything they read or see on television or online. Nonetheless, many are highly influenced by a popular culture that's largely controlled and manipulated by newspapers and magazines, television, movies, music, and ads (refer to Chapter 4). These **mass media**, or forms of communication designed to reach large numbers of people, have enormous power in shaping public attitudes and behavior.

iStock.com/Holger Mette

American fast-food restaurants are popular in many Asian countries. Because of greater competition from local fast-food chains, U.S. companies now offer a large number of items geared specifically to Asian tastes. At one of China's Pizza Huts, the menu includes shrimp pizza, fried squid, and green tea ice cream. In Latin America, McDonald's sells "McMollettes," English muffins filled with refried beans, cheese, and salsa. In India, McDonald's offers an entirely meat-free menu.

Many of my students who claim they "don't pay any attention to ads" come to class wearing branded apparel: Budweiser caps, Old Navy t-shirts, and Adidas footwear. Advertising affects our self-image and self-esteem. As you'll learn in later chapters, many people spend considerable time and money to meet ideal standards of femininity and masculinity—generated by ads—that are unrealistic and impossible to attain. (We'll examine fashion and fads, two important components of popular culture, in Chapter 16.)

Because companies are driven by the pursuit of profit, considerable mass media content is basically marketing. Many shows have become "a kind of sophisticated infomercial" (O'Donnell, 2007, p. 30). Product placement, which is the practice of featuring products in scenes and storylines of movies and shows, is now a 23-billion-dollar industry (Haigney, 2022). Infomercials are another approach to marketing products through shows. The *Dr. Phil Show* is crowded with infomercials that promote the host's numerous products, including

cultural capital resources and assets that give a group advantages.

mass media forms of communication designed to reach large numbers of people.

his diet and relationship books, his wife's skin care line, and his and his son's TV show, publishing company, and "Doctor on Demand" app. U.S. mass media also exert a powerful influence abroad.

3-5b The Globalization of Popular Culture

Because the United States is the biggest producer of popular culture, many countries have become inundated with America's movies, music, television shows, satellite broadcasts, fast food, clothing, and other consumer goods. Overseas corporations contribute to the spread of U.S. popular culture because they profit by distributing the products of popular musicians, authors, and filmmakers (The Levin Institute, 2013).

Some observers have described the spread of U.S. popular culture as **cultural imperialism**, a process by which the cultural values and products of one society influence or dominate those of another. A great deal of cultural imperialism is voluntary. For example, much of England's population has embraced American pop music, films, programs, and architecture and is now experiencing an "obesity epidemic" because of the "bread-potatoes-and-lard-based fast food on which America nourishes itself" (Heffer, 2010).

Other countries complain that U.S. cultural imperialism displaces authentic local culture and results in cultural loss. Iran's government has denounced Batman, Spider-Man, and Harry Potter toys as a "cultural invasion" that challenges the country's conservative and religious values. The curvaceous and often scantily clad Barbie dolls with peroxide-blond hair have been especially singled out as "destructive culturally and a social danger" (Peterson, 2008, p. 4).

Only 42 percent of Americans, compared with 79 percent in 2002, believe that the world at large thinks about the United States favorably. This belief may be a major reason for many Americans' wariness of U.S. global involvement. A large majority (70 percent) say the president should focus on domestic rather than foreign policy, 57 percent believe that other countries should deal with their own problems, and 43 percent think that the United States should "mind its own business" internationally ("Public Uncertain...," 2016; Swift, 2017).

3-6 Cultural Change and Technology

Why do cultures persist? How and why do they change? And what happens when technology changes faster than cultural norms, values, and laws?

3-6a Cultural Persistence: Why Cultures Are Stable

In many ways, culture is a conservative force. As you learned earlier, values, norms, and language are transmitted from generation to generation. Such **cultural integration**, or the consistency of various aspects of society, promotes order and stability. Even when new behaviors and beliefs emerge, they commonly adapt to existing ones. Recent immigrants may speak their native language at home and celebrate their own holy days, but they're expected to gradually absorb the host country's values, to obey its civil and criminal laws, and to adopt its national language. Life would be chaotic and unpredictable without such cultural integration.

3-6b Cultural Dynamics: Why Cultures Change

Cultural stability is important, but all societies change over time. Some of the major reasons for cultural change include diffusion, invention and innovation, discovery, and external pressures.

Diffusion

A society may change because of *diffusion*, the spread of cultural beliefs and activities from one group to another. The influences may have occurred so long ago that the members of a society consider their culture to be entirely their own creation. However, anthropologist Ralph Linton (1964) has estimated that 90 percent of the elements of any culture are a result of diffusion (**Figure 3.4**).

Diffusion can be direct and interpersonal, occurring through trade, tourism, immigration, intermarriage, or the invasion of one country by another. Diffusion can also be indirect and largely impersonal, as with the Internet transmissions that zip around the world.

Invention and Innovation

Cultures change because people are continually finding new ways of doing things. *Invention*, the process of creating new things, brought about products such as toothpaste (invented in 3000 B.C.E.), eyeglasses (262 C.E.), flushable toilets (the sixteenth century), fax machines (1843—that's right, invented in 1843!), credit cards (1920s), computer mouses (1964), the Internet (1960s and 1970s), and Post-it™ notes (1980).

cultural imperialism the cultural values and products of one society influence or dominate those of another.

cultural integration the consistency of various aspects of society that promotes order and stability.

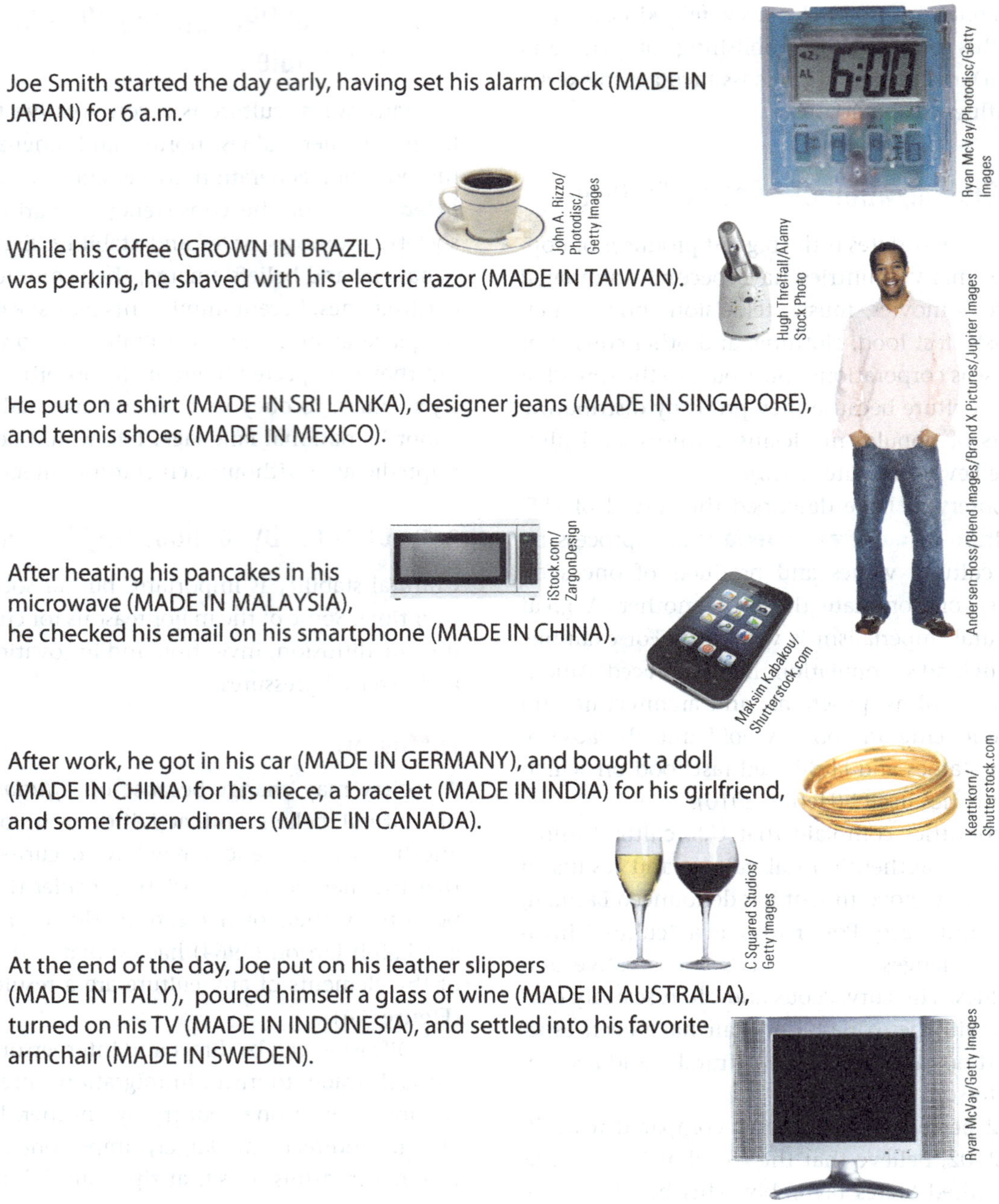

Figure 3.4 The 100 Percent American

Innovation—turning inventions into mass-market products—also sparks cultural changes. An innovator is someone who markets an invention, even if it's someone else's idea. For example, Henry Ford invented nothing new but "assembled into a car the discoveries of other men behind whom were centuries of work," an innovation that changed people's lives (Evans et al., 2006, p. 465).

Inventors and innovators are rarely isolated geniuses who work alone. Instead, they depend on collaborative relationships and social networks for ideas and help. Thomas Edison's laboratory, which had a staff of 35 inventors, produced the light bulb. The Human Genome Project involved 110 scientists around the world from 20 universities, research centers, institutes, and hospitals. Even Michelangelo used 12 assistants to paint the renowned Sistine Chapel in Rome (Dahlin, 2011). The work of inventors and innovators also lays the foundation for others to expand the technology or products. For instance, Dr. Shirley Jackson "conducted breakthrough basic scientific research that enabled others to invent the portable fax, touch tone telephone, solar cells, fiber optic cables, and the technology behind caller ID and call waiting" (Famous Black Inventors, 2019).

Discovery

Like invention, *discovery* involves exploration and investigation and results in new products, insights, ideas, or behavior. The discovery of penicillin prolonged lives, which, in turn, means that more grandparents (as well as great-grandparents) and grandchildren get to know each other. However, longer life spans also mean that many adult children and grandchildren now care for older family members over many years (refer to Chapter 12).

Discovery usually requires dedicated work and years of commitment, but some discoveries occur by chance, called the *serendipity effect*. George de Mestral, a Swiss electrical engineer, was hiking through the woods and was annoyed by burrs that clung to his clothing. Why were they so difficult to remove? A closer examination showed that the burrs had hooklike arms that locked into the open weave of his clothes. The discovery led de Mestral to invent a hook-and-loop fastener. His invention, Velcro—derived from the French words *velour* (velvet) and *crochet* (hooks)—can now be found on everything from wallets to spacecraft.

External Pressures

External pressure for cultural change can take various forms. In its most direct form—war, conquest, or colonization—the dominant group uses force or the threat of force to change an existing culture. When the Soviet Union invaded and occupied many small countries (e.g., Lithuania, Latvia, Estonia, Ukraine, and Armenia) after World War II, it forbade citizens to speak their native languages, banned traditions and customs, and turned churches into warehouses.

Pressures that are indirect can also change culture and a society's support for its cultural practices. The COVID-19 pandemic sparked the worst global recession in nearly 100 years and impacted the cultural economy, especially for nations in the Global South. As a result, "many heritage sites suffer neglect and illicit trafficking of cultural artefacts continues unabated" (de Vries, 2022).

3-6c Technology and Cultural Lag

Some parts of culture change more rapidly than others. **Cultural lag** refers to the gap that occurs when material culture changes faster than nonmaterial culture.

There are numerous examples of cultural lag in modern society because nonmaterial culture (e.g., norms, values, and laws) hasn't kept up with material culture (e.g., technological advances). Texting while driving has resulted in thousands of fatal car crashes (refer to Chapters 1 and 2), but many states have neither banned this practice nor enforced the laws they've passed.

Because technology is moving faster than the development of laws to regulate and ensure Internet and phone-based communication privacy, there are many breaches. In 2013, Edward Snowden, a contractor for the NSA, called out the organization's mass surveillance program of American phone records. Seven years later, a U. S. appeals court found NSA's mass surveillance program to be unlawful (Satter, 2020).

Cultural lags can create confusion, ambiguity about what's right and wrong, conflict, and a feeling of helplessness. They also expose contradictory values and behavior. U.S. technology executives have railed against the NSA's mass compilation of data, but the companies themselves collect information on users that they sell to advertisers (Miller, 2013; Sengupta, 2013). Some consumers complain that retailers track their behavior but have no problem with cookies, profiles, apps, and other online tools that let e-commerce sites know who they are, how they shop, and what they purchase (Clifford and Hardy, 2013). And as you'll learn in several later chapters, even when mores, values, and laws catch up with technological innovations and inventions, they can create unexpected problems such as pollution, drug shortages, and high medical costs.

3-7 Sociological Perspectives on Culture

What is the role of culture in modern society? And how does culture help us understand ourselves and the world around us? Functionalist, conflict, feminist, and symbolic interaction scholars offer important insights in answering these and other questions. **Table 3.3** summarizes the perspectives.

3-7a Functionalism

Functionalists focus on society as a system of interrelated parts (refer to Chapter 1). Similarly, in their analysis of culture, functionalists emphasize the social bonds that attach people to society.

cultural lag the gap that occurs when material culture changes faster than nonmaterial culture.

Table 3.3 Sociological Perspectives of Culture

Theoretical Perspective	Functionalist	Conflict	Feminist	Symbolic Interactionist
Level of Analysis	**Macro**	**Macro**	**Macro and Micro**	**Micro**
Key Points	• Similar beliefs bind people together and create stability. • Sharing core values unifies a society and promotes cultural solidarity.	• Culture benefits some groups at the expense of others. • As powerful economic monopolies increase worldwide, the rich get richer and the rest of us earn less.	• Women and men often experience culture differently. • Cultural values and norms can increase inequality because of sex, race/ethnicity, and social class.	• Cultural symbols forge identities (that change over time). • Culture (like norms and values) helps people merge into a society despite their differences.
Examples	• Speaking the same language (English in the United States) binds people together because they can communicate with one another, express their feelings, and influence one another's attitudes and behaviors.	• Much of the English language reinforces negative images about gender ("slut"), race ("cracker"), ethnicity ("jap"), and age ("boomer") that create inequality and foster ethnocentrism.	• Using male language (e.g., "congressman," "fireman," and "chairman") conveys the idea that men are superior to and dominant over women, even when women have the same jobs.	• People can change the language they create as they interact with others. Many Americans now use "police officer" instead of "policeman," and "single person" instead of "bachelor" or "old maid."

Key Points

For functionalists, culture is the cement that binds people together. As you learned earlier, norms and values shape our lives, provide guideposts for our everyday behavior, and promote cultural integration and societal stability. Especially in developed countries like the United States that have high immigration rates, cultural norms and values help newcomers adjust to the host society.

All societies have similar strategies for meeting human needs. Cultural universals, like religious rituals, may play out differently in different countries, but all known societies have religious rituals. Critics often blame popular culture for societal problems, including violence and crime, but popular culture also unites people with similar interests and produces new technologies and software (e.g., LinkedIn, YouTube, Instagram).

Functionalists also note that culture can be dysfunctional. Some countercultures, particularly right-wing extremist groups, can create chaos by bombing federal buildings and killing or injuring hundreds of people.

Critical Evaluation

In emphasizing culture's role in meeting people's daily needs, functionalism often overlooks diversity and social change. A number of influential functionalists have proposed, for instance, that immigration should be restricted because it dilutes shared U.S. values. Such proposals overlook the many contributions that newcomers make to society refer to Chapter 10).

Another issue is how much culture really binds people together. Only 38 percent of adults in the U.S. say they're "extremely proud" to be Americans, down from 70 percent in 2003. Intertwining factors that may have reduced national pride may include American's struggles with the highest inflation rates in more than 40 years that were brought on in part by the COVID-19 pandemic. In addition, recent mass shootings in Buffalo, New York, and Uvalde, Texas may have influenced people's pride in the nation (Brenan, 2022).

3-7b Conflict Theory

Unlike functionalists, conflict theorists argue that culture can generate considerable inequality instead of unifying society. Because the rich and powerful determine economic, political, educational, and legal policies for their own benefit and control the mass media, the average American has little power to change the culture—whether it's low wages, unpopular wars, or corporate corruption (refer to Chapters 8 and 11).

Key Points

Conflict theorists maintain that many cultural values and norms benefit some members of society more than others. For example, people are to work hard but who

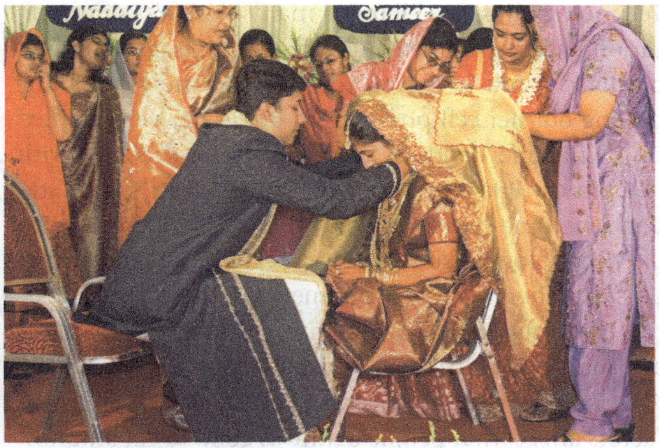

India's high-income families usually have lavish weddings that follow centuries-long traditions. For functionalists, the customs and rituals reinforce cultural identity and bind people. For conflict theorists, India's inequality also creates a large population of low-incoming-earning people who are excluded from the traditions and ceremonies that reinforce social bonds.

profits from the average worker's efforts? U.S. taxpayers paid billions of dollars to bail out financial industries that collapsed because of greedy top executives. The executives then used the money to give themselves and their staff higher salaries and bonuses than ever before. Very rarely, moreover, are wealthy people incarcerated for crime (refer to Chapters 7 and 8). Thus, those at the top of the socioeconomic ladder can violate laws and mores, including honesty, that other Americans are expected to observe.

Conflict theorists also point out that technology serves primarily the rich. Desperate low-income women use online sites to sell their hair, breast milk, and ova. Selling kidneys is illegal in the United States, but one kidney sells for more than $100,000 on the Internet black market (Weller, 2015).

Critical Evaluation

According to some critics, conflict theorists place too much emphasis on societal discord and downplay a culture's benefits. For instance, cultural values integrate members of a society and decrease divisiveness.

Conflict theorists place responsibility with entities such as mass media conglomerates—companies that own several smaller businesses with varying product and service lines. Conflict theorists maintain that conglomerates control popular culture and promote goods and services that most people don't need and/or can't afford (e.g., big screen TVs, the most recent smartphones, and designer clothes). Critics counter that people aren't mindless puppets, instead, they can make choices, including not living on credit. Thus, according to some conflict theory critics, conglomerates shouldn't be blamed for people's irrational behavior.

3-7c Feminist Theories

Feminist scholars agree with conflict theorists that culture creates considerable inequality, but they focus on gender differences. Feminists are also more likely than other theorists to examine multicultural variations across groups.

Key Points

Culture affects gender roles. When media portrayals of women are absent or stereotypical, we get a distorted view of reality (Neuhaus, 2010). Feminist scholars also emphasize that subcultures (e.g., female students or single mothers) may experience culture differently than their male counterparts do because women typically have fewer resources, particularly income and power (refer to Chapters 9 and 11).

Unequal access to resources often results in women having fewer choices than men and living under laws and customs that subordinate women. In the United States, a nonmarital birth typically impoverishes women but not men. Only 10 states or jurisdictions have specific laws that prevent or punish forced marriages of girls under age 16 (Reiss, 2015). In 29 countries and territories in Latin America and the Caribbean, at least 4,473 women were victims of femicide in 2021, and approximately one-quarter of women reported previously or presently being the victim of physical or sexual violence by a partner (Economic Commission for Latin America and the Caribbean, 2022).

In South Sudan, 73 percent of girls do not attend school for primary education, which has contributed to a high rate of illiteracy among girls and women (Suliman, 2017). Factors deterring girls from attending

Which sociological perspectives help explain how children's birthday parties reproduce culture?

school include cultural practices that encourage an early age of marriage and gendered division of labor in addition to the distance to schools and poverty (Yongoro, 2022).

Critical Evaluation

Feminist analyses expand our understanding of cultural components that other theoretical perspectives ignore or gloss over. Like conflict theorists, however, feminist theorists often emphasize discord rather than how culture integrates all people into society. Another weakness is that feminist scholars tend to focus on the cultural experiences of women who have lower incomes and may be racial, ethnic, or religious minorities, but they do not focus on the experiences of their male counterparts.

3-7d Symbolic Interaction

Interactionists examine culture through micro lenses. They're most interested in understanding how individuals—in contrast to groups, organizations, or societies—create, maintain, and modify culture.

Key Points

Interactionists study how culture influences our everyday lives. Language, you may recall, shapes people's views and behavior. People within and across societies create a variety of symbols that may change over time. Technology has also changed many people's communication patterns. Some complain that texting dominates our lives. A much-viewed YouTube clip shows the bride pulling out her cell phone from under her veil and texting with one hand as she walks down the aisle with her father (Pflaumer, 2011). On the other hand, students become more engaged when they're required to use technology, like tweeting weekly observations that illustrate their reading assignments (Lang, 2013).

For interactionists, many cultural components are socially constructed. Consider food consumption. Americans may think it's odd to eat insects, but cultural factors affect what people define as good to eat and how food is prepared. Thus, some visitors consider many U.S. foods strange or "nasty" (e.g., Cheez Whiz, Velveeta, corn dogs, supermarket white bread and rolls, supersweet iced tea, grits). According to some international travelers, many of our sugary breakfast cereals, particularly those with different colors and marshmallows, are "disgusting" (Martin, 2013; Govender, 2014).

Symbolic interactionists also note that our values and norms, like other components of culture, aren't superimposed by some unknown external force. Instead, as people construct their perception of reality, they create, change, and reinterpret values and norms through interactions with others. Peer pressure, for example, can either encourage or discourage inappropriate language or behavior.

Critical Evaluation

Micro approaches help us understand what culture means to people and how these meanings differ across societies. However, symbolic interactionists don't offer a systematic framework that explains *how* people create and shape culture or develop shared meanings of reality. Why, for instance, are some of us less polite than others even though we share the same cultural values and norms? Another weakness is that interactionists don't address the linkages between culture and subcultures. Language bonds people, but interactionists say little about how organized groups (like the English-only movement) try to maintain control over language to promote their own beliefs.

Chapter Review

Key Takeaways

3-1 What is culture and the main types of culture? *Culture* is learned, transmitted from one generation to another, adaptive, and always changing. A *society* shares a culture and considers itself to be a social unit. People construct a *material culture* (e.g., buildings) and *nonmaterial culture* (e.g., rules for behavior) that influence each other (such as forbidding smoking in public buildings).

3-2 What are the building blocks of culture? The following are some of the fundamental building blocks of culture:

- *Symbols* take many forms, distinguish one culture from another, can change over time, and can unify or divide a society.
- *Language* makes us human; it can change over time and affects perceptions about gender, race, social class, and ethnicity.
- *Values* provide general guidelines for behavior; they sometimes conflict, are usually emotion laden, vary across cultures, and change over time.
- *Norms*—whether folkways, mores, or laws—regulate our behavior; they vary across cultures and sanctions range from mild to severe.
- *Rituals* transmit and reinforce norms that unite people and strengthen their relationships.

Example: Sanctions for Violating the Dead

Sanctions are more severe for violating laws than folkways. Legacy.com, which carries a death notice or obituary for almost all of the roughly 2.4 million Americans who die each year, dedicates at least 30 percent of its budget to weeding out comments (a relatively mild punishment) that "diss the dead" (Urbina, 2006). In contrast, in many states, vandalizing a tombstone is a property crime that can result in a fine of up to $1,000, a jail sentence of up to a year, or both.

3-3 What cultural similarities exist across societies? Customs and specific behaviors vary across countries, but *cultural universals*, such as food taboos, are common to all societies. *Ideal culture* often differs from *real culture*. *Ethnocentrism* has its benefits but can also lead to conflict and discrimination. *Cultural relativism*, the opposite of ethnocentrism, encourages cross-cultural understanding and respect.

3-4 How does culture vary within societies? *Subcultures* and *countercultures* account for some of the complexity within a society. Subcultures differ from the larger society in some ways, while countercultures oppose or reject some of the dominant culture's basic beliefs, values, and norms. In *multiculturalism*, several cultures coexist without trying to dominate one another. People who encounter an unfamiliar way of life or environment may experience *culture shock* because everyone is culture bound to some degree.

3-5 What are the differences between high and popular culture? *High culture* refers to the cultural expression of a society's highest social classes, while *popular culture*—which is widely shared among a population—includes music, social media, sports, hobbies, fashions, the food we eat, the people with whom we spend time, the gossip we share, and the jokes we tell. *Cultural capital* affects social class boundaries, but our everyday life can contain elements of both popular and high culture. Popular culture is typically spread through *mass media*, including television and the Internet, and has enormous power in shaping our attitudes and behavior.

3-6 How does culture change and how does technology affect these changes? Some societies are relatively stable because of *cultural integration*, but all societies change over time because of diffusion, innovation, invention, discovery, and external pressures. *Cultural lags* can create confusion, ambiguity about what's right and wrong, conflict, and a feeling of helplessness. They also expose contradictory values and behavior.

3-7 What are the differences among the main sociological perspectives on culture? Refer to Table 3.3 on page 60.

Key Terms

Culture (p. 41)
Society (p. 41)
Material culture (p. 42)
Nonmaterial culture (p. 42)
Symbol (p. 43)
Language (p. 44)
Values (p. 47)
Norms (p. 48)
Folkways (p. 49)
Mores (p. 50)
Laws (p. 50)
Sanctions (p. 50)
Rituals (p. 51)
Cultural universals (p. 51)
Ideal culture (p. 51)
Real culture (p. 52)
Ethnocentrism (p. 52)
Cultural relativism (p. 53)
Subculture (p. 53)
Counterculture (p. 53)
Multiculturalism (p. 54)
Culture shock (p. 55)
High culture (p. 55)
Popular culture (p. 55)
Cultural capital (p. 56)
Mass media (p. 56)
Cultural imperialism (p. 57)
Cultural integration (p. 57)
Cultural lag (p. 59)

Discussion and Critical Thinking Questions

1. In 2011, a tsunami triggered by an earthquake devastated many towns and villages along Japan's northeastern coast. Journalists who had covered similar natural disasters elsewhere were surprised that there was no looting, even though shops and homes stood unguarded. And instead of complaining, victims waited patiently in orderly lines for water, food, gas, blankets, or whatever was offered (Ford, 2011).
   a. Does this example illustrate a cultural variation or a cultural universal?
   b. Unlike in Japan, looting is common after many natural disasters in the United States. Why?
2. The U.S. describes itself as a bastion of free speech that welcomes debate and differences in views. In recent years, however, the practice of shutting down or cancelling people or organizations who express contrary views has become more frequent. "Cancel culture" can include mass online shunning or boycotting, for instance, of products. Supreme Court Justice Clarence Thomas experienced cancel culture when students at George Washington Law School demanded that he be removed from his academic position following the Supreme Court's decision to overturn *Roe v. Wade*. Thomas opted to step down (Turley, 2022). Cancel culture has had a wide reach, affecting people and organizations who express ideas aligning with liberal viewpoints as well as more conservative ones. Is cancel culture teaching people that free speech should only apply to people with whom they agree?
3. What norms did you learn growing up? If you're raising children or plan to do so in the future, are you teaching or would you teach them norms, particularly folkways and mores, that differ from those that you learned? If so, which ones? If not, why not?
4. Does multiculturalism strengthen or weaken America's cultural identity? Explain why.

True or False?

How Much Do You Know about Culture?

The answers to 2 and 5 are true. The answers to 1, 3, 4, and 6 are false.

1. Culture influences many aspects of life including what a person eats; how they were raised and may raise their own children (if they have them); if, when, and whom a person may marry; how a person makes and spends money; and even what they read.
2. Nonmaterial culture consists of the ideas that people create to interpret and understand the world. Rules of behavior are examples of nonmaterial culture.
3. Ethnocentrism is the belief that one's own culture, society, or group is inherently superior to others. It can be functional; for instance, appreciating one's own country and heritage promotes loyalty and cultural unity, reinforces conformity, and maintains stability. Yet, ethnocentrism is also dysfunctional because viewing others as inferior generates hatred, discrimination, and conflict.
4. There's considerable cultural variation not only *across* societies but also *within* the same society. Some of the variation within a culture can be attributed to subcultures and countercultures.
5. Mass media, or forms of communication designed to reach large numbers of people, have enormous power in shaping public attitudes and behavior. Many people are highly influenced by this form of popular culture that's largely controlled and manipulated by newspapers and magazines, television, movies, music, and ads.
6. Sometimes there is a gap (cultural lag) between the development of material culture and the related nonmaterial culture. There are many examples of cultural lag in modern society; for example, the lack of nonmaterial culture (norms and laws) to address texting while driving and regulation of Internet and phone-based communication privacy.

4 Socialization

Comstock Images/Getty Images

Learning Objectives

After studying this chapter, you will be able to…

4-1 Explain socialization and its importance.

4-2 Describe the nature versus nurture debate.

4-3 Compare social learning and symbolic interaction theories of socialization.

4-4 Describe the primary socialization agents.

4-5 Explain how socialization changes throughout life.

4-6 Explain when and how resocialization occurs.

True or False?

How Much Do You Know about Socialization?

1. Socialization only occurs during childhood.
2. Experiencing social isolation has little effect on a person's development.
3. Studies of twins demonstrate that nature (genetic makeup) is the main factor affecting human behavior.
4. Family is the most important primary agent of socialization.
5. Popular culture and the media are types of socialization agents.
6. New parents may experience resocialization with the dramatic life changes prompted by the birth of a firstborn child.

The answers are at the end of the chapter.

16 and Pregnant, a popular reality show that aired between 2009 and 2014, spawned three equally popular *Teen Mom* series. The producers say the goal of the programs is to decrease teen pregnancy by showing the harmful effects of not using protection or birth control and the resulting struggles in raising a baby. Critics argue that the shows glamorize teen pregnancy, make the teen moms instant celebrities, and ignore the long-term negative impact on the mothers' and children's healthy social development (Thompson, 2010). Both sides are talking about **socialization**, the lifelong process through which people learn culture and become functioning members of society.

Be sure to take the True or False quiz in the chapter opener to better understand how much you know about these topics.

What do you think?

I'm always the same person, no matter where I am or who's around.

1	2	3	4	5	6	7
	strongly agree				strongly disagree	

4-1 Socialization: Its Purpose and Importance

Socialization is critical in all societies. To understand why, let's begin by discussing the purpose of socialization.

4-1a What Is the Purpose of Socialization?

Socialization—from childhood to old age—can be relatively smooth or bumpy, depending on factors such as age, gender, race/ethnicity, and social class. Generally, however, socialization has four key functions that range from providing us with a social identity to transmitting culture to the next generation.

Socialization Establishes Our Social Identity

Have you ever thought about how you became the person you are today? Sociology professors sometimes ask their students to give 20 answers to the question "Who am I?" (Kuhn and McPartland, 1954). How would you respond? You would probably include a variety of descriptions like college student, single or married, your gender, and your nationality. Your answers would show a sense of your *self* (a concept we'll examine shortly). You are who you are largely because of socialization.

Socialization Teaches Us Role Taking

Why do you act differently in class than with your friends? Because we play different roles in different settings. A *role* is the behavior expected of a person in a particular social position (refer to Chapter 5). The way we interact with a parent is typically very different from the way we talk to an employer, a child, or a professor. We all learn appropriate roles through socialization.

Socialization Controls Our Behavior

In learning appropriate roles, we absorb values and a variety of rules about how we should (and shouldn't) interact in everyday situations. If we follow the rules, we're usually rewarded. If we break the rules, we may be punished. Socialization maintains social order and controls our behavior by teaching us to conform to social expectations.

We conform because we've internalized societal norms and values. **Internalization** is the process of learning cultural behaviors and expectations so deeply that we accept them without question. Obeying laws, paying bills on time, and respecting teachers are examples of internalized behaviors.

Socialization Transmits Culture to the Next Generation

Each generation passes its culture on to the next generation. The culture that is transmitted, as you learned in Chapter 3, includes language, beliefs, values, norms, and symbols.

socialization the lifelong process through which people learn culture and become functioning members of society.

internalization the process of learning cultural behaviors and expectations so deeply that we accept them without question.

The Harlow Studies and Emotional Attachment

In the early 1960s, psychologists Margaret and Harry Harlow (1962) conducted several studies on infant monkeys. In one group, a "mother" made of terry cloth provided no food, while a "mother" made of wire did so through an attached baby bottle containing milk. In another group, the cloth mother provided food, but the wire mother didn't. Regardless of which mother provided milk, when both groups of monkeys were frightened, they clung to the cloth mother. The Harlows concluded that physical contact and comfort were more important to the infant monkeys than nourishment.

Since then, some sociologists have cited the Harlow studies to argue that emotional attachment may be even more critical than food for human infants. Do you think there are any problems with sociologists' generalizing the results of animal studies to humans?

Nina Leen/Time Life Pictures/Getty Images

4-1b Why Is Socialization Important?

Social isolation can be devastating. An example of its negative effects is Genie. When Genie was 20 months old, her father decided she was impaired, put her in a wire mesh cage, and locked her away in a back room with the curtains drawn and the door shut. Genie's father frequently beat her with a wooden stick, and no one in the house was allowed to speak to her. When she was discovered in 1970 at age 13, a psychiatrist described Genie as "unsocialized, primitive, and hardly human." Except for high-pitched whimpers, she never spoke. Genie had little bowel control, experienced rages, and tried to hurt herself. After living in a rehabilitation ward and a foster home, she learned to eat normally, was toilet-trained, and gradually developed a vocabulary, but her language use never progressed beyond that of a 3- or 4-year-old (Curtiss, 1977).

The research on children who are isolated (such as Genie) demonstrates that socialization is critical to our development. Behaviors such as talking, eating with utensils, and controlling our bowel movements don't come naturally. Instead, we learn to do all of these things beginning in infancy. When children are deprived of social interaction, they don't develop the characteristics that most of us interpret as typical and human.

4-2 Nature and Nurture

Biologists focus on the role of heredity (or genetics) in human development. In contrast, most social scientists, including sociologists, underscore the importance of learning, socialization, and culture. This difference of opinion is often called the *nature–nurture debate* (**Table 4.1**).

4-2a How Important Is Nature?

Those who argue that nature (biology) shapes behavior point to examples such as the developmental and health differences between males and females.

Developmental and Health Differences

Boys mature more slowly than girls, get sick more often, and are less likely to have mastered the self-control

Table 4.1 The Nature–Nurture Debate

Nature	Nurture
Human development is . . .	**Human development is . . .**
Innate	Learned
Biological, physiological	Psychological, social, cultural
Due largely to heredity	Due largely to environment
Fairly fixed	Fairly changeable

and fine motor skills necessary for a successful start in school. Boys are also at greater risk than girls for most of the major learning and developmental disorders; they are as much as four times more likely to suffer from autism, attention deficit disorder, and dyslexia (Eliot, 2012). Girls, however, report higher rates of depression symptoms (57 percent) compared to boys (29 percent), and girls are more likely to experience suicidal ideation (30 percent) than boys (14 percent) (Donegan, 2023).

Among adults, the senses of smell and taste are more acute in women than in men, and hearing is better and lasts longer in women than in men. Some conditions (migraine headaches and breast cancer) are more common in women, while others (hemophilia and skin cancer) are more common in men (McDonald, 1999; Kreeger, 2002). Men are also more likely to be diagnosed with type 2 diabetes (Medical News Today, 2023).

4-2b How Important Is Nurture?

Most sociologists maintain that nurture is more significant than nature because socialization and culture shape human behavior. They point to studies showing cross-cultural variations in male violence and the environment's effect on biology.

Cross-Cultural Variations in Male Violence

We are often told that males are "naturally" more aggressive than females because their glands produce more testosterone, the dominant male hormone. If men were innately aggressive due to biology, they would be equally violent across all societies. This isn't the case. The proportion of women who have ever suffered physical violence by a male partner varies considerably across nations: 12 percent in Switzerland, 14 percent in Cuba, 26 percent in the United States, 41 percent Zambia, and 42 percent in Bolivia (World Health Organization, 2021). Nearly all mass shooters in the United States (those who have killed a large number of people during one incident) have been men (98 percent). Yet, the numbers of mass shootings vary across countries. Seventy-three percent of mass shootings occurring in higher-income countries have been in the United States (The Violence Project, 2021; Silva, 2021). Such variations reflect cultural laws and practices and other environmental factors (nurture) rather than biology or genetics (nature) (Chesney-Lind and Pasko, 2004).

How Environment Affects Biology

Much research shows that environment (nurture) influences children's genetic makeup (nature). For example, birth defects associated with prenatal alcohol exposure can occur in the first 3 to 8 weeks of pregnancy, before a woman even knows she's pregnant. A pregnant woman's single drinking "binge"—lasting four hours or more—can permanently damage an unborn child's brain. A man who drinks heavily may have genetically damaged sperm that also leads to birth defects. Thus, alcohol misuse (an environmental factor) by either a woman or a man can have a devastating, irreversible, and lifelong negative biological impact on a child (Denny et al., 2009; Warren, 2012; Finelli et al., 2022).

In a growing field known as "fetal origins," scientists are finding that the nine months of pregnancy constitute the most consequential period of our lives, permanently influencing the wiring of the brain, the functioning of organs like the heart and liver, and behavior. That is, the kind and quantity of nutrition you received in the womb; the pollutants, drugs, and infections you were exposed to during gestation; your parents' health and eating habits; and your mother's stress level during pregnancy—all these factors have lasting effects in infancy, childhood, and adulthood (Begley, 2010; Paul, 2010).

Adult behavior can influence a child's biological makeup in other ways. Physical, psychological, or sexual maltreatment can blunt a child's biological development and lead to behavioral and emotional problems throughout life (refer to Chapter 12). Siblings share much of their genetic makeup, but the quality of the schools they attend and the teachers they have affect their interests, social skills, academic outcomes, and earnings in adulthood (Whitehurst, 2016).

4-2c What Can We Conclude About the Nature–Nurture Debate?

Humans have hundreds of traits—like height, metabolism, aggression, leadership traits, and cognitive ability—that are partly inherited (Freese, 2008). A team of researchers analyzed 50 years of twin studies and more than 300 characteristics. They found that, on average, 49 percent of the individual differences were genetic, and 51 percent were environmental, but some traits

Ezra Shaw/Getty Images

Three members of the Manning family (father and two sons) played football at the collegiate and professional levels. Is children's athleticism due to nature, nurture, or both?

Twin Studies, A Mix of Nature and Nurture

Identical twins, those who share the same genetic makeup at birth (monozygotic) provide a useful way to explore the influences of nature and nurture on human behavior and health. The Minnesota Study (1979 through 1999) focused on 81 pairs of identical twins who had been raised apart, which allowed researchers to better isolate biological and social influences on behavior. The study found some interesting similarities about the adult twins, including one pair who shared a tendency to bite their nails, drove the same car, and preferred to smoke the same brand of cigarettes (Lewis, 2014).

Yet, it is not a simple as thinking, "all identical twins will come out the same." Additional research on twins and social factors has demonstrated some of the differences emerge through the interplay of environment and genetics (both nature and nurture). The study of epigenetics is helping us better understand how the genes that people are born with tell only part of the story and how, through the course of life, different environmental triggers that people encounter can "turn on" or "turn off" genes that affect the development of certain diseases such as various types of cancer.

For more information on twin studies, refer to the work of Nancy Segal, https://www.apa.org/news/podcasts/speaking-of-psychology/studying-twins.

(e.g., a cleft lip) were more likely to be inherited than others (e.g., antisocial behavior) (Polderman et al., 2015).

Our social environment can enhance or dampen biological characteristics. Children with biological parents who are skilled athletes may have a genetic predisposition (genetic markers) for enhanced athletic skill (Ahmetov, 2016). Social influences can also shape a child's experience with athletics. Parents can help kids excel at sports by encouraging them to try a range of sports and allowing them to have fun with the activities (Spiker, 2015).

Genetic intelligence thrives in an enriching environment but is stifled in poor and disadvantaged conditions. Children of average intelligence who are adopted by parents with high incomes tend to grow up to be rich adults; genetically talented children raised by biological parents who are economically disadvantaged are much less likely to enjoy economic success as adults (Bryant, 2014; Black et al., 2015; Tucker-Drob and Bates, 2016; refer also to Chapters 8 and 13). Biological factors play an important role in human behavior, but sociology's larger concern is how socialization affects people's development, even overriding some genetic predispositions and influences.

4-3 Sociological Explanations of Socialization

Functionalism provides a foundation for understanding the purposes of socialization described at the beginning of this chapter but doesn't tell us *how* socialization works. There are well-known psychological and psychosocial theories of human development (e.g., Sigmund Freud, Erik Erikson, Jean Piaget, and Lawrence Kohlberg). In sociology, two influential micro approaches that explain socialization are *social learning theories* and *symbolic interaction theories*. (**Table 4.2** summarizes these perspectives).

4-3a Social Learning Theories

The central notion of **social learning theories** is that people learn new attitudes, beliefs, and behaviors through social interaction, especially during childhood. The learning is direct and indirect and a result of observation, reinforcement, and imitation (Bandura and Walters, 1963; Mischel, 1966; Lynn, 1969).

SelectStock/Getty Images

social learning theories people learn new attitudes, beliefs, and behaviors through social interaction.

Table 4.2 Key Elements of Socialization Theories

Social Learning Theories	Symbolic Interaction Theories
• Social interaction is important in learning appropriate and inappropriate behavior. • Socialization relies on direct and indirect reinforcement.	• The self emerges through social interaction with significant others. • Socialization includes role taking and controlling the impression we give to others.
Example: Children learn how to behave when they are scolded or praised for specific behaviors.	*Example:* Children who are praised are more likely to develop a strong self-image than those who are always criticized.

Direct and Indirect Learning

Reinforcement refers to direct or indirect rewards or punishments for particular behaviors. Consider gender roles. A little girl who puts on her mother's makeup may be told that she's cute, but her brother will be scolded ("boys don't wear makeup"). Children also learn through indirect reinforcement. If a little boy's male friends are punished for crying, he'll learn that "boys don't cry."

Children also learn through *observation* and *imitation*. Even when children aren't directly rewarded or punished for "behaving like a boy" or "behaving like a girl," they learn about gender by watching who does what in their families. A father who is rarely at home because he's always working sends the message that men are supposed to earn money with less focus on the family. A mother who regularly expresses concern about her own food consumption and exercises excessively sends the message that women are supposed to be focused on maintaining a thin physique.

Because parents are emotionally important to their children, they're typically a child's most powerful **role models**, people we admire and whose behavior we imitate. Other role models, as you'll learn shortly, include siblings, grandparents, teachers, friends, and even celebrities.

Parents and other role models reinforce particular behaviors, but much of our learning is informal and occurs in a variety of social contexts. A study analyzed nearly 6,000 children's picture books published from 1900 to 2000. Females were central characters in no more than 33 percent of the books published in any given year (McCabe et al., 2011). Such underrepresentation teaches children, though informally, that males occupy more important roles in society than women.

Learning and Performing

Social learning theorists also distinguish between *learning* and *performing* behavior. Children and adults learn norms and roles through observation, but they don't always imitate the behavior. For example, children may know their friends cheat in school but don't do so themselves. Adults, similarly, may be aware that their coworkers steal office supplies but choose to buy their own. We behave as we do, then, because our society teaches us what's appropriate and inappropriate.

Critical Evaluation

Social learning theories help us understand why we behave as we do, but much of the emphasis is on early socialization rather than what occurs throughout life. Social learning theories don't explain why reinforcement and modeling affect some children more than others, especially those in the same family. There may be personality differences, but if learning is as effective as social learning theorists maintain, siblings' attitudes and behavior should be more similar than different. This is often not the case, even with identical twins.

Social learning theories often ignore factors such as birth order, which brings different advantages and disadvantages. Except for affluent families, larger families have fewer resources for the second, third, and later children. The first child may be disciplined and monitored more but also has higher grades than their younger siblings and is more likely to attend college. In contrast, later-born children often have fewer rules, less parental supervision, lower grades, and are less likely to receive financial support to attend college (Conley, 2004; Hotz and Pantano, 2013).

Another limitation is that social learning theories disregard the interconnections between social structures and socialization. As family forms become more complex, children's socialization becomes more intricate. In the last 50 years or so, children have experienced greater family instability due to a biological parent's divorce, remarriage, cohabitations, and having children with more than one partner. One outcome is that by age 9,

role models people we admire and whose behavior we imitate.

Pat Byrnes The New Yorker Collection/The Cartoon Bank

children who have undergone such disruptions are more likely to demonstrate behavioral problems that include delinquency (e.g., stealing, vandalizing property) and classroom disruptions (e.g., fighting, bullying, having temper tantrums) (Fomby and Osborne, 2017).

4-3b Symbolic Interaction Theories

Symbolic interaction theories have had a major impact in explaining social development. Sociologists Charles Horton Cooley (1864–1929), George Herbert Mead (1863–1931), and Erving Goffman (1922–1982) were especially influential in showing how social interaction shapes socialization.

Charles Horton Cooley: Emergence of the Self and the Looking-Glass Self

Newborn infants lack a sense of **self**, an awareness of their social identity. Gradually, they begin to differentiate themselves from their environment and develop a sense of self.

After carefully observing the development of his young daughter, Charles Horton Cooley (1909/1983) concluded that children acquire a sense of who they are through their interactions with others, especially by imagining how others view them. The sense of self, then, isn't innate but emerges out of social relationships.

Cooley called this social self the *reflected self*, or the **looking-glass self**, a self-image based on how we think others view us. He proposed that the looking-glass self develops in an ongoing process of three phases:

- **Phase 1: Perception.** We imagine how we appear to other people and how they *perceive* us ("She thinks I'm attractive" or "I bet he thinks I'm fat").
- **Phase 2: Interpretation of the perception.** We imagine how others *judge* us ("She's impressed with me" or "He's disgusted with the way I look").
- **Phase 3: Response.** We experience *self-feelings* based on how we think other people judge us. If we think they feel we are favorable light, we may feel proud, happy, or self-confident ("I'm terrific"). If we think others interpret our actions, behaviors, or characteristics in a negative light, we may feel angry, embarrassed, or insecure ("I'm pathetic").

Our interpretation of others' perceptions (phase 2) may be totally wrong. The looking-glass self, remember, refers to how we *think* others perceive us rather than how they *actually* perceive us. Our perceptions of other people's views—whether we're right or wrong—affect our self-image, which then affects our behavior.

Cooley focused on how children acquire a sense of who they are through their interactions with others, but he noted that the process of forming a looking-glass self doesn't end in childhood. Instead, our self-concept may

Photodisc/Getty Images

Parents who encourage their children to express themselves help them develop a positive self-image.

self an awareness of one's social identity.

looking-glass self a self-image based on how we think others perceive us.

change over time because we reimagine ourselves as we think others perceive us—attractive or ugly, interesting or boring, intelligent or stupid, graceful or awkward, and so on.

George Herbert Mead: Development of the Self and Role Taking

Cooley described how an individual's sense of self emerges, but not how it develops. George Herbert Mead, one of Cooley's colleagues, took up this task. For Mead (1934), the most critical social interaction occurs in the family, the foundation of socialization.

Our self develops, according to Mead, when we learn to differentiate the *me* from the *I*—two parts of the self. The *I* is creative, imaginative, impulsive, spontaneous, self-centered, and sometimes unpredictable. The *me* that has been successfully socialized is aware of other people's attitudes, has self-control, and has internalized social roles. Instead of impulsively and selfishly grabbing another child's toys, for example, as the *I* would do, the *me* asks for permission to use someone's toys and shares them with others.

For Mead, the *me* forms as children engage in **role taking**, learning to take the perspective of others. Children gradually acquire this ability early in the socialization process through three sequential stages:

1. **Preparatory stage (roughly birth to 2 years).** An infant doesn't distinguish between the self and others. The *I* is dominant, while the *me* is forming in the background. In this stage, children learn through imitation. They may mimic daddy's shaving or mommy's angry tone of voice without really understanding the parent's behavior. In this exploratory stage, children engage in behavior that they rarely associate with words or symbols, but they begin to understand cause and effect (e.g., crying leads to being picked up). Gradually, as the child begins to recognize others' reactions and forms a looking-glass self, they develop a self.
2. **Play stage (roughly 2 to 6 years).** The child begins to use language and to understand that words (like *dog* and *cat*) have a shared cultural meaning. Through play, children begin to learn role taking in two ways. First, they emulate significant others. A **significant other** is someone whose opinions we value and who influences our thinking, especially about ourselves. A significant other can be anyone: usually a parent, grandparent, or sibling in early childhood; a teacher, romantic partner, spouse, best friend, or employer in later childhood and adulthood. The child learns that they have a self that's distinct from others, that others behave in many different ways, and that others expect them to behave in specific ways. In other words, the child learns social norms (refer to Chapter 3).

   In the play stage, the child moves beyond imitation and acts out imagined roles ("I'll be the mommy and you be the daddy"). The play stage involves relatively simple role taking because the child plays one role at a time but doesn't yet understand the relationships between roles. This stage is crucial, according to Mead, because the child *is learning to take the role of the other.* For the first time, the child tries to imagine how others behave or feel. The *me* grows stronger because the child is concerned about the judgments of significant others.

   Also in the play stage, children experience **anticipatory socialization**, learning how to perform a role they'll occupy in the future. By playing "mommy" or "daddy," children prepare themselves for eventually becoming parents. Anticipatory socialization continues in later years when, for instance, expectant parents attend childbirth classes, job seekers practice their skills in mock interviews, and high school students visit campuses and attend orientation meetings to prepare for college life.
3. **Game stage (roughly 6 years and older).** This stage involves acquiring the ability to understand connections between roles. The child must "not only take the role of the other . . . but must assume the various roles of all participants in the game, and govern his action accordingly" (Mead, 1964, p. 285).

Mead used baseball to illustrate this stage. In baseball (or other organized games and activities), the child plays one role at a time (batter) but understands and anticipates the actions of other players (pitcher, shortstop, runners on bases) on both teams. To successfully participate in a game of baseball, the individual must be able to anticipate the roles of others. The same is true of participating in society. Thus, as children grow older and interact with a wider range of people, they learn to anticipate, respond to, and fulfill a variety of social roles.

At the game stage, children connect to societal roles through the **generalized other**, the norms, values, and expectations of society that affect a person's behavior.

---

role taking learning to take the perspective of others.

significant other someone whose opinions we value and who influences our thinking, especially about ourselves.

anticipatory socialization learning how to perform a role that a person will occupy in the future.

generalized other the norms, values, and expectations of society that affect a person's behavior.

STAGE 1: **Preparatory Stage** (under age 2)

No distinction between self and others; the child is self-centered and self-absorbed

Learns through observation

STAGE 2: **Play Stage** (aged 2 to about 6)

Distinguishes between self and others

Imitates **significant others** (usually parents)

Learns role taking, assuming one role at a time in "let's pretend" and other play that teaches **anticipatory socialization**

STAGE 3: **Game Stage** (aged 6 and older)

Understands and anticipates multiple roles

Connects to societal roles through the **generalized other**

Figure 4.1 Mead's Three Stages in Developing a Sense of Self

This concept suggests that when an individual acts, they take into account people in general. For example, a child is honest not merely to please a particular parent but because parents, teachers, and others endorse honesty as an important social value. For sociologists, the development of the generalized other is a central feature of socialization because the *me* becomes an integral part of the self (**Figure 4.1**).

After the generalized other has developed, the *me* never fully controls the *I*, even in adulthood. We sometimes break rules or act impulsively, even though we know better. When we're frustrated, sick, tired, or angry, we may lash out against other people even though they haven't done anything wrong.

Erving Goffman: Staging the Self in Everyday Life

Cooley and Mead described how the self and role taking emerge and develop during early socialization. Erving Goffman (1959, 1969) extended these analyses by showing that we interact differently in different settings throughout adulthood. Goffman proposed that social life mirrors the theater because we're like actors: We engage in "role performances," want to influence an "audience," and can have considerable control over the image that we project while we're "on stage." (We'll explore these concepts in greater detail in Chapter 5.)

In a process that Goffman called **impression management**, we provide information and cues to others to present ourselves in a favorable light while downplaying or concealing our less appealing characteristics. Being successful in this presentation of the self requires managing three types of expressive resources. First, we try to control the *setting*—the physical space, or "scene," where the interaction takes place. In the classroom, "a professor may use items such as chalk, lecture notes, computers, videos, and desks to facilitate a class and show that he or she is an excellent teacher" (Sandstrom et al., 2006, p. 105).

A second expressive resource is controlling *appearance*, such as clothing and titles, that convey information about our social status. When physicians or professors use the title "Doctor," for example, they're telling the "audience" that they expect respect.

The third expressive resource is *manner*—the mood or style of behavior we display that sends important messages to the audience. Faculty members regularly manage their manner when interacting with students. When you email or text a professor a question about your course paper, the response is usually encouraging ("Have you also thought about...") even though the professor may actually be annoyed ("You haven't thought about this very much; stop wasting my time").

As we move from one situation to another, according to Goffman, we maintain self-control, for example, by avoiding emotional outbursts and altering our facial expressions and verbal tones. Thus, all of us engage in impression management practically every day.

AndreyPopov/iStock/Getty Images

Who's engaging in impression management? How? Why?

impression management providing information and cues to others to present oneself in a favorable light while downplaying or concealing one's less appealing characteristics.

Critical Evaluation

Symbolic interactionists have provided major insights about socialization, particularly of young children, and have shown that fitting into our social world is a complex process that continues through adulthood. Like other theories, however, symbolic interaction has its limitations.

First, it's not clear why some children have a positive looking-glass self and are successful later in life even when the cues are consistently negative. Over the years, some of my best students have included those who grew up in households where they were mistreated or attended lower-quality schools.

Second, essential concepts such as *self, me,* and *I* are vague. Because the concepts are imprecise, it's difficult, if not impossible, to measure them.

Third, stage theories, like Mead's, maintain that children automatically pass through play and game stages as they get older. Some critics argue that the stages aren't as rigid as symbolic interactionists claim. The stages may overlap or vary, for example, depending on whether parents and other primary caregivers are actively involved in the child's upbringing and provide enriching interaction. Even then, some children never develop the role of the generalized other (refer to Ritzer, 1992).

Fourth, some have questioned the value of the concept of the generalized other in understanding early socialization processes. Because children interact in many social contexts (home, preschool, play groups), they don't simply assume the role of one generalized other, but many. That is, as children get older, they may have several **reference groups**—people who shape an individual's self-image, behavior, values, and attitudes in different contexts (Merton and Rossi, 1950; Shibutani, 1986).

Fifth, interactionists credit people with more free will than they have. Individuals don't always have the power or ability to affect others' reactions or their own lives. For example, impression management is less common among lower than higher socioeconomic groups, whose members have internalized such skills to be successful in jobs. Also, how people react to us may be due to characteristics (e.g., gender, age, race/ethnicity) that we can't change (Powers, 2004).

Finally, a major criticism of Cooley, Mead, Goffman, and other interactionists is that they tend to downplay or ignore macro-level factors that affect people's development. As the next two sections show, institutions such as the family, education, economy, and popular culture have a major impact on socialization. Let's begin by discussing socialization agents who play important roles in our lives.

4-4 Primary Socialization Agents

By fretting, crying, and whining, infants teach adults when to feed, change, and pick them up. Babies are actively engaged in the socialization process, but family, friends, peer groups, teachers, and the media are some of the primary **socialization agents**—the individuals, groups, or institutions that teach us how to participate effectively in society.

4-4a Family

Parents, siblings, grandparents, and other family members play a critical role in our socialization. Parents, however, are the first and most influential socialization agents.

How Parents Socialize Children

The purpose of socialization is to enable children to regulate their behavior and to make responsible decisions. In teaching their children social rules and roles, parents rely on several learning techniques, such as reinforcement, that you've just read about. Parents also manage many aspects of the environment that influence a child's social development: They choose the neighborhood (which often determines what school a child attends), may decorate the child's room in a gender-based style, provide them with particular toys and books, and arrange enriching social activities (e.g., sports, art, music).

Parenting Styles

Four common parenting styles affect socialization (**Table 4.3**). In *authoritarian parenting,* parents tend to be harsh, unresponsive, and rigid, using their power to control a child's behavior. *Authoritative parenting* is warm, responsive, and involved, but parents set limits and expect appropriately mature behavior from their children. *Permissive parenting* is lax: Parents set few rules but are usually warm and responsive. *Uninvolved parenting* is indifferent and neglectful: Parents focus on their own needs, spend little time interacting with the children, and know little about their interests or whereabouts (Baumrind, 1968, 1989; Maccoby and Martin, 1983; Aunola and Nurmi, 2005).

Healthy child development is most likely in authoritative homes because parents are consistent in combining warmth, monitoring, and discipline. Authoritative parenting tends to produce children who are self-reliant, achievement-oriented, and successful in school.

---

reference groups people who shape an individual's self-image, behavior, values, and attitudes in different contexts.

socialization agents the individuals, groups, or institutions that teach us how to participate effectively in society.

Table 4.3 Parenting Styles

Parenting Style	Characteristics	Example
Authoritarian	Very demanding, controlling, punitive	"You can't borrow the car because I said so."
Authoritative	Demanding, controlling, warm, supportive	"You can borrow the car, but be home by curfew."
Permissive	Not demanding, warm, indulgent, set few rules	"Borrow the car whenever you want."
Uninvolved	Neither supportive nor controlling	"I don't care what you do; I'm busy."

Adolescents in households with authoritarian, permissive, or uninvolved parents tend to have poorer psychosocial development, lower school grades, lower self-reliance, higher levels of delinquent behavior, and are more likely to be swayed by harmful peer pressure (e.g., to use drugs and alcohol) (Hillaker et al., 2008; Meteyer and Perry-Jenkins, 2009).

Such findings vary by social class and race/ethnicity, however. For many Latina/o/x and Asian immigrants, authoritarian parenting produces positive outcomes, like respectful behavior and academic success. This parenting style is also effective in safeguarding children who are growing up in low-income neighborhoods with high crime rates and drug peddling (Pong et al., 2005).

Across all social classes, Black parents—particularly mothers—engage in *racial socialization*, a process that teaches children to be proud of their race and how to recognize and diminish the negative effects of discrimination. For example, many middle- and upper-middle-class Black parents spend considerable time teaching their sons how to navigate negative race-based stereotypes. Some of the strategies include suppressing their anger, frustration, or excitement when interacting with authority figures, like teachers and police officers, and monitoring how they dress (Dow, 2016).

Siblings

Siblings, like parents, are important socialization agents. Older siblings may transmit beliefs that having sex at an early age, smoking, drinking, and marijuana use "aren't a big deal" (Altonji et al., 2010). Yet, siblings may influence each other regardless of birth order. One study found that younger siblings who participate in risky behaviors, such as excessive alcohol consumption, may act as role models and create opportunities for their older siblings to do the same (Whiteman et al. 2017). Siblings are also positive role models when they encourage each other to do well in school, to stay away from friends who get in trouble, and to get along with people. Siblings can have a positive impact by helping each other with homework and protecting them from neighborhood bullies as well as reducing the chances of risky sexual behavior and teen pregnancy. During adulthood, siblings often act as confidantes and offer financial and emotional assistance during stressful times (McHale et al., 2012; Killoren et al., 2014; Killoren and Roach, 2014).

Grandparents

In many ways, grandparents are the glue that keeps a family close. Some grandparents and grandchildren live far apart and rarely spend in-person time together, but the majority spend in-person time with their grandchildren often, frequently do things with them, and offer them emotional and instrumental support (e.g.,

wavebreakmedia/Shutterstock

Numerous camps exist to resocialize youth with behavioral, emotional, or substance misuse issues. These efforts may involve year-round boarding schools with therapeutic programs and classes (Golightley, 2020). Others are shorter-term bootcamp formats. In some cases, parents have support in forcibly removing the teens from their homes and placing them in the resocialization programs. Behavior modification results from these experiences are mixed. Some youth in these programs are negatively affected by physical or sexual maltreatment and a lack of necessary physical or mental health treatment (Krebs, 2021).

listening and giving them money). Many recent Asian immigrants live in extended families and are more likely to do so than any other group, including Latinos/as/x. In such co-residence, grandparents are often "historians" who transmit values and cultural traditions to their grandchildren even if there are language barriers (Benokraitis, 2015; refer also to Chapter 12).

Grandparents are sometimes *surrogates* because they provide regular care or raise the grandchildren. Whether custody is legal or informal, about 10 percent of all U.S. children live with a grandparent, up from 5 percent in 1990. About 2.3 million grandparents (32.7 percent) are living with and raising their grandchildren; approximately half of this group has been doing so for five or more years (U.S. Census Bureau, 2021)

The most common pattern is *living-with* grandparents who have the grandchild in their own home or, less often, live in the home of a grandchild's parents. Living-with grandparents take on child-rearing responsibilities either because their children haven't moved out of the house or because teenage or adult parents can't afford to live on their own. These living arrangements have increased the number of **multigenerational households**, homes in which three or more generations live together. In effect, then, grandparents are important socialization agents.

Pressmaster/Shutterstock

Grandparents often provide stability in family relationships and a continuity of family rituals and values. When a teacher asked her class of 8-year-olds what they liked best about a grandparent, the essays included comments like "They don't say 'Hurry up'" and "When they take us for a walk, they slow down for things like pretty leaves and caterpillars."

4-4b Play, Peer Groups, and Friends

As you learned earlier, Mead believed that play and games are important in children's development. They help us become more skilled in using language and symbols, learn role playing, and internalize roles that we don't necessarily enact (the generalized other). A **peer group** consists of people who are similar in age, social status, and interests. All of us are members of peer groups, but such groups are especially influential until about our mid-20s. After that, coworkers, spouses, children, and close friends are usually more important than peer groups in our everyday lives.

Play

Play serves several important functions. First, it promotes cognitive development. Whether it's doing simple five-piece puzzles or tackling complex video games like *SimCity* and *RollerCoaster Tycoon*, play encourages children to think, formulate strategies, and budget and manage resources. From an early age, however, play has generally been gender typed. "Girls'" sections of online and toy stores are swamped with cosmetics, dolls and accessories, arts and crafts kits, and housekeeping toys. "Boys'" sections feature sports supplies, building sets, workbenches, construction equipment, and toy guns. Such gender stereotyping sends girls the message that their most important goals should be looking pretty, nurturing others, and developing cooking and housework skills, often to the exclusion of exploration and invention. Boys are getting the message that they should be adventurous, aggressive, competitive, and athletic, and that "caring for a baby or the home is girly and ... undesirable" (Filipovic, 2016, p. 51).

About 92 percent of American girls aged 3 to 12 own at least one Barbie doll. After Barbie sales plummeted 20 percent because of greater competition by other toy manufacturers, in 2016, Mattel introduced three new Barbie body types—tall, short, and curvy—in a greater number of skin tones, eye colors, and hairstyles (Abrams, 2016; Dockterman, 2016).

Some toymakers have responded to parents' demands for more gender-neutral toys. For example, GoldieBlox and Roominate, created by female engineers, offer construction toys aimed at girls that encourage interest in math, science, and engineering, and Amazon has stopped using gender-based categories for children's toys (Li, 2015; Tabuchi, 2015). Mattel expanded the Barbie line to showcase a greater variety of careers such as video game developer, presidential and vice-presidential candidates, race car driver, and police officer (Mattel Inc., 2022).

---

multigenerational households homes in which three or more generations live together.

peer group people who are similar in age, social status, and interests.

AP Images/Nick Ut

Godong/BSIP SA/Alamy Stock Photo

When Target announced that it would stop labeling girls' and boys' toys in their stores, hundreds of parents flocked to its website to complain because "Boys will always be boys and girls will always be girls!" (Steinmetz, 2015, p. 25). Others argue that toys are more gender typed than in the past (refer to Tabuchi, 2015, and Chapter 9). What do *you* think?

The second function of play—especially when it's structured—is to enhance children's social development. These activities help kids build interests, skills, and confidence as well as support mental health. Sports may also have some positive benefits for improved academic performance. In effect, sports and hobbies provide children with constructive ways to channel their energy and intelligence (Brennan, 2021; Wretman, 2017; McHale, 2001).

Third, play can strengthen peer relationships. Beginning in elementary school, few things are more important to most children than being accepted by their peers. Even if children aren't popular, belonging to a friendship group enhances their psychological well-being and ability to cope with stress (Rubin et al., 1998; Scarlett et al., 2005). Through extracurriculars, children can develop a sense of belonging (being part of the group) (Brennan, 2021).

Peers and Friends

Peer influence usually increases as young children get older. Especially during the early teen years, friends often reinforce desirable behavior or skills that enhance a child's self-image ("Wow, you're really good in math!"). Thus, to apply Cooley's concept, peers can help each other develop a positive looking-glass self.

Peers also serve as good role models. Children acquire a wide array of information and knowledge by observing their peers. Even during the first days of school, children learn to imitate their peers by standing in line, raising their hands in class, and being quiet while the teacher is speaking.

Among teens and young adults who are LGBTQIA+, heterosexual friends can be especially supportive in accepting one's sexual orientation and disclosing it to family. Peers can also have a significant impact in teen dating violence by encouraging their friends to leave a relationship, report the abuse, and seek protection orders (Shilo and Savaya, 2011; Oudekerk et al., 2014). In college, close friends often offer academic support and motivation to succeed: They study together, quiz each other before exams, and proofread each other's papers. Close college friends can also provide social support, a sense of belonging, and relationships that last beyond college (McCabe, 2016).

Not all peer or friend influence is positive, however. Among seventh- to twelfth-graders, having a best friend who engages in sexual intercourse, is truant, joins a gang, and uses tobacco and marijuana increases the probability of imitating such behavior. If close college friends are low achievers and unmotivated, they can interfere in studying, discourage class attendance, and offer little, if any, emotional or academic support (Haas and Schaefer, 2014; McCabe, 2016).

4-4c Teachers and Schools

Oprah Winfrey often praised her fourth-grade teacher, who recognized her abilities, encouraged her to read, and inspired her to excel academically. Like family and peer groups, teachers and schools are important socialization agents.

The School's Role in Socialization

By the time most children are 4 or 5, school fills an increasingly large portion of their lives. The school's primary purpose is to instruct children and enhance their cognitive development. Schools don't simply transmit knowledge; they also teach children to think about the world in different ways. Because of the emphasis on multiculturalism, for example, children often learn about other societies and customs. Even outside of classes, schools affect children's daily activities through homework assignments and participation in clubs and other extracurricular activities.

Because many parents are employed, schools have had to devote more time and resources to topics—such as sex education, abusive adolescent relationships, and drug misuse prevention—that were once the sole responsibility of families (Miller et al., 2015). In many ways, then, schools play an increasingly important role in socialization.

Teachers' Impact on Children's Development

Teachers are among the most important socialization agents. From kindergarten through high school, teachers play numerous roles in the classroom—instructor, role model, evaluator, moral guide, and disciplinarian, to name just a few. Well-trained teachers can motivate students and increase their academic achievement regardless of their innate ability, talents, behaviors, or home circumstances (Gershenson, 2015; refer also to Chapter 13).

Many educators have criticized students' increasing disrespect toward adults and each other, bad manners, and bullying. Instead of complaining, some schools are using programs to enhance students' skills in communication, etiquette and leadership. Programs such as Toastmasters instruct students on public speaking skills as well as constructive interpersonal skills (Toastmasters International, 2023). According to some critics, teachers are wasting valuable time and resources on behavior that should be taught at home, but such programs are growing in popularity across the country.

4-4d Popular Culture and the Media

Because of tablets, smartphones, YouTube, and social networking sites, young people are constantly connected to electronic media (refer to Chapter 5). How does such technology affect socialization?

Electronic Media

The American Academy of Pediatrics (AAP) advises parents to avoid the use of screen media, other than video-chatting, for children younger than 18 months. It also recommends limiting screen use for children ages 2 to 5 to 1 hour per day and only of high-quality programs. The controls would encourage more interactive activities such as talking, playing, singing, and reading together "that will promote proper brain development" (American Academy of Pediatrics, 2016). Despite the AAP's advice, 97 percent of children under age 4—regardless of family income—use mobile devices like smartphones and tablets. Parents are especially likely to let their youngsters use mobile devices when driving in the car (84 percent), and some allow the use of devices when dining at a restaurant (34 percent). Still, 45 percent of parents say they do not spend enough time with their children (Auxier et al., 2020).

Moodboard-Mike Watson Images/Getty Images

Class field trips enhance children's socialization. Among other benefits, the field trips expand learning beyond the classroom, enrich students' understanding of subject matter, and encourage them to think about occupational choices and careers they haven't considered.

The average young American now spends more time with electronic media than in school. The typical 8- to 10-year-old spends nearly 8 hours a day using some type of electronic media (which is more time than many adults spend in a full-time job), and 13- to 18-year-olds spend almost 9 hours a day (Rideout, 2015). Too much time on the Internet has been linked with violence, cyberbullying, trouble in school, obesity, lack of sleep, and lower grades (Strasburger and Hogan, 2013). Screen media usage itself isn't necessarily harmful, but it typically involves "passive consumption," such as watching and listening to content produced by someone else, rather than writing and creating one's own art and music or interacting face-to-face with family and friends (Rideout, 2015).

There's also a growing concern that violent video games make violence seem normal. Playing violent video games (e.g., *Grand Theft Auto, Call of Duty,* and *Halo*) can increase a person's aggressive thoughts, feelings, and behavior both in laboratory settings and in real life. Violent video games also encourage male-to-female aggression because much of the violence is directed at women (Carnagey and Anderson, 2005; Brody, 2015).

What? Adolescents Can Buy Ultraviolent Video Games but Not a Magazine with an Image of a Nude Woman?!

In 2011, the Supreme Court ruled (7–2 in *Brown v. Entertainment Merchants Association*) that video games, even ultraviolent ones, that are sold to minors are protected by the First Amendment's guarantee of free speech. The majority said that none of the scientific studies *prove* that violent games *cause* minors to act aggressively, and that parents, not the government, should decide what is appropriate for their children (Schiesel, 2011; Walls, 2011).

Not surprisingly, the video game industry was jubilant over the decision. Many parents and lawmakers, on the other hand, agreed with one of the dissenting justices that it makes no sense to forbid selling a 13-year-old boy a magazine with an image of a nude woman but not an interactive video game in which the same boy "actively . . . binds and gags the woman, then tortures and kills her" (Barnes, 2011, p. A1).

Hot Property/Shutterstock.com

Still, it's not clear why violent video games affect people differently. Many young males who enjoy playing such games aren't any more aggressive, vicious, or destructive than those who aren't violent video game enthusiasts (Ferguson and Olson, 2013). Instead, factors such as angry and aggressive personalities, indifference to other people's feelings, and the quantity of violent gaming increase the likelihood of violent behavior (refer to Markey et al., 2014; Przybylski and Mishkin, 2015).

Gillian Charters-Barnes/Shutterstock.com

Many men's health and fitness magazines routinely feature models who have undergone several months of extreme regimens, including starvation and dehydration, to tighten their skin and make their muscles "pop." The magazines also use lighting tricks and Photoshop to project an idealized image of hypermasculinity that, in reality, is impossible to attain (Christina, 2011; refer also to Ricciardelli et al., 2010).

Advertising

In 2006, dozens of companies promised to limit child-directed TV ads that used cartoons and other characters to promote unhealthy food. Some companies have run shorter commercials, but the content hasn't changed very much (Kunkel et al., 2015; refer also to Chapter 3).

Advertisers are targeting children at increasingly younger ages. A new form of advertising called *advergaming* combines free online games with advertising. Advergaming is growing rapidly. Sites like Wonka.com, operated by Nestlé, and Barbie.com, operated by Mattel, attract millions of young children and provide marketers with an inexpensive way to "draw attention to their brand in a playful way, and for an extended period of time" (Moore, 2006, p. 5).

Young people, on average, are exposed to more than 1,000 alcohol advertisements per year, which is an average of 3 per day. The amount of exposure to these ads varies by group. Black and Latina/o/x youths have contact with, on average, 4.1 and 3.4 alcohol ads per day, while non-Hispanic White peers are exposed to an average of 2. Girls are exposed to 30 percent more alcohol advertisements than boys (Martino, 2018).

What effect do ads have on girls' and women's alcohol consumption? Over the years, teen boys were more likely to drink and binge drink compared to girls. Now, more teen girls consume alcohol and binge drink. Alcohol use and misuse is also increasing

Marcos Mesa Sam Wordley/Shutterstock

Susan Stewart, in her book *On the Rocks: Straight Talk About Women and Drinking* (2022), highlights deeper explanations for why women are drinking more alcohol and its effects. Beyond alcohol marketing geared towards women, higher levels of stress, due in part to the Covid-19 pandemic, are reasons women are turning to alcohol. Also, the integration of available alcohol into everyday events enjoyed by women has contributed to consumption (e.g., alcohol availability at running events and movie theaters). Because of society's gender-based lenses, women who consume alcohol face more challenges navigating others' perceptions of them (Cramer, 2022).

among women (National Institute on Alcohol Abuse and Alcoholism, 2022a; National Institute on Alcohol Abuse and Alcoholism, 2022b)

4-5 Socialization Throughout Life

Biological aging comes naturally, but social aging is a different matter. As we progress through the life course—from infancy to death—we are expected to learn culturally approved norms, values, and roles.

4-5a Infancy and Toddlerhood

Infancy (between birth and 12 months) and toddlerhood (ages 1 to 3) encompass only a small fraction of the average person's life span but are periods of both extreme helplessness and considerable physical and cognitive growth. Some scientists describe healthy infants' brains as "small computers" because of their enormous capacity for learning (Gopnik et al., 2001, p. 142).

The quality of relationships with adults and other caregivers has a profound effect on infants' and toddlers' development. Engaging infants in talk increases their vocabulary. By age 2, toddlers who know more words have better language skills that, in turn, help them control their behavior at age 3 and later (Vallotton and Ayoub, 2011).

On average, family members spend about 30 minutes a day reading to children under the age of 2 (Rideout, 2014). On the other hand, very young children spend a good deal of time interacting with media and using electronic devices. Parents report that children under the age of 2 interact with various forms of media and electronic devices: television (74 percent), smartphone (49 percent), tablet computer (35 percent), and gaming device (9 percent). In addition, 57 percent of children under the age of 2 watch videos on YouTube (Auxier et al., 2020).

You learned earlier that the AAP has advised parents to limit their infants' and toddlers' screen time. Only 28 percent of parents say that media decreases face-to-face family time, but they often underestimate the amount of time that children under age 2 spend with electronic devices (Rideout, 2013). Electronic media usage increases digital literacy. According to a pediatrics professor, however, young children "need laps more than apps" (quoted in Louis, 2015, p. A16).

4-5b Childhood Through Adolescence

Physical, emotional, and other development continues through childhood and adolescence (roughly ages 4 to 12). Children acquire cognitive skills that include processing thoughts, speaking, reasoning, memorizing and recalling information, analyzing problems, and evaluating options. The quantity and quality of time that parents and teachers invest in cognitive development affect children's long-term outcomes.

That's Not All ...

There are many socialization agents other than family, play and peer groups, teachers and schools, and the media. If you think about the last few years, which groups or organizations have influenced who you are? What about the military, a company you've worked for, college and athletic clubs, religious organizations, support groups, athletic memberships, community groups, or others? All affect our socialization throughout life.

Childhood—viewed as a distinct stage of development—is a fairly recent phenomenon (Ariès, 1962). Until 1938, children made up a large segment of the U.S. labor force, especially in factories. They worked 6 days a week, 12 or more hours a day, received very low wages, and were often injured because of dangerous equipment.

Social class enhances or limits cognitive growth. Middle-class parents, regardless of race or ethnicity, engage in "concerted cultivation" child-rearing processes that deliberately try to improve their children's cognitive and social skills. In contrast, working-class and lower-income parents—because of a lack of free time or resources—facilitate "natural growth" by granting their children more autonomy in everyday life (Lareau, 2011). Especially since the early 1990s, the gap between high- and low-educated parents' investment in children's development has increased (Altintas, 2016).

Besides reading to their children, helping them with homework, and attending their children's school-based events, higher-income parents are much more likely than lower-income parents to enroll children in extracurricular activities Participation in the type of activity varies by income level. With sports, there are more similar levels of participation between kids below poverty level (23 percent) or at or above poverty level (28 percent). For activities including clubs, kids who are at or above the poverty level are more likely to participate, even though participation numbers overall are declining (**Figure 4.2**) (Mayol-Garcia, 2022).

Self-help books and self-proclaimed child experts routinely instruct parents to praise their kids to bolster their self-esteem. Is such advice misguided? A recent study of parents with children ranging from 7 to 12 years old found that constantly praising kids as "special" or "exceptional" and applauding every little achievement results in narcissism. The children often demand special treatment, can't tolerate setbacks, are more violent than other people, and are at high risk for drug addiction and depression (Brummelman et al., 2015). A review of 41 studies concluded that, in the long run, consistent and well-deserved (but not constant) praise strengthens the parent–child relationship, teaches children to overcome setbacks, and improves their problem-solving skills (Owen et al., 2012; Petersen, 2012).

Most U.S. children enjoy happy and healthy lives, but many don't. Almost 1.5 million children aged 17 or younger had at least one parent incarcerated in a state or federal prison (Maruschak et al., 2021). Another 407,000 children are in the foster care system (Administration for Children and Families, 2021). Nearly 1 out of 8 children are living with at least one parent who has a substance use disorder (Lipari and Van Horn, 2017). And, every year, one in seven children experience maltreatment or neglect (U.S. Centers for Disease Control and Prevention, 2022). Thus, for millions of U.S. children, the socialization process is shaky at best.

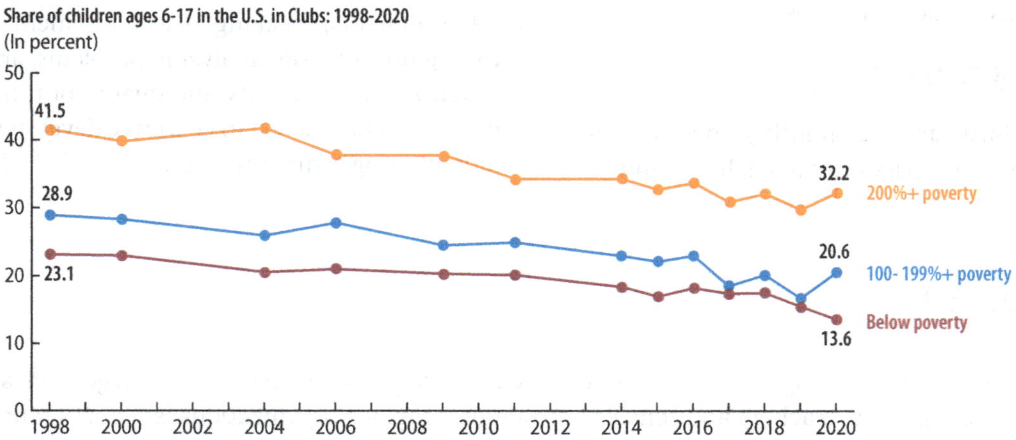

Figure 4.2 Children from Higher-Income Families Are More Likely to Be in Extracurricular Activities

Source: Mayol-Garcia, U.S. Census Bureau, 2022.

Tim Pannell/Corbis

Nationally, 57 percent of 6- to 17-year-olds participate in at least one after-school extracurricular activity. The benefits include learning time management, exploring diverse interests, and building social skills, but some family practitioners believe that today's children are too overbooked and overscheduled (Laughlin, 2014; Lawrence, 2015). Do you agree?

4-5c The Teenage Years

Ages 13 to 19 is another period of tremendous change. Teenagers establish their own identities and become more independent as they mature and break away from parental supervision, a healthy development process. The most difficult part of raising adolescents, according to some parents, is dealing with their changing moods and behavior.

For many years, people attributed such dramatic changes to "raging hormones." Scientists now believe there's a link between a teen's baffling behavior and the fact that their brain may be changing far more than was thought previously. During adolescence, the brain matures at different rates. Areas involved in basic functions such as processing information and controlling physical movement mature first. The parts of the brain responsible for controlling impulses, avoiding risky behavior, and planning ahead—the hallmarks of adult behavior—are among the last to mature (National Institute of Mental Health, 2011).

Parental involvement is usually beneficial in a child's development, but "helicopter parents," who hover over their kids, are hyperinvolved, intrusive, and overcontrolling. Anecdotal examples include parents verbally attacking teachers about their children's low grades, demanding that their child be moved to another class before the school year has even begun, completing their homework assignments, showing up in the guidance counselor's office with college applications they have filled out for their children, and haggling with college professors over the student's grade on an exam or paper (Krache, 2008; Weintraub, 2010; Rochman, 2013).

Helicopter parenting diminishes teens' and young adults' ability to develop decision-making and problem-solving skills. However well-intentioned, helicopter parents increase their own stress and decrease their children's feelings of autonomy and competence. In turn, not believing in one's ability to achieve goals on one's own leads to teenagers' and young adults' feeling anxious, depressed, and more dissatisfied with life (Shellenbarger, 2013; Schiffrin et al., 2014).

4-5d Adulthood

Socialization continues throughout adulthood. People adopt a series of new roles that may include work, marriage, parenthood, divorce, remarriage, buying a house, and experiencing the death of a child, parent, or grandparent. We'll examine these transitions in later chapters. Two of the most important roles in adulthood are work and parenthood, but let's begin with young adults who are reluctant to leave their parents' nest.

The Crowded Empty Nest

During the 1960s and 1970s, sociologists almost always included the "empty nest" in describing the family life course. This is the stage in which parents, typically in their 50s, find themselves alone after their children have married, gone to college, or found jobs and moved out.

Today, young adults are living at home longer than they did in the past. The terms *boomerang children* and *boomerang generation* refer to young adults who move back in with their parents, often with a spouse, girlfriend or boyfriend, and children in tow.

Most young adults leave the parental nest by age 23, but the proportion of adults aged 18 to 34 who are living with parents increased from 20 percent in 1960 to 32 percent in 2014 (Fry, 2016). Among young adults aged 18 to 34, men are more likely than women to live with their parents (**Figure 4.3**).

Macro-level factors have contributed to many Americans' delayed transition to adulthood, especially living independently. Financial insecurity, student loan debt, low wages, divorce, credit card debt, and going from job to job until they find work they enjoy have made it harder for young middle-class adults to maintain the lifestyles that their parents created. And the transition to independence gets tougher the lower a person's occupation and education. In 2014, for instance, only 19 percent of 18- to 34-year-olds who were college graduates were living with their parents, compared with 36 percent who had not completed a college degree (Fry, 2016).

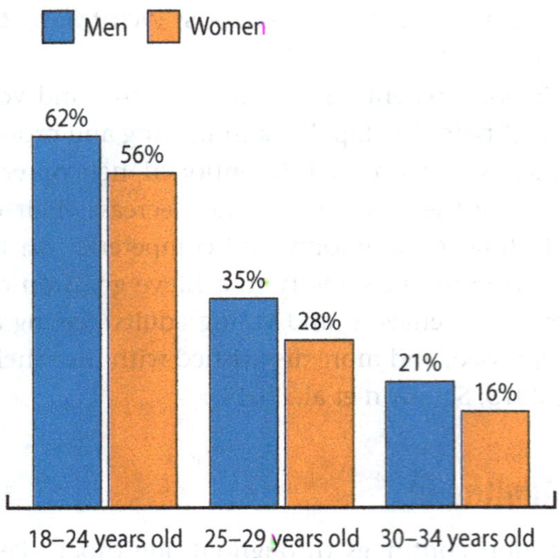

Figure 4.3 Who's Not Leaving the Nest?

Source: Based on Fry, 2016.

On a micro level, almost 33 percent of today's parents, compared with 19 percent in 1993, say that children shouldn't be expected to be on their own financially until age 25 or later (Taylor et al., 2012). Among those aged 18 to 24, 87 percent live at home because parents "make it easy for me to stay" (Payne and Copp, 2013). They enjoy the comforts of the pampering parental nest. According to one of my male students in his late 20s, for example, "My mom enjoys cooking, cleaning my room, and just having me around. I don't pitch in for any of the expenses, but we get along great because she doesn't hassle me about my comings and goings" (Benokraitis, 2015, p. 354).

Many young adults, as you'll learn in Chapter 12, are postponing marriage and don't feel a need to establish their own homes. Perhaps most importantly, the stigma traditionally linked to young adults living at home has faded. Among those aged 25 to 34 who live at home, 61 percent say they have friends or family members who have moved back in with their parents (Parker, 2012). The implication (and perhaps justification) of moving back home or not leaving in the first place is that "a lot of people are doing it."

Work Roles

The average American who was born between 1957 and 1964 has held 11 jobs from the ages of 18 to 46 alone; 26 percent have held 15 jobs or more ("Number of Jobs Held . . . ," 2012). Such job changing means that our occupational socialization is an ongoing process.

Learning work roles can be difficult because each job has different demands and expectations. Even when there's formal job training, we must also learn the subtle rules that are implicit in many job settings. A supervisor who says, proudly and loudly, "I have an open-door policy. Come in and chat whenever you want," may *really* mean "I have an open-door policy if you're not going to complain about something." Even if we remain in the same job for many years, we must often acquire new skills, especially technological ones, or risk being laid off.

Being laid off can be stressful at any age. In the case of midlife men (generally defined as those ages 45 to 60), such changes are especially traumatic: They must learn new job-hunting skills and often adjust to as much as a 70 percent decrease in their earnings (refer to Chapter 11). Instead of enjoying the security of a job in their midlife years, these men must undergo occupational socialization that assaults their self-image as good employees and family providers.

Parenting Roles

Like workplace roles, parenting does *not* come naturally. Most first-time parents muddle through by trial and error. Family sociologists often point out that we get more training for driving a car or a fishing license than for parenting.

Most couples don't realize that children are expensive. Middle-income couples, with an average income of $81,700 a year, spend about 16 percent of their earnings on a child during the first 2 years (**Figure 4.4**). These couples will spend almost $234,000 for each child

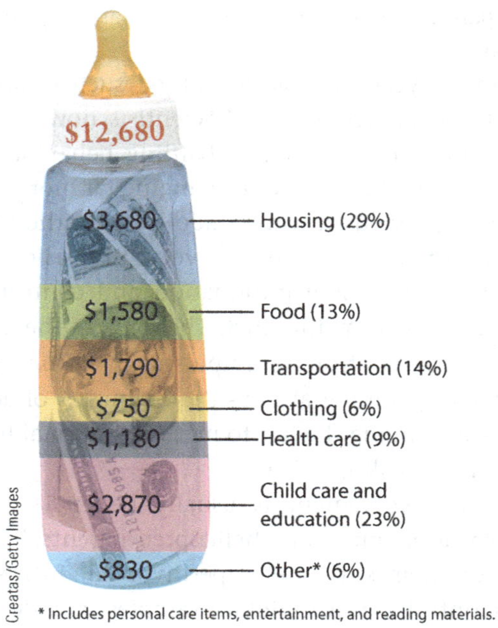

Figure 4.4 How Much Does It Cost to Raise a Child?

In 2015, middle-income married couples—those with before-tax income between $59,200 and $107,400—spent $12,680 per year for a child under age 2. This amount doesn't include the costs of prenatal care or delivery.

Source: Based on Lino et al., 2017, Table 1.

from birth through high school. Child-rearing costs are much higher for single parents and especially for low-income families if a child is disabled, chronically ill, or needs specialized care that welfare benefits don't cover. Expenses are also much higher for parents whose children attend private schools (Lino et al., 2017).

4-5e Later Life

Socialization continues in later life. People aged 65 and older often remarry and must learn to play new roles such as stepparent or step-grandparent. Many others undergo unwanted rites of passage such as widowhood, divorce or re-divorce, becoming a custodial parent, and dealing with the death of an adult child or grandchild.

Because we live longer, we may spend 20 percent of our adult lives in retirement. When retired people are unhappy, it's usually because of health or income problems rather than loss of the worker role (refer to Chapter 14). If retirement benefits don't keep up with inflation or if a retiree isn't covered by a pension plan, people can plunge into poverty soon after retirement. Poverty is devastating at any age, but sliding down the economic ladder during one's 60s and 70s is especially traumatic because there's little chance of recouping one's income (refer to Chapters 12 and 14).

Older people who are in poor health and experience continuous pain sometimes welcome death. Among those who are healthy, some reenter the labor force or volunteer in various organizations, many are involved with their grandchildren, and others forge new relationships after widowhood (refer to Chapter 12). Thus, even in the later years, many people continue to learn new roles.

Approximately 23 percent of Americans are "sandwiched" between their children and their parents. That is, they're caring for at least one parent aged 65 or older, raising a child under the age of 18, and/or providing economic assistance to a grown child. More than half of Americans in their 40s are balancing care for their older and younger relatives. Pulled in many directions, these midlife adults must learn to deal with the stress of juggling a full-time job and caregiving responsibilities (Horowitz, 2022; refer also to Chapter 12).

4-6 Resocialization and Total Institutions

Throughout life, many people undergo **resocialization**, the process of unlearning old ways of doing things and adopting new attitudes, values, norms, and behavior. Resocialization ranges from mild to intense and can be voluntary or involuntary.

4-6a Resocialization

Much resocialization is voluntary, as when an American wife moves to the Middle East with her Iranian husband and follows local customs about veiling, women's submissive roles, and staying out of public life. Other examples include entering a religious order, seeking treatment in a drug rehabilitation facility, or serving in the military.

Voluntary resocialization is usually mild. First-year college students, new employees, and members of support groups, for instance, must learn new rules and live up to an individual's, group's, or organization's expectations. Such changes, however, are usually short term and may even be enjoyable.

On the other hand, some voluntary resocialization can be long, difficult, and intense. New parents' lives change dramatically as they learn parenting roles.

resocialization unlearning old ways of doing things and adopting new attitudes, values, norms, and behavior.

iStock.com/Davincidig; iStock.com/Bridget McGill

Most resocialization occurs in institutions ranging from monastic orders to the military.

In divorce, people may experience a sense of relief and liberation but must also cope with emotional distress, financial changes, and child custody disputes. People who experience a permanent physical disability must cope with physical limitations, psychological issues, and stigma. Soldiers who return from wars, even if they haven't experienced physical injuries, describe coming home and living a normal life as daunting. They must reconnect with their children, adjust to spouses who have become more independent, and deal with posttraumatic stress disorder (PTSD) symptoms that include depression and nightmares (Jones, 2013; van Agtmael, 2013).

Resocialization can also be involuntary when it occurs against a person's will. Examples include children sent to a foster home or older people persuaded to move to an assisted-living facility. Whether voluntary or not, total institutions are the most intense forms of resocialization.

4-6b Total Institutions

A **total institution** is an isolated and enclosed social system that controls most aspects of its participants' lives. Prisons, mental hospitals, nursing homes, juvenile detention camps, convents and monasteries, boarding schools, and military training centers are all total institutions to varying degrees. Because a total institution's goal is resocialization, people are isolated from the rest of society, stripped of their former identities, and required to conform to new rules and behavior (Goffman, 1961).

Resocialization in total institutions is an intense two-step process. First, the staff uses *degradation ceremonies*, humiliating rituals that publicly stigmatize people and try to erode their identities and independence (Garfinkel, 1956). Incarcerated persons are strip-searched, deloused, and fingerprinted; their heads are shaved; they're issued a uniform and are given a serial number; they have no privacy, limited access to family and friends, and little communication with the outside world; and they're told what to do and when. In the military, soldiers get short haircuts, wear matching uniforms, follow rigid schedules, and make few personal decisions. The purpose of these practices and restrictions is to destroy any sense of individualism, autonomy, and past identity and to produce a more compliant person (Goffman, 1961; refer also to Chapter 7).

In the second resocialization step, the staff tries to build a new identity that conforms to the institution's expectations. Incarcerated persons who conform are rewarded with simple privileges (e.g., making phone calls, more visits with family) and an earlier release. Soldiers often bond with one another, become more self-disciplined, may use their new skills in civilian jobs, and are proud of having served in the military ("Once a Marine, always a Marine").

Total institutions don't always succeed in resocializing people. Incarcerated persons, for instance, find ways to make money, to barter, to smuggle in banned items (including drugs and smartphones), and to make and enforce their own rules through a system of rewards and punishments (Kilgore, 2015; Biunno, 2016). Netflix's *Orange Is the New Black* series dramatizes female incarcerated persons' making and breaking formal and informal rules to cope with imprisonment.

As this chapter shows, socialization is a powerful force in shaping who we are. It doesn't produce robots, however, because people are creative, adapt to new environments, and change as they interact with others.

---

total institutions isolated and enclosed social systems that control most aspects of the participants' lives.

Chapter Review

Key Takeaways

4-1 What is the importance of socialization? Socialization is a lifelong process that fulfills four key purposes: It establishes our social identity, teaches us roles, controls our behavior (through internalization), and transmits culture to the next generation. The research on children who are isolated from social interaction shows that socialization is critical to our social and physical development.

4-2 How do perspectives based on nature and those based on nurture differ in their explanations regarding human experience? Biologists tend to focus on the role of heredity (or genetics) in human development. In contrast, most social scientists, including sociologists, underscore the role of learning, socialization, and culture. This difference of opinion is often called the nature–nurture debate. Sociobiologists argue that genetics (nature) can explain much of our behavior, while most sociologists maintain that socialization and culture (nurture) shape even biological inputs.

4-3 What are the key components of the different sociological explanations of socialization? Sociologists have offered many explanations of socialization, but two of the most influential, both at the micro level, have been the social learning and symbolic interaction theories (refer to **Table 4.2** on page 71). These theories help us understand social development processes, but they also have limitations.

4-4 What socialization agents have the most impact on the socialization process? Parents are the first and most important socialization agents, but siblings, grandparents, and other family members also play important roles. Other significant socialization agents include play and peer groups, teachers and schools, popular culture, and the media. Electronic media and advertising are especially powerful socialization sources.

4-5 How does socialization occur throughout the life course? As we progress through the life course, we learn culturally approved norms, values, and roles. Infants are born with an enormous capacity for learning that parents and other caregivers can enrich and shape. In adolescence, these and other socialization agents teach children how to form relationships, get along with others, and develop their social identity through play and peer groups. In adulthood, people must learn new roles that include singlehood, marriage, parenthood, divorce, work, and experiencing the death of a loved one. Socialization continues in later life when many people learn new roles such as grandparents, retirees, older workers, and being widowed.

4-6 How does resocialization occur? Resocialization can be voluntary or involuntary. Voluntary resocialization includes entering a religious order, joining a religious cult, seeking treatment in a drug misuse rehabilitation facility, or serving in the military. The changes can be short-term and pleasant or long, difficult, and intense. Most involuntary resocialization takes place in total institutions, where people are isolated from the rest of society, stripped of their former identities, and required to conform to new rules and behavior.

Key Terms

Socialization (p. 67)
Internalization (p. 67)
Social learning theories (p. 70)
Role models (p. 71)
Self (p. 72)
Looking glass self (p. 72)
Role taking (p. 73)
Significant other (p. 73)
Anticipatory socialization (p. 73)

Generalized other (p. 73)
Impression management (p. 74)
Reference groups (p. 75)
Socialization agents (p. 75)
Multigenerational households (p. 77)
Peer group (p. 77)
Resocialization (p. 85)
Total institution (p. 86)

Discussion and Critical Thinking Questions

1. The Parents Television Council (parentstv.org) monitors children's TV programs. In its ratings of the most popular prime-time programs every week, the council rarely endorses a show as family-friendly if the content includes sex, violence, and/or obscene language. In 2011, in contrast, the U.S. Supreme ruled (Brown v. Entertainment Merchants Association) in a 7–2 opinion that violent video games sold to minors are protected by free speech.
   a. Does the Supreme Court ruling undermine the Parents Television Council's efforts to discourage children from viewing violence? Are both groups interfering with parents' responsibilities as primary socialization agents?
   b. Should parents try to protect their children from television images of sex and violence? Or should they watch the programs with their children and discuss the content?
2. With whom did you hang out in middle school and high school? What did you learn from your peers that was harmful or beneficial? Who are your friends now? Why?
3. Do you think that Cooley's concept of the looking-glass self shaped your self-identity and behavior when you were growing up? What about now?
4. How do you manage people's impressions of you? How does your impression management vary across situations, as when you're with friends, in class, or at work?
5. Wearing clothes protects the skin against harmful ultraviolet rays, but are there other reasons for covering our bodies almost all of the time? Especially in states with hot climates (e.g., Texas, Florida), wouldn't being nude save people money, closet space, and time spent shopping? Tie these questions into a discussion about how free we think we are and the powerful effect of socialization on community standards that are passed down from generation to generation.

True or False?

How Much Do You Know about Socialization?

The answers to 4, 5, and 6 are true. The answers to 1, 2, and 3 are false.

1. Socialization is a lifelong process.
2. Research on children who are isolated demonstrates that socialization is critical to our development. Behaviors such as talking, eating with utensils, and controlling our bowel movements don't come naturally.
3. Exploring various studies of twins shows that both nature and nurture affect human behavior. The Minnesota Study (1979 through 1999), that focused on identical twins who had been raised apart, showed interesting behavioral similarities about the adult twins (Lewis, 2014). Yet, additional research on twins and social factors has demonstrated some of the differences emerge through the interplay of environment and genetics (both nature and nurture).
4. Parents, siblings, grandparents, and other family members play a critical role in our socialization. Parents, however, are the first and most influential socialization agents.
5. New parents experience voluntary resocialization as their lives change dramatically through the process of learning parenting roles. This, and some other experiences with voluntary resocialization, can be long, difficult, and intense.
6. There are several socialization agents including popular culture and media as well as family, peers, schools and teachers.

5 Social Interaction in Everyday Life

Chip Somodevilla/Getty Images News/Getty Images

Learning Objectives

After studying this chapter, you will be able to…

5-1 Explain the importance of social interaction and its relationship to social structure.

5-2 Define status set, ascribed and achieved statuses, master status, and status inconsistency.

5-3 Explain how and why social roles differ and how people cope with role conflict and role strain.

5-4 Compare symbolic interaction, social exchange, and feminist explanations of social interaction.

5-5 Describe nonverbal communication, its importance, and its cross-cultural variations.

5-6 Summarize the benefits and costs of online interaction.

True or False?

How Much Do You Know about Social Interaction in Everyday Life?

1. The existence of social structure typically supports order and predictability in social life.
2. Religious affiliation is a type of ascribed status.
3. Planning lectures and grading exams are typical roles of a professor.
4. Nonverbal communication does not have as strong of an impact on human interaction as does verbal communication.
5. Dramaturgical analysis is a research method that analyzes messages in plays and shows to help sociologists understand human interaction.
6. Social media has had a mixed effect on relationships between parents and children.

The answers are at the end of the chapter.

What, if anything, annoys you about airplane passengers? Among U.S. adults, the top offenders are parents who don't control their whining or misbehaving children, people behind them who kick the seat, and travelers who talk loudly or listen to loud music, usurp several overhead bins, become drunk or disruptive, and never stop talking (Expedia, 2017).

Such complaints illustrate **social interaction**, the process of acting toward and reacting to people around us. Social interaction—verbal, nonverbal, face-to-face, or on social media—is central to all human social activity, can be cooperative or competitive, and can be interpreted in various ways across cultures and groups over time. As people interact, they create social structure.

Be sure to take the True or False quiz in the chapter opener to better understand how much you know about these topics.

What do you think?

I spend more time on social media than I do talking face-to-face with my family and friends.

1	2	3	4	5	6	7
	strongly agree				strongly disagree	

5-1 Social Structure

Social structure is an organized pattern of behavior that governs people's relationships (Smelser, 1988). Because social structure shapes our actions, life is typically orderly and predictable rather than chaotic or random. We're often not aware of the impact of social structure until we violate the cultural rules, formal or informal, that dictate our daily behavior—whether in class, at home, at work, or on an airplane.

Every society has a social structure that encompasses statuses, roles, groups, organizations, and institutions (Smelser, 1988). We'll examine groups, organizations, and institutions in later chapters. Here, let's dig deeper into statuses and roles, which are two building blocks of social structure.

social interaction acting toward and reacting to people around us.

social structure an organized pattern of behavior that governs people's relationships.

status a social position that a person occupies in a society.

status set a collection of social statuses that a person occupies at a given time.

5-2 Status

For most people, the word *status* signifies prestige. An executive, for example, has more status than a receptionist, and a physician has a higher status than a nurse. For sociologists, **status** is a social position that a person occupies in a society (Linton, 1936). Thus, executive, teacher, physician, and nurse are all social statuses. Other examples of statuses are musician, voter, sister, parent, police officer, and friend.

Sociologists don't assume that one position is more important than another. A mother, for example, isn't more important than a father, and an adult isn't more important than a child. Instead, all statuses are important because they determine social identity, or who we are.

5-2a Status Set

Every person has many statuses (**Figure 5.1**). Together, they form their **status set**, a collection of social statuses that a person occupies at a given time (Merton, 1968). Dionne, one of my students, is female, Black, 42 years old, divorced, mother of two, daughter, aunt, Baptist, voter, bank supervisor, volunteer at a soup kitchen, and country music fan. All of these socially defined positions (and others) make up Dionne's status set.

Status sets change throughout the life course. Because she'll graduate next year and is considering remarrying and starting an after-school program, Dionne will add at least three more statuses to her status set and will also lose the statuses of divorced and college student. As Dionne ages, she'll continue to gain new statuses (perhaps grandmother, retiree) and may lose others (supervisor at a bank, or wife, if she becomes widowed).

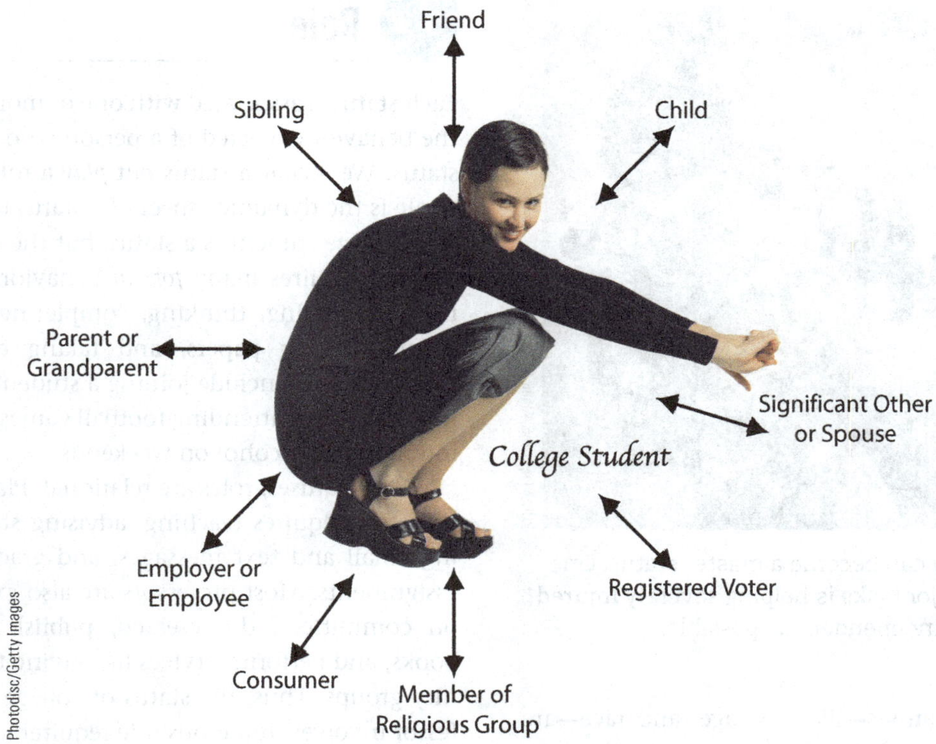

Figure 5.1 Is This Your Status Set?

Being a college student is only one of your current statuses. What other statuses comprise your status set?

Statuses are *relational*, or complementary, because they're connected to other statuses: A *real-estate agent* has *customers*, a *teacher* has *students*, *and a parent has a child or children*. No matter how many statuses you occupy, every status is linked to those of other people. These connections between statuses influence our behavior and relationships.

5-2b Ascribed and Achieved Status

Status sets include both ascribed and achieved statuses. An **ascribed status** is a social position that a person is born into. We typically can't control, change, or choose our ascribed statuses, which include sex category, age, race, ethnicity, and relationships to our family of origin (born or raised in). Your ascribed statuses, for example, might include male, Latino, and brother.

An **achieved status**, in contrast, is a social position that a person attains through personal effort or assumes voluntarily. Your achieved statuses might include college graduate, mother, and employee. Unlike with our ascribed statuses, we have greater ability to change our achieved statuses. We have little choice about being a son or daughter (an ascribed status), but we typically have more ability to choose whether to become a parent (an achieved status).

Students sometimes think that religion and social class are ascribed rather than achieved statuses. It's true that someone may be born into a family that practices a certain religion or one that is lower, middle, or higher income. Because we can change these statuses through our own actions, however, neither religion nor social class is an ascribed status. A Catholic might convert to Judaism (or vice versa), and thousands of Americans born into lower-income families have become millionaires (refer to Chapter 8).

5-2c Master Status

A **master status** overrides other statuses and forms an important part of a person's social identity (Hughes, 1945; Becker, 1963). A master status is usually immediately apparent, makes the biggest impression, affects others' perceptions, and, consequently, often shapes a person's entire life.

ascribed status a social position that a person is born into.

achieved status a social position that a person attains through personal effort or assumes voluntarily.

master status overrides other statuses and forms an important part of a person's social identity.

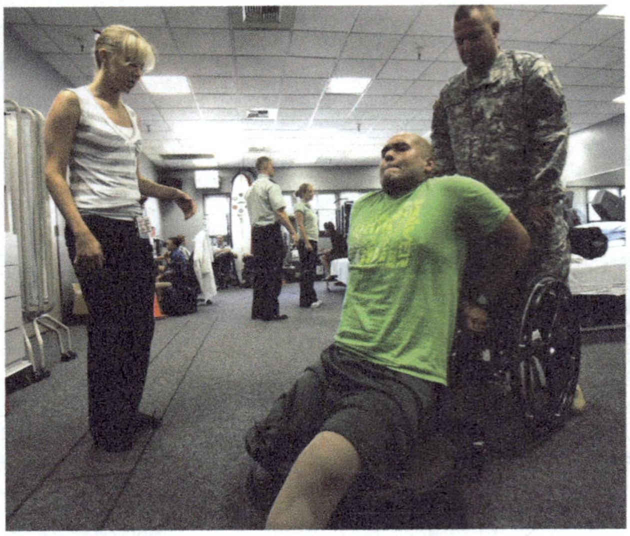

ZUMA Press, Inc./Alamy Stock Photo

A physical disability can become a master status. One of the military's major tasks is helping severely injured soldiers become as independent as possible.

Some master statuses—like sex, age, and race—are ascribed. Whatever you do, for example, people will make assumptions about you based on your age group, child, teen, young adult, middle-aged adult, or older adult and react accordingly. Other master statuses are achieved. For many people, their occupation is their master status: Occupation tends to be the first thing we ask about when we meet someone; conveys much information about income, education, skills, and achievements; and lingers after we've left our jobs (e.g., someone is introduced as a former CEO or ex-military officer). Other achieved master statuses can be based on religion, community leadership, family wealth, or deviant behavior.

5-2d Status Inconsistency

Because we hold many statuses, some clash. **Status inconsistency** refers to the conflict that arises from occupying social positions that are ranked differently. Examples include a teacher who works as a bartender or a skilled welder who stocks shelves at a grocery store because neither can find a better job in a stagnant economy. We'll cover status inconsistency in more detail in Chapter 7. For now, you should be aware that you occupy many statuses, and some of them may clash now or in the future.

status inconsistency the conflict that arises from occupying social positions that are ranked differently.

role the behavior expected of a person who has a particular status.

role performance the actual behavior of a person who occupies a status.

5-3 Role

Each status is associated with one or more roles. A **role** is the behavior expected of a person who has a particular status. We *occupy* a status but *play* a role. In this sense, a role is the dynamic aspect of a status (Linton, 1936).

College student is a status, but the role of a college student requires many *formal* behaviors such as going to class, reading, thinking, completing weekly assignments, writing papers, and taking exams. *Informal* behaviors may include joining a student club, befriending classmates, attending football games, and even overindulging on alcohol on weekends.

Like statuses, roles are relational. Playing the role of professor requires teaching, advising students, answering email and text messages, and grading exams and assignments. Most professors are also expected to serve on committees, do research, publish articles and/or books, and perform services like giving talks to community groups. Thus, the status of college student or professor involves numerous role requirements that govern who does what, where, when, and how.

Roles can be rigid or flexible. An administrative assistant typically plays a role that's defined by rules about when to come to work, how to answer the phone, when to submit the necessary work, and in what format. A boss, on the other hand, usually enjoys considerable flexibility: They have more freedom to come in late or leave early, to determine which projects should be completed first, and to decide when to hire or fire employees.

Because roles are based on mutual obligations, they ensure that social relations are fairly orderly. We know what we're supposed to do and what others expect of us. If professors fail to meet their role obligations by missing classes or coming to classes unprepared, students may respond by studying very little, cutting classes, and submitting negative course evaluations. If students miss classes, don't turn in the required work, or cheat, professors can fail them or assign low grades.

5-3a Role Performance

Roles define how we're *expected* to behave in a particular status, but people vary considerably in fulfilling the responsibilities associated with their roles. Many college students succeed, while others fail; some professors inspire their students, while others put them to sleep. These differences reflect **role performance**, the *actual* behavior of a person who occupies a status. For example, a professor may demand more of graduate students than of undergraduates. An instructor may also interact differently with a 19-year-old student than a 40-year-old student who's anxious about returning to college.

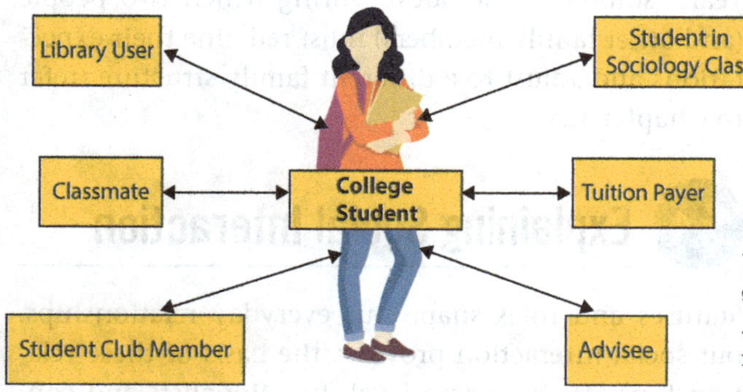

rodrigofoca/Getty Images

Figure 5.2 Role Set of a Typical College Student

What other roles comprise your college student role set?

5-3b Role Set

We occupy many statuses and play many roles associated with each status. A **role set** is the array of roles attached to a particular status. The status of a college student, for example, may include at least six different roles (**Figure 5.2**). Each of these roles is linked to people with different statuses (e.g., professor, registrar, librarian). Because a different set of norms governs each of these relationships, the student interacts differently with a classmate than with a reference librarian or a professor.

Figure 5.2 shows only one status—that of college student. If you think about a college student's other statuses (employee, sibling, parent, friend, teammate, or spouse), you can understand that meeting the expectations of numerous role sets can create considerable role conflict and role strain.

5-3c Role Conflict and Role Strain

Playing many roles often leads to **role conflict**, difficulties in playing two or more contradictory roles. College students who have a job, especially if it's full-time, often experience role conflict because both employers and professors expect them to excel, but it's very difficult to meet these expectations. The role conflict increases if the student has young children or cares for an aging parent or grandparent.

While role conflict arises because of the expectations of contradictory roles, **role strain** is due to the conflicting demands within the same role. Students experience role strain when several exams are scheduled on the same day, and they're also involved in time-consuming extracurricular activities. Faculty members experience role strain when the requirements of being a professor—teaching, research, and community service—sap their energy and time. Military chaplains report role strain in preaching about peace while blessing those about to go to war.

Almost all of us experience role strain because many inconsistencies are built into our roles. **Table 5.1** gives examples of some of the reasons for role conflict and role strain.

5-3d Coping with Role Conflict and Role Strain

Role conflict and role strain can produce tension, hostility, aggression, and stress-related physical problems that include insomnia, headaches, ulcers, eating disorders, anxiety attacks, chronic fatigue, nausea, weight loss or gain, and drug and alcohol misuse (Weber et al., 1997). Some role conflicts and role strain may last only a few weeks, but others are long-lived (e.g., working in a stressful or low-paying job).

To deal with role conflict and role strain, some people *deny that there's a problem,* while others find strategies to address the issues. As a result of the COVID-19 pandemic, many parents juggled remote work along with childcare and sometimes children's remote schooling. Parental burnout was quite common. Sixty-six percent of parents reported these feelings (Gawlik and Melnyk, 2021). Parents experiencing burnout may feel

Table 5.1 Why Do We Experience Role Conflict and Role Strain?

Reason	Example
Because many people are overextended, some roles are bound to conflict with others.	Students may study less than they want because employers demand that they work overtime or on weekends.
People have little or no training for many roles.	Parents are expected to turn out "perfect" kids, even though they receive training for driving a car but none for parenting.
Some role expectations are unclear or contradictory.	Some employers pride themselves on having family-friendly policies but expect employees to work 12 hours a day, travel on weekends, and use vacation days to care for a sick child.
Highly demanding jobs often create difficulties at home.	Some jobs (being in the military, policing, and firefighting) require people to be away from their families for extended periods of time or during crises.

role set array of roles attached to a particular status.

role conflict difficulties in playing two or more contradictory roles.

role strain difficulties due to conflicting demands within the same role.

isolated or ashamed, which makes it harder to acknowledge that they're struggling (Abramson, 2021).

There are five more effective ways to minimize role conflict and role strain. First, we can reduce role conflict through *compromise* or *negotiation*. To decrease the conflict between work and family roles, many couples draw up schedules that require both partners to take responsibility for housework and child rearing (refer to Chapter 12).

Second, we can *set priorities*. If children's extracurricular activities interfere with parents helping with children's homework or completing household chores, which is more important? Burned out parents may decrease stressors by reducing extracurricular activities (Gawlik and Melnyk, 2022).

Third, we can *compartmentalize* our roles. Parents may find ways to keep responsibilities separate, such as working in the morning and afternoon, helping children with schoolwork in the evening, and saving weekends for housework and extracurriculars. It's not always easy, but it may be helpful to separate our various roles.

Fourth, we can decide *not to take on more roles*. One of the most effective ways to avoid role conflict is to just say "no" to requests to do volunteer work, pressure from family or friends to take on unwanted tasks (e.g., babysitting others' children), or pleas to become involved in community activities.

Finally, we can *exit* a role or status. Withdrawing from community activities, for example, can decrease role conflict and role strain considerably. There's always pressure to remain in a role, but none of us is indispensable, and there are many people who are eager to replace us. Although necessary, some role exits are painful and long-lived. Divorce, for instance, isn't a quick event but a process that may go on for many years (sometimes decades), during which two people (and other family members) must redefine their expectations and adjust to a different family structure (refer to Chapter 12).

Geo Martinez/Shutterstock.com

Like other people, college students can reduce their role conflict and role strain by setting priorities and compartmentalizing their roles.

5-4 Explaining Social Interaction

Statuses and roles shape our everyday relationships, but social interaction provides the basis of these relationships. At the macro level, functionalists and conflict theorists study interaction among and between social structures, institutions, and societies (refer to Chapters 1 and 6). However, three micro perspectives—symbolic interaction, exchange theory, and some feminist theories—are especially useful in understanding why we interact as we do, and why people sometimes interpret the same words differently. The explanations offer different insights, but each shows how people interact in their daily lives. **Table 5.2** summarizes these approaches.

5-4a Symbolic Interaction

For interactionists, the most significant characteristic of all human communication is that people take each other and the context into account (Blumer, 1969). When your professors ask, "How are you?" they expect a "Fine, thanks" and probably barely look at you. A doctor, on the other hand, usually looks you in the eye when asking "How are you?" and takes notes as soon as you start

Table 5.2 Sociological Explanations of Social Interaction

Perspective	Key Points
Symbolic Interactionist	• People create and define their reality through social interaction. • Our definitions of reality, which vary according to context, can lead to self-fulfilling prophecies.
Social Exchange	• Social interaction is based on a balancing of benefits and costs. • Relationships involve trading a variety of resources, such as money, youth, and good looks.
Feminist	• Men and women act similarly in many interactions but may differ in communication styles and speech patterns. • Men are more likely to use speech that's assertive (to achieve dominance and goals), while women are more likely to use language that connects with others.

to reply. Thus, "How are you?" has different meanings in different social contexts and elicits different responses ("I'm doing okay" vs. "I've been having a lot of headaches lately"). In this sense, people construct reality.

Social Construction of Reality

"Human reality is socially constructed reality" (Berger and Luckmann, 1966, p. 172). That is, we produce, interpret, and share the reality of everyday life with others. This social construction of reality typically evolves through direct, face-to-face interaction, but the interaction can also be indirect, as in watching television or participating in social media.

Businesspeople, advertisers, politicians, educators, and even social scientists use words intentionally to shape or change our perceptions of reality. For example, doublespeak is "language that pretends to communicate but really doesn't. [It] makes the bad seem good, the negative appear positive, the unpleasant appear attractive or at least tolerable" (Lutz, 1989, p. 1).

There are several kinds of doublespeak: *Euphemisms* are words or phrases that avoid a harsh, unpleasant, or distasteful reality; *gobbledygook* (or *bureaucratese*) overwhelms the listener with big words and long sentences; and *inflated language* makes simple, everyday things seem complex (Lutz, 1989; refer also to Beard and Cerf, 2015). Take the Doublespeak quiz to check how much doublespeak you understand or use.

Social Interaction and Self-Fulfilling Prophecies

Our perceptions of reality shape our behavior. In an oft-cited statement, known as the *Thomas Theorem,* sociologists W. I. Thomas and Dorothy Thomas (1928, p. 572) observed, "If [people] define situations as real, they are real in their consequences." That is, our behavior is a result of how we interpret a situation ("My mother-in-law hates me, so I go shopping when she drops in").

Carrying this idea further, sociologist Robert Merton (1948/1996) proposed that our definitions of reality can result in a **self-fulfilling prophecy**: If we define something as real and act on it, it can, in fact, become real. For example, physical education teachers who publicly humiliate students may turn them off physical fitness for good. One man said, "To this day I feel totally inadequate in team-related activities and have a natural reflex to AVOID THEM AT ALL COSTS" (Strean, 2009, p. 217; capitalization in original). Thus, teachers' negative comments may make students feel inadequate, regardless of their ability, and change their behavior during adulthood.

Our perceptions of reality shape our behavior, but how do people define that reality? Interactionists use two research tools—*ethnomethodology* and *dramaturgical analysis*—to help answer this question.

Ethnomethodology

A term coined by sociologist Harold Garfinkel (1967), **ethnomethodology** is the study of how people construct and learn to share definitions of reality that make everyday interactions possible. That is, we base our interactions on common assumptions about what makes sense in specific situations (Schutz, 1967; Hilbert, 1992).

People make sense of their everyday lives in two ways. First, by observing conversations, people discover the general rules that we all use to interact. Second, people can understand interaction rules by breaking them. Over a number of years, Garfinkel instructed his students to purposely violate everyday interaction rules and then to analyze the results.

In these exercises, some of his students went to a grocery store and insisted on paying more than was asked for a product. Others were instructed, in the course of an ordinary conversation "to bring their faces up to the subject's until their noses were almost touching" (Garfinkel, 1967, p. 72).

When students violated interaction norms, they were sanctioned. Grocery clerks became hostile when the students insisted on paying more than the marked price for a product, and people backed off when their noses were almost touching: "Reports were filled with accounts of astonishment, bewilderment, shock, anxiety, embarrassment, and anger" (Garfinkel, 1967, p. 47).

Violating interaction rules, even unspoken ones, can trigger anger, hostility, and frustration. College students become upset if professors are sarcastic or disrespectful, are unprepared or consistently late, constantly read from the book, or discourage students' comments and questions (Berkos et al., 2001).

Dramaturgical Analysis

Dramaturgical analysis is a research approach that examines social interaction as if occurring on a stage where people play different roles and act out scenes for the "audiences" with whom they interact. According to

self-fulfilling prophecy if we define something as real and act on it, it can, in fact, become real.

ethnomethodology the study of how people construct and learn to share definitions of reality that make everyday interactions possible.

dramaturgical analysis examines social interaction as if occurring on a stage where people play different roles and act out scenes for the audiences with whom they interact.

Doublespeak

1. automotive internist
2. reducing costs
3. alternative fact
4. auto dismantler and recycler
5. air support
6. previously owned
7. genuine imitation
8. bereavement care expert
9. equity retreat
10. revenue enhancement

A. a lie
B. bombing
C. undertaker
D. cutting salaries and/or firing people
E. fake
F. car mechanic
G. junk dealer
H. stock market crash
I. used
J. tax increase

Answers: 1-F; 2-D; 3-A; 4-G; 5-B; 6-I; 7-E; 8-C; 9-H; 10-J

sociologist Erving Goffman (1959, 1967), life is similar to a play in which each of us is an actor, and our social interaction is much like theatre because we're always on stage and always performing. In our everyday "performances," we present different versions of ourselves to people in different settings (audiences).

Because most of us try to present a positive image of ourselves, much social interaction involves *impression management*, a process of suppressing unfavorable traits and stressing favorable ones (refer to Chapter 4). To control information about ourselves, we often rely on props to convey or reinforce a particular image. For example, physicians, lawyers, and college professors may line their office walls with framed diplomas, medical certificates, or community awards to give the impression that they're competent, respected, and successful.

For Goffman, a performance involves front- and back-stage behaviors. The *front stage* is an area where an actual performance takes place. In front-stage areas, such as living rooms or restaurants, the setting is clean, and the servers or hosts are typically polite and deferential to guests. The *back stage,* an area concealed from the audience, is where people can relax. Bedrooms and restaurant kitchens are examples of back stages. After guests have left, the host and hostess may kick off their shoes and gossip about their company. In restaurants, cooks and servers may criticize the guests, use vulgar language, and yell at each other. Thus, the civility and decorum of the front stage may change to rudeness in the back stage.

Another example of front- and back-stage behavior involves faculty and students. Professors want to create the impression that they're well prepared, knowledgeable, and hardworking. In back-stage areas like their offices, however, faculty may complain to their colleagues that they're tired of teaching a particular course or grading terrible exams and papers, or they may confess that they don't always prepare for classes as well as they should. Students often plead with instructors for

Luca Campedri/Nonstock/Jupiter Images

Thinkstock Images/Getty Images

The chaotic and crowded back-stage conditions in many restaurant kitchens differ markedly from the efficient and relaxed front-stage behavior of its servers. Can you think of a situation where your front- and back-stage behaviors differ dramatically?

higher grades: "But I studied very hard" or "I'm an A student in my other courses." In back-stage areas like dorms, student lounges, or libraries, however, students may admit to their friends that they barely studied or that they have a low grade-point average.

5-4b Social Exchange Theory

Social exchange theory proposes that individuals seek through their interactions to maximize their rewards and minimize their costs. An interaction that elicits a reward, such as approval or a smile, is more likely to be repeated than an interaction that evokes disapproval or criticism (Thibaut and Kelley, 1959; Homans, 1974; Blau, 1986). Because interactions are most satisfying when there's a balance between giving and taking, individuals must decide whether and which resources to exchange.

People bring various tangible and intangible resources to a relationship (money, status, intelligence, good looks, youth, power, affection). Any of a person's resources can be traded for more, better, or different resources. Marriages between older men and younger women often reflect an exchange of the man's power, money, and/or fame for the woman's youth, physical attractiveness, and ability to bear children.

Many of our cost–reward decisions are conscious and deliberate, but others are passive or based on long-term negative interactions. Careers in science, technology, engineering, and math (STEM) provide good professional opportunities and higher than average incomes. Still, girls and women are less likely to enter these fields and make up only 27 percent of STEM workforce. Stereotypes that boys are better at math may contribute to girls' assessment that their math skills aren't as good and lead to them earning lower math scores. Over the long term, this messaging and experiences with teachers may decrease girls' desire to pursue STEM study (Martinez and Christnacht, 2021; Hill et al., 2010).

Tyler Olson/Shutterstock.com

In predominantly White organizations, many Black professionals experience being held to different interaction standards than their White coworkers. Consequently, they manage their emotions to avoid being labeled as "too sensitive," "angry," "irritating," or "unpleasant" (Wingfield, 2010).

5-4c Feminist Theories

Cultural norms and gender role expectations shape our interaction patterns. Generally, women are socialized to be more comfortable talking about their feelings, while men are often socialized to be dominant and take charge, especially in the workplace. Women often ask questions that probe about feelings ("Were you glad it happened?"). Women are also more likely than men to do conversational "maintenance work" that encourages conversation ("What happened at the meeting?") (Lakoff, 1990; Farley et al., 2010).

Women's speech tends to be supportive ("You must've felt terrible") and tentative ("I kind of feel that . . . "). "Uptalk" is a style of speech in which every sentence rises in pitch, as if the speaker is always asking a question ("My name is Jennifer?"). Women are much more likely than men to uptalk, which shows uncertainty and a lack of self-confidence. And, in many social situations, women—particularly those ages 40 to 60—often apologize for the way they look (weight, hair, wrinkles). Men don't do so, "even when they look terrible" (Wolven, 2014; Wood, 2015).

Compared with women's speech, men's speech often reflects *conversational dominance*, speaking more frequently and for longer periods. Men also show dominance by interrupting or ignoring others, reinterpreting the speaker's meaning, changing the topic, or giving unsolicited and unwanted advice (Tannen, 1990; Toth, 2011; Gurkoff and Ranieri, 2012).

Such interaction differences aren't innate because of one's sex. Instead, much interaction is contextual—that is, it varies depending on the situation. For example, fathers who do much of the parenting have communication styles that are similar to those of mothers. In the workplace, women and men who occupy high-level decision-making positions have similar interaction styles with superiors and subordinates (Cameron, 2007).

You've learned, so far, that much of our interaction is verbal and face-to-face. Nonverbal communication also affects our relationships.

---

social exchange theory proposes that individuals seek through their interactions to maximize their rewards and minimize their costs.

5-5 Nonverbal Communication

Nonverbal communication refers to messages sent without using words. This silent language, a "language of behavior" that conveys our real feelings, can be more potent than our words (Hall, 1959, p. 15). For example, sobbing sends a much stronger message than saying "I feel very sad." Some of the most common nonverbal messages are silence, visual cues, touch, and personal space.

5-5a Silence

Silence expresses a variety of emotions: agreement, apathy, awe, confusion, disagreement, embarrassment, regret, respect, sadness, thoughtfulness, and fear, to name just a few. In various contexts and at particular points in a conversation, silence means different things to different people. Sometimes silence saves us from embarrassing ourselves. Think, for example, about the times you fired off an angry email or text message, regretted doing so an hour later, and then dreaded getting a fuming response.

Many of us have experienced the pain of getting "the silent treatment" from friends, family members, coworkers, or lovers. Not talking to people who are important to us builds up anger and hostility. Initially, the "offender" may work hard to make the closemouthed person discuss a problem. Eventually, the target of the silent treatment gets fed up, gives up, or ends a relationship. In intimate relationships, silence often signals emotional detachment and the beginning of a breakup.

5-5b Visual Cues

Visual cues, another form of nonverbal communication, include gestures, facial expressions, and eye contact.

nonverbal communication messages that are sent without using words.

"This concludes my lecture on non-verbal communication. Any comments or questions?"

Chris Wildt/CartoonStock.com

Let's consider a few examples of such body language in our daily interactions.

Gestures

Most of us think we know what certain gestures mean—folding your arms across your chest indicates a closed, defensive attitude; leaning forward often shows interest; shrugging your shoulders signals indifference; and knitted eyebrows show worry or concentration. Constantly rolling one's eyes, which conveys contempt and superiority, can be an early predictor of divorce (Weisman, 2016).

Finger-pointing usually directs attention outward, placing blame or responsibility on someone else. The common "talk to the hand" gesture sends a stronger message: "Go away!" or "I'm not listening to you." Gestures, however, should be interpreted in context. Because of habit or hearing problems, for instance, a coworker may always lean forward when listening, regardless of interest.

Hemera Technologies/AbleStock.com/Getty Images/Jupiter Images

The same gesture may have different meanings in different countries:

- Tapping one's elbow several times with the palm of one's hand indicates that someone is sneaky (in Holland), stupid (in Germany and Austria), or cheap or stingy (in Mexico).
- Twisting an index finger into one cheek means "She's beautiful" in Libya, and that something tastes good in Italy. The same gesture in southern Spain means that a man is effeminate, and "You're crazy!" in Germany.
- In many Middle Eastern countries, people view the shoes and the soles of one's feet as unclean. Thus, stretching out one's legs with the feet pointing at someone or crossing one's legs so that a sole faces another person is considered rude (Morris, 1994; Lynch and Hanson, 1999; Jandt, 2001; Donadio, 2013).

In a remarkable tarmac ritual, and regardless of weather, the ground crews at Japanese airports use many gestures as a jet pushes back from the gate. First, the crew members line up, snap to attention, and then, "in perfect unison," they bow. This gesture is directed toward the passengers ("Thank you for visiting Japan"), the plane's flight crew ("We respect your expertise and dedication"), and, ultimately, as recognition of the ground crew's commitment to service ("We have fulfilled our duties to the best of our abilities"). Then they straighten up, smile broadly, and wave goodbye to show the passengers that they had been welcome to Japan (Gottlieb, 2011, p. 44).

Facial Expressions

Facial expressions reveal emotions, but they can be deceptive. First, our facial expressions don't always show our true feelings. Parents may tell their children "Don't you roll your eyes at me!" or "Look happy when Aunt Minnie hugs you." Thus, children learn that displaying their real feelings—especially when they're negative—is unacceptable.

Sankei/Getty Images

Departing jets get a wave from the ground crew in Tokyo.

Digital Vision/Thinkstock

Is she being friendly? Or flirting?

Second, faces can lie about feelings. Parents often know when children are lying because they avoid eye contact, cry, or blink and swallow frequently. Some doctors say they can tell if their patients are lying because of signs such as avoiding eye contact, pausing, or voice inflections, but many adults can control their facial expressions. They can deceive, successfully and over many years, because they've rehearsed the lies in their heads, are smooth talkers with a reputation for being trustworthy, or have gotten away with lying in the past. In nonverbal interaction, people can correctly distinguish truth from lies only about 54 percent of the time (Reddy, 2013; Shea, 2014).

Third, facial expressions can sometimes be misleading because of cultural variations. Although 70 percent of the time, people from various cultures use similar facial expressions in response to social and emotional situations, there are differences (Anwar, 2021). American businesspeople have grumbled that Germans are cool and aloof, while many German businesspeople have complained that Americans are excessively friendly and hide their true feelings with grins and smiles. The Japanese, who believe it's rude to display negative feelings in public, smile more than Americans do to disguise embarrassment, anger, and other negative emotions (Jandt, 2001).

The only consistent research finding, across cultures, is that women smile more than men. This difference may be due to cultural norms that socialize women to hide negative feelings that might make people uncomfortable and to show deference to men (Szarota, 2010; McDuff et al., 2017). Men are more likely than women to mistake smiles and friendliness as being flirtatious and sexually seductive (La France et al., 2009; Rutter and Schwartz, 2012).

Eye Contact

Eye contact serves several social purposes (Eisenberg and Smith, 1971; Ekman and Friesen, 1984; Siegman and Feldstein, 1987). First, we get much information about other people by looking at their eyes. Eyes open wide show surprise, fear, or a flicker of interest. When we're angry, we stare in an unflinching manner. When we're sad or ashamed, our eyes may be cast down.

Second, appropriate eye contact depends on the social context. Especially during job interviews, eye contact conveys attentiveness, confidence, and respect. Managers expect eye contact from subordinates, but subordinates often complain that a manager's prolonged eye contact is domineering or intimidating. Looking at a coworker when speaking shows trust and interest, but an overly intense gaze can be disturbing (Chen et al., 2013). When two people like each other, they establish eye contact more often and for longer periods than when there's tension in the relationship.

Finally, cultural norms affect eye contact. In many Asian cultures, including Japan, meeting another's eyes can be rude. Asians are more likely than Westerners to regard a person who makes frequent eye contact as angry, unapproachable, and unpleasant (Akechi et al., 2013). In effect, then, cross-cultural communication breaks down when we don't understand cultural rules about acceptable body language.

5-5c Touch

Touching, another important form of nonverbal communication, sends powerful messages about our feelings and attitudes. Parents worldwide communicate with their infants through touch—stroking, holding, patting, rubbing, and cuddling.

Touching can be positive (hugging, embracing, kissing, and holding hands) or negative (hitting, shoving, pushing, spanking). Other forms of touching are controlling. One of my students left a boyfriend because "He said he loved and trusted me, but gripped my arm tightly every time I talked to another guy and pulled me away." Between intimate partners, a decline in the amount of touching may signal that feelings are cooling off.

Gendered Touching

Whether touching is viewed as positive or negative depends on the situation and one's gender. Generally, women are more likely than men to initiate hugs and touches that express support, affection, and comfort. In contrast, men more often use touching to assert power or show sexual interest (Wood, 2015). Platonic (nonsexual) touch between American men is typically acceptable in limited interactions such as violence (physical fighting) or contact sports (martial arts). Men are less likely to use physical touch (hugging) with other men to express friendly intentions (Reiner, 2017).

Rod Aydelotte-Pool/Getty Images

Pictured are former President Bush and Saudi Crown Prince Abdullah at Bush's Texas ranch in 2002. Many U.S. journalists raised questions about two men holding hands. In response, Arab Americans were quick to point out that Arab society views the outward display of affection between men as an expression of respect and trust. Thus, government officials and military officers often hold hands as they walk together or converse with one another.

Cross-Cultural Variations in Touching

As with other forms of nonverbal communication, the interpretation of touching varies from culture to culture. A firm handshake is the norm in the United States and Brazil. A light hand grip with slight bow is more appropriate in China. A longer handshake is better in the United Arab Emirates. And in Thailand, instead of shaking hands, it's better to place the hands together near the chest and bow (Gould and Cheng, 2018).

Cultural variations intertwine with gender-based expectations as well. In Morocco, only shake hands with people of the same gender, while in Australia, women do not shake hands with other women. And platonic, physical touch between men is more widely accepted in Italy than in the United States (Gould and Cheng, 2018; Reiner, 2017; Payne, 2004).

Health Influences on Touching

Touching and nonverbal communication can also be affected by health considerations. For instance, during the height of the COVID-19 pandemic, people were less likely to hug due to fear of virus transmission. In 2021, when more people were vaccinated and new virus cases slowed, the return to hugging came with a new set of medical suggestions including: shorter time, faces turned away, and hugging outside is best (Gillett, 2021).

5-5d Personal Space

In the example of airlines at the beginning of this chapter, you read that the distance that people establish between themselves is an important aspect of nonverbal communication. Americans are more annoyed by a passenger who invades their seat space (47 percent) than skimpy airline leg room (39 percent), long security lines (37 percent), or weather delays (20 percent) (Aguila, 2014). Personal space plays a significant role in our everyday nonverbal interactions, reflects power and status, and varies across societies.

When Is Our Personal Space Violated?

Our space is public or private. In the public sphere, which is usually formal, we have clearly delineated spaces: "This is your locker," "That's her office," or "You're parking in my spot." We usually decorate our public spaces with businesslike artifacts such as awards or framed photos of a group's accomplishments.

After a snowstorm, people in densely populated cities who shovel parking spots "mark" their personal spaces with lawn chairs, construction cones, strollers, and even a table set for two, complete with a bottle of wine. One resident, who was new to Boston, moved an orange road cone to park in someone else's "saved" spot. The next day, she discovered a note written on her windshield, "'Where is my cone?'... 'Did you take it as well?'" (Holt, 2018; Goodnough, 2010).

Private spaces send a different message. They convey informality and a relaxed feeling. Private spaces—homes, apartments, and dorm rooms—are often mini-museums that reflect people's interests, hobbies, hygiene, and personalities.

People, particularly men, sometimes invade others' space by standing too close in a line, leaning against others, using all of the shared armrest in tight airline seats, or plopping their feet on the chair next to us in a classroom, airport, or movie theatre. Not all space intrusions are physical encroachments. Loud cell phone conversations are a good example of an auditory intrusion into our personal space.

In intimate relationships (as between family members), our personal space is often 2 feet or less because we're at ease with close physical proximity. In other

dpa picture alliance/Alamy Stock Photo

Why do many men use more public space than women? Because they're larger and have longer legs? Or because men, but not women, feel "totally empowered to take up a lot of public space," even if it inconveniences someone else (refer to Bahadur, 2013)?

social situations, such as small classroom settings, the appropriate social distance may be 4 to 10 feet, which allows for some separation yet promotes interaction with each other. And people may prefer more distance in public spaces or those intended for larger groups. Therefore, these spaces may be designed to allow for more distance between individuals (Hall, 1966; Varnum, 2023).

Space and Power

Space usage signifies who has privilege, status, and power. People with higher incomes can afford enormous apartments in the city or houses with large yards in the suburbs, while the people with lower incomes may be crowded into the most undesirable sections of a city or town or in trailer parks. Executives, including college presidents, usually have huge offices (even entire suites), while faculty members often share office space, even though many of their discussions with students are confidential. Generally, the higher the socioeconomic status, the greater the consumption of space, including large cars, reserved parking spaces, private dining areas,

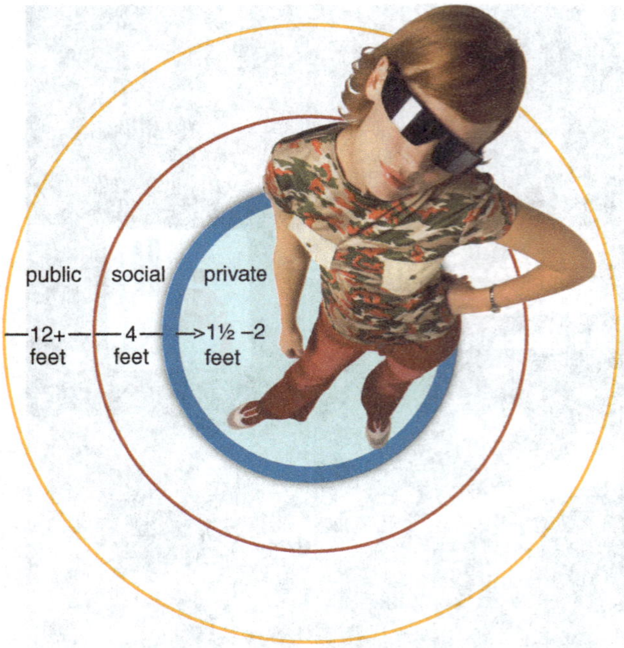

George Doyle and Ciaran Griffin/Stockbyte/Getty Images

larger-sized first-class airline seats, and luxurious skyboxes at sports stadiums.

Some wealthy men feel entitled to violate women's personal space. In a leaked recording, Donald Trump boasted about "grabbing women by the pussy." "When you're a star," he crowed, "you can do anything." Trump denied such sexual assaults, but more than a dozen women have accused him of groping them, grabbing their buttocks, or kissing them against their will. Almost a third of the men who voted for Donald Trump in 2016 said that grabbing a woman by her genitalia without consent isn't a sexual assault (Nelson and Crockett, 2017; *New York Times*, 2016; PerryUndem, 2017).

Cross-Cultural Variations and Space

Cultural norms and values determine how we use space. Americans not only stand in line but also have strict queuing rules: "Hey, the end of the line is back there" and "That guy is trying to cut into the line." In contrast, "along with Italians and Spaniards, the French are among the least queue-conscious in Europe" (Jandt, 2001, p. 109). In these countries and others, people routinely push into the front of a group waiting for taxis, food, and tickets, and nobody objects. Getting people from different cultures to use a shared set of queuing norms can require guidance. The Wimbledon tennis tournament, which draws fans from around the world, has a guide on queuing expectations (Spinks, 2018).

Americans maintain personal space in an elevator by moving to the corners or to the back. In contrast, an Arab male may stand right next to another man even when no one else is in the elevator. Because most Arab men don't share American concepts of personal space in public places and in private conversations, they consider it offensive if the other man steps or leans away. In many Middle Eastern countries, however, there are strict rules about women and men separating themselves spatially during religious ceremonies and about women avoiding any physical contact with men in public places (Office of the Deputy Chief . . . , 2006; Commisceo Global, 2016; refer also to Chapter 3).

5-6 Online Interaction

Many of us now interact in *cyberspace*, an online world of computer networks. In cyberspace, **social media** are websites that enable users to create, share, and/or exchange information and ideas. Social media include social networking sites (e.g., Instagram, LinkedIn, and Twitter), gaming sites (e.g., *Second Life*), video sites (e.g., YouTube), and blogs (websites maintained by people who provide commentary and other material). Who's online and why? And what are the benefits and costs of online interaction?

5-6a Who's Online and Why?

The percentage of Americans aged 18 and older who are connected to the Internet increased from practically zero in 1994 to 93 percent in 2021. The first iPhone was released in 2007. By 2021, 85 percent of American adults owned a smartphone, giving them online access all the time ("Driven to Distraction," 2017; Smith, 2017; Pew Research Center, 2021; Perrin, 2021).

Demographic Variations

About 93 percent of women and 94 percent of men use the Internet. Only 75 percent of Americans aged 65 and older are online compared with 99 percent of those aged 30 to 49 and 99 percent of 18- to 29-year-olds (Pew Research Center, 2021).

Race, ethnicity, and social class also affect a person's likelihood of being online. As **Figure 5.3** shows, Asian Americans are the most connected group in the United States. Their greater "connectivity" is due primarily to high education and income levels. Many Asian American parents are professionals who can afford computers and online service, and they encourage their children to use technology for education, a major avenue of upward mobility. Asian Americans also tend to live in urban and suburban areas where high-speed Internet connections

social media websites that enable users to create, share, and/or exchange information and ideas.

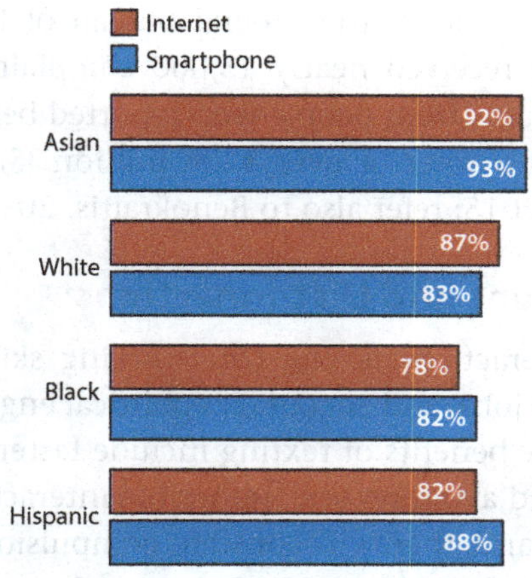

Figure 5.3 Internet and Smartphone Use by Race and Ethnicity

Note: The data are for households.

Source: Based on Martin, Michael. "Computer and Internet Use in the United States: 2018." U.S. Census Bureau, April 2021. https://www.census.gov/content/dam/Census/library/publications/2021/acs/acs-49.pdf

are readily available (File and Ryan, 2014; refer also to Chapters 10 and 13).

The U.S. offline population has declined substantially—from 48 percent in 2000 to only 7 percent in 2021 (Pew Research Center, 2021)—but there's still a digital divide between social classes. The higher a person's family income, the greater the likelihood of being online (**Figure 5.4**). Thus, children from families with lower incomes are the most likely to lack technological skills.

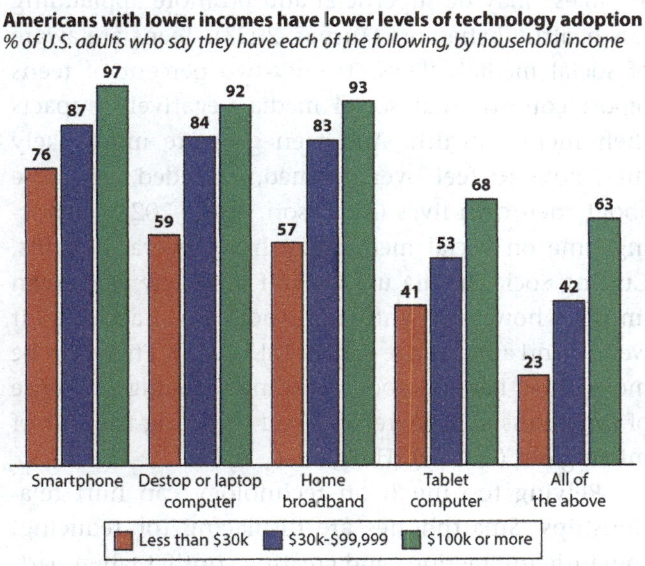

Note: Respondents who did not give an answer are not shown.

Figure 5.4 Social Class and Internet Use, 2016

Source: Based on Smith, 2017.

What We Do Online

The average American spends almost 11 hours a day staring at a screen (Nielsen, 2018). Among our many online activities, the most popular are *social networking sites* that connect people who have similar personal or professional interests. About 64 percent of all smartphone users are on social media each day. For 18–34-year-olds, that number is higher at 72 percent. We spend more time on social media and email than on radio and news sites (refer to **Table 5.3**).

Among online adults aged 18 and older, 81 percent say they use YouTube, and 69 percent use Facebook, which is more than those who use Instagram (40 percent), Pinterest (31 percent), LinkedIn (28 percent), Snapchat (25 percent), Twitter (23 percent), and Tik Tok (21 percent) (Auxier and Anderson, 2021). Besides social networking, more than half of online users access healthcare information, seek parenting advice, do banking, learn about community events or activities, and use navigation devices (Greenwood et al., 2016). Thus, people go online for a variety of reasons.

5-6b What Are the Benefits and Costs of Online Interaction?

You'll learn in Chapter 16 that technology brings both benefits and costs. The same is true of online interaction, ranging from family ties to privacy issues.

Family Ties

Social media has had a mixed effect on families. By age 3, many children have learned technological skills (e.g., creating a password, navigating some websites, and using tablets). Video games like *Minecraft* encourage school age children to build, explore, collaborate, and improve hand-eye coordination and problem solving (Dockterman, 2013; Lewin, 2013). On the other hand, screen time erodes the quality and quantity of interactions between parents and young children: "Some parents are perpetually tuned into their own devices, responding to every ping of their cellphones and tablets" (Brody, 2015, p. D7).

Table 5.3 How Americans Spend Their Online Time

Average minutes per day spent on . . .	
Email	56
Social media	50
Search	26
Online radio and news	14 minutes each

Source: Based on Statista, 2016.

Cell phones have increased the frequency of interaction between parents, especially if both are employed, to coordinate schedules and to chat with their children. Also, parents report spending more time with their children by playing home video games. On the other hand, some children complain that they rarely receive their parents' full attention because a parent is often immersed in email, texting, or being online even when pushing a swing, driving, or eating meals.

A team of researchers observed 55 parents and their young children at fast-food restaurants. Forty parents immediately took out mobile devices and used them throughout most of the meal (Radesky et al., 2014). By paying more attention to their electronic gadgets than their children, parents are sending their kids a powerful message: "My cell phone is more important than you are."

Among adults who are married or in committed relationships, 74 percent say that the Internet, smartphones, and social media have strengthened their communication and emotional closeness. On the other hand, 20 percent believe that the technology has had a negative effect on their relationships (e.g., a cell phone distracts a spouse or partner when the couple is together, they argue about the amount of time one of them spends online) (Lenhart and Duggan, 2014).

Online Relationships

An increasing number of people use social media to discuss important matters, get social support, and keep up or revive dormant relationships. Some ethnic groups, especially Asian Americans, who are underrepresented on television, have become successful on YouTube. They've found "millions of eager fans" who follow their comedy sketches and discuss topics, particularly sex and race, that are taboo among older generations (Considine, 2011, p. ST6; refer also to Pipkin, 2013).

Millions of Americans turn to the Web to find romance, using sites such as Match.com, Tinder, and Hinge. About 3 in 10 American adults say they have used online dating sites or apps. And 12 percent say they have either been in a committed relationship or married someone they met online. The use of online dating apps varies both by age (higher use among 18- to 29-year-olds) and sexual orientation (higher use among those identifying as LGB compared to heterosexual) (Anderson et al., 2020). Seventy one percent of those using dating sites say it was very or somewhat easy to find someone attractive or with shared interests. Younger women (18 to 34 years old) report the highest levels of negative interactions (unwanted persistence, name calling, and threats of physical harm). And, in 2015 alone, the Federal Bureau of Investigation (2016) received nearly 13,000 complaints about romance scams from people who reported being swindled out of a total of over $204 million (Smith and Anderson, 2015; refer also to Benokraitis, 2015).

Connections and Disconnections

Online interaction develops networking skills, helps people find jobs, and encourages political engagement. Some of the benefits of texting include faster communication and avoiding uncomfortable interactions. On the other hand, there's a "constant compulsion to connect" (Suddath, 2013, p. 80). For example:

- 85 percent of Americans are online each day, and 31 percent of them are online almost constantly (Perrin and Atske, 2021). On average, 18- to 24-year-olds use their smartphones more than 80 times a day ("Driven to Distraction," 2017).
- Nearly half of adults say they couldn't live without their smartphones, 41 percent check them several times an hour, and 11 percent do so every few minutes (Anderson, 2015; Newport, 2015).

Most teens report that social media gives them a connection to others, outlets for creativity, and a source of support during rough times. Teens also report that social media boosts their sense of being accepted either a lot (15 percent) or a little (42 percent). Still, only 32 percent feel social media has a positive effect on their lives, and 38 percent say that the drama on these platforms makes them feel overwhelmed (Anderson et al., 2022).

Social media fosters connection, but gestures such as "likes" may be superficial and promote applauding less healthy behaviors (Turner, 2023). Teens are aware of social media's flaws. Twenty-two percent of teens report concern that social media negatively impacts their mental health. And teen girls are more likely than boys to feel overwhelmed, excluded, or worse about their own lives (Anderson, et al., 2022). Reducing time on social media may have several benefits. Cutting social media use by half for a few weeks can improve how teens and young adults feel about their weight and appearance (Thai et al., 2023). There can be more broad benefits too, including reducing the sense of loneliness and depression and lessening the fear of missing out (FOMO) (Hunt et al., 2018).

Relying too much on technology can hurt relationships. Smartphones are displacing (or reducing) romantic interactions and creating conflict when partners don't give each other their full attention. Instead of mingling and conversing with coworkers during lunch or a break, many are texting or checking their email (Turkle, 2015; Roberts and David, 2016). The next time you go to a restaurant, watch how often—instead of chatting with each other—people check messages and social networks even before scanning the menu and how often they check their cell phones throughout a meal.

Online Harassment

Nationally, 94 percent of parents say they've talked with their children about proper online behavior, but 23 percent of 13- to 17-year-olds experienced cyberbullying in a single month. *Cyberbullying* refers to deliberately using digital media to communicate false, embarrassing, or hostile information about someone. Cyberbullying is less common than traditional bullying but has more profound negative outcomes that include depression, academic and mental health problems, severe isolation, and, most tragically, suicide (Anderson, 2016; Hinduja, 2021; Cyberbullying Research Center, 2016).

Harassment—from name calling to threatening behavior—is becoming a common part of online life. Fully 72 percent of adult Internet users have witnessed someone harassed online, and 47 percent have experienced it personally. Half of the latter didn't know the person who had most recently attacked them. Men are more likely to experience name-calling and embarrassment, while women are targets of more intense and serious forms of online assaults like sexual harassment and cyberstalking (Lenhart et al., 2016).

Cyberstalking is the repeated use of electronic communications to harass, threaten, or frighten someone. Nationally, 8 percent of Internet users have been cyberstalked to the point of feeling unsafe or afraid. Young people, particularly women under 30 and LGBTQIA+ individuals, are especially likely to be targets of cyberstalking. Cyberstalking may evolve into offline stalking, including abusive or harassing phone calls, vandalism, threatening or obscene mail, trespassing, and physical assault (Lenhart et al., 2016).

Tweeters can be engaging, funny, and supportive. Many of these faceless strangers can also be racist, sexist, abusive, and hateful. Because tweets are short and easy to read, public accusations and vile, ugly things said about anyone can quickly get a huge audience's attention.

Privacy Issues

A major cost of online interaction is jeopardizing our privacy because email and text messages are neither anonymous nor confidential. Many companies monitor and preserve their employees' email messages and can use them to discipline or fire people. Companies can also monitor online activities such as web browsing. This form of surveillance became a greater area of interest when more people shifted to remote work spurred by the COVID-19 pandemic. Monitoring of online activity is also becoming increasingly common for college admissions officers and employers to search social media sites before deciding whether to accept a student or make a job offer (Belkin and Porter, 2012; Abril and Harwell, 2021; Singer, 2013).

People often text and say things in emails they'd never say in person, especially when they're flirting, angry, frustrated, or tired. Emails, text messages, Snapchat photos, and videos don't disappear after being deleted but may last indefinitely in cyberspace. Divorce lawyers have successfully retrieved deleted emails, Facebook posts, tweets, and Instagram photos to convince judges that a divorcing spouse shouldn't get financial support or custody of the children (Hillin, 2014).

Americans have broad distrust for how social media companies will protect their personal data. This includes how tech companies manage users' personal information and information about their online activities. Around 70 percent distrust Facebook and around 60 percent distrust TikTok and Instagram (Kelly and Guskin, 2021).

It takes a private investigator just a few clicks to get a composite picture of someone from public records, email messages, and social media sites.

Tech giants like Apple, Meta (owner of Facebook, Instagram, and WhatsApp), and Google know more about their users than many people realize—where they go, what they search for, what they buy, which restaurants they frequent, what they do for fun, and when they go to bed. Besides collecting data on its own users, Meta has partnerships with several companies that collect behavioral data—from store transactions and customer email lists to divorce and Web browsing records—that it then sells to advertisers (Angwin et al., 2016). Neither state nor federal laws prohibit the collection or sharing of such data by third parties.

Chapter Review

Key Takeaways

5-1 **What is the purpose of social interaction and social structure?** *Social interaction*, central to all social activity, affects people's behavior. Our interaction is part of the *social structure*, which guides our actions and gives us a feeling that life is orderly and predictable. Every society has a social structure that encompasses statuses and roles.

5-2 **How do different types of statuses affect social experience?** A *status* is a social position that a person occupies in a society. Every person has many statuses that form their *status set*, which includes both ascribed and achieved statuses. An *ascribed status* is a social position that a person is born into and can't control, change, or choose (e.g., age, race, and ethnicity). An *achieved status* is a social position that a person attains through personal effort or assumes voluntarily (e.g., college student or wife). A *master status* is usually immediately apparent, makes the biggest impression, affects others' perceptions, and, consequently, often shapes a person's entire life.

Because we hold many statuses, some clash. People experience *status inconsistency* when they occupy social positions that are ranked differently (such as being a low-paid college professor).

5-3 **What are roles and complications related to roles?** A *role* defines how we're expected to behave in a particular status, but people vary considerably in fulfilling the responsibilities associated with their roles. These differences reflect *role performance*, the actual behavior of a person who occupies a status. A *role set* encompasses different roles attached to a single status (e.g., a parent who's a teacher, volunteer, and PTA member). Playing many roles often leads to *role conflict* because it's difficult to meet the requirements of two or more statuses and to *role strain*, the stress that arises because of incompatible demands among roles within a single status.

To deal with role conflict and role strain, some people deny that there's a problem. More effective ways to minimize role conflict and role strain include compromising, negotiating, setting priorities, compartmentalizing our roles, refusing to take on more roles, and exiting one or more current roles.

Example: Exiting a Marriage

Divorce is a good example of role exit, but it often involves a long process of five stages that may last several decades (Bohannon, 1971):

- The "emotional divorce" begins when one or both partners feel disillusioned or unhappy.
- The "legal divorce" is the formal dissolution of the marriage during which the partner who doesn't want the divorce may try to stall the end of the marriage.
- During the "economic divorce" stage, the partners may argue about who should pay past debts and unforeseen expenses (like moving costs).
- The "coparental divorce" stage involves parents' agreeing on issues such as child support and visitation rights.
- During the "community divorce" stage, partners inform friends, family, and others that they are no longer married.
- Finally, the couple goes through a "psychic divorce" in which the partners separate from each other emotionally. One or both spouses never complete this stage because they can't let go of their pain and anger—even if they remarry.

5-4 **What are the key components of the different sociological explanations of social interaction?** Refer to Table 5.2 on page 94.

5-5 What does nonverbal communication add to social interaction? Our *nonverbal communication* includes gestures, facial expressions, eye contact, and silence. Touching and how we use space are also important forms of nonverbal communication because they send powerful messages about our feelings and power. Nonverbal communication, which varies from society to society, can lead to cross-cultural misinterpretation and misunderstanding.

5-6 What new contributions to social interaction have been made through online technologies? Many people interact in *cyberspace*, an online world of computer networks that includes *social media* and *social networking sites*. Internet usage varies by sex, age, race, ethnicity, and social class. Online interaction can be impersonal and socially isolating and can jeopardize our privacy, but it can also save time, foster closer ties among family members and friends, and facilitate working from home.

Key Terms

Social interaction (p. 90)
Social structure (p. 90)
Status (p. 90)
Status set (p. 90)
Ascribed status (p. 91)
Achieved status (p. 91)
Master status (p. 91)
Status inconsistency (p. 92)
Role (p. 92)
Role performance (p. 92)
Role set (p. 93)
Role conflict (p. 93)
Role strain (p. 93)
Self-fulfilling prophecy (p. 95)
Ethnomethodology (p. 95)
Dramaturgical analysis (p. 95)
Social exchange theory (p. 97)
Nonverbal communication (p. 98)
Social media (p. 102)

Discussion and Critical Thinking Questions

1. What is social structure? How does social structure affect our interaction in settings such as the family and workplace? How does social structure shape (and even dictate) a situation's norms, statuses, and roles?
2. Why is age a master status? What are other examples of a master status? How can a master status be stigmatizing or empowering?
3. Have you (or have people you know) ever unknowingly engaged in a self-fulfilling prophecy? Were the results positive or negative?
4. Think about your interactions with someone who identifies with a gender that differs from your own. How have they indicated gender differences in your nonverbal communication (e.g., visual cues, touching, personal space)?
5. How do your front-stage and back-stage performances differ? Why? Have you ever been sanctioned when your back-stage behavior occurred in public? In what kinds of situations?
6. Has technology increased rudeness? Are you annoyed, for instance, by people texting while you're talking to them? Or have smartphones and texting increased our daily connections with others?

True or False?

How Much Do You Know about Social Interaction in Everyday Life?

The answers for 1, 3, and 6 are true. The answers to 2, 4, and 5 are false.

1. Social structure shapes people' actions and typically supports order and predictability in life. This helps to ensure that social life is not chaotic or random.
2. Although a person may be born into a family that practices a certain religion, people can change these statuses through their own actions. Therefore, religion is not an ascribed status. For instance, a person who is born Christian might convert to be Muslim later in life (or vice versa).
3. Professor is a status that has some typical roles such as planning lectures and grading exams. Roles are the behaviors expected of a person who has a particular status.
4. Nonverbal communication can be more impactful than spoken words. For example, a big smile sends a much stronger message than saying "I feel very happy." Some of the most common nonverbal messages are silence, visual cues, touch, and personal space.
5. Dramaturgical analysis is a research approach that examines social interaction in everyday life. It explores interactions as if they are occurring on a stage where people play different roles and act out scenes for the "audiences" with whom they interact.
6. Social media has had a mixed effect on parent-child relationships. For example, some children complain that they rarely receive their parents' full attention because a parent is immersed in online activities. On the other hand, parents report spending more time with their children by playing home video games.

6 Groups, Organizations, and Institutions

PG Alpha/Alamy Stock Photo

Learning Objectives

After studying this chapter, you will be able to…

6-1 Describe the different types of social groups, why people conform to group pressure, and the impact of social networks.

6-2 Describe three types of formal organizations and the strengths and shortcomings of bureaucracies.

6-3 Compare the theoretical explanations of groups and organizations, including their contributions and limitations.

6-4 Explain why social institutions are important and how they are interconnected.

True or False?

How Much Do You Know about Groups, Organizations, and Institutions?

1. Primary groups fill people's emotional needs while secondary groups are task-orientated.
2. Larger groups have greater cohesion and closeness than smaller groups.
3. Conformity is always a benefit to individual members of groups.
4. One critique of bureaucracies is that they create less freedom for workers.
5. People join utilitarian organizations because of shared interests and to pursue goals that they consider personally worthwhile or rewarding.
6. According to sociologists, family and religion are the only two organized and established social systems that meet one or more of a society's basic needs.

The answers are at the end of the chapter.

Every year, about a dozen of our neighbors, all fervent Ravens football fans, get together to buy season tickets and plan elaborate tailgating parties. They wear Ravens caps and purple jerseys, and most of the drivers attach pennants to their car antennas. These fervent football fans are a social group.

Be sure to take the True or False quiz in the chapter opener to better understand how much you know about these topics."

What do you think?

It's a good idea for companies to use video surveillance, including in restrooms.

1	2	3	4	5	6	7
	strongly agree				strongly disagree	

6-1 Social Groups

A **social group** is two or more people who share some attribute and interact with one another. They have a sense of belonging or "we-ness." Friends, families, work groups, religious congregations, clubs, athletic teams, and military veterans are all examples of social groups. Each of us is a member of many social groups, but we identify more closely with some groups than others.

6-1a Types of Social Groups

Some groups are small and personal (families); others are large and impersonal (financial organizations). Some are highly organized and stable (political parties); others are fluid and temporary (high school classmates). The most basic types of social groups are primary and secondary groups.

Primary and Secondary Groups

A **primary group** is a a relatively small group of people who engage in intimate face-to-face interaction over an extended period (Cooley, 1909/1983). Primary groups, such as families and close friends, are our emotional glue. They have a powerful influence on our social identity because we interact with them on a regular basis over many years, usually throughout our lives. Because primary group members genuinely care about each other, they contribute to one another's personal development, security, and well-being. Our family and close friends, in particular, stick with us through good and bad, and we feel comfortable being ourselves with them.

In contrast to primary groups, a **secondary group** is a large, usually formal, impersonal, and temporary collection of people who pursue a specific goal or activity. Your sociology class is a good example of a secondary group. You might have a few friends in class, but students typically interact infrequently and impersonally. When the semester (or quarter) is over and you've accomplished your goal of passing the course, you may not see each other again (especially if you're attending a large college or university). Other examples of secondary groups include sports teams, labor unions, and a company's employees.

Unlike primary groups, secondary groups are usually highly structured: There are many rules and regulations, people know (or care) little about each other personally, relationships are formal, and members are expected to accomplish specific tasks. While primary groups meet our *expressive* (emotional) needs, secondary groups fulfill *instrumental* (task-oriented) needs. Once a task or activity is completed—whether it's earning a grade, turning in a committee report, or building a bridge—secondary groups split up and become members of other secondary groups.

Table 6.1 summarizes the characteristics of primary and secondary groups. These characteristics are **ideal types**—general traits that describe a social phenomenon rather than every case. Ideal types provide composite pictures of how structures, events, and behaviors differ rather than specific descriptions of reality. Because primary and secondary groups are ideal types, their characteristics can vary. Thus, primary group members may sometimes devote themselves to meeting instrumental

social group two or more people who share some attribute and interact with one another.

primary group a small group of people who engage in intimate face-to-face interaction over an extended period.

secondary group a large, usually formal, impersonal, and temporary collection of people who pursue a specific goal or activity.

ideal types general traits that describe a social phenomenon rather than every case.

Table 6.1 Characteristics of Primary and Secondary Groups

	Characteristics of a Primary Group	Characteristics of a Secondary Group
Interaction	• Face-to-face • Usually small	• Face-to-face or indirect • Usually large
Communication	• Emotional, personal, and satisfying	• Emotionally neutral and impersonal
Relationships	• Intimate, warm, and informal • Usually long-term • Valued for their own sake (expressive)	• Typically remote, cool, and formal • Usually short-term • Goal-oriented (instrumental)
Individual Conformity	• Relatively free to stray from norms and rules	• Expected to adhere to rules and regulations
Membership	• Members aren't easily replaced	• Members are easily replaced
Examples	• Family, close friends, romantic partners, self-help groups, street gangs	• College classes, political parties, professional associations, religious organizations

needs (running a family-owned business), and secondary group members (military units and athletic teams) can develop lasting ties.

In-Groups and Out-Groups

The rivalry between the American baseball teams the New York Yankees and the Boston Red Sox (dating back to the 1919 trade of Babe Ruth) demonstrates how group identification affects people's experience (players and fans). Members of an **in-group** (same team) share a sense of identity and belonging that typically excludes and devalues outsiders. **Out-group** members are people who are viewed and treated negatively because they're seen as having values, beliefs, and other characteristics different from those of an in-group. For example, "we" vegetarians are healthier than "you" meat eaters, "we" computer geeks are smarter than "you" fraternity and sorority "types," and so on.

AF archive/Alamy Stock Photo

Is the cast of characters on the television series NCIS an example of a primary group? A secondary group? Both? Neither?

Almost everyone sees others as members of in-groups and out-groups. From ancient times to the present, people in various parts of the world have made "we" and "they" distinctions based on race, ethnicity, gender, sexual orientation, religion, age, social class, and other social and biological characteristics (Tajfel, 1982; Hinkle and Schopler, 1986). Such distinctions can promote in-group solidarity and cohesion. They can also create conflict and provoke inhumane actions, including wars, massacres of out-group members, and civil wars (refer to Chapter 3).

One person's in-group may be another person's out-group. The general public and the criminal justice system treat gangs as dangerous out-groups. Many young people who join gangs, however, experience in-group benefits. Attractions include making money, although illegally; close relationships with family members and friends who are already involved in gang life; protection against violent family or community members and rival gangs; a sense of self-worth, support, and belonging; and the status of being an "outlaw" that the entertainment industry often glamorizes as sexy and exciting (Taylor and Smith, 2013).

in-group people who share a sense of identity and belonging that typically excludes and devalues outsiders.

out-group people who are viewed and treated negatively because they're seen as having values, beliefs, and other characteristics different from those of an in-group.

Reference Groups

An in-group or an out-group can become a **reference group**, people who shape our behavior, values, and attitudes (Merton and Rossi, 1950). Reference groups influence who we are, what we do, and who we'd like to be in the future. Unlike primary groups, however, reference groups rarely provide personal support or face-to-face interaction over time.

Reference groups might be people with whom we already associate (a college club or community volunteer group). They can also be groups that we admire and want to be part of (teachers or doctors). Each person has many reference groups. Your sociology professor, for example, may be a member of several professional associations, a golf enthusiast, a parent, and a homeowner. Identification with each of these groups influences their everyday attitudes and actions.

Like in-groups and primary groups, reference groups can have a strong impact on our self-identity, self-esteem, and sense of belonging because they shape our current and future attitudes and behavior. We typically add or drop reference groups throughout our life course. If you aspire to move up the occupational ladder, your reference group may change from entry-level employees to managers, vice presidents, and CEOs.

6-1b Group Size and Structure

Two important characteristics of a social group are its size and leadership. Both affect a group's interaction and dynamics.

Group Size: Dyads, Triads, and Beyond

German sociologist Georg Simmel (1858–1918) pioneered the study of **dyads**, groups with two members, and **triads**, groups with three members. A dyad (a parent and a child, two spouses, two close friends) is the most cohesive of all groups because its members tend to have a personal relationship and interact more intensely than people in larger groups. Dyads can be unstable, however, because both persons must cooperate. If either member doesn't fulfill their responsibilities, the dyad collapses (Simmel, 1902; Mills, 1958).

Adding only one member to a dyad has important consequences because it changes the group's dynamics (behavior over time). A triad is more stable than a dyad because the group can continue if a person drops out, and the third member may patch up an argument that erupts between the other two. On the other hand, a third person may intentionally create conflict to break up a relationship or attain a dominant position. In any triad, two of the members can take sides against or leave out the third (Simmel, 1902; Mills, 1958). Triads also tend to interact less often and intensely than dyads (think about a third roommate who moves in, another person who joins you and a coworker at lunch, or someone who tags along on a date).

As more members are added—a fourth, fifth, and so on—interaction changes even more and alliance possibilities increase. At about seven or eight members, the group may break down into separate dyads and triads because it's difficult to have a single conversation about the task at hand (Becker and Useem, 1942). As a result, larger groups often require some kind of leadership to keep the members focused.

Group Leadership

Several classic studies (Lewin et al., 1939; Bales, 1950) identified three basic types of group leaders. An **authoritarian leader** gives orders, assigns tasks, and makes all major decisions. There's a clear division between leaders and followers, and leaders focus on completing *instrumental* tasks (e.g., marketing a new smartphone). Authoritarian leadership is most effective when there's little time for group decision making and the leader is the most knowledgeable member of the group. On the other hand, authoritarian leaders tend to stifle creativity and may be controlling or even oppressive.

Sam Edwards/Media Bakery

This triad studies together for sociology exams. What might happen if one student drops out of the study group? What if three more students join the group?

---

reference group people who shape our behavior, values, and attitudes.

dyad a group with two members.

triad a group with three members.

authoritarian leader gives orders, assigns tasks, and makes all major decisions.

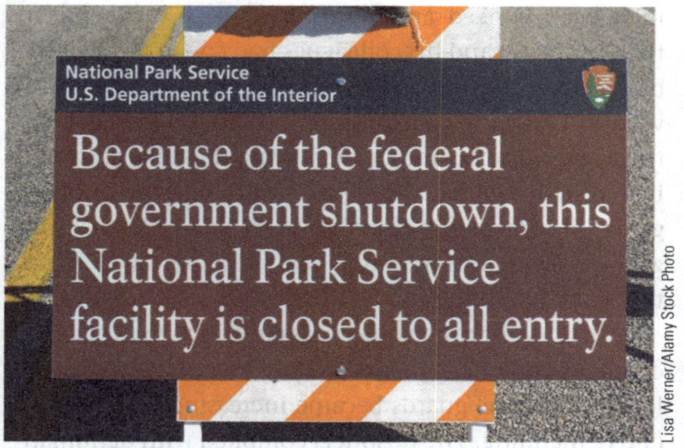

Lisa Werner/Alamy Stock Photo

U.S. government shutdowns occur when political leaders don't compromise on legislation that funds federal agencies and operations.

By contrast, a **democratic leader** encourages group discussion and includes everyone in the decision-making process. Democratic leaders have the final say, but they tend to be concerned about meeting a group's *expressive* needs (e.g., maintaining harmony, high morale, cohesiveness). Democratic leaders engage group members and encourage their creativity, but they have also been blamed for not taking charge in times of crisis and being less productive than groups with authoritarian leaders.

A **laissez-faire** (pronounced les-ey-fair) **leader** offers little or no guidance to group members and allows them to make their own decisions. Laissez-faire leaders are similar to the permissive parents you met in Chapter 4: There are few rules, few demands, and people do pretty much what they want. Laissez-faire leadership is effective in situations where group members are highly qualified in an area of expertise. However, this leadership style often leads to poorly defined roles, low productivity, a lack of motivation, little cooperation between group members, and not meeting a group's goals.

It's difficult for a leader to meet both a group's instrumental and expressive needs because involving everyone in decision making may result in not accomplishing a group's objectives. Also, leadership styles are more effective in some settings than others. Imagine, for example, if your college professors were laissez-faire ("Take the exams whenever you want.") or if your employers were always democratic ("Let's vote on who'll do what today.").

6-1c Group Conformity

Most Americans view themselves as rugged individualists who have minds of their own (refer to the discussion of values in Chapter 3). A number of studies have shown, however, that many people are profoundly influenced by group pressure. Four of the best known of these studies are by Solomon Asch, Stanley Milgram, Philip Zimbardo, and Irving Janis.

Asch's Research

In a now classic study of group influence, social psychologist Solomon Asch (1952) told subjects that they were taking part in an experiment on visual judgment. After seating six to eight male undergraduates around a table, Asch showed them the line drawn on card 1 and asked them to match the line to one of three lines on card 2 (**Figure 6.1**). The correct answer, clearly, is line C.

All but one of the subjects—who usually sat in the last chair—were Asch's confederates, or accomplices. In the first test, all the confederates selected the correct matching line. In the other tests, each of them, one by one, deliberately chose an incorrect line. Thus, the nonconfederates faced a situation in which seven other group members had unanimously agreed on a wrong answer. Averaged over all of the trials, 37 percent of the nonconfederate subjects ended up agreeing with the group's incorrect answers. When they were asked to judge the length of the lines alone, away from the influence of the group, they made errors only 1 percent of the time.

Asch's research demonstrated the power of groups over individuals. Even when we know that something is wrong, we may go along with the group to avoid ridicule or exclusion. Remember that these experiments were done in a laboratory and with people who didn't

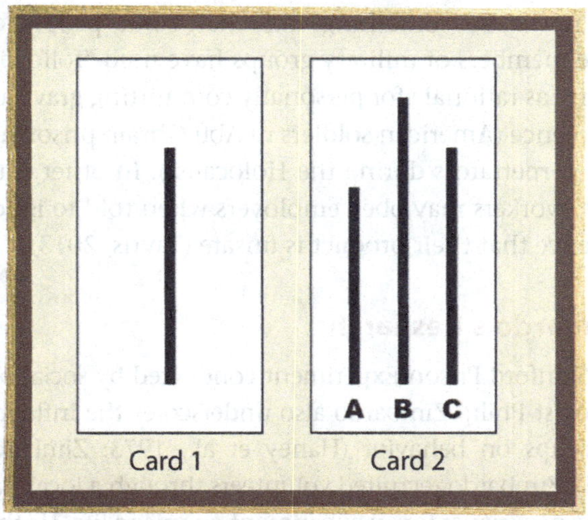

Figure 6.1 Cards in Asch's Experiment

democratic leader encourages group discussion and includes everyone in the decision-making process.

laissez-faire leader offers little or no guidance to group members and allows them to make their own decisions.

know each other. A group's influence on a person's attitudes and behavior can be even stronger when it's a real-life situation.

Milgram's Research

In a well-known laboratory experiment on obedience, psychologist Stanley Milgram (1963, 1965) asked 40 male volunteers to administer electric shocks to other study participants. In each experimental trial, one participant was a "teacher" and the other a "learner," one of Milgram's accomplices. The teachers came from a variety of professions including business professionals and blue-collar workers.

The learner was strapped into a realistic-looking chair that supposedly regulated electric currents. The teacher read aloud pairs of words that the learner had to memorize. Whenever the learner gave a wrong answer, the teacher was told to apply an electric shock from a low of 15 volts to a high of 450 volts. The learner didn't actually receive a shock but was told to fake pain and fear. When the learners shrieked in agony, the majority of the teachers, although distressed, obeyed the study supervisor and administered the shocks when told to do so.

Milgram's study was controversial. Directing participants in the teacher role to order electric shocks raised numerous ethical questions about them suffering extreme emotional stress (refer to Chapter 2 and Perry, 2013, for a recent critique). However, the results showed that an astonishingly large proportion of ordinary people obeyed an authority figure's instructions to inflict pain on others. It's easy to sit back and say, "I'd never do something like that," but people do. Some members of military groups have used "following orders" as rationale for personally committing grave acts of violence (American soldiers in Abu Ghraib prison and Nazi perpetrators during the Holocaust). In other situations, workers may obey employers when told to ignore evidence that their product is unsafe (Tavris, 2013).

Zimbardo's Research

The Stanford Prison Experiment conducted by social psychologist Philip Zimbardo also underscores the influence of groups on behavior (Haney et al., 1973; Zimbardo, 1975). Zimbardo recruited volunteers through a local California newspaper for an experiment on prison life. He then selected 24 young men, most of them college students.

On a Sunday morning, nine of the men were "arrested" at their homes as neighbors watched. The men were booked and transported to a mock prison that Zimbardo and his colleagues had constructed in the basement of the psychology building at Stanford University. The "prisoners" were searched, issued an identification number, and outfitted in a dresslike shirt and heavy ankle chains. Those assigned to be "guards" were given uniforms, billy clubs, whistles, and reflective sunglasses. The guards were told that their job was to maintain control of the prisoners but not to use violence.

All the young men quickly assumed the roles of either obedient and docile prisoners or autocratic and controlling guards. The guards became increasingly more cruel and demanding. The prisoners complied with dehumanizing demands (such as eating filthy sausages) to gain the guards' approval and bowed to their authority.

Zimbardo's study was scheduled to run for 2 weeks but was stopped after 6 days because the guards became increasingly aggressive. Among other things, they forced the prisoners to clean out toilet bowls with their bare hands, locked them in a closet, and made them stand at attention for hours. Instead of simply walking out or rebelling, the prisoners became withdrawn and depressed. Zimbardo ended the experiment because of the prisoners' stressed-out reactions.

The experiment raised numerous ethical questions about the harmful treatment of participants. It demonstrated, however, the powerful effect of group conformity: People exercise authority, even to the point of hurting others, or submit to authority if there's group pressure to conform (Zimbardo et al., 2000).

Janis's Research

Sometimes, intelligent people, including those in highly responsible positions, make disastrous and irrational decisions. Why? Social psychologist Irving Janis (1972, 1982) cautioned presidents and other heads of state to be wary of **groupthink**—a situation in which in-group members make faulty decisions because of group

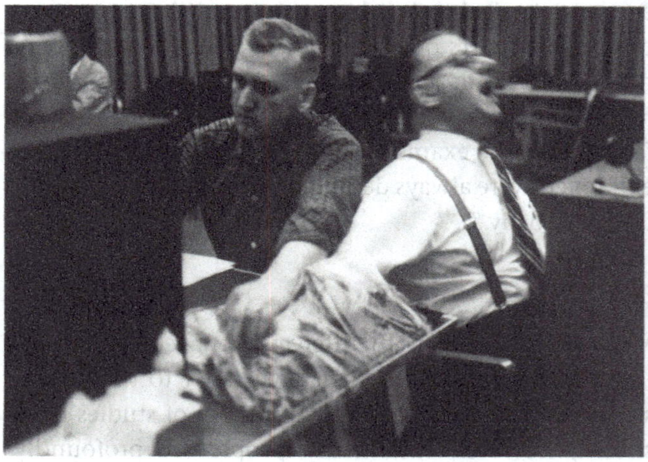

© Stanley Milgram/Alexandra Milgram

groupthink in-group members make faulty decisions because of group pressure rather than critically testing, analyzing, and evaluating ideas and evidence.

Wavebreakmedia/Shutterstock.com

Recommendations to successfully counter bullying have been made based on Milgram's findings about obedience and resistance to authority. Suggestions include diversifying resistance techniques, starting early, and sustaining resistance even when countered (Hollander, 2015).

pressure, rather than critically testing, analyzing, and evaluating ideas and evidence. To demonstrate group loyalty, Janis argues, individuals don't raise controversial issues, don't question weak arguments, and don't probe "soft-headed thinking." As a result, influential leaders often make decisions based on their advisors' consensus that turn out to be political and economic disasters.

> Sometimes intelligent people, even those in highly responsible positions, make disastrous and irrational decisions.

Examples of groupthink "fiascoes" that Janis studied included the U.S. failure to anticipate the attack on Pearl Harbor in 1941 and the escalation of the Vietnam War during the 1960s. More recently, a U.S. Senate study (2004) concluded that U.S. leaders' decision to invade Iraq in 2003 was based on a groupthink dynamic that relied on unproven and inaccurate assumptions, inadequate or misleading sources, and a dismissal of conflicting information showing that Iraq had no weapons of mass destruction.

Janis and other researchers have focused on high-level decision making, but groupthink is common in all kinds of groups—student clubs, PTAs, search committees, juries, religious groups, and college administrators (Hansen, 2010; Cohen and DeBenedet, 2012). People

Jim Barber/Shutterstock.com

Do social networks affect divorce rates? A research team that analyzed three decades of divorce data found that people were 75 percent more likely to get divorced if a friend was divorced and 33 percent more likely to end a marriage if a friend's friend was divorced (McDermott et al., 2013; Chapter 12 discusses some of the micro- and macro-level reasons for divorce).

can avoid groupthink by working through disagreements and seeking advice from informed and objective people outside the group.

6-1d Social Networks

A **social network** is a web of social ties that links individuals or groups to one another. It may involve as few as three people or as many as millions.

Network links between people or groups can be strong or weak. Some of our social networks, like our primary and secondary groups, may be tightly knit, involve interactions on a daily basis, and have clear boundaries about who belongs and who doesn't. In other cases, our social networks connect us to large numbers of people whom we don't know personally and with whom we interact only rarely or indirectly, and the boundaries change as people come and go. Examples of distant networks include members of the American Sociological Association and LinkedIn subscribers.

Electronic communities (also called *online communities* and *virtual communities*) are social networks whose members interact via the Internet. They exchange information, give advice, and engage in public discussions. Some electronic communities—ranging from art to yoga enthusiasts—have thousands of members. Others are small, focused on a single interest, and geographically close. One of my friends used meetup.com to find a local group of equally fervent cyclists who ride on Saturday mornings.

---

social network a web of social ties that links individuals or groups to one another.

6-2 Formal Organizations

A **formal organization** is a complex and structured secondary group designed to achieve specific goals in an efficient manner. We depend on a variety of formal organizations to provide goods and services in a stable and predictable way. Examples include water departments, food producers and grocery chains that stock our favorite items, and garment industries and retailers that manufacture and sell the clothes we wear.

6-2a Characteristics of Formal Organizations

Formal organizations share some common characteristics:

- Social statuses and roles are organized around shared expectations and goals.
- Norms governing social relationships among members specify rights, duties, and sanctions.
- A formal hierarchy includes leaders or individuals who are in charge.

Modern, complex societies rely on formal organizations, including businesses and schools, to accomplish complex tasks. Organizations differ, however, in their goals, membership, and degree of hierarchy.

6-2b Types of Formal Organizations

Sociologist Amitai Etzioni (1975) identified three types of formal organizations—utilitarian, normative, and coercive—based on their purpose and people's reasons for participating.

Utilitarian Organizations

Most people belong to at least one *utilitarian organization*, one that provides an income or other specific current or future material reward. Examples include government agencies, factories, corporations, and schools. Colleges are workplaces, but they also offer students the opportunity to earn degrees and future income.

Normative Organizations

People join *normative organizations* (also called *voluntary organizations* or *voluntary associations*) because of shared interests and to pursue goals that they consider personally worthwhile or rewarding. Examples include political parties, civic organizations (Habitat for Humanity), religious organizations, choirs, cultural groups, occupational groups (American Accounting Association), and groups that have a recreational focus (Beer Collectors of America).

Stan Rohrer/Alamy Stock Photo

Is this a utilitarian, normative, or coercive organization? (If your immediate response was "coercive," try again.)

Helping others is a strong American value (refer to Chapter 3). Thus, every year, millions of Americans—many of whom have full-time jobs—are volunteers in a variety of local, regional, and national organizations, saving government agencies billions of dollars by providing needed services (**Table 6.2**). Charitable donations, more than $390 billion, set a record in 2016. Individual contributions made up 72 percent of all giving, compared with only 5 percent from corporations (Giving USA Foundation, 2017).

Coercive Organizations

Membership in *coercive organizations* is largely involuntary. People are pushed or forced to join these organizations because of punishment (prisons) or treatment (psychiatric hospitals or drug treatment centers). Most coercive organizations are total institutions characterized by strict rules, the members' isolation, and resocialization (refer to Chapter 4).

Table 6.2 Formal Volunteering in the United States, September 2020 through September 2021

Percent of Americans who volunteer	23
Total number of volunteers	61 million
Total number of volunteer hours	4 billion
Total dollar value of volunteer time	$123 billion

Sources: Based on AmeriCorps. "Volunteering and Civic Life in America: Research Summary," 2021.

formal organization a complex and structured secondary group designed to achieve specific goals in an efficient manner.

A single formal organization can fall into all three categories. For example, the military is a coercive organization because of its rigid rules, resocialization, and members' isolation from the rest of the population. It's a utilitarian organization that provides millions of jobs for both soldiers and civilians. The military is also a normative organization: Some people join during wartime because of civic responsibility; many volunteer their time and resources to support soldiers and their families; and others help veterans find jobs, housing, and health care for war-related injuries.

Dave Carpenter/Cartoonstock.com

6-2c Bureaucracies

A **bureaucracy** is a formal organization designed to accomplish goals and tasks in an efficient and rational way. Bureaucracies aren't a modern invention but existed thousands of years ago in ancient Egypt, China, and Africa. Some bureaucracies function more smoothly than others, and some formal organizations are more bureaucratic than others. Whether they're relatively small (a 100-bed hospital) or huge (the Social Security Administration), bureaucracies have some common characteristics.

Ideal Characteristics of Bureaucracies

Max Weber (1925/1947) identified six key characteristics of the ideal type of bureaucracy. In Weber's model, the following characteristics describe what an efficient and productive bureaucracy *should* be like:

- **High degree of division of labor and specialization.** People perform very specific tasks.
- **Hierarchy of authority**. Workers are arranged in a hierarchy in which each person is supervised by someone in a higher position. The resulting pyramids—often presented in organizational charts—show who has authority over and is responsible to whom. Thus, there's a chain of command, stretching from top to bottom, that coordinates decision making.
- **Explicit written rules and regulations.** Detailed written rules and regulations cover almost every possible kind of situation and problem that might arise, including hiring, firing, salary scales, rules for sick pay and absences, and everyday operations.
- **Impersonality.** There's no place for personal likes, dislikes, or tantrums. Workers are expected to behave professionally. An impersonal workplace in which all employees are treated equally minimizes conflict and favoritism and increases efficiency.
- **Qualifications-based employment**. People are hired based on objective criteria such as skills, education, experience, and standardized test scores. If workers perform well and have the necessary credentials and technical competence, they'll move up the career ladder.
- **Separation of work and ownership**. Neither managers nor employees own the offices they work in, the desks they sit at, the technology they use, or the products that they assemble, invent, or design.

A "rational matter-of-factness," Weber maintained, makes bureaucracies more productive by "eliminating from official business love, hatred, and all purely personal, irrational, and emotional elements" (Weber, 1946, p. 216). Weber viewed bureaucracies as superior to other forms of organization because they're more efficient and predictable. He worried, however, that bureaucracies could become "iron cages" because people become trapped in them, "their basic humanity denied."

Bureaucratic Dysfunctions

Weber described the ideal characteristics of a productive and efficient bureaucracy, but what's the reality? As you read through the following list of problems, think about the ones you've experienced while working in a bureaucracy or dealing with one.

bureaucracy a formal organization designed to accomplish goals and tasks in an efficient and rational way.

Health Care on Aisle 7!

To avoid medical and health insurance bureaucracies, more people are turning to *retail clinics* (also called *convenient care* or *quick care clinics*). The first one opened in 2000, and they have grown in popularity. By 2019, 29 percent of adults used these services at least once per year (Bachrach and Frohlich, 2016; Statista, 2017; Black and Adjaye-Gbewonyo, 2021).

The clinics—usually staffed by nurse practitioners—offer low-cost treatment for more than 25 common conditions (e.g., strep throat and ear infections) and provide health screening tests, immunizations, and physicals. The clinics are located in pharmacies, grocery stores, and "big box" stores like Target and Walmart. There's no need to make an appointment. Waiting time, if any, is minimal.

Brand X Pictures/Stockbyte/Getty Images

Patients are in and out in about 15 minutes, and most of the clinics are open evenings and weekends ("Health Care in America," 2015).

- *Weak reward systems* reduce the motivation to do a good job and are thus a major source of inefficiency and lack of innovation (Barton, 1980). Besides low wages or salaries, weak reward systems include few or no health benefits, little recognition, unsafe equipment and work environments, and few incentives to be creative.
- *Rigid rules* squelch creativity and can lead to **goal displacement**, a preoccupation with rules and regulations rather than achieving the organization's objectives (Merton, 1968). A global management consultant criticizes companies for creating numerous "nonsense rules" that hobble the best and most talented workers: "The more rules, the less passion, which means less motivation" (Daskal, 2016, p. 19).
- Rigid rules and goal displacement often lead to **alienation**, a feeling of isolation, meaninglessness, and powerlessness. Alienation—at all levels—may result in high turnover, tardiness, absenteeism, stealing, sabotage, stress, health problems, and in some cases, whistle-blowing (reporting organizational misconduct to legal authorities).
- *Communication problems* are common in bureaucracies. Because communication typically flows down rather than up the hierarchy, employees (and many managers below the highest levels) rarely know what's going on. Supervisors and their subordinates may be reluctant to discuss problems or offer suggestions because they fear being criticized, demoted, or fired (Blau and Meyer, 1987; Patel, 2017). Communication problems also waste time and resources. An employee

goal displacement a preoccupation with rules and regulations rather than with achieving the organization's objectives.

alienation a feeling of isolation, meaninglessness, and powerlessness.

PzAxe/Shutterstock.com

About 63 percent of U.S. employees are disengaged from the job. Reasons include poor leadership, lack of upward mobility, boredom, and similar problems. Such alienation can lead to poor physical and mental health, low productivity, inferior customer service, and high absenteeism—all of which jeopardize a team's performance and the organization's financial vitality (Harter, 2022; Harter and Adkins, 2017).

prepared a monthly report for nearly three years because no one ever mentioned that the company no longer needed it (Harnish, 2014).

- The **iron law of oligarchy** is the tendency of a bureaucracy to become increasingly dominated by a small group of people (Michels, 1911/1949). A handful of people can control and rule a bureaucracy because the top officials and leaders monopolize information and resources. As a result, those at the top maintain their power and privilege.
- The cumulative effect of these and other bureaucratic dysfunctions can result in a *dehumanization* that stifles organizational creativity and freedom. As a result, work becomes more automated and impersonal (refer to "The McDonaldization of Society").

Bureaucracies are plagued by other problems like favoritism and dishonest employee evaluations (Kilmann, 2011). They function, however, largely because of the internal development of informal groups and collaborative work teams.

6-2d Getting Around Bureaucratic Dysfunctions

Weber's model overlooked the many ways that workers get around rigid bureaucratic regulations and impersonality. Since around 2000, self-managing work teams have become more common. Let's begin with a brief examination of some of the early studies' missteps and contributions regarding the importance of informal group networks.

Historic Studies of Informal Group Networks

For many years, and well into the 1930s, experts concentrated on organizational efficiency based on the principles of *scientific management* developed by Frederick Winslow Taylor, a mechanical engineer who wanted to improve industrial productivity. Taylor (1911/1967) considered workers—especially those in factories and on assembly lines—as mere adjuncts to machines, assumed that people don't like to work, maintained that employees need close supervision, and believed that management had to enforce cooperation because most workers aren't capable of handling even the simplest tasks.

Studies conducted by industrial psychologists and sociologists between 1927 and 1932 shook up Taylor's views. The research teams studied employees at the Western Electric Company's Hawthorne plant in Chicago. The *Hawthorne studies*, as they're often called, found that informal groups were critical to the organization's functioning (Roethlisberger and Dickson, 1939/1942; Mayo, 1945; Landsberger, 1958).

Informal social groups can promote an organization's goals if they collaborate, are cohesive, and motivate their members. In one of the Hawthorne studies, researchers spent 5 years observing a group of six women who were taken out of the main work area and placed in special test room where they could be observed while they assembled telephone relay switches. Over the years, productivity increased, and it was not related to any other changes made to the working conditions (such as different lighting). What mattered was the small group being together. These workers formed bonds with each other with one stating, "'We've been the best friends since the day we were in the test room.'" This experiment led the researchers to point to informal social relationships as central to both productivity and satisfaction with the job (Anteby and Khurana, n.d.).

Another Hawthorne experiment highlighted that informal social groups can support workers in resisting an organization's goals and formal rules. In the "bank-wiring room" a work group consisting of nine wiremen, three solderers, and two inspectors were observed for 6 months. Management offered financial incentives for higher productivity, but the work group pressured people to limit output. The wireworkers developed the following norms that controlled the group's behavior:

1. **You shouldn't turn out too much work.** If you do, you're a "rate-buster" and a "speed king."
2. **You shouldn't turn out too little work.** If you do, you're a "chiseler."
3. **You shouldn't tell a supervisor anything that will be detrimental to a coworker.** If you do, you're a "squealer."
4. **You shouldn't "act officious."** If you're an inspector, for example, you shouldn't act like one.

If individuals violated any of these norms, the other workers used "binging" (striking a person on the shoulder) to punish them.

Why did the employees control productivity rather than take advantage of the management's promise of higher wages? They feared that if they produced at a high level, they'd be required to do so regularly. They also worried that high productivity would lead to layoffs because supervisors would conclude that fewer workers could achieve the same output. In addition, the bank-wiring workers experienced high morale and job satisfaction because they felt they had some control over

iron law of oligarchy the tendency of a bureaucracy to become increasingly dominated by a small group of people.

The McDonaldization of Society

According to sociologist George Ritzer (1996, p. 1), the organizational principles that underlie McDonald's, the well-known fast-food chain, are beginning to dominate "more and more sectors of American society as well as of the rest of the world." McDonaldization has four components: efficiency, calculability, predictability, and control.

1. *Efficiency* means that consumers have a quick way of getting meals. In a society where people rush from one place to another and where both parents are likely to work or single parents are pressed for time, McDonald's (and similar franchises) offers an efficient way to satisfy hunger and avoid much "fuss and mess" (refer also to Jargon, 2013). Like their customers, McDonald's workers function efficiently: The menu is limited, the registers are automated, and employees perform their tasks rapidly and easily.
2. *Calculability* emphasizes the quantitative aspects of the products sold (portion size and cost) and the time it takes to get the products. Customers often feel that they're getting a lot of food for what appears to be a nominal sum of money and that a trip to a fast-food restaurant will take less time than eating at home. In reality, soft drinks are sold at a 600 percent markup because most of the space in the cup is taken up by ice. It costs at least twice as much and takes twice as long to get to the restaurant, stand in line, and pick up the food than to prepare a similar meal at home.
3. *Predictability* means that products and services will be the same over time and in all locales. "There's great comfort in knowing that McDonald's offers no surprises" (p. 10). The Quarter Pounder is the same regardless of where customers order. Outside the United States, "homesick American tourists in far-off countries can take comfort in the knowledge that they'll likely run into those familiar golden arches and the restaurant they have become so accustomed to" (p. 81). Workers, like customers, behave in predictable ways: "There are, for example, six steps to window service: greet the customer, take the order, assemble the order, present the order, receive payment, thank the customer and ask for repeat business" (p. 81).
4. *Control* means that technology shapes behavior. McDonald's controls customers subtly by offering limited menus and uncomfortable seats that encourage diners to eat quickly and leave. "Consumers know that they're supposed to line up, move to the counter, order their food, pay, carry the food to an available table, eat, gather up their debris, deposit it in the trash receptacle, and return to their cars. People are moved along in this system not by a conveyor belt, but by the unwritten, but universally known, norms for eating in a fast-food restaurant" (p. 105).

   McDonald's controls employees more openly and directly. Workers are trained to do a limited number of things in precisely the same way, managers and inspectors make sure that subordinates toe the line, and McDonald's has steadily replaced human beings with technologies that include soft-drink dispensers that shut off when the cup is full and a machine that rings and lifts the French fries out of the oil when they're crisp. Such technology increases the corporation's control over workers because employees don't have to use their own judgment or need many skills to prepare and serve the food.

Efficiency, calculability, predictability, and control reflect a rational system (remember Weber?) that increases a bureaucracy's efficiency. On the other hand, Ritzer contends, McDonaldization reflects the "irrationality of rationality" because the results can be harmful. The huge farms that now produce "uniform potatoes to create those predictable French fries" of the same size rely on the extensive use of chemicals that then pollute water supplies. And the enormous amount of nonbiodegradable trash that McDonald's produces wastes our money because we—and not McDonald's—pay for landfills.

McDonald's is such a powerful model that many businesses have cloned its four dimensions of efficiency, calculability, predictability, and control. Examples include "McDentists" and "McDoctors," drive-in clinics designed to deal quickly with minor dental and medical problems; "McChild" care centers like KinderCare; and "McPaper" newspapers, such as *USA Today* (Ritzer, 2008).

iStock.com/EllenMoran

their work. In contradiction to Taylor's scientific management perspective, then, the Hawthorne studies concluded that there's an important relationship between formal and informal organization (Roethlisberger and Dickson, 1939/1942).

Modern Work Teams

Since the Hawthorne studies, many companies have recognized that informal groups affect workers' productivity. As a result, numerous organizations have implemented alternative management strategies.

Today, *self-managing work teams* can be found in many large organizations (Cloke and Goldsmith, 2002; Neider and Schriesheim, 2005). They function best in organizations with a supportive culture and when individuals are willing to take on responsibility (McPheat, 2019). Contrary to Taylor's notion that workers do and managers think, self-managing work teams, sometimes referred to as *postbureaucratic organizations*, involve groups of 10 to 15 people who take on the duties of their former supervisors. They gather, interpret, and act on the information and take collective responsibility for their actions—whether it's designing a new refrigerator or handling a university's food service. Not held back by unresponsive managers, effective self-managing groups focus on their goals and are committed to the organization's success, and the team members themselves can come up with creative ideas (Colvin, 2012). Yet, this team style can also suffer from disadvantages such as groupthink and disconnect with management (McPheat, 2019).

Collaboration

Many organizations emphasize collaboration. The benefits include a diverse pool of ideas, greater worker participation and job satisfaction, a sense of belonging, and speedier completion of tasks because people can achieve more when they combine their skills.

On the other hand, "collaborative overload" can result in interruptions that increase the total time needed to complete a task. The most vulnerable are *knowledge workers*, people who acquire, organize, analyze, and use information (programmers, data and systems analysts, product developers). Some researchers estimate that knowledge workers "waste" 70 to 85 percent of their time attending meetings (virtual or face-to-face), answering emails or phone calls, and dealing with an avalanche of requests (Drucker, 2001; "The Collaboration Curse," 2016; Cross et al., 2016).

6-3 Sociological Perspectives on Social Groups and Organizations

How do organizations operate? And how do social groups affect massive bureaucracies? Functionalist, conflict, feminist, and interactionist perspectives offer different insights, providing a multifaceted understanding of groups and organizations. **Table 6.3** summarizes these perspectives.

Table 6.3 Sociological Perspectives on Groups and Organizations

Theoretical Perspective	Level of Analysis	Main Points	Key Questions
Functionalist	Macro	Organizations are made up of interrelated parts and rules and regulations that produce cooperation in meeting a common goal.	• Why are some organizations more effective than others? • How do dysfunctions prevent organizations from being rational and effective?
Conflict	Macro	Organizations promote inequality that benefits elites, not workers.	• Who controls an organization's resources and decision making? • How do those with power protect their interests and privileges?
Feminist	Macro and micro	Organizations tend not to recognize or reward women and regularly exclude them from decision-making processes.	• Why do many women hit a glass ceiling? • How do gender stereotypes affect women in groups and organizations?
Symbolic Interactionist	Micro	People aren't puppets but can affect what goes on in a group or organization.	• Why do people ignore or change an organization's rules? • How do members of social groups influence workplace behavior?

6-3a Functionalism: Groups and Organizations Benefit Society

Functionalist perspectives emphasize that social groups and organizations are composed of interrelated, mutually dependent parts. As Weber pointed out, when a bureaucracy operates rationally and efficiently, workers and bosses cooperate to turn out a final product, whether it's an automobile or a can of soup.

Effective leadership is a primary factor in job satisfaction. For instance, leaders make decisions about employee compensation and employment conditions which affect work-life balance. Workers also want their leaders to be honest, have follow-through, hear their ideas and concerns, and show them they are valued (Koch, 2023; Wigert, 2023). Based on the leadership research of Rosabeth Kanter (Director of Harvard University Advanced Leadership Initiative), there are six recommendations for leaders to foster positive change and provide employee support (Peek, 2023):

1. **Show up:** Leaders should be present with their teams.
2. **Speak up:** Leaders should share information and talk through processes with teams.
3. **Look up:** Leaders who are focused on goals should "look up" and take a step back to understand their employees' processes and what they can achieve.
4. **Team up:** Leaders should help employees develop relationships because their sense of power comes from their connections with co-workers at all levels. Social events can be a source of connection.
5. **Never give up:** Leaders should be flexible and resilient.
6. **Lift others up:** Employees benefit from feedback. When leaders set goals for their employees, they should also invest in training and resources so people can be successful. Finally, to lift employees up, leaders should provide recognition for work well done.

Groups and organizations can also be dysfunctional. Workers may be alienated because of few rewards, favoritism, and incompetent supervisors. Sixty-three percent of employees report feeling disengaged with their jobs (Harter, 2022). Among employees who feel unchallenged, bored, or dissatisfied, 75 to 80 percent spend more than 2 hours a day surfing the Internet or on Twitter, Facebook, and other social media sites. Wasting time—which results in low productivity, unhappy customers, and lower profits—costs organizations up to $398 billion annually (Adkins, 2015; OfficeTime, 2015).

The Image Party/Shutterstock

Since 1975, the number of employee-owned companies in the United States has grown from 1,600 to more than 11,500 and now represents about 12 percent of the private sector workforce. These firms are profitable because the workers are highly motivated: They have a stake in the business, control the decision making, contribute original ideas and cost-saving solutions, and depend on each other. WinCo Foods is the largest 100 percent employee-owned company in the United States (Koba, 2013; Marikar, 2015; National Center for Employee Ownership, 2022).

Critical Evaluation

For critics, especially conflict theorists, functionalists exaggerate cooperation and tend to gloss over dysfunctions like worker dissatisfaction and alienation. At some of the largest and most successful companies like Amazon, current and former employees have complained of 80-hour work weeks, interrupted vacations, sending negative feedback about coworkers to bosses, and little tolerance for those struggling with life-threatening illnesses or family crises (Kantor and Streitfeld, 2015). Still, many Americans' views of corporations are contradictory. About 63 percent (compared with 48 percent in 2001) are dissatisfied with the size and power of major companies, but only 20 percent want greater regulation (Riffkin, 2016).

Another issue is whether informal social networks improve worker morale and control as much as functionalists claim. After all, a tedious and monotonous job is tedious and monotonous regardless of the degree of informal coworker interaction.

6-3b Conflict Theory: Some Benefit More Than Others

Conflict theorists contend that organizations promote inequality that benefits the top of the hierarchy, not workers. In many companies, those at higher levels are more comfortable hiring and promoting people like themselves (male, White, and middle class or higher).

Phrases like "fitting the mold" and "having common interests" are often code words for belonging to an in-group. In 2020, 82.5 percent of Fortune 500 corporate board seats were filled by people who are White. The representation was lower for people of other races: Black (8.7 percent), Asian/Pacific Islander (4.6 percent), and Latina/o/x (4.1 percent). In addition, women hold a disproportionately lower percentage of these board seats (26.5 percent) compared to men (72.5 percent) (Deloitte and Alliance for Board Diversity, 2023).

Inequality in income, status, and other rewards means that owners and managers can easily exploit workers. Those at the top dictate to those at the middle and the bottom. Because organizations serve elites, conflict theorists argue, they routinely ignore workers' needs and interests.

What about rewarding the best employees? According to a management professor, "I've examined scores of empirical studies since the early 1980s and have not found convincing evidence that performance reviews are fair, accurate or consistent across managers, or that they improve organization effectiveness." Instead, subjective evaluations "are intimidating tools that make employees too scared to speak their minds" (Culbert, 2011, p. A25).

There's also considerable incompetence, waste, and corruption in many government agencies and big businesses. For example:

- In the last two days for spending in the 2018 budget year, 67 Federal agencies spent 97 billion dollars, including $308,994 on alcohol, $673,471 on golf carts, and 4.6 million dollars on lobster tails and crab (U.S. Government Publishing Office, 2020).
- Since at least 2012, the Veterans Administration has ignored the agency's competitive bidding laws, wasting at least $6 billion a year on high prices for medical goods and services (Rein and Wax-Thibodeaux, 2015).
- The Federal government spends more than $1.7 billion each year to keep up thousands of empty buildings. Various regulations create lengthy processes that complicate selling these buildings (Wire, 2016)
- In 2016, the Pentagon buried an internal study that it had wasted $125 billion on accounting, human resources, and property management. At least 40 percent of all U.S. weapons given to the Afghan and Iraqi armies "can't be traced" (Vicens, 2015; Whitlock and Woodward, 2016).
- Since 2000, the federal government has used more than $3 billion taxpayer dollars to build or renovate 36 professional sports stadiums (e.g., for the New York Yankees and Indianapolis Colts). Dozens of studies have shown that such facilities don't increase local economic development or create jobs (Gayer et al., 2016).
- In 2014, Michigan government officials decided to shift from Detroit's water system to the Flint River. It was supposed to save money during a pipeline project. The shift resulted in water contamination, causing over a dozen people to die and more than 80 to become very sick. Contrary to money-saving aims, the project cost 600 million dollars in a legal settlement from a lawsuit against the State of Michigan (among others) (Booker, 2021).
- At least 110 Americans have been injured or killed by an airbag that Ford, Honda, Nissan, and Toyota automakers knew was defective. Volkswagen sold more than 600,000 diesel cars over a decade that the company equipped with illegal devices to evade U.S. emissions rules (refer to Chapters 7 and 15).

Besides wasting billions of taxpayer dollars, according to many federal employees, government agencies hire and promote incompetent people, offer few advancement opportunities and rewards for good performance, and keep increasing the layers of unnecessary executives and managers. One employee complained, for instance, that "It takes 13 steps and five layers to get a signature from our office director" (Rein, 2011, p. B4; U.S. Office of Personnel Management, 2014).

Compared with government agencies, privately owned organizations have more freedom, authority, and control over activities and decision making, but at what cost?

ZUMA Press, Inc./Alamy Stock Photo

When Walmart opened its small Express stores in many rural or isolated towns, local groceries and pharmacies closed because they couldn't compete with Walmart's prices. To increase profits at its Supercenters and some Neighborhood Markets, in 2016, Walmart shut all 102 of its Express stores. Investors were pleased, but some residents had to make 50-minute roundtrip drives to the nearest grocery stores and pharmacies (Pettypiece, 2016).

Because of these and similar problems, many Americans are disillusioned with a number of institutions and organizations (**Figure 6.2**).

Critical Evaluation

Conflict theory has several weaknesses. First, it assumes that greater equality leads to a more successful and productive organization. However, self-managing work teams can fail because of ineffective team leaders or management interference. Employee-owned companies have been bankrupted because of inexperienced management, competition, and low worker morale (Lencioni, 2002; "When Workers Are Owners," 2015).

Second, conflict theorists rarely credit many organizations' support or controversial social issues that promote greater equality. Regardless of motives (improving a tarnished public image, boosting profits, becoming more socially responsible), a number of the nation's most powerful CEOs and businesses—including Walmart, NASCAR, Apple, AT&T, Southwest Airlines, and recently, Disney—have supported same-sex marriage laws and fought discrimination against LGBTQIA+ customers (Green and Higgins, 2016; refer also to Chapters 7, 9, and 11).

Third, do conflict theorists focus too much on organizational deficiencies? There are, after all, many organizations that are efficient, profitable, and that don't exploit their workers (Kaplan, 2010).

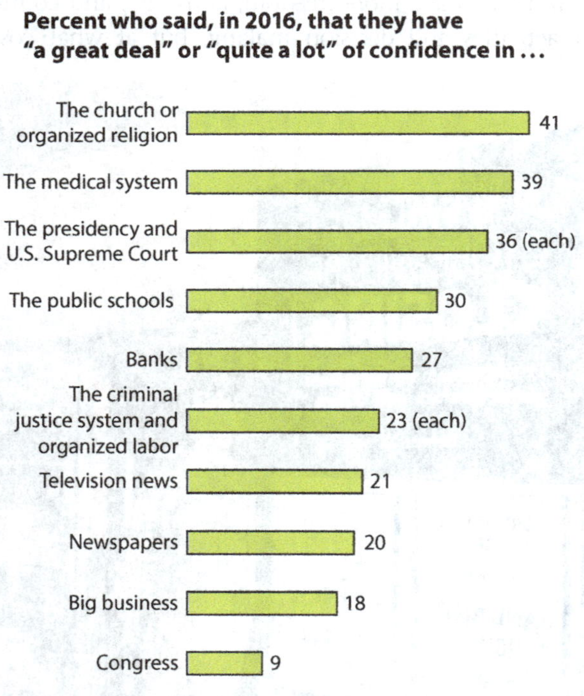

Figure 6.2 Many Americans Have Little Confidence in Organizations

Note: A majority of Americans had high confidence only in the military (73 percent), small business (68 percent), and the police (56 percent).

Source: Based on Norman, 2016.

6-3c Feminist Theories: Men Benefit More Than Women

Feminist scholars point out that across all social classes, women (and especially those who have been systematically excluded) consistently fare worse than men, especially in leadership roles. Women have enjoyed increased success in organizations since the 1980s, but rarely at the highest levels. For example, women hold only 21 (4.2 percent) of the CEO positions at Fortune 500 companies (Merelli, 2017), and there's a huge gender gap in S&P 500 companies between men and women in the labor force and among CEOs (**Figure 6.3**).

Companies benefit from women's leadership. Women who are CEOs are more likely to mentor women workers, to match people to job skills, and to pay wages that reward productivity. On average, women in management roles are more engaged than men in these positions, more likely to encourage their subordinates' development, to recognize and praise their accomplishments, and to make them feel valued (Fitch and Agrawal, 2015; Flabbi et al., 2016).

Note: The Standard & Poor's 500 (S&P 500) is made up of the largest public companies of the leading U.S. industries (as selected by economists).

Figure 6.3 Women in S&P 500 Companies

Source: Catalyst, 2017.

Why, then, does the gender gap in top positions persist? Large numbers of women still hit a **glass ceiling**—workplace attitudes or organizational biases that prevent them from advancing to leadership positions. Examples of obstacles include men's negative attitudes about women in the workplace, women's placement in staff positions that aren't on the career track to the top, a lack of mentoring, biased evaluations by male supervisors, and little or no access to highly visible committees or task forces (Barreto et al., 2009; Auster and Prasad, 2016; refer also to Chapters 9 and 11).

In contrast, many men who enter female-dominated occupations (nursing and teaching) receive higher wages and faster promotions, a phenomenon known as a **glass escalator**. In 2016, for example, men made up only 11 percent of all full-time registered nurses. However, the average female nurse earned $59,436, about 10 percent less than the $65,572 that the average male nurse earned. Men on glass escalators also move up to supervisory positions more quickly than women (Goudreau, 2012; U.S. Bureau of Labor Statistics, 2017).

Critical Evaluation

Feminist theories show that groups and organizations treat many women like outsiders. One weakness, however, is that much of the emphasis is still on White and Black women, even though Latinas and Asian American women comprise a large segment of the labor force.

A second limitation is that even when feminist scholars say that both sexes suffer from organizational stereotypes, they tend to gloss over stereotypes that affect men and how female supervisors reinforce glass ceilings and escalators (refer to Huffman, 2013). Many women experience a double standard, and some fail for some of the same reasons as men: Women who are managers and executives may avoid lateral moves (that can lead to a promotion), turn down promotions and job offers (that involve risk-taking, long hours, and work-related stress), or not network within and outside a company (impacted by exclusion from male colleagues or additional family-related responsibilities) (Reingold, 2016; Bhat, 2022).

6-3d Symbolic Interaction: People Define and Shape Their Situations

While functionalist, conflict, and (some) feminist theorists examine groups and organizations on a macro level, symbolic interactionists focus on micro-level behavior. They emphasize that an individual's perception and definition of a situation shape group dynamics and, consequently, organizations.

Group leaders or members can create or reinforce conformity (as shown in the Asch and Milgram studies). Informal groups can also determine what goes on in an organization by refusing to obey the rules and implementing their own (as in the Hawthorne studies). Thus, according to symbolic interactionists, individuals make choices, change rules, and mold their own identities instead of being manipulated (Kivisto and Pittman, 2001).

Symbolic interactionists also note that people's outcomes depend on how coworkers and bosses interpret the *same* behavior. For example, whether women are supervisors or clerical workers, if they lose their temper, they're overwhelmingly seen as too emotional, incompetent, out of control, weak, and worth less pay. Their angry male counterparts, in contrast, are often viewed as authoritative, tough, and forceful (Brescoll and Uhlmann, 2008).

Critical Evaluation

Symbolic interaction explains how members of groups and organizations interpret the world around them and, as a result, affect what goes on. However, symbolic interaction, a micro-level theory, ignores macro-level factors that affect workers and consumers. About 51 percent of Americans seek out "all natural" products, but much of the food contains chemically processed vitamins and artificial ingredients. Around 70 percent of those seeking healthy foods depend on food labels (Pew Research Center, 2016). Yet, the average consumer has little knowledge of and control over false advertising, a macro-level factor (Esterl, 2013).

In contrast to interactionists' claims, most people can't shape or change their situations. Formal organizations (the U.S. government and large companies) often invade the privacy of citizens, workers, or consumers by collecting information and monitoring people's behavior. Knowledge workers must share large, noisy, open office spaces and do much of their work when they get home at night. A majority (51 percent) of police officers say their job is frustrating because of factors over which they have no control, such as biased media coverage, unfair departmental disciplinary procedures, and a shortage of resources ("The Collaboration Curse," 2016; Morin et al., 2017).

We've examined social groups and formal organizations, both of which are part of larger structures that sociologists call *social institutions*. The final section introduces you to this important concept and lays the foundation for analyses of social institutions in later chapters.

---

glass ceiling workplace attitudes or organizational biases that prevent women from advancing to leadership positions.

glass escalator men who enter female-dominated occupations receive higher wages and faster promotions than women.

6-4 Social Institutions

A **social institution** (or simply, *institution*) is an organized and established social system that meets one or more of a society's basic needs. There are five major social institutions worldwide that ensure a society's survival:

- The *family* replaces a society's members through procreation, socializes children, and provides another important reason for sexual activity between adults.
- The *economy* determines how a society produces, distributes, and consumes goods and services.
- The *government* (which includes the police, military, and courts) maintains law and order, creates and enforces laws, regulates elections, and protects people's civil liberties.
- *Education* helps socialize children, transmits values and knowledge, and teaches work-related skills.
- *Religion* offers meaning and purpose through shared beliefs, values, and practices related to the supernatural.

Besides these core institutions (which we'll examine in Chapters 7 and 11–14), people develop new institutions in response to cultural changes. In the past, for example, the family and religion addressed most health problems. Only in relatively recent times have *medicine* and *health care systems* emerged as distinct institutions that diagnose, treat, and try to prevent illness and disease. Because of technological and economic changes, the *mass media* is another relatively recent institution. It disseminates information to a large audience, creates desires for goods and services, influences our attitudes about social issues, and reinforces a society's values and beliefs.

6-4a Why Social Institutions Are Important

People create social institutions to address basic needs or problems. Institutions have an organized purpose, weave together norms and values, and, consequently, guide behavior. They may differ across countries, but each institution carries out certain tasks in a particular society that contribute to its overall functioning and stability. No two families are exactly alike, but the family institution affirms broadly shared cultural agreements about what a family should be and do. In the same vein, no two grocery stores are exactly alike, but the economy establishes and maintains a variety of rules that make food shopping predictable.

social institution an organized and established social system that meets one or more of a society's basic needs.

Five major social institutions

- family
- economy
- government
- education
- religion

Institutions aren't just "somewhere out there." We create, live inside, change, resist, and reproduce them, and we participate simultaneously in multiple institutions. Understanding institutions and how they're interconnected can tell us a lot about how a society functions.

6-4b How Social Institutions Are Interconnected

Institutions are linked. Consider the connections among seven institutions that affect health related to food intake and exercise and specifically the potential for the development of diabetes or heart disease. *Medicine* tells us that diabetes and heart disease are major (and often preventable) conditions affecting other aspects of health, disability, and death and for steadily rising *health care* costs (refer to Chapter 14). The *mass media* publicizes such information and medical advice about prevention. It also promotes unhealthy food through persuasive marketing techniques (free gifts, celebrity endorsements, false nutrition claims) on television and other electronic devices (Jenkin et al., 2014).

In *education*, many schools now offer healthier meals, but school vending machines still sell high-fat snacks and sugary drinks. When state and local *governments* slash education budgets, schools decrease physical education and sports programs (Evich, 2016; Yeager, 2016). At the federal level, the most recent U.S. food recommendations include "Guideline 3: Focus on meeting food group needs with nutrient-dense foods and beverages, and stay within calorie limits" (U.S. Department of Agriculture, 2020, p. 30). Yet, U.S. food assistance programs often provide cheap, high-calorie, processed foods versus foods that are nutrient-dense (Aubrey, 2022).

Families contribute to the growing health problems associated with food consumption and exercise. By the time children enter school, their parents have already shaped their eating habits and preferences (refer to Chapters 4 and 14). Parents and other family members can encourage physical activity, provide guidance on healthy eating, and limit exposure to mass media marketing that promotes unhealthy habits.

Caspar Benson/fStop/Getty Images

On average, a 12-ounce can of soda contains at least 9 teaspoons of sugar. Over the course of a year, drinking one soda a day can make you 16 pounds heavier (Loumarr, 2015). Which social institutions affect your beverage choices?

The *economy* has far-reaching effects on obesity through its linkages to other institutions. Almost all politicians rely on the food industry to help finance their elections and to hire them as lobbyists after they leave public office (refer to Chapter 11). Beverage giants like Coca-Cola and PepsiCo have donated millions of dollars to nearly 100 prominent health groups (National Institutes of Health, American Diabetes Association) while simultaneously spending millions to defeat public health measures aimed at curbing soda consumption. The mass media relies on food ads even though companies make unhealthy products. And, because of advertising and packaging claims, many families think that fruit and sports drinks and flavored waters—all of which have high amounts of sugar—are healthy (Munsell et al., 2015; Aaron and Siegel, 2017).

Chapter Review

Key Takeaways

6-1 What is the importance of social groups? A *social group* (such as friends or work groups) gives us a common identity and a sense of belonging. Social groups include *primary groups* (such as family members) that shape our social and moral development and *secondary groups* (such as the students in your sociology class) that pursue a specific goal or activity.

Members of an *in-group* share a sense of identity and "we-ness." In contrast, *out-groups* are viewed and treated negatively because they're seen as having values, beliefs, and other characteristics that differ from those of the in-group.

We also have *reference groups* that influence who we are, what we do, and who we'd like to be in the future. Groups often form a *social network* that may be tightly knit and interact every day or may include large numbers of people whom we don't know personally and with whom we interact only rarely or indirectly.

Example: Secondary Groups Can Replace Primary Groups

In 1864, a person with an alcohol use disorder who had ruined a promising career on Wall Street because of his constant alcohol misuse cofounded Alcoholics Anonymous (AA), a program that would enable people to stop drinking by undergoing a spiritual awakening and seeking help from a buddy to stay sober. Initially, AA was a secondary group that encouraged its members to attend regular meetings and discuss their accomplishments in staying sober. Over the years, however, AA became a primary group for many members: It offers support to a relatively small group of people who engage in face-to-face interaction over an extended period, especially when their family and friends have rejected them.

6-2 How are formal organizations described? We depend on a variety of *formal organizations* to provide goods and services in a stable and predictable way. Utilitarian, normative, and coercive organizations differ in their general characteristics and membership, but a single formal organization can fall into all three categories.

Bureaucracies (such as your college) are supposed to accomplish goals and tasks in the most efficient and rational way possible, but many experience shortcomings such as weak reward systems, rigid rules, goal displacement, alienation, communication problems, and dehumanization.

6-3 How do sociological perspectives regarding social groups and organizations differ? Refer to **Table 6.3** on page 121.

6-4 What is the purpose of social institutions? A *social institution* meets a society's basic survival needs. Social institutions are abstractions, but they have an organized purpose, weave together norms and values, and, consequently, guide behavior. Because social institutions are linked to one another, they can tell us a lot about how a society functions and how we're connected to one another.

Key Terms

Social group (p. 110)
Primary group (p. 110)
Secondary group (p. 110)
Ideal types (p. 110)
In-group (p. 111)
Out-group (p. 111)
Reference group (p. 112)
Dyad (p. 112)
Triad (p. 112)
Authoritarian leader (p. 112)
Democratic leader (p. 113)
Laissez-faire leader (p. 113)
Groupthink (p. 114)
Social network (p. 115)
Formal organization (p. 116)
Bureaucracy (p. 117)
Goal displacement (p. 118)
Alienation (p. 118)
Iron law of oligarchy (p. 119)
Glass ceiling (p. 125)
Glass escalator (p. 125)
Social institution (p. 126)

Discussion and Critical Thinking Questions

1. Whether we think reference groups are positive or negative depends on who we are and want to be. Sociologists describe reference groups like the Ku Klux Klan (KKK) as negative because they encourage hatred, divisiveness, and violence. The members of the KKK, however, describe themselves as "delivering a message of hope and deliverance for White Christian America."
   a. How does this example show that whether a reference group is positive or negative depends on a person's attitudes and goals?
   b. How does it also illustrate in-groups and out-groups?
2. Do you think that your college/university has become McDonaldized? If yes, what are some of the benefits and costs? If not, would your college run more smoothly if it were McDonaldized?
3. What is your most important reference group? Why?
4. Asch, Milgram, and Zimbardo published their research in the 1950s and 1970s. Why do faculty still teach about these studies?
5. At many large organizations, employees work in cubicles. Employers typically applaud such office arrangements because they save space. From an employee's perspective, what are some of the functional and dysfunctional aspects of cubicles?

True or False?

How Much Do You Know about Groups, Organizations, and Institutions?

The answers to 1 and 4 are true. The answers to 2, 3, 5, and 6 are false.

1. Primary groups are more likely to meet emotional needs (also referred to as expressive) while secondary groups are more likely to fill task-oriented needs (also called instrumental).
2. A dyad (group of two) is the most cohesive of all groups because its members have a personal relationship and interact more intensely than people in a triads (group of three) or larger size group.
3. Studies, such as Zimbardo's Stanford Prison Experiment, have shown that conformity can lead to harm of group members.
4. Bureaucracies aim for efficiency which is achieved through use of control that often limits workers' independence of activity and creative thought.
5. People seek out utilitarian organizations for income or other material rewards while they join normative organizations (also called voluntary organizations) because of shared interests and goals.
6. There are five main social institutions: family, government, economy, education, and religion. Plus, there are some newer institutions such as health care and mass media.

7 Deviance, Crime, and Social Control

©bikeriderlondon/Shutterstock.com

Learning Objectives

After studying this chapter, you will be able to…

7-1 Describe the key characteristics of deviance and crime.

7-2 After evaluating the two major crime measures, identify the different types of crime.

7-3 Describe functionalist perspectives on deviance.

7-4 Describe conflict perspectives on deviance.

7-5 Describe feminist perspectives on deviance.

7-6 Describe symbolic interaction perspectives on deviance.

7-7 Evaluate the criminal justice system's social control methods.

True or False?

How Much Do You Know about Deviance, Crime, and Social Control?

1. All deviant behavior is criminal.
2. Deviance serves a useful purpose in society.
3. Most crime victims in the United States are women.
4. Crime rates in the United States have increased since 2000.
5. Incarceration rates in the United States have decreased since 2000.
6. Most countries around the world have the death penalty.

The answers are at the end of the chapter.

A group of hackers broke into AshleyMadison.com, a website that arranges extramarital affairs, and released more than 35 million subscribers' names, addresses, and personal information. The online profiles included thousands of American and European military and government personnel and high-level executives. Many people believed that the deceitful spouses got what they deserved, but hundreds of church leaders resigned, some people lost their jobs, and several subscribers died by suicide (Weise and Vanden Brook, 2015; Rosenthal, 2016).

Who was deviant? The hackers? The website owners? The subscribers? And what, exactly, is deviance?

Be sure to take the True or False quiz in the chapter opener to better understand how much you know about these topics.

What do you think?

There would be less crime if the punishments were more severe.

1	2	3	4	5	6	7
	strongly agree				strongly disagree	

7-1 What Is Deviance?

Have you ever driven above the speed limit? Cheated on an exam? Engaged in underage drinking? All are examples of **deviance**—violations of social norms. Unlike the general public, sociologists use the concept *deviance* nonjudgmentally: They're interested in understanding and explaining why all of us violate norms from time to time, and why people judge some acts (hacking) more negatively than others (subscribing to an infidelity site).

Deviance becomes **crime**, a violation of society's formal laws, when it breaks rules that have been written into law and enforced by a political authority. For example, trafficking children under age 18 for sex work, pornography, or sex is deviant. It's not a crime, however, until a country enacts laws that punish offenders.

7-1a Some Key Characteristics of Deviance

Deviance is universal because it exists in every society. Still, deviance can vary quite a bit over time, from situation to situation, from group to group, and from culture to culture (Sumner, 1906; Schur, 1968):

- **Deviance can be a trait, a belief, or a behavior.** People usually do something to be considered deviant. We can also be treated as outsiders simply because of our appearance, skin color, religious beliefs, or sexual orientation (Becker, 1963; refer also to Chapters 9, 10, and 13).
- **Deviance is accompanied by social stigmas.** A **stigma** is a negative label that devalues a person and changes their self-concept and social identity. Stigmatized individuals may react in many ways: They may alter their appearance (such as having cosmetic surgery), associate with others like themselves who accept them (such as in gangs), hide information about some aspect of their deviance (a person convicted of a crime who doesn't reveal that status), or divert attention from a stigma by excelling in some area (such as music or sports) (Goffman, 1963).
- **Deviance varies across and within societies.** What's accepted or tolerated in one society may be deviant in another. The United States comprises only 5 percent of the world's population, but it accounts for 31 percent of global public *mass shootings* (incidents that result in killing four or more people outside homes and that aren't gang-related, hostage situations, or robberies). The number of mass shooting in the United States is on the rise,

---

deviance a violation of social norms.

crime a violation of society's formal laws.

stigma a negative label that devalues a person and changes their self-concept and social identity.

from 273 in 2014 to 647 in 2022. And gun related deaths make up a larger percentage of all homicides in the United States (79 percent) compared to other nations: Canada (37 percent), Australia (13 percent), and United Kingdom (4 percent). Across and within countries, stricter gun control laws and lower firearm ownership rates are correlated with fewer mass shooting incidents (BBC News, 2023; Florida, 2013; Lankford, 2015; Lemieux, 2015; Myers, 2016).

- **Deviance varies across situations.** What's acceptable in one context may be stigmatized in another. Among U.S. adults, between 74 and 77 percent say it's generally okay to use a cell phone on public transportation and while walking down the street or waiting in line. In contrast, only 4 to 5 percent approve of using cell phones at meetings, movie theaters, and religious services (Rainie and Zickuhr, 2015).
- **Deviance is formal or informal.** *Formal deviance* is behavior that violates laws. A leading example is crime, a topic we'll examine shortly. In contrast, *informal deviance* is behavior that disregards accepted social norms, such as picking one's nose or teeth or scratching one's private parts in public, belching loudly, and not dressing appropriately (e.g., wearing jeans to a formal wedding reception or job interview in a professional office setting).
- **Perceptions of deviance can change over time.** Many behaviors that were acceptable in the past are now perceived as deviant. Only during the 1980s and 1990s did U.S. laws define stalking and child abuse as crimes. Smoking—widely accepted in the past—has been banned in most public places. Texas is now one of the eight states that allows students to carry concealed guns in four-year public colleges. In 2017, Nevada became the first state to install vending machines with free needles for people who use intravenous drugs. Laws regarding child safety restraints in automobiles were only implemented beginning in 1978; before then, even babies could roam free in vehicles (Bae et al. 2014 and The Staff at the Henry Ford, 2022).

Markus Matzel/ullstein bild/Getty Images

Present-day accusations of witchcraft in Ghana can cause significant harm. These allegations can prompt dangerous public responses, including mob activity or murder. Most witchcraft accusations are directed towards women who are low-income widows or older. They are blamed for issues outside of their control, such as outbreaks of disease in their villages and relatives' or neighbors' illnesses or deaths. In the northern region of Ghana, which has high rates of poverty, there are approximately 277 women banished to four witch camps (Global Fund Community Foundations, 2022; Küntzle and Blondé, 2013).

On the other hand, most Americans now shrug off other behaviors that were stigmatized in the past. Cohabitation, perceived as sinful and immoral 40 years ago, is now widespread and considered normal. Americans' support for legalizing marijuana for recreational use increased from only 12 percent in 1969 to 68 percent in 2022 (Geiger, 2016; Van Green, 2022). And a third of Americans have at least one tattoo, up from only 16 percent in 2003 (Geiger, 2016; Shannon-Missal, 2016).

7-1b Who Decides What's Deviant?

Because deviance is culturally relative and standards change over time, who decides what's right or wrong? An important group is those who have authority or power. During our early years, parents and teachers specify acceptable and unacceptable behavior (refer to Chapter 4). As we get older, laws also define what's deviant (e.g., in the U.S., being forbidden to drive until age 16 in most states or to purchase or consume alcohol until age 21).

Many employers—especially those in businesses and federal agencies—don't hire people with visible "body art" ("Tattoos in the Workplace . . . ," 2014). Lawmakers pass numerous laws that people must obey but often exempt themselves. Examples include having immunity from employment discrimination and sexual harassment lawsuits and passing laws that allow guns in schools, restaurants, public buildings, and churches but not in state legislative chambers (Skoning, 2013; Deprez and Selway, 2014).

Are any of these people deviant? Why or why not?

Nazar_ab/E+/Getty Images; Jeff J Mitchell/Getty Images News/Getty Images; Dmitry Naumov/Shutterstock.com; iStock.com/belterz

Public attitudes and behavior also affect definitions of deviance. Many people, including health practitioners, now use the word "overweight" instead of "fat" or "obese" to decrease the stigma of being fat. On the other hand, there has been considerable *normalization of deviance*, the gradual process through which unacceptable practices or standards become acceptable. Examples include the growing acceptance of same-sex and interracial marriage, legalized gambling, political corruption, and recreational marijuana (refer to Chapters 3, 10, and 11). Normalizing some types of deviance generates considerable income, particularly at the state level.

7-2 Types of Deviance and Crime

Noncriminal deviance—such as dying by suicide, alcohol misuse, lying, mental illness, and adult pornography—is common, and these behaviors and conditions are labeled as deviant because they violate a society's social norms. Sociologists are especially interested in *criminal deviance*, behavior that violates laws, because most of the transgressions are visible, publicized by the media, and threaten people's lives or property. Measuring criminal deviance may seem straightforward, but it isn't as simple as it appears.

7-2a Measuring Crime

Two of the most important sources of crime statistics are the FBI's *Uniform Crime Report* (UCR) and the U.S. Department of Justice's *National Crime Victimization Survey* (NCVS). Each has its strengths and weaknesses.

The UCR data are crimes reported to the police and arrests made each year. The statistics are useful in examining trends over time, but the UCR doesn't include offenses such as corporate crime, kidnapping, and Internet crimes; simple assaults (those not involving weapons or serious injury); or the 53 percent of all violent crimes (e.g., rape, robbery, and aggravated assault) and 65 percent of property crimes (e.g., burglary and motor vehicle theft) that aren't reported to the police (Gramlich, 2017). Because of such limitations, many sociologists also use victimization surveys to measure the extent of crime.

A **victimization survey** interviews people about being crime victims. Because the response rates are at least 90 percent, the NCVS offers a more accurate picture of many offenses than the UCR. The NCVS includes both reported and unreported crime and isn't affected by police discretion in deciding whether to arrest an offender.

Still, many victims don't report a crime to either the police or interviewers for a variety of reasons: They fear reprisal for getting the offender in trouble, believe that the police can't or won't do anything to help, or the crime, particularly rape, is too painful to discuss. Some people don't want to admit having been victims—especially when the perpetrator is a family member or friend. Others may be too embarrassed to tell an interviewer that they were victimized while under the influence of alcohol or engaged in drug sales (Hagan, 2008; Truman and Langton, 2015).

No one knows the full extent of U.S. crime because even the best sources (the UCR and NCVS) provide only estimates, and data for higher-income lawbreakers are more elusive than for lower-income offenders. Let's begin with the latter, often called *street crimes*, offenses committed by ordinary people, usually in public places.

7-2b Street Crimes

A majority of Americans (80 percent) worry about crime and violence in the United States (Brenan, 2022). The main concern is *violent crimes* (e.g., murder, robbery, aggravated assault) that use force or the threat of force against others. *Property crimes* (e.g., burglary, motor-vehicle theft, arson) involve the destruction or theft of property, but offenders don't use force or the threat of force.

victimization survey interviews people about being crime victims.

Esline/Shutterstock.com

Becky Stares/Shutterstock.com

One of the biggest costs for retail stores is *wardrobing*, returning expensive clothes after wearing them. Such "borrowing" has also become prevalent in fine jewelry, high-priced cameras, men's clothes (like business suits for special occasions), and even household tools. Wardrobing has increased 40 percent since 2009 and cost retailers $9.1 billion in 2015 (Timberlake, 2013; Allen, 2015). Who do you think pays for wardrobing fraud?

Incidence of Violent and Property Crimes

Property crimes are more common than violent crimes, but both have been falling, and property crimes continue to be more prevalent than violent crimes (**Figure 7.1**). In 2000, there were 507 violent crimes reported per 100,000 persons, and by 2019, the number had decreased to 367. Also, in 2000, there were 3,618 property crimes reported per 100,000 persons, and that rate also decreased to 2,110 by 2019 (Federal Bureau of Investigation, 2020). When looking at victimization rates, the average person is almost six times more likely to be a victim of property crime than a violent crime (Gramlich, 2020; Morgan and Thompson, 2022).

About 75 percent of Americans say that neither they nor anyone in their household has experienced a crime.

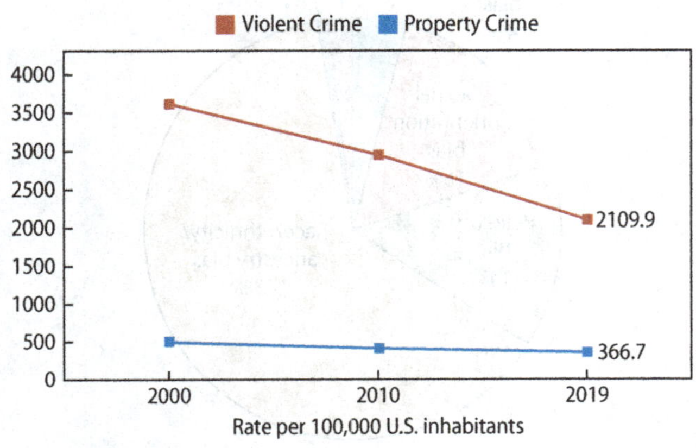

Figure 7.1 U.S. Crime Rates Have Decreased

Source: Based on Morgan and Thompson, 2022.

Nonetheless, 53 percent worry "a great deal" about being a crime victim. Since 1989, up to 89 percent have said that crime had increased compared with the previous year (McCarthy, 2015; Davis, 2016).

Why is there such a disconnect between many Americans' perceptions of crime and official statistics? Among other reasons, the media inspire fear by publicizing violent crimes, television shows and movies exaggerate the prevalence of crime, and many people believe misinformation. Both during and after the 2016 presidential election, Donald Trump claimed that the U.S. murder rate "is the highest it's been in 27 years," that Philadelphia's murder rate was "just terribly increasing," and that "carnage" plagued "inner cities" (Diamond, 2017; Hee Lee, 2017; refer also to Chapter 2). The statements were false, but they sparked public concerns and misperceptions about crime.

Offenders and Victims

Most offenders are never caught, but crimes aren't random. Of those arrested in 2019, for example:

- 16.8 percent were between the ages of 25 and 34. In contrast, only 7 percent were under age 18, and 1.5 percent were 65 or older, and many of the crimes involved substance misuse and driving under the influence of alcohol rather than violent and property crimes.
- 72 percent were men. They accounted for 79 percent of arrests for violent crime and 62 percent for property crime.
- By race, 70 percent were White, 26 percent were Black, and 2.4 percent were Native American (represented as American Indian in the data), 1.3 percent were Asian, and 0.3 percent were Native Hawaiian or Other Pacific Islander. Across all races, 18.8 percent were Latina/o/x (Federal Bureau of Investigation, 2020).

Looking at victimization rates per every 1,000 persons, men (17.5) are more likely than women (15.5) to be crime victims. Victimization rates for men and women do differ by type of crime. Men experience higher rates of homicide, robbery, and aggravated assault, and women are more likely to be victims of rape, sexual assault, and intimate partner violence. Violent victimization rates are lower for White and Latina/o/x individuals (16 per 1,000 persons aged 12 and older) than for Black individuals (18.5). The rates are also much higher for people aged 18 to 24 (23.1) than those aged 65 and older (6.4). People in lower-income households (under $25,000 a year) are approximately three times more likely than higher-income households ($200,000 or more a year) to experience violent victimization (Thompson and Tapp, 2022).

Comstock/Stockbyte/Getty Images

Men are much more likely than women to be victims of violent crime.

Crime rates are higher in lower-income areas, but does this mean that people with lower incomes are more deviant? Because police devote more resources to low-income neighborhoods, those at the lower end of the socioeconomic ladder are more likely to be caught, arrested, prosecuted, and incarcerated. *Intellectual property theft*—which includes offenses like software piracy, bootlegging musical recordings and movies, selling company trade secrets, and copyright violations—is primarily a middle-class crime, but the offenders are difficult to catch.

7-2c Hate Crimes

In 2016, 21-year-old Dylann Roof killed nine Black people in a South Carolina church. Roof said that "Blacks were taking over the world" and that he was rescuing the "White race" (Silverstein, 2015). The massacre is an example of a **hate crime** (also known as a *bias crime*), a criminal offense motivated by the fact or perception that the victims differ from the perpetrator. The attacks, often violent, are based on prejudice toward race, ethnicity, ancestry, religion, disability, sexual orientation, gender, or gender identity (Federal Bureau of Investigation, 2016; refer also to **Figure 3.3** and the related discussion of hate groups on page 147).

The FBI recorded 7,074 single bias hate crimes affecting 8,753 victims in 2021. The number is probably much higher because more than half of victims don't report hate crimes to the police. Most hate crimes are attributed to racial and ethnic bias (63 percent) (**Figure 7.2**). Another 16 percent are based on bias towards a person's sexual orientation. Offenders in hate crimes are overwhelming white (56 percent). Most hate crimes happened near or inside residences (32.2 percent) or on roadways, sidewalks, or alleys (16.9 percent) (U.S. Department of Justice, 2023).

Why do people commit hate crimes? In *thrill hate* crimes, the offenders are "just having fun" by attacking people based upon their race or ethnicity, bashing LGBTQIA+ individuals, or destroying property. Perpetrators of *defensive hate crimes* feel justified, even obligated, to protect their neighborhood against outsiders who are the "wrong" race, ethnicity, and so on. *Retaliatory hate crimes* are motivated by revenge for a real or imagined incident (e.g., attacking Muslims after the 2013 Boston Marathon bombing) (McDevitt et al., 2002; Levin and Nolan, 2017).

7-2d White-Collar Crime

White-collar crime refers to illegal activities for financial gain, which are committed by people with greater power accessed through their professional connections (Sutherland, 1949; Cressey, 1953). The offenses don't involve violence because white-collar criminals use their status and powerful positions to enrich themselves or others. Typical white-collar crimes include embezzlement (stealing an employer's or client's money); identity theft; bribery; insider trading (illegal stock manipulation); and mass marketing, mortgage, and investment fraud (Federal Bureau of Investigation, 2014).

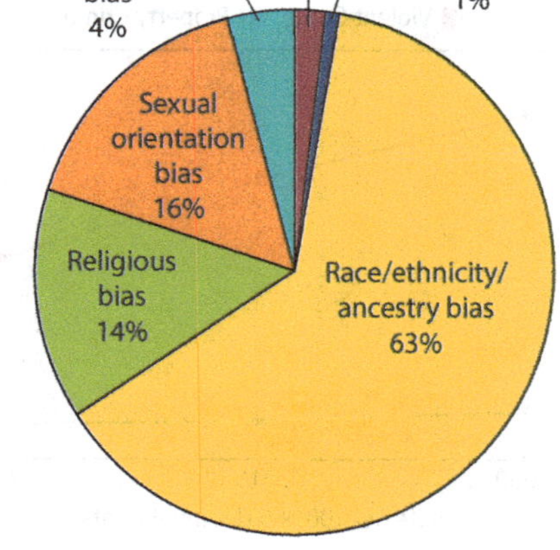

Figure 7.2 Reasons for Hate Crimes, 2020

Source: U.S. Department of Justice, 2023.

---

hate crime (also known as *bias crime*) a criminal offense motivated by the fact or perception that the victims differ from the perpetrator.

white-collar crime illegal activities for financial gain which are committed by people with greater power accessed through their professional connections.

Bank tellers, who have low salaries and easy access to customer information, are particularly susceptible to white-collar crime. Between 2011 and 2015, Wells Fargo employees opened more than 2 million fake accounts to meet sales goals. The fraud resulted in $2.6 million in unnecessary fees for tens of thousands of clients. Wells Fargo fired 5,300 tellers and low-level managers. The executive who supervised the 6,000 retail branches and knew about the swindles retired with a $125 million payout (Clifford and Silver-Greenberg, 2016; Ochs, 2016).

White-collar crimes are lucrative because most offenders aren't caught. According to one estimate, there's only one FBI investigator per almost 4,000 financial services sales agents, personal financial advisers, and financial analysts. Thus, the odds of being arrested are very small (Rohrlich, 2010; refer to fbi.gov/wanted/wcc for annual updates of the FBI's most wanted white-collar criminals).

Even when white-collar criminals are arrested and convicted, few are incarcerated. Of the almost 3,800 suspected embezzlers, for example, only 14 percent have been arrested and sentenced; of those arrested and sentenced, only 7 percent served any prison time (Motivans, 2013). Because so few of the most affluent white-collar offenders are punished, they enjoy lavish lifestyles that include yachts, private jets, and $2 million birthday parties (Lublin, 2013; Spitznagel, 2013).

7-2e Corporate Crimes

Corporate crimes (also known as *organizational crimes*) are illegal acts committed by executives to benefit themselves and their companies. Corporate crimes include a vast array of illicit activities such as conspiracies to stifle free market competition, price-fixing, tax evasion, and false advertising. The target of the crime can be the general public, the environment, or even a company's own workers (refer to Chapters 11 and 15). Often, corporate offenders commit multiple crimes (fraud, insider trading, and perjury).

Corporate crime is common. In the auto industry alone, and during just the last few years, General Motors (GM) knew about a malfunctioning ignition switch as far back as 2001 but only started recalling cars in 2014 after dozens of death and injury lawsuits. At least four automakers knew that Takata's airbags had a deadly defect linked to deaths and injuries but installed them anyway "to save on costs." Well over a year before Volkswagen publicly admitted to cheating on pollution tests, executives knew that the diesel cars were giving off more than 40 times the U.S. legal limit of nitrogen oxide, which can cause respiratory problems (Vlasic, 2015; Krisher and Biesecker, 2016; Tabuchi and Boudette, 2017, p. B1).

AP images/Kathy Willens

In 2009, Bernard Madoff—a highly respected stockbroker and investment advisor—pleaded guilty to defrauding thousands of investors of almost $18 billion between 1991 and 2008. Madoff was sentenced to 150 years in a federal prison. So far, it has cost almost $1 billion to recover 61 percent of the stolen money (Ross et al., 2016).

Corporate crime is common because it's profitable. In the ignition switch problems, the Department of Justice (DOJ) fined GM $900 million, which amounts to less than 1 percent of the company's annual revenue. Education Management Corporation, one of the nation's largest for-profit education companies, received $11 billion from students and the federal government by lying about its programs. The DOJ fine, again, was less than 1 percent of the corporation's revenue but left tens of thousands of students with worthless degrees and huge debts (United States Senate, 2016).

7-2f Cybercrime

Cybercrime (also called *computer crime*) refers to illegal activities that are conducted online. These high-tech crimes include defrauding consumers with bogus financial investments, embezzling, sabotaging computer systems, hacking, and stealing confidential business and personal information.

Since 2005, data breaches have been on the rise with a high point of 1,862 in 2021 and then a slight decrease to 1,802 in 2022. The three industries with the most recorded data beeches were healthcare, financial services, and manufacturing (Petrosyan, 2023). Since 2007, the leading reasons for data breaches have been *hacking*

---

corporate crimes (also called *organizational crimes*) illegal acts committed by executives to benefit themselves and their companies.

cybercrime (also called *computer crime*) illegal activities that are conducted online.

(illegal access to a computer), *phishing* (masquerading as a legitimate person or group to obtain a login name and password), and *skimming* (stealing credit card information). Criminals can use stolen information (e.g., social security numbers, addresses, and names) to file false tax returns, order credit cards, and drain people's bank accounts (Identity Theft Resource Center, 2017; Risk Based Security, 2017).

In 2020, the FBI received almost 792,000 complaints of suspected internet crime—which is 300,000 more than in 2019. The most common offenses were phishing scams, non-payment or non-delivery, and extortion. Over 28,500 complaints were related to COVID-19 related fraud. About 64 percent of Americans have experienced illegal credit card charges (Federal Bureau of Investigation, 2021; Olmstead and Smith, 2017). Stealing personal information has become so routine, according to one observer, "that most such breaches are quickly forgotten" and are "merely an occasional inconvenience" ("The Ashley Madison Hack," 2015, p. 14).

7-2g Organized Crime

Organized crime refers to activities of individuals and groups that supply illegal goods and services for profit. Organized crime includes human and drug trafficking, loan-sharking (lending money at unlawful rates), illegal gambling, pornography, theft rings, hijacking cargo, and laundering money obtained from illicit activities (e.g., sex work) by depositing the money in legitimate enterprises (e.g., banks) (Federal Bureau of Investigation, 2017).

One of the most profitable activities, *organized retail crime*, involves stealing billions of dollars in merchandise that is then sold online, a practice known as "e-fencing." Organized crime costs the retail industry approximately $30 billion each year (National Retail Federation, 2016).

Organized crime rings also orchestrate global cybercrimes, usually targeting financial institutions. In 2013, a hack compromised three billion Yahoo accounts, in 2014, another 500 million Yahoo usernames, passwords, and email addresses were compromised, and in 2016, 200 million Yahoo usernames and passwords were up for sale on the dark web (Sarkar, 2022). Criminal organizations employ highly sophisticated staff who circumvent bank security systems. Russia and China are especially likely to sponsor organized cybercrime, which makes money by stealing other countries' intellectual property and government secrets (IBM Journal Staff, 2016; Kessem, 2016; Rodgers, 2016). Certain conditions can escalate cybercrime. In the early years of the COVID-19 pandemic, three-quarters of banks and insurers experienced a rise in cybercrime (Businesswire, 2021).

Mladenbalinovac/E+/Getty Images

Is this a victimless crime?

7-2h Victimless Crimes

Victimless crimes (also called *public order crimes*) are illegal acts that have no direct victim. Because crimes such as illicit drug use, sex work, illegal gambling, and pornography are voluntary, the people involved don't consider themselves victims or report the offenses. Sex workers argue, for instance, that they're simply providing services that people want, and drug users claim that they're only hurting themselves.

Some contend that this term is misleading because victimless crimes often lead to property and violent crimes, such as when a person with a substance use disorder commits burglary, robs people at gunpoint, and engages in identity theft to get money for drugs. Family members also experience the effects of substance use disorders. Problems can be emotional, psychological, financial, and social. Children with parents experiencing these disorders can experience neglect or abuse. These conditions can also lead to financial problems, including poverty or bankruptcy. Family members may choose to isolate themselves from social relationships out of shame. The condition may also erode trust within family relationships (American Addiction Centers, 2022). In 2017, 46 percent of Americans reported having a friend or family member with a current or past substance use disorder (Gramlich, 2017).

The next sections examine four important sociological perspectives that help us understand why people are deviant. **Table 7.1** summarizes these theories.

---

organized crime activities of individuals and groups that supply illegal goods and services for profit.

victimless crimes (also called *public order crimes*) illegal acts that have no direct victim.

Table 7.1 Sociological Explanations of Deviance

Theoretical Perspective	Level of Analysis	Key Points
Functionalist	Macro	• Deviance is both functional and dysfunctional. • Anomie increases the likelihood of deviance. • People are deviant when they experience blocked opportunities to achieve the culturally approved goal of economic success.
Conflict	Macro	• There's a strong association between capitalism, social inequality, power, and deviance. • The most powerful groups define what's deviant. • Laws rarely punish the illegal activities of the powerful.
Feminist	Macro and micro	• There's a large gender gap in deviant behavior. • Women's deviance reflects their general oppression due to social, economic, and political inequality. • Many women are offenders or victims because of patriarchal beliefs and practices and living in a rape culture.
Symbolic Interactionist	Micro	• Deviance is socially constructed. • People learn deviant behavior from significant others such as parents and friends. • If people are labeled or stigmatized as deviant, they're likely to develop negative self-concepts and engage in criminal behavior.

7-3 Functionalist Perspectives on Deviance

For functionalists, deviance is a normal part of society. They don't endorse undesirable behavior, but they view deviance as both functional and dysfunctional.

7-3a Dysfunctions and Functions

Dysfunctions are the negative consequences of behavior, but what's dysfunctional for one group or individual may be functional for another. For example, gangs are functional because they provide members with a sense of belonging, identity, and protection. Gangs are also dysfunctional because they commit violent and property crimes (Egley and Howell, 2013).

Deviance is *dysfunctional* because it:

- **Creates tension and insecurity.** Any violation of norms—a babysitter who cancels at the last minute or the theft of your smartphone or tablet—makes life unpredictable and increases anxiety.
- **Erodes trust in personal and formal relationships.** Crimes like stalking make many women suspicious of men, and victims of identity theft have had problems obtaining banking services or credit cards because financial organizations don't trust them (refer to Harrell and Langton, 2013).
- **Decreases confidence in institutions.** In the 2008 stock market crash, taxpayers had to pay for the financial industry's corporate fraud and mismanagement. Since then, millions of people, even those who didn't lose money, worry that their retirement funds may disappear in the future (refer to Chapter 12).
- **Is costly.** Besides personal costs to victims (e.g., fear, emotional trauma, physical injury), deviance is expensive. All of us pay higher prices for consumer goods and services (e.g., auto and property insurance) and higher taxes for prosecuting criminals and for building and maintaining prisons. And as you saw earlier, white-collar, corporate, cyber, and organized crime cost the U.S. economy hundreds of billions of dollars every year.

Deviance can also be *functional* because it provides a number of societal benefits (Durkheim, 1893/1964; Erikson, 1966; Sagarin, 1975). For example, deviance:

- **Affirms cultural norms and values.** Negative reactions, like expelling a college student who's caught cheating or incarcerating a robber, reinforce a society's norms and values about being honest and law-abiding.
- **Provides temporary safety valves.** Some deviance is accepted under certain conditions, enabling people to "blow off steam." Typically, a community and police will tolerate noise, underage drinking, loud music, and obnoxious behavior during college students' spring break because it's a short-lived nuisance.

- **Bolsters the economy.** Deviance can benefit a community financially. Some towns welcome new prisons because they generate jobs and stimulate the economy (Semuels, 2010). Unlike a manufacturing plant that might close down or move to another country, prisons are rarely dismantled.
- **Triggers social change.** When auto collisions increased, states started passing laws that banned distracted driving. Since 1999, drunk driving and deadly crashes among persons aged 16 to 25 have declined primarily because of drinking age laws, mass media campaigns, and sobriety checkpoints (Governors Highway Safety Association, 2014; Azofeifa et al., 2015).

7-3b Anomie and Social Strain Theory

Functionalists offer many explanations of deviance, but two of the most influential are anomie and strain theories. Both analyze why so many people engage in deviant behavior even though they share many of the same goals and values as those who conform to social norms.

Durkheim's Concept of Anomie

Émile Durkheim (1893/1964, 1897/1951) introduced the term **anomie** to describe the condition in which people are unsure how to behave because of absent, conflicting, or confusing social norms. During periods of rapid social change, such as industrialization in Durkheim's time, societal rules may break down. As many young people moved to the city to look for jobs in the nineteenth century, norms about proper behavior that existed in the countryside crumbled. Even today, as you'll learn in Chapter 15, many urban newcomers experience anomie and miss the neighborliness that was common at home.

Merton's Concept of Social Strain

Robert Merton (1938) expanded the concept of anomie to explain how social structure helps create deviance. According to Merton, Americans are socialized to believe that anyone can realize the American dream of accumulating wealth and being successful economically. To achieve the *cultural goal* of economic success, society emphasizes legitimate and *institutionalized means* such as education, hard work, saving, starting at the bottom and working one's way up, and making sacrifices instead of seeking instant gratification.

John Lund/Blend Images/Getty Images

In many countries, including the United States, however, people don't always have access to institutionalized means for financial success because of poverty, low wages, or long-term unemployment. How do they respond?

Merton's **strain theory** posits that people may engage in deviant behavior when they experience a conflict between goals and the means available to obtain the goals. Not all people turn to deviance to resolve social strain (**Table 7.2**). Most of us *conform* by working harder and longer to become successful. You're reading this textbook, to illustrate Merton's typology, because you've chosen conformity to achieve success through an institutionalized means (a college education).

Merton's other four modes of adaptation signify deviance. *Innovation* occurs when people endorse the cultural goal of economic success but turn to illegitimate means, especially crime, to achieve the goal. Resorting to crime is functional for people who believe they can't achieve success through hard work, education, and deferred gratification.

In *ritualism,* people have modest aspirations, but follow customs and rules because they enjoy doing so or they want to keep their jobs. Ritualists aren't criminals, but they're deviant because they've given up on becoming financially successful. Instead, they do what they're told and "go along to get along."

In *retreatism,* people reject both the goals and the means for success. They try to escape their failure by withdrawing—mentally, physically, or both—through alcohol and other drugs, becoming drifters, living on the streets, or otherwise dropping out of society.

---

anomie the condition in which people are unsure how to behave because of absent, conflicting, or confusing social norms.

strain theory people may engage in deviant behavior when they experience a conflict between goals and the means available to obtain the goals.

Table 7.2 Merton's Strain Theory of Deviance

Mode of Adaptation	Accept Cultural Goals?	Accept Institutionalized Means to Achieve Goals?	Examples
Conformity	Yes	Yes	College graduate
Innovation	Yes	No	Thief
Ritualism	No	Yes	Low-level bureaucrat
Retreatism	No	No	Person with alcohol misuse disorder
Rebellion	No—seeks to replace goals	No—seeks to replace means for achieving goals	Antigovernment militia member

Source: Based on Merton, 1938.

In *rebellion*, people feel so alienated that they want to change the social structure entirely by substituting new goals and means for the current ones. Contemporary examples are terrorists and U.S. militia groups that oppose the federal government (refer to Chapter 3).

7-3c Critical Evaluation

A major contribution of functionalism is showing how social structure, not just individual attitudes and behavior, produces or reinforces deviance and crime. Functionalist theories also have weaknesses: (1) The concepts of *anomie* and *strain theory* are limited because they overlook the fact that not everyone in the United States embraces financial success as a major goal in life; (2) the theories don't explain why women's crime rates are much lower than men's, especially because women generally have fewer legitimate opportunities for financial success than men; (3) why crime rates have declined since 2000 despite much poverty and unemployment; and (4) why people commit some crimes (setting fires just for kicks or murdering an intimate partner) that have nothing to do with being successful (Anderson and Dyson, 2002; Williams and McShane, 2004).

The most consistent criticism is that functionalism typically focuses on lower-class deviance and crime. Conflict theorists have filled this gap by examining middle- and upper-class crime.

7-4 Conflict Perspectives on Deviance

Functionalists ask, "Why do some people commit crimes and others don't?" Most conflict theorists ask, "Why are some acts defined as criminal while others aren't?" (Akers, 1997).

7-4a Capitalism and Social Inequality

For conflict theorists, there's a strong association between capitalism, social inequality, and deviance. People and groups with economic and political power view deviance as any behavior that threatens their own interests (Vold, 1958; Turk, 1969, 1976; Kraska, 2004). As a result, the mass media, owned by the affluent, routinely focus on "street crimes" rather than "suite crimes."

The United States, like many other countries, is a capitalist economy based on the private ownership of property (refer to Chapter 11). Characterized by personal gain rather than by collective well-being, capitalistic goals include continually maximizing profits, accumulating wealth, and promoting economic self-interest.

From a conflict perspective, capitalism produces crime for several interrelated reasons (Chambliss, 1969; Quinney, 1980; Snider, 1993). First, greed and self-interest perpetuate deviance. A study of 36 countries found that 42 percent of board directors and senior managers—pressured to increase growth and fearing personal pay cuts—"cooked the books" by overstating revenue, encouraging customers to buy unnecessary stock, and bribing powerful people to begin or continue their business (Ernst & Young, 2013).

Second, capitalism is a competitive system that encourages corporate crime. When only a few firms dominate an industry, they can easily eliminate competitors, conspire to fix prices, falsify financial statements, and commit fraud (refer to Eavis, 2015).

Third, because capitalism creates considerable social inequality, those at the top can control and exploit lower-level workers. In the financial services/banking industry, almost half of the workers said that illegal or unethical activity was "part and parcel of succeeding in this highly competitive field," and that their company's confidentiality policies and procedures bar them from reporting wrongdoing (Tenbrunsel and Thomas, 2015).

Orlando Police/The New York Time/Redux

In 2016, 29-year-old Omar Mateen killed 49 people and wounded 53 others at a gay nightclub in Orlando, Florida. Do you think that Merton's strain theory explains Mateen's crime?

The current U.S. tax system allows companies to avoid paying $135 billion each year in taxes. Such "tax dodging" prevents crucial investments in healthcare, infrastructure, education, family homelessness, and poverty reduction (Oxfam America, 2017).

Sam Baird/Investigation Discovery/Everett Collection Inc.

True crime is the most in-demand subgenre of all documentary types—four times more popular than sports documentaries. This includes shows such as *Tiger King*, *Making a Murderer*, and *American Greed* (Sayles, 2021). Why do you think true crime shows are so popular?

7-4b Laws, Power, and Social Control

From a conflict perspective, criminal laws serve the interests of the capitalist ruling class. The most powerful groups in society control the law, which defines what's deviant, and who will be punished (Chambliss and Seidman, 1982; Reiman and Leighton, 2010).

Why do so many people with power commit crimes? Because they can, according to conflict theorists. First, most white-collar and corporate crimes are *not criminalized* (Turk, 1969; Lilly et al., 1995). Generally, the wealthy and powerful make and enforce laws that protect their property. Laws against criminals with higher levels of power are relatively lenient and seldom enforced, while laws against offenders with less power, particularly those who commit property crimes, are harsher and enforced more often (Thio et al., 2012).

Second, the powerful have more *opportunities for deviance*. As you saw earlier, white-collar and corporate criminals have access to resources and the necessary occupational skills to defraud customers and employees.

Third, powerful individuals and corporations are *rarely prosecuted*. Prosperous corporations pay millions, or billions, "to make cases disappear before any public hearing." Early in his presidency, Biden's DOJ policy changes aimed to reinstate a stronger approach on corporate crime (compared to Trump's softer approach). Yet, in 2021, only 90 companies pled guilty or were found guilty of corporate crimes. This was fewer than the number in 2020 (94) and far below the high point in 2000 (296) (Public Citizen, 2022).

Fourth, there are *few penalties* for high-status crimes because lawmakers and defendants share a common cultural background. Of the senior Wall Street bank executives who played leading roles in the 2008 financial crisis, only one has been convicted of fraud. Offshore banks knowingly shelter U.S. tax evaders because there are no punishments for doing so. Fraud by the rich wipes out 40 percent of the world's wealth, but no one goes to jail (Taibbi, 2014; "Financial Crime," 2015).

When U.S. prosecutors bring civil charges, the fines are typically less than half of 1 percent of annual revenue and are passed on to shareholders. Top officials at Goldman Sachs agreed to pay a $550 million fine to settle accusations of securities fraud that made almost $52 billion in profits. The fine represented four days of the bank's revenues (Summers, 2010). If, using a comparable percentage, you made $100,000 a year illegally and someone paid your $1,000 fine, would there be a financial incentive for honesty?

Sun Sentinel/Tribune News Service/Getty Images

Rich Criminals, Lenient Sentences Ryan LeVin (pictured here), a super-rich heir to a family costume jewelry fortune, killed two men while hurtling down a street at more than 100 miles an hour, then fled the scene. He pleaded guilty to vehicular homicide and faced 30 years in prison. LeVin was sentenced to only 2 years of house arrest, which he served at one of his parents' luxury seaside condos. LeVin never apologized to the victims' families. His major concern was getting back his Porsche (Shepherd, 2011; Mayo, 2012).

In another recent case, Ethan Couch, 18, killed four people while driving drunk. The defense argued for leniency because Couch suffered from "affluenza," being so rich and spoiled that he was totally unaware of consequences and, thus, not responsible for his actions. The judge sentenced Couch to only 10 years of probation, which he violated 2 years later (Garza and Williams, 2016).

7-4c Critical Evaluation

Conflict theories explain the linkages between capitalism, laws, power, and social class that benefit those at the top. According to some critics, however, conflict theories (1) exaggerate the importance of capitalism because other industrialized societies (e.g., Japan, Scandinavian countries), which enforce fraud laws, have much lower white-collar and corporate crime rates; (2) downplay street crimes even though victim injuries, incarceration, and crime prevention programs are costly to society; (3) overlook the fact that some affluent people, including executives, don't always get away with crimes; and (4) ignore the ways that crime is functional for society (e.g., provides jobs, affirms law-abiding cultural norms and values) (Leonardsen, 2004; Ernst & Young, 2013; "Jail Bait," 2016).

Finally, some contend, the most influential conflict theories focus almost entirely on men as both victims and offenders (Moyer, 2001; Belknap, 2007). Feminist theories have been filling this gap.

7-5 Feminist Perspectives on Deviance

Until recently, most sociological research on deviance focused almost entirely on men. Nearly all sociologists were men who saw women as not worthy of much analytical attention or assumed that explanations of male behavior were equally applicable to females (Flavin, 2001; Belknap, 2007). Feminist scholars' analyses of female offenders and victims have greatly enriched our understanding of deviance.

7-5a Women as Offenders

Crime is still generally a male activity. Men constitute about half of the U.S. population but make up 72.5 percent of all arrests. Overall, and across 28 offenses, men commit more crimes and more serious crimes than women (Federal Bureau of Investigation, 2021). Among

Table 7.3 Five Top Reasons for Arrest by Sex, 2021

Percentage of Men Arrested for ...		Percentage of Women Arrested for ...	
Rape	96.6	Prostitution and commercialized vice	63.0
Sex offenses (except rape and prostitution)	93.2	Embezzlement	50.2
Weapons (carrying, possessing)	90.4	Larceny-theft	42.6
Murder and nonnegligent manslaughter	88.0	Property crime	37.7
Robbery	84.2	Fraud	35.9

Sources: Based on Federal Bureau of Investigation, 2020 "Uniform Crime Report: 2019".

the top arrests, embezzlement has a fairly even distribution between women and men, and sex work is the only female-dominated crime (**Table 7.3**).

Socialization explains some of the gender gaps in arrest rates. Starting from an early age, girls have more responsibilities compared with boys and are supervised more closely, expected to behave, and discouraged from risk-taking behavior. When women take risks, they're likely to do so to protect loved ones rather than to gain status or financial profits (England, 2005; Steffensmeier et al., 2013; refer also to Chapter 4).

Men are much more likely than women to commit violent crimes (**Table 7.3**), but women's arrest rates have increased for some offenses. Between 2010 and 2019, arrests of men dropped for all offenses except murder, nonnegligent manslaughter (+2.7), and motor vehicle offenses (+14.0), and arrests of women rose across five categories including larceny-theft (+56.4) and drug use violations (+26.3) (Federal Bureau of Investigation, 2021).

Why is embezzlement the only crime category where women are arrested at higher rates than men? More women have white-collar occupations (e.g., bank tellers, accountants, sales managers) that present opportunities for fraud, theft, and embezzlement. Because women, on average, experience pay inequity compared to men in the same positions, some resort to crime—especially shoplifting, petty theft, and sex work—to survive financially or to support a family. Even at managerial levels, women are much more likely than men to commit fraud because of financial difficulties due to divorce or family problems rather than "generally unscrupulous or shrewd behavior" (Abrahams, 2015; Association of Certified Fraud Examiners, 2016).

Almost 7 percent of individuals arrested are under age 18. Like adults, boys account for more arrests (69.5 percent) than girls, but girls' arrests have increased dramatically—from 17 percent in 1980 to 30.5 percent in 2019 (Federal Bureau of Investigation, 2021).

The higher arrest rates are due to a combination of factors: Policy changes, including more aggressive policing, parents' and school officials' greater likelihood of calling the police to deal with girls' (especially Black girls') disruptive behavior, and a substantial increase in girls (18 and under) committing crimes that range from substance misuse to assaulting family members or intimate partners (Levintova, 2016; Morris and Perry, 2017).

7-5b Women as Victims

Many of the serious crimes (murder and robbery) are committed by men against men. Women and girls, however, are almost always the victims of sexual assault, rape, intimate partner violence, stalking, sexual exploitation, sexual harassment, female infanticide, and other crimes that degrade and terrorize women.

Feminist scholars offer several explanations for women's victimization. First, **patriarchy**—a social system in which men control cultural, political, and economic structures—produces violence against women. In patriarchal societies, including the United States, women have less access to power, so they're at a disadvantage in creating and implementing laws that protect women (Price and Sokoloff, 2004; DeKeseredy, 2011). Such inequity diminishes women's control over their lives and increases their invisibility as victims.

Second, gender roles reinforce inequality: "Girls are rewarded for passivity and feminine behavior, whereas boys are rewarded for aggressiveness and masculine behavior" (Belknap, 2007, p. 243). As early as middle school, girls experience physical and sexual harassment in hallways, classrooms, and gym classes. Although the incidents are upsetting, girls often dismiss the boys' unwanted behavior as "only joking" (Espelage et al., 2016). In effect, then, both sexes internalize the belief that the victimization of women is normal. Because patriarchal societies rarely punish men who victimize women, or give them light sentences, many female victims feel trapped.

A third reason for women's abuse is living in a **rape culture**, an environment in which sexual violence

Justin Sullivan/Getty Images News/Getty Images

Elizabeth Holmes was the founder and chief executive officer (CEO) of Theranos, a health technology startup that claimed to have transformed blood testing. The company's profits soared, increasing Holmes' net worth to $4.5 billion. She was found guilty of fraud for deceiving investors about the abilities of the blood-testing technology. Holmes was sentenced to serve 11 years in prison (Hartmans et al., 2023).

patriarchy a social system in which men control cultural, political, and economic structures.

rape culture an environment in which sexual violence is prevalent, pervasive, and perpetuated by the media and popular culture.

is prevalent, pervasive, and perpetuated by the media and popular culture (Parrott and Parrott, 2015). Here are a few examples of the pervasiveness of rape and sexual assault in U.S. culture:

- During their lifetimes, 3 percent of men and 17 percent of women have been victims of an attempted or completed rape.
- One out of every 10 rape victims are men.
- 21 percent of TGQN (transgender, genderqueer, nonconforming) college students have been sexually assaulted, compared to 18 percent of non-TGQN women, and 4 percent of non-TGQN men (RAINN, 2023).

Across 77 countries, the number of rapes reported to the police ranges from fewer than 1 per 100,000 people in Japan to 37 in the United States and 51 in England and Wales. Rape statistics vary considerably depending on the legal definition, recording practices, and other factors. Nonetheless, the variation across countries shows that sexual violence is impacted by societal norms and policies (United Nations Office on Drugs and Crime, 2017; refer also to Roden, 2017).

7-5c Critical Evaluation

Feminist sociologists have been at the forefront of raising awareness of crimes like date rape, intimate partner violence, and international sex trafficking. Some critics contend, however, that some feminist perspectives focus much attention on violence and crime perpetrated by men without providing equivalent attention to crimes and violence committed by women. In addition, early critiques highlighted a need for feminist sociologists to address the interdependent effects of gender, age, social class, and race/ethnicity. In more recent decades, this gap has been addressed with the voices of BIPOC and LGBTQIA+ feminists (Friedrichs, 2009; Hunnicutt, 2009; Dao, 2013; Garcia-Hallett et al., 2022).

7-6 Symbolic Interaction Perspectives on Deviance

For symbolic interactionists, deviance is socially constructed because it's in the eye of the beholder. For instance, a study of eleven different countries found that there are different attitudes towards marital infidelity (Scleza et al., 2020). Interactionists offer many explanations of deviance, but two of the best known are differential association theory and labeling theory.

7-6a Differential Association Theory

Differential association theory asserts that people learn deviance through interaction, especially with significant others like family members and friends. People become deviant, according to sociologist Edwin Sutherland, if they have more contact with significant others who violate laws than with those who are law-abiding. In effect, people learn techniques for committing criminal behavior, along with the values, motives, rationalizations, and attitudes that reinforce such behavior (Sutherland and Cressey, 1970). That is, deviance isn't inherent but is learned through interactions with others.

Sutherland emphasized that differential association doesn't occur overnight. Instead, people are most likely to engage in crime if they're exposed to deviant values (1) early in life, (2) frequently, (3) over a long period of time, and (4) by significant others and reference groups (e.g., parents, siblings, close friends, and business associates).

Considerable research supports differential association theory. Almost 47 percent of people incarcerated in state prisons have a parent or other close relative who has also been incarcerated. Even before age 13, children who associate with peers who smoke, use drugs, or commit crimes are more likely to do so themselves because they learn these behaviors from their friends (Adamson, 2010; Roman et al., 2012; Huizinga et al., 2013). Thus, according to differential association theory, we're products of our socialization.

7-6b Labeling Theories

Have you ever been accused of something wrong you didn't do? What about getting credit for something you and others know you didn't deserve? In either case, did people start treating you differently? The reactions of others are the crux of **labeling theory**, which holds that society's reaction to behavior is a major factor in defining oneself or others as deviant. Some of the earliest sociological studies found that teenagers who misbehaved were tagged as delinquents. Such labeling changed the child's self-concept and resulted in more deviance and criminal behavior (Tannenbaum, 1938). During the 1950s and 1960s, two influential sociologists—Howard Becker and Edwin Lemert—extended labeling theory.

---

differential association theory asserts that people learn deviance through interaction, especially with significant others.

labeling theory posits that society's reaction to behavior is a major factor in defining oneself or others as deviant.

Becker: Deviance Is in the Eyes of the Beholder

According to Howard Becker (1963, p. 9), "The deviant is one to whom that label has successfully been applied; *deviant behavior is behavior that people so label*" (emphasis in original). That is, it's not an act that determines deviance but whether and how others react.

Some people are never caught or prosecuted for crimes they commit and thus aren't labeled deviant. In other cases, people may be falsely accused (e.g., cheating on taxes) and stigmatized. In effect, then, deviance is in the eye of the beholder because societal reaction, rather than an act, labels people as law-abiding or deviant. Moreover, labeling can lead to secondary deviance.

Lemert: Primary and Secondary Deviance

Edwin Lemert (1951, 1967) developed labeling theory by differentiating between primary and secondary deviance. **Primary deviance** is the initial act of breaking a rule. Primary deviance can range from relatively minor offenses, like not attending a family member's funeral, to serious offenses, like rape and murder.

Even if people aren't guilty of primary deviance, labeling can lead to **secondary deviance**, rule-breaking behavior that people adopt in response to others' reactions. A teenager who is caught trying marijuana may be labeled a "druggie." If the individual is rejected by others, they may accept the label, associate with drug users, and become involved in a drug-using subculture. According to Lemert, a single deviant act will rarely result in secondary deviance. The more times a person is labeled, however, the higher the probability that they will accept the label and engage in deviant behavior.

There's considerable evidence that labeling affects people's lives. For example, those who have been jobless for a year or longer report stigmas in job searches: Employers assume they're lazy, don't really want to work, or there's something wrong with them. Some companies have explicitly barred the long-term unemployed from certain job openings, telling them outright in job ads that they need not apply (refer to Chapter 11).

The Medicalization of Deviance

In 1952, the American Psychiatric Association published the first edition of the *Diagnostic and Statistical Manual of Mental Disorders (DSM)*, which describes mental health disorders. Because of continual changes in medical science, the manual has been revised several times. It is now in its sixth edition—*DSM-5-TR*. The number of recognized disorders in the manual have greatly increased with newly included disorders such as "prolonged grief disorder" (Tavris, 2013; American Psychiatric Association, 2023).

The *DSM*'s ever-changing diagnoses and labels are an example of the **medicalization of deviance**, diagnosing and treating a violation of social norms as a medical disorder. The *DSM* has far-reaching effects, according to one of the past editors. "Anything you put in that book ... has huge implications not only for psychiatry but for pharmaceutical marketing, research, for the legal system, for who's considered to be normal or not, for who's considered disabled. *And it has huge implications for stigma because the more disorders you put in, the more people get labels*" (Carey, 2010, p. 1, emphasis added; Chapter 14 discusses the costs of medicalizing deviance).

7-6c Critical Evaluation

Interaction theories help us understand deviance, but critics point to several weaknesses. Differential association theory doesn't explain (1) impulsive crimes triggered by rage or fear; (2) crimes committed by people who have grown up in law-abiding families; (3) why only children from racial and ethnic groups that have historically excluded and who grow up in disadvantaged communities join gangs and/or commit crimes; and (4) why only one sibling raised in the same family environment may commit juvenile or adult crime (Williams and McShane, 2004; Graif and Sampson, 2009; Breining et al., 2017).

Critics have faulted labeling theory for not explaining why (1) people are deviant before labeling occurs; (2) crime rates are higher in some parts of the country or at particular times of the year (such as before holidays);

Jeff Greenberg/Universal Images Group/Getty Images

Do you think that this arrest is the result of primary or secondary deviance? Why?

primary deviance the initial act of breaking a rule.

secondary deviance rule-breaking behavior that people adopt in response to others' reactions.

medicalization of deviance diagnosing and treating a violation of social norms as a medical disorder.

Boston Globe/Getty Images

Can People Change Labels? Yes. An organization called Boston Uncornered offers former and current gang members an opportunity to earn a high school credential and transition into college for opportunities for positive change. Here, people look at the "Uncornered" art display in Boston in 2022. This exhibit, called "Uncornered Project," showcases the personal stories of those who have felt "cornered" but then made positive shifts for themselves and others.

and (3) labeling doesn't necessarily increase deviant behavior (Colvin and Pauly, 1983; Schurman-Kauflin, 2000; Benson, 2002). Conflict theorists, in particular, criticize symbolic interactionists for ignoring structural factors—like poverty and low-paying jobs—that create or reinforce deviance and crime (Currie, 1985).

7-7 Controlling Deviance and Crime

Most of us conform to laws and cultural expectations. We usually conform, however, not because we're innately good but because of **social control**, techniques and strategies that regulate people's behavior in society. The purpose of social control is to eliminate, or at least reduce, deviance.

7-7a Control Theory and Types of Social Control

Control theory proposes that deviant behavior decreases when people have strong social bonds with others. The bonds that strengthen our self-control are based on (1) *attachment* to other people (e.g., family, friends, and coworkers), (2) *commitment* to conformity (e.g., participating in community activities), (3) *involvement* in legitimate activities (e.g., getting an education, holding a job), and (4) *belief* in social norms and values (e.g., being moral, obeying laws) (Hirschi, 1969).

Most conformity is due to the internalization of norms during the powerful socialization process. Because most of us care about the opinions of family members, friends, and teachers, we try to live up to their expectations (refer to Chapter 4). Thus, we conform because of *informal social controls* that we learn and internalize during childhood.

Formal social control, which also regulates social behavior, exists outside of the individual. For example, a college dean might threaten a student with suspension or expulsion, a business might install security cameras, or a police officer might arrest someone.

7-7b The Criminal Justice System and Social Control

Social institutions such as the family, education, and religion try to maintain *social* control over behavior; the criminal justice system has the *legal* power to control crime and punish offenders. The **criminal justice system** is made up of government agencies—including the police, courts, and prisons—that enforce laws, judge offenders, and try to decrease criminal behavior. The criminal justice system relies on three major strategies to control crime: prevention and intervention, punishment, and rehabilitation. Punishment, as you'll learn, is the least effective approach.

Prevention and Intervention

Structural inequality—in employment, housing, education, and other areas—is the best predictor of violence and crime (Basken, 2016; Carpenter et al., 2016; Light and Ulmer, 2016). An exploration of government spending opens the question of national priorities. It costs U.S. taxpayers, annually, about $39,924 per incarcerated person compared with $15,621 per public school K-12

social control the techniques and strategies that regulate people's behavior in society.

control theory proposes that deviant behavior decreases when people have strong social bonds with others.

criminal justice system government agencies that are charged with enforcing laws, judging offenders, and changing criminal behavior.

student (Bureau of Prisons, 2021; National Center for Education Statistics, 2022).

Crime prevention is much less expensive than punishment. Among adults, the average cost for an intensive outpatient 30-day addiction treatment program ranges from $3,000 to $10,000, compared with $39,924 for incarceration (Brown, 2022; Bureau of Prisons, 2021). Nearly 28 percent of state and federal incarcerated persons are aged 50 and older, and many have serious health problems. An older offender's annual medical costs can exceed $100,000 (Johnson and Beiser, 2013; Carson and Anderson, 2016).

Nationally, states with the most firearms and fewest restrictions have the highest gun-related violence, homicide, and suicide rates, but gun control laws are controversial. Many Americans (85 percent) support background checks, but only a slight majority (55 percent) want stricter gun control laws. Advocating for harsher laws often surges after high-profile shooting incidents but then subsides again (Fleegler et al., 2013; Anestis and Anestis, 2015; Doherty, 2015; Newport, 2016; Parsons and Weigand, 2016).

Social Service Agencies and Community Outreach Programs. Many organizations—federal and state agencies, law enforcement professionals, social workers, and nonprofit groups—try to prevent and decrease crime (e.g., CrimeSolutions.gov), but some programs are more successful than others.

Compared with police-only interventions, when police and local social service workers respond together to an incident of intimate partner violence, victims experience significantly less violence immediately after the episode and months later. Why? The collaborative partnership provides victims with more protective strategies and resources (e.g., getting medical attention and protection orders, going someplace where a partner can't find them) (Messing et al., 2014).

Programs like "Scared Straight" are organized visits to prisons and face-to-face interactions with adult incarcerated persons to deter juvenile delinquency. Evaluations of scared-straight programs have concluded, however, that they may increase rather than deter adolescent deviance. The reasons are unclear but may be due to the program participants imitating the incarcerated persons' "brutal and vulgar" language and crimes. Still, the tours continue to be popular (Robinson and Slowikowski, 2011; Petrosino et al., 2013).

Police. The primary role of the police is to enforce society's laws, but can they prevent crimes? Police can deter some deviance by cruising high-risk areas in patrol cars or having more officers on foot patrols who return to the same street several times during a shift ("Policing Philadelphia . . . ," 2013). Concentrating on *hot spots,* areas of high criminal activity, reduces crime only temporarily, however, because criminals simply move to other parts of the city (U.S. Department of Justice, 1996; Braga, 2003).

And do some police actually reduce safety for specific categories of citizens? Negative views of police have spiked since 2014, after White police officers shot and killed unarmed Black men during routine law enforcement encounters, resulting in the deaths of Eric Garner, Michael Brown, and Tamir Rice. The deaths sparked protests in some cities and violence and riots in Baltimore (refer to Chapter 16). Compared with 56 percent of White individuals, only 27 percent of Black individuals have "a great deal" or "quite a lot" of confidence in their local police (Jones, 2021).

People who are systematically disadvantaged, particularly Black individuals, accuse police of discrimination, *racial profiling* (using race to stop and detain someone suspected of having committed a crime), and brutality. Many police officers (72 percent) say that the Black men's deaths are isolated incidents. Only 27 percent have ever fired their weapons while on duty, but they become aggressive because a "well-armed population" makes their job unpredictable and dangerous, and many suspects fight and resist arrest. Racial profiling is justified, according to some officers, because a high proportion of Black and Latina/o/x people commit crimes. An overwhelming 83 percent of Americans say they understand the risks that police face, but only 14 percent of officers agree and believe that they're scapegoated for social problems ("Police Culture," 2015; Morin et al., 2017). Police have no control over macro-level factors—poverty, unemployment, limited educational opportunities, and neighborhood deterioration—that are associated with high crime rates.

Punishment

Those who endorse a **crime control model** believe that crime rates increase when offenders don't fear apprehension or punishment. This model endorses a tough approach in sentencing, imprisonment, and capital punishment.

Sentencing. After someone has been found guilty of a crime or has pleaded guilty, a judge (and sometimes a jury) imposes a *sentence,* or penalty. A sentence can be a fine, *probation* (supervision instead of serving time in jail), incarceration, or the death penalty. Offenders who receive *parole* are released from prison before the end of their full sentence on the condition that they check in regularly with an officer and obey the law.

Of the 6.8 million adults in the corrections system in 2015, 56 percent were on probation, 13 percent were

crime control model proposes that crime rates increase when offenders don't fear apprehension or punishment.

on parole, and the rest were in a local jail or a state or federal prison. Since 2007, the number of people on parole has increased slightly, while the number on probation and in jails and prisons has declined (Kaeble and Glaze, 2016; Kaeble and Bonczar, 2017).

Why? In 2016, only 14 percent of Americans said that the criminal justice system wasn't "tough enough," down from 65 percent in 2003, and 38 percent described drug crime sentences as "too tough" (McCarthy, 2016). It's not clear whether (or how much) people's opinions affect the criminal justice system, but as budgets shrink, even states with reputations for being tough on crime are embracing more lenient policies. Some have begun to shorten the average number of years on probation and parole, which decreases supervision costs. Others are reducing the number of people sent to prison: It costs an average of $79 a day to keep an incarcerated person in prison but only about $3.50 a day to monitor the same person on probation (Richburg, 2009; Porter, 2016).

Incarceration. The number of people incarcerated in the United States has surged since 1987. In 2019, almost 2.1 million people were incarcerated in the U.S., including 1.43 million in federal and state prisons and approximately 735,000 in local jails. The incarceration rate in the U.S. is 810 out of every 100,000 adults aged 18 and older. This rate is actually less than the recent high point of 1,000 out of 100,000 in 2000. Still, the United States incarcerates a greater portion of its population than any other country (**Figure 7.3**), compared with the next highest countries El Salvador (564), Brazil (357), and Turkey (335) (Gramlich, 2021; Walmsley, 2016).

Of all people who are incarcerated in state and federal prisons, 93 percent are men, but imprisonment rates vary by race/ethnicity. As **Figure 7.4** shows, incarceration rates are highest for Black individuals, both women and men. There has also been a broader decline in incarceration rates over the last decade, with the most notable decline from 2019 to 2020 (15 percent). This is the result of pandemic-related delays in court operations and virus mitigation strategies that reduced overall prison populations (U.S. Department of Justice, 2021).

> It's tempting to assume that the more people behind bars, the lower the crime rate.

The imprisonment rate has been dropping since 2007 (Carson and Anderson, 2016). Arrest rates have declined, as you saw earlier, but prison counts have dropped primarily because of economic reasons. Driven by overcrowding and budget crises, many states are releasing low-level offenders before they complete their

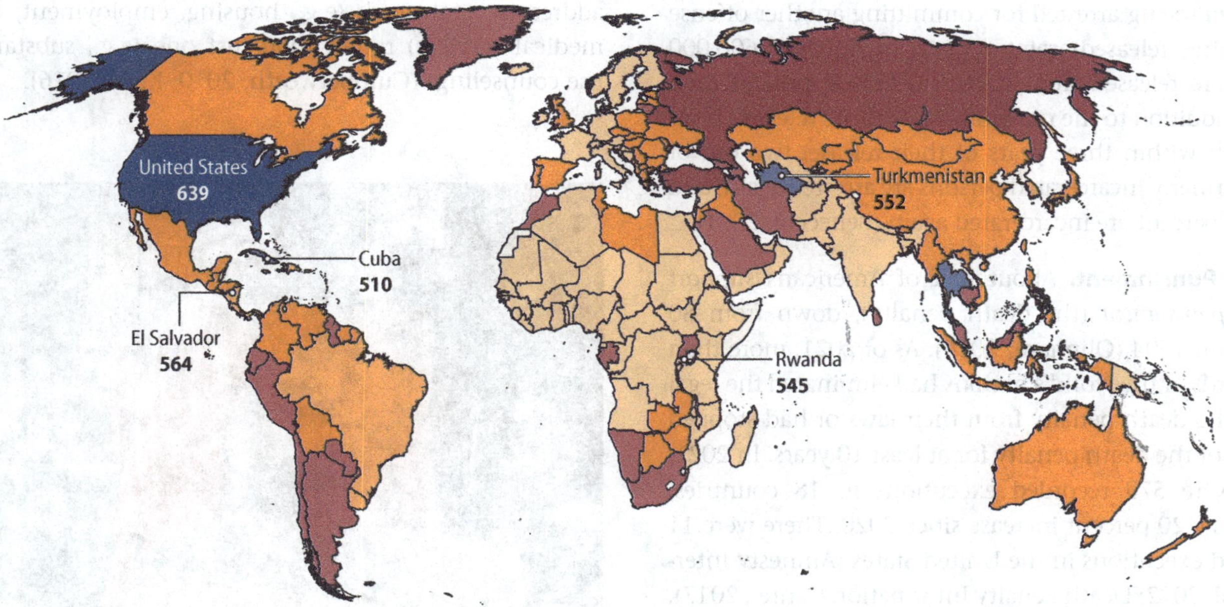

Figure 7.3 The U.S. Incarcerates a Larger Share of Its Population Than Any Other Country

Incarceration rate per 100,000 people of any age

Sources: Based on Gramlich, Pew Research Center, 2021.

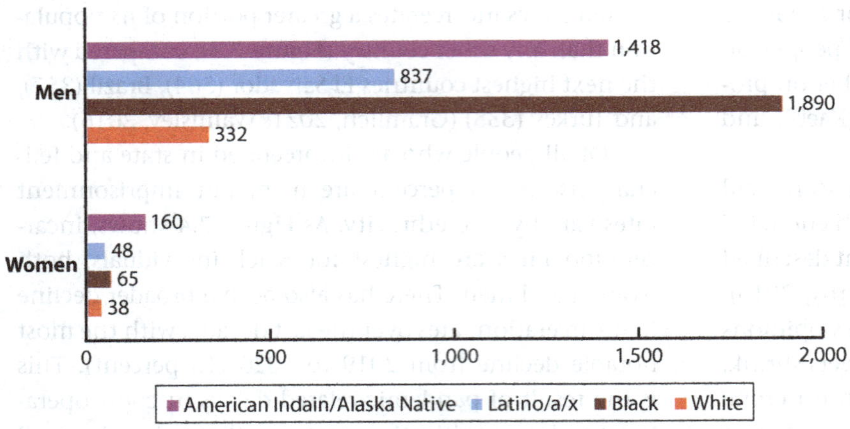

Figure 7.4 U.S. Prisoners by Race/Ethnicity and Sex, 2020

*Includes Native Americans, Alaska Natives, Asians, Native Hawaiians, Pacific Islanders, and persons of two or more races.

Source: Based on Carson, 2021, "Prisoners in 2020—Statistical Tables" Table 1.

sentences. Releasing people early from their sentences has been controversial, but state officials maintain that their budgets can't afford the high incarceration costs (Goode, 2013; Kearney et al., 2014).

It's tempting to assume that the more people that are behind bars, the lower the crime rate will be. State spending on corrections surged from $19 billion in 1977 to $86 billion in 2020 (accounting for inflation), which is 363 percent increase (The Urban Institute, 2023). However, *recidivism* (being arrested for committing another offense after being released) rates remain high. Over 600,000 people are released from state and federal prisons each year in addition to the nine million who are released from jails. Yet, within three years of their release, two out of three formerly incarcerated persons are arrested again, and over 50 percent are incarcerated again (Benecchi, 2021).

Capital Punishment. About half of Americans support *capital punishment* (the death penalty), down from 80 percent in 1994 (Oliphant, 2016). As of 2021, more than two-thirds of the world's nations had eliminated the legal use of the death penalty from their laws or had stopped the use of the death penalty for at least 10 years. In 2021, there were 579 recorded executions in 18 countries, which is a 20 percent increase since 2020. There were 11 recorded executions in the United States (Amnesty International, 2022; Death Penalty Information Center, 2017).

There are several reasons for fewer executions: (1) homicide rates have declined sharply since 1980; (2) the death penalty doesn't deter crime; (3) public support for the death penalty has waned; (4) hundreds of false convictions have been overturned; and (5) most juries can now impose life sentences without the possibility of parole. Some also argue that the death penalty is a waste of money because incarcerated persons can spend 15 to 20 years appealing a sentence. In California, for example, it costs taxpayers $137 million annually for incarcerated persons who appeal a death sentence, compared with $12 million for those serving life sentences (McPhate, 2016; Oliphant, 2016; Amnesty International USA, 2017; Death Penalty Information Center, 2017).

Rehabilitation

Rehabilitation, a third approach to controlling deviance, maintains that appropriate treatment can change offenders into productive, law-abiding citizens. Federal and state incarcerated persons produce more than $2 billion worth of products a year. They make everything from redwood canoes to specialty motorcycles and saddles. They also raise fish, milk cows and goats, teach dogs obedience, and manage vineyards at prison-run facilities (Alsever, 2014; "Prisons," 2017).

Are rehabilitation programs successful? Only if they provide employment after release. Other effective rehabilitation efforts include learning a trade while in prison, earning a high school diploma or college degree while in prison or after release, and receiving services that address several needs (e.g., housing, employment, *and* medical services) rather than just one (e.g., substance use counseling) (Cumberworth, 2010; Kelly, 2016).

Jim West/Alamy Stock Photo

About 700,000 U.S. incarcerated persons have jobs. Whether they work for the government or private businesses that have contracts with local correctional authorities, their earnings range from nothing to as little as 12 to 40 cents an hour (Vongkiatkajorn, 2016; "Prisons," 2017). Would paying incarcerated persons much higher wages decrease recidivism?

Chapter Review

Key Takeaways

7-1 **How is deviance described?** *Deviance*, the violation of social norms, includes *crime*. Perceptions of deviance and crime vary across and within societies and can change over time. People with authority or power decide what's right or wrong, and a *stigma* accompanies deviant behavior.

Example: Deviance and College Drinking

According to many college presidents, alcohol misuse is the most serious problem on campus. It can lead to alcohol poisoning, blackouts, sexual assault, violent behavior, injuries, and academic problems. Because drinking laws are rarely enforced, some college presidents have proposed that the drinking age be lowered from 21 to 18.

Others argue that doing so would increase traffic fatalities and drinking problems. Young people can get a driver's license at 16 and vote and enlist in the military at 18. Should they be the ones, then, to decide whether drinking laws should be changed?

7-2 **What are the different types of deviance and crime?** Sociologists study *noncriminal deviance*—such as suicide, alcohol misuse, lying, mental illness, and adult pornography—and *criminal deviance*, behavior that violates laws. Two of the most important sources of U.S. crime statistics are the FBI's *Uniform Crime Report* and *victimization surveys*.

The media usually focuses on street crimes, but Americans are much more likely to be victimized by *hate crimes*, *white-collar crimes*, *corporate crimes*, *cybercrime*, and *organized crime*. *Victimless crimes* violate laws, but the parties involved don't consider themselves victims.

7-3 **What are the main components of functionalist perspectives on deviance?** Refer to Table 7.1 on page 137.

7-4 **What are the main components of conflict perspectives on deviance?** Refer to Table 7.1 on page 137.

7-5 **What are the main components of feminist perspectives on deviance?** Refer to Table 7.1 on page 137.

7-6 **What are the main components of symbolic interactionist perspectives on deviance?** Refer to Table 7.1 on page 137.

7-7 **How effective are the criminal justice system's social control methods?** The *criminal justice system* is a form of formal social control, and is made up of government agencies—including the police, courts, and prisons—that enforce laws, judge offenders, and try to decrease criminal behavior. The three major strategies to control crime are prevention and intervention, punishment, and rehabilitation. Punishment, such as incarceration, is the least effective approach.

Key Terms

Deviance (p. 130)

Crime (p. 130)

Stigma (p. 130)

Victimization survey (p. 132)

Hate crime (also known as *bias crime*) (p. 134)

White-collar crime (p. 134)

Corporate crimes (also called *organizational crimes*) (p. 135)

Cybercrime (also called *computer crime*) (p. 135)

Organized crime (p. 136)

Victimless crimes (also called *public order crimes*) (p. 136)

Anomie (p. 138)

Strain theory (p. 138)

Patriarchy (p. 142)

Rape culture (p. 142)

Differential association theory (p. 143)

Labeling theory (p. 143)

Primary deviance (p. 144)

Secondary deviance (p. 144)

Medicalization of deviance (p. 144)

Social control (p. 145)

Control theory (p. 145)

Criminal justice system (p. 145)

Crime control model (p. 146)

Discussion and Critical Thinking Questions

1. The United States jails more children and teenagers than any other developed nation in the world. Those who are incarcerated are 13 percent less likely to graduate high school and 22 percent more likely to end up back in prison as adults than those placed under home monitoring or electronic supervision (Aizer and Doyle, 2013).
   a. What's the purpose of incarcerating children? Crime prevention? Punishment? Rehabilitation? Some other reason?
   b. At what age should children who commit serious crimes (like murder) be treated as adults?
   c. Why does juvenile incarceration increase the likelihood of committing crimes in adulthood?
2. In Los Angeles, California, a group of former gang members, now in their 60s, are trying to undo the violence they created. Among other things, they provide drug counseling to inner-city youth and sponsor community activities such as Easter egg hunts and jazz festivals.
   a. What kind of social control is this, formal or informal? Is it aimed at prevention or intervention?
   b. Do you think that such community outreach programs are more or less effective than social service agencies? Why?
3. Many Americans fear street crimes and school violence but not corporate crime, which is much more prevalent and costly. Why the difference?
4. What's your position on incarcerating nonviolent offenders? Should the federal government decriminalize victimless crimes like drug possession and prostitution? Why or why not?
5. Many studies show that capital punishment doesn't deter crime. Why, then, do many people support the death penalty?

True or False?

How Much Do You Know about Deviance, Crime, and Social Control?

The answers to 2 and 5 are true. The answers to 1, 3, 4, and 6 are false.

1. Not all deviant activities and behaviors are also crimes. For instance, subscribing to an infidelity website may be deviant, but it is not criminal.
2. Deviance serves a purpose in society. It affirms cultural norms and values. It provides temporary safety valves for expression. It bolsters the economy. It triggers social change.
3. Men are more likely than women to be crime victims. Victimization rates do differ by type of crime. Men experience higher rates of homicide, robbery, and aggravated assault. Women are more likely to be victims of rape, sexual assault, and intimate partner violence.
4. Rates for both property crimes and violent crimes were lower in 2019 compared with the rates in 2000.
5. The 2019 incarceration rate in the U.S. was 810 out of every 100,000 adults aged 18 and older. This rate is less than the recent high point of 1,000 out of 100,000 in 2000.
6. As of 2021, more than two-thirds of the world's nations had eliminated the legal use of the death penalty from their laws or had stopped the use of the death penalty for at least 10 years.

8 Social Stratification: United States and Global

Learning Objectives

After studying this chapter, you will be able to…

8-1 Explain social stratification systems and bases.

8-2 Explain the U.S. class structure, including how and why social classes differ.

8-3 Explain the concept of poverty and various reasons why people experience lower socioeconomic status.

8-4 Describe the different types of social mobility, recent trends in social mobility, and factors that affect mobility.

8-5 Describe global stratification, its variations and consequences, and the theoretical models that explain why inequality is universal.

8-6 Evaluate the theoretical explanations of social stratification.

True or False?

How Much Do You Know about U.S. Economic Equality?

1. The middle class in the United States is growing.
2. The most prestigious occupations have the highest incomes.
3. Americans aged 65 and older have the highest poverty rates.
4. The wealth gap has been decreasing in the United States.
5. Many Americans are economically disadvantaged, but poverty has decreased since the 1980s.
6. Income inequality in the United States is higher than income inequality in most other countries.

The answers are at the end of the chapter.

In 2012, American multi-billionaire Larry Ellison—founder of Oracle, the world's biggest database company—bought the Hawaiian island of Lanai for $500 million. The same year, 17 percent of Hawaii's residents were living in poverty (*Hawai'i Free Press*, 2012; Kroll and Dolan, 2013). Many people in the United States and around the world are struggling to survive while others enjoy astonishing wealth. This chapter examines social stratification and why there are haves and have-nots.

Be sure to take the True or False quiz in the chapter opener to better understand how much you know about these topics.

What do you think?

Americans who are economically disadvantaged just aren't working hard enough.

1	2	3	4	5	6	7
	strongly agree				strongly disagree	

8-1 Social Stratification Systems and Bases

Social stratification is a society's ranking of people based on their access to valued resources such as wealth, power, and prestige. All societies are stratified, but some more than others. Sociologists distinguish between stratification systems based on the extent to which they're open or closed.

8-1a Closed and Open Stratification Systems

In a *closed stratification system*, movement from one social position to another is limited by ascribed statuses. An *open stratification system* allows movement up or down because people's achievements affect mobility. Closed stratification systems are considerably more fixed than open ones, but neither is completely open or closed.

Two closed stratification systems exist today: slavery and castes. In a **slavery system**, people own others as property and have almost total control over their lives. Enslaved people have been kidnapped, inherited, or given as gifts to pay a debt. They're bought and sold as commodities, sometimes multiple times. The United Nations banned all forms of slavery worldwide in 1948, but approximately 49.6 million people are enslaved across the globe. That is an increase of 10 million enslaved people over the last five years. Out of all those who are enslaved, approximately 27.6 million are in forced labor, and 6.3 million are in forced commercial sexual exploitation (International Labour Organization et al., 2022).

In a **caste system**, a second type of closed stratification system, people's positions are ascribed at birth and largely fixed. These places in the hierarchy are primarily determined by inherited characteristics such as race, skin color, gender, family background, or nationality (refer to Chapter 5). People must marry within their caste; there are severe restrictions in their choice of occupation, residence, and social relationships; and people rarely move from one caste to another.

A good example is India, where a caste system has existed for more than 3,000 years. At the top were the Brahmins. On the bottom were the Dalits (formerly called "untouchables"), who performed the most undesirable, and sometimes unhealthy, jobs, including collecting human waste and cleaning streets. Fearing being "polluted," higher castes wouldn't interact with the Dalits (Mendelsohn and Vicziany, 1998).

India outlawed the caste system in 1949, but social distinctions are deeply entrenched, particularly in rural areas. When surveyed, sixty-two percent of Indians said it is very important to stop marriage across castes for men and sixty-four percent reported this sentiment for women (Saghal et al., 2021). In 1950, India set up a quota system that reserves nearly half of government

social stratification a society's ranking of people based on their access to valued resources such as wealth, power, and prestige.

slavery system people own others as property and have almost total control over their lives.

caste system people's positions are ascribed at birth and largely fixed.

jobs and public college slots for members of the most disadvantaged castes. Some claim that the quota system is unfair because it benefits less qualified candidates. Others maintain that such complaints come primarily from groups who fear losing some of their deep-rooted privileges (Barstow and Raj, 2015; McCarthy, 2015).

In a **class system**—a relatively open stratification structure—people's positions are based on both birth and achievement. Because achieved characteristics (e.g., education, work skills, occupation) can change, people in open stratification systems can move from one social class to another. A **social class** is a group of people who have a similar standing or rank in a society based on wealth, education, power, prestige, and other valued resources. A majority of Americans (more than 60 percent) believe that hard work leads to success (Abrams, 2021), but is this really the case?

iStock.com/JeremyRichards

India's Dalits still perform the most undesirable, and sometimes unhealthy, tasks such as burning corpses and removing garbage and human waste.

8-1b The Bases of Stratification

In 2018, just 24 percent of young adults were financially independent by age 22, which is a decrease from 1980 when 32 percent of young adults were fully supporting themselves financially. In addition "about six-in-ten parents with children ages 18 to 29 (59%) say they have given their kids at least some financial help in the past year" (Barroso et al., 2019). Yet, more than half of young adults think they will become millionaires, with twenty percent expecting that they will achieve this financial goal by the age of 40 (The Economist, 2019). Even if some become multimillionaires, and income is an important stratification factor, it's not the only one. Sociologists use a multidimensional approach that includes wealth, prestige, and power (Weber, 1946).

Wealth

Wealth refers to the economic assets that a person or family owns. It includes money, property (e.g., real estate), stocks and bonds, retirement and savings accounts, personal possessions (e.g., cars and jewelry), and income. **Income** is the money a person receives, usually through wages or salaries, but it can also include rents, interest on savings accounts, stock dividends, royalties, or business proceeds. Income and wealth differ in several important ways:

- Wealth is *cumulative*. It increases over time, especially through investment, while income is usually spent on everyday expenses.
- Because wealth is accumulated over time, much of it can be *passed on to the next generation*. Inheritances and other monetary gifts offer opportunities to start a business, buy a first (or vacation) home, pursue college without debt, and increase assets (e.g., savings and stocks).
- Passing on wealth to the next generation *preserves privilege*. Wealthy people tend to have the greatest amount of *economic capital* (income and other monetary assets like property), *cultural capital* (advanced degrees and assets such as style of speech, table manners, and physical appearance), and *social capital* (networks comprised of influential people). Differential access to all three types of capital reinforces and reproduces the existing class structure and inequality (Bourdieu, 1984).

U.S. wealth and income inequality are staggering. Over the years, the wealth gap has been increasing. As **Figure 8.1** shows, the top 1 percent of Americans own 31 percent of all wealth. The next 9 percent earn 32 percent, while the bottom 90 percent of the people share the remaining 37 percent. Yet, the COVID-19 pandemic did temporarily prompt some change. Stimulus payments helped increase the wealth for the bottom 50 percent of Americans—bringing their wealth higher than pre-pandemic levels. Yet, that benefit was quickly diminished (Board of Governors of the Federal Reserve, 2023; Kochhar and Sechopoulos, 2022; O'Donnell, 2023).

class system people's positions are based on both birth and achievement.

social class people who have a similar standing or rank in a society based on wealth, education, power, prestige, and other valued resources.

wealth economic assets that a person or family owns.

income the money a person receives, usually through wages or salaries, but can also include other earnings.

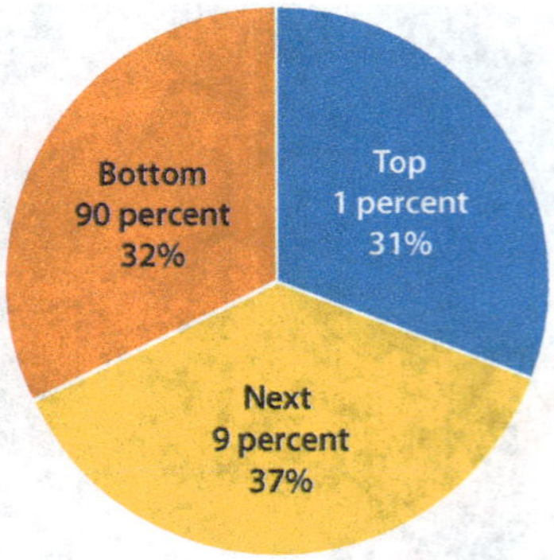

Figure 8.1 Who Has the Biggest Piece of Pie?

Sources: Based on Board of Governors of the Federal Reserve, 2023.

The racial wealth divide remains wide. In 2019, the average White family had $184,000 in wealth compared to Hispanic families with $38,000 and Black families with $23,000. While the White–Hispanic wealth-gap has slightly decreased over time, the gap between White and Black families remains relatively unchanged (**Figure 8.2**) (Harris and Wertz, 2022).

In comparison, Asians' wealth has increased significantly since 1989. Researchers predict that Asian wealth levels will soon surpass those of White wealth: Asian families already have the highest median household income, high educational levels, diversified assets, little or no credit card debt, and save money every year (Emmons and Noeth, 2015; refer also to Chapter 10).

Regardless of how the economy is doing, the rich keep getting richer with the income gap growing as well. From 1970 to 2020, incomes for the top third of the population increased by 69 percent, bringing their

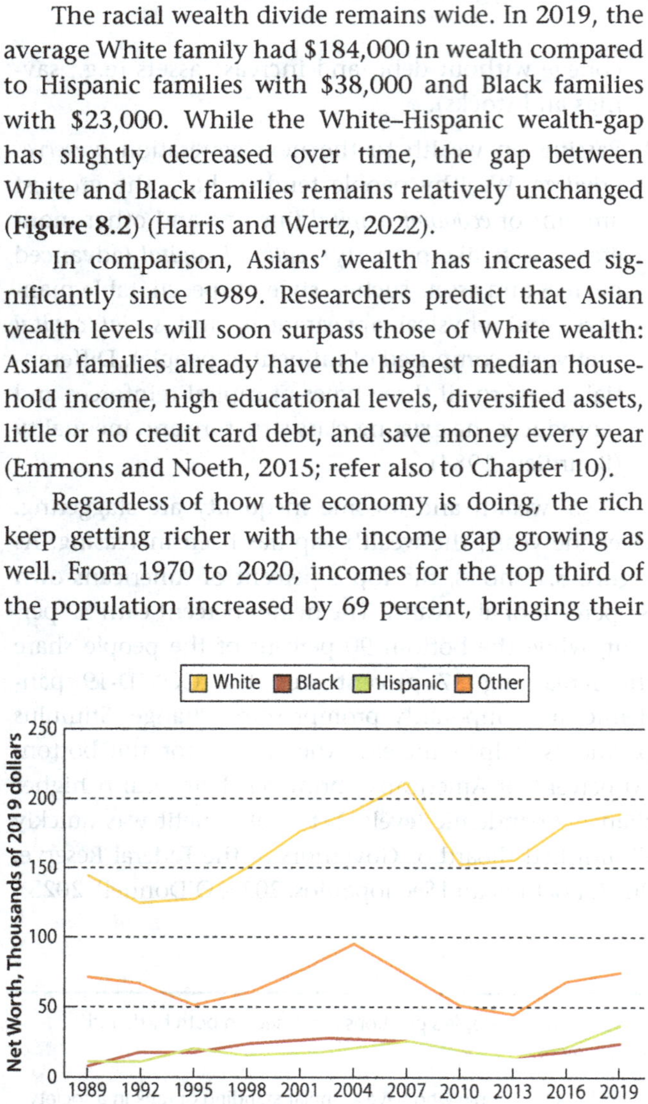

Figure 8.2 Racial Wealth Gap Over Time

Source: Harris and Wertz, 2022.

prestige respect or recognition attached to social positions.

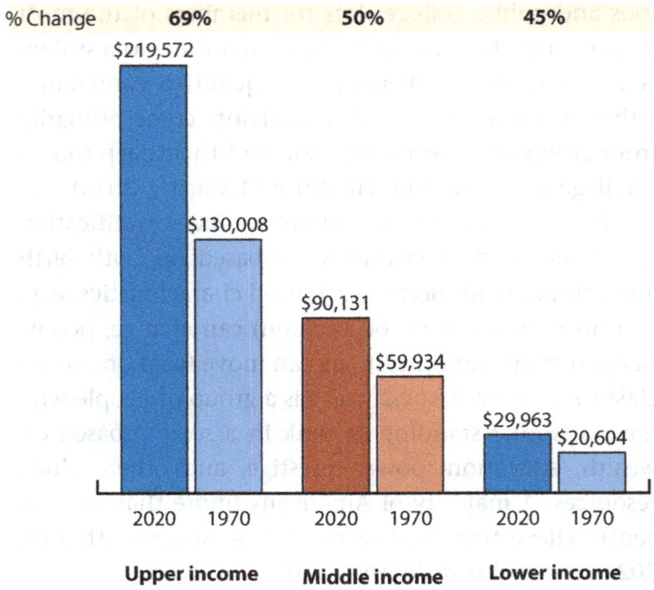

Ink Drop/Shutterstock.com

Figure 8.3 Income Change by Income Level in U.S., from 1970 to 2020.

Source: Based on Kochhar and Sechopoulos, 2022.

average earnings up to $219,572. On the other hand, the average incomes of the middle third made smaller gains, growing by 50 percent (to $90,131), and incomes for the bottom third increased by even less—45 percent ($29,963) (Kochhar and Sechoupolos, 2022) (**Figure 8.3**).

The top 10 percent of income earners bring in $1.3 million income annually, while the average for the remaining 90 percent is approximately $50,000. As of March 2023, *Forbes* recognized Bernard Arnault of France as the richest person in the world with a net worth of nearly $211 billion. Arnault ousted Elon Musk—now the second richest person in the world. Out of the top ten wealthiest people, six are from the United States. And nine out of the top ten are men (Peterson-Withorn, 2023).

Prestige

A second basis of social stratification is **prestige**—respect or recognition attached to social positions. Prestige is based on many criteria, including wealth, family background, power, and accomplishments. Regarding accomplishments, for example, every college convocation acknowledges students who graduate *cum laude*, *magna cum laude*, and *summa cum laude*.

We typically evaluate people according to the kind of work they do. Higher prestige occupations include physicians, lawyers, pharmacists, college professors, architects, dentists, and teachers. Legislators, police officers, actors, librarians, realtors, and firefighters are examples of jobs at a medium prestige level. Occupations at the lower prestige level consist of janitors, carpenters, bartenders, garbage collectors, truck drivers, and food servers (Smith et al., 2011).

The most prestigious occupations have several characteristics:

- They *require more formal education* (college or postgraduate degrees) and/or extensive training. Physicians, for example, must fulfill internship and residency requirements after receiving a medical degree.
- They're primarily *nonmanual and require more abstract thought*. All jobs have some physical activity, but an architect must use more imagination in designing a building than a carpenter, who usually performs very specific tasks.
- They *are paid more*, even though there are some exceptions. A realtor or a truck driver may earn more than a registered nurse, but registered nurses are likely to earn more over a lifetime because they have steady employment, health benefits, and retirement programs.
- They're *viewed as socially more important*. An elementary school teacher may earn less than the school's janitor, but the teacher's job is more prestigious because teachers contribute more to a society's well-being.
- They *involve greater self-expression, autonomy, and freedom from supervision*. A dentist has considerably more freedom in performing their job than a dental hygienist. In effect, then, higher prestige occupations provide more privileges.

Power

A third basis of social stratification is **power**—the ability to influence or control the behavior of others despite opposition. Wealthy people generally have more power than others in dominating top government and economic positions. Even if they don't hold a political office or a corporate position, the most powerful people make the nation's important decisions.

People who experience *status consistency* are about equal in terms of wealth, prestige, and power. A Supreme Court justice, for instance, is usually affluent, enjoys a great deal of prestige, and wields considerable power. In many cases, however, there's *status inconsistency* (refer to Chapter 5) if a person ranks differently on stratification factors. Consider funeral directors. Their prestige is relatively low, but most have higher incomes than college professors, who are among the most educated people in U.S. society and have relatively high prestige.

We've explored stratification systems and their bases, but how, specifically, do people in different social classes act? And how does social class affect our behavior?

8-2 Social Class in America

Social class is a more important divider than race, ethnicity, gender, age, or other characteristics. To measure social class, sociologists typically use **socioeconomic status (SES)**—an overall ranking of a person's position in society based on income, education, and occupation. Some researchers ask people to identify the social classes in their communities (*reputational approach*), some ask people to place themselves in one of a number of classes (*subjective approach*), but most use SES indicators (*objective approach*).

In measuring social class, national surveys tend to use the subjective approach and closed-ended questions (refer to Chapter 2). Because the choices differ, so do respondents' self-identification. In a recent Gallup poll in which two of the five categories included the words "middle class," 62 percent of Americans identified themselves as upper-middle or middle class (Newport, 2017). In another national survey, however, where three of the five categories included the words "middle class," fully 87 percent said that they were upper-middle, middle, or lower-middle class (Doherty and Tyson, 2015).

Because there are different ways of measuring social class, sociologists don't always agree on the number of social classes. There's consensus, however, that there are four general classes: upper, middle, working, and lower. Except for the working class, many sociologists divide these classes further into more specific strata. There aren't rigid dividing lines, but the U.S. social class structure varies by income, education, and occupation, and social classes also differ in values, power, prestige, social networks, and *lifestyles* (ways of living which can include tastes and preferences).

8-2a The Upper Class

Two percent of Americans say they're upper class (Newport, 2017). Their self-identification doesn't necessarily reflect their actual SES, however, because the upper class comprises the *upper-upper class* and the *lower-upper class*.

The Upper-Upper Class

Upper-upper-class people are the *old rich* who have been wealthy for generations. Because they value their privacy, upper-upper-class members rarely appear on the lists of wealthiest individuals published by *Forbes* or other sources.

power the ability to influence or control the behavior of others despite opposition.

socioeconomic status (SES) an overall ranking of a person's position in society based on income, education, and occupation.

An inherited fortune brings power. Upper-upper-class White males, in particular, shape the economic and political climate through a variety of mechanisms: Dominating the upper levels of business and finance, holding top political positions in the federal government, underwriting thousands of think tanks and research institutes that formulate national policies, and shaping public opinion through the mass media (Zweigenhaft and Domhoff, 2006; Hacker and Pierson, 2010).

The Lower-Upper Class

The lower-upper class—which is much more diverse than the upper-upper class—is the *nouveau riche*, those with "new money." Some, like the Kennedys, amassed fortunes several generations ago. Others, like politician and businessman Jeb Bush, are augmenting their vast inheritances through a combination of speaking fees, corporate board memberships, investments, and consulting contracts that pay millions each year (O'Keefe et al., 2015). Many of the new rich—Oprah Winfrey, Beyoncé, and Mark Zuckerberg—worked for their income rather than inherited it.

Besides business entrepreneurs, the lower-upper class also includes high-level managers of international corporations, those who earn at least a million dollars a year, and some highly paid athletes and actors, but the lifestyles vary considerably. Some lower-upper-class members live modestly, but many flaunt their new wealth. Sociologist Thorstein Veblen (1899/1953) coined the term *conspicuous consumption*, a lavish spending on goods and services to display one's social status and enhance one's prestige.

The *nouveau riche* are those who were not born into families with wealth but instead attained wealth themselves. People who are new to the highest economic levels may face challenges. Because they lack the "right" ancestry and have usually made their money by working for it, however, lower-uppers aren't accepted into "old money" circles that have strong feelings of in-group solidarity. Still, they engage in lifestyles and rituals that parallel those of the upper-upper class, such as having personal chefs, taking exotic vacations (often in private planes), and joining country clubs (Sherwood, 2010).

Hemera Technologies/Getty Images

Conspicuous consumption isn't limited to the super-rich. Many upper-middle-class people, in particular, use status symbols to show that they've made it by buying "almost rich" cars (like a Mercedes Benz that starts at $32,000), upscale kitchen appliances, designer handbags, expensive jewelry, and by taking lavish vacations.

8-2b The Middle Class

For about 89 percent of Americans, being middle class means having a secure job and being able to save money for the future. This may help explain why—whether they earn $22,000 or $200,000 a year—people identify themselves as middle class (Brown, 2016; Grobart, 2016). There's no agreement on exactly where the middle class begins and ends, but there are some major differences between the *upper-middle class* and the *lower-middle class*.

The Upper-Middle Class

Upper-middle-class members, although rich, live on earned income rather than accumulated or inherited wealth. The occupations of this group usually require an advanced professional degree or Ph.D. People in this class include corporate executives and managers (but not those at the top), high government officials, owners of large businesses, physicians, and successful lawyers and stockbrokers. A higher level of education is associated with greater occupational prestige and autonomy as well as job quality and security, but people in these occupations are three times more likely than those in the general population to work 50 or more hours per week (Dewan and Gebeloff, 2012).

The Lower-Middle Class

The lower-middle class, which is more diverse than the upper-middle class, is composed of people in nonmanual occupations that require some training beyond high school, and professional occupations that require a college degree. Nonmanual jobs include office staff, low-level managers, owners of small businesses, medical and dental technicians, secretaries, police officers, and sales workers (e.g., insurance and real estate agents). Examples of professional occupations are nursing, social work, and teaching. Most families in the lower-middle class rely on two incomes to maintain a comfortable standard of living.

Unlike the upper-middle class, those in the lower-middle class have less autonomy and freedom from supervision, and there's little chance for advancement. Except for some retirement funds, most have only modest savings to cover emergencies. Many buy used or inexpensive late-model cars, eat out at middle-income restaurants, and take occasional vacations, but they rarely have the income to buy luxury products without going deeply into debt.

8-2c The Working Class

About 30 percent of Americans identify themselves as working class (Newport, 2016). This group includes skilled and semiskilled laborers (e.g., construction and assembly-line workers, truck drivers, auto mechanics, electricians). The semiskilled jobs typically require little training, are mechanized, and closely supervised. Most of the occupations are blue-collar, but some—clerks and retail sales workers—are white collar. The jobs have more flexible educational requirements, which makes them more accessible, but the positions offer little or no opportunity for advancement.

Working-class homeowners may experience foreclosure because of delinquent payments. Many use credit cards but then can barely pay the monthly minimum, don't have enough money to cover a $500 repair bill, and all of their savings may amount to only about four months of income. Debts become overwhelming when borrowers suffer setbacks like divorce, illness, or job loss (Bell, 2015; PEW Charitable Trusts, 2015).

8-2d The Lower Class

Only 8 percent of Americans identify themselves as living below the poverty level (Newport, 2016). People in this group are at the bottom of the economic ladder because they have little education, few occupational skills, work in minimum wage jobs, or are often unemployed. Most of the lower class is economically disadvantaged, but sociologists distinguish between the *working poor* and the *underclass*.

The Working Poor

Almost 9 million Americans are considered **working poor**—people who work at least 27 weeks a year but whose wages fall below the official poverty level. Women are more likely than men to be among the working poor. Black and Latina/o/x individuals are more than twice as likely as Asian and White individuals to be in this group. Full-time workers are less likely than part-time workers to be working poor, but almost 4 percent of people usually employed full time are among the working poor, primarily because of low earnings (BLS Reports, 2017).

The Underclass

The **underclass** are people who are persistently economically disadvantaged, residentially segregated, and relatively isolated from the rest of the population. Most are chronically unemployed or drift in and out of jobs. They may have periods of unemployment or work part time, but their narrow skillsets, fewer educational opportunities compared to others, and, in some cases, mental or physical disabilities make it difficult for them to find full-time jobs. Members of the underclass may have some earnings from employment as well as income or benefits from government programs, including Social Security, Medicaid, and veterans' benefits to enhance their quality of life. Some also subsidize their economic security through criminal activities (Gilbert, 2011; refer also to Chapter 7).

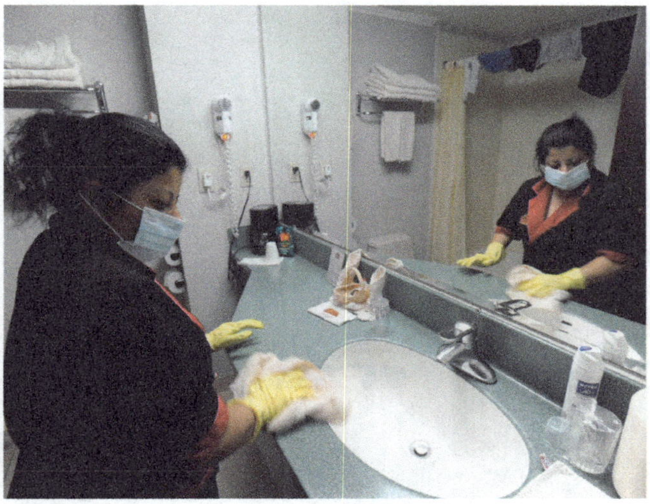
JOSE CABEZAS/Getty Images

Hotel housekeepers are among America's working, economically disadvantaged. On average, those employed even at expensive hotels earn approximately $28,250 a year working full time (Bureau of Labor Statistics, May 2021).

8-2e How Social Class Affects Us

Social class, more than any other single variable, affects practically all aspects of our lives. Max Weber referred to the consequences of social stratification as *life chances*—opportunities to access social and economic resources that improve one's quality of life (e.g., food, housing, education). People's life chances vary across institutions. Regarding health, for example:

- Children living in poverty are two times more likely than children in affluent families to have serious emotional or behavioral problems and are more likely to be unhealthy in adulthood. Unhealthy adults earn less, spend less time in the labor force, and must often retire earlier (Reuben and Pastor, 2015; refer also to Chapter 14).

working poor people who work at least 27 weeks a year but whose wages fall below the official poverty level.

underclass people who are persistently economically disadvantaged and seldom employed, residentially segregated, and relatively isolated from the rest of the population.

Glovatskiy/Shutterstock

Shkvarko/Shutterstock

- Living in a chronically disadvantaged neighborhood and experiencing violence increases anxiety and long-term stress, accelerating the onset of diseases (Harrell et al., 2014; Schanzenbach et al., 2016).
- Compared with the upper class, the lower class is four times more likely to be unhealthy, three times more likely to be unhappy, and twice as likely to be frequently stressed and depressed (Morin and Motel, 2012; Graham, 2015).
- The wealthiest Americans are now living 10 to 15 years longer than the most economically disadvantaged (Dickman et al., 2017).

Americans have an awareness of the issues concerning the gap between those with the highest wealth and income and those with the lowest. In a survey, 61 percent of American adults said there is too much economic inequality in the United States, yet only 41 percent of them felt addressing it should be a top priority for the federal government (Horowitz et al., 2020).

8-3 Poverty

There are two ways to define poverty: absolute and relative. **Absolute poverty** is not having enough money to afford the basic necessities of life, such as food, clothing, and shelter ("what I need"). **Relative poverty** is not having enough money to maintain an average standard of living.

absolute poverty not having enough money to afford the basic necessities of life.

relative poverty not having enough money to maintain an average standard of living.

poverty line the minimal income level that the federal government considers necessary for basic subsistence (also called the *poverty threshold*).

8-3a The Poverty Line

The **poverty line** (also called the *poverty threshold*) is the minimal income level that the federal government considers necessary for basic subsistence. To determine the poverty line, the Department of Agriculture (DOA) estimates the annual cost of food that meets minimum nutritional guidelines and then multiplies this figure by three to cover the minimum cost of clothing, housing, health care, and other necessities. The poverty line, which in 2016 was $24,339 for a family of four (two adults and two children), is adjusted every year to include cost-of-living increases. If a family makes more than the poverty line, it's usually not eligible for public assistance (Semega et al., 2017).

Is the poverty line realistic? Some contend that the poverty line is too high. They argue, for example, that many people aren't as economically disadvantaged as they seem because the poverty threshold doesn't include the value of noncash benefits such as food stamps, medical services (primarily Medicare and Medicaid), public housing subsidies, and unreported income (Eberstadt, 2009; Haskins, 2015).

Others claim that the poverty line is too low because it doesn't include child care and job transportation costs or the cost-of-living expenses, particularly housing, that vary considerably across states, regions, and urban-rural areas. According to some economists, a family of four needs almost $60,000 a year to cover basic necessities, a considerably higher amount than the official poverty line of just over $24,000. Some also argue that poverty estimates exclude millions of Americans who live above the poverty line but rely on food banks, soup kitchens, and clothing thrift stores to survive (Fremstad, 2012; DeGraw, 2014).

8-3b Who Are Most Likely to Experience Poverty?

In 2016, almost 13 percent of Americans (40.6 million people) were living in poverty, down from 15 percent

Six Ways People with Lower Incomes Pay More

Food: Not having a car to get to a supermarket chain or discount store means getting groceries at the corner store where staples are more expensive, produce is limited, and some products are stale.

Transportation: About 45 percent of American households don't have access to public transportation. Old cars break down, can be expensive to fix, use a lot of gas, and can cost people a job.

Doing Business: Most neighborhood stores charge more for products because real estate and insurance are more expensive. Economically disadvantaged people also pay more for financial services. Banks are scarce and "check-cashing stores" charge up to 10 percent of a check's value to cash it.

Credit: Because economically disadvantaged people don't have credit, they must rely on "cash advance" stores if they can prove that they get a regular paycheck. If people with lower incomes qualify for such a loan, they pay an annual percentage rate of about 825 percent.

Everyday Hassles: The economically disadvantaged are hassled almost every day by bill collectors because of overdue payments, deal with laundromat trips, carry groceries long distances, and contend with violence in high-crime areas.

Waiting: Economically disadvantaged people spend much of their time waiting—in food pantry lines, at blood banks to sell their blood, at bus stops to get to work, and for apartments in safer neighborhoods.

Sources: Brown, 2009; Bass and Campbell, 2013; Johnson, 2014 "Tackling Poverty," 2015.

in 2012. Between the ages of 25 and 60, 62 percent of Americans will experience at least one year of poverty (Rank and Hirschl, 2015; Semega et al., 2017). Both historically and currently, people who experience poverty share some of the following characteristics.

Spencer Platt/Getty Images

One in three lower-income Americans live in the suburbs. Compared with central cities, many suburbs offer safer streets, better schools, and a greater number of low-paying jobs (e.g., retail sales, landscaping, restaurants). Rents are rising, however, and many lower income people living in the suburbs worry about the costs of maintaining a car to get to work (Kneebone and Berube, 2014; Kneebone, 2016).

Age

Children under the age of 18 make up only 23 percent of the U.S. population but 33 percent (13.3 million) of the economically disadvantaged, up from a low of 16 percent in 2001. Americans aged 65 and older make up about 15 percent of the total population but 9 percent (4.6 million) of those living in poverty. The poverty rate of older Americans is at an all-time low and lower than that of any other age group because government programs, particularly Medicare and Medicaid, have generally kept up with the rate of inflation. In contrast, many programs for the lowest income children have been reduced or eliminated since 1980 (Children's Defense Fund, 2014; Semega et al., 2017).

Gender and Family Structure

Women's poverty rates are higher than men's (14 and 11 percent, respectively), but family structure is an important factor. In 2016, 5 percent of married-couple families, 13 percent of male-headed families, and 27 percent of female-headed families lived in poverty (Semega et al., 2017).

Feminization of poverty refers to the disproportionate number of economically disadvantaged people who

feminization of poverty the disproportionate number of people experiencing poverty who are women.

are women. Because of increases in divorce, nonmarital childbearing, and low-paying jobs, single-mother families are four to five times more likely to be living in poverty than married-couple families that have two wage earners, and they are more likely to experience extremely poverty over many years (Shriver, 2014; Irving and Loveless, 2015).

Race and Ethnicity

Among those living in poverty, in absolute numbers, there are approximately 11.15 million non-Hispanic Whites experiencing poverty and lower numbers of non-Hispanic Blacks (4.60 million), Hispanics of all races (7.02 million), Asian (1.96 million), and Native American/Alaskan native (512,000). Because each race or ethnic group constitutes a different portion of the total population, absolute numbers do not tell the whole story regarding poverty. It's also important to explore poverty percentages within racial and ethnic groups. Proportionally, non-Hispanic Whites are less likely to live in poverty (5.7 percent). This is compared with higher percentages experiencing poverty for Hispanics (11.2 percent), non-Hispanic Blacks (11.2 percent), Asians (9.5 percent), and Native Americans/Alaskan Natives (12.4 percent) (**Figure 8.4**) (Creamer et al, 2022).

Education

Increasingly, education separates the economically disadvantaged from the economically advantaged. In 2020, just 4 percent of those with at least a four-year college degree were economically disadvantaged compared with 13 percent of high school graduates and 25 percent without a high school diploma (American Association of Community Colleges, 2021).

The more risk factors a person has, the more likely that they will experience poverty. For example, being a young, Black, unmarried woman with young children and little education increases the odds of both poverty and extreme poverty (living below the bottom 10 percent of the income distribution) (Rank and Hirschl, 2015; Annie E. Casey Foundation, 2015).

8-3c Why Do People Experience Poverty?

Social scientists have a number of theories about poverty (refer to Cellini et al., 2008, for a nontechnical summary). Of the two most common explanations, one blames the economically disadvantaged themselves, and the other emphasizes societal factors.

Blaming the Economically Disadvantged: Individual Failings

More than a third of Americans believe that people are economically disadvantaged because they don't work hard enough. Republicans are more likely than Democrats (56 and 19 percent, respectively) to say that poverty is due to lack of effort rather than circumstances beyond people's control (Smith, 2017).

Such attitudes reflect a still influential *culture of poverty* perspective, which contends that people are

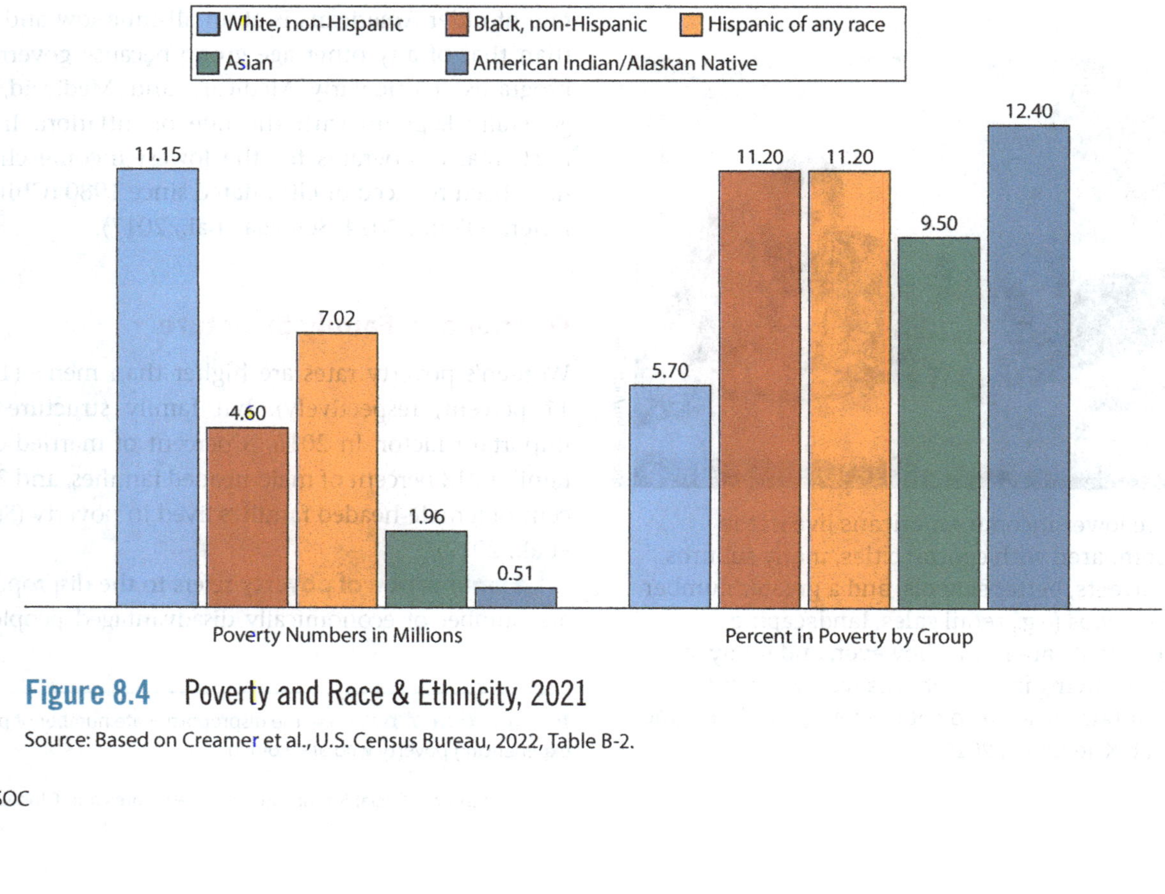

Figure 8.4 Poverty and Race & Ethnicity, 2021

Source: Based on Creamer et al., U.S. Census Bureau, 2022, Table B-2.

In 2021, approximately 326,000 Americans experienced sheltered homelessness on a given night, down by 8 percent since 2020. Sheltered homelessness refers to people in transitional housing, shelters, or others temporary settings (U.S. Department of Housing and Urban Development, 2022).

economically disadvantaged because they're personally "deficient" or "inadequate." If they worked harder, took initiative, were ambitious, and planned for the future instead of seeking immediate gratification, according to this perspective, economically disadvantaged people could succeed. Instead of being responsible and getting jobs, the economically disadvantaged rely on generous government handouts that erode an incentive to work (Lewis, 1966; Banfield, 1974; Murray, 1984, 2012). Thus, the most effective remedy to poverty is to change people's attitudes and behavior.

Blaming Society: Structural Characteristics

In contrast to blaming the economically disadvantaged, most sociologists contend that macro-level structural factors create and sustain poverty. Technology, automation, industry relocation, and the growth of a low-wage service sector produce poverty conditions that are difficult for individuals to overcome. As the number of inner-city manufacturing jobs declined, neighborhood businesses collapsed, middle class people—across all races—moved to the suburbs, and the economically disadvantaged became more isolated and in need of government benefits to meet their needs for survival (Oliver and Shapiro, 1995; Wilson, 1996; Conley, 1999). Thus, the most effective remedy to poverty is changing structural forces like unemployment and very low wages.

Which perspective is more accurate, blaming those who are economically disadvantaged or blaming society? Some people do experience economic hardship because they put in little effort and would rather get a handout than a job. Others "compound their misfortune by self-medicating or engaging in irresponsible, self-destructive behavior" (Kristof, 2016, p. SR1). Most people are economically disadvantaged, however, because of economic factors (low wages, job loss, lack of affordable housing, inability to afford health insurance), which, in turn, can result in acute health and employment problems.

Temporary assistance for the nation's lowest income families has fallen well below the poverty line in every state. In 2022, there were 1 million families and 2.8 million individuals receiving funds through Temporary Assistance for Needy Families (TANF). This is a decrease from the high point of 5.1 million families receiving TANF in 1994. TANF benefits vary state by state. In July 2020, the maximum benefit for a family of two in Mississippi was $146 per month, and in New Hampshire, it was $862. Millions of economically disadvantaged Americans don't seek any help for food, shelter, or utilities because they don't know about the programs or have transportation and childcare difficulties in meeting some of the job-training requirements (Rodrigue and Reeves, 2015; Stanley et al., 2016; Falk and Landers, 2023).

From 2011 until 2021, the poverty rate declined by approximately 3 percent (Creamer et al., 2022). Employers created more jobs, especially in the service sector, and raised minimum wages that attracted low-skill workers. For instance, a 32-year-old man who had survived on odd jobs lifted his family out of poverty when he became an assistant manager at a pizzeria with an annual salary of $40,000 and health benefits (Cohen, 2016).

On the other hand, and despite their challenges, millions of middle-class Americans enhance their quality of life through government aid from programs for individuals with disabilities, college students, older adults, and people who are unemployed. Stimulus payments during the COVID-19 pandemic had a similar effect. Also, higher income farmers receive generous federal subsidies, and for every taxpayer dollar that helps economically disadvantaged families, six dollars benefit the rich who avoid paying taxes (Rampell, 2014; Eckhoff, 2015; Buchheit, 2015; Center on Budget and Policy Priorities, 2016).

8-4 Social Mobility

Social mobility is movement from one social class to another. Is the American dream of moving up the economic ladder a reality? Sixty percent of Americans say that "most people who want to get ahead can make it if they're willing to work hard" (Horowitz, 2020). Are they right?

social mobility movement from one social class to another.

8-4a Types of Social Mobility

Intragenerational mobility is movement up or down a social class over one's lifetime. If Tara begins as a nurse's assistant, becomes a registered nurse, and then a physician's assistant (PA), she experiences intragenerational mobility. **Intergenerational mobility** is movement up or down a social class over two or more generations. If Tara's parents were blue-collar workers, her upward movement to the middle class is an example of intergenerational mobility. Intragenerational and intergenerational mobility can be downward or upward.

8-4b Recent Trends in Social Mobility

Only 50 percent of Americans born in the 1980s, compared with 90 percent born in the 1940s, are economically better off than their parents. About 40 percent of children born in the lowest or highest fifth of the income distribution will still be there as adults. For those born in the bottom 10 percent in the 1980s, 30 percent are worse off than their parents, five times as many as for children born in the 1940s (Chetty, Grusky, et al., 2016; Reeves and Joo, 2016).

The middle class has been shrinking. Since 1971, the upper- and lower-income brackets have expanded, while the middle bracket has dropped from 61 to 50 percent (**Figure 8.5**). Because of such changes, some analysts describe the middle class as "eroding," "squeezed," and "sinking" (Galston, 2013; Porter, 2013; Trumbull, 2015).

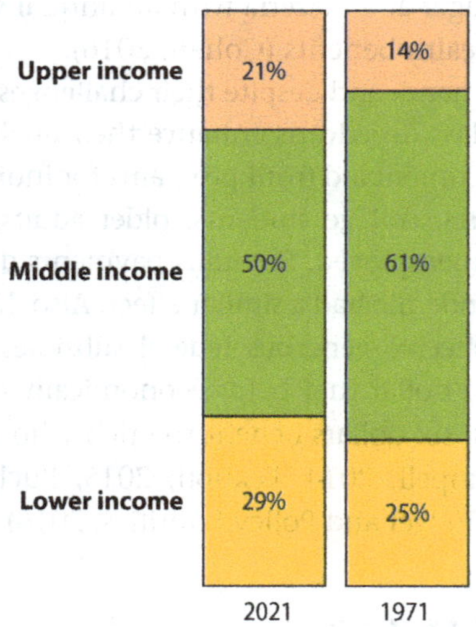

Note: Adults are assigned to income tiers based on thier size-adjusted household incomes in the calendar year prior to the survey year. Shares may not add to 100% due to rounding.

Figure 8.5 Adults in the Middle Class Have Decreased Considerably Since 1971

Source: Kochhar and Sechopoulos, 2022.

Most social mobility moves are short because social classes are fairly rigid (Wysong et al., 2014). A child from a working-class family, for example, is more likely to move up to the lower-middle class than to jump to the lower-upper class. The same is true of downward mobility: Someone from the middle class is more likely to slide into the working class than to drop to the underclass. Upward intergenerational mobility has become more limited because the rungs of the economic ladder have grown further apart (Chetty et al., 2014). As you read earlier, income inequality has increased, making it more difficult for even upper-middle class people to move up.

Across 24 middle- and high-income countries, the United States ranks only 16th in upward generational mobility and only slightly ahead of some lower-income countries, like Brazil and Chile (Corak, 2016). Countries with the greatest income inequality, such as the United States, are the most likely to have limited upward mobility (Greeley, 2013).

8-4c What Affects Social Mobility?

Upward mobility doesn't always reflect people's talents, intelligence, or hard work. Instead, much social mobility depends on structural, demographic, and family background factors, all of which are interrelated.

Structural Factors

Macro-level variables over which people have little or no control affect social mobility in several ways. First, *changes in the economy* spur upward or downward mobility. During an economic boom, the number of jobs increases, and many people have an opportunity to move up. During recessions, such as the Great Recession, long-term unemployment leads to downward mobility (refer to Chapters 11 and 12).

Second, *government policies and programs* affect social mobility. Unlike the United States, countries that have promoted equality (e.g., Canada, Denmark, Finland, Norway) have the highest upward mobility rates. They have universal health care, which reduces the chance of people falling into poverty because of medical emergencies or poor health; provide affordable housing; and fund technical schools where young people learn high-paying trades (Foroohar, 2011; Deparle, 2012).

Third, *immigration* fuels upward mobility. Because many recent immigrants take low-paying jobs, groups that are already in a country often move into higher-paying occupations (Haskins, 2008). In addition, the children of those who immigrate often experience upward mobility. "No matter which country their

intergenerational mobility movement up or down a social class over two or more generations.

intragenerational mobility movement up or down a social class over one's lifetime.

parents came from, children of immigrants are more likely than the children of the U.S.-born to surpass their parents' incomes when they are adults" (Abramitzky and Boustan, 2022).

Demographic Factors

Demographic factors also affect social mobility. Four of the most important are education, gender, race, and ethnicity, but place also has an effect.

Education. Higher education promotes upward mobility. Especially when the economy is slumping, people with a high school education or less often face long and frequent bouts of unemployment, may need to rely on temporary work, and may move down the socioeconomic ladder (Kochhar and Fry, 2015; Chetty et al., 2017).

Gender. Women's massive entry into the labor force since the 1980s has increased family income and many women's upward mobility. You'll learn in later chapters that men's labor participation rates have been dropping, and women's educational attainment has risen faster than men's, but men's mobility is less affected than women's by a divorce, nonmarital children, or widowhood.

Race and ethnicity. Black and Latina/o/x middle classes have grown since the 1970s, but both groups still lag significantly behind White people and Asian Americans (refer to Chapters 10 and 11). Of Black children born into the bottom 20 percent of the income distribution, about half will still be there as adults, compared to less than one-quarter of White children. Economic, cultural, and social capital account for most of this mobility gap. Since the slavery era, White parents could accumulate skills and wealth that, with every generation, gave their children far

Jaguar PS/Shutterstock

Billionaire Oprah Winfrey grew up in an economically disadvantaged family. She focused on school and her media career. Oprah is now one of the most well-known media personalities in the United States and has an approximate net worth of $2.5 billion.

MCT/Getty Images

Children raised in lower income families can be upwardly mobile. Sonia Sotomayor, whose parents were born in Puerto Rico, is the first Latina to serve on the U.S. Supreme Court. Her family first lived in a South Bronx tenement and later a working-class housing project. Sotomayor's father, who had an alcohol misuse disorder, had a third-grade education, and was a tool and die worker. He died when Sotomayor was nine years old. Her mother—a telephone operator and later a practical nurse—stressed the value of education. Sotomayor excelled in school, received a full scholarship to Princeton University, and put in long hours to catch up with her peers' academic knowledge.

better chances than Black children of moving up (Collins and Wanamaker, 2017; Matthew and Reeves, 2017)

Place. Where one lives can stimulate or dampen upward mobility. For example, the probability of a child born into the lowest fifth of the population in San Jose, California, making it to the top fifth is 13 percent compared with only 4 percent for a child born in Charlotte, North Carolina. These upward mobility disparities in different parts of the country are correlated with four factors: residential segregation (whether by income or race), the quality of schooling, how many children live with only one parent, and parents' social capital (Chetty et al., 2014; Berube and Holmes, 2015).

The longer that children live in better environments, the greater their chances of upward mobility. For example, children below age 13 whose parents move to lower-poverty neighborhoods are more likely to attend college, more likely to have substantially higher incomes as adults, and less likely to become single parents (Chetty and Hendren, 2016; Smeeding, 2017).

Family Background Factors

Family wealth affects social mobility. About 60 percent of all U.S. household wealth is inherited; most of it grows, spreads through families and dynasties, and endures for generations. Whether people inherit wealth or amass it themselves, children who grow up in high-income households can experience considerable upward mobility (Alvaredo et al., 2017; Roth, 2017).

In contrast, nearly half of U.S. adults don't have enough money to cover a $400 car repair or any other unexpected expense. Not having savings increases chronic stress. Constantly struggling to solve the crisis of the day makes it difficult to plan for the future, including saving for college (Larimore et al., 2016; Graham, 2016).

Only 7 of the 46 U.S. presidents came from the lower-middle class or below. Although Abraham Lincoln was born in a log cabin, his father was one of the wealthiest people in the community. Former President Donald Trump joined his father's thriving real estate company after college. He started his business with a $14 million loan from his father, received other monetary gifts from him over the years, and relied on his successful father's connections to build his own empire (Kessler, 2016). About a third of the students with low grades at Ivy League universities wouldn't be there if their parents weren't celebrities, well-known politicians, or others who donate millions to the school (refer to Chapters 11 and 13).

Our socialization affects what French sociologist Pierre Bourdieu (1984) called *habitus*—the habits of speech and lifestyle that determine where a person feels comfortable and knowledgeable. Upper-class parents cultivate flexibility, autonomy, and creativity because they expect their children to step into positions that require such characteristics. Economically disadvantaged and working-class parents stress obedience, honesty, and appearance—traits that many employers expect. High-income parents have social connections to jobs and admission to particular schools or colleges. Because of such resources, the children of top earners are likely to grow up to be top earners themselves (Khan, 2012; Corak, 2013).

Many low-SES households are resilient, but meager incomes limit upward mobility. Unlike high-SES parents, low-income parents have less time, health, and job flexibility to read to their children every day, take them to museums, or attend school events. Children in high-earning families also benefit from high-quality schools, mentoring connections, enriching extracurricular activities, and skills for navigating bureaucracies that will help them succeed in schools and workplaces (Mitnik and Grusky, 2015; Putnam, 2015; Corak, 2016; refer also to Chapter 4).

8-5 Global Stratification

Global stratification refers to worldwide inequality patterns that result from differences in wealth, power, and prestige. All societies are stratified, but more than 75 percent of the world's population lives in countries where economic inequality has widened, increasing the possibility of social class tension and conflict (United Nations Development Program, 2013; World Economic Forum, 2013).

8-5a Wealth and Income Inequality

In 2021, global economic inequality increased, even with global wealth reaching $530 trillion. The top 10 percent of people own 76 percent of global wealth, while the bottom 50 percent own only 2 percent of all global wealth (Metcalf, 2022; Chancel et al., 2022). Worldwide, income inequality is also pervasive. The top 10 percent of people worldwide earn an average of $122,100 per year, while the bottom 50 percent average $3,920 (Ventura, 2022).

Recently, the COVID-19 pandemic affected income, wealth, and poverty. For instance, before the COVID-19 pandemic, the number of people worldwide experiencing extreme poverty was declining. The COVID-19 pandemic led to an increased number of people living in extreme poverty and altered poverty projections (Mahler et al., 2021) (**Figure 8.6**).

global stratification worldwide inequality patterns that result from differences in wealth, power, and prestige.

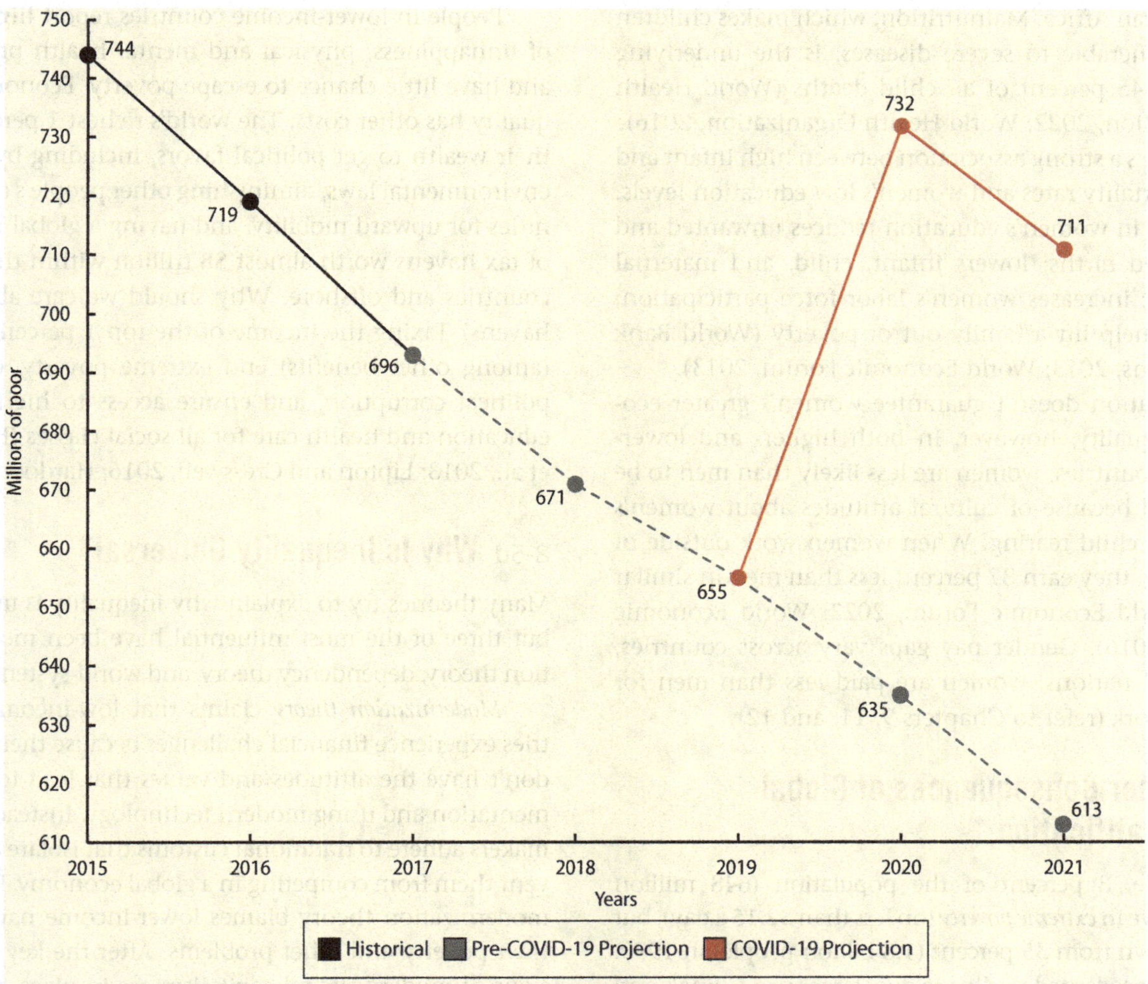

Note: Extreme poverty is measured as the number of people living on less than $1.90 per day. 2017 is the last year with official global poverty estimates. Official poverty estimates are available for East Asia & Pacific, Europe & Central Asia, Latin America & Caribbean, and rest of the world for up to 2019, and for Middle East & North Africa and Sub-Saharan Africa up to 2018. Regions are categorized using PovcalNet definition.

Figure 8.6 Impact of COVID-19 Pandemic on Extreme Poverty

Source: Henry, 2022.

The enormous wealth disparities between nations have declined over the last two decades, yet the inequality within nations has been on the rise. Living in a higher-income country doesn't mean that people enjoy income equality. Among 169 nations, the United States falls behind more than 60 percent of them in equality of incomes. The United States is below many other high-income nations including France, United Kingdom, Sweden, Germany, and Canada. The United States also lags behind many lower-income nations including Sierra Leone, Senegal, and Malaysia (Ventura, 2022).

Even within a nation, there can be differences between levels of income and wealth inequality. For example, in Sweden, which is one of the most egalitarian countries in the world for income, wealth inequality has remained fairly unchanged (Chancel et al., 2022).

8-5b The Plight of Women and Children

Historically, currently, and across *all* nations, women and children experience the greatest inequality. An important measure of a country's health is its **infant mortality rate**, the number of babies under age 1 who die per 1,000 live births in a given year. The infant mortality rate ranges from a high of 97 in some low-income African nations to a low of 3 in affluent Western European countries (Kaneda and Bietsch, 2016).

Children's deaths have declined worldwide from 12.8 million in 1990 to 5 million in 2021. Still, every year, nearly 5 million children die before their fifth birthday; most of the deaths continue to be concentrated in

infant mortality rate the number of babies under age 1 who die per 1,000 live births in a given year.

sub-Saharan Africa. Malnutrition, which makes children more vulnerable to severe diseases, is the underlying factor in 45 percent of all child deaths (World Health Organization, 2022; World Health Organization, 2016).

There's a strong association between high infant and child mortality rates and women's low education levels. Investing in women's education reduces unwanted and unplanned births; lowers infant, child, and maternal mortality; increases women's labor force participation; and can help lift a family out of poverty (World Bank and Collins, 2013; World Economic Forum, 2013).

Education doesn't guarantee women's greater economic equality, however. In both higher- and lower-income countries, women are less likely than men to be employed because of cultural attitudes about women's roles and child rearing. When women work outside of the home, they earn 37 percent less than men in similar jobs (World Economic Forum, 2022; World Economic Forum, 2016). Gender pay gaps vary across countries, but in all nations, women are paid less than men for similar work (refer to Chapters 9, 11, and 12).

8-5c Other Consequences of Global Stratification

Worldwide, 8 percent of the population (648 million people) live in *extreme poverty* (on less than $2.15 a day), but this is down from 35 percent (1.8 billion people) in 1990. This unprecedented decline is due, largely, to China's and India's economic growth, which has benefited those with high and low incomes. Despite such progress, extreme poverty rates have increased in sub-Saharan Africa, which now accounts for half of the world's most economically disadvantaged people (World Bank, 2017; Schoch, 2022).

Because of severe destitution, almost one in ten people worldwide suffer from chronic hunger and undernourishment. China and India have reduced their hunger rates, but in other regions (e.g., sub-Saharan Africa and Southern Asia, which includes Pakistan and Bangladesh), the number of undernourished people has increased. Some of the consequences include recurring illness, reduced work capacity, delayed physical and mental development, and stunted growth. For women, stunted growth can result in low birth weight babies and high child mortality rates (Martins, 2011; FAO, IFAD, and WFP, 2015).

People in lower-income countries report high levels of unhappiness, physical and mental health problems, and have little chance to escape poverty. Economic inequality has other costs: The world's richest 1 percent use their wealth to get political favors, including bypassing environmental laws, diminishing other people's opportunities for upward mobility, and having a global network of tax havens worth almost $8 trillion within their own countries and offshore. Why should we care about tax havens? Taxing the income of the top 1 percent would (among other benefits) end extreme poverty, decrease political corruption, and ensure access to high-quality education and health care for all social classes (Helliwell et al., 2013; Lipton and Cresswell, 2016; Hardoon, 2017).

8-5d Why Is Inequality Universal?

Many theories try to explain why inequality is universal, but three of the most influential have been modernization theory, dependency theory, and world-system theory.

Modernization theory claims that low-income countries experience financial challenges because their leaders don't have the attitudes and values that lead to experimentation and using modern technology. Instead, policy makers adhere to traditional customs that isolate and prevent them from competing in a global economy. In effect, modernization theory blames lower-income nations for their poverty and other problems. After the key foundations of modernity and capitalism are in place, this perspective maintains, low-income countries will prosper.

Dependency theory contends that the main reason why low-income countries are economically challenged is because they're pawns that high-income countries exploit and dominate. Rich nations wield an enormous amount of power by exporting jobs overseas, manipulating foreign aid, draining less

iStock.com/jacus

Aly Song/REUTERS

We hear about the booming economies of India and China, but not everyone has benefited. A tent city in India houses many of the workers who earn about $1.30 a day building new office towers for the affluent. In China, a man in Shanghai begs as wealthier residents pass by.

powerful countries of their resources, penetrating other countries with multinational corporations, and coercing national governments to comply with corporate economic and political interests. In effect, according to dependency theorists, high-income countries benefit because the economically disadvantaged people provide labor for lower wages and collectively don't hold enough power to protest even though they work in hazardous conditions and earn less than $1 a day.

More recently, *world-system theory*, similar to dependency theory, argues that the structure of the world system helps higher income countries maintain their economic advantages while ensuring that lower income countries remain disadvantaged (Bradshaw and Wallace, 1996, p. 44). Countries like the United States that dominate the world economy control the economies of low-income countries because their workers depend on external markets for jobs. High-income countries can extract raw materials (e.g., diamonds and oil) with little cost. They can also set prices, regardless of market values, for the agricultural products that low-income countries export. Doing so forces many small farmers to abandon their fields because they can't pay for labor, fertilizer, and other costs (Carl, 2002; Alvaredo et al., 2013).

8-6 Sociological Explanations: Why There Are Haves and Have-Nots

Why are societies stratified? **Table 8.1** summarizes the key points of the four theoretical perspectives. Let's begin by discussing a long-standing debate between functionalists and conflict theorists on why there are haves and have-nots.

Kanye West was the highest paid musician in 2020, earning approximately $170 million. Do his earnings reflect his contribution to society, especially compared with teachers, physicians, trash collectors, computer scientists, and others?

8-6a Functionalist Perspectives: Stratification Benefits Society

Functionalists view stratification as both necessary and inevitable: Social class provides each individual a place in the social world and motivates people to contribute to society. Without a system of unequal rewards,

Table 8.1 Sociological Explanations of Social Stratification

Perspective	Level of Analysis	Key Points
Functionalist	Macro	• Fills social positions that are necessary for a society's survival • Motivates people to succeed and ensures that the most qualified people will fill the most important positions
Conflict	Macro	• Encourages workers' exploitation and promotes the interests of the rich and powerful • Ignores a wealth of talent among people with lower incomes
Feminist	Macro and micro	• Constructs numerous barriers in patriarchal societies that limit women's ability to achieve wealth, status, and prestige • Requires most women, not men, to juggle domestic and employment responsibilities that impede upward mobility
Symbolic Interactionist	Micro	• Shapes stratification through socialization, everyday interaction, and group membership • Reflects social class identification through symbols, especially products that signify social status

functionalists argue, many important jobs wouldn't be performed.

The Davis–Moore Thesis

Sociologists Kingsley Davis and Wilbert Moore (1945) developed one of the most influential functionalist perspectives on social stratification that persists today. The **Davis–Moore thesis**, as it's commonly called, asserts that social stratification benefits society. The key arguments of the Davis–Moore thesis can be summarized as follows:

1. **Every society must fill a wide variety of positions and ensure that people accomplish important tasks.** Societies need teachers, doctors, farmers, trash collectors, plumbers, police officers, and so on.
2. **Some positions are more crucial than others for a society's survival.** Doctors, for example, provide more critical services to ensure a society's continuation than do lawyers, engineers, or bankers.
3. **The most qualified people must fill the most important positions.** Some jobs require more skill, training, or intelligence than others because they're more demanding, and it's more difficult to replace the workers. Pilots, for example, must have more years of training and aren't replaced as easily as flight attendants.
4. **Society must offer greater rewards to motivate the most qualified people to fill the most important positions.** People won't undergo many years of education or training unless they're rewarded by money, power, status, and/or prestige. If doctors and nurses earned the same salaries, there wouldn't be much incentive for people to spend so many years earning a medical degree.

According to the Davis–Moore thesis and other functionalist perspectives, then, stratification and inequality are necessary to motivate people to work hard and to succeed. For functionalists, social stratification is based on **meritocracy**—people's accomplishments. That is, people are rewarded for what they do and how well, rather than their ascribed status.

Davis–Moore thesis the functionalist view that social stratification benefits a society.

meritocracy a belief that social stratification is based on people's accomplishments.

bourgeoisie those who own and control capital and the means of production.

proletariat workers who sell their labor for wages.

Critical Evaluation

Sociologist Melvin Tumin (1953) challenged the Davis–Moore thesis. First, he argued, societies don't always reward the positions that are the most important for the members' survival. If the multimillionaire professional athletes, actors, and pop musicians went on strike, many of us would probably barely notice. If, on the other hand, those with much lower income—whether they're doctors, garbage collectors, teachers, truck drivers, or mail carriers—refused to work, society would grind to a halt. Thus, according to Tumin, there's little association between earnings and the jobs that keep a society going.

Second, Tumin claimed, Davis and Moore overlook the many ways that stratification limits upward mobility. Where wealth is differentially distributed, access to education, especially higher education, depends on family wealth. As a result, large segments of the population are likely to be deprived of the chance to discover what their talents are, and society loses.

Third, Tumin criticized Davis and Moore for ignoring the critical role of inheritance. Children born into the upper class don't have to work because their inherited wealth guarantees a lifetime income and perpetuates privileges over generations.

Functionalist theories also don't explain why (1) upward social mobility is more limited in the United States than in other economically privileged (and even some economically disadvantaged) countries; (2) so many college graduates can find only low-paying jobs; and (3) racial/ethnic income and wealth gaps persist across all social classes (Legatum Prosperity Index, 2013; McKernan et al., 2013).

8-6b Conflict Perspectives: Stratification Harms Society

Like Tumin, conflict theorists maintain that social stratification is dysfunctional because it hurts individuals and societies. Karl Marx's (1934) analysis of social class and inequality has had a lasting impact on modern sociology, especially conflict theory.

Capitalism Benefits the Rich

Marx was aware that a diversity of classes can exist at any one time, but he predicted that capitalist societies would ultimately be reduced to two social classes: the capitalist class, or bourgeoisie, and the working class, or proletariat. The **bourgeoisie**, those who own and control capital and the means of production (e.g., factories, land, banks, and other sources of income), can amass wealth and power. The **proletariat**, workers who sell their labor for wages, earn barely enough to survive.

For conflict theorists, the economic struggles of the U.S. middle and working classes since the late 1970s weren't primarily the result of globalization and technological changes but, instead, a long series of government policies that overwhelmingly favored the rich. This policy of **corporate welfare** consists of an array of subsidies, tax breaks, and assistance that the government has created for businesses. For example:

- Taxpayers paid $7 trillion to bail out mismanaged financial institutions. The companies' executives still receive multimillion-dollar annual salaries and benefits (Ivry et al., 2015; Kiel and Nguyen, 2015; refer also to Chapter 11).
- Each household pays about $10,000 per year to subsidize corporations. Companies benefit enormously from the taxpayer-supported highways they use, law enforcement and judicial systems that protect their intellectual and physical property, public education that provides their workforce, and military personnel who safeguard corporate assets abroad. Other costs include corporations' unpaid taxes, oil producers' pollution and environmental disaster clean-ups, and taxpayer-funded research that results in patent-protected technology that reaps massive corporate profits (Buchheit, 2015, 2017).
- At least 55 of the largest U.S. corporations paid no taxes in 2020. The corporate tax rate is 21 percent, well below the average employee's 30 percent tax rate. If these corporations had paid, they would have contributed $8.5 billion. The savings, along with $3.5 billion in rebates, saved the companies a total of $12 billion (Gardner and Wamhoff, 2021).

Considerable data support conflict theorists' contention that growing inequality is dysfunctional for society. Among other social problems, ongoing economic inequity intensifies poverty, undermines people's trust in political and economic institutions and, consequently, erodes national solidarity.

Whether their concern is driven by self-interest or a sense of justice, even powerful financiers and conservative economists have begun to worry about economic inequality. They fear that the increasing income and wealth gaps and low wages have made the United States less productive, have narrowed the consumer spending base, and have discouraged many people from entering or staying in the labor force. They also wonder if the huge gulf between the top 1 to 5 percent and the rest of society may lead to "oppressive taxes" on the superrich, widespread social unrest, and even a revolution (Dumaine, 2015; Hightower, 2015).

Critical Evaluation

Conflict theories have their limitations. First, and despite Marx's prediction, even though corporate wealth has surged during the past 100 years while many people have become less financially stable, there have been some protests but no revolutions in capitalist countries. Americans aren't demanding greater economic equality, and many go into debt to purchase luxury items like TVs, jewelry, and pet supplies and to replace expensive smartphones and other high-tech hardware every few years. Thus, according to critics, conflict theorists are exaggerating the existence and effects of economic inequality.

Second, some critics point out that conflict theorists overlook the fact that government programs have cut poverty by 40 percent since the 1960s. About 46 percent of American households pay no federal income taxes (and these tax breaks have nearly doubled since 1975) because they receive public assistance, get deductions for raising children under age 18, or don't report income. Moreover, the United States spends far more, per capita, than do many other wealthy countries on pensions, health care, education, unemployment, housing assistance, and similar benefits (Kirkegaard, 2015; "Inequality," 2017).

8-6c Feminist Perspectives: Stratification Primarily Benefits Men

For feminist scholars, functionalist and conflict theories are limited because they typically focus on men in describing and analyzing stratification and social class. As a result, women are largely invisible.

Gender Stratification and Inequality

Gender stratification—unequal access to wealth, power, status, prestige, and other valued resources because of one's sex—contributes to systemic inequality. At the top rungs, only 13 percent of the world's billionaires are women, and 83 percent have inherited their fortunes. Only 16 percent of the top 1 percent of Americans are women. "The higher up you move in the income distribution," according to an influential economist, "the lower the proportion of women" (Piketty et al., 2016; Frank, 2017, p. BU3; Wealth-X, 2017).

In both high- and low-income countries, gender stratification leads to discrimination in the economy, politics, and access to medical services. Especially in low-income

---

corporate welfare subsidies, tax breaks, and assistance that the government has created for businesses.

gender stratification unequal access to wealth, power, status, prestige, and other valued resources because of one's sex.

countries, almost 4 million girls and women are "missing" each year because of high death rates. About 20 percent are aborted because of a preference for sons, and more than a third die of childbirth and pregnancy complications. Besides excessive deaths, many women do much of the unpaid work (such as household chores and child rearing), earn less than men across all jobs, and hold only a minority of decision-making positions in public and private institutions (United Nations, 2015).

More than 100 countries ban women from certain jobs, accessing financial services, owning businesses, or conducting legal affairs. "Almost four out of five of those in forced commercial sexual exploitation are women or girls" (UN Women, 2014; International Labour Organization et al., 2022).

Feminist theorists contend that, in a patriarchal system, men dominate the stratification system: They control a disproportionate share of wealth, prestige, and power, and the feminization of poverty results in women's downward mobility. Women must often overcome economic inequities as well as juggle domestic and workplace responsibilities (refer to Chapters 11 and 12). The gender gaps in wealth, income, and household burdens make it harder for women to build a strong future for themselves and their families.

Critical Evaluation

Some critics point out that feminist theorists often focus only on economically disadvantaged women in showing how patriarchy affects stratification and social class (refer to Kendall, 2002). Another criticism is that many feminist scholars don't explain why so many women succeed despite patriarchal barriers.

8-6d Symbolic Interaction Perspectives: People Create and Shape Stratification

Symbolic interactionists focus on how people create, change, and reproduce social classes. They address micro-level issues such as how people learn their social positions in everyday life and how such learning affects their attitudes, behavior, and lifestyles.

Social Contexts Shape Social Class

People across social classes interact with and socialize their children differently. By age 2, children from higher-income homes know 30 percent more words than those from lower-income homes. Early vocabularies increase reading comprehension and proficiency as early as kindergarten (Fernald et al., 2013).

Children as young as 4 years old are aware of and enact their social class. Among preschoolers, upper-middle-class children speak, interrupt, ask for help, and argue more often than do working-class children. Such behavior receives more adult attention and gives upper-middle-class children more opportunities to develop their language skills (Streib, 2011).

Social contexts also affect mobility. The social classes into which children are born affect their aspirations, the skills they value and can access, and their networks and resources. Doctors' children, for example, are more likely to become physicians themselves because they're exposed to medical discussions at home, are encouraged to pursue medicine as an occupation, and are embedded in social networks that provide medical school contacts (Jonsson et al., 2009).

You've learned that U.S. economic inequality is at its highest level since the 1930s, yet many Americans are relatively unconcerned. Fully 62 percent are satisfied with their opportunities to get ahead, and only 47 percent say that the high–low income gap is a problem (DeSilver, 2014; Newport, 2016).

Why are so many people content instead of challenging policies that benefit primarily the top 1 percent? The average American knows very little about economic inequality (e.g., whether it's increasing or decreasing), where they are on the income spectrum, and believes (incorrectly) that there's more upward than downward mobility. Also, what people think they know is often wrong (Gimpelson and Treisman, 2015; Kraus and Tan, 2015). Thus, and because popular culture reinforces the idea that inequality is normal and inevitable, many people don't understand the connection between structural inequality and their personal situation.

Many Americans aren't angry about rising inequality, moreover, because they embrace "mobility optimism"—a belief in the American Dream that anyone can succeed, and that they (or their children) will be rich someday (Manza and Brooks, 2016; refer also to Davidai and Gilovich, 2015). Political speeches and social media, particularly, fan beliefs that the American Dream is achievable.

Critical Evaluation

Symbolic interaction theories help us understand the everyday processes that underlie social stratification, but there are several weaknesses. First, interactionism, unlike the other perspectives, doesn't explain *why* stratification exists. Second, the theories don't explain why—despite the same family background, resources, and socialization—some siblings are considerably more upwardly mobile than their brothers and sisters. Third, conflict theorists, especially, fault symbolic interactionists for ignoring structural factors—like the economy, government policies, and educational institutions—that create and reinforce inequality.

Chapter Review

Key Takeaways

8-1 What is social stratification and its main components? *Social stratification* is a society's ranking of people who have different access to valued resources. A *closed stratification system* differs from an *open stratification system* because the latter allows movement from one social class to another. Stratification includes *wealth, prestige,* and *power.* People are more likely to experience status inconsistency if they rank differently on these three dimensions, such as a football player who has great wealth but little power.

8-2 What are the main aspects of social class in the United States? A good indicator of social class is *socioeconomic status (SES),* an overall rank based on a person's income, education, and occupation. Using SES and other variables (such as values, power, and *conspicuous consumption*), most sociologists agree that there are at least four social classes in the United States: upper, middle, working, and lower. These groups can be divided further into upper-upper, lower-upper, upper-middle, lower-middle, and the working class. The lower class includes the *working poor* and the *underclass.* A major outcome of social stratification is *life chances.*

Example: Restaurant Menus and Stratification

Two sociologists—in Iowa and Virginia—asked their students in introductory sociology classes to do a content analysis of 10 menus that represented a sampling of restaurants by social class. The students found that the restaurants that catered to upper-class clientele had higher than average entrée prices, described the entrées in foreign languages, used fancy sauces, recommended expensive wines, and had few illustrations. Menus at restaurants geared towards middle-income diners emphasized "value for the dollar," presented photos of entrées with "bountiful plates overflowing with appetizing food," and popular items such as quesadillas. Menus at restaurants catering to lower-income clientele featured low prices ($3 to $10 entrées), the items were numbered, none of the entrées had "pretentious names," and the typesetting was simple (Wright and Ransom, 2005). In effect, then, even menus denote social class and social status.

8-3 What are the key aspects of sociological explanations of poverty? *Absolute poverty* is a more serious social problem than *relative poverty* because millions of Americans live below the *poverty line*. Poverty varies by age, gender, family structure, race, and ethnicity. One explanation for poverty maintains that people are economically disadvantaged because of individual failings; another contends that societal structures create and sustain poverty.

8-4 How is social mobility described? *Social mobility* can be *horizontal, vertical, intragenerational,* or *intergenerational.* There has been a sharp decline in the share of middle-income adults since 1971 because many have slid into a lower class. Structural, demographic, and individual factors affect a person's social mobility. Much social mobility depends on structural, demographic, and family background factors, all of which are interrelated.

8-5 How is global stratification described? Global inequality is widespread, and all societies are stratified, but some countries are much wealthier than others, and inequality is increasing across the globe. Historically, currently, and across all nations, women and children experience the greatest poverty. Sociologists use *modernization theory, dependency theory,* and *world-system theory* to explain why inequality is universal.

8-6 What are the differences among the main sociological explanations of stratification? Refer to Table 8.1 on page 167.

Key Terms

Social stratification (p. 152)
Slavery system (p. 152)
Caste system (p. 152)
Class system (p. 153)
Social class (p. 153)
Wealth (p. 153)
Income (p. 153)
Prestige (p. 154)
Power (p. 155)
Socioeconomic status (SES) (p. 155)
Working poor (p. 157)
Underclass (p. 157)
Absolute poverty (p. 158)
Relative poverty (p. 158)
Poverty line (p. 158)
Feminization of poverty (p. 159)
Social mobility (p. 161)
Intragenerational mobility (p. 162)
Intergenerational mobility (p. 162)
Global stratification (p. 164)
Infant mortality rate (p. 165)
Davis–Moore thesis (p. 168)
Meritocracy (p. 168)
Bourgeoisie (p. 168)
Proletariat (p. 168)
Corporate welfare (p. 169)
Gender stratification (p. 169)

Discussion and Critical Thinking Questions

1. As Chapter 8 demonstrates, U.S. income and wealth inequality is rising, and the median family income is lower now than 30 years ago. Nonetheless, 58 percent of Americans, compared with 49 percent in 2008, say that the nation is not divided into "haves" and "have-nots" (Jones, 2019).

a. Why do so many Americans deny the existence of income and wealth inequality? Lack of understanding? A belief that they might be rich some day? Other reasons?

b. Some politicians have called for increasing taxes on the rich to help redistribute income. Why have most Americans been mute about such issues?

2. Career counselors, commencement speakers, and others often encourage us to "Follow your dreams!" and "Do what you love!" Is such advice sound? For example, entry level careers in performing arts pay an average annual salary of $39,000, and those in theology pay $36,000 (Picchi, 2023). Parents and others also tell children "You can be anything you want." Is this true? What about economic, social, and cultural capital?

3. Is it possible or desirable to establish a truly classless society? Why or why not?

4. Do you agree or disagree with the Davis–Moore thesis that stratification benefits society? Explain why.

5. Do you expect that, in the future, you'll experience horizontal or vertical mobility? What about intragenerational and intergenerational mobility? Explain why or why not.

True or False?

How Much Do You Know about U.S. Economic Equality?

The answers to 4 and 6 are true and the rest are false.

1. The percent of people in the middle class in the United States has been declining.
2. A higher level of prestige does not always mean higher income. Funeral directors have relatively low prestige, but most have higher incomes than college professors, who have relatively high prestige.
3. Children have the highest poverty rates.
4. The wealth gap in the United States has been growing, and the top 1 percent of Americans own 31 percent of all wealth.
5. Many Americans are economically disadvantaged, but poverty has decreased since the 1980s.
6. The United States falls behind more than 60 percent of nations worldwide for income equality.

9 Gender and Sexuality

Learning Objectives

After studying this chapter, you will be able to…

9-1 Differentiate between sex and gender.

9-2 Describe sexual identity and sexual orientation within the context of society's changing perspectives.

9-3 Explain how gender stratification affects the family, education, workplace, and politics.

9-4 Describe contemporary sexual attitudes and practices, including sexual scripts and double standards.

9-5 Summarize changing viewpoints and policies related to abortion and same-sex marriage trends.

9-6 Describe gender and sexual inequality across cultures.

9-7 Compare the theoretical explanations of gender and sexuality.

Women pay more than men, sometimes twice as much, for many things, including cars, mortgages, health care, high-end jeans from the same designer, and similar grooming products like moisturizers, deodorants, and even razors (Hill, 2015; Ngabirano, 2017). Why? This chapter examines how gender and sexuality affect our lives.

Be sure to take the True or False quiz to better understand how much you know about these topics.

What do you think?

Today, women have more employment advantages than men.

1	2	3	4	5	6	7
	strongly agree				strongly disagree	

True or False?

How Much Do You Know about Gender and Sexuality?

1. Approximately 25 percent of the U.S. adult population identifies as LGBTQIA+.
2. Few American teens use contraceptives.
3. Women make up close to 50 percent of Congress in the United States.
4. Women earn approximately 82 cents for every 1 dollar earned by a man.
5. U.S. abortion rates have increased since 1990.
6. In heterosexual two-parent families, women still do more childcare and housework than men.

The answers are at the end of the chapter.

9-1 Sex, Gender, and Culture

Sex and gender aren't synonymous. Some differences between people are biological; others are social creations. Both biology and social factors shape a person's identity and their experiences related to sex and gender.

9-1a How Sex and Gender Differ

Many people use *sex* and *gender* interchangeably, but they're not the same. **Sex** refers to the biological characteristics with which we are born—chromosomes, anatomy, hormones, and other physical and physiological attributes that are then used to assign us to a specific sex category such as male, female, or both. These attributes influence our behavior but *don't determine* how we think or feel. Gender is a more complex concept than sex.

Gender refers to learned attitudes and behaviors that characterize people based on social and cultural expectations associated with sex categories. Gender, therefore, is not based on physical traits. Thus, most people are *born* with traits that are used to assign them as either male or female at birth, but people *learn* to be women, men, neither, or both through internalizing the behavior patterns expected of each sex. In many societies, for example, women are expected to look young, thin, and attractive, and men are expected to amass as much wealth as possible.

9-1b Sex: Our Biological Component

Physical characteristics like breasts and beards may be indicators of whether someone is male or female, but sex categories aren't always clear-cut and are influenced by cultural ideas and language. Our cultural expectations dictate whether a person is female or male, but these categories are limited and do not adequately encompass all people. For example, people who are **intersex** have sex characteristics that fit those typically associated with both male and female. About 1 in 1,500 or 2,000 children born each year are classified

---

sex the biological characteristics with which we are born that are then used to assign us to a specific sex category such as male, female, or both.

gender learned attitudes and behaviors that characterize people based on social and cultural expectations associated with sex categories.

intersex having sex characteristics that fit with those typically associated with both male and female.

as intersex because they're born with both male and female external genitals or an incomplete development of internal reproductive organs. Some parents seek surgery; others wait until a child is old enough to decide what to do (Interact, 2022; Bendavid, 2013).

9-2 Sexual Identity and Sexual Orientation

Sexual identity is an awareness of our gender and how we express our sexual values, attitudes, and feelings.. Sexual identity incorporates a **sexual orientation**—an emotional or sexual attraction to sexual partners of the same sex, of another sex, of more than one sex, or of neither sex.

- **Heterosexual**, often referred to as being *straight*, is an emotional and sexual attraction to people of the opposite sex—here, "opposite" is in the context of a society where norms are built around a binary (two part) system of male and female.

LGBTQIA+ is an acronym that unifies many categories of sexual orientations and identities for the purpose of addressing shared concerns and for supporting efforts to build more inclusive cultural norms and attain greater rights within society. The acronym stands for lesbian, gay, bisexual, transgender, queer/questioning, intersex, asexual and plus (+) to include many other categories (such as non-binary).

- **Gay** refers to emotional and sexual attraction to people of the same sex. While gay may be used to describe both men and women, some gay men prefer to be called *gay* and some gay women prefer to be called *lesbian(s)*. *Lesbian* refers to a woman (or girl) who is emotionally and sexually attracted to people of the same sex.
- **Bisexual**, sometimes called *bi*, is an emotional and sexual attraction to more than one gender.
- *Pansexual* is an emotional and sexual attraction to a person regardless of gender.
- **Asexual** is having little interest in or desire for sex.

Sexual orientation, like biological sex, isn't as clear-cut as many people believe. Alfred Kinsey (1948) and his associates' classic study found that most people weren't exclusively heterosexual or gay. Instead, they fell somewhere along a continuum in terms of sexual desire, attractions, feelings, fantasies, and experiences. Researchers have recently added *asexual* to Kinsey's classification.

Most people's sexual identity corresponds with their biological sex, sexual attraction, and sexual behavior, but not always. Among Americans ages 15 to 44, for example, 86 percent *identify themselves* as straight, but 23 percent have had *same-sex experiences* (Moore, 2015).

How Many Americans Are LGBTQIA+?

From 2012 until 2022, the share of adults identifying as lesbian, gay, bisexual, or transgender (LGBT) has increased from 3.5 percent to 7.2 percent (**Figure 9.1**). There are generational differences with LGBT identification. In 2022, Generation Z adults (born 1997 through 2013) were the most likely age cohort to identify as LGBT (19.7 percent). In comparison, members of the Silent Generation (born 1945 and before) were least likely to identify as LGBT (1.7 percent) (**Table 9.1**) (Jones, 2023).

The LGBT population may have actually increased since 2012, but there are other explanations for the higher numbers. More Americans (62 percent) now report being satisfied with the nation's acceptance of people identifying as gay or lesbian as compared to in 2020 (56 percent) (Jones, 2022). Greater acceptance may encourage more people to identify as LGBT.

What Determines Our Sexual Orientation?

A Hong Kong billionaire offered $65 million to any man who succeeded in marrying his daughter after she eloped with her female partner. The offer attracted 20,000 suitors, but none were successful (Nichols, 2014). Like this father, 30 percent of Americans (down from 56 percent in 1978) believe that being gay or lesbian is a "personal choice" or due to a person's upbringing (Jones, 2015).

Culture shapes people's sexual attitudes and behavior, but there isn't a singular answer for why people are straight, gay, bisexual, or asexual. Sexual orientation

---

sexual identity an awareness of our gender and how we express our sexual values, attitudes, and feelings.

sexual orientation an emotional or sexual attraction to partners of the same sex, of another sex, of more than one sex, or of neither sex.

heterosexual emotional and sexual attraction to people of the opposite sex in a society where norms are built around a binary system of male and female.

gay emotional and sexual attraction to people of the same sex.

bisexual emotional and sexual attraction to more than one gender.

asexual having little interest in or desire for sex.

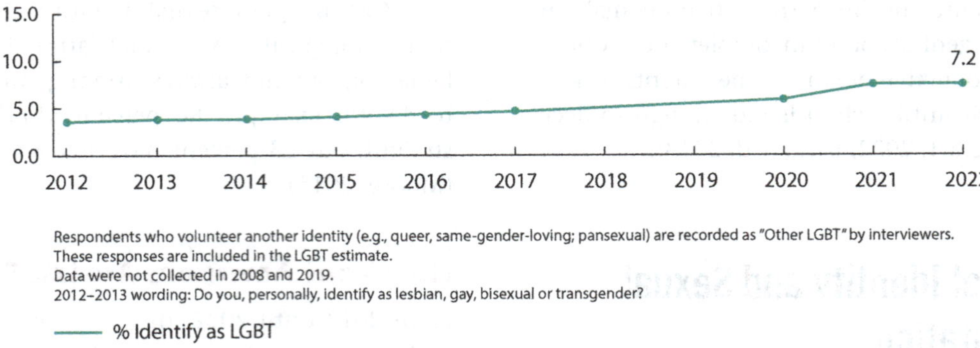

Figure 9.1 Americans' Self-Identification as Lesbian, Gay, Bisexual, Transgender, or Something Other Than Heterosexual, 2012-2022

Source: Jones, 2023 "U.S. LGBT Identification Steady at 7.2%."

must have biological roots, according to some researchers, because gay and lesbian sexuality exists in all societies and, across cultures, the gay population is roughly the same—about 5 percent (Barash, 2012).

There's also growing scientific consensus that biological factors, particularly the early influence of sex hormones after conception and around childbirth, have a strong effect on sexual orientation (refer to LeVay, 2011, for a comprehensive summary of the studies). Other researchers speculate that a combination of genetic and cultural factors influence sexual orientation (Slater, 2013).

9-2a Gender: Our Cultural Component

Gender doesn't occur naturally but is socially constructed. This means that gender aspects may differ across time, cultures, and even groups within a society. Let's begin with gender identity.

Gender Identity

People develop a **gender identity**, a perception of their own gender, early in life. Many Mexican baby girls, but not boys, have pierced ears, for example, and hairstyles and clothing for American toddlers often differ by sex. In many societies, there is an expectation that a feminine gender identity will correspond with a female biological sex and a masculine gender identity will be associated with male biological sex. It is more accurate to note that gender identity may or may not correspond to a person's assigned biological sex. Most importantly, gender identity is part of our self-concept and usually remains relatively fixed throughout life.

Transgender is an umbrella term for having a gender identity that varies from the sex assigned to a person at birth. Approximately 1.3 million people in the United States (0.5% of adults and 1.4% of 13- to 17-year-olds) identify as transgender (Allen, 2022).

Because transgender is independent of sexual orientation, people may also identify as heterosexual, gay, bisexual, or asexual. Facebook users can now choose their gender identity from more than 50 possibilities, but here are some of the most common transgender categories (American Psychological Association, 2014):

- *Transsexual* people have permanently changed, or plan to change, their physical sex through medical intervention. Unlike transgender, transsexual is not an umbrella term to describe a person whose gender identity differs from their assigned sex. Transsexual is largely considered an outdated term and should only be used to describe someone who openly identifies as such.

Table 9.1 U.S. Adults Identifying as LGBT by Birth Year (Age Cohort)

Birth Year (Age Cohort)	% LGBT
1945 and earlier (Silent Generation)	1.7
1946 through 1964 (Baby Boomer)	2.7
1965 through 1980 (Generation X)	3.3
1981 through 1996 (Millennial)	11.2
1997 through 2004 (Generation Z)	19.7

Source: Jones, 2023.

gender identity a perception of one's own gender.

transgender a gender identity that varies from the sex assigned to a person at birth.

Paul J. Richards/AFP/Getty Images

Olivier Douliery/Abaca Press/Washington/USA/Newscom

Kristin Beck, who received multiple military awards and decorations, retired from the elite U.S. Navy SEALs in 2011. A few years later, she became openly transgender, changing her name from Christopher to Kristin Beck.

- *Genderqueer* people identify their gender as falling somewhere on a continuum between female and male or have a combination of gender identities and sexual orientations.

Gender expression is how a person communicates gender identity to others and includes behavior, clothing, hairstyles, voice, and body characteristics. Cross-dressing, girls' frilly dresses, and men's business suits are all examples of gender expression. Even if a person's gender identity is constant, gender expression can vary from situation to situation and change over time. For example, more men are getting cosmetic treatments (surgical and non-surgical), including body contouring and facial procedures. In 2018, more than 1.3 million cosmetic procedures were performed on men, a 29 percent increase from 2000 (American Society of Plastic Surgeons, 2019). In addition, half of American men now routinely use moisturizers, facial creams, or self-tanning lotions and sprays, and some of the National Basketball Association's "toughest players" have promoted products for Dove, La Mer, and other skin-care companies (Boyle, 2013; Holmes, 2013). Another form of gender expression is cross-dressing. People who cross-dress wear clothing that's traditionally or stereotypically worn by another gender in their culture. Those who cross-dress are usually comfortable with their assigned sex and don't wish to change it.

Gender Roles, Gender Stereotypes, and Sexism

Gender roles are the characteristics, attitudes, feelings, and behaviors that society expects of girls/women and boys/men. As you read in Chapter 4, a major purpose of socialization, which begins at birth, is to teach people gender roles. As a result, people learn societal expectations for gender through interactions with family members, teachers, friends, and the larger society.

Americans are more likely now than in the past to pursue jobs and other activities based on their abilities and interests rather than their sex. For the most part, however, most societies still have fairly rigid gender roles and widespread **gender stereotypes**—expectations about how people should look, act, think, and

Mike Theiler/Getty Images News/Getty Images

How Do Gender Roles Differ? While the world scrutinized Michelle Obama's gowns, dresses, shoes, jewelry, and hairdos, President Obama wore the same tuxedo and shoes during their eight years in the White House (Feldman, 2017).

gender expression how a person communicates gender identity to others.

gender roles the characteristics, attitudes, feelings, and behaviors that society expects of girls/women and boys/men.

gender stereotypes expectations about how people will look, act, think, and feel based on their sex.

feel based on their sex. People tend to associate stereotypically female characteristics with weakness and stereotypically male characteristics with strength. Consider, for example, how often people describe the same behavior differently for women and men:

- He's firm; she's stubborn.
- He's good with details; she's picky.
- He's honest; she's opinionated.
- He's raising good points; she's complaining.
- He's experienced; she's "been around."
- He's enthusiastic; she's shouting.

Gender stereotypes fuel **sexism**, an attitude or behavior that discriminates against one sex (usually the female sex) based on the assumed superiority of another sex (typically male sex). In 1967, Kathrine Switzer was the first woman to officially run and complete the Boston Marathon. At that time, women were not allowed in the race based on the idea that they could not manage longer distances. Kathrine was able to sign up because she used her initials (versus first name) on the event registration. Although Kathrine finished the race, during the event, she was physically pushed by a man trying to get her off the course (Fell, 2017).

There is a difference in opinion on equal opportunity regardless of sex. A smaller percentage of women (33 percent), compared with 61 percent of men, believe that "women have equal job opportunities as men." Satisfaction with women's equality in the workforce has been declining in recent years, and it is at its lowest point since 2002 (Brenan, 2021).

Men also experience negative outcomes of gender stereotypes. Here's what one of my students wrote during an online discussion of gender roles:

> Some parents live their dreams through their sons by forcing them to be in sports. I disagree with this but want my [9-year-old] to be "all boy." He's the worst player on the basketball team at school and wanted to take dance lessons, including ballet. I assured him that this was not going to happen. I'm going to enroll him in soccer and see if he does better.

Is this mother suppressing her son's natural dancing talent? We'll never know because she, like many parents, expects her son to fulfill oppressive gender roles that meet with society's approval.

sexism an attitude or behavior that discriminates against one sex (typically women) based on the assumed superiority of another (typically men).

Colin McConnell/Toronto Star/Getty Images

At an early age, sex-associated activities prepare girls and boys for future adult roles. As a result, there are few boys and men who are ballet dancers and few women who are auto mechanics.

9-2b Societal Reactions to LGBTQIA+ People

People's attitudes toward LGBTQIA+ people are mixed. There's been greater acceptance in some countries, but considerable repression in others.

Greater Acceptance, But ...

Australian passports and birth certificates designate male, female, and transgender. In India, the 2011 national census for the first time offered three options: male, female, or the more inclusive category of "third sex." In Thailand, which has the world's biggest transgender population, an airline recruits "third sex" flight attendants. In the United States, in 2017, Oregon was the first state to allow residents to mark their sex as "not specified" on a driver's license. As of 2022, in the United States, there are 22 states and territories that allow an X for nonbinary gender (Migdon, 2022).

In the United States, many jurisdictions, corporations, and small companies now extend more health care and other benefits to gay employees and their partners than to unmarried heterosexual employees who live together. The U.S. Supreme Court and a growing number of states have legalized same-sex marriages, and large numbers of Americans support equal rights for LGBTQIA+ people in the workplace and elsewhere (Von Drehle, 2014).

In 2012, the Army promoted the first openly gay female officer to brigadier general. In 2013, the Pentagon added benefits for same-sex partners, including services on U.S. military bases. Federal workers and Medicare recipients are now eligible for Gender Confirmation Surgery (GCS) (Mach and Cornell, 2013; "Transgender Rights," 2015). And, since the mid-1990s, many LGBT characters have appeared in leading and supporting

Jonathan Daniel/Getty Images Sport/Getty Images

In mid-2013, the Washington Wizards' Jason Collins appeared on the cover of *Sports Illustrated*. "I'm a 34-year-old NBA center, I'm Black, and I'm gay," he announced. Collins said that his teammates were supportive, but he retired a year later. Ten years later, there are still very few openly gay men in professional sports (Lipka, 2014; Hernandez, 2021).

roles in popular TV programs (e.g., *Transparent, Modern Family, Gotham, This Is Us, Empire,* and *Game of Thrones*). As of 2022, almost 12 percent of regular characters on primetime scripted shows were LGBTQ, which is an increase from the previous year (Deerwater, 2022)

There's greater LGBTQIA+ acceptance, yet, as we will explore soon, Americans' views on gender identity and transgender issues are varied. Americans are about evenly divided on two issues: whether wedding-related businesses (like caterers and florists) should be required to serve same-sex couples and whether transgender people should be able to use public restrooms that correspond to their gender identity rather than their birth sex (Masci, 2016; McCarthy, 2017; Parker et al., 2022).

... Also Widespread Intolerance

According to one scholar, what makes gay people different from others is that "we are discriminated against, mistreated, [and] regarded as sick or perverted" (Halperin, 2012: B17). **Heterosexism**, a belief that heterosexuality is the only legitimate sexual orientation, pervades societal practices, laws, and institutions. For example, 13 percent of Americans—including judges, religious leaders, and politicians—want to undo recently achieved rights like gay marriage (McCarthy, 2016). Heterosexism can trigger **homophobia**, a fear and hatred of lesbians, gay men, and other non-heterosexual sexualities.

Homophobia often takes the form of *gay bashing:* threats, assaults, or acts of violence directed at people who are LGBTQIA+. Of the nearly 8,000 single bias hate crimes reported to the police in 2021, 16 percent were based on sexual orientation bias, but a lot of gay bashing isn't reported (refer to Chapter 7).

Younger people who are lesbian, gay, or bisexual are more likely to face many challenges at school. More LGB high school students report being bullied on school property (32 percent) compared with heterosexual students (17.1 percent). LGB students are also more likely to report being cyberbullied (26.6 percent) compared with 14.1 percent of heterosexual students. And more LGB students (13.5 percent) than heterosexual students (7.5 percent) report that safety concerns contributed to not going to school. LGB youth were almost four times more likely to report that they had attempted suicide (23.4 percent) compared to their heterosexual peers (6.4 percent) (Centers for Disease Control and Prevention, 2020).

Without warning, in mid-2017, President Trump tweeted that the government "will not accept or allow transgender individuals to serve in any capacity in the U.S. Military." (LeBlanc, 2017). This policy was reversed by President Biden in January 2021, allowing transgender people to openly serve in the U.S. military (Baldor and Miller, 2021).

9-3 Contemporary Gender Inequality

The World Economic Forum determined that it would take 132 years of change efforts to achieve worldwide gender equality; this considers education, economic and political engagement, and health and safety (Alves, 2023). You learned in Chapter 8 that there's still widespread gender stratification because of sex. *Gendered institutions* are social structures that enable and reinforce gender stratification. Let's begin with the family, remembering that institutions are interrelated (refer to Chapter 6).

9-3a Gender and Family Life

About 56 percent of married adults—with and without children—say that sharing household chores is "very important" to a successful marriage (Geiger, 2016). In heterosexual marriages, men do more at home than in the past, but time devoted to key activities is still varied along gender lines. Even in marriages where women earn roughly the same amount as their male partners, there are differences. Men spend more hours each week in paid work than women, do less

heterosexism belief that heterosexuality is the only legitimate sexual orientation.

homophobia a fear and hatred of lesbian women, gay men, and people with other non-heterosexual sexualities.

caregiving and housework, and spend more time on leisure (**Figure 9.2**). Perceptions of men's and women's value in different areas of life also differ. Fewer Americans say men's contributions at home and work are equally valued (35 percent), while more say this about women (49 percent) (Fry et al., 2023).

About 58 percent of parents say that parenting is "extremely important" to their identity, but parenting tasks are also gendered. Women do less housework than they used to but devote twice as much time as men to child care. In two-parent, heterosexual families, fathers tend to do more of the enjoyable tasks (e.g., reading to children, playing with them, taking them to games) and are much more likely to "join in" with child care than to "take over" from mothers. In contrast, mothers do most of the daily, nonstop tasks like picking children up from school or day care and feeding, bathing, and putting them to bed. Compared with fathers, mothers' greater investments in children result in less happiness, more stress, and greater fatigue (Craig, 2015; Musick et al., 2016; Parker and Livingston, 2016).

Every year, media organizations feature and applaud stay-at-home dads. The numbers are growing but are still negligible. Out of children under 15 years old and living in married-couple families, 24 percent have a stay-at-home mom, while only 1 percent have a stay-at-home dad (U.S. Census Bureau, 2021). A stay-at-home dad is usually a temporary role that is due to unemployment or health problems. In some cases, however, White, college-educated, upper-middle-class men choose to be stay-at-home dads because their partners or wives have high incomes (Kane, 2015; "Father's Day ...," 2017).

For many people, there's nothing sexier than a partner who does housework.

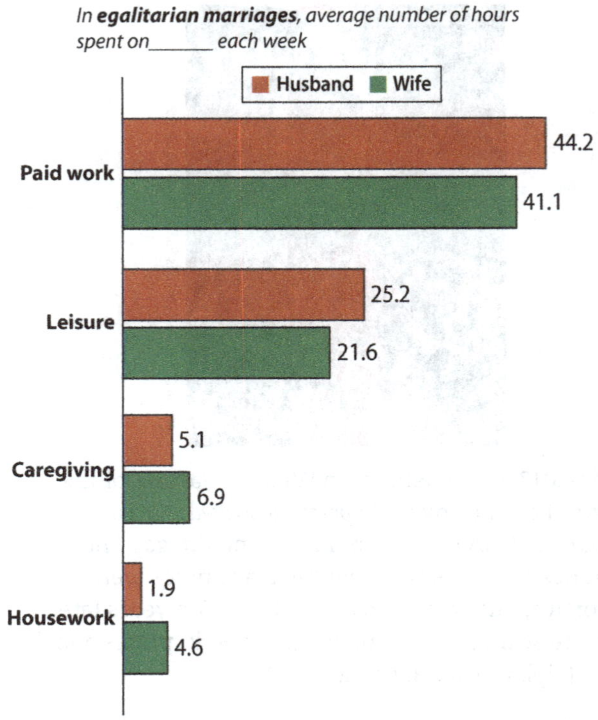

Note: Earnings analysis limited to opposite-sex marriages including spouses ages 25 to 64. In an egalitarian marriage, each spouse earns 40%–60% of the couple's joint earnings. A primary breadwinner earns more than 60% of the couple's joint earnings. A sole breadwinner earns 100% of the couple's earnings.

Figure 9.2 How Men and Women in Heterosexual Marriage Spend Their Time

Source: Fry, et al., 2023.

9-3b Gender and Education

Despite substantial progress, there are gender differences at all educational levels. In public K–12 schools, as rank and pay increase, the number of women decreases. Women fill almost 80 percent of teaching positions but only slightly more than half of principal positions. Although there are now women in school leadership, for similar positions, women earn around $1000 less than their male counterparts (Superville, 2021).

Because women across all racial and ethnic groups are more likely than men to finish college, some observers have described this phenomenon as "the feminization of higher education." For more than a decade, women have been earning a majority of the doctoral degrees awarded at U.S. universities (Perry, 2021). Even when women earn doctoral degrees in male-dominated STEM (science, technology, engineering, and math) fields, they're less likely than men to be hired (refer to Chapter 13). Once hired, women are less likely to be promoted; as the academic rank increases, the number of women faculty decreases. Out of full-time faculty who are men, almost one-third are (full) professors while

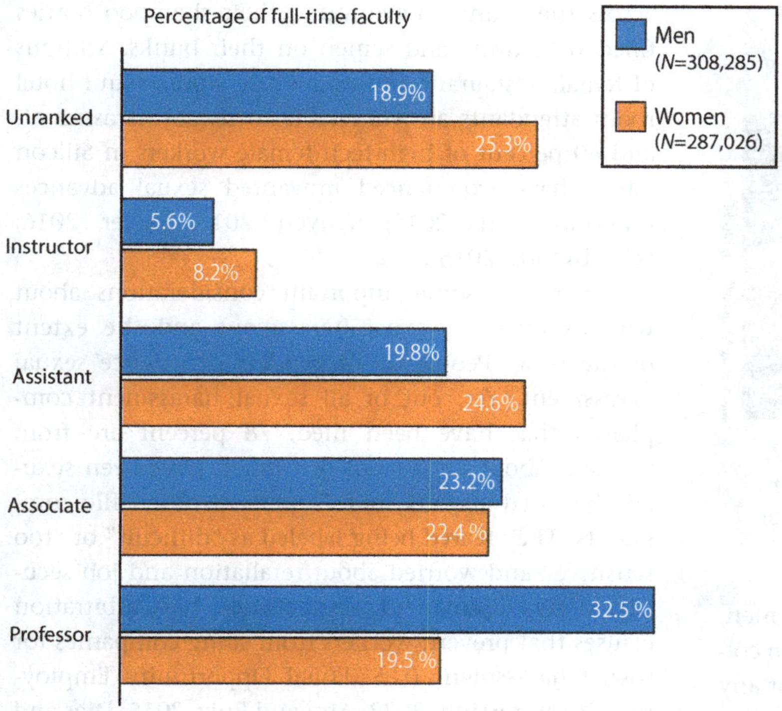

Note: Figures represent full-time nonmedical noninstructional staff in degree-granting nonprofit institutions participating in Title IV federal financial aid programs.

Figure 9.3 Percentage of Men and Women Professors by Faculty Rank, 2020

Source: Based on American Association of University Professors. "The Annual Report on the Economic Status of the Profession, 2021–22." 2022, Figure 10.

only one-fifth of full-time professors who are women earn this title (**Figure 9.3**). In addition, across all levels, professors who are women earn less than men—the average salary for men is $113,000 compared to $93,00 for women (American Association of University Professors, 2022). Such data contradict the description of higher education as feminized.

9-3c Gender and the Workplace

There has been progress toward greater workplace equality, but we still have a long way to go. In the United States (as around the world), many jobs are segregated by sex, there are ongoing gender pay gaps, and numerous women experience sexual harassment.

Occupational Gender Segregation

Occupational gender segregation (sometimes called *occupational sex segregation*) is the process of channeling people into different types of jobs based on gender stereotypes.. As a result, a number of U.S. occupations are filled almost entirely by either women or men. Within education, training, and library occupations for people aged 16 years and older, 73 percent of employees are women. In comparison, within computer and mathematical occupations, almost 73 percent of employees are men (Bureau of Labor Statistics, 2023).

The issue isn't people working in different spaces or locations but that male-dominated occupations usually pay higher wages. And, as in education, women are much less likely than men to move up the occupational ladder (refer to Chapters 8 and 11).

The Gender Pay Gap

On average, full-time, year-round working women earn 82 cents for every dollar men earn (Aragão, 2023). To state this differently, *the average woman must work almost nine extra weeks every year to make the same wages as a man.*

This earnings difference between women and men is the **gender pay gap** (also called the *wage gap, pay gap,* and *gender wage gap*). Among year-round, full-time workers, compared with White men, women from all racial and ethnic groups earn less (82 percent): Asian women, 93 percent; White women, 83 percent; Black women, 70 percent; and Latina (Hispanic women), 65 percent. The pay gap is narrower for younger women aged 25–34. Lower wages and salaries reduce women's savings, purchasing power, and quality of life, and they receive less income from Social Security and pensions after retirement (AAUW, 2017; BLS News Release, 2017; BLS Reports, 2017; Kochhar, 2023; Aragão, 2023).

Why is there a gender pay gap? Women tend to enter fields with lower earnings (e.g., health care and education), while men are more likely to major in higher paying fields (e.g., engineering and computer science). As you learned earlier, women on average also work fewer hours at paid labor than men, primarily to care for children or other family members. However, there's a pay gap after controlling for a number of variables—including occupation, hours worked, GPA, age, experience, and marital status. Even a college degree does not level the playing field. Not only do women earn less than men at all educational levels, the higher the education level, the bigger the gender pay gap. Note that, as a group, women with advanced degrees earn

occupational gender segregation (sometimes called *occupational sex segregation*) the process of channeling people into different types of jobs based on gender stereotypes.

gender pay gap the difference between men's and women's earnings (also called the *wage gap, pay gap,* and *gender wage gap*).

Marmaduke St. John/Alamy Stock Photo

Men constitute around one-fifth of teachers in K–12. A major reason is low salaries compared with other occupations, but gender stereotypes are also a factor (Rich, 2014; Bureau of Labor Statistics, 2017).

less than men with a college degree. Compared to men, women are now more likely to have graduated from college. Still, the gender pay gap continues and is not any narrower than the gap between women and men without college degrees (AAUW, 2013; Blau and Kahn, 2016; Goldin et al., 2017; Kochhar, 2023).

A year after they graduate, women with Ph.D.s in science and engineering earn 31 percent less than men. The pay gap disappears when women receive doctorates in better-paid fields (engineering versus mathematics or chemistry), *and* work in industry rather than government or higher education, *and* don't marry (to avoid moving because of a husband's job), *and* don't have children (Buffington et al., 2016; Goldin et al., 2017).

Sexual Harassment

Sexual harassment is any unwanted sexual advance, request for sexual favors, or other conduct of a sexual nature that makes a person uncomfortable and interferes with their work. It includes *verbal behavior* (e.g., pressure for dates or the threat of sexual violence), *nonverbal behavior* (e.g., indecent gestures, display of sexually explicit posters, photos, or drawings), and *physical contact* (e.g., pinching, touching, or groping).

Sexual harassment occurs at all occupational levels. Women employed at the National Park Service have reported being groped, propositioned, verbally abused, and threatened with retaliation if they refused or reported incidents to supervisors. Female firefighters across the country have found their shampoo bottles filled with urine and semen on their bunks. Millions of female restaurant and home care workers and hotel room attendants are vulnerable to sexual harassment, and 60 percent of high-tech female workers in Silicon Valley have experienced unwanted sexual advances (Vassallo et al., 2015; Nguyen, 2016; Parker, 2016; Schrobsdorff, 2016).

There are some important considerations about who experiences sexual harassment and the extent of the issue. People of all genders experience sexual harassment. Yet, out of all sexual harassment complaints that have been filed, 78 percent are from women. About 25 percent of women have been sexually harassed at work, but 71 percent didn't file complaints. They feared being labeled as "difficult" or "too sensitive" and worried about retaliation and job security. Many employment contracts now have arbitration clauses that prevent workers from suing companies for sexual harassment (U.S. Equal Opportunity Employment Commission, 2022; Ahn and Ruiz, 2015; Dias and Dockterman, 2016).

9-3d Gender and Politics

Much of the existing information about gender and politics focuses on the gender binary of women and men. Hence, in this section, the focus is on the two categories for which data is readily available. The United States has lagged behind some other nations in terms of representation of women in political office. Vice President Kamala Harris holds the highest office ever held by a woman in the United States. Twenty-eight percent of members of the 118th Congress of the United States are women. This is an increase of 59 percent from a decade ago (Leppert and DeSilver, 2023). In other important elective offices (governor, mayor, state legislator), only a handful of the decision makers are women (**Table 9.2**). In 2022, women made up around one-quarter of the mayors in the 100 largest cities in the U.S., which is a slight increase from two years prior (Ceron, 2022). While women are not equally represented in all political positions, there have been improvements, moving the U.S. up in the Women's Power Index ranking (Vogelstein and Fliegauf, 2021).

Women's voting rates in the United States have been higher than men's since 1984. Why, in contrast, are there so few women in political office? There's a combination of reasons: (1) Women run for office at a far lower rate than men with similar credentials because women don't consider themselves qualified; (2) they have to do more than their male counterparts to prove themselves; (3) U.S. presidents appoint

sexual harassment any unwanted sexual advance, request for sexual favors, or other conduct of a sexual nature that makes a person uncomfortable and interferes with their work.

Table 9.2 U.S. Women in Elective Offices, 2022

Political Office	Total Number of Office Holders	Number (Percentage) Who Are Women
Senate	100	25 (25%)
House of Representatives	440	128 (29%)
Governor	50	12 (24%)
State Legislator	7383	2414 (33%)
Lieutenant Governor	43	21 (49%)
Mayor (100 largest cities)	100	33 (33%)

Source: Based on Center for American Women and Politics (CAWP), 2023.

more men than women to important positions; and (4) there's a lingering sexism, among both men and women, that women who are politicians are both less feminine and compassionate than the average woman and lack the leadership traits associated with men who are politicians (e.g., confident, assertive) (Schneider and Bos, 2014; Parker and Horowitz, 2015; refer also to Chapter 11).

9-4 Sexuality

Sexuality is considerably more complex than just having sex because it's a product of our sexual identity, sexual orientation, and sexual scripts, and includes desire, expression, and behavior.

9-4a Contemporary Sexual Attitudes and Practices

Sex doesn't "just happen." It typically progresses through a series of stages such as approaching, flirting, touching, or asking directly for sex. Sexual attitudes and behavior can vary from situation to situation and change over time, including the reasons why we have sex.

Why We Have Sex

In some cases, people have sex to reproduce. People may also have sex to experience physical pleasure or to deepen an emotional connection with another person. In some circumstances, people have sex in exchange for payment or other resources. There are other reasons for sex as well.

Nationwide, 3 percent of male and 10 percent of female high school students have been physically forced to have unwanted sexual intercourse. Teenagers are also more likely to engage in sex at any early age if they use alcohol or other drugs or experience domestic violence; 21 percent of teens reported using drugs or alcohol before their last sexual intercourse (Kann, McManus et al., 2016; Centers for Disease Control and Prevention, 2020). A study of nearly 2,000 college students identified 237 reasons for having sex that ranged from physical (stress reduction) to altruism (to make the other person feel good) to spite (to retaliate against a partner who had cheated) (Meston and Buss, 2007).

Sexuality throughout the Life Course

There are many misconceptions about sex at different stages of the life course (teenage years through older adulthood). We'll address some of these misconceptions here.

Some may worry that teens are having sex too early and are engaging in risky sexual behaviors, but teens' reports of their own sexual practices counter some of these concerns and provide insight into their changing sexual behaviors. According to the Centers for Disease Control, among never-married teenagers (age 15 to 19), the percentage who've had penile-vaginal intercourse has declined from 51 percent in 1988 to 40 percent in 2017 (most recent data). Rates for oral and anal sex with a different sex partner have been declining as well. Three percent of male teens and 15 percent of female teens report having sex with a same-sex partner. Most teens report using contraception, with most reporting the use of condoms. Among 15- to 19-year-olds, 91 percent of females and 94 percent of males report the use of some contraception method. Fifty-one percent of females aged 15 to 19 years old use condoms most or every time they have intercourse, while 68 percent of males do (McCarthy, 2018; Dunleavy, 2020; Centers for Disease Control and Prevention, 2020; Lindberg et al., 2021).

By age 44, 93 percent of Americans have had penile-vaginal intercourse, 87 percent have had oral sex, and 39 percent have had anal sex with a different-sex partner. Among people aged 18 to 44, 17 percent of women and 6 percent of men have had same-sex contact (Copen et al., 2016).

A majority of adults aged 45 and older agree that a satisfying sexual relationship is important. Sexual frequency may decrease because concerns about earning a living, making a home, and raising a family become more pressing than lovemaking. Others may be going through a divorce, dealing with unemployment, helping to raise grandchildren, or caring for aging parents—all of which reduce people's sexual interest (ConsumerReports.org, 2009; Twenge et al., 2017).

Christopher Weyant/The New Yorker Collection/The Cartoon Bank

"Don't you have any sexual fantasies that don't involve me cleaning?"

As people age, many continue to desire sex and engage in sexual activities. A poll of 65- to 80-year-olds showed that two-thirds want to have sex and around half say it's significant for a good quality of life (Steckenrider, 2023). A third of men and women aged 70 and older report having sex at least twice a month. Health-related considerations and challenges finding a partner, rather than just advancing age, are more closely linked to less frequent sexual activity (Lee et al., 2015).

9-4b Sexual Scripts and Double Standards

We like to think that our sexual behavior is spontaneous, but all of us have internalized sexual scripts. A **sexual script** specifies the formal and informal norms for acceptable or unacceptable sexual behavior. Social scripts can change over time and across groups but are highly gendered in two ways—women's increasing hypersexualization and a persistent sexual double standard.

The "Sexy Babes" Trend

Sexualized social messages are reaching ever younger audiences, teaching or reinforcing the idea that girls and women should be valued for how they look rather than their personalities and abilities. For example, there are "bikini onesies" for infant girls and padded bras for 7- and 8-year- olds (that's right, for 7- and 8-year-olds!).

Many girls are obsessed about their looks, and from an early age, for a variety of reasons, including their mothers' role modeling. By age 9, some girls start imitating the clothes, makeup, and behavior of their mothers who dress and act in highly sexualized ways (Starr and Ferguson, 2012).

For many girls, constantly seeking "likes" and attention on social media is like being a contestant in a never-ending beauty pageant. The boom in selfie culture has increased girls' sexualization because validation is only a tap away, and "one of the easiest ways to get that validation is by looking hot. Sex sells, whether you're 13 or 35" (Sales, 2016, p. 26).

Media images also play a large role in girls' hypersexualization. Girls and boys are exposed to cheerleaders (with increasingly sexualized routines) on TV far more than female basketball players or other athletes. Women are now represented in more diverse TV roles—as doctors, lawyers, and criminal investigators—but they're often sexy ("hot"). Also, top female athletes regularly pose naked or semi-naked for men's magazines.

Who benefits from girls' and women's hypersexualization? Marketers, who convince girls (and their parents) that being popular and "sexy" requires the right clothes, makeup, hair style, and accessories, create a young generation of shoppers and consumers who will increase business profits more than ever before (Lamb and Brown, 2007; Levin and Kilbourne, 2009).

The Sexual Double Standard

Some believe that the **sexual double standard**—a code that permits greater sexual freedom for men than women—has faded. Others argue that it persists. Among U.S. adolescents, the higher the number of sexual partners, the greater the boy's popularity. In contrast, girls who have more than eight partners are far less popular than their less-experienced female peers. By age 44, many more men (21 percent) than women (8 percent) report having had at least 15 sexual partners. And, over a lifetime, men are more likely than women to have sex outside of marriage (19 percent and 14 percent, respectively) (Kreager and Staff, 2009; Chandra et al., 2011; Drexler, 2012).

The sexual double standard can also be explored through the experience of *hooking up,* which can mean anything from kissing to sexual intercourse. Hooking up generally suggests that no money is spent, there is no pre-existing romantic relationship, and the event involves genital contact. Some college students are now referring to this as a "date," so the category seems to be fluid. The format of hooking up does allow breaking norms for dating and sex. Many men prefer hookups because they're inexpensive compared with dating. For women, hookups offer sex without becoming involved in time-consuming relationships that compete with

---

sexual script specifies the formal and informal norms for acceptable or unacceptable sexual behavior.

sexual double standard a code that permits greater sexual freedom for men than women.

schoolwork, dealing with boyfriends who become demanding or controlling, and experiencing breakups (Bogle, 2008; Rosin, 2012). Hooking up may also allow the flipping of sexual scripts. Using data from the Online College Social Life Survey, one study found that 90 percent of heterosexual dates (following hookup standards) were initiated by men, yet genital contact more likely occurred on hookup dates initiated by women (63 percent) compared to dates prompted by men (36 percent) (Dolan, 2018; Kendrick and Kepple, 2022).

9-5 Some Current Social Issues about Sexuality

Most Americans view sex as a private act, but others believe that the government should control some sexual behavior and decisions. People disagree about social policies on sex-related topics such as teenagers' birth control, sex work, reproductive technologies (refer to Chapter 16), and teen pregnancy (refer to Chapter 12). Two of the most controversial and politically contested issues continue to be abortion and same-sex marriage.

9-5a Abortion

Abortion is the expulsion of an embryo or fetus from the uterus. It can occur naturally—in *spontaneous abortion* (miscarriage)—or be induced medically. Abortion policies have shifted over the years. It was outlawed in the nineteenth century, but in 1973, the U.S. Supreme Court's *Roe v. Wade* ruling made abortion legal throughout the United States. In 2022, the Supreme Court's decision in *Dobbs v. Jackson Women's Health Organization* again removed abortion rights protection at the federal level. The right is now determined at the state level, and currently, 14 states have total abortion bans (**Figure 9.4**) (Haines et al, 2023).

CBZ/ZOB/WENN/Newscom

Galia Slayen built a life-sized Barbie to show what she would look like if she were a real woman. (Slayen used a toy for the head because she wasn't able to create a proportional head) ("Life Size Barbie...," 2011).

Trends

Two organizations regularly track abortions in the United States: The Guttmacher Institute and the U.S. Centers for Disease Control (CDC). They use different methods for collecting abortion information, so they report different numbers. Guttmacher gathers information from all known abortion providers (hospitals, doctor's offices, and clinics), giving estimates for those that don't respond. Guttmacher also uses health department data. The CDC uses information reported by central health agencies in most states (the CDC's latest totals don't include California, Maryland, or New Hampshire). For 2020, the CDC said there were 620,327 abortions in the U.S. (1.5 percent less than in 2019). Guttmacher reported 930,160 (a 1.5 increase from 2019). Abortion numbers have been declining since the mid-1990s (**Figure 9.5**) (Diamant and Mohamed, 2023).

Why have abortion rates decreased? At least half of the states have restricted access to abortion, but most of the decline has been due to an overall drop in pregnancy rates, delaying child-bearing, more effective usage of contraceptives, and greater access to emergency contraception that prevents pregnancy (Finer and Zolna, 2016; Dreweke, 2017).

Abortion is most common among women who are in their twenties, White, and never married. Most have previously given birth. About 20 percent of women who get abortions have at least a college degree, but most have lower incomes and are financially unable to support a baby. Women with lower incomes are also much more likely than women with higher incomes to have experienced intimate partner violence that included being impregnated against their will. Abortion cuts across all religious groups with 62 percent of abortion patients affiliating with a religion. This includes 24 percent who are Catholic and 30 percent who are Protestant (either mainline or evangelical Protestant) (Jones et al., 2013; Guttmacher Institute, 2014; Reeves and Venator, 2015; Diamant and Mohamed, 2023).

---

abortion expulsion of an embryo or fetus from the uterus.

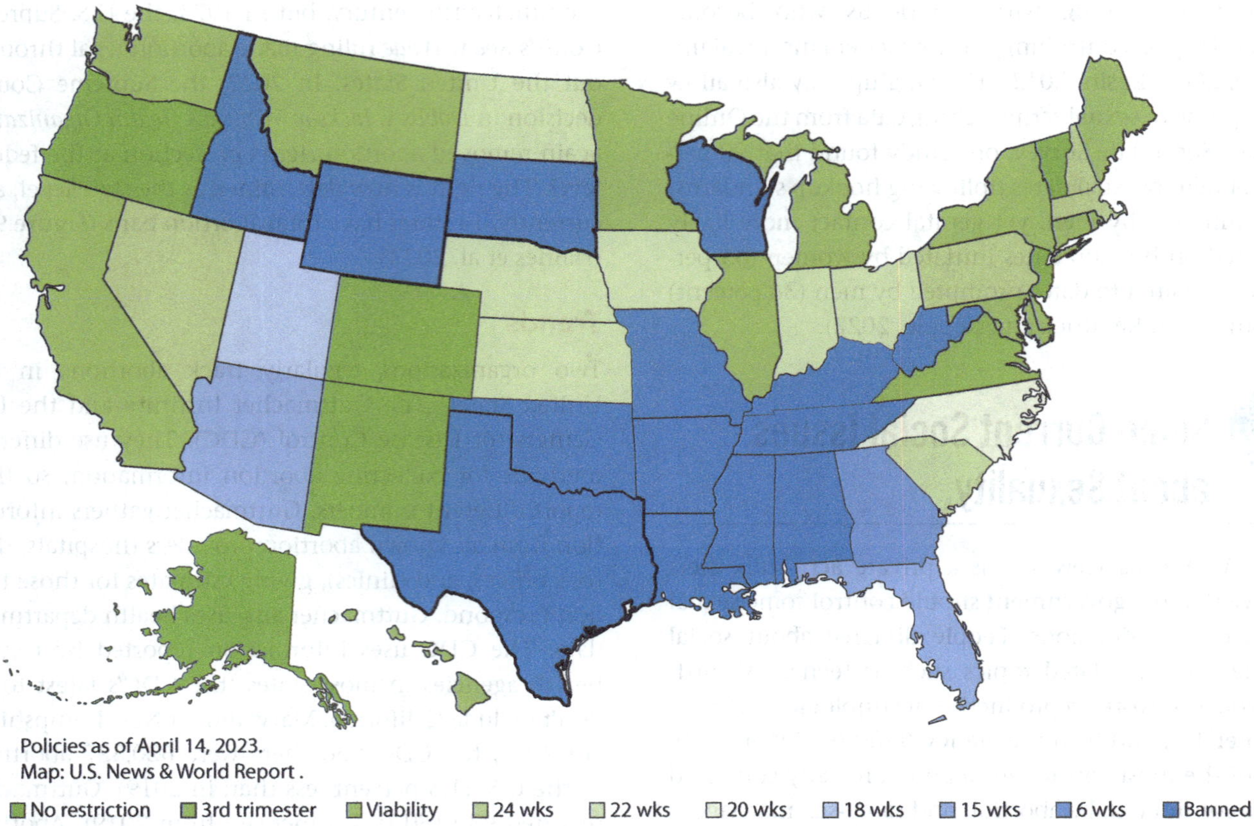

Figure 9.4 Abortion Access, State by State

Source: Haines et al, 2023.

Why Is Abortion Controversial?

Abortion has been one of the most persistently contentious issues in U.S. politics and culture. More Americans describe themselves as "pro-choice" (55 percent) than "pro-life" (39 percent). There is more support among women, Democrats, and younger adults (ages 18 to 34). Most people (73 percent) say abortion should be legal in all or most cases if the pregnancy puts the mother's life at risk. Sixty-nine percent feel it should be legal if the pregnancy was the result of rape (Saad, 2022; Pew Research Center, 2022).

Antiabortion groups believe that the embryo or fetus isn't just a mass of cells but a human being from the time of conception and, therefore, has a right to life. In contrast, abortion rights advocates point out that, at the moment of conception, the organism lacks a brain and other specifically and uniquely human attributes, such as consciousness and reasoning, and that a pregnant woman—not legislators—should decide whether or not to bear children.

Antiabortion groups maintain that abortion is immoral and endangers a woman's physical, mental, and emotional health. Whether abortion is immoral is a religious and philosophical question. On a physical level, a legal abortion in the first trimester (up to 12 weeks) is safer than driving a car, playing football, motorcycling, getting a penicillin shot, or continuing a pregnancy. There's also no evidence that having an abortion increases the risk of breast cancer or causes infertility (Sheppard, 2013; Pazol et al., 2014; Holloway, 2015).

What about mental and emotional health? Antiabortion activists argue that abortion leads to postabortion stress disorders, depression, and even suicide. National studies have consistently found that abortion poses no hazard to an adolescent or adult woman's mental health, doesn't increase emotional problems like depression or low self-esteem, and doesn't lead to drug or alcohol misuse or suicide. An unwanted pregnancy or being denied an abortion, not an abortion, increases the risk of mental health problems (Academy of Medical Royal Colleges, 2011; Steinberg and Finer, 2011; Biggs et al., 2017).

A large majority (61 percent) of Americans want to keep abortion legal in all or most cases, but there are limitations to accessing abortion services. Aside from the 14 states that have total abortion bans, many other states have restrictions on what point in pregnancy a person can have an abortion (**Figure 9.5**). In 2023, Idaho become the first state to restrict travel for abortion,

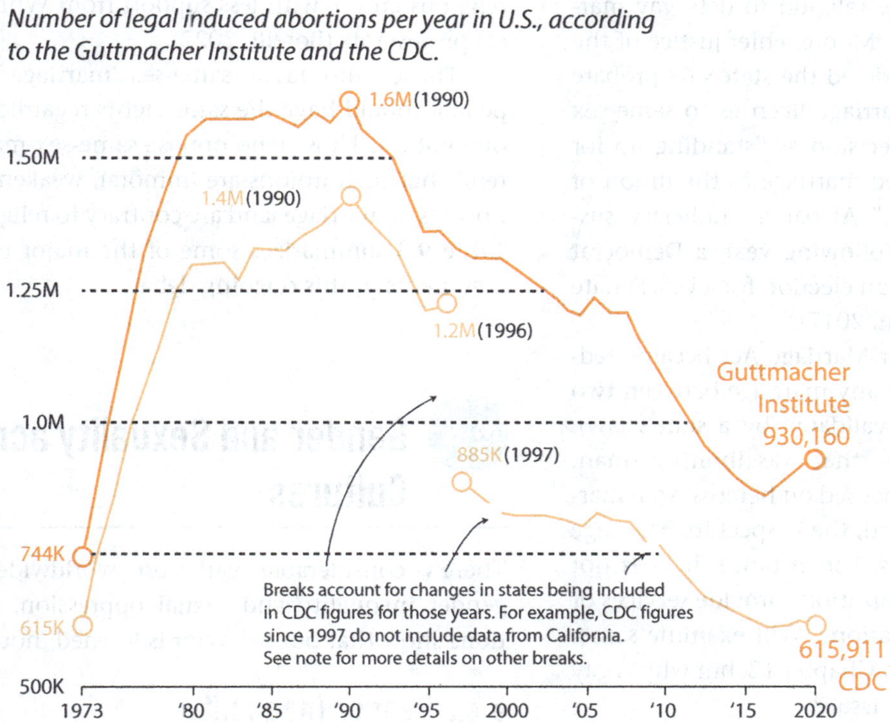

Note: Guttmacher data for 1983, 1986, 1989–90, 1993–94, 1997–98, 2001–03, 2006, 2009, 2012, 2015 and 2018 are estimates from Guttmacher. CDC totals from 1973 to 1996 include all 50 states and District of Columbia. CDC data since 1997 excludes states that did not report data each year of the specified time period, to facilitate comparison. CDC data in this graphic from 1997–1999 excludes six states: Alaska, California, Louisiana, New Hampshire and West Virginia. CDC data in this graphic from 2010–2020 excludes three states–California, Maryland and New Hampshire – as well as the District of Columbia.

Figure 9.5 Changing Numbers for Legal Abortions in the United States since the 1970s

Source: Diamant and Mohamed, 2023.

making it a felony to assist a minor with accessing abortion services or pills without consent from parents (Pew Research Center, 2022; Diamant and Mohamed, 2023; Ziegler, 2023).

9-5b Same-Sex Marriage

Same-sex marriage (also called *gay marriage*) is a legally recognized marriage between two people of the same biological sex and/or gender identity. Although still controversial, same-sex marriage is becoming more acceptable in the United States and some other countries.

Trends

In 2015, the U.S. Supreme Court issued a landmark ruling (*Obergefell v. Hodges*) granting same-sex couples a constitutional right to marry. The 5-4 decision gave gay and lesbian couples nationwide the same legal rights and benefits as heterosexual couples. The United States and 23 other countries (so far) allow same-sex marriage (Human Rights Campaign, 2023).

Prior to the ruling, gay marriage was illegal in 13 states. Some of these states' lawmakers urged their constituents to accept the new law. Others pressed their residents to "stand and fight by seeking a constitutional amendment banning gay marriage" (de Vogue and Diamond, 2015).

same-sex marriage (also called *gay marriage*) a legally recognized marriage between two people of the same biological sex and/or gender identity.

Opponents often invoke religion to defy gay marriage laws. For example, Roy Moore, chief justice of the Alabama Supreme Court, ordered the state's 68 probate judges to refuse to issue marriage licenses to same-sex couples. He defended his decision as "standing up for God" because "God ordained marriage as the union of one man and one woman." Alabama's judiciary suspended Judge Moore. The following year, a Democrat narrowly defeated Moore in an election for a U.S. Senate seat (Robertson, 2016; Cason, 2017).

In 2022, the Respect for Marriage Act became federal law. This act recognizes any marriage between two individuals (that has been validated by a state's law), removing previous language that was limiting (man, woman, opposite sex) and focused on heterosexual marriage. Although a step forward, the Respect for Marriage Act leaves open some issues. For instance, it does not mandate that religious organizations provide services or goods for a marriage celebration. We'll examine same-sex marriages and families in Chapter 12, but why is gay marriage such a contentious issue?

Why Is Same-Sex Marriage Controversial?

Support for same-sex marriage has been growing. In 2022, 71 percent of Americans felt that marriages between same-sex couples should be recognized by the law and have the same rights as heterosexual marriages (up from 27 percent in 1996) (McCarthy, 2022). Also in 2022, a majority (61 percent) of Americans said same-sex marriage is good for society. There are differences based on group affiliation, including political party, with less support from Republicans (especially conservative learning); age, with less support from older Americans; and religious group, with less support from White evangelical protestants (Borelli, 2022).

Those who favor same-sex marriage argue that people should have the same rights regardless of sexual orientation. Those who oppose same-sex marriage contend that such unions are immoral, weaken traditional notions of marriage, and are contrary to religious beliefs. **Table 9.3** summarizes some of the major pro and con arguments in this ongoing debate.

9-6 Gender and Sexuality across Cultures

There's considerable variation worldwide regarding gender inequality and sexual oppression. Such variations show that our behavior is learned, not innate.

9-6a Gender Inequality

A recent United Nations (2020) report concluded that women continue to face discrimination in access to work, economic assets, and participation in private and public decision making. They're also more likely than men to live in poverty, to be illiterate, and to experience violence. And the COVID-19 pandemic created more barriers.

In many countries, women's progress toward equality has been mixed. For example, Saudi Arabia, one of the wealthiest countries in the world, has some of the most educated women in the world (including STEM college and advanced degrees), and ranks near the bottom in women's economic and political participation. In *all countries and regions*, the greatest gender gaps

Table 9.3 Why Do Americans Favor or Oppose Same-Sex Marriages?

What do you think? What other reasons can you add for each side of the debate?

Same-sex marriage should be legal because . . .	Same-sex marriage should be illegal because . . .
• Same-sex marriages strengthen families and long-term unions that already exist. Children are better off with parents who are legally married.	• Children need a mom and a dad, not two dads or two moms.
• There are no scientific studies showing that children raised by gay and lesbian parents are worse off than those raised by heterosexual parents.	• There are no scientific studies showing that children raised by gay and lesbian parents are better off than those raised by heterosexual parents.
• Every person should be able to marry someone that they love.	• People can love each other without getting married.
• Same-sex marriages are good for the economy because they boost businesses such as restaurants, bakeries, hotels, airlines, and florists.	• What's good for the economy isn't necessarily good for society, especially its moral values and religious beliefs.

Sources: Bennett and Ellison, 2010; Olson, 2010; Sullivan, 2011; Sprigg, 2011; Whitehead, 2011; Bogage, 2015; Kaufman, 2015.

are in economic participation and political leadership (World Economic Forum, 2016).

Economic Participation

Worldwide, on average, women have just three-quarters of the legal rights that men do. There are 150 countries that have at least one law that treats women and men differently, and 63 countries have five or more. These laws make it difficult for women to own property, open bank accounts, start businesses, and enter certain professions (World Bank Group, 2021; World Bank, 2017).

Globally, in 2019, women made up 39.4 percent of employed persons. With the onset of the COVID-19 pandemic, women disproportionately faced employment loss—constituting almost 45 percent of job losses in 2020 (United Nations, 2022). Economic rights are also important for increasing women's participation in other areas. "As women's economic rights increase, so, too, does their representation in national lawmaking bodies" (World Bank Group, 2021).

Countries that have closed education gaps and have high levels of women's economic participation—the Scandinavian countries, United States, Canada, New Zealand, and Australia—have strong economic growth (Worley, 2014). However, gender gaps still persist in senior positions, wages, and leadership. For example, Germany is Europe's No. 1 economy, but it also has one of the largest pay gaps in the European Union, with women earning 17.2 percent less than men (European Parliament, 2023). And in 2021, out of the 160 companies that are part of the country's main stock indexes, there were 603 men and only 93 women on their executive boards (Wrede and Pladson, 2021).

Political Leadership

Worldwide, at the beginning of 2022, women filled just over one third of seats in local governments, and 26.2 percent of seats in lower and single houses of national parliaments were held by women (an increase of 3.8 percentage points since 2015) (United Nations, 2022). Worldwide, women are unequally represented in the highest positions of political power. Out of the 193 United Nations member states, just 13 had women serving as their head of government in 2023. And less than one third have ever had a woman as leader (Clancy and Austin, 2023). In 2021, the United States made gains in its global ranking for women's political leadership, going from 128th out of 193 nations to 43rd. This was the result of President Joe Biden's efforts to appoint a gender-balanced cabinet as well as women's gains in Congressional seats (Vogelstein and Fliegauf, 2021).

9-6b Sexual Inequality

Globally, women have fewer rights and opportunities than men. There has been more acceptance of LGBTQIA+ people in some countries, but heterosexism prevails.

Violence against Women

Violence against women is a persistent problem. The World Health Organization (WHO) estimates that 30 percent of women worldwide have experienced physical and/or sexual intimate partner violence or sexual violence perpetrated by a non-partner during their lives. For 19 countries, the estimates are much higher, between 40 and 53 percent (World Health Organization, 2021).

Female genital mutilation/cutting (FGM/C) is the partial or total removal of the female external genitalia. Two hundred million girls and women in 31 countries have had FGM/C including half of the women in Egypt, Ethiopia, and Indonesia. The mutilation typically occurs between 3 and 12 years old. The operator is often an elderly village woman who uses a knife or other sharp object and doesn't administer an anesthetic. A more recent trend in some countries is to medicalize the practice with the procedure conducted by a health-care provider. Approximately one quarter of FGM survivors had their procedure in this medicalized format. Countries justify FGM/C on the grounds that it controls a girl's sexual desires and preserves her virginity, a prerequisite for marriage. Although Nigeria, Egypt, and other countries have outlawed FGM/C, the practice remains widespread and widely accepted ("Female Genital Cutting," 2016; UNICEF, 2023).

Creastista/Shutterstock.com

Violence against Men

You learned earlier that Americans are more accepting of gay and lesbian sexuality and that more nations are legalizing gay marriage. In contrast, many countries in Asia, Africa, and the Middle East are less tolerant of people who are LGBTQIA+. On average, across 27 countries (from various regions or the world and with different income levels), 54 percent of people surveyed support the right of LGBT people to legally marry and 16 percent support some other legal recognition (Ipsos, 2021). Sixty-four countries have laws that criminalize being gay or lesbian—almost half are in Africa. In Brunei, Iran, Mauritania, Saudi Arabia, Yemen, and some areas of Nigeria, the death penalty is the legal punishment for same-sex sexual acts. In Afghanistan, Pakistan, Qatar, Somalia, and the United Arab Emirates, the death penalty could be used since the laws are not clear (Reality Check Team, 2023).

While women and girls disproportionately face sexual violence worldwide, men and boys are victims as well. For example, rape, genital violence, and castration have been used against Sudanese refugee men and boys (Women's Refugee Commission, 2019). There are about 10 million transgender people in Asia and the Pacific. Governments in Bangladesh, India, Nepal, and Pakistan have recognized transgender people as a legal category that has rights, but many—often labelled "mentally ill" by the public—experience discrimination and violence. In China, the attackers are often the victim's relatives. In Fiji, 40 percent of *trans women* (male-to-female) have been raped. In Australia, 60 percent of *trans men* (female-to-male) suffer abuse from their partners ("Knife-edge Lives," 2016).

9-7 Sociological Explanations of Gender and Sexuality

Gender and sexuality affect all people's lives, but why is there so much variation over time and across cultural groups? The four sociological perspectives answer this and other questions somewhat differently. **Table 9.4** summarizes these theories.

9-7a Functionalism

Functionalists view women and men as having distinct roles that ensure a family's and society's survival. These roles help society operate smoothly and have an impact on the types of work that people do.

Division of Gender Roles and Human Capital

Some of the most influential functionalist theories, developed during the 1950s, proposed that gender roles differ because women and men have distinct roles and responsibilities. A man (as a husband and father) plays the *instrumental role* of economic provider; he's competitive and works hard. A woman (as a wife and mother) plays an *expressive role*; she provides the emotional nurturance that sustains the family unit and supports the father/husband (Parsons and Bales, 1955; Betcher and Pollack, 1993).

Instrumental and expressive roles are complementary, and each person knows what's expected: If the house is clean, she's a "good wife"; if the bills are paid, he's a "good husband." The duties are specialized, but both roles are equally important in meeting a family's needs and ensuring a society's survival.

Table 9.4 Sociological Explanations of Gender and Sexuality

Theoretical Perspective	Level of Analysis	Key Points
Functionalist	Macro	• Gender roles are complementary, equally important for a society's survival, and affect human capital. • Agreed-on sexual norms contribute to a society's order and stability.
Conflict	Macro	• Gender roles give men power to control women's lives. • Most societies regulate women's, but not men's, sexual behavior.
Feminist	Macro and micro	• Women's inequality reflects their historical and current domination by men, especially in the workplace. • Many men use violence—including sexual harassment, rape, and global sex trafficking—to control women's sexuality.
Symbolic Interactionist	Micro	• Gender is a social construction that emerges and is reinforced through everyday interactions. • The social construction of sexuality varies across cultures because of societal norms and values.

Arthur Greenberg/Alamy Stock Photo

Such traditional gender roles help explain occupational gender segregation because people differ in the amount of human capital they bring to the labor market. *Human capital* is the array of competencies—including education, job training, skills, and experience—that have economic value and increase productivity.

From a functionalist perspective, what individuals earn is the result of the choices they make and, consequently, the human capital that they accumulate to meet labor market demands. Following a functionalist viewpoint, women may diminish their human capital because they choose lower paying occupations (social work rather than computer science) and postpone or leave the workforce for childbearing and child care. When they return to work, women have lower earnings than men because, even in higher paying occupations, their human capital has deteriorated or become obsolete (Kemp, 1994).

Why Is Sexuality Important?

For functionalists, sexuality is critical for reproduction, but people should limit sex to marriage and forming families. Functionalists view sex outside of marriage as dysfunctional because children from single parent households are more likely to be economically disadvantaged (refer to Chapter 8)

You might be tempted to dismiss the functionalist view of limiting sex to marriage as outdated. However, sex outside of marriage is prohibited in some countries, including Qatar, Iran, and Afghanistan. And *arranged marriages*—in which parents or relatives choose their children's future mates—are the norm in some countries, including India, China, and Pakistan. Most children agree to arranged marriages because of social custom and out of respect for their parents' wishes. The matches solidify relationships with other families and ensure that the woman's sexual behavior will be confined to her husband, avoiding any doubt about the offspring's parentage (refer to Benokraitis, 2015).

Some functionalists encourage marrying during one's mid-to-late twenties instead of delaying marriage. The benefits may include enjoying more frequent sex, having an easier time getting pregnant, and being able to have more than one child compared to people who marry in their thirties or later (Wilcox, 2015).

Critical Evaluation

Critics fault functionalist gender role perspectives on three counts. First, even during the 1950s, White middle-class male sociologists ignored almost a third of the labor force that was composed of working-class, immigrant, and systematically disadvantaged women who played *both* instrumental and expressive roles. Second, functionalists tend to overlook the fact that many people don't have a choice of filling only instrumental or expressive roles because most families rely on two incomes for economic survival. Third, the human capital model assumes that women have lower earnings than men because they "choose" lower paying occupations. As you read earlier, however, there's a gender pay gap across *all* occupations, even those that require advanced degrees.

Functionalists tend to reject sexual relationships outside of marriage. Compared with married couples, for example, those who cohabit have lower quality relationships and lower happiness levels. As you'll learn in Chapter 12, however, marriage doesn't guarantee long or happy relationships.

Antigay discrimination, which is legal in 28 states, is dysfunctional. Companies don't attract young and talented LGBTQIA+ workers, and forbidding transgender people from using a bathroom that corresponds to their gender identity has sparked considerable interpersonal and group conflict (Green, 2016; McCarthy, 2017).

9-7b Conflict Theory

For conflict theorists, gender inequality is built into the social structure. In both developing and industrialized countries, men control most of a society's resources and dominate women. Like functionalists, conflict theorists discuss sexuality as a key component of a society's organization, but they view sexuality as reflecting and perpetuating sexism and discrimination.

Capitalism and Gender Inequality

Conflict theorists maintain that capitalism, not complementary roles, explains gender roles and men's social and economic advantages. Women's inequality is largely due to economic exploitation—both as underpaid workers in the labor force and unpaid domestic workers who care for children and aging family members. In effect, gender roles are profitable for business. Companies can require their male employees to work long hours or make numerous business trips and not worry about workers demanding payment for child care services. Costs can average $15,400 for an infant and $9,200 for a school age child in a care center (Landivar, 2023).

Women comprise 51 percent of the U.S. population, but Congress "looks much like the face of corporate America—overwhelmingly wealthy, white, and male" (Weathers, 2015). Women's underrepresentation in political institutions can help explain why occupational gender segregation, gender pay gaps, and sexual harassment persist.

Is Gender Inequality Linked to Sexual Inequality?

From a conflict perspective, gender inequality gives men economic, political, and/or interpersonal power to control or dominate women's sexual lives. Most of the domestic violence and rape victims, both in the United States and around the world, are women and girls. In workplace sexual harassment cases, the offender is typically a male supervisor. For example, 40 percent of women in the fast food industry have experienced sexual harassment by a manager, owner, or supervisor at least once a month. As noted in Chapter 8, women and girls are disproportionately (80 percent) the victims of forced commercial sexual exploitation (International Labour Organization, 2022).

In some countries, men dictate how women should dress and whether they can access a broad array of rights. For instance, in Qatar, women and girls fall under male guardianship—a father, uncle, brother, or husband. The guardian determines whether women and girls can exercise their rights for government scholarships for higher education, reproductive health care, and travel out of the country (Human Rights Watch, 2021). Many women have internalized such sexism. For example, 25 to 38 percent of women in some African countries, Egypt, Palestine, and Indonesia believe that wives sometimes deserve beatings (e.g., for leaving the house without their husband's permission) (Kaneda and Bietsch, 2015; "The State of Arab Men," 2017).

Critical Evaluation

Critics point out several limitations of conflict theory. First, women aren't as powerless as some conflict theorists claim. Like men, women often barter to increase their economic and political power. Second, some conflict theorists ignore women's exploitation of other women. Both within the United States and globally, women comprise around one-third of perpetrators of the sex trafficking industry (Jeglic, 2021). The thriving underground sex economy relies heavily on nannies, secretaries, and escort services and brothels owned or operated by females (Dank et al., 2014).

Third, capitalism can discourage sexism. For years, Fox News and Bill O'Reilly, the station's popular political commentator, paid up to $13 million to settle several sexual harassment lawsuits. Fox fired O'Reilly only after more than 70 corporations dropped their ads—largely in response to social media campaigns—because the companies "couldn't afford to alienate women and their considerable purchasing power" (Abbey-Lambertz, 2017; Chira, 2017, p. B7).

9-7c Feminist Theories

Feminist scholars agree with conflict theorists that gender stratification benefits men and capitalism but emphasize that women's subordination also includes their daily

AP Images/Ben Curtis

In 2014, a group of men surrounded a woman at a busy bus station in Nairobi, Kenya. They violently tore off her clothes because she was wearing a miniskirt, accused her of being a jezebel, and left her naked on the street. Bystanders watched. A week later, about 1,000 women (some dressed in miniskirts) protested the rising violence against girls and women, vowing to wear whatever clothes they want. Some men joined the protestors; others threw rocks or shouted lewd comments.

vulnerability to male violence (Katz, 2006). Feminist scholars, more than any other group of theorists, are especially concerned about men controlling women's sexual lives.

Living in a Gendered World

Men comprise 48 percent of the U.S. population age 18 and older, but they account for 98 percent of the directors of the 700 top-grossing films since 2007, 97 percent of the televised news media's and ESPN's sports broadcasts, 96 percent of Fortune 500 CEOs, 83 percent of the largest private law firm partners, 67 percent of college/university presidents, 74 percent of federal and state judges, and 73 percent of generals and admirals (Cooky et al., 2015; Smith et al., 2015; Johnson, 2016; American Bar Association, 2017; Kidder et al., 2017; Merelli, 2017; American Council on Education, 2023). In many of these sectors, women's representation has decreased or remained about the same for at least a decade.

Women often experience harsher sanctions than men for workplace offenses. In the financial services industry, for instance, fraud and forgery are punished by demotion or dismissal. Compared with advisers who are women, men engage in almost three times more fraud and forgery and are more likely to be repeat offenders. Nonetheless, women are 20 percent more likely to be fired and 30 percent less likely to find new jobs (Egan et al., 2017).

Sexism is prevalent in almost all workplaces and online environments. Social media has amplified feminist voices but also silenced them. Once a writer is singled out by an anti-feminist men's group, they are deluged with hateful and threatening messages, both public and private. Men who are the most disadvantaged are the most likely to be threatened by and hostile toward an influx of talented women. Regardless of the reasons, some women no longer participate online because of the harassment (Goldberg, 2015; Kasumovic and Kuznekoff, 2015).

Feminist theorists emphasize that gender, race, and social class intersect to form a hierarchical stratification system that shapes people's experiences and behavior. Privileged women have less status than privileged men, but upper-class White men *and* women subordinate women *and* men of lower economic statuses who are systematically disadvantaged (Andersen and Collins, 2010). Understanding the interconnections between gender, race/ethnicity, and social class provides a more comprehensive picture of living in a gendered world than does any single variable.

Sexuality, Social Control, and Commercializing Sex

Men assert their power and control—across cultures and over time—through sexual violence, intimate partner violence, sexual harassment, exploiting women through forced sex work and pornography, forcing women to marry against their will, honor killings, FGM/C, and punishing women, but not men, for seeking a divorce or committing adultery (World Health Organization, 2013).

Sex is big business, and pharmaceutical industry has tapped into this. Men's testosterone levels decline naturally with aging but also due to obesity and alcohol. Pharmaceutical companies "have seized on the decline in testosterone levels as pathological and applicable to every man." As a result, sales of testosterone-boosting drugs are surging, even though they increase the risk of heart attacks (La Puma, 2014, p. A21; O'Connor, 2014).

In 2015, the Food and Drug Administration approved the drug flibanserin, marketed as Addyi, to increase premenopausal women's sexual desire. Some women's groups lobbied for the "female Viagra," but a study found that flibanserin resulted in "one-half additional satisfying sexual event per month." The medical researchers didn't define "one-half" of a "satisfying sexual event" but concluded that the drug significantly increased the risk of dizziness, sleepiness, nausea, and fatigue. They recommended an "integrative approach" that includes medical, psychiatric, psychological, and couple-relationship treatment (Jaspers et al., 2016, p. 457, 461).

Pornography, "sex drugs," and the earlier discussion of the "sexy babes" trend are just a few examples of the increasing *commercialization of sex*, making sexuality a commodity that can be sold for financial gain. Commercializing sex demeans both women and men, but the "products" are usually women.

Critical Evaluation

Feminist explanations are limited for several reasons. First, they pay little attention to parenting barriers s that many men face. For instance, President Obama signed the Bathrooms Accessible in Every Situation Act (BABIES Act), requiring diaper-changing spaces in restrooms in public federal buildings, but stations are still limited in men's restrooms in other locations (Chapman, 2018). Second, feminist analyses are inclusive, but this strength can also be a weakness. Regarding the gender pay gap, for example, should race, ethnicity, or social class be given priority in implementing change? Third, are feminist scholars overstating women's underrepresentation in some sectors? Since 2010, for instance, both cable and network shows have featured women in strong leading roles (e.g., *Girls, How to Get Away with Murder, Scandal, Madam Secretary, The Marvelous Mrs. Maisel*) (O'Keefe, 2014; Blay, 2015). And, like conflict theorists, feminist scholars are sometimes accused of glossing over women's exploitation of others; women are perpetrators in sex trafficking.

9-7d Symbolic Interaction

While functionalist, conflict, and some feminist theories are macro level, symbolic interactionists focus on the everyday processes that produce and reinforce gender roles. We "do" gender, sometimes consciously and sometimes unconsciously, by adjusting our behavior and our perceptions depending on the sex of the person with whom we're interacting (West and Zimmerman, 2009). Our sexual expression, similarly, isn't inborn but is a product of socialization and what families and other societal groups deem as appropriate and inappropriate behavior (Hubbard, 1990).

Gender Is a Social Construction

For interactionists, gender is a social creation, and we learn gender roles through everyday social interaction. For example, when teachers tell girls and boys that both are equally capable in math and science, "the difference in performance essentially disappears" (Hill et al., 2010, p. 2). When, on the other hand, teachers discourage girls from pursuing math and science, girls fare worse than boys on exams, enroll in fewer advanced math and science courses in high school and college, and, consequently, choose lower-paying careers (Lavy and Sand, 2015). Thus, teachers' gender biases can affect occupational choices and earnings in adulthood.

Believing in gender differences can actually *produce* differences. Because men are self-confident about their worth, they're four times more likely than women to ask for a raise. When women do ask for a raise, they ask for 30 percent less than men. Men in higher-level positions who aren't promoted often threaten to quit and are offered retention bonuses to stay. Women typically decide to work harder and try again next year (Lipman, 2015).

Sexuality Is Also a Social Construction

For interactionists, sexuality is also socially constructed. As with gender roles, we *learn* to be sexual and to express our sexuality differently over time and across groups because the people around us affect our attitudes and behavior. For example, students who attend conservative religious high schools are less likely than their public school counterparts to report same-sex attraction or to identify as LGBT in adolescence or young adulthood (Wilkinson and Pearson, 2013).

In Iran, men may have premarital sex, but some still expect women to produce a virginity certificate before marriage (Akbarian and Bettiza, 2022). In the United States, as you learned earlier, men who have casual sex are "studs," while women are "sluts." Thus, sexual double standards are socially constructed.

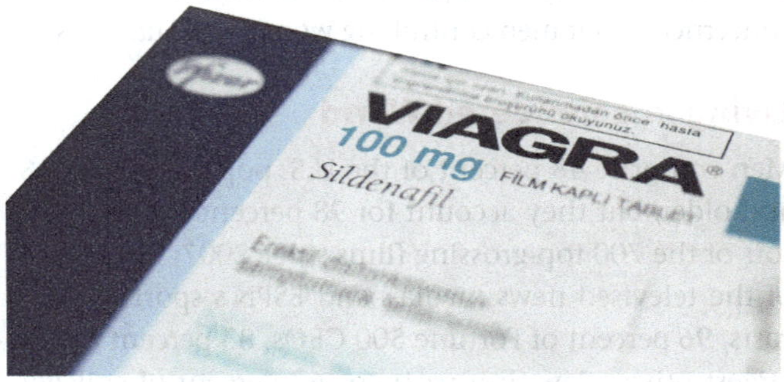

Sean Nel/Shutterstock.com

Many sexually healthy people in their 20s to 60s now take drugs like Viagra and Addyi. Are the drugs an example of commercializing sex?

In China, a university that planned to publicly shame students who engaged in "uncivilized behavior" (e.g., hugging or kissing in public) withdrew the policy after widespread protests. In much of India, people may be beaten for kissing on the street or even holding hands. To protest such moral policing by right-wing groups, two college students started a "Kiss of Love" campaign on Facebook. Despite arrests, hundreds of people joined the demonstration. Similar protests quickly spread to other cities (Ming, 2011; Bhardwaj, 2014).

Critical Evaluation

A common criticism is that interactionists ignore the social structures that create, maintain, or change gender roles and gender inequality. Many 18- to 32-year-olds plan to share earnings and household/child care responsibilities equally with their future partners. An equal division of labor is unlikely, however, because current workplace and government policies don't support the ability of women and men to balance work and family

AP Images/Altaf Qadri

Activists supporting the "kiss of love" campaign in New Delhi, India.

life. Thus, people don't have as much ability to shape their lives as interactionists claim.

Interactionists emphasize that language, erotic images, and other symbols evoke sexual interest or desire, but they neglect the relationship between biological factors and sexual orientation. Thus, interactionism doesn't explain why siblings, even identical twins—who are socialized similarly—may have different sexual orientations. A third limitation is that interactionists don't explain why, historically and currently, women around the world are considerably more likely than men to be controlled and sexually exploited. Such analyses require macro-level analyses that examine religious, political, and economic institutions.

Chapter Review

Key Takeaways

9-1 **How do sex and gender differ?** *Sex* refers to the biological characteristics with which we are born that are then used to assign us to a specific sex category such as male, female, or both , while *gender* refers to learned attitudes and behaviors. *Gender roles* differ depending on whether people perceive themselves as masculine or feminine and because a society expects women and men to think and behave differently. Many Americans still have *gender stereotypes* about how people will look, act, think, and feel based on their sex.

9-2 **What are the differences between sexual identity and sexual orientation?** A person's sexual identity incorporates a *sexual orientation* which is emotional or sexual attraction to partners of the same sex, of another sex, of more than one sex, or of neither sex. Many biological theories maintain that sexual orientation has a strong genetic basis, but social constructionists argue that culture, not biology, plays a large role in forming people's *gender identity*.

9-3 **How does gender stratification affect education, work, and family?** *Sexism* is widespread because of *gender stratification*, which can lead to inequality in the family (particularly childcare and housework), education, the workplace (as in *occupational gender segregation* and a *gender pay gap*), and politics. *Sexual harassment* and workplace bullying are also common in the workplace.

9-4 **What are some contemporary sexual attitudes and practices?** Sexual attitudes and behavior can vary from situation to situation and change over time, including why we have sex. Contrary to some stereotypes, adolescents aren't sexually promiscuous and older people aren't asexual. All of us have internalized *sexual scripts*, and *sexual double standards* persist.

Example: Gender Roles and Hooking Up—Are Women the Losers?

Hooking up refers to physical encounters, no strings attached, and can mean anything from kissing and genital fondling to oral sex and sexual intercourse. The prevalence of hooking up has increased only slightly since the late 1980s but is now more common than dating at many high schools and colleges. Hooking up has its advantages. It's much cheaper than dating and requires no commitment of time or emotion. In addition, hookups remove the stigma from those who can't get dates but can experience sexual pleasure, and they make people feel sexy and desirable. Hooking up also has disadvantages, especially for women, because they generally get a bad reputation as being "easy" (refer to Chapter 9). In effect, then, and despite the advantages of hooking up, the sexual double standard persists.

9-5 **What are Americans' viewpoints on abortion and same-sex marriage?** Two of the most controversial and politically contested issues continue to be *abortion* and *same-sex marriage.* Almost equal percentages of Americans support or condemn abortion. Those who favor same-sex marriage argue that people should have the same legal rights regardless of sexual orientation; those who oppose same-sex marriage contend that such unions are immoral and contrary to religious beliefs.

9-6 How do experiences with gender and sexuality vary across cultures? Worldwide, women have fewer rights and opportunities than men. In all countries and regions, the greatest gender gaps are in economic opportunity and participation in political leadership. People in some countries are now more accepting of gay, lesbian, and other sexualities, but *heterosexism* prevails.

9-7 How do various sociological explanations of gender and sexuality compare? Refer to Table 9.4 on page 190.

Key Terms

Sex (p. 174)
Gender (p. 174)
Intersex (p. 174)
Sexual identity (p. 175)
Sexual orientation (p. 175)
Gay (p. 175)
Heterosexual (p. 175)
Bisexual (p. 175)
Asexual (p. 175)
Gender identity (p. 176)
Transgender (p. 176)
Gender expression (p. 177)
Gender roles (p. 177)
Gender stereotypes (p. 177)
Sexism (p. 178)
Heterosexism (p. 179)
Homophobia (p. 179)
Occupational gender segregation (p. 181)
Gender pay gap (p. 181)
Sexual harassment (p. 182)
Sexual script (p. 184)
Sexual double standard (p. 184)
Abortion (p. 185)
Same-sex marriage (p. 187)

Discusssion and Critical Thinking Questions

1. What's the difference between homophobia and heterosexism? How do both affect people who are of different sexual orientations?
2. Do you think the United States is too restrictive or too permissive about sexual behavior? Why?
3. How has your gender role been shaped by your family, peers, school, and the mass media? Which seem to have had the most impact on you?
4. Which of the sociological perspectives on gender inequality do you find the most convincing? Why?

True or False?

How Much Do You Know about Gender and Sexuality?

The answers to 4 and 6 are true. The answers to 1, 2, 3, and 5 are false.

1. Approximately 7.2 percent of the adult population identifies as LGBTQIA+.
2. Over 90 percent of teens (ages 15 to 19) report using some contraception method, with most opting for condoms.
3. Women make up about 28 percent of the 118th U.S. Congress.
4. The gender pay gap is at 82 percent and varies by racial and ethnic group.
5. U.S. abortion rates have declined since 1990, when there were 1.6 million.
6. Men spend more hours in paid work while women do more caregiving and housework.

10 Race and Ethnicity

Stanislaw Tokarski/Shutterstock.com

Learning Objectives

After studying this chapter, you will be able to...

10-1 Describe the current racial and ethnic composition of the United States and how it is projected to change.

10-2 Define race, ethnicity, and racial-ethnic group.

10-3 Explain how immigration patterns have changed, and describe Americans' reactions to documented and undocumented immigrants.

10-4 Describe dominant group, minority groups, and the most common patterns of dominant-minority group relations.

10-5 Describe the most common sources of racial-ethnic friction.

10-6 Compare the five major minority groups in the United States in terms of origins, social class, and achievements.

10-7 Evaluate the main theoretical explanations of race and ethnicity.

10-8 Explain changes with interracial and interethnic relationships.

In 2016, NFL player Colin Kaepernick knelt during the American national anthem. His action aimed to highlight the United States' role in oppression of people of color. Later, it was mentioned that protesting players should be fired, and in response, 200 NFL players joined to kneel in protest. These forms of protest have become more widespread in professional sports. Some feel professional sports are useful, highly visible platforms to protest racial oppression and violence. Others feel these kinds of actions are disrespectful, and there should be a separation between race, politics, and sports.

Such situations illustrate our diversity and divisions. This chapter examines the impact of race and ethnicity on our lives, why racial-ethnic inequality is still widespread, and the growth of interracial and interethnic relationships.

Be sure to take the True or False quiz to better understand how much you know about these topics.

What do you think?

The United States is a melting pot.

1	2	3	4	5	6	7
	strongly agree				strongly disagree	

True or False?

How Much Do You Know about U.S. Racial and Ethnic Groups?

1. Race is determined biologically.
2. The United States has the most residents from other nations than any country in the world.
3. People who are prejudiced also discriminate.
4. The U.S. Census counts race but not ethnicity.
5. Asian Americans have the highest median household income of all racial groups in the United States.
6. More Americans now say there is equality of opportunity, regardless of race, in the United States.

The answers are at the end of the chapter.

10-1 U.S. Racial and Ethnic Diversity

The United States is one of the most multiracial and multiethnic countries in the world, a magnet that draws people from hundreds of nations and is home to millions of Americans who are bilingual or multilingual. Approximately 13.7 percent of the U.S. population is from other nations (up from only 5 percent in 1965), and that number is expected to reach a historic 19 percent (78.2 million people) in 2065. America's multicultural umbrella includes at least 150 ethnic or racial groups and 350 languages (U.S. Census Bureau, 2021; U.S. Census Bureau News, 2015; López and Radford, 2017).

The population of people who are White and non-Hispanic is expected to decrease from 198 million in 2022 (around 60 percent of the population) to 179 million in 2060 (around 44 percent). This is because the White population, including both Hispanic and non-Hispanic White individuals, is expected to increase from 253 million to 275 million. The population of those who are two or more races is expected to grow the most, with large increases for the Hispanic and Asian populations (Vespa et al., 2020) (**Figure 10.1**).

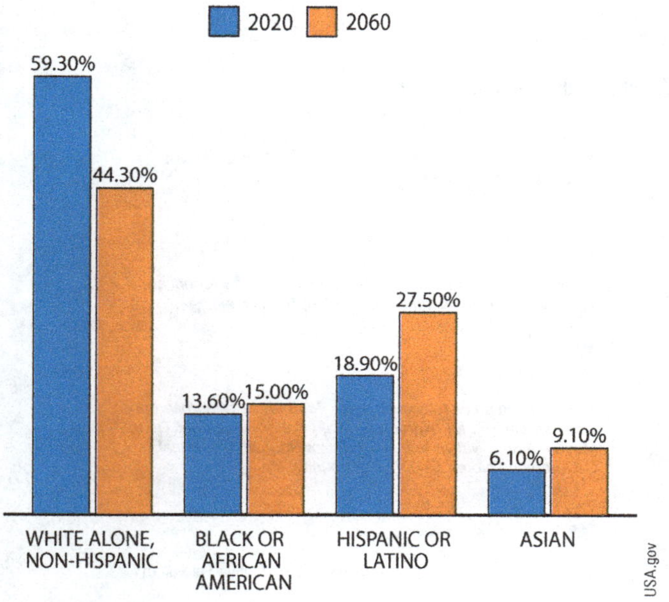

Figure 10.1 Racial and Ethnic Composition of the U.S. Population, 2020-2060

Note: Represents the four most populous racial-ethnic groups

Sources: Based on U.S. Census Bureau, Population Census, April 1, 2020; Vespa et al., 2020.

10-2 The Social Significance of Race and Ethnicity

All of us identify with some groups in terms of sex, age, social class, and other factors. Two of the most common and important sources of self-identification and labeling by others are race and ethnicity.

10-2a Race

A **racial group** refers to people who share visible physical characteristics, such as skin color and facial features, that members of a society consider socially important. Contrary to the popular belief that race is determined biologically, it's a *social construction*, a societal invention that labels people based on physical appearance, social class, or other characteristics (refer to Daniel, 2014).

Possibly only six of the human body's estimated 35,000 genes determine the color of a person's skin. Because all human beings carry 99.9 percent of the same genetic material (DNA), the "racial" genes that make us look different are minuscule compared with the genes that make us similar (Graves, 2001; Pittz, 2005).

If our DNA is practically identical, why are we so focused on race? People react to the physical characteristics of others, and those reactions have consequences. Skin color, hair texture, and eye shape, for example, are easily observed and mark groups for unequal treatment. As long as we act on the basis of these characteristics, our life experiences will differ in access to jobs and other resources, how we treat people, and how they treat us.

10-2b Ethnicity

An **ethnic group** (from the Greek word *ethnos*, meaning "nation") refers to people who identify with a common national origin or cultural heritage. Cultural heritage includes language, geographic roots, food, customs, traditions, and religion. Ethnic groups in the United States include Puerto Rican, Chinese, Serbian, Palestinian, Swedish, Hungarian, Jewish, and many others. Like race, ethnicity can be a basis for unequal treatment, as you'll read shortly.

10-2c Racial-Ethnic Group

People who have distinctive physical and cultural characteristics are a **racial-ethnic group**. Some people use the terms *racial* and *ethnic* interchangeably, but remember that *race* refers to a category based on social perceptions of the physical characteristics with which we're born, and *ethnicity* refers to cultural characteristics that we learn. The term *racial-ethnic* includes both physical and cultural traits.

Describing racial-ethnic groups has become more complex. Beginning with the 2000 Census, people could choose more than one race. And with the 2020 Census, people could write in their race response as "Some other race." The U.S. government now allows people to identify themselves in terms of both race and ethnicity, with separate questions for each. For example, someone who is Latina/o/x could check off "Black" or "White" for race and "Cuban" for ethnic origin. Such choices generate dozens of racial-ethnic categories. People prefer some "labels" to others, and racial and ethnic self-identification varies (Marks, 2021).

---

racial group people who share visible physical characteristics that members of a society consider socially important.

ethnic group people who identify with a common national origin or cultural heritage.

racial-ethnic group people who have distinctive physical and cultural characteristics.

Figure 1.
2020 Census Hispanic Origin Question

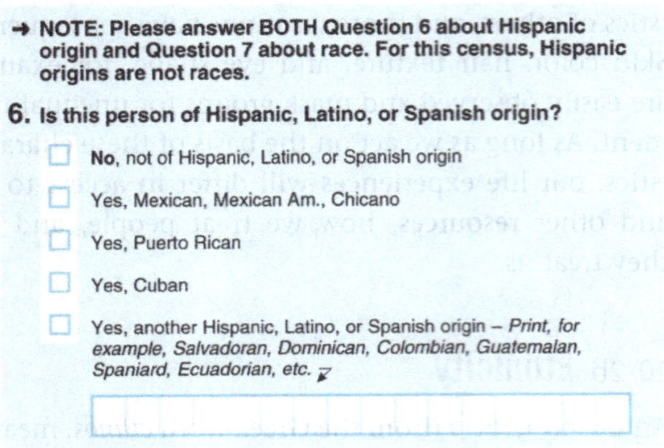
➜ **NOTE: Please answer BOTH Question 6 about Hispanic origin and Question 7 about race. For this census, Hispanic origins are not races.**

6. Is this person of Hispanic, Latino, or Spanish origin?

- ☐ **No**, not of Hispanic, Latino, or Spanish origin
- ☐ Yes, Mexican, Mexican Am., Chicano
- ☐ Yes, Puerto Rican
- ☐ Yes, Cuban
- ☐ Yes, another Hispanic, Latino, or Spanish origin – *Print, for example, Salvadoran, Dominican, Colombian, Guatemalan, Spaniard, Ecuadorian, etc.*

Figure 2.
2020 Census Race Question

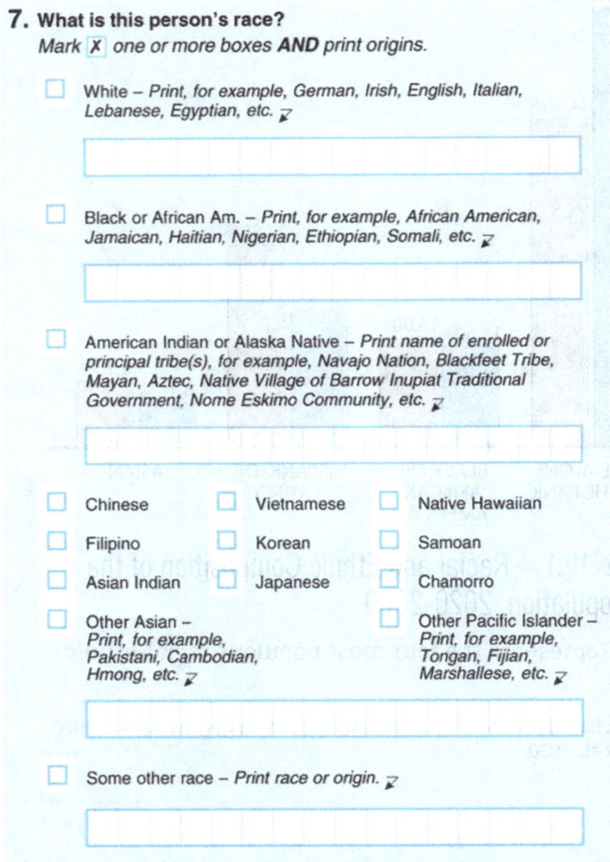
7. What is this person's race?
Mark ☒ *one or more boxes* ***AND*** *print origins.*

- ☐ White – *Print, for example, German, Irish, English, Italian, Lebanese, Egyptian, etc.*
- ☐ Black or African Am. – *Print, for example, African American, Jamaican, Haitian, Nigerian, Ethiopian, Somali, etc.*
- ☐ American Indian or Alaska Native – *Print name of enrolled or principal tribe(s), for example, Navajo Nation, Blackfeet Tribe, Mayan, Aztec, Native Village of Barrow Inupiat Traditional Government, Nome Eskimo Community, etc.*
- ☐ Chinese
- ☐ Vietnamese
- ☐ Native Hawaiian
- ☐ Filipino
- ☐ Korean
- ☐ Samoan
- ☐ Asian Indian
- ☐ Japanese
- ☐ Chamorro
- ☐ Other Asian – *Print, for example, Pakistani, Cambodian, Hmong, etc.*
- ☐ Other Pacific Islander – *Print, for example, Tongan, Fijian, Marshallese, etc.*
- ☐ Some other race – *Print race or origin.*

Jstaley401/Getty Images

10-2d What We Call Ourselves

In 1976, the U.S. government began to use the word *Hispanic* or *Latino* to categorize Americans who trace their roots to Spanish-speaking countries. To be more inclusive of gender, some are now using the term Latinx. About 61 percent prefer Hispanic, 29 percent prefer Latino, 4 percent prefer Latinx, 5 percent prefer something else, and others say it doesn't matter if they're referred to as *Latino* or *Hispanic*. If given a choice, however, 51 percent prefer to identify themselves by their family's country of origin (e.g., Mexican, Cuban, Salvadoran) (Noe-Bustamante et. al, 2020; Pew Hispanic Center, 2012; Jones, 2013).

The Census Bureau uses different terms for specific races—most recently *Black* and *African American*. Currently, 64 percent of people in this group say it doesn't matter whether they're called *Black or African American*, 18 percent prefer African American, and 17 percent prefer Black. These preferences haven't changed much since 1991 (Saad, 2020). Many people, including African American scholars, use *Black* and *African American* interchangeably. Yet, Black is preferred if it is uncertain whether the person is African American.

For *Native Americans, people in* these groups may prefer their tribal identities (e.g., Cherokee, Apache, Lumbi) to being lumped together under a single term. Among White Americans, in contrast, recent immigrants rarely refer to themselves by their country of origin (e.g., Armenian or Armenian American). Third and later generations may even be unsure of their ancestors' roots ("My grandparents on one side of the family were Polish, but I don't know much about my grandparents on the other side").

Jupiterimages/Getty Images

There's much variation in skin color across and within groups. People of African descent have at least 35 different shades of skin tone (Taylor, 2003). So, can you determine someone's race simply by looking at them?

10-3 Our Changing Immigration Mosaic

The United States is a nation of immigrants and has historically and currently both welcomed immigration and feared its consequences. Continued immigration and changes in the population, along with divisions in political opinions, keeps immigration in the spotlight for public debate.

10-3a The Population of People from Other Nations

When counting people who were born in other nations, the Census Bureau refers to this group as *foreign born*, and the definition includes naturalized citizens, documented permanent residents (immigrants), temporary migrants (e.g., foreign students), refugees, and undocumented migrants (e.g., people crossing the U.S. border without documentation) ("Foreign Born," 2016). Among all countries, the United States has one of the highest populations of people from other nations, but there's been a significant shift in their countries of origin. In 1960, more than 70 percent of the people from other nations came from Europe, compared with around 10 percent in 2019 (**Figure 10.2**). Today, most immigrants come from Mexico (24 percent), India (6 percent), China (5 percent), Philippines (4 percent), El Salvador, Vietnam, Cuba and the Dominican Republic (3 percent each), and Guatemala and Korea (2 percent each) (Ward and Batalova, 2023).

The United States has more migrants than any other nation. While it has around 5 percent of the total world population, it is home to almost 20 percent of all migrants (Ward and Batalova, 2023). The number of *undocumented immigrants* has changed over the years, from 180,000 in the early 1980s to a high point of 12 million in 2007 and now approximately 11 million. They make up 26 percent of all residents from other nations, and nearly 4 percent of the nation's population. Almost 53 percent are from Mexico, 21 percent from Central and Latin America, 13 percent from Asia, and 13 percent from other countries, including Canada, Europe, and Africa (Passel and Cohn, 2016; Ward and Batalova, 2023). Most immigrants, documented and undocumented, come to the United States for the same reasons as in the past—religious and political freedom, economic and educational opportunities, and escape from wars and natural disasters (refer to Chapter 15).

Americans have usually opposed admitting large numbers of *refugees*, people who flee their country to escape war, persecution, or death. In 1990, refugee numbers reached a high point at 122,066. Since then, the numbers have been declining. In 2020, 11,840 refugees were admitted to the United States. Most came from the Democratic Republic of the Congo, Burma, and Ukraine. Refugee admissions have decreased drastically since 2015 (Bolter, 2022; Ward and Batalova, 2023).

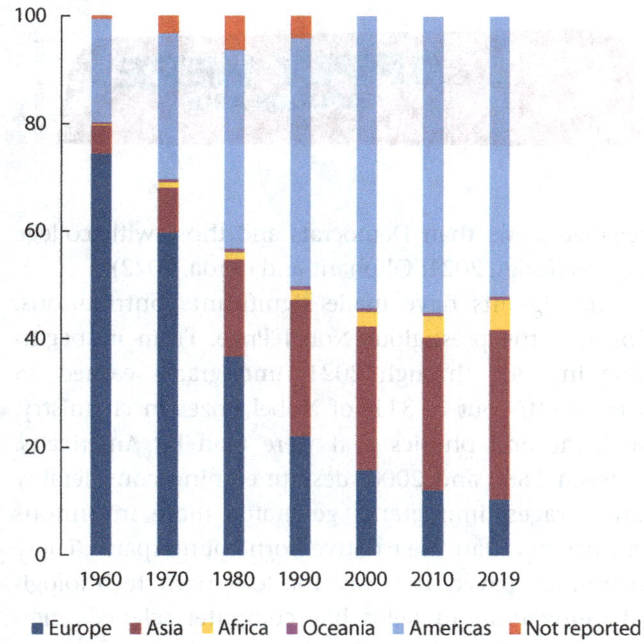

Figure 10.2 Origins of U.S. Population from Other Nations: 1960 to 2019

Source: Based on Migration Policy Institute (MPI), "U.S. Immigrant Population by World Region of Birth, 1960-2019," MPI Migration Data Hub, 2021.

10-3b Americans' Reactions to Immigrants

American attitudes toward immigrants have grown more positive since the mid-1990s, with 75 percent saying immigration is good for the country (Jones, 2021). But views are still mixed: 69 percent say immigration contributes to more economic growth over the long term, and 58 percent say the U.S. is strengthened because of immigrants' hard work and talents (Ekins and Kemp, 2021). And 27 percent believe that immigrants come to the United States to use government services, including welfare. Americans are more divided about immigration levels. More people in 2022, compared to 2020, felt that immigration levels should decrease (28 percent in 2020 to 38 percent in 2022), and fewer people felt those levels should increase (34 percent in 2020 to 27 percent in 2022) (Gallup, 2023). On both issues—whether immigrants are more of a benefit or burden and immigration levels—Republicans and those without college degrees have more

Bloomberg/Getty Images

negative views than Democrats and those with college degrees (Jones, 2021; Oliphant and Cerda, 2022).

Immigrants have made significant contributions. Consider the prestigious Nobel Prize. From its beginning in 1901 through 2021, immigrants earned 35 percent (109 out of 311) of Nobel Prizes in chemistry, medicine and physics that were won by Americans. Between 1880 and 2000, despite earning considerably lower wages, immigrants generated more inventions and patents than their native-born counterparts. These inventions paved the way for long-term technological innovations in fields like computer science, engineering, and medicine. Foreign-born Americans have founded or co-founded about 25 percent of U.S. tech companies (including Intel, Yahoo, and Google) and more than 52 percent of the Silicon Valley start-ups between 1995 and 2005 (National Foundation for American Policy, 2021; Najam, 2016; Agress, 2017; Akcigit et al., 2017).

10-3c Two Big Questions

Does immigration redirect jobs needed by native-born Americans to immigrants? And does immigration burden government budgets?

Of the nation's 161 million people in the labor force, 83 percent are U.S.-born, 15 percent are documented immigrants, and 5 percent are undocumented. Regardless of legal status, immigrants work in a variety of jobs and don't make up the majority of workers in any one industry, but employment patterns differ markedly. Documented immigrants are most likely to be in professional, management, or business and finance jobs (37 percent) or service jobs (22 percent). Undocumented immigrants, by contrast, are most likely to be in service (32 percent) or construction jobs (16 percent) (López and Bialik, 2017).

Some Americans (34 percent) believe that immigration hurts wages. Except for people who did not earn a high school diploma or equivalent and who have few skills that employers seek, there's little evidence that immigration has had a negative impact on native-born workers' wages or employment levels. Whether documented or undocumented, unskilled immigrants are more willing than their native-born counterparts to take grueling but low-paying jobs in service occupations (e.g., child care, housekeepers, cooks), agriculture, and the seafood industry. ("The State of American Jobs," 2016; Borjas, 2017; Gee et al., 2017).

Does immigration burden government budgets? The answer is mixed. President Biden's most recent budget includes funding to increase and expedite legal pathways for migration (focusing on Cuba, Haiti, Nicaragua, and Venezuela) as well as more border security funding. First-generation immigrants generally cost state and local governments about $57 billion a year, mainly because of health care and education services. The second generation, better educated and with higher wages and salaries, adds about $30 billion a year to government finances. The third generation pays almost $224 billion a year in taxes. Undocumented immigrants pay approximately $12 billion a year in state and local taxes—about 8 percent of their income compared with just 5 percent that the wealthiest 1 percent pays. Each immigrant, documented or undocumented, generates 1.2 local jobs because businesses need more people to meet the increased demand for goods and services (White House, 2023; Hong and McLaren, 2015; Blau and Mackie, 2016; Gee et al., 2017).

On a number of measures, immigrants and their children—both documented and undocumented—outperform native-born Americans. Children of immigrants are more likely than their U.S.-born peers to live in a two-parent household, to have a parent with a secure job, to earn a college degree, to attend religious services regularly, and to be "prodigious job creators." While immigrants and native-born populations have similar

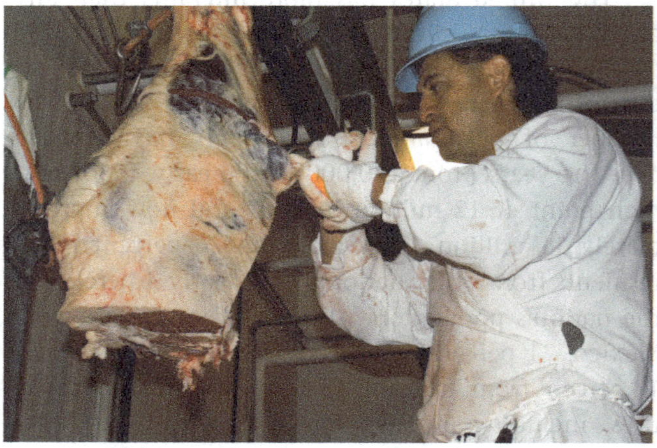
Design Pics Inc/Alamy Stock Photo

In Colorado, Iowa, and other states, refugees and immigrants from Asia, Africa, Mexico, and Central America fill $15-an-hour jobs that native-born workers spurn. According to a local resident who worked at a pork processing plant for nearly 40 years, "Even if pay were raised to $20 or $25 an hour, I don't think you could get White guys" (Cohen, 2017, p. A12; Etter and Singh, 2017).

rates of college completion (34 and 35 percent), a higher percentage (47 percent) of recently arrived immigrants have college degrees (Ewing et al., 2015; Waters and Pineau, 2015; Landgrave and Nowrasteh, 2017; Livingston, 2016; Ward and Batalova, 2023).

10-4 Dominant and Minority Groups

Race and ethnicity often make people feel like outsiders, or "the other." Otherness is being different or having characteristics that set you apart from the dominant group (Thorpe-Moscon and Pollack, 2014).

10-4a What Is a Dominant Group?

A **dominant group** is any physically or culturally distinctive group that has the most economic and political power, the greatest privileges, and the highest social status. As a result, the dominant group can treat other groups as subordinate. In most societies, for instance, men are a dominant group because they have more status, resources, and power than women (refer to Chapters 8 and 9).

Dominant groups don't necessarily have to be the largest in number of members. From the seventeenth century until 1994, about 10 percent of South Africans were White and had almost complete control of the Black population. Because of *apartheid*, a formal system of racial segregation, the Black residents couldn't vote, lost their property, and had minimal access to education and politics. Apartheid ended in 1994, but most Black South Africans are still a minority because White people "hold the best jobs, live in the most expensive homes, and control the bulk of the country's capital" (Murphy, 2004. p. A4). South Africa remains one of the nations with the highest wealth inequality in the world (Chancel et al., 2022), and race plays a role.

10-4b What Is a Minority?

Sociologists describe Latina/o/x American, Black Americans, Asian Americans, Middle Eastern Americans, and Native Americans as minorities. A **minority** is any group that may be treated differently and unequally because of their physical, cultural, or other characteristics. The characteristics include gender, age, sexual orientation, religion, ethnicity, or skin color. Because the term minority can refer to categories outside of race or ethnicity, it is most accurate to use *people of color* or *racial-ethnic minority* when referring to those specific groups. Minority groups may have more members than a dominant group, but they have less power, privilege, and social status. For example, middle-class Black people are much more likely than their White counterparts to be called by debt collectors, even though both groups have similar debt levels and repayment rates (Ruetschlin and Asante-Muhammad, 2013). **Table 10.1** offers other examples of everyday privileges associated with skin color.

dominant group a physically or culturally distinctive group that has the most economic and political power, the greatest privileges, and the highest social status.

minority people who may be treated differently and unequally because of their physical, cultural, or other characteristics.

Table 10.1 Am I Privileged?

Most White people don't feel privileged because they don't have high levels of income or wealth. Nonetheless, they enjoy everyday benefits and take them for granted simply because they're members of the dominant group. Would you add any other advantages of being White?
1. I can go shopping and feel fairly sure that I won't be followed or harassed by store detectives.
2. If a traffic cop pulls me over, I can be sure that I haven't been singled out because of my race.
3. I can be late to a meeting without having the lateness reflect on my race.
4. I can turn on the television and see people who look like me represented in positive ways and in a wide range of roles.
5. I live in a safe neighborhood with good schools.
6. I never think twice about calling the police when trouble occurs.
7. I can wear a hoodie and not worry about others thinking that I'm up to no good.
8. I don't get dirty looks if I listen to loud music at a gas station.

Sources: McIntosh, 1995; Independent Television Service, 2003; Williams, 2014.

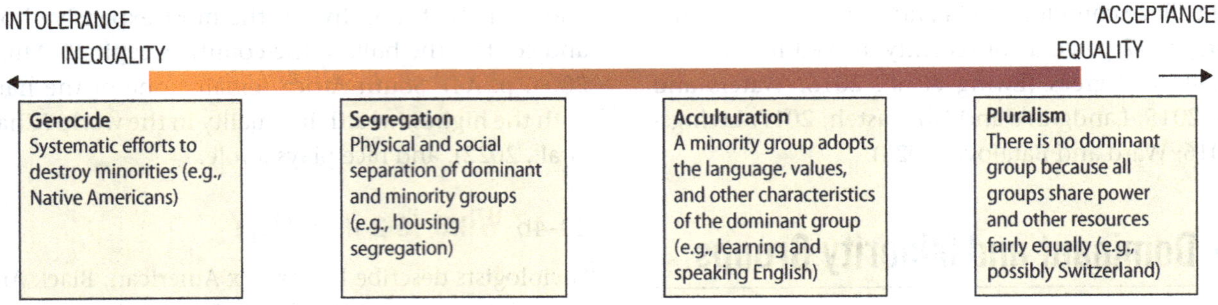

Figure 10.3 Continuum of Some Dominant-Minority Group Relations

10-4c Some Patterns of Dominant-Minority Group Relations

To understand some of the complexity of dominant-minority group relations, think of a continuum. At one end of the continuum is genocide; at the other end is pluralism (**Figure 10.3**).

Genocide

Genocide is the systematic effort to kill all members of a particular ethnic, religious, political, racial, or national group. By 1710, the colonists in America had killed thousands of Native Americans, poisoned others, and promoted scalp bounties. In 1851, the governor of California officially called for the extermination of all Native Americans in the state (de las Casas, 1992; Churchill, 1997). Worldwide, in the 20th and 21st centuries, genocides have taken the lives of many, including those in Darfur (400,000), the Tutsi of Rwanda (800,000), Bosnian Muslims of Bosnia-Herzegovina (200,000), the Cultural Revolution of China (27 million), and the Holocaust (6 million). These numbers are estimates since it is difficult to accurately determine the total number of lives lost (The Genocide Education Project, 2014; United States Holocaust Memorial Museum, 2020).

Segregation

Segregation is the physical and social separation of dominant and minority groups. In 1954, the Supreme Court ruling in *Brown v. Board of Education* declared *de jure*, or legal, segregation unconstitutional. A variety of federal laws then prohibited racial segregation in public schools and discrimination in employment, voting, and housing.

De facto, or informal, segregation has replaced *de jure* segregation in the United States and many other countries. Some *de facto* segregation may be voluntary, such as when minorities prefer to live among their own racial or ethnic group. In most cases, however, *de facto* segregation is due to discrimination. In the case of equally qualified homeseekers, for instance, realtors sometimes show people of color fewer homes and apartments than they do White people, restricting their housing options, including access to neighborhoods with higher-performing schools, public playgrounds, and lower crime rates (Turner et al., 2013; refer also to Chapter 15 on residential segregation).

Acculturation and Assimilation

Many racial-ethnic minority group members blend into U.S. society through **acculturation**, the process of adopting the language, values, beliefs, and other characteristics of the host culture (e.g., learning English, celebrating Thanksgiving). Acculturation doesn't include intermarriage, but the newcomers merge into the host culture

Moviestore collection Ltd/Alamy Stock Photo

Based on true events, *Hidden Figures* depicts the lives of three Black women who helped pioneer space travel and desegregation at NASA.

---

genocide the systematic effort to kill all members of a particular ethnic, religious, political, racial, or national group.

segregation physical and social separation of dominant and minority groups.

acculturation the process of adopting the language, values, beliefs, and other characteristics of the host culture.

in most other ways. **Assimilation** involves conforming to the dominant group's culture, adopting its language and values, and intermarrying with that group. Some may use *assimilation* and *acculturation* interchangeably, but the former absorbs, rather than just changes, a group culturally.

Pluralism

In **pluralism**, sometimes called *multiculturalism*, minority groups maintain many aspects of their original culture—including using their own language and marrying within their own racial or ethnic group—while living peacefully with the host culture. Pluralism is especially evident in urban areas with large racial and ethnic communities (e.g., "Little Italy," "Greek Town," "Little Korea," "Spanish Harlem"). U.S. racial and ethnic minorities have numerous ethnic newspapers and radio stations and the same constitutional rights (such as freedom of speech) as the dominant group. Nonetheless, people of various races and ethnicities don't always experience the same social standing, and there can be considerable racial-ethnic friction.

A number of colleges and professional teams have replaced their logos, nicknames, and mascots because many Native Americans have denounced them as racist. Most recently, Cleveland's MLB team became the Guardians (formerly Indians), and the Washington D.C. NFL team became the Commanders (dropping the former name Redskins).

10-5 Some Sources of Racial-Ethnic Friction

During a trip to Switzerland, Oprah Winfrey asked a clerk at an expensive shop to show her a $37,000 purse. The clerk refused: "That one will cost too much, you won't be able to afford that." Winfrey didn't know for sure whether or not the snub was motivated by racism, but she said that the incident was what "people with black or brown skin experience every day" (Goyette, 2013). Ahmed Mohamed, a 14-year-old Sudanese-American student in Texas, was arrested and suspended after teachers mistook his homemade digital clock for a bomb (Fantz et al., 2015). Would people have reacted differently if Winfrey and Mohamed had been White? And were the reactions racist?

10-5a Racism

Racism refers to beliefs that one's own racial group is inherently superior to other groups. Using this definition, anyone can be racist if they believe that another group is inferior. It's a way of thinking about racial and ethnic differences that justifies and preserves the social, economic, and political interests of dominant groups.

Many White people believe that racism has been pretty much "solved." After all, Americans elected a Black president twice and a Black Vice President. And there are 133 people of color in the 118th Congress as well as one Latina and two Black Supreme Court justices. In fact, racism is widespread. Consider that 63 percent of hate crimes are motivated by race, ethnicity, or ancestry bias (Schaeffer, 2023; U.S. Department of Justice, 2023) (refer to Chapter 7). Racism fuels both prejudice and discrimination.

10-5b Prejudice

Prejudice is an *attitude* that prejudges people, usually in a negative way, who are different from "us" in race, ethnicity, religion, or other ways. If an employer assumes, for example, that White workers will be more productive than Black or Latina/o/x workers, they are prejudiced. Prejudice isn't one-sided because *anyone* can be prejudiced ("White people can't be trusted" or "Black women can't afford expensive purses"). Prejudice is most evident in stereotypes and scapegoating.

---

assimilation conforming to the dominant group's culture, adopting its language and values, and intermarrying with that group.

pluralism minority groups maintain many aspects of their original culture while living peacefully with the host culture.

racism beliefs that one's own racial group is inherently superior to other groups.

prejudice an attitude that prejudges people, usually in a negative way.

A **stereotype** is an oversimplified or exaggerated generalization about a group of people (refer to, for example, www.stuffwhitepeoplelike.com). Stereotypes can have a positive spin ("Black people are great athletes") or a more negative one ("Black people break laws"). Whether positive or negative, stereotypes distort reality. Some Black people have exhibited great athletic talents while others have broken laws, just like people in other groups. Whatever the intention, stereotypes lump people together and reinforce the belief that many traits are biological, fixed, and cannot be changed. Once established, stereotypes are difficult to change because people often dismiss any evidence to the contrary as an exception ("For an Asian, Jeremy Lin is a terrific basketball player").

Stereotypes can lead to a displacement of anger and aggression on **scapegoats**, individuals or groups whom people blame for their own problems or shortcomings ("They didn't hire me because the company wants Black people" or "I didn't get into that college because Asian students are at the top of the list"). Racial and ethnic minorities are easy targets because they typically differ in physical appearance and are usually too powerless to strike back.

At one time or another, almost all newcomers to the United States have been scapegoats, with some facing more backlash than others. Recent immigrants can be especially targeted for scapegoating in times of economic hardship ("Latina/o/x immigrants are replacing Americans in construction jobs"). Prejudice, stereotypes, and scapegoating are attitudes, but they often lead to discrimination.

10-5c Discrimination

Discrimination is *behavior* that treats people unequally because of some characteristic (e.g., race, ethnicity, gender, age, religion, sexual orientation). It encompasses all sorts of actions, ranging from social slights (e.g., not inviting coworkers to lunch because of their race or ethnicity) to the rejection of job applications and racially motivated hate crimes. Discrimination can be subtle (e.g., not sitting next to someone) or blatant (e.g., racial slurs), and it occurs at individual and institutional levels.

---

stereotype an oversimplified or exaggerated generalization about a group of people.

scapegoats individuals or groups whom people blame for their own problems or shortcomings.

discrimination behavior that treats people unequally because of some characteristic.

individual discrimination unequal treatment on a one-to-one basis.

institutional discrimination unequal treatment because of a society's everyday laws, policies, practices, and customs.

Individual discrimination is unequal treatment on a one-on-one basis, usually by a dominant group member against someone in a minority group. Americans have become less optimistic that people have equal opportunities regardless of race. Forty percent feel people of color have equal job opportunities with Whites, which is a less than the 53 percent who felt this way in 2008 (Saad, 2021). A study of Uber and Lyft drivers in Boston and Seattle found that Black people had to wait as much as 35 percent longer for some rides, and drivers were more than twice as likely to cancel the rides of passengers with Black-sounding names than passengers with White-sounding names (Ge et al., 2016; "On Views of Race...," 2016). And Uber and Lyft drivers themselves are more likely to face discrimination. Sixty-nine percent of drivers of color said their driving privileges had been deactivated compared with only 57 percent of White drivers having this experience (Bienasz, 2023).

In **institutional discrimination** (also called *institutionalized discrimination*, *systemic discrimination*, and *structural discrimination*), minority group members experience unequal treatment because of a society's everyday laws, policies, practices, and customs. For example, sickle cell disease (SCD) affects three times as many Americans as cystic fibrosis (CF), but government spending on CF is four times higher than SCD. Why? About 90 percent of SCD patients are Black, while CF affects mostly White people. The CF community has more wealthy and powerful parents and advocates who pressure the government to do research and provide treatment and develop and fund private foundations, which, in turn, underwrite pharmaceutical companies' creation of drugs, even though a new drug only works for a handful of patients (Butler, 2015). Thus, although individuals may be unprejudiced, institutional policies and processes contribute to racially unequal access to health care.

Discrimination, both individual and institutional, also occurs *within* racial-ethnic groups. A study of New York City's Korean-owned nail salons found an "ethnic caste system." Among Asian manicurists, Koreans earned twice as much as their non-Korean counterparts. The latter frequently worked longer hours, were often forced into less desirable salons that had fewer customers and paltry tips, were told not to chat during their entire 12-hour shifts, and were assigned male customers that manicurists often dread because of the men's "thick toenails and hair-covered knuckles" (Nir, 2015).

10-5d Relationship Between Prejudice and Discrimination

Sociologist Robert Merton (1949) created a model showing how the relationship between prejudice and

Table 10.2 Relationship Between Prejudice and Discrimination

Is the Person Prejudiced?	Does the Person Discriminate? Yes	Does the Person Discriminate? No
Yes	Prejudiced discriminator (e.g., a prejudiced person who attacks minority group members verbally or physically)	Prejudiced nondiscriminator (e.g., a prejudiced person who goes along with equal employment opportunity policies)
No	Unprejudiced discriminator (e.g., an unprejudiced person who joins a club that excludes minorities)	Unprejudiced nondiscriminator (e.g., an unprejudiced employer who hires minorities)

Source: Based on Merton, 1949.

discrimination can vary. His model includes four types of people and their possible response patterns (**Table 10.2**).

Unprejudiced nondiscriminators aren't prejudiced and don't discriminate. They believe in the American creed of freedom and equality for all and cherish egalitarian values. They may not do much, however, individually or collectively, to change discrimination. In contrast, but equally consistent in attitude and action, are *prejudiced discriminators* who are both prejudiced and who discriminate. They're willing to defy laws, such as not renting to minorities, because of their beliefs.

Unprejudiced discriminators aren't prejudiced but discriminate because it's expedient or in their own self-interest to do so. If, for example, an insurance company charges higher automobile insurance premiums to people with low occupational and educational levels (often people of color), agents will implement these policies even though they themselves aren't prejudiced.

Prejudiced nondiscriminators are prejudiced, but they don't discriminate. Despite their negative attitudes, they hire people of color and are civil in everyday interactions because they believe they must conform to antidiscrimination laws or situational norms. If, for example, most of their neighbors or coworkers don't discriminate, prejudiced nondiscriminators will go along with them.

David Sipress/The New Yorker Collection/The Cartoon Bank

"You look like this sketch of someone who's thinking about committing a crime."

10-6 Major U.S. Racial and Ethnic Groups

Levels of diversity vary around the United States. The diversity index measures the chance that two randomly selected people would be from different racial and ethnic groups. The states with the most diversity are Hawaii, California, Nevada, Maryland, and Texas (Hubbard, 2021; U.S. Census Bureau, 2021) (**Figure 10.4**). Of the major racial-ethnic groups, some encounter more barriers than others, but all have numerous strengths that enhance U.S. society. *Let's* begin with White ethnic groups whose ancestors are from Europe.

10-6a White Non-Hispanic Americans: A Declining Majority

During the seventeenth century, English immigrants settled the first colonies in Massachusetts and Virginia. Other immigrants included people from Wales and Scotland. Most of these groups spoke English. Some of the immigrants were affluent, but many were economically underprivileged or had criminal backgrounds.

Diversity

About 59 percent of Americans identify as White, non-Hispanic, with about 53 percent of Americans (170 million people) identifying their ancestry as European. The largest groups have ancestors from Germany, Ireland, England, Italy, Poland, France, and the Scandinavian countries.

Characteristics and Changes

Subsequent generations of these early White immigrants generally looked down on later waves of immigrants from southern and eastern Europe. They viewed the newcomers as inferior, dirty, lazy, and uncivilized because they differed in language, religion, and customs. New England, which was 90 percent Protestant, was particularly hostile to Irish Catholics, characterizing them as irresponsible and shiftless (Feagin and Feagin, 2008).

All the later waves of European immigrants faced varying degrees of hardship in adjusting to the new land because the first English settlers had a great deal of power in shaping economic and educational institutions. In

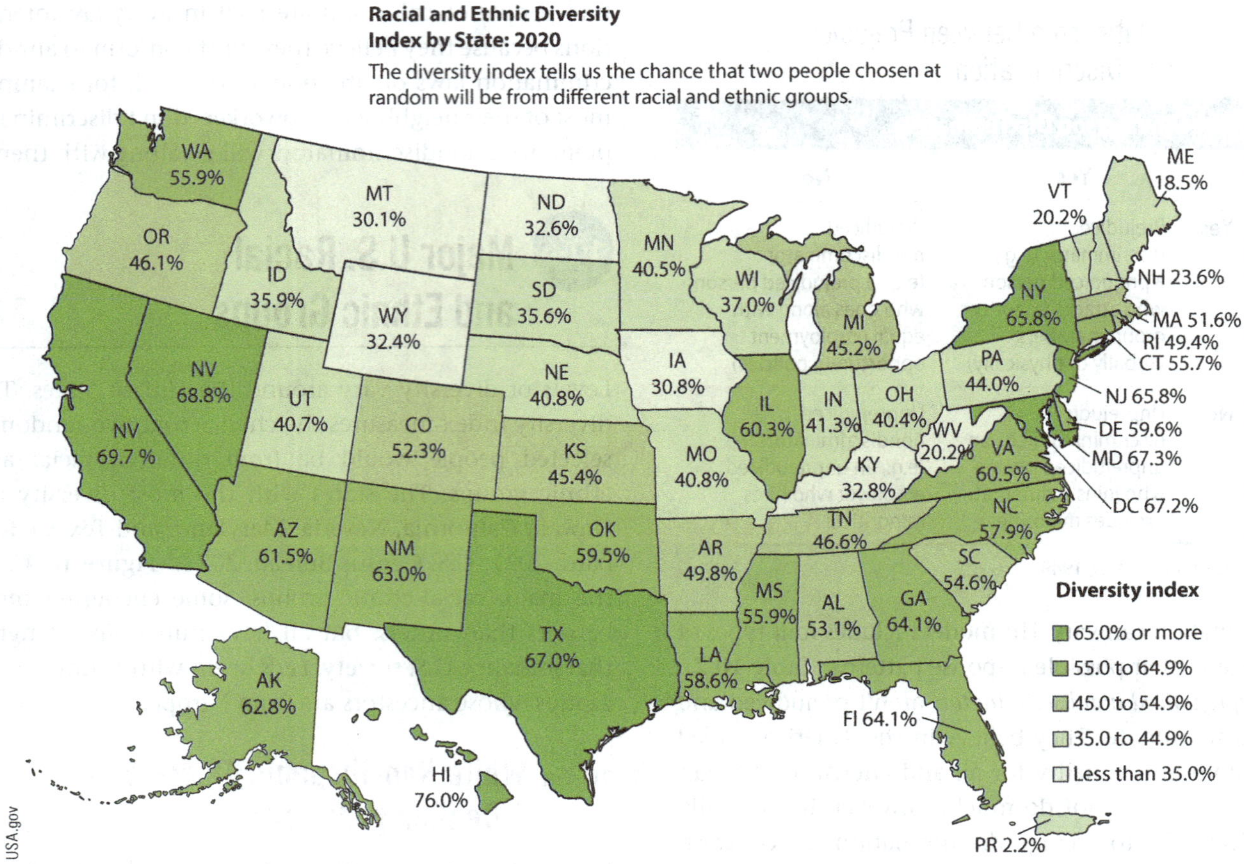

Figure 10.4 U.S. Racial and Ethnic Diversity Index by State, 2020

Source: U.S. Census Bureau, 2021. "2020 Census: Racial and Ethnic Diversity Index by State".

response to prejudice and discrimination, many of the immigrants founded churches, schools, and recreational activities that maintained their languages and traditions (Myers, 2007).

Despite stereotypes, prejudice, and discrimination, European immigrants began to prosper within a few generations. They surmounted numerous obstacles and became influential in all sectors. Overall, they now fare much better financially than most other groups. This doesn't mean that all are rich. In fact, in absolute numbers, economically underprivileged White people outnumber those of other racial-ethnic groups (refer to Chapter 8).

10-6b Latina/o/x Americans: The Largest Minority

Latina/o/x is the youngest and largest racial or ethnic group in the United States, with nearly 19 percent of the nation's population identifying with this group (**Figure 10.5**). The size and growth of the Latina/o/x population in this century is due mainly to births in the United States, not recent immigration. On average, Latinas have three births each, one more than White, Black, and Asian or Pacific Islander women.

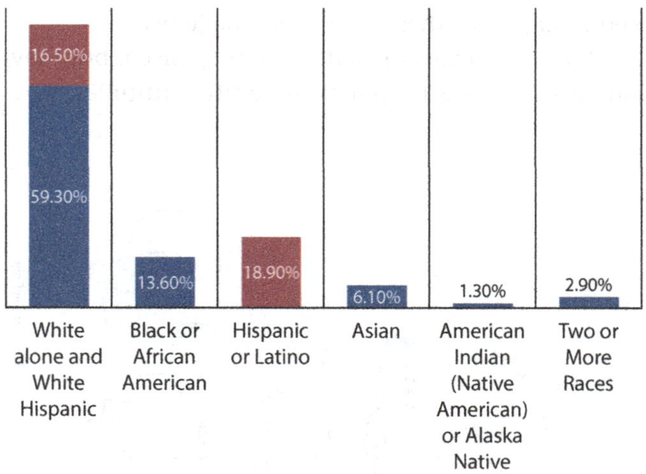

Figure 10.5 Major U.S. Racial and Ethnic Groups, 2021

Source: Based on U.S. Census Bureau, Population Estimates, July 1, 2021 (V2021)..

Jeffrey Macmillan for U.S News & World Report

Dominican-born Alfredo Rodriguez is one of numerous successful Latino businessmen who are rebuilding neglected inner-city neighborhoods. In 1985, Rodriguez bought his first grocery store in Queens, New York, with the $25 a week his mother had been setting aside for him for a decade. In 2002, he purchased a supermarket in Newark, New Jersey, to meet the needs of local Latino shoppers. Five years later, his Xtra Supermarket had annual sales of $9 million (Rayasam, 2007).

Since the Great Recession in 2007, however, Latina/o/x immigration and birth rates have declined ("Hispanic Heritage Month...," 2016; Patten, 2016; Stepler and Lopez, 2016).

Diversity

Worldwide, only Mexico has a larger Latina/o/x population than the United States. More persons of Puerto Rican origin now live in the 50 states and the District of Columbia than in Puerto Rico (Cohn et al., 2014; "Hispanic Heritage Month...," 2016).

Some Latina/o/x people trace their roots to the Spanish and Mexican settlers who established homes and founded cities in the Southwest before the arrival of the first English settlers on the East Coast. Others are recent immigrants or children of the immigrants who arrived in large numbers at the beginning of the twentieth century. Of the Latina/o/x people living in the United States, most are from Mexico, but people identifying with this group come from different nations and vary widely in their customs, cuisines, and cultural practices (Figure 10.6).

Characteristics and Changes

As with other groups, Latina/o/x people's socioeconomic status reflects a number of interrelated factors, particularly education, occupation, English language proficiency, and recency of immigration

Speaking a language other than English at home has become more common. In 2019, there were 68 million people who spoke a language other than English at home, which is three times as many as in 1980. Out of those who speak at language other than English at home, nearly 62 percent speak Spanish or Spanish Creole, followed by 5 percent who speak Chinese. The number who speak English has grown too—from 187 million in 1980 to 241 million in 1990 (Dietrich and Hernandez, 2022).

Latina/o/x median household income is lower than the national median but higher than that of Native American and Black households (**Figure 10.7**). Over 57 percent of Latina/o/x families earn $50,000 or more each year, up considerably from only 7 percent in 1972, but 11 percent live below the supplemental poverty measure (SPM) threshold (Semega and Kollar, 2022; Creamer et al., 2022).

Latina/o/x people have some of the lowest education levels (**Figure 10.8**), but there's considerable variation across subgroups. For instance, around 57 percent of Venezuelan immigrants (ages 25 and older) had bachelor's degrees or a higher degree, which is much higher than all U.S. adults and Latina/o/x individuals as a whole (Hoffman and Batalova, 2023). Yet, other Latina/o/x immigrants have lower levels of bachelor's degree attainment: Mexican (17 percent), Honduran

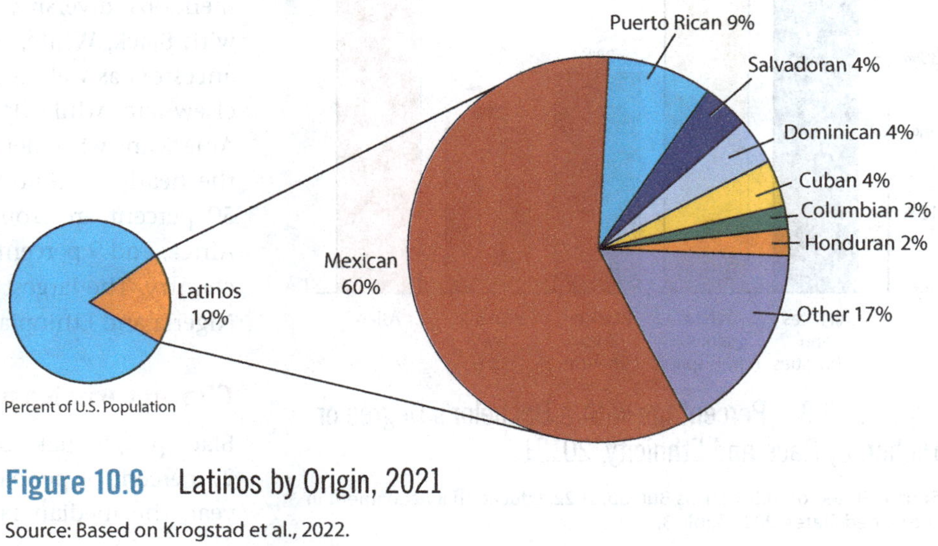

Figure 10.6 Latinos by Origin, 2021

Source: Based on Krogstad et al., 2022.

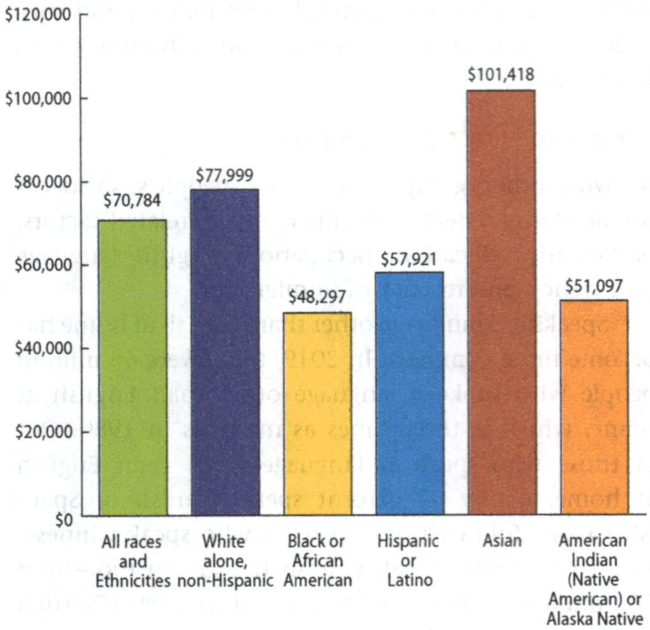

Figure 10.7 U.S. Median Household Income, by Race and Ethnicity, 2021

Source: Based on U.S. Census Bureau, 2021, Table A-2. Households by Total Money Income, Race, and Hispanic Origin of Householder: 1967 to 2021.

(12 percent), Salvadoran (8 percent), and Guatemalan (6 percent) (Noe-Bustamante et al., 2020).

Despite numerous economic, legal, and social barriers, 77 percent of Latina/o/x people, compared with

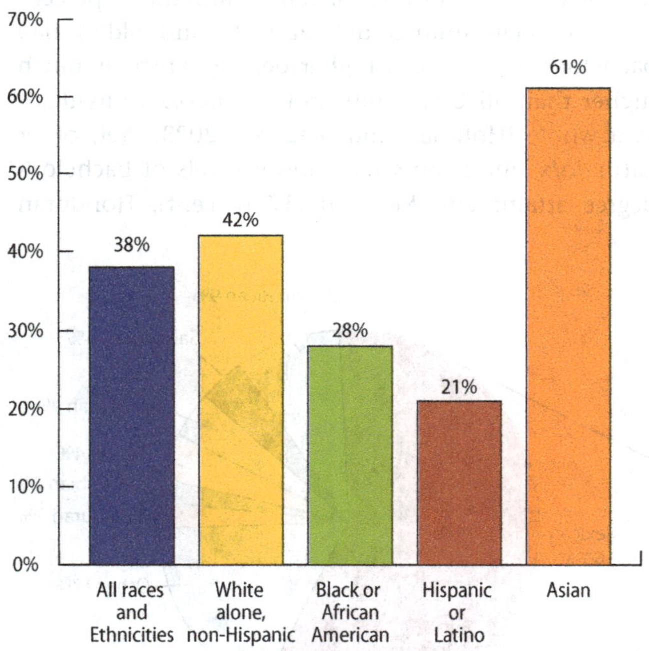

Figure 10.8 Percentage with a Bachelor's Degree or Higher, by Race and Ethnicity, 20121

Source: Based on U.S. Census Bureau, 2022, Educational Attainment in the United States, 2021, Table 3.

62 percent of all U.S. adults, believe that most people can get ahead if they're willing to work hard. Such attitudes help explain why many Latina/o/x people are successful. Across 10 Latino groups, the second generation is working in better jobs with higher occupational status than their first-generation parents. Fully 72 percent expect their children will be better off financially than they themselves are right now. Latina/o/x people are making gains in economic areas, including home ownership. Latina/o/x home ownership has increased to nearly 49 percent and has been growing for the last eight years (National Association of Hispanic Real Estate Professionals, 2022).

From 2007 to 2012, Latina/o/x people launched over 60 times more new small businesses than any other group. Hispanic-owned businesses increased by 8.6 percent, making up 6.5 percent of all businesses in the United States (Lee, 2023). Since 2007, the number of Latina-owned businesses has grown by 137 percent, outpacing all other categories of minority women ("2012 Survey of Business Owners," 2015; Berenson, 2016; Lopez et al., 2018; Tran, 2016).

10-6c Black and African Americans: A Changing Minority

The 46 million Black and African American people in the United States are the second largest racial-ethnic group, making up 13.6 percent of the population (Figure 10.5). Over the past 40 years, Black people have made progress on many fronts, but racial gaps persist.

Diversity

Most Black Americans share a common characteristic: They're members of the only group ever brought to the United States involuntarily and legally enslaved. The term *Black American* can be used to encompass tremendous diversity, including native-born Americans with Black, White, Native American, and/or Latina/o/x ancestors as well as recent immigrants from Africa and elsewhere, while *African American* specifically denotes Americans who identify as being of African descent. Of the nearly 4 million Black people from other nations, 50 percent are from the Caribbean, 36 percent from Africa, and 9 percent from a Central or South American country. The largest African-born populations are from Nigeria and Ethiopia (Anderson, 2015).

Characteristics and Changes

Black people make up 13 percent of the population but 23 percent of people living in poverty. At $48.000 per year, the median family income of Black households

Dappled History/Alamy Stock Photo

Science History Images/Alamy Stock Photo

Few people know that Madame C. J. Walker (1867–1919), who manufactured hair care products for Black women, was one of the first American female millionaires or that Dr. Charles R. Drew (1904–1950) was a renowned surgeon, teacher, and researcher. He founded two of the world's largest blood banks, saving untold lives during and since World War II.

is the lowest of all racial-ethnic groups (**Figure 10.7**). Between 2010 and 2021, the number of Black households with annual incomes of $50,000 or more rose from 36 to 49 percent (Semega and Kollar, 2022).

A growing share of Black people are completing high school and college but lag behind White and Asian people in getting a bachelor's or advanced degree (**Figure 10.8**). Since the mid-1990s, Black women have outpaced Black men (and Latinos) in educational attainment. In 2021, for example, 31 percent of Black women, compared with 25 percent of Black men, had completed four years of college or more. Black women have higher labor force participation rates than other minority or White women and are far more likely to work after motherhood, but they experience the highest gender wage gaps (Holloway, 2016; U.S. Census Bureau, "Educational Attainment…," 2022; refer also to Chapter 9).

The wage gap between college-educated Black and White Americans was larger in 2015 than in 1979, primarily because wage disparities build up over time. Some economists estimate that at least 30 percent of the Black–White wage gap is due to differential treatment in hiring, firing, and pay rather than differences in formal schooling or specific job skills (Fryer et al., 2011; Wilson and Rodgers, 2016).

Economic factors alone don't explain race inequality. For example, White mothers with less than a high school education have lower infant mortality rates than well-educated middle-class Black mothers. Affluent Black families, with annual incomes of more than $100,000, are four times more likely to live in economically underprivileged neighborhoods than comparable White families. Some Black people prefer lower-income neighborhoods because of family ties, but most live in lower-income areas because of segregation and White hostility in many suburbs (Eligon and Gebeloff, 2016; Matthew and Reeves, 2016).

Despite discrimination, many Black people are successful. The share of Black people in Congress is at an all-time high, increasing from 2 percent in 1971 to 11 percent in 2022, which is 60 out of 534 seats in the 118th Congress. Also, the college completion gap between White and Black people has narrowed by 20 percent since 1976 ("2012 Survey of Business Owners," 2015; Bialik and Cilluffo, 2017; Schaeffer, 2023). In 2020, Black-owned businesses represented 3 percent of all businesses in the United States. This is a 14 percent increase from 2017 (Leppert, 2023).

10-6d Asian Americans: A Model Minority?

Asian Americans comprise just over 6 percent of the population (Figure 10.5). In 2012, Asians became the fastest-growing racial or ethnic group in the country; over 60 percent of the growth came from international migration. Hawaii is the only state with an Asian majority of 56 percent ("Asians Fastest-Growing…," 2013; "Asian-American and Pacific…," 2017).

Diversity

Asian Americans encompass a broad swath of cultural groups. They come from over 26 countries in East and Southeast Asia (e.g., China, Korea, Vietnam, Cambodia, the Philippines) and South Asia (especially India, Pakistan, and Sri Lanka) and speak at least 19 languages in the United States (Ryan, 2013; "Asian-American and Pacific…," 2017).

These diverse origins mean that there are huge differences in languages and dialects (and even alphabets), religions, cuisines, and customs. The largest Asian American group is Chinese, followed by Indian and Filipino (**Figure 10.9**). At least 85 percent of Asian Americans

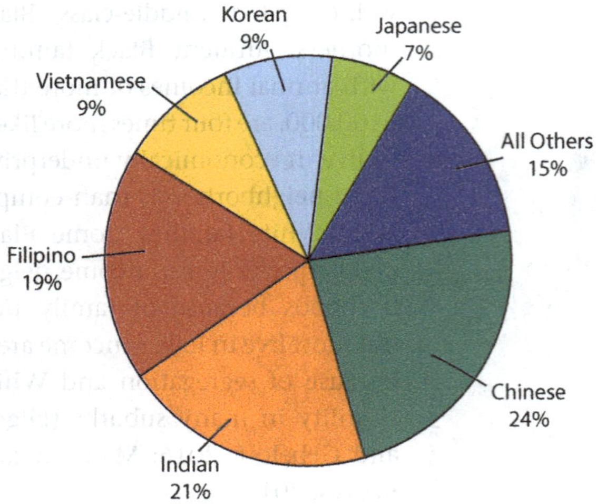

Figure 10.9 Asian Americans by Origin

Source: Based on Budiman and Ruiz, 2021.

trace their roots to only six countries—China, India, Japan, Korea, the Philippines, and Vietnam (Budiman and Ruiz, 2021).

Characteristics and Changes

Compared with White people and other racial-ethnic groups, many Asian Americans are doing well. They have the highest education levels (**Figure 10.8**), and almost 31 percent have a graduate or professional degree, compared with only 14 percent of the general population.

Alex Wong/Getty Images

Indian American students have won the Scripps National Spelling Bee 21 out of the last 23 competitions. The winning words in recent Scripps competitions were *murraya* and *moorhen.* (Am I the only one who had to look them up?) Indian Americans' extraordinary performance has been attributed to factors like hard work, sacrifice, pursuing a championship over many years, and growing up in households where one or both parents are well-educated professionals who set high academic goals (Dhingra, 2022; Heim, 2015).

Because of their high education levels, almost half of Asian Americans are in highly skilled and well-paying occupations such as information technology, science, engineering, and medicine ("Asian-American and Pacific...," 2017). Consequently, they have the highest median household income (**Figure 10.7**).

Because of their educational and economic success, Asian Americans are often hailed as a "model minority." Such labels are misleading, however, because there's considerable variation across subgroups. For example, more people who are Indian (75 percent), Malaysian (65 percent), Mongolian (60 percent), and Sri Lankan (60 percent) have at least a bachelor's degree compared with people who are Laotian (18 percent) and Bhutanese (15 percent) (Budiman and Ruiz, 2021).

Many Asian college students succeed, despite discrimination, because they work hard to achieve their goals. Stereotypes of Asian Americans as hard-working and high-achieving can enhance performance: Teachers and guidance counselors place the students in advanced classes, the honors track, and give them extra help with homework and college applications. On the other hand, the model minority stereotype pigeonholes Asian students into specific careers and doesn't give them the chance "to see themselves as something other than doctors, engineers, or accountants," assumes that all minorities can overcome institutionalized discrimination, and holds Asian Americans to unrealistically high standards (Kay et al., 2013; Leung, 2013; Lee and Zhou, 2015).

Applauding Asian educational attainment obscures other barriers. Compared with White people, for example, Asians are nearly twice as likely to hold PhDs, law degrees, MBAs, or MDs, yet they earn 5 percent less. Among technology companies—including those in Silicon Valley that hire a disproportionately high number of Asian professionals—Asians are vastly underrepresented in upper management: "Whites are the only group whose proportion increases as they rise through the ranks" (Chin, 2016: 70; Hilger, 2016).

10-6e Native Americans: A Growing Nation

Native Americans used to be called the "vanishing Americans" but have "staged a surprising comeback" due to higher birth rates, a longer life expectancy, and better health services (Snipp, 1996, p. 4). There are about 4 million Native Americans and Alaska Natives (not identifying with another race), and they make up around 1.3 percent of the population. This group is expected to increase to nearly 2.5 percent (10 million people) by 2060 (U.S. Census Bureau, 2022; Vespa, 2020). There are actually more people who identify as Native American because census counts for Native Americans are lower than actual numbers. "Census Bureau has identified twelve factors that are associated with census undercounts, including unemployment, poverty, linguistic isolation, lack of a high school diploma, and lack of a telephone" (Bureau of Indian Affairs, 2023).

Diversity

Like Asian Americans, Native Americans and Alaska Natives are a heterogeneous group. Of the 574 federally recognized tribes, the Cherokee and Navajo are the two largest in the United States, each with nearly 400,000 members. There are, in total, nearly one-half million Cherokee tribal citizens around the world and 141,000 Cherokee Nation citizens living on the tribe's reservation. The Navajo tribe had an increase in membership with the COVID-19 pandemic. This may have been due in part to Coronavirus Aid, Relief and Economic Security Act funds associated with tribal membership (U.S. Census, 2022 "Facts for Features: American Indian and Alaska Native Heritage Month: November 2022"; Cherokee Nation, 2023; Lonas, 2021).

The various tribes speak 150 native languages, although many are quickly vanishing. For example, only about 10 people nationwide speak Comanche. Tribes also vary widely in their religious beliefs and cultural practices. Thus, a Comanche-Kiowa educator cautions against placing all Native Americans into one group because "tribes ... are sovereign nations and are as different from another tribe as Italians are from Swedes" (Pewewardy, 1998, p. 71; Mangan, 2013; U.S. Census Bureau News, 2015; Ticeahkie, 2022).

Characteristics and Changes

Native Americans and Alaskan Natives are a unique racial-ethnic group because they're not immigrants and have been in what is now the United States longer than any other group. They have experienced centuries of subjugation, exploitation, and political exclusion (e.g., not having the right to vote until 1924). Some tribes are still trying to reclaim billions of dollars that the federal government has squandered or mismanaged (Wilkinson, 2006; Volz, 2012; Kindy, 2013).

Many Native Americans and Alaskan Natives are better off today than they were a decade ago, but long-term institutional discrimination has been difficult to shake. For example, Native Americans have the highest percentage of people living in poverty of all racial-ethnic groups—11.8 percent following the supplemental poverty measure (SPM) threshold. This is twice as many as the White, non-Hispanic group. The median household income for Native Americans and Alaskan Natives is slightly higher than that of Black households but lower than that of other racial-ethnic groups (**Figure 10.7**).

The educational levels of Native Americans and Alaskan Natives have increased, but only 14 percent have a bachelor's or advanced degree compared with 36 percent of the general population. One of four civilian-employed Native Americans or Alaskan Natives works in a management or professional occupation. Even when they are similar to White people in age, gender, education level, marital status, state of residence, and other factors, their odds of being employed are 31 percent lower than those of White people. Such data suggest that Native Americans and Alaskan Natives are experiencing discrimination in the labor market (Austin, 2013; "American Indian and Alaska Native...," 2016).

Despite numerous obstacles, Native Americans and Alaskan Natives have made considerable economic progress by insisting on self-determination and the rights of tribes to run their own affairs. For example, 244 tribes now generate about 43 percent of all U.S. casino gaming revenue. The Justice Department allows tribes to grow or sell marijuana on their lands, even in states that ban the practice. Some tribes, "mindful of the painful legacy of alcohol abuse in their communities," are opposed to selling or using marijuana on their territory. Others view marijuana sales as a huge source of future income, similar to cigarette sales and casino gambling (Cano, 2015; "2013 Indian Gaming Industry Report," 2015).

10-6f Middle Eastern Americans: An Emerging Minority

The Middle East is "one of the most diverse and complex combinations of geographic, historical, religious, linguistic, and even racial places on Earth" (Sharifzadeh, 1997, p. 442). It encompasses about 30 countries, including Armenia, Turkey, Israel, Iran, Afghanistan, Pakistan, Algeria, Iraq, Kuwait, Saudi Arabia, and the United Arab Emirates.

Diversity

Because of the continued immigration from the Middle East and North Africa, Arabic is the fastest-growing and seventh most commonly spoken non-English language in the United States (Brown, 2016). Of the 68 million people in the United States who speak a language other than English at home, about 1.2 million speak Arabic, which is the seventh most spoken second language in the United States. Another 450,000 speak Persian

iStock.com/dcdebs

AP Images/Jeff Robbins

Mohegan Sun in southern Connecticut is one of the largest casinos in the United States. It has spent some of its profits on college scholarships, a $15 million senior center, and health insurance for tribal members. In contrast, some of the economically underprivileged tribes, such as the Navajo and Hopi, have rejected gaming for religious reasons and have many members who live in poverty without kitchen facilities (like stoves or refrigerators) or indoor plumbing.

(including Farsi and Dari), and over 200,000 speak Hebrew (Dietrich and Hernandez, 2022).

There are around 3.7 million people in the U.S. who trace their ancestry to a Middle Eastern country. Yet, the census does not currently have a category for people who identify as Middle Eastern or North African (Wintersmith, 2022). Those who identify themselves as Arab Americans come from many different countries (**Figure 10.10**). As in the case of Asian American families, Middle Eastern families make up a heterogeneous population that is a "multicultural, multiracial, and multiethnic mosaic" (Abudabbeh, 1996, p. 333).

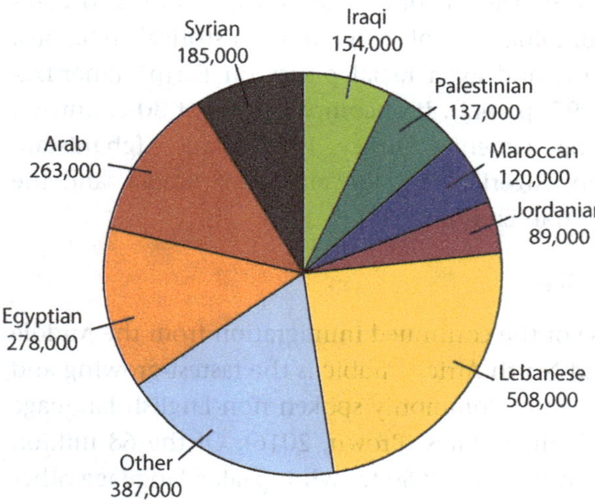

Figure 10.10 Origins of Middle Eastern Americans

Source: Based on Alsarif and Tensley, 2022.

Characteristics and Changes

Middle Eastern Americans tend to be better educated and wealthier than other Americans. About 46 percent

Justin Sullivan/Getty Images

Steve Jobs—the late co-founder, chairman, and CEO of Apple Inc.—was born to unmarried university students. His father was a Syrian-born Muslim, and his mother was a U.S.-born Catholic of Swiss descent. The father put the baby up for adoption because his girlfriend's family wouldn't allow her to marry a person from the Middle East ("Steve Jobs . . . ," 2011).

have a college or advanced degree, compared with 36 percent of the general population. In 2010, the median income of Arab American households was almost $57,000, compared with $52,000 for all U.S. households. As in other groups, however, there are wide variations. Lebanese have higher median family incomes ($67,300) than Iraqis ($32,000) (Asi and Beaulieu, 2013; Motel and Patten, 2013).

Michigan has the country's second highest number of residents with Middle Eastern roots and many with Arabic origins (278,000). As the auto industry recovers, companies are hiring immigrants and refugees from Syria, Iraq, Somalia, Yemen, and Myanmar. According to a manufacturing recruiter: "They work really hard, and that's what companies are looking for" (Green, 2016, p. 18). California is the state with the largest number of Arab Americans (373,000), most identifying their roots as being in Iraq or Egypt (Arab American Institute, 2021).

10-7 Sociological Explanations of Racial-Ethnic Inequality

Four major sociological theories help us understand racial-ethnic relations. (**Table 10.3** summarizes the key points of each perspective.)

10-7a Functionalism

People who criticize immigrants for not becoming Americanized quickly enough reflect a functionalist view of racial-ethnic relations. That is, if a society is to work harmoniously, newcomers must adopt the dominant group's values, goals, and particularly the language. Doing so increases a society's cultural solidarity.

Stability and Cohesion

Immigration is functional for the host nation if it gains needed workers. Highly educated and skilled immigrants fill important positions in medicine, science, and business. Many employers also rely on immigrants to work in fields, orchards, and vineyards at low wages; others are actively recruiting immigrants for decent-paying but difficult and often less desirable jobs such as at meat and fish factories that native-born Americans avoid (Newkirk and Douban, 2012; Davey, 2014).

Functionalists view racial-ethnic inequality as dysfunctional but attribute much of the inequity to individual failings and a lack of acculturation. There are incentives for not acculturating, however. For example, hundreds of Puerto Ricans get federal disability benefits regardless of work experience or level of education because they're considered less employable if they can't speak English (Office of the Inspector General, 2015).

Critical Evaluation

Functionalist explanations are limited in several ways. First, acculturation and assimilation increase social solidarity but can also have negative outcomes. For example, second-generation immigrants are more likely than their peers from other nations to join gangs, commit crimes, and experience obesity and other health problems (Bersani, 2014; Akbulut-Yuksel and Kugler, 2016; refer also to Chapter 14).

Second, by focusing on order and stability, functionalists ignore racial-ethnic inequalities that often spawn tension and discord. Instead of implementing an immigration policy, President Trump tried to prevent refugees and people from seven majority-Muslim countries from entering the United States. His action polarized many Americans—particularly the 64 percent who think the

Table 10.3 Sociological Explanations of Racial-Ethnic Inequality

Theoretical Perspective	Level of Analysis	Key Points
Functionalist	Macro	Immigration provides needed workers; acculturation and assimilation increase social solidarity; racial-ethnic inequality can be dysfunctional but benefits dominant groups.
Conflict	Macro	There's ongoing strife between dominant and minority groups; powerful groups maintain their advantages primarily through economic exploitation; race is a more important factor than social class in perpetuating racial-ethnic inequality.
Feminist	Macro and micro	Minority women suffer from the combined effects of racism and sexism; gendered racism occurs within and across racial-ethnic groups.
Symbolic Interactionist	Micro	Because race and ethnicity are socially constructed, social interaction can increase or reduce racial and ethnic hostility; antagonistic attitudes toward minorities, which are learned, can be lessened through cooperative interracial and interethnic contacts.

nation's growing racial and ethnic diversity makes the country a better place ("In First Month...," 2017).

Third, functionalism doesn't explain why, despite acculturation and assimilation, minorities experience exclusion. For instance, 61 percent of Black people and 45 percent of White people say that U.S. race relations are "generally bad," and 43 percent of Black people believe that America will never make the changes needed for Black people to achieve equal rights with White people (Geiger, 2016; Bialik and Cilluffo, 2017). Functionalists acknowledge that inequality is dysfunctional, but this isn't their major focus. Thus, they seem to accept discrimination as inevitable (Chasin, 2004).

10-7b Conflict Theory

Conflict theorists observe the ongoing strife between dominant and minority groups. Dominant groups try to protect their power and privilege, while subordinate groups struggle to gain a larger share of societal resources.

Economic and Social Class Inequality

For conflict theorists, capitalism creates and sustains racial-ethnic inequality. According to a classic explanation, there's a "split labor market." Jobs in the *primary labor market*, held primarily by White workers, provide better wages, health and pension benefits, and some measure of job security. In contrast, workers in the *secondary labor market* (e.g., fast-food employees) are largely minorities and easily replaced. Their wages are low, there are few fringe benefits, and working conditions are generally lower (Doeringer and Piore, 1971; Bonacich, 1972).

Such economic stratification pits minorities against each other and low-income White people. Because these groups compete with each other instead of uniting against exploitation, capitalists don't have to worry about increasing wages or providing safer work environments.

Social class doesn't always protect minorities from economic inequality. Because of residential segregation, for instance, middle-income Black and Latina/o/x households are much more likely than White ones to live in lower-income neighborhoods, exposing children to inadequate schools and more crime. Racial-ethnic minorities are still underrepresented in upper-level corporate positions. For instance, there are now six Black CEOs of Fortune 500 companies, which is around 1 percent of these businesses (Andrews, 2015; Kang, 2015; Reardon et al., 2015; Makoni, 2022).

© 2001 Brian Fairrington and PoliticalCartoons.com/Cagle Cartoons, Inc.

Critical Evaluation

Conflict theories have several drawbacks. First, discrimination isn't always as conscious and deliberate as some conflict theorists claim. To increase diversity, Google and Facebook have, for some time, recruited at historically Black colleges, trained employees about unconscious bias, and rewarded recruiters and managers for hiring Black and Latina/o/x engineers (McGirt, 2017; Zarya, 2017).

Second, conflict theories are better at explaining racial-ethnic competition than cooperation. In 2021, 48 percent of Americans worried a "great deal" about U.S. race relations. This is up from only 31 percent in 2020 and only 13 percent in 2010 (Brenan, 2021). This is likely due to more public focus on racial injustice following several police-involved deaths of Black people, including George Floyd. Despite such tragic and divisive events, Americans generally work together and cooperate to improve race relations, especially at the local level. Finally, economic inequality reinforces racial-ethnic inequality, but racism existed hundreds of years before the rise of capitalism.

10-7c Feminist Theories

Walk through almost any hotel, large discount store, nursing home, or fast-food restaurant in the United States, and you'll notice two things: Most of the low-paid employees are women and predominantly minority women. For feminist scholars, such segregation of minority women is due to gendered racism.

Gendered Racism

Gendered racism refers to the overlapping and cumulative effects of inequality due to racism *and* sexism. Many White women encounter discrimination on a

---

gendered racism the overlapping and cumulative effects of inequality due to racism *and* sexism.

Jason LaVeris/Getty Images

In 2015, Viola Davis became the first Black woman to win an Emmy for best actress in a drama series (*How to Get Away with Murder*). In her acceptance speech, Davis said, "The only thing that separates women of color from anyone else is opportunity."

daily basis (refer to Chapter 9). Women of color, however, are also members of a racial-ethnic group, bringing them a double dose of inequality. If social class is included, some minority women experience *triple oppression*. Many affluent women, in particular, have no qualms about exploiting recent immigrants, especially Latinas, who perform demanding housework at very low wages (Hondagneu-Sotelo, 2001).

Gendered racism also occurs *within* racial-ethnic groups. According to a Black male sociologist, scholars rarely discuss Black male privilege, which is characterized by having advantages over Black women. Examples include being promoted more often and getting higher pay than their Black female counterparts who are equally skilled and educated (National Public Radio, 2010).

Women of color are eager to reach high-level positions: 48 percent, compared with 37 percent of White women, aspire to be a top executive. Minority women make up one-third of the workforce but less than 4 percent of executives and only 0.4 percent of CEOs at the largest companies. Some label roadblocks to women of color as a "concrete ceiling." In contrast to a "glass ceiling" that White women might shatter, concrete ceilings are more difficult to break through (Piazza, 2016; Calacal, 2017; refer also to Chapter 6 on glass ceilings).

Critical Evaluation

Feminist perspectives can be faulted on two counts. First, because all of us have internalized institutional discrimination, minority group members are also guilty of reinforcing gendered racism in schools, workplaces, and elsewhere. Black girls as young as 15 have to cope with Black men's sexual harassment, stereotypes that Black girls and women are sexually promiscuous, and many Black men's preference for women with lighter complexions and facial features that are closer to European standards of beauty (Friedman, 2011; Thomas et al., 2011). Second, feminist scholars seldom explore women's contribution to gendered racism (e.g., affluent Black women and Latinas who exploit domestic and farm workers).

AP Images/Dave Martin

After Hurricane Katrina in 2005, the caption on a photo of a young Black man wading through water described him as "looting."

Chris Graythen/Getty Images news/Getty Images

A caption described a White man and a light-skinned woman as "finding" goods at a flooded local grocery store.

For symbolic interactionists, images shape our perceptions of racial and ethnic groups.

10-7d Symbolic Interaction

According to symbolic interactionists, we learn attitudes, norms, and values throughout the life course. Because race and ethnicity are constructed socially by thoughts and conversations, labeling, selective perception, and social contact can have powerful effects on everyday intergroup relations.

Labeling, Selective Perception, and the Contact Hypothesis

Labeling can increase racial tension and conflict because people attach meaning to symbols and act according to their subjective interpretation of the symbols. Students on some college campuses have demanded the removal of paintings and Civil War statues and the renaming of buildings that, according to students, honor slave owners, segregationists, and white supremacists. These protests antagonize many alumni whose donations support academic and sports programs, scholarships, and libraries ("Universities," 2015).

Selective perception also affects racial attitudes and behavior. When White people get proof of being privileged, they may insist that *other* people, not they themselves, benefit personally from privilege (Phillips and Lowery, 2015). People who observe what they want and ignore any evidence that challenges their perceptions are unlikely to support policies to reduce racial inequality.

Negative images create and reinforce racial and ethnic stereotypes, but people can decrease labeling and selective perception. The **contact hypothesis** posits that the more people get to know minority group members personally, the less likely they are to be prejudiced against that group. Such contacts are most effective when dominant and minority group members have approximately the same ability and status (e.g., both are coworkers or bosses), when they share common goals (e.g., work on a project), when they cooperate rather than compete, and if an authority figure supports intergroup interaction (e.g., an employer encourages White supervisors to mentor co-workers who are racial-ethnic minorities) (Allport, 1954; Kalev et al., 2006; Carrell et al., 2015).

Critical Evaluation

Interactionism is limited for three reasons. First, it's not clear why labeling, selective perception, and racial bias are more common among some people than others, especially when they're similar on a number of variables such as social class, gender, age, religion, race, and ethnicity (Dovidio, 2009).

Second, symbolic interaction tells us little about the social structures that create and maintain racial-ethnic inequality. For instance, people who aren't prejudiced can foster discrimination by simply going along with the inequitable policies that have been institutionalized in education, the workplace, and other settings (i.e., the unprejudiced discriminators described earlier).

A third criticism is that contact between and among racial-ethnic groups doesn't necessarily increase acceptance and understanding. For instance, darker- and lighter-skinned Black people report more everyday discrimination even from Black people than those with a medium tone (Monk, 2015; Jones et al., 2016).

contact hypothesis posits that the more people get to know members of a minority group personally, the less likely they are to be prejudiced against that group.

10-8 Interracial and Interethnic Relationships

President Obama, the son of a Black father from Kenya and a White mother from Kansas, identified himself as "Black" on the 2010 census questionnaire, although he could have checked off both Black and White or "other." The president is just one of the growing number of Americans who are biracial or multiracial.

10-8a Growing Multiracial Diversity

The 2000 U.S. Census, for the first time, allowed people to mark more than one race, which generated about 126 categories and combinations. In 2020, 97 percent

Francis Miller/The LIFE Picture Collection/Shutterstock.com

Mildred and Richard Loving's Supreme Court win against the state of Virginia made interracial marriage legal throughout the United States.

of Americans reported being only one race, but almost 3 percent (nearly 9.5 million people) self-identified as being two or more races, up from 2.4 percent in 2000. Every state saw its multiple-race population jump by at least 8 percent, but Native Hawaiians/Pacific Islanders were the most likely to identify themselves as belonging to two or more races.

A recent survey found that 17 percent of U.S. adults could be considered multiracial based on how people describe themselves as well as their parents' and grandparents' racial backgrounds. The percentage may be even higher because 61 percent of adults with a mixed racial background don't consider themselves "multiracial" (Parker et al., 2015; Patten, 2016).

10-8b Interracial Dating and Marriage

Laws against **miscegenation**, marriage or sexual relations between people of different races, existed in America as early as 1661. It wasn't until 1967 that the U.S. Supreme Court's *Loving v. Virginia* decision overturned antimiscegenation laws nationally.

Racial-ethnic intermarriages have increased slowly—from only 0.7 percent in 1970 to 10 percent (11 million people) in 2015. In that year, 17 percent of newlyweds had a spouse of a different race or ethnicity. Latinos and Asians have higher intermarriage rates than Black and White people, but recent overall increases have been fueled by rising intermarriage rates among Black and White newlyweds (**Figure 10.11**).

The increase in intermarriage is due to many interrelated factors—both micro and macro level—that include everyday contact and changing attitudes. Public approval for interracial marriages has been increasing—recently at 94 percent, up from 87 percent in 2013 and only 4 percent in 1958 (McCarthy, 2021).

Proximity and education also affect intermarriage rates. We tend to date and marry people we interact with on a regular basis. The higher the educational level, the greater the potential for intermarriage because educated minority group members often attend integrated colleges, and their workplaces and neighborhoods are more racially mixed than in the past (Chen and Takeuchi, 2011; Wang, 2012).

People may also marry outside of their racial-ethnic groups because of a shortage of potential

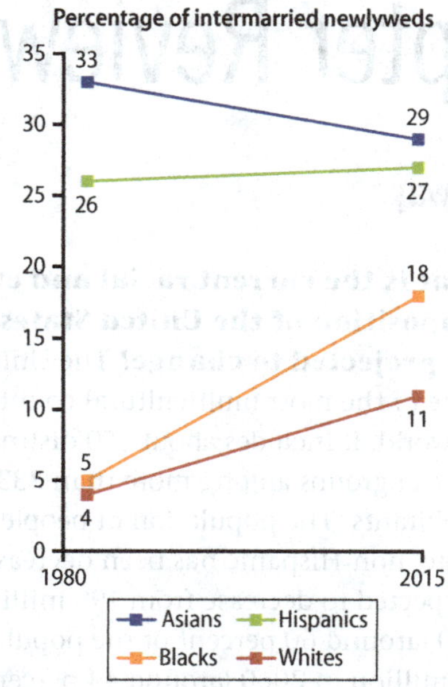

Figure 10.11 U.S. Intermarriage Rates Vary by Race and Ethnicity

Source: Based on Livingston and Brown, 2017, p. 11.

spouses within their own group. Because the Middle Eastern American population is so small, 80 percent of U.S.-born Arabs have non-Arab spouses. Acculturation also affects intermarriage rates. Intermarriage is more common among second-generation immigrants (15 percent) than those in the first generation (8 percent). By the third generation, 31 percent of Latinos and Asian Americans have a spouse of a different race or ethnicity (Kulczycki and Lobo, 2002; "Second-Generation Americans," 2013).

The White majority is shrinking as multiracial marriages surge. According to demographer William Frey (2015, p. 191), the rise in intermarriages and multiracial children shows "an unmistakable trend toward a softening of racial boundaries" that should lead to greater equality in education, politics, the economy, and other institutions.

miscegenation marriage or sexual relations between people of different races.

Chapter Review

Key Takeaways

10-1 What is the current racial and ethnic composition of the United States? How is it projected to change? The United States is one of the most multicultural countries in the world. It includes about 150 distinct ethnic or racial groups among more than 333 million inhabitants. The population of people who are White, non-Hispanic has been decreasing and is expected to decrease from 199 million in 2020 (around 60 percent of the population) to 179 million in 2060 (around 44 percent).

10-2 What is meant by the concepts of race, ethnicity, and racial-ethnic group? *Race* refers to physical characteristics, while an *ethnic group* identifies with a common national origin or cultural heritage. A *racial-ethnic group* has both distinctive physical and cultural characteristics.

10-3 How have immigration patterns and attitudes changed in the United States? In 1900, almost 85 percent of immigrants came from Europe; now, immigrants come primarily from Asia and Latin America. Many Americans are ambivalent about immigrants, especially those who are in the country and undocumented. Many scholars argue, however, that, in the long run, both documented and undocumented immigrants bring more benefits than costs.

10-4 What is the difference between dominant and minority groups? What are the main patterns of dominant-minority group relations? A *dominant group* has more economic and political power than a *minority group*. The latter may be larger in number than a dominant group but is often subject to differential and unequal treatment because of its physical, cultural, or other characteristics. Patterns of dominant-minority group relations include *genocide*, *segregation*, *acculturation*, *assimilation*, and *pluralism*.

10-5 What are the main sources of racial-ethnic friction? *Racism* justifies and preserves the social, economic, and political interests of dominant groups. *Prejudice* is an attitude; *discrimination* is an act that occurs at both the individual and institutional level. All of us can be prejudiced, but minorities are typically targets of *stereotypes* and *ethnocentrism* that often lead to *scapegoating*.

Example: Stuff White People Like

The popular blog *Stuff White People Like* has generated clones (e.g., *Stuff Educated Black People Like* and *Stuff Asian People Like*). Why are these sites so popular? Many fans say that the descriptions are funny because they're true. According to some critics, however, by poking fun at privileged upper-middle-class white people, the sites fuel stereotypes instead of having painfully frank discussions about U.S. race and racism (Sternbergh, 2008). Do you agree or disagree?

10-6 What are key points about the major U.S. racial and ethnic groups in the United States? European Americans, who settled the first colonies, are declining in population, and Latina/o/xs now comprise 19 percent of the population. Other large racial and ethnic populations are African Americans (13 percent), Asian Americans (6 percent), and Native Americans (almost 2 percent). Middle Eastern Americans, who comprise less than 0.5 percent of the population, come from more than 30 countries. All of these groups have experienced prejudice and discrimination, but they have enhanced U.S. society and culture.

10-7 What are the similarities and differences among the sociological theories of racial-ethnic inequality? Refer to Table 10.3 on page 215.

10-8 How have interracial and interethnic relationships in the United Stated changed? About 97 percent of Americans report being only one race, but the number of biracial children is increasing because of interracial dating and marriage. The rise of racial-ethnic intermarriage reflects many micro and macro factors, such as greater interethnic and interracial contact, changing attitudes, and acculturation.

Key Terms

Racial group (p. 199)
Ethnic group (p. 199)
Racial-ethnic group (p. 199)
Dominant group (p. 203)
Minority (p. 203)
Genocide (p. 204)
Segregation (p. 204)
Acculturation (p. 204)
Assimilation (p. 205)
Pluralism (p. 205)
Racism (p. 205)
Prejudice (p. 205)
Stereotype (p. 206)
Scapegoats (p. 206)
Discrimination (p. 206)
Individual discrimination (p. 206)
Institutional discrimination (p. 206)
Gendered racism (p. 216)
Contact hypothesis (p. 218)
Miscegenation (p. 219)

Discusssion and Critical Thinking Questions

1. Piyush "Bobby" Jindal was born in 1971 in Baton Rouge, Louisiana. Shortly before his birth, Jindal's parents came to the United States from India so his mother could pursue a graduate degree in nuclear physics. His father, a civil engineer, was the only one of nine siblings who attended high school. In 2007, Jindal was elected governor of Louisiana, becoming the nation's youngest governor and the first Asian American to occupy a governor's mansion.
   a. Jindal's success illustrates the belief that anyone can "make it" in America, but what kind of resources did his parents have that many recent immigrants lack?
   b. In addition to Jindal's success, President Obama is biracial, many of the largest cities' mayors are African Americans, and a U.S. Supreme Court justice is a Latina. Does this mean that race and ethnicity are becoming less important in the United States?
2. Even though the Washington D.C. NFL team changed its name in 2022 to the Commanders (dropping Redskins), there are 37 high schools that still use "Redskins" in their team names. Many people, including Native American leaders, denounce "Redskins" as a racist slur, but the schools' students and parents dismiss such complaints (ESPN, 2022; Shapira, 2014). If "Redskins" is derogatory, what about the National Association for the Advancement of Colored People? Isn't "colored" offensive?
3. When you are sitting in a restaurant, what do you think people first notice about you? Is it your race, ethnicity, or something else? Why do you think this is the first thing that they notice?
4. How do individual and institutional discrimination differ? Do you think that one is more harmful than the other?
5. Have you ever engaged in unprejudiced discrimination? If so, why? If you've ever been a prejudiced nondiscriminator, what factors influenced your behavior?

True or False?

How Much Do You Know about U.S. Racial and Ethnic Groups?

The answers to 2 and 5 are true. The answers to 1, 3, 4, and 6 are false.

1. Race is a *social construction*, a societal invention that labels people based on physical appearance, social class, or other characteristics.
2. The United States has more migrants than any other nation. It is home to almost 20 percent of all migrants worldwide.
3. Prejudice is an *attitude* that prejudges people. It does not necessarily mean that a person with prejudice will actually carry out the action of discrimination. This can be the case with *prejudiced nondiscriminators*.
4. The U.S. Census allows people to identity as one or more races and select their ethnicity.
5. The U.S. Census allows people to identity as one or more races and select their ethnicity.
6. Americans have become less optimistic that people have equal opportunities regardless of race. Only 40 percent feel people of color have equal job opportunities compared to those who are White; 53 percent felt this way in 2008.

11 The Economy and Politics

Learning Objectives

After studying this chapter, you will be able to...

11-1 Compare the different types of global economic systems.

11-2 Describe corporations and how corporate and political power are interwoven.

11-3 Explain how and why macro-level variables have changed the U.S. economy and jobs.

11-4 Compare the theoretical explanations of work and the economy.

11-5 Compare the different types of global political systems.

11-6 Explain how power and authority differ and the different types of authority.

11-7 Describe the U.S. political system, including who votes, who doesn't, and why.

11-8 Evaluate the theoretical explanations of politics and power.

Recently, 43 percent of Americans said current economic conditions are bad, and 72 percent said the economy is getting worse. Americans' confidence in the economy is still lower than pre-COVID-19 pandemic levels. Also recently, only 21 percent of Americans say they have confidence in government to do what is right (Saad, 2023; Pew Research Center, 2022). In this chapter, we explore how the economy and government affect people's everyday lives.

The **economy** determines how a society produces, distributes, and consumes goods and services. In **politics,** individuals and groups acquire and exercise power and authority and make decisions. Societies worldwide differ in the kinds of economic and political systems they develop because of factors such as globalization and technology. Let's begin by discussing global economic systems.

Be sure to take the True or False quiz in the chapter opener to better understand how much you know about these topics.

What do you think?

The country would be better off if there were more gender diversity in political office.

1	2	3	4	5	6	7
	strongly agree				strongly disagree	

True or False?

How Much Do You Know about the Economy and Politics?

1. Countries follow the economic models of capitalism, socialism, or communism.
2. Unemployment rates have been decreasing in recent years (2022 and 2023).
3. American employees take fewer vacation days than workers in most other countries.
4. Many Americans are dissatisfied with the two major political parties in the United States.
5. Worldwide, democracy is decreasing.
6. Voter turnout for the 2020 U.S. presidential election was the lowest ever.

The answers are at the end of the chapter.

11-1 Global Economic Systems

The two major economic systems around the world are capitalism and socialism. A handful of countries endorse communism, but in actual practice, economies are usually some mixture of economic systems.

11-1a Capitalism

Ideally, *capitalism*, an economic system based on the private ownership of property and the means of production, has four essential characteristics (Smith, 1776/1937; Heilbroner and Thurow, 1998):

- **Private ownership of the means of production.** Property (e.g., real estate, banks, utilities) belongs to individuals or organizations rather than the state or the community.
- **Market competition.** Owners of production compete in deciding what goods and services to produce and setting prices that offer consumers the greatest value in price and quality.
- **Profit.** Selling something for more than it costs to produce generates an accumulation of wealth for individuals and companies.

economy determines how a society produces, distributes, and consumes goods and services.

politics individuals and groups acquire and exercise power and authority and make decisions.

- **Investment.** By investing profits, capitalists can increase their own wealth. Workers, too, can save and invest their money.

In reality, capitalism doesn't function ideally because of abuses, greed, and worker exploitation. It usually results in monopolies and oligopolies rather than a free market that encourages competition. Federal laws prohibit the establishment of a **monopoly**, the domination of a particular market or industry by one person or company. With little or no competition, a company would be able to dictate prices and lower the quality of goods and services.

An **oligopoly**, which is legal, is the domination of a market by a few large producers or suppliers. In the mid-1980s, 50 companies controlled 90 percent of all U.S. media. Today, it's six companies—including CBS, Time Warner, and the Walt Disney Corporation. There were hundreds of airlines in the mid-1980s, but four giant airlines (United, Delta, American, and Southwest) now control 80 percent of the U.S. market. CEOs maintain that merging companies increases efficiency and safety. Critics argue that oligopolies raise prices, stifle competition, and reduce consumer choices. Companies that dominate industries make huge profits, but workers' wages and salaries tend to decline (Yglesias, 2013; Koenig and Mayerowitz, 2015; Coy, 2016; Semeuls, 2023). Moreover, entrepreneurs have little chance of breaking into an industry dominated by an oligopoly.

11-1b Socialism

Socialism is an economic system based on the public ownership of the production of goods and services. Ideally, socialism is the opposite of capitalism and has the following characteristics:

- **Collective ownership of property.** The community, rather than the individual, owns property. The state owns utilities, factories, land, and equipment but distributes them equally among all members of a society.
- **Cooperation.** Working together and providing social services to all people are more important than competition.
- **No profit motive.** Private profits fueled by greed and the exploitation of workers are forbidden.
- **Collective goals.** The state is responsible for all economic planning and programs. People are discouraged from accumulating individual profits and investments and are expected to work for the greater good.

There have been many socialist governments during the last 150 years, but none has reflected pure socialism. Competition and individual profits are officially forbidden, but government officials, top athletes, and high-ranking party members enjoy more freedom, larger apartments, higher incomes, and greater access to education and other resources than the average person.

11-1c Communism

Communism is a political and economic system in which property is communally owned and all people are considered equal. While socialism allows some free market economy and people receive resources according to how hard they work, communism demands that all production be owned by the public and that people receive resources according to their need.

Walt Disney Company is an example of an oligopoly. Among other holdings, it owns publishing companies, web portals, 19 major television and cable stations, at least 60 radio stations, numerous magazines, and 14 theme parks and resorts around the world. All of these outlets promote and sell Disney products, increasing the corporation's profits.

Pictorial Press Ltd/Alamy Stock Photo; Alexey Stiop/Shutterstock; iStock.com/CribbVisuals; Nlizer/Dreamstime.com; iStock.com/Viafilms

---

monopoly domination of a particular market or industry by one person or company.

oligopoly domination of a market by a few large producers or suppliers.

socialism an economic system based on the public ownership of the production of goods and services.

communism a political and economic system in which property is communally owned and all people are considered equal.

- **Shared ownership**. A system based on the collective ownership of the means of production (think factories), distribution (how products are dispersed), and exchange (system to ensure people have the variety of products and services they need). Everyone is entitled to products and services based on their needs.
- **No private property.** Individual people don't have the sole possession of means of production (businesses).
- **No social classes**. This system aims to eliminate hierarchy among people.
- **No money.** Currency is not needed for transactions.
- **No state.** The system may use a form of self-governance, but its exact meaning does not have widespread agreement.

Karl Marx (1867/1967) believed that socialist states would evolve into communist societies. This hasn't happened, and most formerly communist countries have collapsed. Today, only China, Cuba, Laos, North Korea, and Vietnam practice some form of communism, but they have also incorporated capitalism. In contrast to its principles, communism has failed, according to historians, because of widespread corruption and mismanagement, oppression, the imprisonment or execution of people who questioned coercive policies, economic inefficiency that resulted in shortages of food and other goods, and few rewards for working hard (Brown, 2010; Johnson, 2010; Pollick, 2014).

11-1d Mixed Economies

State capitalism (also called by the less preferred term *welfare capitalism*) is an economic system that combines private ownership of property, market competition, and governmental regulation of many programs and services. Many countries, including the United States, have some form of state capitalism. Most industry is private, but the government owns or operates some of the largest industries and services (e.g., education, transportation, postal services) that benefit the entire population (Esping-Andersen, 1990; Hacker, 2002).

- *Crony capitalism* is an economy based on close relationships between business people and the government. In this system, the government gives wealthy people and large corporations preferential treatment through special tax breaks (called *tax credits* and *deductions*), direct payments or loans (called *subsidies*, not welfare), and grants that rarely require competition in an open market. As you learned in Chapter 8, at least 55 of the largest U.S. corporations paid no taxes in 2020. Their tax savings, along with $3.5 billion in rebates, saved the companies $12 billion (Gardner and Wamhoff, 2021). In addition, the federal Cares Act, to off-set the economic impacts of the COVID-19 pandemic, provided $560 billion to help individuals. It also gave another $500 billion to businesses as loans (Andrzejewski, 2014; Snell, 2020; refer also to the discussion of corporate welfare in Chapter 8).

Nir Elias/REUTERS

China is an example of a country with a mixed economy: It espouses communism, practices socialism, and has endorsed many aspects of capitalism. China's economic boom hasn't benefited all of its citizens. Many urban centers are thriving—offering those in upper and middle classes high-rise apartments, stores, and restaurants. In contrast, millions of Chinese, including many older people, like the one pictured here, survive by scouring trash bins for plastic bottles to recycle.

Large companies drive both capitalism and mixed economies. Corporations, in particular, wield enormous influence both at home and abroad.

11-2 Corporations and the Economy

In 2005, Hamdi Ulukaya, a Turkish immigrant, founded Chobani, a yogurt company that's now valued at about $5 billion. In 2016, Ulukaya told his 2,000 full-time employees that they'd receive shares worth up to 10 percent of the company if it went public or is sold. Many workers would become millionaires (Strom, 2016). Unlike Ulukaya, most corporate heads have luxurious lifestyles, eliminate employees' pensions, retire as billionaires, and pass on their massive wealth to their heirs

state capitalism (also called by the non-preferred term *welfare capitalism*) an economic system that combines private ownership of property, market competition, and a government's regulation of many programs and services.

(refer to Chapter 8). Corporate and political power are interwoven, but let's begin by exploring some of the characteristics of corporations.

11-2a Corporations

A **corporation** is an organization that has legal rights, privileges, and liabilities apart from those of its members. Until the 1890s, there were only a few U.S. corporations—in the textile, railroad, and oil and steel industries. The number of corporations has grown, as have corporate profits. Consider that in one quarter (three-month period) in 1951, corporate profits were around $14.5 billion. In 2022, they reached a high point of $2.5 trillion in a single quarter. Many industrialized nations have closed corporate tax loopholes, but the United States has expanded them. As a result, many corporations exercise more power than governments and have gained great wealth. A number of U.S. corporations use legal tax loopholes to avoid paying tens of billions of dollars every year on income earned overseas (U.S. Census Bureau, 2012; Durden, 2011; Niquette and Rubin, 2014; Trading Economics, 2023; refer also to Chapter 8).

11-2b Conglomerates

A **conglomerate** is a corporation that owns a collection of companies in different industries. Conglomerates emerged during the 1960s and grow by acquiring companies through mergers. Mergers might increase the value of shareholders' stock, but they typically make chief executives "truly, titanically, stupefyingly rich" (Morgenson, 2004, p. C1). An example is Kraft Heinz (after the merger of Kraft and Heinz in 2015), which owns companies that produce snacks, beverages, and pet foods, owns restaurants (e.g., Taco Bell), and has ties with other corporations such as Starbucks (refer to www.kraft.com/brands).

Conglomerates diversify business risk by participating in a number of different markets and can ward off emerging competition. Conglomerates waned after the 1980s but are making a comeback. Google, for example, has set up companies to create driverless cars, extend human lifespans, and develop online education ("From Alpha to Omega," 2015).

11-2c Interlocking Directorates

In an **interlocking directorate**, the same people serve on the boards of directors of several companies or corporations. An estimated 40 percent of corporate directors sit on two or more corporate boards (Rockel, 2022). Some interlocking directorates are especially powerful because they include past U.S. presidents and past members of Congress who provide connections in Washington, D.C. Interlocking directors develop similar economic perspectives; those who sit on two or more corporate boards are especially likely to receive appointments to influential government advisory committees (Domhoff, 2013). Thus, a very small group of people has the power to shape the national agenda.

11-2d Transnational Corporations and Conglomerates

Interlocking directorates have become more influential than ever because of the proliferation of transnational corporations. A **transnational corporation** (also called a *multinational corporation* or an *international corporation*) is a large company that is based in one country but operates across international boundaries. By moving production plants abroad, large U.S. corporations can avoid trade tariffs, bypass environmental regulations, and pay low wages. On the positive side, many corporate giants (e.g., Johnson & Johnson, Microsoft, Apple) invest heavily in local and global education, medical research, and developing economic opportunities (Preston, 2016; Gray, 2017).

The most powerful are **transnational conglomerates** (also called *multinational conglomerates*). These corporations own a collection of different companies in various industries in a number of countries. Amazon is the world's biggest transnational conglomerate. It owns over 40 subsidiaries, including IMDb, Whole Foods, Amazon Web Services, and China-based Joya (Haqqi, 2022).

11-3 Work in U.S. Society Today

Economic problems vary by social class, but much financial hardship has to do with changes in **work**, a physical or mental activity that produces goods or services. Even those who work hard have lost economic ground because

---

corporation an organization that has legal rights, privileges, and liabilities apart from those of its members.

conglomerate a corporation that owns a collection of companies in different industries.

interlocking directorate the same people serve on the boards of directors of several companies or corporations.

transnational corporation (also called a *multinational corporation* or an *international corporation*) a large company that's based in one country but operates across international boundaries.

transnational conglomerate (also called a *multinational conglomerate*) a corporation that owns a collection of different companies in various industries in a number of countries.

work a physical or mental activity that produces goods or services.

of macro-level variables such as deindustrialization, globalization, offshoring, and weakened labor unions.

11-3a Deindustrialization and Globalization

Many Americans have been casualties of **deindustrialization**, a process of social and economic change that reduces industrial activity, especially manufacturing. The number of manufacturing jobs was at its peak in 1979 with 20 million workers, equaling 22 percent of the nonfarm workforce. Forty years later, in 2019, there were about 13 million manufacturing jobs—around 9 percent of workforce. Now, other areas have greater demand for more workers. There are more workers in education and health services (24 million), professional and business services (21 million), and leisure and hospitality (17 million) (Harris, 2020).

A major reason for deindustrialization is that, beginning in the early 1960s, employers easily replaced workers with the lowest skill levels, usually those on assembly lines, with automation. Machines and robots have also replaced many blue-collar jobs that provided a comfortable standard of living in the past. Robots are commonplace in manufacturing jobs, but employers report a shortage of workers with science, technology, engineering, or math degrees who can program robots, repair sophisticated hardware, and design the software that runs a factory's operations (Gregory, 2017; Manyika et al., 2017).

The manufacturing industry is also facing a major shortage of specialized workers like welders, ironworkers, and brick masons. An estimated two million such jobs will go unfilled over the next decade because employers can't find enough people with the required skills, despite above-average wages. Entry-level welders, for example, can move quickly from $17 to $30 an hour (Cohen, 2015; Payne and Somerville, 2015).

Avatar_023/Shutterstock

Many U.S. factories today need only a few workers in specialized areas to manage the high-tech machines that have replaced people.

Deindustrialization was hastened by **globalization**, the growth and spread of investment, trade, production, communication, and new technology around the world. One example of globalization is a car that is assembled in the United States with practically all of its parts manufactured and produced in Germany, Japan, South Korea, or industrializing countries.

Supporters of globalization argue that it creates millions of jobs and brings affordable goods and services (e.g., cell phones) to billions of people around the world. Critics contend that globalization has reduced the number of jobs in Western economies. They also maintain that globalization primarily benefits the world's most powerful transnational corporations: It increases their profits by expanding their worldwide base of consumers, exploits economically disadvantaged people in industrializing countries, creates greater income inequality between workers and the global elite, and gives corporations unprecedented political power (Schaeffer, 2003; Porter, 2014).

11-3b Offshoring and Labor Unions

Offshoring refers to sending work or jobs to another country to cut a company's costs at home. Sometimes called *international outsourcing* or *offshore outsourcing*, the transfer of manufacturing jobs overseas has been going on since at least the 1970s.

Between 2001 and 2011, U.S. companies moved more than 2.7 million jobs to China, 77 percent of them in manufacturing and primarily blue-collar jobs (Scott, 2012). During the same period, U.S. firms also offshored 28 percent of high-level, well-paid information technology (IT) jobs, including those in accounting, computer science, and engineering (National Science Board, 2012). Most of the offshored jobs go to India and China, but many have also moved to Canada, Hungary, Mexico, Poland, Russia, Egypt, Venezuela, Vietnam, and South Africa.

Companies can get computer programming services in India that cost much less than in the United States (**Figure 11.1**). Wage differences exist in other industries, such as manufacturing. In Mexico, the average General Motors worker earns much less than GM workers in the U.S. Employees at one factory in Mexico "earn less than

deindustrialization social and economic change that reduces industrial activity, especially manufacturing.

globalization the growth and spread of investment, trade, production, communication, and new technology around the world.

offshoring sending work or jobs to another country to cut a company's costs at home.

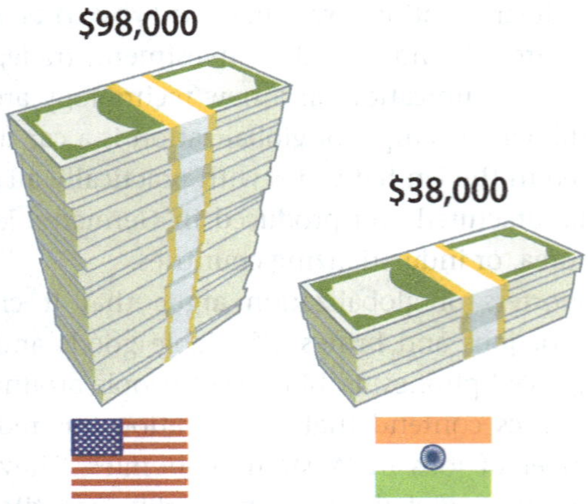

Figure 11.1 Median Annual Earnings of a Computer Programmer in the United States and India

Sources: U.S. Bureau of Labor Statistics, 2022 and Ahmed, 2020

$25 a day, compared with a range of $18 to $32 an hour at GM plants in the US," and even with a raise of 8.5 percent, there is still a large difference in compensation (Navarro, 2022). Because of these wage differences, American consumers can purchase goods and services at low prices, and the majority don't care where products are made (Newport et al., 2014).

Deindustrialization, globalization, and offshoring have weakened *labor unions*, organized groups that seek to improve wages, benefits, and working conditions. Union membership has dropped sharply—from 35 percent of the workforce in the mid-1950s to around 10 percent in 2022. Five states prohibit unionization; 27 states have significantly restricted collective bargaining rights; and anti-union groups have helped defeat efforts to unionize autoworkers in Tennessee and other states (Whitesides, 2014; Bookman and Neeley, 2017; U.S. Bureau of Labor Statistics, 2023).

Do we still need unions? Opponents argue that union members are overpaid and that unions drain state resources because of the high pensions and salaries of public sector employees (e.g., teachers, nurses, sanitation workers, police). Critics also contend that unions have limited employers' flexibility in hiring and firing decisions and that ever-increasing labor and health care costs have forced some employers to move their operations overseas to remain competitive (McKinnon, 2011; Schlesinger, 2011).

Supporters argue that states have experienced budget deficits because of the housing crisis and a recession that was due to Wall Street greed and not overpaid union members. Others point out that union members have made numerous concessions, like decreasing their wages and benefits. Most important, historically, unions have benefited almost all workers by insisting on paid holidays and vacations, greater workplace safety, and overtime pay (Stepan-Norris, 2015).

11-3c How Americans' Work Has Changed

The COVID-19 pandemic caused a spike in unemployment. Since then, the U.S. has focused on pandemic recovery and added jobs and brought back workers. In 2021, the U.S. economy added 6.4. million jobs, which was the strongest year for job growth in decades. This was followed by another 4.5 million jobs in 2022 (Pickert, 2022; Mast and Perez, 2023). Still, millions of Americans have only low-paying and part-time jobs. If these strategies fail, they find themselves among the unemployed.

Low-Wage Jobs

About one-third of U.S. jobs (almost 52 million workers) pay less than $15 an hour. After taxes, the wages of a full-time year-round worker who earns $15 an hour is only about $26,000 a year, and many jobs pay much less. The federal minimum wage, which rose from $6.55 to $7.25 an hour in 2009, increased the wages of less than 4 percent of the workforce. The District of Columbia and 29 states now have minimum wages that exceed the federal minimum wage. If the wage had been raised since 1968 at the same rate as the average worker's growth in productivity, the federal minimum wage would be nearly $18.50 an hour (Cooper, 2015; DeSilver, 2017; Hoxie, 2017; Ceron, 2022).

Almost 3 percent (2.2 million) of workers 18 and older are at or below the federal minimum wage. Another 30 percent (20.6 million) are "near-minimum" workers who earn less than $10.10 an hour (DeSilver, 2017). Thus, most jobs don't pay enough to allow workers to save money or get ahead. In 2016, Walmart, the nation's largest low-wage employer, increased their hourly wage to $10 an hour. If Walmart paid its workers $12 an hour and passed every penny of the costs to consumers, the average customer would pay just 46 cents more per shopping trip, or about $12 a year (Cooper, 2015; Montlake, 2016). Thus, many Americans are being paid less for doing more.

Except for registered nurses, whose median annual salary is about $69,000, eight of the occupations expected to have the most jobs added through 2024 are low paying, under $24,000 a year. These occupations include home health aides and personal care aides who help older adults and people with disabilities. Other fast-growing but low-paying jobs—from about $22,000 to $25,000 a year—include retail salespersons, cooks, stock clerks, janitors, housekeeping cleaners, and fast-food workers ("Employment Projections," 2017).

Part-Time, Temporary, and Gig Work

Of the almost 26 million part-timers (those who work less than 35 hours a week), 20 percent are part time involuntarily because they can't find suitable full-time employment or employers have reduced their hours. Around 22 million people work part-time, versus full-time, for non-economic reasons, which may include school, childcare, or other family commitments. The "alternative workforce"—which includes working for temporary help agencies, as freelancers, or as independent contractors—rose from 11 percent in 2005 to 16 percent in 2015 (15 million and nearly 24 million, respectively) (Katz and Krueger, 2016; "Economic News Release," 2017; Bureau of Labor Statistics, 2023).

Rapid technological advances have spurred a *gig economy* (also called a sharing, on-demand, platform, collaborative, or digital economy). Although there's no official definition, *gig work* is a project or task for which an individual, often through a digital marketplace, receives payment (e.g., Uber, Airbnb, TaskRabbit, Postmates, Etsy). One survey showed that 36 percent of employed people (equivalent to 56 million Americans) identify as "independent workers," including gig workers, freelancers, and those with "side hustles." Around 25 percent report doing this work out of a need for basic necessities, while others point to the flexibility, enjoyment, or ability to add extra income. Gig work also has few formal education or experience requirements. On the other hand, people generally earn less than in traditional occupations, must cover their own health care costs, and may face unpredictable hours and financial instability (Smith, 2016; Hicks, 2017; Morgan, 2017; Katz and Krueger, 2016; Torpey and Hogan, 2016; McKinsey and Company, 2022).

Unemployment and Discouraged Workers

Although the U.S. unemployment rate had been declining, with the onset of the COVID-19 pandemic, it surged to a high of 14.7 percent in April 2020. It has since dropped to a low of 3.4 percent in January 2023 (**Figure 11.2**).

Approximately 63 percent of Americans aged 16 and older have a job or are looking for one. Even excluding the labor force changes resulting from the COVID-19 pandemic, among men aged 25 to 54, who are in the prime of working life, the percentage of those who are working has been declining since 1965. Currently, 86.5 percent of men in this group are employed. As a result, almost 13 percent (7 million men) aren't employed or looking for work. Large numbers are Black, veterans, or were formerly incarcerated; have a high school diploma as their highest academic credential; or receive disability benefits (Eberstadt, 2016; Krueger, 2016; White House Council of Economic Advisers, 2016; Bureau of Labor Statistics, 2023 "Table A-13").

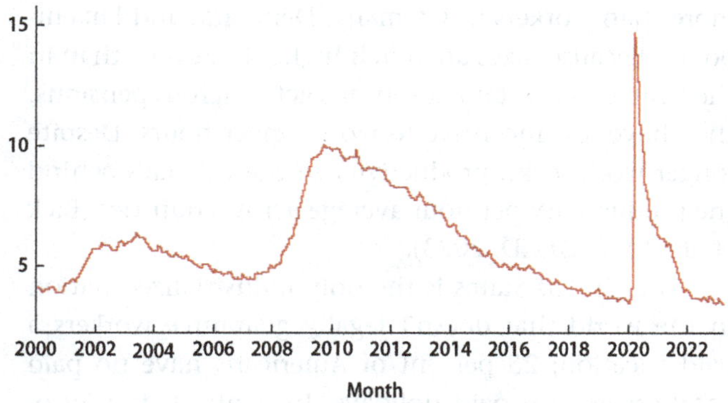

Figure 11.2 U.S. Unemployment Rate, 2000 to 2023

Source: Bureau of Labor Statistics, 2023, "Labor Force Statistics from the Current Population Survey." https://data.bls.gov/pdq/SurveyOutputServlet

About 21 percent of the jobless are *long-term unemployed*, people who have searched for but not found a job for 27 weeks or longer. The longer that people are unemployed, the harder it is to find a job: Only 11 percent find steady, full-time employment a year later. Many of the long-term unemployed become **discouraged workers**, people who stop looking for work because they believe that job hunting is futile. As of April 2023, there were 363,000 discouraged workers, which is about 20 percent lower than the year before (Center for Economic and Policy Research, 2015; U.S. Bureau of Labor Statistics, 2023).

Job Satisfaction and Productivity

Only 32 percent of Americans report being engaged with their jobs, and another 18 percent are actively disengaged. Those who are disengaged are more likely to be uninvolved, unhappy that their needs aren't being met, and put in as little time as possible. Younger workers and women have experienced the most decline in engagement recently. Some reasons include not feeling cared about and not having support at the workplace. Also, those who are working on-site but could work remotely have high rates of disengagement (Harter, 2023). Job type matters too. Service workers are less engaged than managers and executives (28 percent and 40 percent, respectively), but 60 percent of the latter aren't engaged or are actively disengaged (Adkins, 2015).

In most European countries, the typical workweek is 35 hours—comparable with the U.S. definition of part-time work. Americans work nearly 258 more hours per year compared to most Europeans and almost 400 hours

discouraged workers people who stop looking for work because they believe that job hunting is futile.

more than workers in Germany, Denmark, and Luxembourg. Because taxes are much higher in Europe than in the United States, and Europeans get generous pensions, they have less incentive to work longer hours. Despite longer workweeks, productivity in the U.S. lags behind the productivity per hour average across countries (Bick et al., 2017; OECD, 2023).

The United States is the only industrialized nation in the world that doesn't legally guarantee workers a paid vacation; 25 percent of Americans have no paid vacation *and* no paid holidays. In contrast, the European Union requires 20 vacation days, and the United Kingdom mandates 28. Beyond this, the typical worker, especially in Western Europe, has at least 6 weeks of paid vacation, regardless of job seniority or the number of years worked (Ray et al., 2013; Greenfield, 2015). U.S. workers have the least vacation days and don't use all of them (**Table 11.1**). Workers in France have and use more vacation days compared to those in Mexico. Yet, nearly the same percentage of workers in France (71 percent) and Mexico (77 percent) report that they're "vacation deprived." Sixty-three percent of Americans feel this way (Cuadra, 2022; Expedia, 2023).

Clearly, even when workers have paid time off, they don't use all of it. But why? Around half say they don't need to take additional days off. Around 50 percent also are concerned that more time away from the office could lead to falling behind. Other reasons include feeling bad about co-workers having to take on work in their absence and concerns over risks to job advancement or

Table 11.1 Vacation Days, on Average, that Workers Receive and Use

	Receives	Uses
France	28.5	23.0
Germany	28.5	24.5
United Kingdom	26	22
New Zealand	25	21.5
Japan	22	26
Australia	21.5	18
Hong Kong	21.5	25.5
Canada	19	17
Singapore	17	14.5
Mexico	13.5	12.5
United States	12.5	11.0

Source: Based on Expedia's Vacation Deprivation Report, 2023.

Jeff Zelevansky/The New York Times/Redux

In good times and in bad, Black men have the highest unemployment rates. And compared with White workers, Black workers are unemployed longer after their unemployment insurance benefits end. Pictured here are laid-off workers examining listings at a job fair.

possibly losing their jobs (Horowitz and Parker, 2023). Some reduce their vacations to a few days or long weekends because they couldn't afford a vacation, had too much to do on the job, wanted to impress their boss with their dedication, or feared being replaced while away. Employees who forfeit paid time off don't get more raises or bonuses and report higher stress than those who take all of their vacation time, but employers save almost $53 billion a year when people work for free (U.S. Travel Association, 2014; Dickey, 2015; Mudallal, 2015).

11-3d Women and Racial-Ethnic Minorities in the Workplace

One of the most dramatic changes in the United States during the twentieth century was the increase of women in the labor force (**Table 11.2**). Many factors have contributed to the surge in women's

Table 11.2 Women in the Labor Force, 1960–2020

Percentage of Women by Age Cohort in the Labor Force			
Year	Women 16 and over	Women 25 to 54	Women 55 to 64
1960	37.7	42.9	37.2
1980	51.5	64.0	41.3
2000	59.9	76.7	51.9
2010	58.6	75.2	60.2
2020	56.2	75.1	59.1

Source: Based on U.S. Bureau of Labor Statistics, Current Population Survey, 1948-2022. https://www.dol.gov/agencies/wb/data/lfp/women-by-age

employment, especially since the 1970s, including the growth in the number of college-educated women (who, consequently, had more job opportunities), an increase in the number of working single mothers, and the higher costs of homeownership that require two incomes.

Largely because of their higher educational attainment, in heterosexual marriages where both spouses work, around 16 percent have women as the sole or primary income earner, and in another 29 percent of these marriages, both spouses have nearly equal incomes (Fry et al., 2023). As a group, however, women have lower earnings than men in both the highest and lowest-paying occupations, and the wage gaps are greater in high-income jobs (**Figure 11.3**) (refer to Chapter 9).

There are earnings disparities across racial-ethnic groups, but the differences are especially striking by sex. As you examine **Figure 11.4**, note two general characteristics. First, earnings increase across all racial-ethnic groups and for both sexes as people shift into higher-paid positions. But across all occupations, men have higher earnings than women of the same racial-ethnic group. In entry-level and less specialized positions, Black and Latina women fare worse than any of the other groups, with Latina women faring the worst. Thus, both gender *and* race-ethnicity affect earnings (refer to Hegewisch et al., 2015, for a detailed analysis of the gender wage gap by race and ethnicity since 1955.).

11-4 Sociological Explanations of Work and the Economy

Sociological theories offer different insights on the economy and its impact on society. **Table 11.3** summarizes each perspective's key points.

11-4a Functionalism

Functionalists typically emphasize the benefits of work and the economy. They also describe capitalism as bringing prosperity to society as a whole.

Some Key Issues

For functionalists, work is necessary for a society's survival and to define its members' roles. Work is also important because it connects people to each other: As

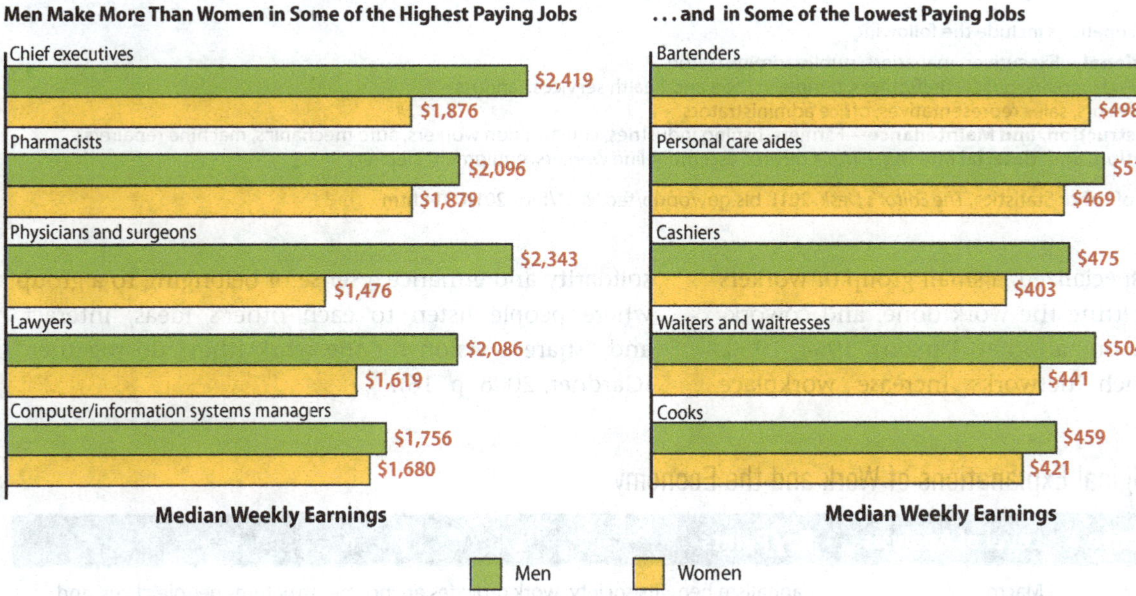

Figure 11.3 Women Earn Less Than Men Whether They're CEOs or Cooks

These are five of the highest and lowest paid occupations of full-time, year-round U.S. workers in 2016. How might you explain why the earnings differ by sex, especially in the highest paid jobs?

Note: Some of the differences between men's and women's median weekly earnings may seem small, but multiply each number by 52 weeks. Thus, in annual earnings, male physicians and surgeons average more than $121,836 compared with only $76,752 for women.

Source: Based on Labor Force Statistics, Current Population Survey, 2017, Table 39, "Median Weekly Earnings of Full-Time Wage and Salary Workers by Detailed Occupation and Sex." Accessed July 1, 2017 (bls.gov/cps).

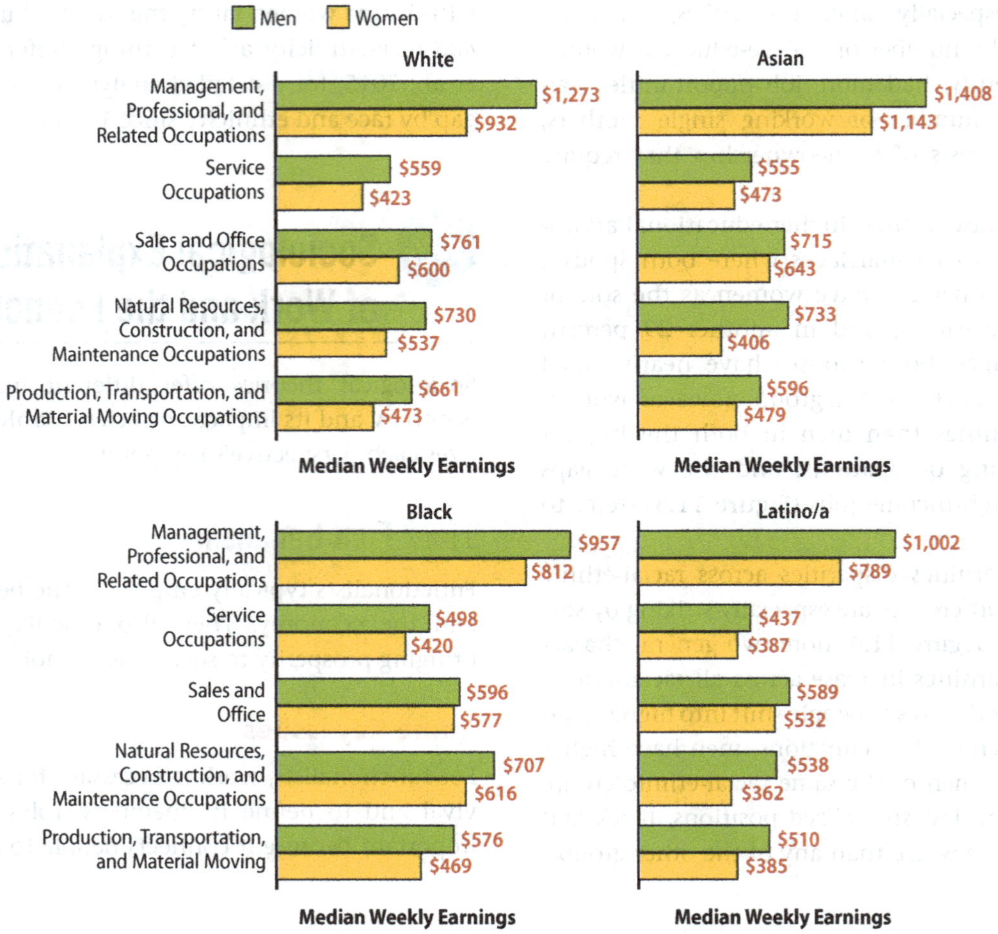

Figure 11.4 Median Weekly Earnings of Full-Time Workers by Occupation, Sex, Race, and Ethnicity

Note: Examples of the occupations include the following:

Managerial and Professional—Executives, managers, public administrators
Service—Private household workers, police, firefighters, people in food and health services, janitors
Sales and Office—Supervisors, sales representatives, office administrators
Natural Resources, Construction, and Maintenance—Farmers, fishing industries, construction workers, auto mechanics, machine repairers
Production, Transportation, and Material Moving—Truck drivers, assembly-line workers, equipment cleaners

Source: Based on Bureau of Labor Statistics, *The Editor's Desk*, 2011. bls.gov/opub/ted/2011/ted_20110914.htm

jobs become more specialized, a small group of workers is responsible for getting the work done, and coworkers can get to know one another (Parsons, 1954, 1960; Merton, 1968). Such networks increase workplace solidarity and enhance a sense of belonging to a group where people listen to each other's ideas, interact, and "share a vision for the work [they] do together" (Gardner, 2008, p. 16).

Table 11.3 Sociological Explanations of Work and the Economy

Theoretical Perspective	Level of Analysis	Key Points
Functionalist	Macro	Capitalism benefits society; work provides an income, structures people's lives, and gives them a sense of accomplishment.
Conflict	Macro	Capitalism enables the rich to exploit other groups; most jobs are low-paying, monotonous, and alienating; productivity isn't always rewarded.
Feminist	Macro and micro	Gender roles structure work experiences differently and inequitably across genders
Symbolic Interactionist	Micro	How people define and experience work in their everyday lives affects their workplace behavior and relationships with coworkers and employers.

Besides providing income, work has social meaning. For many Americans, jobs offer a sense of accomplishment and value; stability, order, and a daily rhythm; and a network of interesting social and professional contacts (Katzenbach, 2003; Brooks, 2007). Functionalists also maintain that wage inequities spur people to persevere and to set higher goals. Low-paying jobs, for example, can motivate people to work harder or obtain more education to move up the economic ladder (refer to Chapter 8).

Critical Evaluation

Critics point out that work is a very or somewhat significant source of stress for 66 percent of Americans and often leads to myriad health problems (American Psychological Association, 2019). Many people have unstimulating and lower-paying jobs that require repetitive and routine tasks, instead of providing a connection to a product or a group (refer to the discussion of McDonaldization in Chapter 6). Yet, many types of jobs have become stressful. From 2019 to 2022, worker engagement declined the most among healthcare workers, while white-collar workers saw less of a decline. For both categories of jobs, most of the downturn is connected to "quiet quitting" when workers are not actively engaged with work activities (Harter, 2023).

Functionalists also gloss over many U.S. corporations' disinterest in workers' well-being. In the private sector, 21 percent of workers don't have paid sick days. And these paid sick days are not evenly dispersed. Among the highest paid workers, 90 percent have this benefit. In comparison, only 43 percent of the lowest 10 percent of income earners have paid sick days to use (U.S. Bureau of Labor Statistics, 2021). Even when business improves, many employers convert full-time jobs to part-time or temporary positions and cut health benefits (Schultz, 2011).

AP Images/Connecticut Post, Ned Gerard

For functionalists, the economy provides order and stability. Many unemployed men—including recent Asian and Latina/o/x immigrants—join the military because it offers steady work, college education benefits, and an opportunity for career advancement.

11-4b Conflict Theory

While functionalists emphasize the economy's benefits, conflict theorists argue that capitalism creates social problems. They contend, for example, that globalization has led to job insecurity and that a handful of transnational conglomerates have enormous power.

Some Key Issues

For conflict theorists, low wages alienate employees rather than motivate them to work harder, as functionalists claim. Many people are stuck in low-paying, monotonous jobs and don't always get promotions, regardless of their skills, attitudes, perseverance, or productivity (Clements, 2012).

Conflict theorists also argue that capitalism enables the rich and powerful to exploit other groups, resulting in huge wealth disparities. In 2021, CEO pay was 399 times more than the average worker. In some of the largest U.S. companies, CEOs earn 1,000 times more than their employees, and the gap has grown over the decades. From 1978 to 2021, CEO pay increased by a whopping 1,460 percent, yet pay for the average worker went up by only 18.1 percent. It is estimated that CEOs at the top 350 companies were paid around 28 million dollars in 2021. And, unlike other industrialized countries, the U.S. government doesn't cap corporate pay (Bivens and Kandra, 2022; Krantz, 2015; Weaver, 2016).

> **Are conflict theorists too quick to blame capitalism for economic problems?**

Instead of higher wages, annual raises, and health benefits, more businesses are rewarding employees with perks (e.g., free gym memberships, $100 gift cards, pet health insurance). Working-class families' eroding job opportunities and wages have led to personal problems such as relationship discord and mental health problems. As economic problems accumulate, "deaths of despair"—death by drugs, alcohol, or suicide—increase, especially among middle-aged White men with low education levels (Mui, 2015; Case and Deaton, 2017; refer also to Chapter 14).

Critical Evaluation

Critics fault conflict theorists for emphasizing economic constraints rather than choices. In 2021, 61 percent of Americans were living paycheck-to-paycheck, and by 2022, it increased to 64 percent. Out of that group of 166 million, 8 million of them earn $100,000 or more (Tanzi, 2023).

Also, according to functionalists, conflict theorists underestimate people's ability to get ahead. Many young people are thronging schools that teach high-tech skills in welding and other trades, fast-food workers are unionizing to raise the industry's minimum wage to $15.00 an hour, and 63 percent of small-business owners find qualified new employees through word-of-mouth referrals (Berfield, 2013; Jacobe, 2013; Phillips, 2014). Thus, many Americans are using several strategies to succeed economically.

ivector/Shutterstock.com

In 2017, the median CEO compensation was nearly $16 million. In contrast, the median annual wages for full-time workers was only $44,980—a CEO-to-worker pay ratio of 356 to 1 (Economic News Release, 2017; Marcec, 2017).

motherhood penalty (also called *motherhood wage penalty* or *mommy penalty*) a pay gap between women who are and aren't mothers.

11-4c Feminist Theories

Feminist scholars agree with conflict theorists that there's widespread workplace inequality. They emphasize, however, that gender is a critical factor in explaining the inequity.

Some Key Issues

At all income levels and in all occupations, women, especially Latina and Black women, earn less than men (**Figure 11.4**). There's a 40 percent gender wage gap—after accounting for race, ethnicity, education, occupation, workplace experience, and other factors—that's presumably due to gender discrimination (Blau and Kahn, 2016). Both historically and currently, as you learned in Chapters 6 and 9, women have lower earnings than men because of occupational sex segregation, glass ceilings, and glass escalators.

Employment rates increased after the recession, but the gains for women have been largely in low-paying jobs, particularly waitressing, in-home health care, food preparation, and housekeeping. From 2009 to 2012, 60 percent of the increase in women's employment, versus 20 percent for men, was in jobs that pay less than $10.10 an hour. Women make up less than half of the labor force but hold more than two-thirds of all low-wage jobs (e.g., personal and home health aides, fast-food workers, lowest-paid retail positions). Following the pandemic, recovery for women's employment is behind that of men, even in industries that have been typically dominated by women—retail, leisure, and hospitality (National Women's Law Center, 2014; Robbins and Vogtman, 2016; Almeida and Salas-Betsch, 2023).

Many women's earnings suffer from a **motherhood penalty** (also called the *motherhood wage penalty*), a pay gap between women who are and aren't mothers. On average, women without children make 90 cents to a man's dollar, married mothers make 73 cents to a man's dollar, and single mothers make only 60 cents to a man's dollar. This motherhood penalty accounts for most of the overall wage gap between men and women. On the other hand, there is a *fatherhood bonus*, where men with children earn more than men without children. More recently, the fatherhood bonus is declining (Rowe-Finkbeiner, 2012; Entmacher et al., 2014; Whiting, 2022; Hymowitz, 2020).

After declining for several decades, the share of stay-at-home mothers began increasing around 2000.

There are several reasons for this increase, but low wages and motherhood penalties are at the top of the list. In 2022, following the COVID-19 pandemic, women with children under 18 made employment leaps, supported by the reopening of schools, availability of vaccines, and expansion of child care. Still, during the post-pandemic recovery, the lack of affordable childcare has been a barrier to over a million women returning to work (Cohn et al., 2014; Kaplan, 2022; Almeida and Salas-Betsch, 2023).

Critical Evaluation

Critics note that men don't always dominate women in the workplace. There are many situations where female supervisors have power and authority over women and men, and top female managers aren't always concerned about gender wage gaps nor support qualified women's promotions (Huffman, 2013). Second, many feminist scholars maintain that capitalism exploits women by crowding them into lower-paying occupations. However, there's considerable economic gender inequality in socialist, communist, and mixed economies in both industrialized and industrializing nations (Cudd and Holstrom, 2010).

Some people also question whether the economy reinforces gender-based discrimination or simply reflects cultural sexism. For example, 51 percent of Americans, including women, say that children are better off if the mother is at home and doesn't work; only 8 percent say the same about stay-at-home dads (Cohn et al., 2014). Thus, patriarchal values may be as important as discrimination in explaining many women's economic inequality.

11-4d Symbolic Interaction

Symbolic interactionists rely on micro-level approaches to explain the day-to-day meaning of work. They're especially interested in how work shapes people's self-identity.

Some Key Issues

Symbolic interaction has provided numerous insights into how people define and experience work. Regardless of income level, 63 percent of U.S. workers who are engaged in their jobs, and even 42 percent of those who aren't engaged say they would continue to work in their current job if they won a $10 million lottery. All of them might quit, of course, but such responses show that work provides many people with "a source of identity, purpose, and satisfaction that money alone may not replace" (Harter and Agrawal, 2013).

Interactionists also study how people are socialized into their jobs and the informal rules that shape behavior. In medicine, surgeons teach their interns and residents that it's normal to make some mistakes, but that carelessness and continued errors are unprofessional (Bosk, 1979). Assembly-line workers also control

iStock.com/CatLane

Would you work if you won a $10 million lottery? Why or why not?

coworkers who overproduce or underproduce by punishing or rewarding them (refer to Chapter 6).

Critical Evaluation

The most common criticism is that symbolic interaction, although providing in-depth analyses, sacrifices scope. Studies of small work groups can tell us much about people's interaction. We don't know, however, if the findings are applicable to large companies or different parts of the country because the research is based on small and nonrepresentative samples.

Another limitation is that interactionism neglects macro-level social forces that affect people's work and choices. Consider government: When it allows monopolies, consumers must pay high prices for low-quality goods and services because there are no options. Low wages and discrimination, which most politicians ignore, alienate workers and increase personal and family stress. Thus, structural obstacles can thwart individual choices and efforts.

11-5 Global Political Systems

Every society has a **government**, a formal organization that has the authority to make and enforce laws. Governments are expected to maintain order, provide social services, regulate the economy, establish educational systems, create armed forces to discourage (real or imagined) attacks by other countries, and ensure their residents' safety.

Worldwide, governments vary from democratic to totalitarian, but there are also authoritarian

government a formal organization that has the authority to make and enforce laws.

governments and monarchies. Let's begin by looking at democracy.

11-5a Democracy

A **democracy** is a political system in which, ideally, citizens have control over the state and its actions. Democracies are based on several principles:

- Individuals participate in decisions and select leaders who are responsive to the majority of the people's wishes.
- Suffrage (the right to vote) is universal, and elections are frequent, free, fair, and secret.
- The government recognizes individual rights, such as the freedoms of speech (including dissent), press, and assembly and the right to organize political parties whose members compete for public office.
- The "rule of law" requires that everyone obey the law and be held accountable if they violate it.

Worldwide, 36 percent of the world's population (2.7 billion people) live in countries with repressive governments that deny basic political and civil rights. People in sub-Saharan and North Africa, Middle East, China, Eurasia, and Russia are the least likely to have political freedom (**Figure 11.5**). Despite some countries' attempts to establish democracies, democracy is declining in nearly half of the world's existing democracies (International Institute for Democracy and Electoral Assistance, 2022).

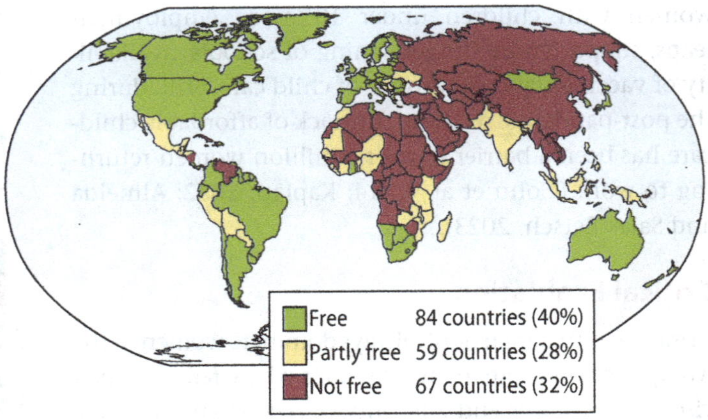

Figure 11.5 Global Freedom Around the World, 2023

This map shows the degree of global freedom for 195 countries and 15 territories. Global freedom includes both political rights and civil liberties.

Source: Freedom House, 2023.

11-5b Totalitarianism and Dictatorships

At the opposite end of the continuum from democracy is **totalitarianism**, a political system in which the government controls almost every aspect of people's lives. Totalitarianism has several distinctive characteristics (Taylor, 1993; Tormey, 1995; Arendt, 2004):

- A pervasive ideology that legitimizes state control and instructs people how to act in their public and private lives.
- A single political party controlled by one person, a *dictator*—a supreme, sometimes idolized leader—who stays in office indefinitely.
- A system of terror that relies on secret police and the military to intimidate people into conformity and punish dissenters.
- Total control by the government over other institutions, including the military, education, family, religion, economy, media, and all cultural activities, including the arts and sports.

Contemporary examples of the most repressive totalitarian governments include Afghanistan, Myanmar, North Korea, Central African Republic, and Syria. In these countries and territories, state control over daily

Timofeev Sergey/Shutterstock.com

Some political analysts believe that Russia is moving from authoritarianism to totalitarianism. Vladimir Putin, Russia's prime minister/president since 1999, censors much of the Internet content, crushes gay and other civil rights activists, represses his regime's critics, and is resurrecting the KGB—a secret police force and spy network involved in all aspects of people's everyday lives (Ryzhkov, 2015; Soldatov, 2016).

democracy a political system in which, ideally, citizens have control over the state and its actions.

totalitarianism the government controls almost every aspect of people's lives.

In 1995, at age 3, Oyo Nyimba Kabamba Iguru became the world's youngest monarch when he was crowned as king of the Ugandan Kingdom of Toro. At age 18, he officially took control of the kingdom.

life is pervasive and wide-ranging, independent organizations and political opposition are banned or suppressed, and those who criticize the government are imprisoned, tortured, or killed (Puddington and Roylance, 2017).

11-5c Authoritarianism and Monarchies

Many nations have some version of democracy or totalitarianism. However, a number of countries are characterized by **authoritarianism**, a political system in which the state controls the lives of citizens but permits some degree of individual freedom. Authoritarian governments, like Russia, may permit elections, often controlled, but prohibit free speech, punish those who publicly question state policies, and imprison people without giving them a trial. Russia has recently moved more towards totalitarianism (Kolesnikov, 2022).

In a **monarchy**, the oldest type of authoritarian regime, power is based on heredity and passes from generation to generation. A member of a royal family, usually a king or queen, reigns over a kingdom. A monarch's power and authority are legitimized by tradition and religion. There are more than 40 monarchies around the world. In some countries—especially in the Middle East and parts of Africa, including North Africa—monarchs have absolute control. Those in Norway, Denmark, Belgium, Britain, Japan, and many other countries have little political power because they're limited by democratic constitutions and serve primarily ceremonial roles (Glauber, 2011).

In sum, political systems vary worldwide. Despite the variation, political leaders wield considerable power and authority.

11-6 Politics, Power, and Authority

Power and authority are important concepts in sociological analyses of politics. The use and misuse of power and authority affect the quality of our everyday lives.

11-6a Power

Power is the ability of a person or group to influence others, even if they resist. For example, a government can quash a peaceful protest or, conversely, protect people's right to protest. *Legitimate power* comes from having a role, position, or title that people accept as legal and appropriate (e.g., Congress has the right to pass laws, the police have the right to enforce laws). *Coercive power* relies on force or the threat of force to impose one's will on others (e.g., a dictator who imprisons or executes protestors).

Power is relational. Lawmakers affect their constituents' lives because of the laws they pass on taxes, the environment, and other issues. Voters can protest, refuse to follow laws, and vote politicians out of office.

Whether legitimate or coercive, power—especially political power—is about controlling others. Because people may revolt against sheer force, many governments depend on authority to establish order, shape people's attitudes, and control their behavior.

11-6b Authority

Authority, the legitimate use of power, has three characteristics. First, people *consent* to authority because they believe their obedience is for the greater good (e.g., following traffic rules). Second, people think of the authority as *legitimate*—valid, justifiable, and necessary (e.g., paying taxes to support public education, increase national security, and remove trash and snow). Third, people accept authority because it's *institutionalized*, accepted by a large number of people, in organizations (e.g., police departments and government agencies).

Max Weber (1925/1978) described three ideal types of legitimate authority: traditional, charismatic, and rational-legal (**Table 11.4**). Remember that ideal types are models that describe the basic characteristics of any phenomenon (refer to Chapters 1 and 5). In reality, these types of authority often overlap.

---

authoritarianism the state controls the lives of citizens but permits some degree of individual freedom.

monarchy power is based on heredity and passes from generation to generation.

power the ability of a person or group to influence others, even if they resist.

authority the legitimate use of power.

Table 11.4 Weber's Three Types of Authority

Type of Authority	Description	Source of Power	Examples
Traditional	Power is based on customs, traditions, and/or religious beliefs.	Personal	Medieval kings and queens, emperors, tribal chiefs
Charismatic	Power is based on exceptional personal abilities or a calling.	Personal	Adolf Hitler, Gandhi, Martin Luther King, Jr.
Rational-legal	Power is based on the rules and laws that are inherent in an elected or appointed office.	Formal	U.S. presidents, members of Congress, state officials, police, judges

Traditional Authority

Traditional authority is power based on customs that justify the ruler's position. The source of power is personal because the ruler inherits authority on the basis of longstanding customs, traditions, or religious beliefs. Traditional authority is most common in nonindustrialized societies where power resides in kinship groups, tribes, and clans. In many African, Middle Eastern, and Asian countries, power is passed down among men within a family line (Tétreault, 2001; refer also to Chapter 8).

Charismatic Authority

Charismatic authority is power based on exceptional individual abilities and characteristics that inspire devotion, trust, and obedience. Like traditional authority, charismatic authority is personal and reflects extraordinary deeds or even a belief that a leader has been chosen by God, but the leaders don't pass their power down to their offspring.

Charismatic leaders can inspire loyalty and passion whether they're heroes or tyrants. Examples of the former include Aung San Suu Kyi, Mahatma Ghandhi, and Martin Luther King Jr.. Charismatic leaders who are more well known for heavy control of those they led include Adolf Hitler, Napoleon, Ayatollah Khomeini, and Fidel Castro. Whether inspiring people to improve the world or harm it, charismatic leaders are often spellbinding orators who radiate magnetism, dynamism, and tremendous self-confidence and often promise to improve a nation's future (Taylor, 1993).

traditional authority power based on customs that justify the ruler's position.

charismatic authority power based on exceptional individual abilities and characteristics that inspire devotion, trust, and obedience.

rational-legal authority power based on the belief that laws and appointed or elected political leaders are legitimate.

Rational-Legal Authority

Rational-legal authority is power based on the belief that laws and appointed or elected political leaders are legitimate. Unlike traditional and charismatic authority, rational-legal authority comes from rules and regulations that pertain to an office rather than to a person. For example, anyone running for mayor must have specific qualifications (e.g., U.S. citizenship). When a new mayor (or governor or other politician) is elected, the rules don't change because power is vested in the office rather than the person currently holding the office.

Mixed Authority Forms

People may enjoy more than one type of authority. Some historians believe that presidents Abraham Lincoln, Theodore Roosevelt, John F. Kennedy, Ronald Reagan, and Barack Obama enjoyed charismatic appeal beyond their rational-legal authority. In some countries, including Japan and England, emperors, kings, and queens have traditional authority and perform symbolic state functions (e.g., attending ceremonies). Both of these

Getty Images/Getty Images

First Lady Michelle Obama is an example of a charismatic political leader. She worked to end childhood obesity, helped military families, and visited other countries as a goodwill ambassador.

countries also have parliaments that exercise legal-rational authority in determining laws and policies.

11-7 Politics and Power in U.S. Society

Politics plays a critical role in our everyday lives. Three important components of the political system are political parties, special-interest groups, and voters.

11-7a Political Parties

A **political party** is an organization that tries to influence and control government by recruiting, nominating, and electing its members to public office. Political parties are avenues for citizens to shape public policy at the local, state, and national levels.

Functions of U.S. Political Parties

Political parties engage in a range of activities—from designing social media messaging, mailing flyers, and calling voters to drafting laws if a candidate is elected. Parties perform a number of vital political functions, especially recruiting candidates for public office, organizing elections, and, if elected, running the government.

Ideally, that's how political parties *should* work. In reality, parties perform some of these functions more effectively than others. For example, parties typically concentrate on winning elections rather than passing laws that benefit the general population once their candidates are in office.

The Two-Party System

Many democracies around the world have a number of major political parties: 4 in Australia, 9 in Spain, and 7 each in Germany, Romania, and the United Kingdom. Thus, compared with many other countries, the two-party system of Democrats and Republicans in the United States is uncommon. Political parties base their activities on an *ideology*, a set of ideas that constitutes a person's or group's beliefs, goals, expectations, and actions. **Table 11.5** gives some examples of major ideological differences between Democrats and Republicans.

As of March 2023, equal percentages of Americans identify with the two major parties (Democrat and Republican)—25 percent each. Because of dissatisfaction with the two parties, another 49 percent identify as independent (Gallup, 2023). Fifty-six percent of Americans feel the United States needs a third party (Saad, 2022). Because there's no national Independent Party, however, independents usually vote for Democrats, Republicans, or don't vote at all.

political party an organization that tries to influence and control government by recruiting, nominating, and electing its members to public office.

Table 11.5 How Do Democrats and Republicans Differ?

Issue	Many Democrats Believe That . . .	Many Republicans Believe That . . .
Economy	The government should raise taxes on the wealthy and corporations and expand programs for the economically disadvantaged.	The government should lower taxes on the wealthy and corporations to encourage investment and economic growth; economically disadvantaged people are too dependent on the government.
Abortion	Women should have the right to choose abortion in practically all cases; the government should pay for abortions for low-income women.	The unborn should be protected in all cases; there should be no public money for abortions; *Roe v. Wade* should remain overturned.
Women's Rights	Significant obstacles still make it harder for women to get ahead than men; women should have paid child care, family leave, and equal pay.	The obstacles that made it harder for women to get ahead are largely gone; there are as many or more women than men in powerful positions; women have equal or more financial stability than men.
Education	The government should strengthen public schools by raising salaries for teachers and decreasing classroom sizes. It shouldn't use tax dollars to help students attend private schools.	The government should increase state and local control of schools. It should use tax dollars to help students attend private schools of their choice if their local school is underperforming.
Immigration	Immigrants help the economy by creating small businesses, taking jobs Americans don't want, and accepting lower wages that keep the costs of goods and services down; immigrants have enriched U.S. culture through food, music, and the arts.	Immigrants hurt the economy by driving down wages and decreasing the incomes of Americans who work in the lowest-paying jobs, increasing crime rates, and paying few (if any) taxes.

Sources: Doherty et al., 2014; Wilke and Newport, 2014; Dugan, 2015; Fingerhut, 2016, 2017; McCarthy, 2017; PerryUndem, 2017; Swift, 2017.

SHAWN THEW/EPA-EFE/Shutterstock.com

In 5 to 4 decisions in 2010 and 2014, the Supreme Court overturned previous limits on federal campaign donations. Now, super PACs can contribute as much as they want.

Especially at the national level, political candidates routinely promise that, if elected, they'll reform campaign spending, curb lobbying, and stop wasting taxpayer money. Instead, special-interest groups play an important role in political outcomes.

11-7b Special-Interest Groups

A **special-interest group** (also called an *interest group*) is made up of people who seek or receive benefits or special treatment. While political parties include diverse individuals, special-interest groups are usually made up of people who are similar in social class and political objectives. There are thousands of U.S. special-interest groups that include everything from agribusiness to unions. Special-interest groups use many tactics to influence the government, but the most effective are campaign contributions and lobbying.

Campaign Contributions

Eighty-four percent of Americans agree that money has too much influence in our political campaigns, but it's often the most important factor in winning elections. Money doesn't guarantee a victory, but candidates with little financing usually have little public visibility. In the 24 months leading up to the 2020 election, candidates for both the presidency and Congress raised and spent a combined 8 billion dollars. Congressional candidates brought in and spent nearly 4 billion dollars. The spending continues to increase from 6.5 billion in 2016 (Montanaro et al., 2014; Kamp, 2016; Center for Responsive Politics, 2017; Federal Election Commission, 2021).

Contributions come from individuals and national, state, and local party committees, but the most powerful are **political action committees (PACs)**, special-interest groups that raise money to elect one or more candidates to public office. PACs can contribute directly to candidates but have donation limits. Super PACs can't contribute directly to political candidates but may raise and spend unlimited funds to advocate for or against candidates.

PACs have mushroomed. They have grown in number—from 608 in 1974 to almost 15,700 in 2014 (35 percent represent corporate interests). In the 24 months before the 2020 elections, PACs raised and spent about $13 billion. Leading up to the 2022 midterm elections, PACs brought in around $7.4 billion and spent around $6.5 billion (Federal Election Commission, 2021; Federal Election Commission, 2023). Some PACs, like the International Brotherhood of Electrical Workers, support primarily Democrats, while the National Rifle Association contributes only to Republicans. A number of PACs, however, contribute to both parties to cover their bases regardless of who's elected. PACs and super PACs provide access to politicians and obligate lawmakers to pass the legislation a PAC wants.

Lobbyists

A **lobbyist** is someone hired by a special-interest group to influence legislation on the group's behalf. At the federal level, the number of lobbyists rose from 10,400 in 1998 to almost 11,200 in 2016, an average of 21 lobbyists per congressional member. Some of the best-paid lobbyists include former members of Congress, former congressional aides, and lawmakers' relatives. Specific industries also lobby more aggressively. In 2021, Apple, Amazon, Google, and Facebook spent over $55 million to lobby the federal government, more than the $34 million they spent in 2020. Big technology directed more money to lobbying than the defense and pharmaceutical industries (Drutman and Furnas, 2014; "Lobbying Database," 2017; Birnbaum, 2022).

In 2016, lobbyists spent $3.2 billion to send members of Congress and their family members on expensive vacations and, most importantly, made generous campaign contributions. In return, corporations and other organizations receive favorable legislation, such as Congress passing lax gun and banking laws, approving corporate mergers, and not investigating monopolies (Schweizer, 2013; "Top Spenders," 2017).

Lobbying also generates *earmarks* (known commonly as "pork")—funding requests by Congress to provide

---

special-interest group (also called an *interest group*) a group of people that seeks or receives benefits or special treatment.

political action committee (PAC) a special-interest group that raises money to elect one or more candidates to public office.

lobbyist someone hired by a special-interest group to influence legislation on the group's behalf.

federal money to companies, projects, groups, and organizations in their district—that don't require external competitive bidding. After considerable public pressure, Congress banned earmarks in 2010, but pork spending continues. In 2023, there were 7,396 earmarks, almost 44 percent more than in 2022. These 2023 earmarks cost $26 billion (Citizens Against Government Waste, 2023).

11-7c Who Votes, Who Doesn't, and Why

In 2020, almost 67 percent of Americans voted in the presidential election, 5 points higher than in 2016. The 2020 turnout was actually the highest since the 1900 presidential election (nearly 74 percent). Typically, voter turnout is between 50 and 60 percent (Fabina, 2021; Colarossi, 2020). While 2020 was an improvement for Americans voting, fewer people vote in U.S. elections compared to other countries. Of 49 other nations that recently had elections, U.S. voter turnout ranks 31st. In their 2018 and 2019 elections, three countries had very high rates: Turkey (89 percent), Sweden (82 percent), and Australia (81 percent) (DeSilver, 2022; Alas, 2020).

What are the challenges to voting in the United States? Who votes, who doesn't, and why reflect demographic characteristics, attitudes about politics, and situational and structural factors.

Demographic Factors. Many demographic factors (e.g., marital status, geographic region, religion) affect registration and voting. The most important, however, are sex, age, social class, and race and ethnicity.

Sex. In every presidential election since 1996, women, across all racial/ethnic groups, have voted at higher rates than men. Following this pattern, in the 2020 presidential election, more women (68 percent) than men (65 percent) voted. Also, in the 2020 election, the biggest gender voting gap was among Black people—Black women voted at higher rates than Black men by 8 percentage points. The reasons for this difference are unclear but may reflect higher turnouts by women who believe that social services, abortion rights, and health care coverage are threatened (U.S. Census Bureau, 2021).

Age. Historically, and across all elections, voting rates increase with age. In 2020, those aged 65 to 74 had the highest turnout (76 percent) compared to those aged 18 to 24, who had the lowest (51 percent) (**Table 11.6A**). A key factor is registration. Only 60 percent of young people (aged 18–24) were registered to vote, compared with 79 percent of those aged 65–74 (U.S. Census Bureau, 2021).

Young people have lower registration rates because they're more geographically mobile than older people and are less likely to reregister after a move. Moreover, they often feel uninformed about politics, believe that elections aren't relevant in their lives, and may be preoccupied with major life events such as going to college and finding jobs (Schachter, 2004; Child Trends Data Bank, 2013). In effect, young people's low voting rates means they have little effect on political processes.

Social Class. Social class has a significant impact on voting behavior. At each successive level of educational attainment, the voting rate increases, and those

Table 11.6 Voter Rates in the 2020 Presidential Election, by Selected Characteristics

A. Age	
18–24	51.4%
25–34	60.3%
35–44	65.1%
45–54	69.0%
55–64	72.7%
65–74	76.0%
75 and older	72.3%
B. Education	
High school, no diploma	41.5%
High school graduate	55.5%
Some college or associate degree	69.6%
Bachelor's degree	77.9%
Advanced degree	83.0%
C. Annual Family Income	
Under $10,000	47.1%
$10,000 to $14,999	50.4%
$15,000 to $19,999	49.8%
$20,000 to $29,999	56.1%
$30,000 to $39,999	63.1%
$40,000 to $49,999	68.6%
$50,000 to $74,999	72.0%
$75,000 to $99,999	77.0%
$100,000 to $149,999	80.5%
$150,000 and over	84.8%

Source: Based on U.S. Census Bureau, 2021, Tables 5 and 7.

How Would You Describe Congress?

In 2010, the Pew Research Center asked Americans to use one word to describe Congress. The responses were put into a "word cloud." The larger the word, the more times it was used. What one word would *you* use to describe Congress?

word cloud graphic, created using http://wordle.net, from "Congress in a Wordle," Mar. 22, 2010, The Pew Research Center For the People & the Press, a project of the Pew Research Center

with advanced degrees are the most likely to vote (**Table 11.6B**). People with higher educational levels are usually more informed about and interested in the political process and more likely to believe that their vote counts (U.S. Census Bureau, 2021; "The Politics of Financial Insecurity," 2015).

Voting rates also increase with income levels (**Table 11.6C**). For instance, 81 percent of people with incomes of $100,000–$149,999 voted, compared to 63.6% of those earning $30,000–$39,999. People with higher incomes are more likely to be employed and to have assets (e.g., houses, stock) and, therefore, are more likely to vote to protect or increase their resources. In contrast, low-income people usually have few assets and may be too disillusioned with the political system to vote ("Who Votes . . . ," 2006; File and Crissey, 2010).

Race and Ethnicity. Across racial-ethnic groups, there are more native-born or naturalized White people who are eligible to vote. Around 70 percent of Black people are eligible to vote, which is only slightly less than the overall population (72 percent). They're also the most likely to register and to vote. Racial-ethnic minorities, especially Black people, played a pivotal role in Barack Obama's 2008 and 2012 presidential wins. In 2020, in contrast, 64 percent of White people versus 63 percent of Black people voted. The 2020 Black voter turnout was higher than the 2016 election (Moslimani, 2022; U.S. Census Bureau, 2021; Morris and Grange, 2021)

In the past and currently, Black voting rates have been higher than those of Latina/o/x people and Asians. In 2020, the Latina/o/x and Asian/other percentage of the vote grew, while the Black percentage remained the same (**Figure 11.6**). About 75 percent of Asians and Latina/o/xs were born in the United States, but Asian Americans and Latina/o/x Americans had lower voter registration rates (64 and 61 percent) than Black people (69 percent) and White people (77 percent) (Frey, 2021; U.S. Census Bureau, 2021).

The racial-ethnic voting gaps may close in the future as immigrants become eligible for citizenship and as Latina/o/xs and Asians become more familiar with registration and voting practices. Low Asian voter turnout is puzzling, however: They have the highest education and income levels (refer to Chapters 8 and 10), but even among the college-educated, voter turnout has lagged behind White and Black individuals (Krogstad, 2014; Frey, 2015).

Attitudes

Millions of Americans have a low opinion of the government, and it has been that way for most of the last two decades. In 2022, around 21 percent of Americans trust the federal government to "do what is right" almost always or most of the time. This is a 3 percent decline in trust from the year before (Pew Research Center, 2022). Trust in government aligns with party affiliation. The party with the greatest control of the government has more support from members of its own party than members

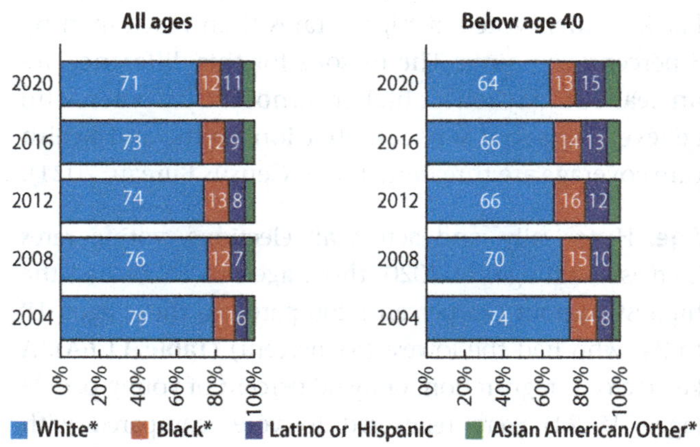

Figure 11.6 Shares of U.S. Voters in Presidential Elections by Race and Ethnicity: 2004 to 2020

Source: Based on Frey, 2021.

of the other party. Trust also varies by race, with Asian Americans having greater trust (37 percent), followed by Latina/o/x Americans (29 percent), Black Americans (24 percent), and finally, White Americans (16 percent). Americans' trust in their local governments is higher than their trust in the federal government (Brenan, 2021).

Such distrust and dissatisfaction explains some of the reason why many people don't vote: They believe their lives won't improve regardless of who's elected. While only 21 percent of registered voters say that members of Congress should be re-elected, few incumbents are voted out of office election after election. Yet, there is turnover. In the 2021 Congress, 72 percent of House members and 65 percent of senators had been there for 12 years or less. And counter to public messaging about voter distrust with the 2020 election count, when it comes to how elections are run and votes are counted, most people trust that elections are done well at the local level (90 percent) and federal level (70 percent) (Pew Research Center, 2022; Brenan, 2021; Jones, 2022; DeSilver, 2022).

Access

When it comes to voter participation, it is also important to consider access. Not all people have the same ability to register or cast their ballots. For the 2020 election, some non-voters mentioned a lack of transportation (2.4 percent) or inconvenient polling places (2.6 percent). Consider rural areas where it may be challenging to find large, public buildings that are reasonably located for voters. And travel distance may be challenging for both voters and for potential poll workers. Other reasons that people did not vote in 2020 were health-related issues, with 4.3 percent citing concerns about the COVID-19 pandemic and 13 percent indicating illness or disability as barriers. People with disabilities may have to navigate inaccessible polling places or a lack of accessible voting machines (Levine et al., 2020; U.S. Census Bureau, 2021; Dimmick and Aksosz, 2022).

Situational and Structural Factors

In the 2020 presidential election, 14.5 percent of Americans didn't vote because they didn't like the candidates, and another 17.6 weren't interested, but situational and structural factors also affect elections. Some of the high turnout rates in other countries are partly due to *compulsory voting*: The government imposes fines for not voting, may cut off public assistance benefits, or requires evidence of voting to get a passport or a driver's license. Such sanctions send the message that voting isn't a privilege but a civic responsibility. In many European and other countries, voters are registered automatically when they pay taxes or receive public services, elections are held on weekends, and people vote electronically (Seward, 2012; Samuelson, 2016; U.S. Census Bureau, 2021).

Instead of making voting easier, 34 states (particularly those controlled by Republican legislatures) have passed or are considering laws that will make voter registration more difficult. Some of the new rules require government-issued photo identification (e.g., a driver's license or passport) and a birth certificate or proof of citizenship. Some states have reduced early voting, limited the time polls are open, done away with same-day voter registration, and moved polling locations outside of areas with large racial-ethnic populations. Proponents contend that the new laws will crack down on voter fraud; opponents argue that the new regulations will stifle voter turnout, especially among students, people who have lower incomes, and racial-ethnic minorities, who tend to vote for Democrats. Although there is some concern over voter fraud, in 2020, there were less than 475 potential cases of voter fraud out of more than 25 million votes cast (Whiteaker et al., 2014; Rios, 2016; Underhill, 2017; PBS, 2021).

Because many Americans don't vote, who has the most power? And do government leaders represent the average citizen? Such questions have generated considerable debate among sociologists and other social scientists.

11-8 Sociological Perspectives on Politics and Power

According to symbolic interaction, power is socially constructed through interactions with others; people learn to be loyal to a political system—whether it's a democracy or a monarchy—and to show respect for its symbols and leaders. For the most part, however, sociologists rely on functionalist, conflict, and feminist theories to explain political power. **Table 11.7** summarizes these perspectives.

11-8a Functionalism: A Pluralist Model

For functionalists, the people rule through **pluralism**, a political system in which power is distributed among a variety of competing groups in a society (Riesman, 1953; Polsby, 1959; Dahl, 1961).

Some Key Issues

For functionalists, the government is an important institution that creates and enforces laws, regulates elections, and protects people's civil liberties. In the pluralist model,

pluralism a political system in which power is distributed among a variety of competing groups in a society.

Table 11.7 Sociological Explanations of Political Power

	Functionalism: A Pluralist Model	Conflict Theory: A Power Elite Model	Feminist Theories: A Patriarchal Model
Who has political power?	The people	Rich upper-class people—especially those at top levels in business, government, and the military	White men in Western countries; most men in traditional societies
How is power distributed?	Very broadly	Very narrowly	Very narrowly
What is the source of political power?	Citizens' participation	Wealthy people in government, business corporations, the military, and the media	Being White, male, and very rich
Does one group dominate politics?	No	Yes	Yes
Do political leaders represent the average person?	Yes, the leaders speak for a majority of the people.	No, the leaders are most concerned with keeping or increasing their personal wealth and power.	No, the leaders are rarely women who have decision-making power.

individuals have little direct power over political decision making but can influence government policies through special-interest groups—unions, professional organizations, and so on (**Figure 11.7**). The various groups rarely join ranks because they concentrate on single issues such as health care, pollution, or education. This focus on different issues fragments groups but also results in a broad representation of interests and a distribution of power.

Because there are a number of single-issue groups, functionalists maintain, there are multiple leaderships (Dye and Ziegler, 2003). As a result, many leaders, not just a few, can shape decisions that represent numerous groups and issues. Pluralists note that people also have power outside of interest groups: They can vote, run for office, contact lawmakers, and put specific issues on a ballot. Therefore, there are continuous checks and balances as individuals and groups vie for power and try to influence laws and policies.

Pluralists believe that groups resolve differences by bargaining and compromising. As this process unfolds, the government is a neutral referee: It reduces outright hostility, helps groups achieve their goals to at least some degree, and ensures that no particular group has absolute authority. Thus, group competition benefits society as a whole.

Critical Evaluation

Does pluralism work as democratically as functionalists maintain? Critics argue that interest groups have unequal resources. People who are economically disadvantaged and historically marginalized rarely have the skills and

PLURALIST MODEL

Power is dispersed among multiple groups that influence the government. For example,

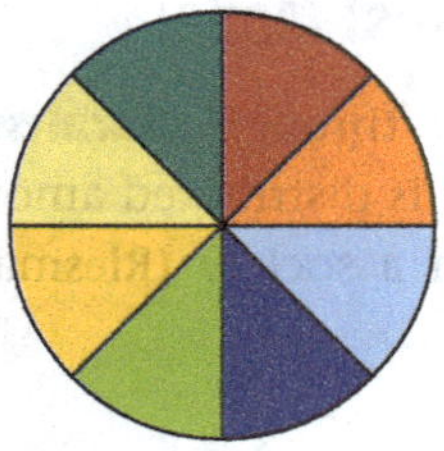

- government employees
- victims' rights groups
- labor unions
- banks and other financial institutions
- realtors and home builders
- teachers
- environmental groups
- women's rights groups

POWER ELITE MODEL

Power is concentrated in a very small group of people who make all the key decisions. For example,

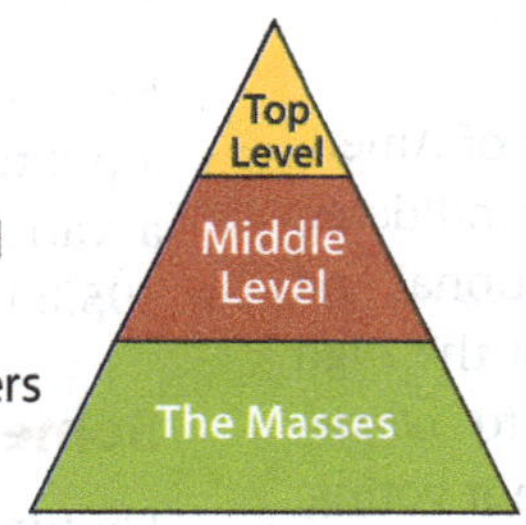

Top level (1 percent)—CEOs of large corporations, high-ranking lawmakers in the executive branch, and top military leaders

Middle level (8 percent)—most members of Congress, lobbyists, entrepreneurs of small businesses, leaders of labor unions and other interest groups (in law, education, medicine, etc.), and influential media commentators

The masses (91 percent)—people who are systemically exploited and disempowered, which reduces their abilities to organize for change.

Figure 11.7 Pluralist and Power Elite Perspectives of Political Power educational backgrounds to organize or promote their interests. In contrast, wealthy individuals and organizations can influence the government through lobbyists, political contributions, and personal connections.

Critics also maintain that pluralists aren't realistic about the emergence of multiple leaderships because the affluent control the government. For example:

- President Trump's cabinet was the wealthiest (at $6.2 billion) in U.S. history. President Biden's cabinet has less money and more women and more racial-ethnic diversity. Still, Biden's group has a lot of money, considering Biden's own worth around $8 million, Vice President Harris' at $7 million, and the entire cabinet's worth of $115 million (Alexander and Tindera, 2021).
- Supreme Court justices also have high salaries and notable wealth. In 2023, associate Supreme Court justices earned salaries of $285,400, and the chief justice earned $298,500. They also have income from other sources. For example, Sonya Sotomayor earned $3.3 million in book payments (United States Courts, 2023; Carlisle, 2022).
- Cabinet memberships and cabinet-level positions are dominated by business executives who have been abolishing or watering down pollution and banking regulations that conflict with corporate interests (Flesher and Biesecker, 2017; Walsh, 2017).

Because the wealthy control the political system, according to critics, the government isn't a neutral referee that benefits the general population.

11-8b Conflict Theory: A Power Elite Model

Conflict theorists contend that the United States is ruled not by pluralism but by a **power elite**, a small group of influential people who make the nation's major political decisions. Sociologist C. Wright Mills (1956) coined the term *power elite* to describe a pyramid of power he believed characterized American democracy.

Key Characteristics

According to Mills, the power elite is made up of three small but dominant groups of people at the top level who run the country: political leaders, corporate heads, and military chiefs (**Figure 11.7**). Practically all of the members of these groups are White Protestant men who form an inner circle of power. Those at the middle power level include members of Congress, lobbyists, influential media commentators, and others. The bottom level, and the largest and least powerful group, the masses, is composed of everyone else.

The power elite tolerates the masses—including their elections and laws—but in the end, it simply does what it wants. In 2013, almost 80 percent of Americans wanted limits on the amount of money that congressional candidates could spend on their campaigns (Saad, 2013). The following year, the Supreme Court did just the opposite by striking down laws that limited contributions to political candidates, parties, and PACs.

Like Mills, contemporary conflict theorists maintain that the United States isn't a democracy because of the close ties and "revolving door" between the power elite in politics, business, and the military. For example, fewer than 31,390 people—less than 0.01 percent of the nation's population—contributed more than $1.6 billion to political campaigns in 2012, and nobody was elected to Congress without the economic elite's money. In return, politicians pass favorable legislation, such as $83 billion in taxpayer-financed subsidies to the country's banks, award earmarks and lucrative military and other contracts, and may later serve as corporate lobbyists (Gilson, 2014; Kristof, 2014; Winship, 2015).

Critical Evaluation

Functionalists accuse conflict theorists of exaggerating the power elite's influence. They point out, for example, that "the masses" aren't just puppets but vote, mobilize, support particular interest groups, and protest current political and corporate policies.

Also, according to critics, the power elite perspective assumes, incorrectly, that the ruling class acts as a unified force in protecting its interests. Because the Democratic and Republican parties and their top officials endorse very different agendas, the power elite are rarely unified.

Some also question whether elites are always as self-serving and self-interested as conflict theorists contend. For example, past administrations have reduced smoking, pollution, and unsafe automobiles despite considerable lobbying and resistance by wealthy and influential corporations.

11-8c Feminist Theories: A Patriarchal Model

In a recent national poll, 75 percent of women and men said that both sexes are equally good political leaders (Associated Press and NORC, 2016). Why, then, are there so few female officeholders?

Some Key Issues

Feminist theorists maintain that women are generally shut out of the most important political positions because, in a patriarchal society, men rule. A "gender gap in political ambition" begins early in the socialization process.

---

power elite a small group of influential people who make the nation's major political decisions.

Parents often encourage their sons, but not their children of other genders, to think about politics as a career and to join political clubs in college. In adulthood, family, friends, and party leaders are more likely to urge men and not other genders to promote themselves and to run for a political office. Thus, despite comparable educational and occupational backgrounds, in 2021, around 60 percent of men have considered running for office while almost the same percentage of women have not. There has not been much improvement over the years. Today's gender gap in political ambition is almost as large as it was 20 years ago (Lawless and Fox, 2021). Potential candidates who are LGBTQIA+ may also find less support for running for office. Around one-third of registered voters say they would be "less excited" to support an LGB candidate for president, and over 40 percent would be less supportive of a transgender candidate (Associated Press, 2019).

When women run for political office, they're almost as likely to be elected as men, but they must overcome stereotypes and double standards. According to 40 percent of Americans, women who seek the highest political offices are held to higher standards than men. Some voters believe that mothers with young children who run for office won't be able to balance work and family life. In contrast, men routinely include their children in political advertising to project an image of being a father and family man, a traditional gender role. Voters are also more likely to expect female than male candidates to be likeable, honest, and ethical (Hayes and Lawless, 2016; Barbara Lee Family Foundation, 2017; Ditmar, 2017).

During the 2020 vice presidential debate, Kamala Harris was expected to follow debate rules and "remain calm and collected in the face of disrespect, to resist emotionality, and as a Black woman, to especially resist anger." Although Harris had to navigate extra expectations for gender and race, "Mike Pence was not tasked with walking that tightrope. He broke even the most basic rules" (Dastagir, 2020). Similar issues occurred during the 2016 presidential campaign. Some political observers rebuked both Democrat Hillary Clinton and Republican Carly Fiorina for appearing too stern and not smiling enough. No one criticized Donald Trump's or Bernie Sanders' smiles or scowls. Talk show commentators (particularly Fox News) and male journalists accused Clinton of being "shrill," "angry," and "grating." Sanders, who often shouted, was described as "enthusiastic" and "tough." Except for a handful of female journalists, few news correspondents or talk show hosts objected to Trump's continuous lewd, degrading, and sexist comments about women or his multiple sexual assaults (Christina, 2016; Holloway, 2016; Ditmar, 2017).

Critical Evaluation

Critics fault feminist explanations for several reasons. First, there are powerful women on some of the most influential congressional committees. Women have attained top leadership posts—president, prime minister, or other equivalent positions. Out of the 193 United Nations member states, just under one third have had a woman as their leader. Thirteen nations currently have women leading them, and Prime Minister Sheik Hasina of Bangladesh has been in office for 19 years (Clancy and Austin, 2023).

Second, a larger number of female officeholders doesn't guarantee greater gender equality or other rights for women. For example, some Republican women in Congress have opposed equal pay for equal work laws because (they maintain) well-qualified women are already paid the same as men and because such legislation would increase the number of civil lawsuits and hurt businesses. Recently, "Eight women were among the 50 Republican senators who voted to oppose abortion becoming enshrined in federal law" (Lowery, 2014; McDonough, 2014; Carbonaro, 2022).

One might also question whether patriarchy is the root of political power differences between women and men. Some of the world's most patriarchal societies (e.g., in parts of Africa, Latin America, and the Middle East) have more women in high-ranking political positions than does the United States, which professes gender equality. Women have the fewest rights in societies where male political leaders are religious zealots who don't believe in gender equality (refer to Jones, 2014). Thus, religion may be more important than patriarchy in quashing women's political leadership.

The White House

Shortly after the 2016 election, Vice President Mike Pence convened an influential group of Republicans to discuss, among other things, removing Obamacare insurance companies' requirements covering maternity, newborn, and pregnancy care (Filipovic, 2017). *Where are the women!*

Chapter Review

Key Takeaways

11-1 What are similarities and differences among the different global economic systems? The *economy* determines how a society produces, distributes, and consumes goods and services. *Capitalism* frequently spawns *monopolies* and *oligopolies* that dominate the market and discourage competition. *Socialism* and *communism* promise cooperation and collective ownership of property, but all countries have mixed economies that allow private profits.

11-2 How are corporate and political power interwoven? A *corporation*, usually created for profit, often forms a *conglomerate*. Both corporations and conglomerates are governed by *interlocking directorates* that have become more powerful than ever because of the growth of *transnational corporations* and *transnational conglomerates*.

11-3 How have macro-level variables changed the U.S. economy and jobs? *Work* produces goods or services. Many Americans have been casualties of *deindustrialization* and *globalization*. Others have lost their jobs to *offshoring*. Widespread *downsizing* has resulted in unemployment, *underemployment*, discouraged workers, and low-paid jobs.

11-4 What are similarities and differences among the various sociological explanations of work and the economy? Refer to Table 11.3 on page 232.

11-5 What are similarities and differences among the types of global political systems? In a *democracy*, ideally, citizens have a high degree of control over the state. *Totalitarianism* controls people's lives; *authoritarianism* generally permits some degree of individual freedom. A *monarchy* is the oldest type of authoritarian regime.

11-6 How do power and authority differ? Politics includes *power* and *authority*. *Legitimate power* comes from having a role, position, or title that people accept as legal and appropriate; *coercive power* relies on force or threat of force to impose one's will on others. Authority can be based on *tradition*, *charisma*, *rational-legal power*, or a combination of these sources.

11-7 What are the major factors in the U.S. political system? A *political party* tries to influence and control government. While political parties include diverse individuals, *special-interest groups* are usually made up of people who are similar in social class and political objectives. The most powerful special-interest groups are *political action committees* and *lobbyists*. The voting rate is much higher among some groups than others. Situational and structural factors, such as voter registration, requiring proof of citizenship, and convenience can also encourage or discourage voting.

11-8 How do the sociological perspectives on politics and power differ? Refer to Table 11.7 on page 244.

Key Terms

Economy (p. 223)
Politics (p. 223)
Monopoly (p. 224)
Oligopoly (p. 224)
Socialism (p. 224)
Communism (p. 224)
State capitalism (p. 225)
Corporation (p. 226)
Conglomerate (p. 226)
Interlocking directorate (p. 226)
Transnational corporation (also called a multinational corporation or an international corporation) (p. 226)
Transnational conglomerate (also called a multinational conglomerate) (p. 226)
Work (p. 226)
Deindustrialization (p. 227)
Globalization (p. 227)
Offshoring (p. 227)

Discouraged workers (p. 229)
Motherhood penalty (also called *motherhood wage penalty* or *mommy penalty*) (p. 234)
Government (p. 235)
Democracy (p. 236)
Totalitarianism p. 236)
Authoritarianism (p. 237)
Monarchy (p. 237)
Power (p. 237)
Authority (p. 237)
Traditional authority (p. 238)
Charismatic authority (p. 238)
Rational-legal authority (p. 238)
Political party (p. 239)
Special-interest group (also called *an interest group*) (p. 240)
Political action committee (PAC) (p. 240)
Lobbyist (p. 240)
Pluralism (p. 243)
Power elite (p. 245)

Discussion and Critical Thinking Questions

1. The national gender wage gap has narrowed slightly since 1990 but varies considerably across states, ranging from a low of 69 cents in Louisiana to a high of 92 cents in California. About 15 percent of the gender wage gap is due to discrimination, particularly employer hiring decisions and unequal pay for the same job.
   a. How does the persistent gender pay gap weaken the economy?
   b. Should all employed women automatically receive a 15 percent bonus every year to decrease the gender pay gap?
2. Do American workers need unions? Or are unions outdated because of offshoring and other factors?
3. Are your political views similar to those of other family members? If not, how do they differ and why?
4. Would the United States be better off if it had several major political parties instead of just two? Why or why not?
5. "My vote won't change anything." Do you agree or disagree? Why?

True or False?

How Much Do You Know about the Economy and Politics?

The answers to 2, 3, 4, and 5 are true. The answers to 1 and 6 are false.

1. Most countries have economic systems that are a blend of the system types.
2. Unemployment rates in January 2023 were at the lowest levels ever at 3.4 percent.
3. Using an average of 11 vacation days, American employees fall behind workers in most other countries such as France, where workers use an average of 23 days.
4. Because of dissatisfaction with the two parties, 49 percent of Americans identify as independent.
5. Democracy is declining in nearly half of the world's existing democracies.
6. The 2020 voter turnout was the highest since the 1900 U.S. presidential election.

12 Families and Aging

Learning Objectives

After studying this chapter, you will be able to…

12-1 Describe how families are similar and different in the United States and worldwide.

12-2 Describe how and why U.S. families are changing.

12-3 Explain why intimate partner violence, child maltreatment, and older adult maltreatment occur.

12-4 Illustrate how the U.S. older adult population is changing and its impact on society.

12-5 Evaluate the theoretical explanations of families and aging.

True or False?

How Much Do You Know about Contemporary U.S. Families and Aging?

1. Nonmarital teenage births have decreased over the past 20 years.
2. Women with college degrees are more likely than those with a high school diploma to cohabitate.
3. Divorce rates have increased since the 1990s.
4. Less than half of children live with both parents.
5. The percentage of the population born between 1946 and 1964 is growing in the United States.
6. Children under the age of 1 year old are most likely to be victims of child maltreatment.

The answers are at the end of the chapter.

American families vary greatly in structure, dynamics, and racial-ethnic diversity. This chapter examines the ways that families and aging are changing, both in the United States and globally.

Be sure to take the True or False quiz in the chapter opener to better understand how much you know about these topics.

What do you think?

I would undergo medical treatments to live to age 120 or longer.

1	2	3	4	5	6	7
	strongly agree				strongly disagree	

12-1 What Is a Family?

Ask five of your friends to define *family*. Their definitions will probably differ not only from each other's but also from yours. For our purposes, a **family** is an intimate group consisting of two or more people who (1) have a committed relationship, (2) care for one another, and (3) share activities and close emotional ties. This definition includes households and families of various types (e.g., foster families) whose members aren't related by birth, marriage, or adoption.

Contemporary households are varied and complex; family structures vary across cultures and have changed over time. In some societies, a family includes grandparents, cousins, and other relatives. In others, only parents and their children are viewed as a family.

12-1a How Families Are Similar Worldwide

The family, a social institution, exists in some form in all societies. Worldwide, families are similar in fulfilling some functions that have persisted over time.

family an intimate group consisting of two or more people who (1) have a committed relationship, (2) care for one another and any children, and (3) share activities and close emotional ties.

incest taboo cultural norms and laws that forbid sexual intercourse between close blood relatives.

Family Functions

Families vary considerably in the United States and globally but fulfill five important functions that ensure a society's survival (Parsons and Bales, 1955):

- **Sexual activity.** Every society has norms regarding who may engage in sexual relations, with whom, and under what circumstances. U.S. laws ban sexual intercourse with someone younger than 18 (or 16 in some states), but several states allow 12- and 13-year-old girls and 14-year-old boys to marry with parental permission, and some countries permit marriage with girls as young as 8 (McClendon and Sandstrom, 2016; Sandstrom and Theodorou, 2016; refer also to Chapter 9).

  One of the oldest rules that regulates sexual behavior is the **incest taboo**, cultural norms and laws that forbid sexual intercourse between close blood (biological) relatives (e.g., siblings, parent and child, or other family members who have close biological relationship).
- **Procreation and socialization.** Procreation is an essential family function because it replenishes a country's population. Through socialization, children acquire language; absorb the accumulated knowledge, attitudes, beliefs, and values of their culture; and learn the social and interpersonal skills they need to function effectively in society (refer to Chapters 3 and 4).
- **Economic security.** Families provide food, shelter, clothing, and other material resources for their members.
- **Emotional support.** Families supply the nurturance, love, and emotional sustenance that people need to be happy, healthy, and secure. Our friends may come and go, but our family is usually our emotional anchor.
- **Social class placement.** Social class affects all aspects of family life. Initially, our social position is based on our parents' social class, but we can move up or down the social hierarchy in adulthood (refer to Chapter 8).

Some sociologists include *recreation* as a basic U.S. family function. This function isn't critical for survival, but since the 1950s, many parents have spent much more time with their children on leisure activities (e.g., visiting amusement parks, playing video games together) that strengthen interpersonal bonds.

Marriage

Marriage, a socially approved mating relationship that people expect to be stable and enduring, is also universal. Countries vary in their specific norms and laws dictating who can marry whom and at what age, but marriage everywhere is an important rite of passage that marks adulthood and its related responsibilities, especially providing for a family.

Endogamy and Exogamy

All societies have formal or informal rules about the "right" partner for marriage or a committed relationship. **Endogamy** (often used interchangeably with *homogamy*) is a cultural practice of marrying or forming a committed partnership within one's group. The partners are similar in religion (e.g., Catholics marrying Catholics), race or ethnicity (e.g., Black people marrying Black people), social class (e.g., a college-educated person forming a committed relationship with another college-educated person), and/or age (e.g., younger people partnering with younger people). Across Southwest Asia and Africa, marrying a first or second cousin is not only common but desirable. The highest rates of cousin marriage are in Pakistan and India. The benefits include knowing a lot about shared relatives and ensuring that property stays in the family, but the practice increases the chance of genetic diseases (such as cystic fibrosis and thalassemia, a blood disorder) in their children (Iqbal et al., 2022; "Consanguineous Marriage," 2016).

Exogamy (often used interchangeably with *heterogamy*) is a cultural practice of marrying or forming a committed partnership with a person outside one's group, such as not partnering with one's own relatives. In the United States, 25 states prohibit marriage between first cousins, but violations are rarely prosecuted (National Conference of State Legislatures, 2014). Even when there are no formal laws, cultural norms and values (as well as social pressure) usually limit our marital partner choices.

12-1b How Families Differ Worldwide

Families also differ around the world. Some variations affect the family's structure, while others regulate where people reside and who has the most household power and authority.

Cartoon Resource/Shutterstock

"The kids are in college ... what are we going to do with ourselves?"

Nuclear and Extended Families

In some societies, such as American and European ones, a **nuclear family** composed of married parents and their biological or adopted children is quite typical. In much of the world, however, the most common family form is the **extended family**, composed of parents, children, and other kin (e.g., grandparents, cousins, and other adult relatives).

As the number of single-parent families increases in industrialized countries, extended families are becoming more common. By helping out with household tasks and child care, other adult members make it easier for a single parent to work outside the home.

Many Americans assume that the nuclear family, with opposite-sex parents, is the most common arrangement, but such families have declined in number. In 2014, just 14 percent of children lived with a stay-at-home mother and a working father, down from 50 percent in 1960 (Livingston, 2015).

---

marriage a socially approved mating relationship.

endogamy (often used interchangeably with *homogamy*) cultural practice of marrying within one's group.

exogamy (often used interchangeably with *heterogamy*) cultural practice of marrying outside one's group.

nuclear family a family form composed of married parents and their biological or adopted children.

extended family a family form composed of parents, children, and other kin.

age fotostock/Alamy Stock Photo

The Mosuo are a matriarchal society in China's Himalayas. A family consists of a woman, her children, and the daughters' offspring. An adult man will join a lover for the night and then return to his mother's or grandmother's house in the morning. Children resulting from these unions belong to the woman and take her last name, she and her relatives raise them, and daughters are preferred to sons. The Mosuo population is decreasing, however, as more young people marry outside the group or move to cities for work (Qin, 2015).

Residence Patterns

In a **patrilocal residence pattern**, newly married opposite-sex couples live with, or near, the husband's family. In a **matrilocal residence pattern**, they live with, or near, the wife's family. In a **neolocal residence pattern**, the couple sets up their own residence.

Around the world, the most common residence pattern is patrilocal. In industrialized societies, married couples are typically neolocal. Since the early 1990s, however, the tendency for young married adults to live with the parents of either spouse—or sometimes with the grandparents of one of the partners—has increased. The number of Americans living in multigenerational households has more than doubled from 7 percent in 1971 to 18 percent in 2021 (Parker, 2012; Cohn et al., 2022).

One result is a **boomerang generation**, young adults who move back into their parents' home after living independently for a while or who never leave home in the first place. These numbers increased with the 2007–2009 recession and then spiked with the COVID-19 pandemic, reaching a rate of 52 percent (**Figure 12.1**). Two years later, almost two-thirds of adults who moved back in with their parents were still there (Dickler, 2022).

You learned in Chapter 4 that many younger adults don't feel a need to set up their own homes because, among other reasons, the stigma of living with parents has faded. The growth in boomerangers may be connected to increased "helicopter parenting," where parents excessively monitor all of their children's activities. This can include hovering, overprotecting and intervening on their kids' behalf with education, recreational activities, or even jobs. With the increase of remote learning during the COVID-19 pandemic, parents had another way to helicopter parent via platforms such as Zoom. While some stepped in to assist their kids out of necessity by helping their kids navigate technology, others were overly monitoring and assisting children with their classes. Helicopter parenting can cause children to have a lessened sense of independence and develop difficulty dealing with challenges of life. When launched into the adult world, like boomerangs, some of these young adults keep flying back to their parents (Braff, 2020; Travers, 2022; West and Lewis, 2018).

Authority and Power

Residence patterns often reflect who has authority and power in the family. In a **matriarchal family system**, the oldest women (usually grandmothers and mothers) control cultural, political, and economic resources and, consequently, have power over men and boys. Some Native American tribes were matriarchal, and in some African countries, the oldest women have considerable authority and influence. For the most part, however, matriarchal societies are rare.

Worldwide, a more typical pattern is a **patriarchal family system**, in which the oldest men (grandfathers, fathers, and uncles) control cultural, political, and economic resources and, consequently, have power over women and girls. In some patriarchal societies, women have few rights within or outside the family; they may not be permitted to work outside the home or attend college. In other patriarchal societies, women may have

---

patrilocal residence pattern newly married couples live with the husband's family.

matrilocal residence pattern newly married couples live with the wife's family.

neolocal residence pattern a newly married couple sets up their own residence.

boomerang generation young adults who move back into their parents' home or never leave it in the first place.

matriarchal family system the oldest females control cultural, political, and economic resources and, consequently, have power over males.

patriarchal family system the oldest males control cultural, political, and economic resources and, consequently, have power over females.

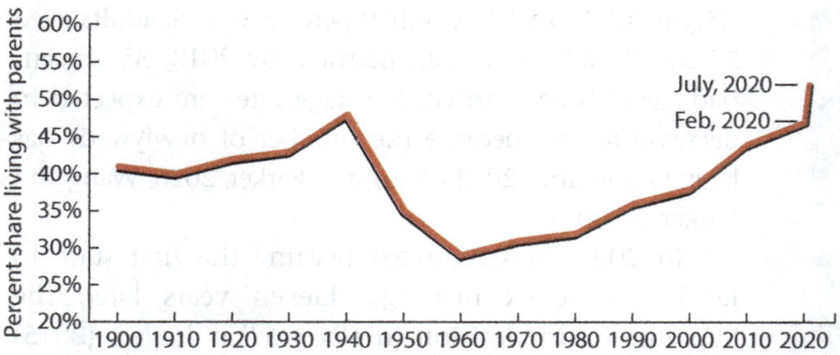

Figure 12.1 Young Adults Living with Parents

Since the pandemic began, the share of young adults living with their parents has risen to levels not seen since the Great Depression. (Percent of 18- to 29-year-olds in the U.S. residing with at least one parent in the household.)

Source: Pew Research Center analysis of decennial census 1900–1990; Current Population Survey annual averages 2000–2019; Current Population Survey monthly files (PUMS).

considerable decision-making power in the home but few legal or political rights, such as getting a divorce or running for political office (refer to Chapter 9).

In an **egalitarian family system**, both partners share power and authority fairly equally. Many Americans think they have completely egalitarian families, but these families tend to retain some elements of patriarchal families. For example, employed women shoulder almost twice as much housework and child care as men and are more likely than men to provide caregiving to aging family members (refer to Chapter 9).

Dating and Mate Selection

Dating fulfills several important functions: recreation and companionship, a socially acceptable way of pursuing love, and opportunities for sexual intimacy and experimentation. Sociologists often describe U.S. dating as a **marriage market**, a courtship process in which prospective spouses compare the assets and liabilities of eligible partners and choose the best available mate. The concept of marriage markets centering around open dating may not sound very romantic, but they serve several important functions (Benokraitis, 2015).

Many societies, in contrast, discourage open dating and have **arranged marriages** in which parents or relatives recommend or select the spouses for the next generation. An arranged marriage is a family decision rather than an individual one. It increases solidarity between families and preserves endogamy. Those receiving marriage guidance or direction from family may have veto power. That power may be affected by several factors, including the person's age (child, teen, or adult). Arranged marriages also follow the belief that if partners are compatible, love will result. In some of India's urban areas, arranged marriages also rely on nontraditional methods (e.g., online dating services) to find prospective spouses (Cullen and Masters, 2008; refer also to Chapter 9).

Monogamy and Polygamy

In **monogamy**, one person is partnered exclusively with another person. Where divorce and remarriage rates are high, as in the United States, people engage in **serial monogamy**. That is, they marry several people but only one at a time—they marry, divorce, remarry, divorce, and so on.

Polygamy, in which a person has two or more spouses, is subdivided into *polygyny*—one man married to two or more women—and *polyandry*—one woman married to two or more men. Polygamy is rare, with around 2 percent of people worldwide living in polygamous households. It is most common in countries in sub-Saharan Africa, including Burkina Faso and Mali, where small but sizeable minorities of people live in households with the marriage type (Kramer, 2020).

Although rarer, there are pockets of polyandry in some remote and isolated parts of India and among the Pimbwe in western Tanzania, Africa. Polyandry serves several functions: The family is more likely to survive in harsh environments if there's more than one husband to help provide food, and if one husband dies, the others care for the widow (Borgerhoff Mulder, 2009; Polgreen, 2010).

In contrast to polyandry, polygyny is common in many societies, especially in some regions of Africa, South America, and the Middle East. Men benefit from polygyny in several ways: They can have many legal or socially approved sexual partners, more chances to have male heirs, more income if some of the wives are employed, and higher social status because they can support multiple wives and children (Al-Jassem, 2011; Nossiter, 2011).

Most industrialized societies forbid polygamy, but there are isolated polygynous groups in the United States, Europe, and Canada. The Church of Jesus Christ of Latter-Day Saints (Mormons) banned polygamy

---

egalitarian family both partners share power and authority fairly equally.

marriage market prospective spouses compare the assets and liabilities of eligible partners and choose the best available mate.

arranged marriage parents or relatives choose the children's spouses.

monogamy one person is married exclusively to another person.

serial monogamy individuals marry several people, but one at a time.

polygamy a man or woman has two or more spouses.

India's Ziona Chana has the world's largest polygynous family—39 wives, 94 children, and 33 grandchildren (so far). They all live together in a 100-room mansion (Sykes, 2015).

in 1890 and excommunicates members who follow such beliefs. Still, it is practiced by men of the Fundamentalist Church of Jesus Christ of Latter-Day Saints (FLDS), a polygynous sect that broke away from the mainstream Mormon Church more than a century ago and mainly live in Texas, Arizona, Utah, and Canada. Women who have escaped from these groups have reported forced marriage between men in their 60s and girls as young as 10 years old, sexual abuse, and incest (Janofsky, 2003).

12-2 How U.S. Families Are Changing

American families have changed considerably since the 1950s. Some of the most important changes are related to marriage, divorce, cohabitation, childbearing, and single-parent families.

12-2a Marriage and Divorce

The United States has one of the highest marriage and divorce rates in the world. By age 60, around 90 percent of Americans have been married at least once; over a lifetime, about half of first marriages end in divorce (Gurrentz and Mayol-Garcia, 2021; Vespa et al., 2013). Despite the possibility for divorce, most people aren't disillusioned about marriage. Indeed, nearly 85 percent of those who divorce remarry, half of them within 4 years, and 8 percent have been married three times or more (Kreider and Ellis, 2011; Livingston, 2014).

Trends

U.S. marriage and divorce rates rose steadily during the twentieth century but have declined since 1990 (**Figure 12.2**). In 1970, only 9 percent of U.S. adults aged 25 to 50 had never been married. By 2018, 35 percent had never been married. Marriage rates are expected to decrease further because the number of newlyweds has been falling since 2012 (Wang and Parker, 2020; Wang and Parker, 2014).

In 2004, Massachusetts became the first state to legalize same-sex marriage. Eleven years later, the Supreme Court decision in *Obergefell v. Hodges* (2015) required states to provide marriage licenses to same-sex couples. It also required that all states recognize marriages validated in other places. Now, around 9 percent of LGBTQIA+ adults are married to a same-sex spouse, and another 7 percent are living with a same-sex partner. Around 11 percent are married to a spouse of the opposite sex (Jones, 2021). About half of people who self-identify as LGBTQIA+ are bisexual, helping explain the high proportion who are married to different-sex partners. In 2019, over half (58 percent) of couples in same-sex households were married, while around 42 percent were unmarried (U.S. Census Bureau, 2021). Out of all same-sex married couples, there is a slightly higher percentage that are women (53 percent) than men (47 percent) (Walker and Taylor, 2021). Thus, despite the Supreme Court's legalization of same-sex marriage in 2015, LGBTQIA+ marriages haven't surged. Like heterosexuals, LGBTQIA+ people may be postponing marriage or deciding not to marry.

In 1970, California was the first state to pass a *no-fault divorce* law; neither partner needs to prove guilt or wrongdoing (e.g., adultery, desertion). Today, in all states, couples can simply give "irreconcilable differences" or "incompatibility" as a valid reason for divorce. As these laws changed, marital dissolutions became

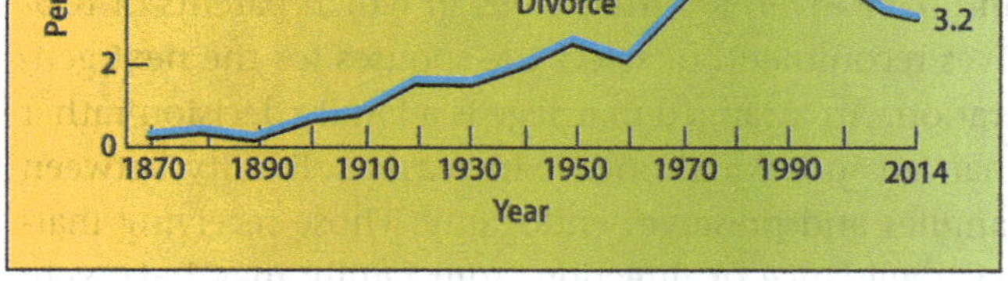

Figure 12.2 U.S. Marriage and Divorce Rates, 1870–2014

Sources: Based on Plateris 1973, Table 1; and CDC/NCHS National Vital Statistics System, 2017.

quick and cheap, and divorce rates rose to historically high levels, particularly in the 1980s and 1990s. Divorce rates are still high, but they are *lower* today than they were between 1980 and 2009 (**Figure 12.2**). Still, people are becoming more accepting of divorce. In 2022, 81 percent of Americans said that divorce was morally acceptable, compared with only 53 percent in 1954 (Gallup, 2023; Dugan, 2017).

Why Are Marriage and Divorce Rates Falling?

There are several macro-level reasons for declining marriage and divorce rates. First, U.S. values have been changing. In 2019, around 30 percent of Americans felt that marriage was not important for a fulfilling life (Barroso, 2020). In addition, 65 percent of Americans are accepting of sex between unmarried adults (Brown, 2020). And less than half (40 percent) say it's very important for couples with children to legally marry (Jones, 2020; Cohn, 2013; Jones, 2017; Vespa, 2017). Thus, the traditional reasons for marriage have waned.

Second, the economy affects marriage and divorce rates. Economic depressions, recessions, and unemployment tend to delay marriage, especially for men. Because of the 2007–2009 recession, 20 percent of 18- to 34-year-olds postponed marriage. With the recent COVID-19 pandemic, marriage rates in 2020 declined with the fewest marriages since 1963. Moreover, when incomes plummet and people are insecure about their jobs, unhappy married couples tend to stay together. They can't afford to divorce and risk the possibility of not being able to maintain separate households. This provides insight into the decline in divorce rates during the COVID-19 pandemic (Cohn, 2012; Martin et al., 2014; Stobbe, 2022; Steverman, 2021).

Demographic variables also affect marriage and divorce rates. In 2022, the estimated median age for first time marriage was 30 years for men and 28 years for women. This was up from median ages of 24 (men) and 21 (women) in 1947 (U.S. Census Bureau, 2022; "Marital Status," 2017).

Marriage has become more of an option instead of a necessity, particularly for women. In 2021, 52 percent of Americans lived with a spouse, which is two percent lower than 2011. Marriage has become an option rather than a goal in part because women are freer now than in the past to pursue a higher education and have a career. Plus, views on cohabitation have become more accepting. Thus, many women have become choosier about whether and when to marry (U.S. Census Bureau, "Living Arrangements Over the Decades", 2021; Traister, 2016; "Unmarried and Single . . .," 2017).

Education is a factor affecting marriage and divorce rates. About 64 percent of Americans with college degrees are married, compared with 47 percent of those with a high school diploma or less. The education–marriage relationship holds even at higher levels. Among women in their early 40s, for example, 80 percent with a Ph.D. or professional degree are married, compared to 63 percent with a bachelor's degree. Also, the highly educated tend to marry other highly educated people. Eighty-one percent of those with college degrees have spouses with college degrees. In effect, then, there's a growing "marriage gap" between the most and least educated (Wang, 2015; Reeves et al., 2016; Fry, 2022). Such data also challenge stereotypes of highly educated women as sad and lonely "old maids" who lavish their attention on a brood of cats.

Americans with a bachelor's or graduate degree also have lower divorce rates than those without college degrees. By age 46, 30 percent of people with a college degree or higher are divorced compared with 59 percent with less than a high school diploma (Aughinbaugh et al., 2013). College graduates have lower divorce rates in part because going to college postpones marriage. As a result, couples with more educational credentials are often more mature and capable of dealing with personal crises. They also have higher incomes and health care benefits, both of which lessen marital stress over financial problems.

There are also *micro-level (individual) reasons* for falling marriage rates and delays in marriage. About 90 percent of Americans say that love is a "very important" reason to marry compared with 73 percent who say it is an important reason to live with a partner. Never-married adults who may marry at some point give different reasons for not being wed. About 59 percent say a major reason is they "haven't found the right person." Forty-one percent say financial stability is a barrier. And another 24 percent point to not being ready to marry. As a result, there are people who are still waiting for their "ideal mate" or "one true love" (Horowitz et al., 2019; Wang and Parker, 2014; Parker and Stepler, 2017).

The most common micro-level reasons for divorce include infidelity, communication and financial problems, substance misuse and spousal violence, premarital doubts, continuous disagreements about how to raise and discipline children, and expecting to change a partner after marriage (refer to Benokraitis, 2015, for a discussion of these studies). Because many people are delaying marriage, they're usually more mature in handling the challenges of married life, which decreases the likelihood of divorce.

12-2b Cohabitation

Cohabitation is an arrangement in which two unrelated and unmarried people live together and are in a sexual relationship. Because it's based on emotional rather than legal ties, "cohabitation is a distinct family form, neither singlehood nor marriage" (Brown, 2005, p. 33).

Trends

Married couples comprise 49 percent of all households, a sharp decline from 78 percent in 1950. The decline is due to falling marriage rates and rising cohabitation. The number of adults in heterosexual cohabiting relationships surged from 430,000 in 1960 to almost 18 million in 2016. This number climbs by another 860,000 if we include same-sex persons who are cohabitating. Despite the high numbers, only 7 to 9 percent of the population is cohabiting in any given year ("Characteristics of Same-Sex Households," 2017; Stepler, 2017).

About half of people who cohabitate are younger than 35. Since 2007, however, the number of cohabiters aged 50 and older has increased faster than other groups (**Figure 12.3**). Because of high divorce rates and a growing share of people who have never been married in this age group, more people are available for cohabitation (Stepler, 2017).

By age 44, 65 percent of women—compared with 33 percent in 1987—have cohabited. Cohabitation is a common experience at all education levels, but the likelihood of cohabiting decreases as women's educational levels increase (e.g., 76 percent of women with less than a high school diploma compared with 58 percent of women with a bachelor's or advanced degree). On average, college-educated women cohabit for the shortest period (17 months) and are more likely than those without college degrees to transition to marriage (Copen, 2013; Vanorman and Scommegna, 2016).

Black women are less likely to have ever cohabited (59 percent) than White women (67 percent) or Latina women (64 percent). Among all cohabiting adults, significantly more Asian people (46 percent) live with a partner of a different race or ethnicity than people who are Latina/o/x (24 percent), Black (20 percent), or White (12 percent) (Vanorman and Scommegna, 2016; Livingston, 2017). The high rate of Asian interracial cohabitation, like intermarriage, may be partially due to this group's small population size and fewer available partners of the same race (refer to Chapter 10).

cohabitation two unrelated and unmarried people live together and are in a sexual relationship.

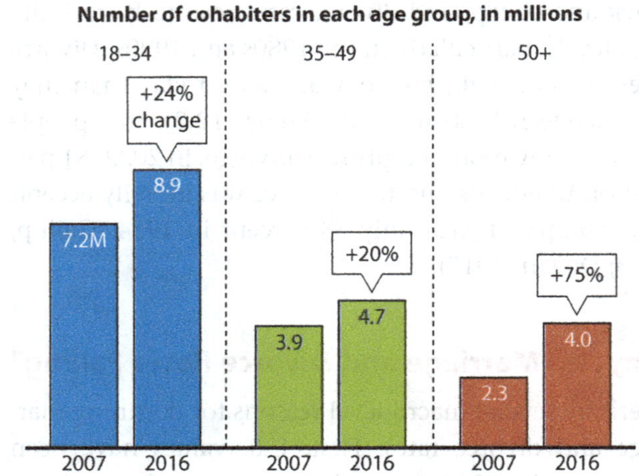

Figure 12.3 Cohabitation, by Age Groups, 2007 and 2016

Source: Stepler, 2017.

Why Has Cohabitation Increased?

On a micro level, some people drift gradually into *dating cohabitation*, when a couple that spends a great deal of time together decides to move in together. Dating cohabitation is essentially an alternative to singlehood because the decision may be based on a combination of reasons. Most people who cohabitate cite love as a reason (73 percent) or companionship (63 percent). A smaller percentage say convenience (37 percent) or finances (38 percent). In this type of cohabitation, and especially among young adults, there's considerable *serial cohabitation*, living with different sexual partners over time (Manning and Smock, 2005; Wartik, 2005; Horowitz, et al., 2019).

For many people, premarital cohabitation is a step between dating and marriage. In *premarital cohabitation*, the couple lives together before getting married. They may or may not be engaged but plan to marry. Such "almost-married" cohabitation may be especially attractive to partners who wonder if they can deal successfully with problems that arise from differences in personalities, interests, finances, ethnicity, religion, or other issues.

Finances are a factor for cohabitation and marriage delay. Thirty-eight percent of those who cohabitate say they do so because it "makes financial sense" compared with only 13 percent of married people. For never-married adults who may get married some day, income affects whether financial instability is viewed as a reason to not get married. Forty-seven percent with family incomes of $30,000 or below say financial insecurity is a major reason for not being married compared with only 21 percent of those with incomes over $75,000. Across all

Do you think that women or men benefit more from cohabitation? Why?

ethnic groups, 3 out of 4 people who cohabit say they're delaying marriage because "Everything's there except money" (Xie et al., 2003; Horowitz et al., 2019; Parker and Stepler, 2017).

Does cohabitation beforehand affect marriages? Perhaps. Those who cohabitate before marriage are 48 percent more likely than those who did not to get divorced. Among those who moved in together before getting engaged, 34 percent divorced compared with 23 percent of those who moved in after engagement. Couples who decide on a commitment to the relationship before living together (married or in a committed relationship) may be less likely to divorce. Marital success also depends on the cohabitors' ages, socioeconomic status, commitment, and attitudes toward marriage (Reinhold, 2010; Manning and Cohen, 2012; Thies, 2023). Cohabitation, like any other relationship, has benefits and costs (**Table 12.1**).

12-2c Childbearing with or without Partners

Americans have different perspectives on, and experiences with, whether or not to have children. In 2021, around 55 percent of adults under age 50 without children say it's likely they will have children some day, while 45 percent say it's unlikely. The percentage who feel having children is unlikely has grown by 8 percent since 2018. Almost 17 percent of adults 55 and over do not have biological children (Brown, 2021; Valerio et al., 2021).

As you will learn in this section, many people have children on their own, and having children does not need to be intertwined with marriage or a committed partnership. Becoming a parent on one's own, separate from marriage, is also referred to *nonmarital childbearing*. Social factors connected to having children on one's own and the rates of these births have both changed. In 1940, the birthrate for unmarried women was 7.1 for every 1,000 women. By 1990, that rate increased nearly 6 times to 43.8. This can be partially explained by the later ages at which women were getting married. The nonmarital birth rate has been declining since 2008, likely affected by the change in the overall birth rate (which has been dropping) and the economic recession of 2007–2009. Since then, birth rates have continued to decline for many reasons, including increased contraceptive use. More recently, the COVID-19 pandemic played a role in the birth rate shifting from 38.4 in 2020 to 37.8 in 2021 (Venture and Bachrach, 2000; Osterman et al., 2023; Livingston and Brown, 2014).

Table 12.1 Some Benefits and Costs of Cohabitation

Benefits	Costs
• Couples can pool their resources instead of paying for separate housing, utilities, and so on. They can also have the emotional security of an intimate relationship but maintain their independence by spending time with their friends and family members separately (McRae, 1999; Fry and Cohn, 2011).	• U.S. laws don't specify a cohabitant's rights and responsibilities. For example, there's no automatic inheritance if a partner dies without a will, and it's more difficult to collect child support from a cohabiting partner than an ex-spouse (Silverman, 2003; Grall, 2013).
• Couples who postpone marriage have a lower likelihood of divorce because being older is one of the best predictors of a stable marriage (Copen et al., 2012).	• Compared with married couples, cohabitants have a lower quality of relationship and lower levels of happiness and satisfaction (Sassler et al., 2012; Wiik et al., 2012).
• Couples find out how much they really care about each other when they have to cope with unpleasant realities (e.g., a partner who doesn't pay bills or rarely showers).	• Cohabitation dilutes intergenerational ties. Compared with their married peers, the longer people live together, the less likely they are to give or receive help from their parents and to be involved in extended family activities (Eggebeen, 2005).
• Children in cohabiting households reap economic advantages by living with two adult earners instead of a single mother (Lundberg et al., 2016).	• Because cohabiting parents are more than twice as likely as married parents to break up, children's academic, emotional, behavioral, and financial problems often increase (Fomby and Estacion, 2011; Rackin and Gibson-Davis, 2012).

Trends

In 2007, the nonmarital birth rate in the United States reached a high point of 51.8 births per 1,000 women ages 15 to 44. The nonmarital birth rate has been declining, and in 2021, the rate was 37.8 which was just under 1.5 million nonmarital births. In 2021, nonmarital births accounted for 40 percent of all births (Osterman et al., 2023). This way of forming a family has become more mainstream.

Births to unmarried women vary widely across racial-ethnic groups. White women have more nonmarital babies than other groups. However, nonmarital birth rates are highest for Latina women and lowest for Asian American women (**Figure 12.4**).

As with marriage and cohabitation, nonmarital births vary by social class. Among college graduates, only 13 percent of recent births were outside of marriage compared with 57 percent of those who have not completed high school (**Table 12.2A**). The percentage of nonmarital births decreases as household income goes up (**Table 12.2B**). Consequently, women who are systematically disadvantaged are the most likely to have babies outside of marriage (Sawhill and Venator, 2014).

Age is also a factor to consider in the rate of unmarried births. Unmarried teenage births peaked in the early 1990s, have declined steadily since then, and are now at their lowest level since 1968. The birth rate for unmarried teens (13.0) is lower than that of the overall population of unmarried women. There is variation in the birth rate even by the age of teens, with a birth rate of 5.6 for those ages 15 to 17 and a rate of 24.4 for ages 18 and 19. The age group with the highest unmarried birth rate is actually women ages 25 to 29. The birth rate for unmarried women in this age group is 58.7, which is 5 times greater than that for teens aged 15 to 19 and 10 times greater than the rate for 15- to 17-year-olds. Still, 89 percent of teenage births are nonmarital, compared with 52 percent for women in their 20s and 22 percent for women in their 30s. Moreover, 20 percent of unmarried teenagers have a repeat birth; 86 percent have a second child, and 15 percent have 3 to 6 children (Gavin et al., 2013; Martin et al., 2017; Osterman et al., 2023).

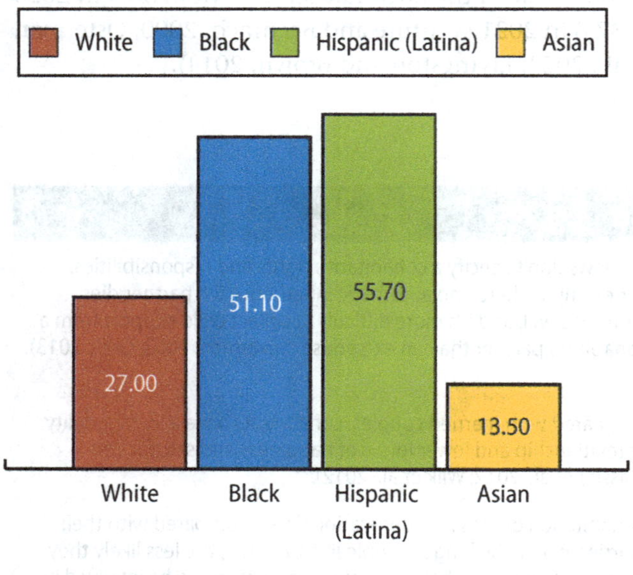

Figure 12.4 Birthrates for Unmarried Women by Race and Ethnicity in U.S., 2021

Source: Based on Osterman et al., "Births: Final Data for 2021", 2023.

Why Has Childbearing on One's Own Increased?

On a *micro (individual) level,* many teens and young adults use contraception (refer to Chapter 9). Yet, contraception may be used inconsistently or incorrectly because they think they or their partners are sterile,

Table 12.2 Births to Unmarried Mothers, by Education and Household Income

A. Educational Attainment	
Percentage of births to unmarried mothers in each educational level	
Less than high school	57%
High school graduate	52%
Some college	40%
Bachelor's degree	13%
Graduate or professional degree	7%

B. Annual Household Income	
Percentage of births to unmarried mothers in each income level	
Under $10,000	77%
$10,000 to $24,999	62%
$25,000 to $49,999	45%
$50,000 to $99,999	30%
$100,000 to $199,999	20%
$200,000 and above	13%

Source: Based on U.S. Census Bureau, 2015 American Community Survey 1-Year Estimates, Table S1301, and "Fertility of Women in the United States: 2016," Current Population Survey, Table 7. Accessed July 10, 2017 (factfinder.census.gov and census.gov/hhes/fertility/data).

they believe that not taking the pill a few times "doesn't really matter," or the partners don't want to use condoms. Others think that unplanned pregnancies are predetermined or controlled by outside forces like a higher power or fate ("It was meant to happen" or "It's God's plan") (Frohwirth et al., 2013; Lindberg et al., 2016).

> The number of births to unmarried women aged 25 to 29 is four times greater than for teenagers.

Unintended pregnancies have decreased by 18 percent since 2008. Nonetheless, 81 percent of pregnancies among never married non-cohabiters, 75 percent of teen pregnancies, and 56 percent of pregnancies among cohabiting women are unintended. Even if the pregnancies are unwanted, women may not get abortions because of personal or religious beliefs (Curtin et al., 2014; Finer and Zolna, 2016). A lack of access to abortion services may be a factor as well.

Demographic variables also affect nonmarital childbearing. Social class and educational attainment are key factors in nonmarital births (**Table 12.2**). Women with bachelor's degrees and graduate degrees are more likely than those with lower levels of educational attainment to postpone parenthood until marriage. They may also have more resources for family planning and have more decision-making power in their relationships, such as regarding contraceptive use (Livingston and Cohn, 2013).

Macro-level variables increase nonmarital birth rates in several ways. Until the early 1970s, nonmarital births were rare and kept secret; young people were often forced into "shotgun" marriages if the girl was pregnant, and young women—especially White women—were pressured to give their babies up for adoption. Now, almost 75 percent of Americans aged 15 to 44 believe that "it's okay for an unmarried female to have and raise a child." Because attitudes have changed, childbearing outside of a committed relationship or on one's own has become more typical in the United States. In contrast, only 2 percent of children in Japan and South Korea are born outside of marriage because unmarried mothers are stigmatized (Daugherty and Copen, 2016; "Marriage in Japan," 2016; Chamie, 2017).

Fewer teens, particularly in nonmetropolitan areas, are receiving formal sex education in school than they

Denise Hager/Catchlight Visual Services/Alamy Stock Photo

A study found that the reality TV show *16 and Pregnant* reduced U.S. teen births by almost 6 percent in the 18 months following its release (Kearney and Levine, 2014). Should middle schools require that adolescents watch such programs?

did 25 years ago. And there are some gaps regarding particular topics that they are taught. For instance, only 64 percent of teen girls and 63 percent of teen boys learned about methods of birth control. Yet even fewer, 48 percent of girls and 45 percent of boys, received information on how to get birth control. STDs were more frequently covered, with over 90 percent of teens learning about them (Lindberg and Kantor, 2022).

Government policies also affect birth rates. Publicly funded family planning services that offer counseling and contraceptives, particularly for teenagers and young adults, reduce unintended pregnancies and births. Many states, however, fund only sex-abstinence education and can now deny federal money to organizations that include contraception services. In 2022, the Supreme Court's decision in *Dobbs v. Jackson Women's Health Organization* eliminated abortion rights protection at the federal level. Within 100 days of this decision, 66 clinics in 15 states stopped providing abortion care. Reducing access to family planning clinics and abortion services can increase nonmarital births, especially among teens. The longer-term effects of the *Dobbs* decision on the birth rate are yet to be determined (Cunningham et al., 2017; Superville, 2017; Kirstein et al., 2022; refer also to Chapter 9).

12-2d Single-Parent Households

Most children live with two parents, but single-parent families are increasingly common. As of 2020, over one quarter of children lived with one parent. And over 80 percent of single parent households are headed by women (Hemez and Washington, 2021).

Trends

In 2020, 70 percent of children under age 18 lived with both parents, down from 85 percent in 1968. The number of one-parent families, on the other hand, has more than doubled from 12 percent in 1968 to 25 percent in 2020. Compared with other families with children under 18, mother-only families:

- make up nearly 21 percent of all families and 82 percent of one-parent families;
- comprise 51 percent of Black, 28 percent of Latina/o/x, 89 percent of White, and 10 percent of Asian families; and
- are twice as likely as father-only families and more than three times more likely than two parent families to be living below the poverty level (Hemez and Washington, 2021; U.S. Department of Justice, 2021).

Asian American children are the most likely to grow up in two-parent homes (**Figure 12.5**). Until 1980, married couple families were the norm in Black families. Since then, Black children have been more likely than children in other racial-ethnic groups to grow up with only one parent, usually the mother. Among Latina/o/x families, 62 percent of children live in two-parent families, down from 78 percent in 1970 (Lugaila, 1998; Hemez and Washington, 2021) (**Figure 12.5**).

Nationally, 15 percent of the same-sex couples had at least one child under the age of 18 living with them, while 38 percent of different-sex couples did. About 200,000 of the children have married or cohabiting same-sex parents; more than a million single LGB adults are raising approximately 2 million children. Even though same-sex individuals and couples are less likely than different-sex couples to be raising children, they're four times more likely to adopt children or have stepchildren. Same-sex couples are also more than twice as likely to foster a child (Taylor, 2020). Thus, large numbers of LGBTQIA+ adults—single, partnered, or married—are parents.

Why Have Women-Headed Households Increased?

On a *micro level*, having sex at an early age and not using contraception at all, incorrectly, or sporadically increases the likelihood of unintended pregnancies. This also helps explain the formation of some mother-only families, especially those among teenagers. Teen pregnancies can lead to challenges for families. For instance, adolescent fathers may face hurdles to getting and keeping regular employment because of their age, lower levels of educational attainment, and less specialized skill sets. In a society that has gendered expectations that fathers bear most of the financial responsibility for children,

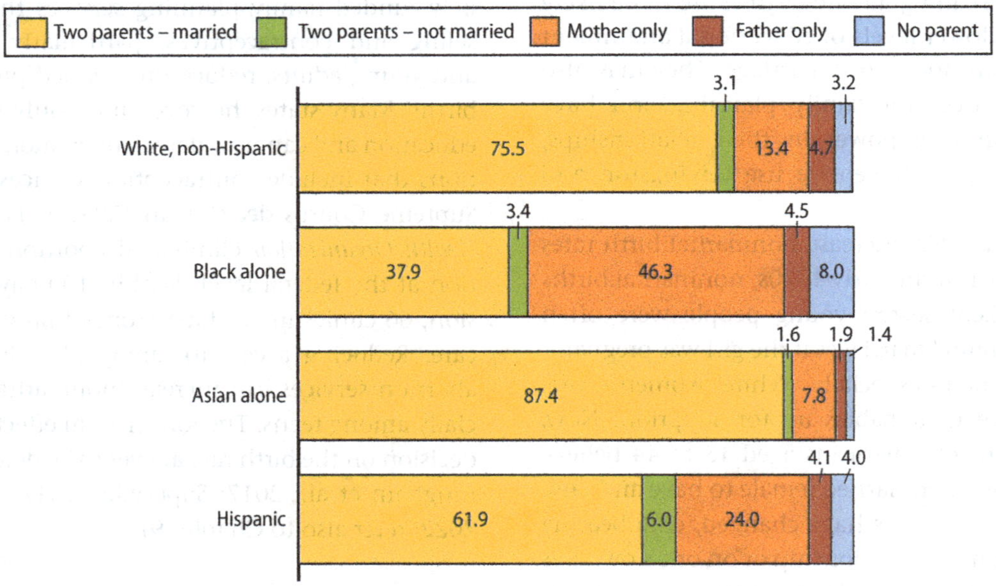

Figure 12.5 Living Arrangement of Children Under 19 by Race and Ethnicity, 2020 (in Percent)

Note: Percents may not sum to 100 due to rounding.

Note: White, non-Hispanic indicates children whose race was reported as White alone, and who were reported as non-Hispanic, Black alone and Asian alone indicate children whose race was reported as a single race group.

Source: U.S. Census Bureau, Current Population Survey, 2020 Annual Social and Economic Supplement" their lower wages can cause obstacles for parenting and partnerships (Weber, 2020).

Mass media routinely feature the "single mother by choice" (SMC). Whether in movies, in newspapers, or on TV shows, SMCs are typically White, well-educated, and successful women in their 30s and 40s who decide to have and raise a baby without a partner. According to some sociologists, however, SMCs are "a relatively rare phenomenon," more common in the media than in reality, because they make up less than 5 percent of women, most of whom are older and well-educated (Hayford and Guzzo, 2015; Braff, 2016).

Demographic variables, especially social class, also affect the rise of female-headed households. College-educated people, as you've learned, tend to postpone marriage, marry rather than cohabit, and delay parenthood—factors that decrease the likelihood of divorce, one-parent households, and, consequently, children experiencing poverty (Redd et al., 2011).

Not all women-headed households have lower incomes, of course. Women-headed households have lower median incomes ($39,000) than those headed by men ($48,000). Yet, they also are more likely to have bachelor's degrees or beyond (40 percent) compared with men-headed households (30 percent). And women-headed households are making gains in home ownership. In 1990, only 51 percent of women heads of home were homeowners, but by 2021, that percentage had increased to 63 percent. This percentage still lags behind that of men-headed households (68 percent) (Choi, 2023). Instead of waiting for "Mr. Right," some women with economic resources are deciding to raise children on their own.

In many Black and Latina/o/x families, men and boys (e.g., sons, brothers, uncles, grandfathers) often provide emotional and financial support to female kin who head households (Sudarkasa, 2007). Many Black families also welcome **fictive kin**, nonrelatives who they accept as part of the family. Fictive kin have strong bonds with biological family members and provide important services (e.g., caring for children when young mothers are employed or negligent) (Billingsley, 1992; Dilworth-Anderson et al., 1993). A variation of fictive kin involves single mothers—many of whom are White, single, and college educated—who turn to one another for companionship and child care help (Bazelon, 2009).

At the *macro level*, values and views about single parenthood have shifted. In 2018, 40 percent of Americans viewed single mother households as a bad thing, and this percentage increased to 47 percent in 2021 (Hurst, 2022). It is possible that economic strains from the COVID-19 pandemic played a role in this view shift.

Karen Struthers/Shutterstock

Is the absent Black father a stereotype? A national study of White, Black, and Latino men found that, among fathers who don't live with their children, Black fathers were the *most* likely to parent their children from birth to age 18. Parenting included, on a daily basis, feeding, bathing, dressing, playing with and reading to children, taking them to and from activities, and helping them with or checking homework (Jones and Mosher, 2013).

The economy also affects the number of women-headed households. As in the case of cohabitation and childbearing on one's own, women with lower incomes often enter parenthood with men who are low-income or unemployed. Men may embrace parenthood and resolve to be good fathers, but already weak bonds with the child's mother can deteriorate further. Seeking financial security and stability, women gain little or nothing by marrying the babies' fathers or other men with lower incomes (Edin and Nelson, 2013).

12-3 Family Conflict and Violence

Conflict is a normal part of family life, but violence is *not* normal. Over a lifetime, we're much more likely to be assaulted or killed by a family member, current/former spouse, or romantic partner than by a stranger (Truman and Langton, 2014).

12-3a Intimate Partner Violence

Intimate partner violence (IPV) is abuse that occurs between people in a close relationship. The term *intimate partner* refers to current and former spouses, couples who live together, and current and former boyfriends

fictive kin nonrelatives who are accepted as part of a family.

intimate partner violence (IPV) abuse that occurs between people in a close relationship.

or girlfriends. IPV, which ranges from a single episode to ongoing abuse, includes three prevalent types of behavior:

- *Physical abuse* is threatening, trying to hurt, or hurting someone using physical force (e.g., throwing objects, pushing, grabbing, slapping, biting, choking, beating).
- *Sexual abuse* is threatening or forcing a partner to take part in a sex act without their consent (e.g., unwanted anal, oral, or vaginal sex).
- *Psychological abuse* (also called *emotional abuse*) is aggressive behavior that threatens, humiliates, manipulates, or controls another person (e.g., criticism, rejection, isolation, name-calling, bullying, stalking).

IPV, which is pervasive in U.S. society (**Figure 12.6**), begins early in life. Among 12- to 18-year-olds in a recent or current dating relationship, 69 percent have been victims of IPV, and 63 percent have perpetrated abuse. Among adults, 36 percent of women and 29 percent of men have been IPV victims at some time in their lives, and nearly half have experienced psychological abuse (Black et al., 2011; Copen et al., 2013; Taylor and Mumford, 2016).

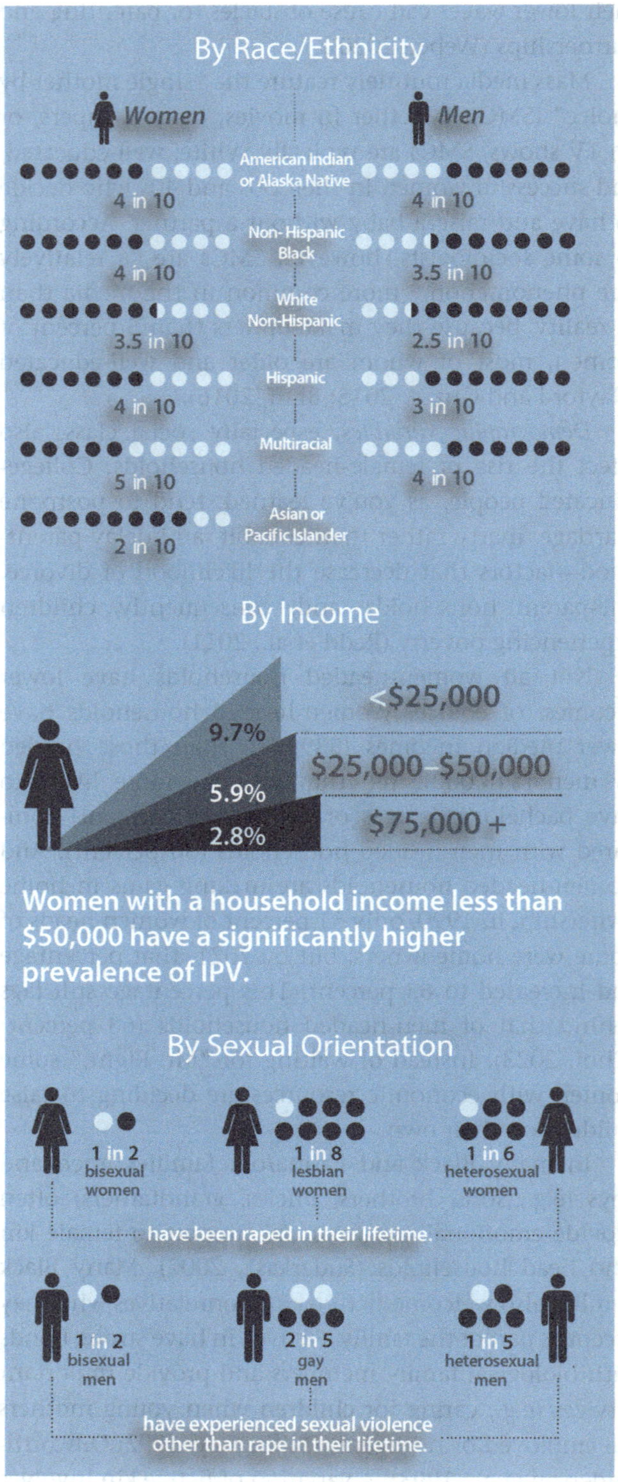

Figure 12.6 Intimate Partner Violence, Sexual Violence, and Stalking in Lifetime

Source: Centers for Disease Control and Prevention, Infographic, 2014. Accessed October 8, 2015 (cdc.gov/violenceprevention/nisvs/infographic.html).

Of the nearly 5,000 women and girls who were victims of murder and nonnegligent manslaughter, nearly one third (34 percent) of them were attributed to IPV (Smith, 2022). When victims survive assaults, 40 percent of girls and women, compared with 5 percent of boys and men, suffer a serious physical injury (e.g., gunshot or knife wounds, internal injuries, broken bones) (Catalano, 2013; Federal Bureau of Investigation, 2016). Women are also much more likely than men to experience IPV over a lifetime, regardless of age, race or ethnicity, or social class (Black et al., 2011; Truman and Morgan, 2014).

There are nearly 1.3 million intimate partner victimizations every year, but this number is conservative. An estimated 44 percent of victims don't report IPV to the police because they're ashamed, believe that no one can help, or fear reprisal (Reaves, 2017).

Why Does Intimate Partner Violence Occur?

The reasons for IPV are interrelated individual, demographic, and societal factors. On a *micro level*, violence escalates if one or both partners use alcohol or other drugs, if they have more children than they can afford (which intensifies financial problems), or if either partner has been raised in a violent household, has low self-esteem, and has a controlling personality. The most common disagreements that can escalate into abuse and violence are gender role expectations (who does what housework), money (saving and spending), children (especially discipline), and infidelity, both personally and online (Spivak et al., 2014; Benokraitis, 2015).

Regarding *demographic variables*, IPV often begins at a young age: 22 percent of girls and 15 percent of boys between 11 and 17 years old), and 47 percent of young women s and 39 percent of young men (between 18 and 24 years old) (Black et al., 2011). IPV cuts across all social classes, but women living in households with an annual income of less than $7,500 are nearly four times more likely to be abused than those living in households with an annual income of $75,000 or more (**Figure 12.6**).

Macro-level variables, particularly unemployment and poverty, increase the likelihood of financial stress and violence. The absence of legal or social sanctions against IPV and a scarcity of shelters discourage victims from trying to escape abuse (Matjasko et al., 2013). A study of girls aged 11 to 16 who had experienced sexual assault concluded that girls (and later women) often didn't report the incidents because such violence had been "normalized in their communities." The girls were ashamed and feared retribution, but also believed that men "can't help it," perceived everyday harassment and abuse as "normal male behavior," and assumed that male authority figures (e.g., police officers, judges) would accuse the girls of overreacting (Hlavka, 2014).

12-3b Child Maltreatment

Child maltreatment (also called *child abuse*) includes a broad range of behaviors that can result in serious harm, including physical and sexual abuse, neglect, and emotional mistreatment. The victims often experience several types of maltreatment.

Trends

U.S. child maltreatment rates fluctuate but have decreased since 2011. In 2020, at least 14 percent of children in the U.S. experienced child abuse or neglect, and 1,750 children died from maltreatment.

Children living below the poverty threshold experience more maltreatment. Experiencing poverty can place a lot of stress on families, which may increase the risk for child maltreatment. Rates of child abuse and neglect are five times higher for children in families with lower incomes. Because only a fraction of the cases are reported, however, federal agencies assume that millions of children experience abuse and neglect every day (Centers for Disease Control and Prevention, 2022).

Neglect is the most common type of child maltreatment (**Figure 12.7**). The most vulnerable children are those younger than 1 year old, with a victimization rate of 25 for every 1,000 children. Child maltreatment rates vary by race and ethnicity: 13.1 for Black children,

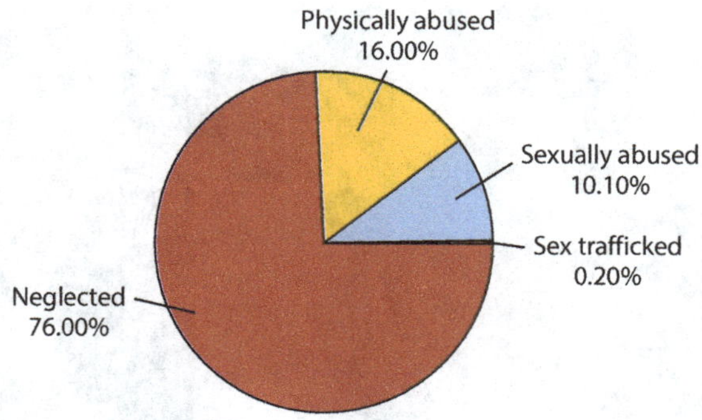

Figure 12.7 Types of Child Maltreatment, 2021

Source: Based on U.S. Department of Health and Human Services, 2021.

15.2 for Native American/Alaska Native children, 7.7 for Latina/o/x children, 10.3 for multiracial children, 8.5 for Pacific Islander children, 7.1 for White children, and 1.4 for Asian children (U.S. Department of Health and Human Services, 2021).

Around 77 percent of the perpetrators are one or both parents. An additional 7 percent of assailants are relatives (U.S. Department of Health and Human Services, 2021).

Why Does Maltreatment of Children Occur?

On a *micro level*, child maltreatment is most common in households where there is parental substance misuse or mental illness, adult or sibling violence, and one or more children who have emotional, developmental, or physical disabilities (Finkelhor et al., 2011). *Demographic variables* that increase the likelihood of child maltreatment include living with a young parent, living in one-parent or stepparent households—particularly those with low socioeconomic status—and experiencing parental conflict before, during, and after a hostile divorce (Truman and Smith, 2012; White and Lauritsen, 2012). On a *macro level*, economic hardship, unemployment, and poverty increase the likelihood of stress that leads to child maltreatment, including infant deaths due to severe head traumas (Eckenrode et al., 2014).

Almost 26 percent of children live in homes where parents or other adults are violent (Hamby et al., 2011). Whether children are targets of maltreatment or are aware of it, violent childhood experiences are associated with lifelong developmental problems, including depression, delinquency, suicide, alcohol misuse, low academic achievement, unemployment, and medical problems in adulthood (Hibbard et al., 2012; Sacks et al., 2014).

---

child maltreatment (also called *child abuse*) a broad range of behaviors that can result in serious emotional or physical harm.

Anne Marie Fox/Lions Gate/Everett Collection

The movie *Precious* (2009) received many awards, but some critics contended that the film stereotyped Black teenage girls—especially those in low-income households—as obese, illiterate, and living with abusive mothers (Amusa, 2010). If you've seen the movie, do you agree with such criticism or not? Why?

12-3c Older Adult Maltreatment

Older adult maltreatment (also called *elder abuse*) is any knowing, intentional, or negligent act by a caregiver or other person that causes harm to people aged 65 or older. This term includes physical, psychological, and sexual abuse; neglect; isolation from family and friends; deprivation of basic necessities such as food and heat; not providing needed medications; and financial exploitation.

Trends

Around one out of ten people aged 60 and over experience some type of older adult maltreatment. The types of maltreatment include: psychological (4.6 percent), physical (1.6 percent), sexual (0.6 percent), financial fraud and exploitation (5.2 percent), and neglect (5.1 percent) The actual rates are probably much higher because only an estimated 20 percent of all abuse and neglect of older adults is reported (National Council on Aging, 2021; Department of Justice, 2021).

Who are the victims? About 83 percent are White; the average age is 76; 76 percent are women; 84 percent live in their own homes; 86 percent have a chronic disease or other health condition; 57 percent are married or cohabiting; 53 percent haven't graduated from high school; 50 percent suffer from dementia, Alzheimer's, or other mental illness; 46 percent feel socially isolated; and the average combined household income is less than $35,000 a year (Acierno et al., 2010; Jackson and Hafemeister, 2011). These data suggest that the most likely victims of older adult maltreatment are those who are the most vulnerable physically, mentally, socially, and financially.

Who are the perpetrators? Most are adult children, spouses or cohabiting partners, or other family members. Less than a third are acquaintances, neighbors, or nonfamily service providers. The average age of the abuser is 45, 77 percent are White, 61 percent are men, 82 percent have a high school diploma or less, 50 percent misuse alcohol and/or other drugs, 46 percent have a criminal record, 42 percent are financially dependent on the older person, 37 percent live with the older person, 29 percent are chronically unemployed, and 25 percent have mental health problems (Jackson and Hafemeister, 2011).

Why Does Maltreatment of Older Adults Occur?

On a *micro level*, misuse of alcohol and other drugs is more than twice as likely among family caregivers who abuse older adults as among those who don't. Both victims and offenders often report a childhood history of witnessing or experiencing family violence, lower-quality family relationships in the past and currently, and communication problems. In addition, older people with cognitive impairment—due to dementia (deteriorated mental condition) after a stroke or the onset of Alzheimer's disease—are abused at higher rates than those without such disabilities (Heisler, 2012; National Center on Elder Abuse, 2012).

On a *macro level*, a shared residence is a major risk factor for the maltreatment of older adults because the caregiver(s) may depend on the older person for housing, while the older person may be dependent on the caregiver(s) for physical help. These situations compound the likelihood of everyday tensions and conflict. Financial stress may also increase the risk of maltreatment. Unlike low-income families, those in the middle class aren't eligible for admission to public facilities, yet few can pay for the in-home nursing care, high-quality nursing homes, and other services that upper-class families can afford (Acierno et al., 2010; Jackson and Hafemeister, 2011).

12-4 Our Aging Society

One of our friends, 55, was shocked and insulted when he ordered a meal at a fast-food restaurant, and the young worker called out "one senior meal!" The *Baltimore Sun* (2012) reported on the death of an "elderly" man in a house fire. He was 62.

older adult maltreatment (sometimes called *elder abuse*) any knowing, intentional, or negligent act by a caregiver or other person that causes harm to people aged 65 or older.

12-4a When Is "Old"?

What images come to mind when you think of the word *old*? In a national survey, people in their 40s said that a person is old at age 63, those in their 70s said age 75, and a 90-year-old woman said that a woman isn't old "until she hits 95" (*AARP Magazine*, 2014, p. 40). So, how old is "old"?

It depends on whom you ask because "old" is a social construction. In some African nations, where people rarely live past 50 because of diseases and civil wars, 40 is old. In industrialized societies, where the average person lives to age 78, 40 is considered young. Still, regardless of how we feel physically, society usually defines old in chronological age. In the United States, for instance, people are deemed old at age 65, 66, or 67 because they can retire and become eligible for Medicare and Social Security benefits.

Many older Americans are vigorous and productive, but others experience physical and mental limitations as they age. There are significant differences between the *young-old* (65–74 years old), the *old-old* (75–84 years old), and the *oldest-old* (85 years and older) in health, living independently, and working. Generally, for example, a 75-year-old is much less likely than an 85-year-old to need caretaking.

Between the ages of 65 and 91, avid mountaineer Hulda Crooks scaled Mount Whitney (the highest mountain in the continental United States) 23 times. She died at the age of 101 in 1997.

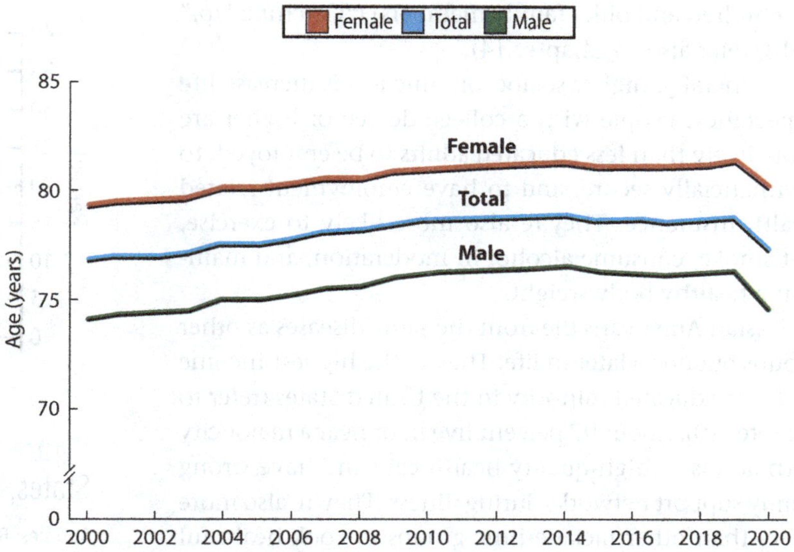

Figure 12.8 U.S. Life Expectancy at Birth, by Sex, 2000-2020

Notes: life expectancies for 2019 by Hispanic origin and race are not final estimates. Estimates are based on provisional data for 2020. Provisional data are subject to change as additional data are received.

Source: Arias et al., 2021 "Provisional Life Expectancy Estimates" Figure 1.

12-4b Life Expectancy

Life expectancy is the average expected number of years of life remaining at a given age. American children born in 2020 have a life expectancy of almost 77 years (compared with 71 in 1970 and 47 in 1900). Despite the United States' enormous wealth, the nation ranks just 42nd globally in life expectancy (Arias et al., 2020; Schwartz, 2017).

Life expectancy varies by sex, social class, and race-ethnicity. Latina/o/x Americans have a longer life expectancy at 79 years compared with White Americans at 78 years and Black Americans at 72 years. Historically, currently, and across all racial-ethnic groups, women live longer than men. While life expectancy had been rising, it has recently begun to decline for both men and women (Arias et al., 2021) (**Figure 12.8**).

The higher rates of cigarette smoking, heavy alcohol consumption, gun-related violence, employment in hazardous occupations, and risk-taking in recreation and driving are responsible for shorter lifespans for many boys and men. The life expectancy gender gap has narrowed since the 1990s, however, because there's been an increase in women's smoking, use of alcohol and other drugs, obesity (which increases the risk of hypertension and heart disease), and stresses due to multiple roles (e.g., juggling employment while caring

life expectancy the average expected number of years of life remaining at a given age.

for children and older family members) ("Catching Up," 2013; refer also to Chapter 14).

Generally, higher socioeconomic levels increase life expectancy. People with a college degree or higher are more likely than less educated adults to be employed, to be financially secure, and to have employment-related health insurance. They're also more likely to exercise, not smoke, consume alcohol in moderation, and maintain a healthy body weight.

Asian Americans die from the same diseases as other groups but do so later in life. They're the highest-income and best-educated minority in the United States (refer to Chapter 10). About 97 percent live in or near a major city with access to high-quality health care and have strong family support networks during illness. They're also more likely than other racial-ethnic groups to cook healthful meals, smoke less, and maintain a healthy body weight (Acciai et al., 2015). There are no definitive reasons for Asians' longevity, but these factors prolong life.

12-4c How Our Aging Nation Is Changing

The 65 and older population is booming and becoming more ethnically and racially diverse. For example,

- 49 million Americans (15 percent of the total population) are 65 or older, a dramatic increase from just 4 percent in 1900, and will comprise 21 percent of U.S. residents by 2030.
- One of the fastest-growing groups is the oldest-old, whose numbers increased from 100,000 in 1900 to 6.3 million in 2015. By 2030, this group will comprise almost 3 percent of the population.
- By 2030, 28 percent of the 65-and-older population will be racial-ethnic minorities, up from 20 percent in 2010; 23 percent of the 85-and-older population will be racial-ethnic minorities, up from 15 percent in 2010 (Colby and Ortman, 2015; "The Nation's Older Population...," 2017; "Older Americans Month...," 2017).

The aging population has four important (and interrelated) implications. First, as you'll read shortly, health care and caregiving needs, services, and costs will surge. Second, the number of older Americans is increasing, and the proportion of young people is falling (**Figure 12.9**). As a result, many adult children—even in their 60s—will be caring for aging parents, grandparents, and other older relatives for more years than in the past.

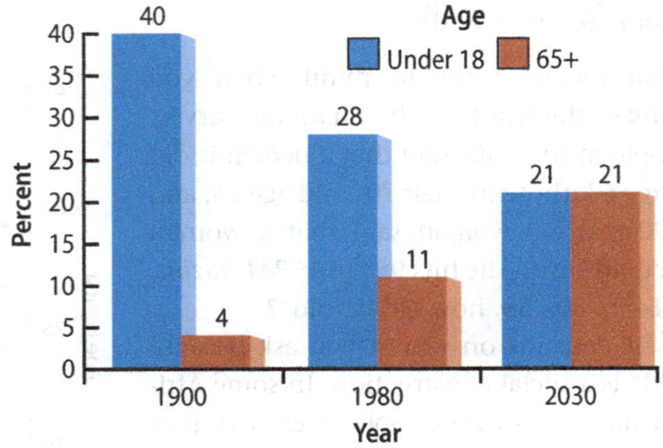

Figure 12.9 The Young and the Old in the United States, 1900–2030

Sources: Based on U.S. Senate Special Committee on Aging et al., 1991, p. 9, and Colby and Ortman, 2015, Figure 4.

Third, disability rates among older Americans have increased, particularly among **baby boomers** (people born between 1946 and 1964) and those 10 years younger than baby boomers. The higher disability rates are due to better medical diagnoses, but baby boomers are more likely than the previous generation to be obese and to have diabetes and high blood pressure (Scommegna, 2013). Thus, the baby boomer generation might have more health concerns and for more years than their predecessors who had lower life expectancies.

Fourth, as the number of older people increases, so does the **old-age dependency ratio** (also called the *elderly support ratio*)—the number of working-age (18 to 64) adults for every person aged 65 and older. The ratio is an approximation because some people work into their late 60s and 70s, but it's a useful measure of the burden of workers who support the older population. The old-age dependency ratio was 14 workers per older person in 1900. By 2060, the ratio of working adults supporting each older person may be two to one (Ortman et al., 2014; Mather et al., 2015). Thus, many people will have to pay much higher federal and state taxes to support the ever-growing older population.

12-4d Some Current Aging Issues

The *sandwich generation* is composed of midlife people who care for aging parents, are raising a child under age 18, or supporting an adult child. Around 23 percent of Americans fall into this category (Kelly, 2023). In an aging society, the sandwich generation and older people experience many changes, including the rise of multigenerational households, work and retirement options, right-to-die issues, and competition for scarce resources.

baby boomers people born between 1946 and 1964.

old-age dependency ratio (also called the *elderly support ratio*) the number of working-age (18 to 64) adults for every person aged 65 and older.

Multigenerational Households

The share of the U.S. population living in multigenerational households rose from 12 to 16 percent between 1980 and 2010. Asians (28 percent) are more likely to live in multigenerational households than Black (26 percent), Latina/o/x (25 percent), and White (14 percent) households.

Why are several generations living under one roof? Aging baby boomers may move in with their adult children and grandchildren to avoid poverty. More grandparents are raising children because drug-addicted parents neglect them, abandon them, or die from an overdose. Adult children may move back home with their parents (often with children and romantic partners in tow) because of unemployment, low wages, or divorce. Extended family living provides intergenerational support, including child care for employed parents. Cultural values, particularly among recent immigrants, reflect long-standing practices of caring for aging parents and grandparents in one's own home (Ellis, 2013; Fry and Passel, 2014; Wiltz, 2016).

The effects of multigenerational households are mixed. Grandparents who live with their adult children and grandchildren may provide crucial economic and child care support. On the other hand, if the multigenerational household is struggling financially, if the grandparents are also raising their own children, and if there's parent–grandparent conflict, grandparents experience emotional problems, including unhappiness, stress, worry, and anger (Deaton and Stone, 2013).

Work and Retirement

Many older Americans are postponing retirement. In 2017, almost 19 percent of people aged 65 and older were working—up from 14 percent in 2000 (Kromer and Howard, 2013; Bureau of Labor Statistics, 2017). The labor force participation rates have increased for both men and women, even for people aged 70 or over (**Figure 12.10**).

Most work out of necessity. A consequence of longer life expectancy is that people need to finance more years of retirement, but are they doing so? Almost a third of employed adults with $10,000 or more in investments give financial assistance to an adult child, a parent, or both (Saad, 2017). The contributions increase family stability and strengthen family ties but also deplete retirement savings.

Since the early 1990s, many large companies (especially in the steel, airline, and auto industries) have greatly reduced or eliminated employee pension plans, but only about half of private sector workers participate in a 401(k) or similar retirement savings program. Among those with savings, the median account value

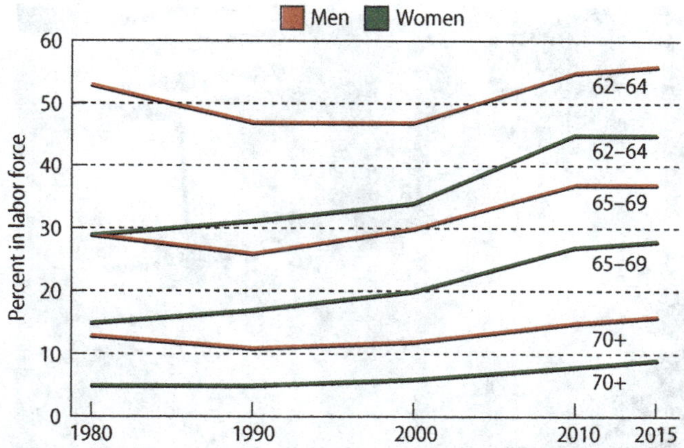

Figure 12.10 The U.S. Labor Force Is Growing Older

Source: Based on Federal Interagency Forum on Aging-Related Statistics, 2016, Table 12.

is just $148,000 for households aged 65 and older. In 2015, the average Social Security payment for people aged 65 and older was only $1,342 a month, but 60 percent rely on Social Security for half or more of their family income ("Who Pays the Bill?" 2013; Center on Budget and Policy Priorities, 2016; Federal Interagency Forum on Aging-Related Statistics, 2016).

Social class is a key factor in living well after retirement. People in higher social classes have more savings, pensions, and property (including homes with paid-off mortgages), higher Social Security income because they had higher lifetime earnings, and 401(k)s that have increased in value over the years. Because of such assets, 64 percent of nonretirees with annual household incomes of $75,000 or more expect to live comfortably in retirement compared with only 32 percent with annual incomes below $30,000 (Norman, 2016). The age for receiving Social Security has recently been increased from 65 to 67 (Brandon, 2022).

For older women, particularly, working is often a necessity rather than a choice. Did you notice in **Figure 12.10** that larger numbers of women in their 60s and 70s are in the labor force? You'll read shortly that divorce rates after age 50 have been increasing. Divorced women, especially if they don't retain their home or a favorable legal settlement, may have to enter (or re-enter) the labor force because they have few economic resources. In other cases, women in their 60s and 70s are staying in the workforce longer because, compared with prior generations, they have more education, work experience, and access to better jobs (Goldin and Katz, 2016).

The Right to Die

Living longer and with more years of chronic diseases and disabilities raises questions about later-life choices.

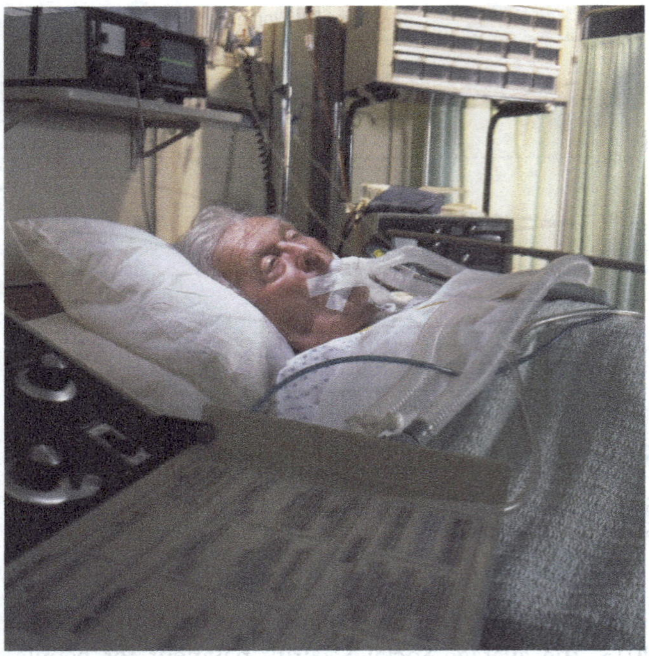

Ken Tannenbaum/Shutterstock.com

Medical care for patients aged 65 and older in the last year of life accounts for about 25 percent of annual Medicare costs (Cubanski et al., 2016). Is this a good investment of our resources?

Both historically and currently, one of the most controversial issues involves end-of-life decisions. Oregon's Death with Dignity Act, which took effect in 1997, authorized lethal prescriptions when two doctors agreed that a patient would die within six months and freely chose *physician-assisted suicide* (also called *physician-aid-in-dying*). Modeling Oregon's law, six states and the District of Columbia have passed similar legislation.

In a recent Gallup poll, 74 percent of Americans supported euthanasia, and 61 percent supported physician-assisted suicide (Jones, 2020). Advocates argue that terminally ill people who face a long and painful death and huge medical costs should have the legal right to die with dignity on their own terms instead of lingering in a nursing home, hospital, or hospice. Opponents maintain that actively ending a life, regardless of a person's frailty or suffering, is a moral violation and that patients might be pushed to die early for the caregiver's convenience (Eckholm, 2014).

Competition for Scarce Resources

When Congress passed the Social Security Act in 1935, life expectancy was about 62 years compared with almost 79 years today. In the future, the growing older population will put a significant strain on the nation's health care services and retirement income programs.

The older population is about half as large as the 18 and younger population. Nonetheless, the federal government spends nearly $6 on older people for every $1 it spends on children (Hahn et al., 2014). The older people get, the higher the medical costs. For example,

- Among Medicare recipients, the average annual health care costs are almost $17,000 for someone aged 65 to 74, compared with almost $26,000 for people aged 85 and over (Federal Interagency Forum on Aging-Related Statistics, 2016).
- A study of 1.8 million Medicare recipients who died in 2008 found that nearly 20 percent had surgery in the last month of life, and nearly 10 percent had surgery in the last week of life. Among those undergoing end-of-life surgery, almost 60 percent were 80 and older (Kwok et al., 2011).
- About 60 percent of prostate cancer is diagnosed in men aged 65 or older. It costs $93,000 per patient to treat advanced prostate cancer; the treatment prolongs life by an average of four months. Although rates have been declining for new cases of prostate cancer, an estimated 288,000 new cases will occur in 2023 (Beil, 2012; National Cancer Institute, 2023).

Many older people maintain that they deserve all the medical benefits they can get because they paid taxes over many years. In fact, an average two-earner couple that retired in 2015 will draw about $212,000 more from Medicare and Social Security than they paid in taxes to support these programs (Steuerle and Quakenbush, 2013).

Younger generations can't count on federally financed health care and retirement benefits in the future. Medicare funds are supposed to last until 2026 and Social Security through 2033 (The Board of Trustees..., 2013), but there are no guarantees. As a result, the age at which a person can draw Social Security benefits was recently increased to 67. As baby boomers age, the costs of these programs will mushroom. Because baby boomers have higher educational levels than previous generations, they're likely to demand more and more expensive health care services (refer to Chapter 14).

Despite rising health care costs, 20 percent of Americans want to live to age 100—about 21 years longer than the current average U.S. life expectancy. Moreover, 38 percent support medical treatments that would allow them to live to at least age 120 (Lipka and Stencel, 2013; Lugo et al., 2013). Who'll pay for such radical life extensions?

12-5 Sociological Explanations of Family and Aging

The four sociological perspectives are useful in understanding families and aging. **Table 12.3** summarizes the key points of these theories.

12-5a Functionalism

We began this chapter by exploring the vital functions that families perform—such as procreation and economic security—that promote societal stability and individual well-being. Functionalists recognize that families differ in structure (e.g., nuclear vs. extended) but believe that their similar functions ensure a society's continuity.

Stability and Activity

For functionalists, marriage, followed by procreation, is critical in fostering social order and cohesion. Parenthood is a crucial social role. One of its tasks, socialization, is essential to maintaining any culture. Problems arise, for instance, when parents can't or don't provide their children with the necessary financial and emotional support or when they divorce.

Kinship ties involve responsibilities, but they also help keep families together, especially in times of trouble. Extended family members offer financial assistance, emotional support, and help care for young children and aging adults. Nearly 1.5 million employed grandparents are responsible for most of the basic care of grandchildren who live with them ("National Grandparents Day...," 2017).

Retirement benefits society because it provides younger people with jobs. Even in retirement, however, the **activity theory** proposes that many older people remain engaged in numerous roles and activities, including work. Moreover, those who are active adjust better to aging and are more satisfied with their lives (refer to Atchley and Arusch, 2004, for a summary of some of this research).

Critical Evaluation

Functionalism has several weaknesses. First, it's questionable whether some family functions are as universal or necessary as functionalists claim. Procreation often occurs outside of marriage, and divorce rates have tripled among people aged 65 and older since 1990. Also, the government has assumed some of the family's functions, including caring for some children (as in foster homes) and older adults (through Medicare, for example) (Lundberg and Pollak, 2007; Stepler, 2017).

Second, is activity theory as representative of older people as some functionalists claim? People may continue to work even in their 80s, but usually not by choice. They can't afford to retire, even though they have health concerns and are unhappy in their lower-income jobs. As their health deteriorates, many older people become less active and more isolated (Lee, 2009).

Third, there is not an orderly progression of when people will retire. For instance, the number of baby boomers who retire has fluctuated up and down year to year. Most recently, in 2020, there was a larger than usual increase (Fry, 2020). Therefore, there isn't always an orderly progression to retirement and the opening of job opportunities for the younger generation.

activity theory proposes that many older people remain engaged in numerous roles and activities, including work.

Table 12.3 Sociological Perspectives on Families and Aging

Theoretical Perspective	Level of Analysis	Key Points
Functionalist	Macro	Families are important in maintaining societal stability and meeting family members' needs. Older people who are active and engaged are more satisfied with life.
Conflict	Macro	Families promote social inequality because of social class differences. Many corporations view older workers as disposable.
Feminist	Macro and micro	Families both mirror and perpetuate patriarchy and gender inequality. Women have an unequal burden in caring for children as well as older family members and relatives.
Symbolic Interactionist	Micro	Families construct their everyday lives through interaction and subjective interpretations of family roles. Many older family members adapt to aging and often maintain previous activities.

12-5b Conflict Theory

Conflict theorists agree that families serve important functions but point out that some groups benefit more than others. Families are sources of social inequality that mirror the larger society, and the inequities persist from birth to older age (Cruikshank, 2009).

Inequality, Social Class, and Power

For conflict theorists, families in higher income brackets have the greatest share of capital, including wealth that they can pass down to the next generation. Differences in inheritance and wealth create an uneven playing field where not all families have equal opportunities for resources such as quality education, housing, and health care (refer to Chapter 8).

Economic inequality affects all aspects of family life. Child abuse and neglect are highest in U.S. counties with the greatest gaps between those with higher and lower incomes. Moreover, the more distance between these social classes, the less likely that economically disadvantaged people (both children and adults) will receive the needed services and support to reduce child maltreatment rates (Eckenrode et al., 2014).

Unequal access to resources continues into older age. Most employers insist that they value older workers' loyalty, work ethic, reliability, and experience, but many are less likely to hire or retain older people because they're usually more expensive than younger workers. Because many large companies have cut their pension plans, numerous older workers must work long after they expected to retire (Kromer and Howard, 2013; Winerip, 2013).

Critical Evaluation

Conflict theory is limited for several reasons. First, conflict theorists tend to overlook the fact that workers of all income levels fund public assistance programs such as Medicaid (a social health care program for individuals and families with lower incomes). They also pay billions each year to cover the costs of unintended pregnancies and the resulting infant care expenses for economically disadvantaged parents (Sonfield and Kost, 2013).

Second, conflict theory links family inequality to capitalism and social class, but there's also considerable family inequality in countries that aren't capitalist (refer to Chapters 8 and 11). Thus, social class affects but doesn't determine whether a child will succeed economically.

Third, many older people don't have to struggle for resources. As you learned earlier, the U.S. government's spending on children has declined, but spending has increased for people aged 65 and older. Older people, regardless of social class, have generous health care coverage through Medicare or Medicaid. They vote in large numbers, have influential lobbyists in Congress, and flood lawmakers with hundreds of thousands of letters and email messages when there's a threat of cutting Medicare or Social Security payments (Johnson, 2011; refer also to Chapter 11).

Alistair Berg/DigitalVision/Getty Images

The average 65-year-old can now expect to live for another 20 years, half of them free of disability. For functionalists, older adults are happiest when they stay active and maintain social relationships.

12-5c Feminist Theories

Feminist scholars agree with conflict theorists that there's considerable inequality between families with lower incomes and higher incomes in accessing necessary resources. However, feminist theorists emphasize the inequality of gender roles in families, especially in patriarchal societies (including the United States).

Gender Roles and Patriarchy

For feminist scholars, families both mirror and perpetuate patriarchy and gender inequality. In most countries, men pass laws about property and inheritance rights, marriage and divorce, and many other regulations that give men authority over women. The United States is similar to many other nations where men (particularly lawmakers and religious officials) control or affect women's decisions about reproduction and access to abortion. In turn, this increases unintended pregnancies and nonmarital birth rates.

Male aggression against women and children is common in patriarchal societies where men hold most of the power, status, and privilege. Females, on the other hand, are marginalized and socialized to accept male domination. In many cases, legal, political, and

In most Latina/o/x communities, the *quinceañera* (pronounced "keen-say-ah-NYAIR-ah") is an important coming-of-age ritual that celebrates a girl's entrance into adulthood on her fifteenth birthday. It's an elaborate and dignified religious and social event that reinforces strong ties with family, relatives, and friends.

religious institutions don't take violence against women and children seriously (Lindsey, 2005). Some judges still view domestic violence charges as frivolous.

Out of 190 nations, the United States, Papua New Guinea, and Swaziland are the only countries that have no national paid parental leave policy. As a result, caregivers, who are predominantly women, must often leave their jobs or work only part-time to care for children and aging parents. Working part-time or stepping out of the workforce reduces women's earnings, career advancement, retirement income, and their families' living standards. Among those who take time off from work following the birth or adoption of a child, women are nearly twice as likely as men (25 percent and 13 percent, respectively) to say that doing so had a negative impact on their job or career (Fry, 2017; Horowitz, 2017; Rossin-Slater, 2017).

Critical Evaluation

Feminist explanations of families and aging have several limitations. First, according to critics, feminist scholars overstate women's domination by men. As you learned earlier in this chapter, the number of women-headed households has been increasing as well as their rates of educational attainment and homeownership (Choi, 2023).

A second criticism is that feminist theories tend to gloss over data that show that intimate partner violence is often mutual. Several dozen studies have found that women initiate from 30 to 73 percent of violent incidents (Straus, 2011, 2014). You'll recall that mothers mistreat their children, and almost 40 percent of the perpetrators of older adult maltreatment are women (Jackson and Hafemeister, 2011).

Third, there can limitations caused by narrow interpretation of feminist theory. Specially, feminist theory often applies an intersectional approach that considers social class and race in addition to sex and gender. For example, the lowest income older adults are minority women, but social class and marital status are important factors. Older married women and those in higher socioeconomic levels are less likely to experience poverty in later life than women who are single (never married, divorced, or widowed) and from lower socioeconomic levels (Hokayem and Heggeness, 2014; Hunter, 2014). Thus, readers must understand that social class, race, and gender (among other identities) are significant factors considered by feminist theorists for how they shape women's experiences in later life.

12-5d Symbolic Interaction

For symbolic interactionists, people create subjective meanings of what a family is and does. Thus, people learn, through interaction with others, how to act as a parent, a grandparent, a teenager, a stepchild, and so on throughout the life course.

Learning Family and Aging Roles

Throughout the socialization process, family members establish trust and develop emotional bonds. Between 90 and 97 percent of Americans believe that it's important for different-sex parents to *both* provide discipline and emotional support and to teach values ("The New American Father," 2013).

Interactionists often use exchange theory to explain mate selection and family roles. The fundamental premise of **exchange theory** is that people seek through their social interactions to maximize their rewards and minimize their costs. In mate selection, people trade their resources (e.g., wealth, good looks, youth, and/or status) for more, better, or different assets. People may stay in unhappy marriages and other intimate relationships

exchange theory people seek through their social interactions to maximize their rewards and minimize their costs.

because the rewards seem equal to the costs. Many women tolerate abuse because they fear loneliness or losing the economic benefits that a man provides (Khaw and Hardesty, 2009; refer also to **Table 12.1** on the benefits and costs of cohabitation).

A well-known psychologist who interviewed more than 200 couples over a 20-year period found that the difference between lasting marriages and those that split up was a "magic ratio" of 5 to 1. That is, if partners have five positive interactions for every negative one, the marriage is likely to be stable over time (Gottman, 1994). Thus, we can improve our family relationships by learning to interact in positive ways.

Stereotypes about older people are deeply rooted in U.S. society. About 84 percent of Americans aged 60 and older encounter **ageism**, discrimination against older people, including insulting jokes, disrespect, and patronizing behavior (Roscigno, 2010).

Language is full of ageist words and phrases that stereotype and disparage older people (e.g., *biddy, old bat, old fart, old fogey, geezer, over the hill, doddering*). Because it's legal to ask about age (but not sex or race), age discrimination is a major barrier in getting a job. If employers assume that older workers are less productive, "technophobic," or inflexible, older applicants—even those applying for low-skilled jobs—are half as likely as people aged 29 to 31 to get a callback. Ageism, especially in hiring practices, helps explain why people try to look younger. In 2016, for example, people aged 40 to 54 had the most plastic surgery (American Society of Plastic Surgeons, 2017; EEOC, 2017; Newmark et al., 2017; Rosenblatt, 2017).

ageism discrimination against older people.

continuity theory older adults can substitute satisfying new roles for those they've lost.

Continuity theory posits that older adults usually maintain the same activities, behaviors, social roles, personalities, and past relationships. The theory also proposes that older people adapt to changes by substituting satisfying new roles for those they've lost (Atchley and Barusch, 2004). For example, a retired music teacher can join a local chorus or orchestra. Thus, developing new roles may lessen some of the emotional distress due to ageism.

Critical Evaluation

A common criticism is that symbolic interaction, a micro-level perspective, doesn't address macro-level constraints. For example, U.S. nonmarital teen births have fallen primarily because of structural factors (e.g., greater access to family planning services, expanded educational opportunities for young women with economic disadvantages) rather than teen attitudes about childbearing (Kearney and Levine, 2014).

Second, exchange theory is limited because people don't always calculate the potential costs and rewards of every decision. In the case of women who care for older family members, genuine love, concern, and a sense of obligation override cost–benefit decisions, even when the person receiving care is abusive (Jackson and Hafemeister, 2013).

Third, continuity theory doesn't distinguish normal aging from disease and disability. Older people who experience chronic illness may be too sick to continue their activities and relationships. Continuity theory also overlooks structural obstacles that discourage older people from pursuing current or new roles. Poverty, for example, can affect physical and mental health, which, in turn, weakens social networks and limits housing, medical, and recreational options.

Chapter Review

Key Takeaways

12-1 **What is a family?** Among other activities, the members of a *family* care for one another and any children. Worldwide, however, families vary in characteristics such as structure (a *nuclear* or an *extended* family), living arrangements (*patrilocal, matrilocal,* or *neolocal*), who has authority (*matriarchal, patriarchal,* or *egalitarian*), and how many marriage mates a person can have (*monogamy* or *polygamy*).

12-2 **How are U.S. families changing?** The United States has one of the highest marriage and divorce rates in the world. *Divorce* is easier to obtain than in the past because all states now have *no-fault divorce* laws, but divorce rates vary by race, ethnicity, and social class. The number of single people has risen greatly, primarily because many people are postponing marriage. There has also been a striking increase in *cohabitation* and nonmarital births. Women-headed single-parent families are increasingly common.

12-3 **Why does family conflict and violence occur?** We're more likely to experience *intimate partner violence (IPV)* than assault by a stranger. IPV causes are complex and often the product of multiple individual, demographic, and societal factors. *Child maltreatment* is widespread, and parents are the most common offenders. Similarly, in *older adult maltreatment*, most of the offenders are adult children, spouses, or other family members. Both micro- and macro-level reasons help explain IPV and maltreatment of children and older adults.

12-4 **How is the aging population changing?** How people define "old" varies across societies. There are also significant differences between the *young-old*, the *old-old*, and the *oldest-old* in their health, ability to live independently, and to work. Historically and currently, women live longer than men, but life expectancy varies by race, ethnicity, and social class. Our *old-age dependency ratio* has increased, placing a larger burden on the working-age population. Some important aging issues include the rise of *multigenerational households*, physician-assisted suicide, and competition for scarce resources.

Example: "He Gets Prettier; I Get Older" When comparing her own public image with that of her actor husband, the late Paul Newman, actress Joanne Woodward once remarked, "He gets prettier; I get older." Was she right? If aging gracefully is acceptable, why do many companies tout antiaging products? And why are the products targeted primarily at women?

12-5 **How do the different sociological explanations of family and aging compare?** Refer to **Table 12.3** on page 269.

Key Terms

Family (p. 250)
Incest taboo (p. 250)
Marriage (p. 251)
Endogamy (often used interchangeably with *homogamy*) (p. 251)
Exogamy (often used interchangeably with *heterogamy*) (p. 251)
Nuclear family (p. 251)
Extended family (p. 251)
Patrilocal residence pattern (p. 252)
Matrilocal residence pattern (p. 252)
Neolocal residence pattern (p. 252)
Boomerang generation (p. 252)
Matriarchal family system (p. 252)
Patriarchal family system (p. 252)
Egalitarian family (p. 253)
Arranged marriage (p. 253)
Marriage market (p. 253)
Monogamy (p. 253)
Serial monogamy (p. 253)
Polygamy (p. 253)
Cohabitation (p. 256)
Fictive kin (p. 261)
Intimate partner violence (IPV) (p. 261)
Child maltreatment (also called *child abuse*) (p. 263)
Older adult maltreatment (also called *elder abuse*) (p. 264)
Life expectancy (p. 265)
Baby boomers (p. 266)
Old-age dependency ratio (also called the *elderly support ratio*) (p. 266)
Activity theory (p. 269)
Exchange theory (p. 271)
Ageism (p. 272)
Continuity theory (p. 272)

Discusssion and Critical Thinking Questions

1. U.S. divorce rates have declined since 1975, but over a lifetime, around half of marriages end in divorce. There are many reasons for divorce, but financial problems are always at the top of the list. Often, couples go into debt even before they marry because they spend, on average, about $30,000 for a wedding

instead of paying off debts, including college loans, or making a down payment on a house (Gillespie and Bennet, 2023; Reaney, 2012).

a. We are often told that people dream about having the "perfect wedding." Is this true? Or is it media hype that benefits jewelers, florists, hair salons, bridal shops, caterers, photographers, and other businesses?

b. Should couples go into debt because a wedding or commitment ceremony is an important rite of passage?

2. Why is it difficult to define family?

3. Do you think that serial monogamy in the United States is a modern version of polygamy?

4. If you live with a romantic partner or have done so in the past, why? Would you advise others to cohabit? If you haven't cohabited and don't plan to do so, why not?

5. What do you think your life will be like at age 80? Do you think your options may be limited because of your gender, sexual orientation, social class, and race or ethnicity?

True or False?

How Much Do You Know about Contemporary U.S. Families and Aging?

The answers to 1 and 6 are true. The answers to 2, 3, 4, and 5 are false.

1. Nonmarital births for people of all ages, including teens, have been decreasing.
2. Cohabitation rates go down as education levels go up. Women with bachelor's degrees or beyond are less likely than those with a high school diploma to cohabitate.
3. Divorce rates have decreased since their high point in the 1990s.
4. In 2020, around 70 percent of children lived with both parents.
5. The percentage of the population born between 1946 and 1964 is growing in the United States and is estimated to make up nearly one-fifth of the population by 2030.
6. Children under the age of 1 year have the highest rate of child maltreatment at 25 per 1,000.

13 Education and Religion

FatCamera/Getty Images

Learning Objectives

After studying this chapter, you will be able to…

13-1 Differentiate between education and schooling.

13-2 Compare the sociological perspectives on education.

13-3 Describe some of the contemporary issues in U.S. education.

13-4 Explain how religion, religiosity, and spirituality differ.

13-5 Describe religious organization and some of the major world religions.

13-6 Explain how and why U.S. religion is diverse, and whether or not secularization is increasing.

13-7 Compare the sociological perspectives on religion.

True or False?

How Much Do You Know about Education and Religion?

1. More women than men are earning bachelor's degrees.
2. Teacher retention rates have improved in the United States.
3. The amount of educational loan debt per college student has remained the same for two decades.
4. Americans' religiosity is decreasing.
5. The percentage of followers of Christianity worldwide is expected to grow.
6. Because of separation of church and state, religion has no presence in the U.S. educational system.

The answers are at the end of the chapter.

Education and religion, two important social institutions, teach values, shape attitudes, maintain traditions, bring people together, control behavior, and grapple with social change. The second half of this chapter examines U.S. and global religions. Let's begin with education.

Be sure to take the True or False quiz in the chapter opener to better understand how much you know about these topics.

What do you think?

I'm optimistic about getting a good job after graduating from college.

1	2	3	4	5	6	7
	strongly agree				strongly disagree	

13-1 Education and Society

Education transmits attitudes, knowledge, beliefs, values, norms, and skills to a society's members. Education can be formal or informal and can occur in a variety of settings. **Schooling**, a narrower term, is formal training and instruction provided in a classroom setting.

U.S. education and schooling have undergone four significant changes since the beginning of the twentieth century: Universal education has expanded, community colleges have flourished, public higher education has burgeoned, and student diversity has increased as more women and racial-ethnic groups enrolled at colleges and universities. As a result of these and other changes, an increasing percentage of Americans have completed high school (91 percent) and obtained a bachelor's or higher degree (37 percent) (U.S. Census Bureau, 2022).

13-2 Sociological Perspectives on Education

Sociologists agree that education and schooling are important but offer different insights into their purpose and outcomes. **Table 13.1** summarizes the four major perspectives.

education transmits attitudes, knowledge, beliefs, values, norms, and skills.

schooling formal training and instruction provided in a classroom setting.

13-2a Functionalism: What Are the Benefits of Education?

For functionalists, education contributes to society's stability, solidarity, and well-being and provides people with an opportunity for upward mobility. In their analyses, functionalists distinguish between manifest and latent functions.

Manifest Functions of Education

Some of the functions of education are *manifest*; that is, they're open, intended, and visible. For example:

- Schools are *socialization agencies* that teach children how to get along with others and prepare them for adult economic roles (Durkheim, 1898/1956; Parsons, 1959).
- Education *transmits knowledge and culture*. Schools teach skills like reading, writing, and counting; they also instill cultural values that encourage competition, achievement, and democracy (refer to Chapter 3).
- Similar values increase *cultural integration*, the social bonds that people have with each other and with the community at large, and *societal cohesion*.
- Education promotes *cultural innovation*. Faculty at research universities receive billions of dollars every year to develop computer technology, treatments for diseases, and programs to address social problems.
- Education *benefits taxpayers* because more highly educated people tend to pay higher taxes, are less likely to rely on public assistance programs, and lead healthier lifestyles, reducing state and federal health care costs (Baum et al., 2013).

For many decades, a large percentage of Americans agreed with functionalists that a college degree is "very important"; in 2013, 70 percent of Americans felt this way (Marken, 2019). For instance, some view a college degree as a path to a better quality of life. In a survey of incoming first-year college students, 88 percent said that the most important reason for going to college is to

Table 13.1 Sociological Explanations of Education

Theoretical Perspective	Level of Analysis	Key Points
Functionalist	Macro	Contributes to society's stability, solidarity, and cohesion and provides opportunities for upward mobility
Conflict	Macro	Reproduces and reinforces inequality and maintains a rigid socioeconomic structure
Feminist	Macro and micro	Produces inequality based on gender
Symbolic Interactionist	Micro	Teaches roles and values through everyday face-to-face interaction and behavior

get a good job (Schneider, 2013; Calderon and Sorenson, 2014; Jones, 2016). Among college graduates, 74 percent said their education helped them grow intellectually (Parker et al., 2011).

Besides expanding a person's intellectual horizons, education increases earnings. On average, college graduates earn 60 to 80 percent more per year than high school graduates. Those with a high school diploma on average earn $22,000 per year while those with a bachelor's degree earn $52,000. Even academically marginal students who manage to get a four-year degree have earnings that are as high, on average, as those of their peers with higher grades (Oreopoulos and Petronijevic, 2013; Hardy, 2022).

As peoples' educational attainment rises, their earnings increase (**Figure 13.1**). Over a lifetime, education affects earnings five times more than other demographic factors like sex, race, ethnicity, and age (Schramm et al., 2013; Daly and Bengali, 2014). By age 64, people with a bachelor's degree earn six times more than those with a high school diploma as their highest educational credential; over a lifetime, this is around $800,000 more in earnings (Selingo, 2013; Hardy, 2022).

Regardless of race, ethnicity, family background, and marital status, people with a bachelor's degree typically live longer, report better physical and emotional health, and are better able to handle stress because of their economic and social positions. They're also more likely than non-college graduates to hold jobs that offer a greater sense of accomplishment, more autonomy, more opportunities for creativity, and more social interaction with coworkers (Rheault and McGeeney, 2011; Oreopoulos and Petronijevic, 2013).

What people study affects their financial payoff, but lifetime earnings vary by occupation, even within the same major field of study. Someone with a degree in the social sciences who works in management can earn $3.4 million versus $1.9 million in education (Julian, 2012; Carnevale and Cheah, 2013). Thus, the combination of what people study in college *and* the careers they pursue after graduation can make a big difference in lifetime earnings.

Latent Functions of Education

Education also has *latent functions*—hidden, unstated, and sometimes unintended consequences. For example:

- Schools *provide child care*, particularly after-school programs, for the growing number of single-parent and two-income earning families.

Median annual earnings for full-time, year-round workers by educational attainment, 2016

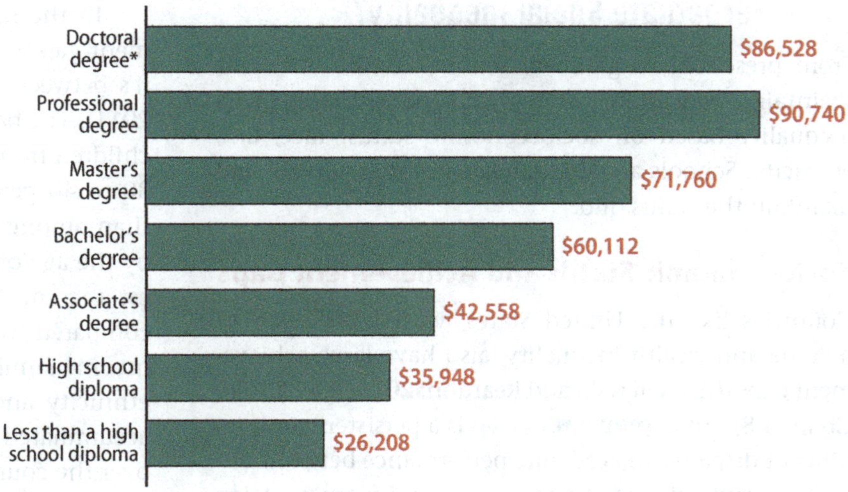

*Includes Ph.D., Ed.D., D.M.A., D.Sc., D.A., and D.M.

Figure 13.1 Education Pays Off

Source: Based on Bureau of Labor Statistics, 2017, "More Education: Lower Unemployment, Higher Earnings," April. Accessed July 25, 2017 (bls.gov/emp).

- High schools and colleges are *matchmaking institutions* that bring together unmarried people.
- Education *decreases job competition*; when young adults are in school longer, current workers are to maintain their jobs for more years.
- Educational institutions *create social networks* that can lead to jobs or business opportunities.
- Education is *good for business*. Thousands of companies offer services that tutor and test students and produce textbooks and related materials.

Critical Evaluation

Do functionalists exaggerate education's benefits? About 35 percent of the world's billionaires don't have a bachelor's degree, and some did not earn high school diplomas. In the 118th Congress, six members of the House of Representatives do not have a bachelor's degree (Wealth-X and UBS, 2014; Schaeffer, 2023). Moreover, some of the highest paying jobs (e.g., occupational therapy assistants, respiratory therapists, aircraft technicians, dental hygienists, web developers and designers, paralegals) that will be in high demand in the coming decades require only a two-year degree or vocational training (Ogunro, 2012; Kowarski and Claybourn, 2023).

Also, according to some critics, functionalists tend to gloss over education's dysfunctions, including rising college costs and student loan debt, topics we'll consider shortly. Conflict theorists, especially, argue that educational institutions produce and reproduce inequality.

13-2b Conflict Theory: Does Education Perpetuate Social Inequality?

From preschool to graduate school, conflict theorists maintain, education creates and perpetuates social inequality based on socioeconomic status, race, and ethnicity. Schools are also gatekeepers that control and maintain the status quo.

Socioeconomic Status and Achievement Gaps

Countries like the United States, with high levels of income and wealth inequality, also have large achievement gaps (Chmielewski and Reardon, 2016; refer also to Chapter 8). An **achievement gap** is a persistent and significant disparity in academic performance between different groups of students (e.g., men and women, Asian students and other racial-ethnic minorities, students with higher incomes and students with lower incomes). The most common measures of achievement gaps are standardized test scores, GPAs, high school dropout rates, and college enrollment and completion rates.

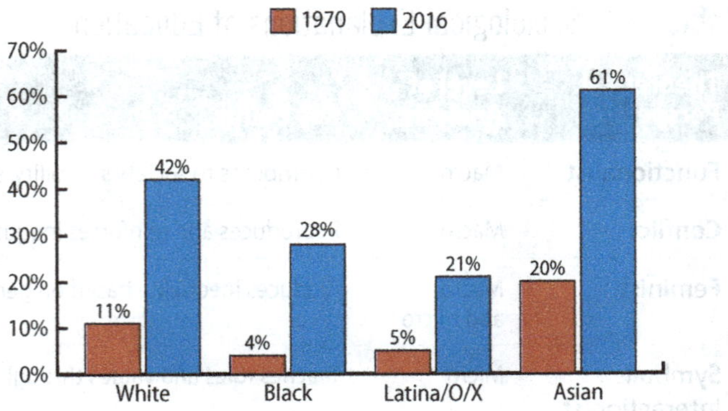

Figure 13.2 Attainment of Bachelor's Degree or Higher, by Race and Ethnicity, 1970 and 2021

Sources: Based on U.S. Census Bureau, 2012, Table 229; U.S. Census Bureau, 2022, Educational Attainment in the United States, 2021, Table 3.

College enrollments have surged since 1990, especially among students from systematically disadvantaged racial-ethnic groups, but who graduates? Asian Americans have made the largest gains (**Figure 13.2**), but there's considerable variation across subgroups. More people who are Indian, Malaysian, Mongolian, and Sri Lankan have at least a bachelor's degree compared with people who are Laotian and Bhutanese (Budiman and Ruiz, 2021) (refer also to Chapter 10). Such variations, as with other systematically disadvantaged subgroups, are due to many factors, including English language proficiency, but the best predictor of educational attainment is socioeconomic status.

In the mid-twentieth century, the largest achievement gap was between Black and White people; today it's between different socioeconomic groups (Reardon, 2011). The bachelor's degree "achievement gap between children from high- and low-income families is roughly 30 to 40 percent larger among children born in 2001 than among those born twenty-five years earlier," and by the age of 24, 79 percent of students from the highest income families have earned a bachelor's degree compared with 11 percent of students from the lowest income families (Education Trust, 2023). A mix of race/ethnicity and socioeconomic status variables produce accumulated educational advantages or disadvantages over the course of a person's life.

Cumulative Reasons for Achievement Gaps

When the 2007–2009 recession began, a number of schools experienced federal, state, and local budget cuts. Five years after the recession, class sizes at many public

achievement gap a persistent and significant disparity in academic performance between different groups of students.

Ariel Skelley/Getty Images

Low- and middle-income students who receive free help from tax professionals to fill out the complicated federal financial aid forms are substantially more likely than students who receive only financial information to submit a college application, to enroll the following fall, and to be awarded financial aid (Bettinger et al., 2012).

schools increased, but staff positions—including teachers, reading and math specialists, and guidance counselors—continued to decrease. Wealthier communities, in contrast, increased property taxes to accommodate higher enrollments (Rich, 2013). During the COVID-19 pandemic, high school graduation rates did not decrease much, but there were socioeconomic-based learning losses. The lowest income students were more likely "to experience food insecurity, homelessness and the loss of a loved one to COVID-19" which caused disruptions to learning (Turner, 2022).

Many of the nation's schools now have a high level of "double segregation" because students are increasingly separated not only by race but also by income. Nationally, 40 percent of Black students and only 6 percent of White students attend a public school with a high percentage of students below the poverty threshold. Educational problems linked to poverty *and* racial segregation include less-experienced and less-qualified teachers, high teacher turnover, inadequate facilities (e.g., computers, science labs), outdated textbooks, high dropout rates, and school buildings that are falling apart (Reich, 2014; Carnoy and García, 2017).

Poverty and lower quality school resources are significant academic roadblocks for lower-income students. They're less likely to attend high schools that offer higher-level courses in mathematics and science. This can mean the difference between passing or failing required courses during the first few years of college. Even when top students from lower-income public high schools enter selective colleges, they may struggle to maintain passing grades. ("Selective" schools are those that admit a relatively low percentage of applicants.) Many college youths from low- and middle-income families must also work part-time or full-time and, consequently, have more stress and less time for academic work (Ross et al., 2012; Black et al., 2014).

Education and Social Control

Functionalists view education as an avenue for upward mobility. Conflict theorists maintain that education restricts upward mobility because of a hidden curriculum, credentialism, and privilege.

Hidden Curriculum. Every school has a formal curriculum that includes reading, writing, and learning other skills. Schools also have a **hidden curriculum**, practices that transmit nonacademic knowledge, values, attitudes, norms, and beliefs that legitimate "economic inequality and the staffing of unequal work roles" (Bowles and Gintis, 1977, p. 108).

Schools in lower-income and working-class neighborhoods tend to stress obedience, following directions, and punctuality so that students can fill jobs that require these characteristics; these jobs are often associated with lower pay (e.g., restaurants, nursing homes, hospitals) (Kozol, 2005). Schools in middle-income neighborhoods emphasize behavior and appearance often expected in jobs with more specialized skill sets, cooperation, conforming to rules, and deference to authority because many of these students will go to college and work in bureaucracies that require such attributes (Hedges, 2011). In contrast, selective schools encourage leadership, creativity, independence, and people skills—all prized traits in elite circles (Persell and Cookson, 1985; refer also to Chapter 8). In effect, then, the hidden curriculum reproduces the existing class structure and provides workers for jobs and occupations in the stratification hierarchy.

Credentialism. Have you noticed that many faculty, doctors', lawyers', and dentists' offices are usually wallpapered with framed degrees? Such tangible symbols of achievement reflect **credentialism**, an emphasis on certificates or degrees to show that people have certain skills, educational attainment levels, or job qualifications.

Functionalists maintain that credentialism rewards people for their accomplishments, sorts out those who are the most qualified for jobs, and stimulates upward social mobility. Conflict theorists contend, however, that for many positions, people can gain skills on the

hidden curriculum school practices that transmit nonacademic knowledge, values, attitudes, norms, and beliefs.

credentialism an emphasis on certificates or degrees to show that people have certain skills, educational attainment levels, or job qualifications.

Fotos International/Getty Images

How important is graduating from a prestigious college or university? Nationally, 84 percent of business leaders say that the amount of knowledge a job candidate has in a particular field is "very important." Only 9 percent say that where a person attended school is very important (Calderon and Sidhu, 2014). Steven Spielberg, the well-known movie producer and director, graduated from California State University at Long Beach after being rejected by the more prestigious University of Southern California and University of California at Los Angeles film schools.

job with a few weeks of training or succeed because of ability or other factors.

Because of a large supply of high school graduates, employers can demand higher levels of education even though some jobs (e.g., retail sales, law enforcement) don't require a college degree for competent performance, a process called *credential inflation.* As more people obtain a college degree, its value diminishes, and students from lower-income families, who are the least likely to have access to a college education, fall further behind (Bollag, 2007).

Largely because of rising college costs and student loan debt, postsecondary certificates have become increasingly popular. A *certificate* is a credential showing that a person has completed educational courses in a specific field (e.g., information systems security, business and office management). Certificates take anywhere from a few months to two years to complete, match skills employers want with a job candidate's qualifications, can be "stacked" as students move in and out of college or a shifting job market, are high paying in some fields, and can be stepping stones to further education, including a college or advanced degree (Blumenstyk, 2015; Mangan, 2015).

Bob Daemmrich/Alamy Stock Photo

Privilege. Beyond the typical challenges of academic coursework and adjusting to college life (roommates, new environment, etc.), students from lower income families may have other hurdles to face. They may experience housing insecurity (perhaps sleeping in their car) or not have funds to meet their nutritional needs (perhaps needing resources from the college food pantry) (Gross, 2022). Many colleges claim that they're committed to admitting talented students with lower-incomes, but award more financial aid and merit-based scholarships to students from high-income families. For instance, Students in the highest income levels are given an average of $11,300 in non-federal financial aid while students at other income levels are granted an average of $7,500 (Helhoski, 2020).

Among the nation's 342 selective four-year public colleges, almost a third of all students come from wealthy households compared with just 8 percent from low-income families. A mere 4 percent of all colleges and universities (138 institutions) hold 75 percent of all postsecondary endowment wealth. Instead of supporting low-income students, "many super wealthy colleges are playgrounds for the children of the wealthiest in our country and the world" (Nichols and Santos, 2016, p. 1; Halikias and Reeves, 2017).

Another privileged group is *legacies*, the children of alumni who have "reserved seats" regardless of their accomplishments or ability. For example, President George W. Bush—who had mediocre high school grades and standardized test scores—was admitted as a legacy at Yale University, which his father and grandfather had attended (Golden, 2006). Legacies also include students with less qualified academic records whose parents, including celebrities, make million-dollar donations. Of the top 250 colleges and universities in the United States over half (56 percent) use legacy as a consideration in their admissions decisions (Massey, 2007; Hurwitz, 2011; Nietzel, 2022).

Critical Evaluation

Conflict theories have several weaknesses. First, a hidden curriculum doesn't necessarily determine job placement because many students aren't passive recipients of educational systems. There are many adults in the labor force who have achieved credentials for employment in a wide range of career fields- without earning a bachelor's degree. Out of workers, who do not hold a bachelor's, 56 percent have a license, 43 percent have certificate, and 21 percent have a certificate (National Center for Education Statistics, 2016). These and other "alternative credentials" have labor market value that increases employment and earnings (Ewert and Kominski, 2014). Conflict theorists may denounce credentialism and credential inflation, but employers often view credentials as useful initial screening tools that signal job candidates' discipline and drive.

Second, do conflict theorists overstate the importance of socioeconomic status achievement gaps? A study tracked more than 5,200 Asian American and White students from kindergarten through high school and found that the former, regardless of income, performed better than their White counterparts. The researchers attributed the Asian American students' greater success to cultural beliefs that achievement is learned rather than innate, parental pressures to succeed, and, especially among recent immigrants, ethnic community resources such as private tutoring and vital information about navigating the education system (Hsin and Xie 2014; refer also to Chapter 10).

13-2c Feminist Theories: Is There a Gender Gap in Education?

Since 1982, women have been graduating from college at higher rates than men. Some view this trend as an indicator of greater gender equality; others are alarmed about the supposed "male crisis" in and "feminization" of higher education (refer to Chapter 9). Has women's progress changed education for other genders? Or are there still gender gaps, particularly in higher education?

Who's Getting Degrees

A higher percentage of girls (88 percent) are graduating high school in four years compared with 82 percent of boys (Reeves et al., 2021). Women earn more associate's, bachelor's, master's, and doctoral degrees than men (**Figure 13.3**). Why are women now more likely to achieve degrees than men?

In both high school and college, women spend more time than men studying, earn higher grades, hold more leadership positions, and are more involved in student clubs and community volunteer work. In high school, boys are twice as likely as girls to describe school as a "waste of time." This may be one reason why teenage girls and women are less likely to drop out of high school or college. In addition, women's enrollment in college

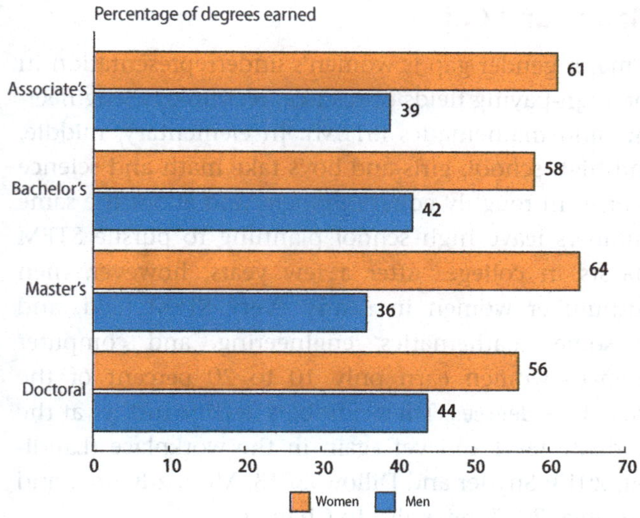

Figure 13.3 Post-Secondary Degrees Award by Sex, 2018–19

Source: Based on National Center for Education Statistics, "Degrees Conferred by Race/Ethnicity and Sex," 2022.

has been increasing alongside their labor force participation rates and structural changes in economy and family (Parker, 2021) (refer also to Chapter 11 and 12). Regardless of the reasons for women's success, many colleges have been giving men preferential treatment in admissions to avoid large gender imbalances in their student bodies ("Gender, Education, and Work," 2015; West, 2021).

Gina Kelly/Alamy Stock Photo

A survey of 27 prominent U.S. universities (including all but one of the Ivy League schools) found that 23 percent of undergraduate women and 5 percent of undergraduate men were victims of non-consensual sexual contact—ranging from penetration to sexual touching. Even in the most serious assaults that included penetration, 72 percent of victims didn't report the assaults because they were "...embarrassed, ashamed or thought it would be too emotionally difficult" or "...didn't think anything would be done about it" (Cantor et al., 2015).

Majors and Careers

A major gender gap is women's underrepresentation in the high-paying fields of science, technology, engineering, and mathematics (STEM). In elementary, middle, and high school, girls and boys take math and science courses in roughly equal numbers, and about the same numbers leave high school planning to pursue STEM majors in college. After a few years, however, men outnumber women in nearly every STEM field, and in some—mathematics, engineering, and computer science—women earn only 10 to 20 percent of the bachelor's degrees. Their numbers decline further at the graduate level and yet again in the workplace (Landivar, 2013; Snyder and Dillow, 2013; Munoz-Boudet and Ravenga, 2017; refer also to Chapter 9).

Why is there gradual attrition? In college, women are initially as persistent as men in a STEM major and earn higher grades. However, they're less satisfied than men with the core courses and more likely to doubt their ability to succeed in a discipline dominated by men. As a result, women's self-confidence falters and they change majors (Shapiro and Williams, 2012; Jagacinski, 2013).

The exit from a STEM major is also associated with having few women faculty role models, and some science professors' beliefs that women students won't benefit from mentoring because they're less competent than men. "Chilly climates"—uncomfortable work environments for women—are common in jobs with higher percentage of men, including STEM occupations. Chilly climates decrease women's job satisfaction and increase the likelihood of leaving a job (Moss-Racusin et al., 2012; Kahn and Ginter, 2017).

Critical Evaluation

A common criticism is that feminist scholars address women's education barriers but not their choices. Socialization, gender stereotypes, and teachers' expectations do affect people's behavior (refer to Chapters 4 and 9), Women, who are empowered with information about systematic biases, often weigh potential benefits of academic and career benefits against possible challenges. Through this process, many women still choose fields of study that they know are on the lower end of the pay scale (e.g., health, education).

Some fault feminist theorists for being more interested in women's than men's educational attainment gaps. There's also the question of why feminist scholars devote little attention to issues such as why boys are more disruptive in elementary school. Boys have greater challenges with behavioral engagement compared to girls; this includes following classroom rules, respecting personal boundaries of others, and turning in assignments on time (Pyne, 2020). The types of disruptions can lead to suspensions and decrease the chance of attending college by at least 16 percent for each suspension (Bertrand and Pan, 2011).

13-2d Symbolic Interaction: How Do Social Contexts Affect Education?

None of us is born a student or a teacher. Instead, these roles, like others, are socially constructed (refer to Chapters 3–5). For symbolic interactionists, education is an active *process* that involves students, teachers, peers, and parents and which includes tracking, labeling, and student engagement.

Tracking

Beginning in kindergarten, practically all schools sort students by aptitude. Such sorting results in **tracking**, assigning students to specific educational programs and classes on the basis of test scores, previous grades, or perceived ability.

Some educators believe that tracking is beneficial because students learn better in groups with others like themselves, and it allows teachers to develop curricula for students with similar ability. Many interactionists maintain, however, that tracking creates and reinforces inequality.

- Students placed in accelerated tracks take classes that involve more critical thinking, problem solving, and creativity that align with occupations that require specialized skillsets and come with higher pay. Students in tracks with slower paces take classes that are limited to simple skills (e.g., punctuality and conformity) that usually fit better with jobs that don't require specialized skill sets and have lower pay.
- Students in the accelerated tracks have more homework, better quality instruction, and more enthusiastic teachers. One result is that students in these tracks are more likely to see themselves as "bright." In comparison, the experiences provided to students in the slower-paced tracks may lead them to view themselves as "slow" or "less capable."
- The effects of tracking are usually cumulative and lasting. Teachers tend to set lower expectations for students in the slower-paced tracks, who therefore fall further behind every year in reading, mathematics, and interaction skills (Oakes, 1985; Hanushek and Woessman, 2005).

---

tracking assigning students to specific educational programs and classes on the basis of test scores, previous grades, or perceived ability.

Middle school and high school become even more stratified as accelerated-track students are sorted into gifted, honors, and advanced courses. In college, students continue to be tracked and sorted into honors programs and accelerated undergraduate courses.

Labeling

Tracking often leads to labeling, a serious problem because "there's a widespread culture of disbelief in the learning capacities of many of our children, especially children of color and the economically disadvantaged" (Howard, 2003, p. 83). Labeling, in turn, can result in a *self-fulfilling prophecy*. That is, students live up or down to teachers' expectations and evaluations that are influenced by a student's income level, skin color, dress, accent, and test scores (refer to Chapter 5).

Implicit bias—unconscious prejudices or stereotypes that affect our attitudes, actions, and decisions—influences teachers' expectations and interactions with students. Both Black and White preschool teachers watch Black children, particularly boys, more closely than White children for misbehavior. When these very young children take other students' toys or hit classmates or instructors, for example, teachers punish Black boys the most severely. One result is that Black children, especially boys, are almost four times more likely than White children to be suspended from preschool (Gilliam et al., 2016; U.S. Department of Education, 2016).

Implicit bias and labeling also affect girls' education. Among seventh- to twelfth-grade students, for example, 46 percent of teenage girls, compared with 62 percent of teenage boys, are confident that they can learn computer science. Only 4 to 10 percent of parents and teachers think that girls are interested in computer science. In contrast, 64 percent say that boys are interested in computer science and are more likely to succeed at it than girls. Viewing computer science as a male field helps explain why the share of women earning a bachelor's degree in computer science slipped from 37 percent in the mid-80s to 25 percent in 2019. This drop in women computer science majors is due to many factors, including chilly classroom and work environments (U.S. Department of Education, 2012; Keating and English, 2015; Sherman, 2015; Fry et al., 2021). Teachers' and parents' implicit biases, however unintentional, shape students' choices and outcomes.

Student Engagement

Many schools assess performance not only through tests but also through *student engagement*, how involved students are in their own learning. A team of researchers who studied more than 2,500 elementary school classrooms concluded that the typical U.S. child has only a 1 in 14 chance of being in a school that encourages their engagement. Because of standardized tests, teachers in public elementary schools spend most of the day on basic reading and math drills and little time on problem solving, reasoning, science, and social studies (Pianta et al., 2007).

By the time they reach middle school, students from racial-ethnic groups that have been systemically disadvantaged—particularly Black and Latina/o/x students—believe that teachers discipline them more harshly than their White peers for similar behavior (e.g., not having a hall pass, disobedience). Students' perceptions that teachers are biased often lead to distrust, defiance, disengagement, underperformance, acting out, and suspensions or expulsions. Teachers can counteract such self-fulfilling prophecies by sending the message that their students are capable, valued, and respected (Yeager et al., 2017).

U.S. high school students aren't as engaged in their education as they could be; 26 percent admit that they usually don't do their homework. The students who are most likely to be disengaged are from economically disadvantaged families, Black or Latina/o/x, don't live with both biological parents, attend financially strapped public high schools that are typically overcrowded and understaffed, and have fewer resources such as computers and even textbooks (Planty et al., 2008; U.S. Department of Education, 2016).

Dean Drobot/Shutterstock.com

Large shares of full-time undergraduates—46 percent at community colleges and 42 percent at 4-year colleges—are employed (McFarland et al., 2017). Is working while in college a necessity or a preference? And, how does balancing work and academic obligations affect student engagement?

Implicit bias unconscious prejudices or stereotypes that affect our attitudes, actions, and decisions.

About 20 percent of first-year college students have difficulty learning course material and getting help with coursework. Students who are the most likely to be engaged in their coursework have mentors, participate in faculty projects, have internships, and are involved in several extracurricular activities (Busteed, 2015; National Survey of Student Engagement, 2016). Such opportunities encourage engagement and decrease attrition.

On average, even full-time college students study only 15 to 16 hours each week compared with 24 hours a week in 1961 (Babcock and Marks, 2011; National Survey of Student Engagement, 2022). This is well below the 24 to 30 hours faculty members say students should be spending on class preparation if they're taking three courses.

According to one college instructor, "Education is the only business in which the clients want the least for their money" (Perlmutter, 2001, p. B1). However, some analysts also blame professors for students' not studying. Many faculty have watered down their required readings, say nothing when students don't prepare for classes or do assignments, and often give easy exams to avoid complaints about low grades. Only 46 percent of students say their courses "challenged them to do their best work" (National Survey of Student Engagement, 2021). When faculty dilute their course requirements and tests, some students don't study because they're bored by courses that don't stimulate them (Benton, 2011; Glenn, 2011).

Critical Evaluation

One weakness is that interactionists presume that once a person (or group) is labelled, a self-fulfilling prophecy will follow. People can overcome labeling effects by changing a situation. In middle school, for example, placing higher-achieving children from racial-ethnic groups that have been systemically disadvantaged into accelerated tracks and programs significantly boosts their academic performance—even in the lowest income neighborhoods of large urban school districts—because students have high-quality teachers and textbooks, as well as positive peer pressure to learn and succeed (Card and Giuliano, 2016).

Another limitation is that interactionists, because of their micro-level analysis, overlook the macro-level structural constraints that affect education. Educational success varies greatly depending on factors such as the availability of high-quality preschools and kindergartens, neighborhood resources (like property taxes) devoted to education, and well-funded and well-staffed high schools. Even more importantly (as you'll learn shortly), politicians—with little or no experience in education—sometimes establish policies that perpetuate racial/ethnic and economic inequality.

13-3 Some Current Issues in U.S. Education

In a survey question about "the biggest problems facing the country today," only 39 percent of Americans expressed concern that the quality of public K-12 schools is a very big problem. More Americans identify inflation (70 percent) and affordable healthcare (55 percent) as concerning issues (Doherty and Gomez, 2022).

13-3a Elementary and Middle Schools

High-income earning economies, including the United States, rely not on physical labor but on "cognitive labor" that requires formal analytical abilities, written communications, and specific technical knowledge—skills that people acquire and cultivate beginning in preschool (Autor, 2014). Three of the ongoing issues regarding elementary and middle schools are low test scores, achievement gaps, and effective teaching.

Low Test Scores and Achievement Gaps

Children who read proficiently by the end of the third grade are more likely to do well in other subjects, including mathematics, and to graduate from high school. Since 1990, eighth-grade reading and math scores have increased, but there are racial and ethnic gaps. On a scale of 0 to 500, Asian students have the highest scores in both subjects, and Black students the lowest, but both Black and Latina/o/x students have narrowed achievement gaps (**Figure 13.4**).

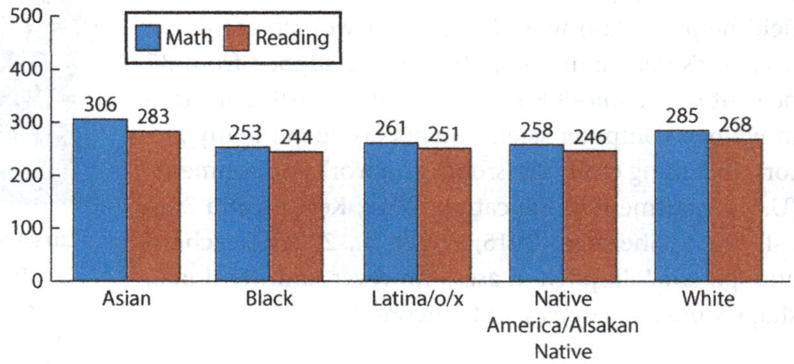

Figure 13.4 Average Reading and Mathematics Scores of U.S. 8th-Grade Students, by Race and Ethnicity

Source: Based on National Center for Education Statistics 2023, "NAEP Report Card: Reading" and "NAEP Report Card: Math".

Socioeconomic status explains some of the variations. In 2021, 57 percent of 3- and 4-year-olds and 90 percent of 5-year-olds whose parents had either a graduate or professional degree, were enrolled in kindergarten, preschool, and nursery school programs that provided enriching educational experiences. This is compared with only 41 percent of 3- and 4-year-olds and 84 percent of 5-year-olds whose parents had a high school diploma as a highest credential, Also, nearly 80 percent of fourth graders in economically disadvantaged families are below proficiency in reading compared with 49 percent of higher-income children (National Center for Education Statistics, 2023 "Enrollment Rates of Young Children"; McFarland et al., 2017).

Socioeconomic status alone doesn't explain student performance, however. Homework also affects test scores. Many elementary and middle school children aren't getting or doing their homework, but parents and the public often blame teachers for the students' academic performance when it falls below expectations (Loveless, 2014).

Effective Teaching

A multi-year project concluded that the most accurate way to evaluate elementary school teachers is to use a three-pronged approach that includes student test scores, multiple classroom observations, and student evaluations (Cantrell and Kane, 2013), but 40 states use only student test scores to evaluate teachers. Much research shows that test scores are invalid and unreliable measures of teacher effectiveness. Other factors affect test scores, including a teacher's quality of undergraduate education, years of experience, being assigned to low-track or high-track classrooms, whether or not students study, and school funding (Kalogrides et al., 2013; Amrein-Beardsley, 2014; Polikoff and Porter, 2014).

Education and Politics

Many teachers believe that politics threaten public education, yet politics can also help. Under the Trump administration *vouchers* for private school were expanded. (Vouchers allow parents to use taxpayer money to pay for all or some of a child's private school fees.) With the current, Biden administration, the American Rescue Plan added $1.9 trillion to offset educational issues that resulted from COVID-19 pandemic. For instance, there is funding to assist with learning loss and mental health (The White House, 2022).

13-3b High Schools

High schools, like elementary and middle schools, must grapple with low achievement scores and effective teaching. In addition, standardized tests (ACTS, SAT) for college admissions are becoming an increasingly controversial issue. The number of colleges and universities using test optional college admissions increased dramatically with the COVID-19 pandemic and have continued at many institutions (Einhorn, 2022).

International Comparison

Among 71 countries, U.S. 15-year-olds rank 38th in math and 24th in both science and reading. U.S. students are roughly in the middle in reading and science, but math performance, which is below average, has declined since 2009. In Singapore, which consistently has the highest achievement, average high school students are roughly three years ahead of their American peers in math. Compared with 35 other wealthy industrial nations, the United States ranks only 30th in math and 19th in science ("Education," 2016; Sparks, 2016; DeSilver, 2017).

The United States is one of the wealthiest countries in the world; spends similar amounts on primary and secondary education compared with 30 other industrialized countries; ranks second among 30 nations for spending on college/university education; and ranks above average in the share of high school students with high socioeconomic status (Carnoy and Rothstein, 2013; OECD, 2022). Why, then, are U.S. students not performing as well? A major difference between the United States and other high-income nations is teacher preparation and quality.

kali9/Getty Images

About 6 percent of Americans ages 16 to 24 have discontinued high school study, down from 11 percent in 2000. The dropout rate is slightly higher for teenage boys than for teenage girls, and is higher for Native Americans, Latina/o/xs from another country of origin and youth whose families are in the bottom 25 percent of all family incomes. Students may discontinue study for a variety of reasons including "pull factors" that include needing to financially support family or care for a family member (Snyder et al., 2016; McFarland et al., 2017).

Teacher Preparation and Quality

At the primary and secondary levels, American teachers spend, on average, more time teaching than their counterparts in most other OECD countries (OECD, 2022). A combination of structural factors helps explain why, despite their individual efforts, many U.S. teachers aren't as effective as they could be.

First, the top-performing countries accept only the top applicants for education programs. Finland, Singapore, and South Korea recruit the top 5 to 30 percent of high school graduates but accept only 10 percent who have high standardized test scores in science, math, and reading (Auguste et al., 2010; Sawchuck, 2012).

Several reports described U.S. teacher preparation programs as "an industry of mediocrity" that has low or no academic entry standards, easy coursework, and gives out many easy A's (Greenberg et al., 2013; Putnam et al., 2014). In top-performing countries, 100 percent of teachers are in the top third of their college graduating classes. In the United States, 47 percent of kindergarten through twelfth-grade teachers come from the bottom third. "In other words, we hire lots of our lowest performers to teach, and then we scream when our kids don't excel" (Cloud, 2010. p. 48).

Second, compared with teachers in top-performing countries, a higher number of U.S. teachers don't have a bachelor's degree or higher in the subjects they teach. For example, 30 percent of public high school teachers who teach math and 54 percent who teach chemistry didn't major in these subjects (Hill et al., 2015). To fill teacher shortages, many states are now lessening requirements to teach; some do not require formal teacher training or certificates (Richman and Crain, 2022).

Third, the top-performing countries offer teachers competitive salaries. In South Korea and Singapore, teachers on average earn more than lawyers and engineers. In these countries, Finland, and many others, teachers are highly regarded and enjoy the same prestige as physicians and other high-status professions and have lifelong careers, resulting in very low teacher attrition. In the United States, teachers' salaries are much lower, compared with the earnings of other similarly educated workers; this puts the U.S. ranking lower than most other OECD nations (OECD, 2022). Teacher job satisfaction in the U.S. has also hit a 50-year low (Ross, 2023). Recently, the teacher turnover rate was 10 percent which is higher than pre-pandemic levels. Forty to 50 percent of new U.S. teachers leave the profession within their first 5 years on the job. High turnover rates are especially harmful to students in low-performing schools: Replacing a teacher—especially a good one—is costly, disruptive, and demoralizing for both students and parents (Ronfeldt et al., 2011; "Schools in Finland," 2016; Startz, 2016; Dilberti and Schwartz, 2023).

Fourth, high-performing countries endorse egalitarian, student-directed policies and learning environments. Unlike the United States, they (1) enroll most children in high-quality preschools; (2) direct more resources to the most disadvantaged pupils; (3) delay steering children into academic or vocational programs until they're 15 or 16 years old, which decreases SES gaps due to tracking; (4) have small class sizes (one teacher for every 12 high school students in Estonia, one of the top-performing countries); (5) apply rigorous, consistent standards across all classrooms; and (6) give principals (not politicians) considerable autonomy over resources, curricula, and other school policies ("Education," 2016; OECD, 2016; Ripley, 2016).

The Growing Controversy over SATs, ACTs, and AP Courses

In 2013, about 3.5 million high school students took either the SAT or the ACT, standardized college entrance examinations. The number of students taking these assessments has declined. In the class of 2022, 3.4 million high school students took one of the tests. The administrators/owners of both exams say that the scores indicate where students are falling behind in college readiness, measure students' capability to do college-level work, and predict college success (College Board, 2013; ACT, Inc., 2014; Claybourn, 2022).

A growing number of critics, including college officials, contend that the tests are little more than gatekeeping tools. They exclude lower socioeconomic students from higher education, don't predict college success, and the test scores reflect accumulated life advantages and disadvantages rather than the ability to do college-level work. As a result, for fall 2023 admissions, 1,750 colleges did not require applicants to submit SAT or ACT scores; that's 80 percent of colleges (Botstein, 2014; Fraire, 2014; Hiss and Franks, 2014; Reeves, 2017; Nietzel, 2022).

Advanced Placement (AP) offers high school students college-level curricula and exams. About 78 percent of degree-granting colleges and universities give course credit to students who get high scores on the exams (Berrett, 2014). Thirty-four percent of the high school class of 2022 took at least one AP Exam which is an increase from class of 2012 (30 percent). The percentage of high school graduates with a passing score of has increased as well (College Board, 2022).

Like the ACT and SAT, AP is becoming more controversial. First, AP courses are rarely as demanding

Alexander Raths/Shutterstock.com

Finnish high school students routinely score at or near the top on international reading, science, and math tests. Why?

as college courses and don't offer the same content. Second, to increase a high school's prestige, superintendents and principals may pressure teachers to offer AP courses and students to take them. Both groups may be unprepared to do so, and failure can make students and teachers feel inferior. Finally, a growing number of public high schools offer AP courses, but 30 percent of schools still do not. Schools that offer the exams are typically in high-income and predominantly White neighborhoods (Simon, 2013; Berrett, 2014; The Feed, 2021).

13-3c Colleges and Universities

Public postsecondary institutions often feel embattled by numerous constituencies and external pressures. Many colleges and universities are grappling with affirmative action issues, grade inflation and cheating, low graduation rates, and rising student loan debt.

Affirmative Action

In higher education, *affirmative action* refers to admission policies that provide equal access for groups, particularly women and racial-ethnic minorities, that have been historically excluded or underrepresented. The enrollment rates of these groups has increased steadily since the 1970s, but using race and ethnicity in college admission decisions has been, and continues to be, a hotly debated topic (refer to "The Affirmative Action Debate in Higher Education").

In 2006, Michigan voters approved an amendment to the state's constitution to ban using race as a factor in deciding who's admitted to the state's public universities. In 2014, the U.S. Supreme Court (*Schuette v. Coalition to Defend Affirmation Action*) ruled that Michigan's ban was constitutional. The court's decision legitimated similar laws in seven other states (Schmidt, 2014). Affirmative action opponents praised the decision for reinforcing **meritocracy**, a system that rewards people because of their individual accomplishments. We say things like "work hard and you'll get ahead" and "pull yourself up by your bootstraps," but is that what happens? As you've learned, the best predictor of who goes to college isn't ability but income.

Grade Inflation and Cheating

Grade inflation is widespread across all education levels, and some children start cheating on tests as early as the third grade. The consequences of grade inflation and cheating in postsecondary education are more serious, however, because someone who cheated their way through college or graduate school may be your dentist, lawyer, doctor, or tax accountant.

Grade Inflation. Has an A replaced the C of the 1970s? In 1971, only 26 percent of first-year college students expected to earn at least a B average in college, compared with 52 percent in 1995, and a whopping 70 percent in 2010 (Pryor, 2011), and even though students are studying less than in the past.

Many faculty give high grades because it decreases student complaints, involves less time and thought in grading exams and papers, and reduces the chances of students' challenging a grade. Some faculty believe that they can get favorable course evaluations from students by handing out high grades, and others accept students' view of high grades as a reward for simply showing up in class. Inflating grades decreases student attrition and satisfies administrators, especially when state legislators base funding on graduation rates (Bartlett and Wasley, 2008; Gillespie, 2014).

> Would you want a military officer, a financial planner, a dentist, or a doctor who cheated her or his way through college or graduate school?

Cheating. Despite grade inflation, cheating is common in college and graduate school. Many studies have uncovered cheating at the Air Force Academy and in

meritocracy a system that rewards people because of their individual accomplishments.

many graduate programs, including business administration and dentistry. Concerns over cheating increase with remote learning, where students are not monitored in a physical classroom, In one online course, over 200 students (one-quarter of the class) "were referred for disciplinary action for using 'tutor-provided solutions' to exam questions" (Newton, 2020). A variety of internet-based tools are causing concern about academic integrity. Faculty are just now considering the impacts of Chat GPT as a tool that students can use to complete assignments without actually doing the work themselves.

Policy analyst David Callahan (2004) contends that dishonesty is prevalent because we live in a "cheating culture," which rationalizes deceit (e.g., "everybody's doing it"), and rewards cutting corners to make money and to succeed. As competition increases and more people become afraid of falling behind, even usually honest people don't feel guilty for cheating at tax time, at work, or in higher education. For many college students, cheating "isn't a big deal," perhaps because they see it everywhere—on Wall Street, by sports stars, and by celebrities. Political (and particularly White House) dishonesty is so common, *Time* magazine offers a weekly list of "Facts vs. Alternative Facts," meaning lies (refer to Chapter 7 on widespread corporate, white collar, and consumer retail fraud).

More than two-thirds of college students admit having cheated at least once on tests and assignments: "To them, higher education is just another transaction, less about learning than about obtaining a credential" (Wolverton, 2016, p. A42). Students cheat because they can. At least 41 percent of faculty don't take disciplinary action against students they know are cheating (Coren, 2011). Students are rarely caught and, if caught, seldom punished. Despite widespread grade inflation and cheating, millions of students don't graduate from college.

Graduation Rates

The United States has a similar 4-year college graduation rate (38 percent) compared with other OECD countries (average 39 percent) (National Center for Education Statistics, 2019). Still, there are specific issues creating challenges at American colleges and universities, and not all groups of students have similar experiences and outcomes.

Once on campus, 40 percent of students must take one or more remedial courses. This percentage is higher at community colleges (65 percent). Students placed in remedial courses tend to be Black and Latina/o/x students, and students from families with lower incomes (Ran and Lin, 2022). Dropout rates are higher for men than women, part-time than full-time students, and students from families with lower incomes. At four-year colleges, the more selective the institution, the higher the graduation rate: 35 percent at colleges with open

The Affirmative Action Debate in Higher Education

Supporters argue that affirmative action programs...

- give disadvantaged and underrepresented students a needed boost in overcoming past and current discrimination
- have doubled or tripled the number of minority applications, resulting in student populations that are more representative of the surrounding community
- have increased minorities' upward social and economic mobility
- promote diversity and encourage people to work together effectively in a multicultural society

Opponents argue that affirmative action programs...

- are discriminatory because they're based on race rather than academic achievement
- benefit primarily middle- and upper-class minorities
- are condescending because they imply that minorities can't succeed without preferential treatment
- admit students who often can't live up to an institution's academic standards, resulting in high failure rates

What else would you add to either side of the debate?

Sources: Messerli, 2012; Krishnamurthy and Edlin, 2014.

Willie B. Thomas/Getty Images

admission policies, 61 percent at colleges that accept 50 percent or more of applicants, and 85 percent at colleges that accept only 25 percent of applicants (Snyder et al., 2016).

Higher graduation rates at more selective institutions don't necessarily mean that the students are more intelligent than those at open admission colleges. Instead, selective colleges require higher GPAs and, when required, standardized test scores, both of which screen out students who may be less prepared for college. Selective colleges also have higher tuition and fees that, in turn, generate more financial aid, on-campus resources and support services, and enriching extracurricular activities—all of which increase the likelihood of earning a degree.

Socioeconomic status and family relationships also affect college completion. Students whose family income is in the top 25 percent are nearly five times more likely to attain a bachelor's degree than those from the lowest 25 percent (58 percent vs. 12 percent). Fully three-quarters of undergraduates are managing some combination of family responsibilities, jobs, and commuting to class. Working while in college, living at home, and experiencing family-related stresses (e.g., parents' divorce, job loss, illness) decrease the odds of graduating from college (Princiotta et al., 2014; Wilbur and Roscigno, 2016; Cahalan et al., 2017). Students also drop out because they get sick, accept attractive job offers, or decide that college isn't for them, but cost is a major reason for attrition.

Student Loan Debt

Student loan debt, which has more than tripled from 2006 (1/2 trillion dollars) to 2022 ($1.75 trillion). There are 45 million Americans with student loan debt. From 1995 to 2019, the average student loan debt at graduation increased by 70 percent. Students can take out federal loans, private loans, or both. Per student borrower, federal loan debt average is around $37,000 while the average for private loan debt is around $55,000. These debts can be hard to pay off. Two decades after starting college studies, around half of borrowers still have $20,000 left to pay on each type of loan. Some groups have more loan debt than others; women have 64 percent of these debts and Black women have a much higher average debt amount. Also, students who identify as LGBTQ typically hold $16,000 more in outstanding debt (Welding, 2022; Hanson, 2023).

The rising student loan debt depends on factors like an institution's endowment, in-state and out-of-state tuition prices, whether the college is public or private, and the growing number and costs of full-time administrators (*Chronicle of Higher Education*, 2014). In the early part of this century, public spending on higher education dropped 30 percent, even though enrollment at public colleges jumped 34 percent (Kena et al., 2014; The Urban Institute, 2023). As state funding decreased, tuition increased, and so did student borrowing to cover tuition, fees, and the costs for room and board. With the COVID-19 pandemic, additional public funds were directed to colleges and universities. Yet, full-time college enrollments have been dropping, down nearly 12 percent from 2011 to 2022. This means fewer students to contribute tuition dollars towards college and university budgets. With these changing factors, tuition dollars continue to rise (**Figure 13.5**) (State Higher Education Finance, 2022; Kerr and Wood, 2022).

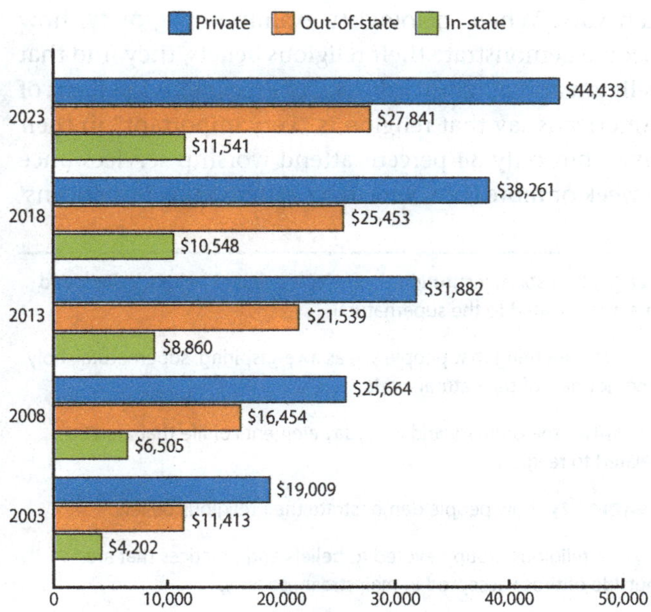

Figure 13.5 Average Tuition Among National Universities, 2003-2023

Source: Kerr and Wood, 2022.

13-4 Religion and Society

Like education, religion has a profound effect on society. Some form of religion exists in all societies and cultures. Throughout history, religion has been a central part of the human experience, affecting the family, economy, and other social institutions. Why is religion important to many people? And does it always benefit society? The four theoretical perspectives offer different insights in answering these and other questions.

13-4a What Is Religion?

Religion is a social institution that involves shared beliefs, values, and practices related to the supernatural. It unites believers into a community, but customs and practices differ across cultures and groups. For example, being Catholic requires confessing one's sins to a priest, whereas silent confessions directly to God are more common among people who are Protestant, Jewish, Muslim, and members of other religious groups.

Émile Durkheim (1961) distinguished between the sacred and the profane. **Sacred** refers to anything that people see as awe-inspiring, supernatural, holy, and not part of the physical world. In contrast, **profane** refers to the ordinary and everyday elements of life that aren't related to religion. Sociologists differentiate religion from religiosity and spirituality.

13-4b Religion, Religiosity, and Spirituality

Religion is a belief system, but religious expression can vary. When sociologists examine **religiosity**, how people demonstrate their religious beliefs, they find that religion and religiosity differ. For example, 41 percent of Americans say that religion is "very important" in their lives, but only 34 percent attend worship services once a week or more (Pew Research Center, 2021). Americans' belief in God is dropping too. From Gallup surveys of Americans, in 1945, 98 percent said they believed in God, but by 2022 it was down to 81 percent (Jones, 2022 "Belief in God").

Spirituality is a personal quest to feel connected to a reality greater than oneself. About 29 percent of Americans describe themselves as "spiritual but not religious." Out of those who are nonreligious, 30 percent have "a slight connection to a higher power," 26 percent pray, 30 percent meditate, and 19 percent say religious is "somewhat important" to them (Brady, 2021). Thus, some people who are nonreligious still engage in activities that are spiritual, but not necessarily religious.

religion a social institution that involves shared beliefs, values, and practices related to the supernatural.

sacred anything that people see as awe-inspiring, supernatural, holy, and not part of the natural world.

profane the ordinary and everyday elements of life that aren't related to religion.

religiosity how people demonstrate their religious beliefs.

cult a religious group devoted to beliefs and practices that are outside of those accepted in mainstream society.

new religious movement (NRM) term used instead of *cult* by most sociologists.

charismatic leader someone that followers see as having exceptional or superhuman powers and qualities.

13-5 Religious Organization and Major World Religions

People manifest their religious beliefs most commonly through organized groups, including cults, sects, denominations, churches, and ecclesia. The major world religions differ in their membership and beliefs.

13-5a New Religious Movements (Cults)

A **cult** is a religious group devoted to beliefs and practices that are outside of those accepted in mainstream society. Many sociologists use **new religious movement (NRM)** rather than *cult* because the media have used the latter term in derogatory ways to describe any unfamiliar, new, or seemingly bizarre religious group (Roberts, 2004).

NRMs usually organize around a **charismatic leader** whom followers see as having exceptional or

joyfull/Shutterstock

Hare Krishna is the popular name for the International Society of Krishna Consciousness (ISKCON), a new religious movement based in Hinduism. Established in the United States in 1965, its practices include an austere and simple life, vegetarianism, chanting, evangelism, and abstinence from drugs and alcohol.

superhuman powers and qualities (refer to Chapter 11). Some NRMs have become established religions. Early Christians were a renegade group, with the charismatic leader Jesus, that broke away from Judaism. Islam began as an NRM built around the charismatic leader Muhammad. Both Christianity and Islam became engrained within particular societies and have endured over time. Most contemporary NRMs are fragmentary, loosely organized, and temporary, but others (e.g., the Church of Scientology) have developed into lasting and highly bureaucratic international organizations.

13-5b Sects

A **sect** is a religious group that has broken away from an established religion. People who begin sects are usually dissatisfied members who believe that the parent religion has become too secular and has abandoned key original doctrines. Like NRMs, some sects are small and disappear after a time, whereas others become established and persist. Examples of sects that have persisted include Amish, Jewish Hasidics, Jehovah's Witnesses, Quakers, and Seventh-Day Adventists (Bainbridge, 1997). Some sects develop into denominations.

13-5c Denominations

A **denomination** is a subgroup within a religion that shares its name and traditions and is generally on good terms with the main group. Denominations can form slowly or develop rapidly, depending on factors like geography, immigration, and a country's birth rate. Some scholars describe a denomination as somewhere between a sect and a church. Like sects, denominations have a professional ministry. Unlike sects, denominations view other religious groups as valid and don't make claims that only they possess the truth (Hamilton, 2001).

Denominations typically accommodate themselves to the larger society instead of trying to dominate or change it. As a result, people may belong to the same denomination as their grandparents or great-grandparents. Denominations exist in all religions, including Christianity, Judaism, and Islam. For example, the main branches, or denominations, of Islam are Shi'ah and Sunni, the latter of which is followed by 85 percent of Muslims worldwide. In the United States, the many Protestant denominations include Episcopalians, Baptists, Lutherans, Methodists, and Evangelicals.

13-5d Churches

A **church** is a large established religious group that has strong ties to mainstream society. Because leadership is attached to an office rather than a specific leader, new generations of believers replace previous ones, and members follow tradition or authority rather than a charismatic leader. As in a denomination, people are usually born into a church, but may later decide to leave it.

Churches (e.g., Roman Catholic Church, Greek Orthodox Church) are typically bureaucratically organized, have formal worship services and trained clergy, and often maintain some degree of control over political or educational institutions (refer to Chapter 11). Because churches are an integral part of the social order, they often become dependent on, rather than critical of, the ruling classes (Hamilton, 2001).

13-5e Ecclesiae

An **ecclesia** (also called a *state religion*) is an official religious organization that claims everyone in society as its members. Membership is automatic at birth, the religion is part of the country's cultural identity, and there's a close association between religious and state officials. Examples include the Anglican Church in England, the Lutheran Church in Norway and Sweden, and Islam in Iran, Iraq, Afghanistan, and Saudi Arabia.

An ecclesia has some of the same characteristics as a church, such as a formal structure and professionally trained and designated clergy. A major difference is an ecclesia's power. In some societies, particularly in the Middle East, ecclesiae have immense authority because there's little or no separation between church and state. For example, 23 nations have Islam built into their constitutions (U.S. Commission on International Religious Freedom, 2023). In others (e.g., England, Sweden, Norway), ecclesiae have little or no influence on government laws and regulations or most cultural practices.

13-5f Some Major World Religions

Worldwide, the largest religious group is Christianity, followed by Islam. The religious affiliation of the world's population is approximately:

- 31.1 percent Christians
- 24.9 percent Muslims

sect a religious group that has broken away from an established religion.

denomination a subgroup within a religion that shares its name and traditions and is generally on good terms with the main group.

church a large established religious group that has strong ties to mainstream society.

ecclesia (also called a *state religion*) an official religious organization that claims everyone in society as its members.

- 15.6 percent Unaffiliated (people who may or may not believe in God or a universal spirit but don't identify with any particular religious group)
- 15.2 percent Hindus
- 6.6 percent Buddhists
- 5.6 percent Folk religionists (includes followers of African traditional religions, Chinese folk traditions, Native American religions, and Australian aboriginal religions)
- Less than 1 percent Other (includes Jews, Baha'is, Sikhs, Jains, Shintoists, and many others) (based on CIA World Factbook, 2022).

Note that no religious group comes close to being a global majority, that the third largest group is religiously unaffiliated, and that non-Christians outnumber Christians 2 to 1. By 2050, however, Islam is expected to surpass Christianity as the world's largest religious group and will make up nearly one-third of the world's population. The main reasons for Islam's growth is that Muslims have more children than members of the other major religious groups. Also, because Muslims have the youngest median age (20 vs. 30 for non-Muslims), a larger share will soon be having children, accelerating the growth of the Muslim population (Lipka, 2017).

Five religious groups in particular have had a worldwide impact on economic, political, and social issues. **Table 13.2** provides a brief overview of these groups.

Table 13.2 Characteristics of Five Major World Religions

Religion	Date of Origin	Founder	Percentage of Followers Worldwide
Christianity	0 C.E.	Jesus Christ	31.1%
 iStock.com/Onfokus	Jesus, the son of God, sacrificed his life to redeem humankind. Those who follow Christ's teachings will enter the Kingdom of Heaven. Sinners who don't repent will burn in hell for eternity.		
Islam	600 C.E.	Muhammad	24.9%
 iStock.com/Kickstand	God is creator of the universe, omnipotent, omniscient, just, forgiving, and merciful. Those who sincerely repent and submit (the literal meaning of Islam) to God will attain salvation; the wicked will burn in hell.		
Hinduism	Between 4000 and 1500 B.C.E.	No specific founder	15.2%
 iStock.com/Ferenc Cegledi	Life in all its forms is an aspect of the divine. The aim of every Hindu is to use pure acts, thoughts, and devotion to escape a cycle of birth and rebirth (samsara) determined by the purity or impurity of past deeds (karma).		
Buddhism	525 B.C.E.	Siddhartha Gautama	6.6%
 iStock.com/Pablo Salgado Barrientos	Life is misery and decay with no ultimate reality. Meditation and good deeds will end the cycle of endless birth and rebirth, and the person will achieve nirvana, a state of liberation and bliss.		
Judaism	2000 B.C.E.	Abraham	<1%
 iStock.com/Howard Sandler	God is the creator and the absolute ruler of the universe. By obeying the divine law God gave them, Jewish bear special witness to God's mercy and justice.		

Note: C.E. (Common Era) is the nondenominational abbreviation for A.D. (Anno Domini, Latin for "In the year of our Lord") and B.C.E. (Before the Common Era) is the nondenominational abbreviation for B.C. (Before Christ).

Sources: Based on a number of sources, including the Center for the Study of Global Christianity, 2007; "Religions of the World . . . ," 2007; and "Global Christianity . . . ," 2011; CIA World Factbook, 2020

13-6 Religion in the United States

Among U.S. adults, a declining number (81 percent) believe in God (Jones, 2022). For sociologists, religiosity is a better measure of being religious than simply asking people whether they believe in God (or a universal spirit) and which religion they follow. Religiosity includes a number of variables, but the most common are religious belief, affiliation, and participation.

13-6a Religious Belief

Some 41 percent of Americans say that religion is very important in their lives, down from 70 percent in 1965 (Newport, 2016; Pew Research Center, 2021). Not surprisingly, religion isn't important for *agnostics* (people who say that it's impossible to know whether there's a God), *atheists* (those who believe that there's no God), or others who are skeptics. Nonetheless, most Americans (six in ten), including some atheists and agnostics, embrace at least one tenet associated with some Eastern religions or elements of New Age spirituality, including reincarnation, psychics, astrology, or spiritual energy in physical objects (such as in nature) (Gecewicz, 2018).

13-6b Religious Affiliation

Americans' religious identity is shifting. For the last three decades, the share of Christians has declined now at 63 percent compared to 78 percent in 2007. In comparison, the number of adults who say they have no religious preference or affiliation has grown to 29 percent (**Table 13.3**). Nowadays, 35 percent of Americans who were raised Christian later become religiously unaffiliated. The religiously unaffiliated, also referred to as "nones," were raised in a specific religion but no longer believe the teachings, dislike formal religion, have lost respect for religious leaders who focus on power and money, consider many religious people to be hypocritical and judgmental, and are more willing to admit not being religious than in the past. People who shift to religious unaffiliation include a broad swath of people but tend to be younger, men, Democrats or political independents, college graduates, or live in the Western areas of the United States (Taylor et al., 2014; Lipka, 2016; Pew Research Center, 2016; Pew Research Center, 2022 "How U.S. Religious Composition Has Changed"; Diamant, 2023).

Table 13.3 Religious Affiliation in the United States, 2021

Religious Affiliation	Percentage of U.S. Population
CHRISTIAN	**63%**
Protestant	40%
Catholic	21%
Mormon	2%
JEWISH	**2%**
MUSLIM	**1%**
HINDU	**1%**
UNAFFILIATED	**29%**
OTHER	**4%**

Source: Based on Smith, 2021.

13-6c Religious Participation

Turning to the third measure of religiosity, religious participation, among Americans there was a slight a decline in weekly attendance at religious services from 2007 to 2014. Recently, fewer people participated in religious services because of the COVID-19 pandemic. Although many groups experienced declines in religious service participation (attendance one to two times per week), the decreases were most notable for younger adults, ages 18 to 34 (down 12 percent), Black (down 15 percent), and married without children (down 10 percent). Since the reopening of many places, including religious buildings, participation in religious services has not returned to pre-pandemic levels (Pew Research Center, 2014; Wang, 2022; Witt-Swanson et al., 2023).

13-6d Some Characteristics of Religious Participants

Americans differ in their beliefs and affiliations. Religious participation also varies by sex, age, race and ethnicity, and socioeconomic status.

Sex

American women are generally more religious than men. They're more likely than men to say that religion is "very important" in their lives (60 percent vs. 47 percent), to pray every day (64 percent vs. 47 percent), and to attend religious services at least once a week (40 percent vs. 32 percent), and they are less likely to be religiously unaffiliated (19 percent vs. 27 percent) ("The Gender Gap...," 2016).

The religious gender gap is due to a mix of factors. One explanation is gender-based social expectations that encourage women to be more pious and spiritual because, especially as nurturers, they should transmit religious values to their children (refer Chapter 9). Because of lower labor force participation rates, women have more time for religious activities, where they may forge friendships. Women give higher priority to religion because it offers a sense of self-identity and well-being, especially for women who experience poverty, poor health, or domestic violence ("The Gender Gap...," 2016). Also, women, on average, live longer than men, and religiosity increases with age.

Age

Generally, Americans aged 65 and older are more likely than younger people to describe themselves as religious, to say that religion is very important in their lives, to pray, and to attend services at least weekly. These age-related differences may reflect several factors: Older Americans grew up decades ago when church attendance was higher, they seek spiritual comfort as elderly friends and relatives die, they want to lessen a sense of isolation or loneliness, and they're preparing for death (Taylor et al., 2009).

Although three in ten Americans are religiously unaffiliated, this is more common for Americans who are younger adults. Forty-five percent of Americans ages 20 to 24 years old are religiously unaffiliated compared with only 10 percent of Americans ages 90 to 94. Young adults are also much more likely to be unaffiliated than earlier generations were at a similar stage in their lives. Among other reasons, many baby boomers taught their children to be independent, think for themselves, and "find their own moral compass." As a result, perhaps, millennials are less trusting of major institutions such as religion, the economy, and the government (Smith, 2021; Pew Research Center, 2022; Masci, 2016).

Race and Ethnicity

Black individuals are more religious than any other U.S. racial or ethnic group. They're the most likely to report a religious affiliation, to belong to a church, to pray every day, and to attend religious services every week. The vast majority (71 percent) are Protestant, primarily Baptist and Evangelical; 5 percent are Catholic; and 18 percent are unreligious (Pew Research Center, 2015).

Latina/o/xs' religious landscape has been shifting. Of the largest affiliations, 48 percent are Catholic (down from 67 percent in 2010), 19 percent are evangelical Protestant, and 20 percent are nonreligious (up from 10 percent in 2010). Some Catholic and evangelical Protestant churches have been especially successful in attracting recent Latina/o/x immigrants because the churches offer services in Spanish, the ceremonies are expressive rather than formal, and the clergy are more responsive to their members' social and economic needs ("Select-a-Faith," 2014; Pew Research Center, 2017).

Of all racial-ethnic groups in the United States, Asian Americans may have the most diversity of religious affiliation. Nationally, 17 percent are Catholic, 16 percent are Hindu, 16 percent are mainline or evangelical Protestant, 6 percent are Buddhist, and 6 percent are Muslim. A large share, 31 percent, are unaffiliated with a religion (Pew Research Center, 2015).

secularization religion loses its social and cultural influence.

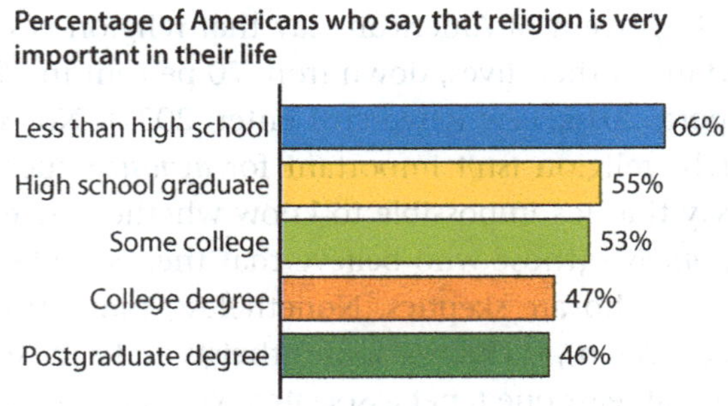

Figure 13.6 Education and Religion

Source: Based on Pew Research Center, 2017, 7.

Socioeconomic Status

As education increases, the importance of religion decreases (**Figure 13.6**). Generally, college graduates are less likely than people with less schooling to believe in God, to pray, and to attend religious services. They're also more likely (11 percent) than adults with a high school education (4 percent) to describe themselves as atheists or agnostics. Among the religiously unaffiliated, 30 percent of atheists and agnostics have an annual family income of $100,000 or more, compared with 17 percent of Christians (Pew Research Center, 2017).

13-6e Secularization: Is Religion Declining or Thriving?

Industrialized nations have been experiencing **secularization**, a process in which religion loses its social and cultural influence. Is secularization endangering religion? Or is religion flourishing?

Is Secularization Increasing?

You've learned that religious unaffiliation is growing, attendance at religious services has dropped, and fewer Americans say that religion is "very important" in their lives. There's other evidence that secularization is rising in the United States and globally. For example:

- 41 percent of Americans have a "great deal" of confidence in the church and organized religion, down from 68 percent in 1975 (Newport, 2017).
- 50 percent of adults belonged to a religious congregation in 2018 compared with 70 percent in 1999 and 73 percent in 1937 (Neuman, 2021).
- 68 percent of people worldwide describe themselves as religious, down from 77 percent in 2005 (WIN-Gallup International, 2013).

Pascal Deloche/Corbis Documentary/Getty Images

- The European Court of Justice, which passes laws for 28 European countries, ruled that private employers can ban workers from wearing religious clothing on the job (Bilefsky, 2017; Margolis, 2021).

There's also evidence of greater "forced secularization" worldwide. About 77 percent of the world's population—up from 58 percent in 2007—lives in countries with high government restrictions on religion. Some of the laws and policies ban particular faiths, prohibit conversions, or limit preaching. Others allow religion-related conflict or terrorism, mob violence against religious minorities, and harassment over religious attire (Grim and Cooperman, 2014; Henne, 2015).

Is Religion Thriving?

Many sociologists contend that the prevalence of secularization has been greatly exaggerated. For example:

- 62 percent of Americans say that it's important for a U.S. president to have strong religious beliefs (Smith, 2017).
- 72 percent endorse Christmas displays on government property (Pew Research Center, 2014).
- Only 1 of the elected 534 members of the 118th U.S. Congress is religiously unaffiliated (Diamant, 2023).
- Nearly all states allow exemptions if vaccinations conflict with parents' religious beliefs; so far, 34 states and the District of Columbia don't prosecute parents for child abuse or manslaughter, even if denying medical treatment for religious reasons results in a child's death or lifelong disability (Sandstrom, 2015, 2016).
- The U.S. Constitution never mentions God or the divine. In contrast, all 50 state constitutions use "God," "Creator," "divine," "Lord," "Supreme Being," and similar religious references at least once (Sandstrom, 2017).

Fundamentalism, a belief in the literal meaning of a sacred text (e.g., Christian Bible, Muslim Qur'an, Jewish Torah), is widespread in practically all countries except Europe. About 25 percent of Americans believe that the Bible should be taken literally because it's the word of God (Saad, 2017).

The U.S. Constitution established a separation of church and state. Over the years, that separation has become fuzzier. Some southern states have passed laws that allow public schools and government buildings to display traditional Christmas scenes and symbols; many public elementary and high schools allow prayer before class, permit students to leave classes for religious instruction, support religious student clubs that faculty sponsor, and offer after-school evangelical programs. And public colleges and universities have approved building privately funded "religious dorms" (Bruinius, 2013; Campo-Flores, 2013; Lawrence, 2013).

fundamentalism belief in the literal meaning of a sacred text.

AP Images/Harry Cabluck

BRENDAN SMIALOWSKI/Getty Images

In 2005, the U.S. Supreme Court ruled that a 6-foot-high monument containing the Ten Commandments outside the Texas state capitol was constitutional, even though it's on government property. In 2014, the court ruled that prayer in civic life, including legislative bodies, is not only constitutional, but can use explicit Christian or other religious language. Do such decisions illustrate greater secularization or religiosity?

> "We establish no religion in this country. We command no worship. We mandate no belief, nor will we ever. Church and state are and must remain separate." Ronald Reagan, Republican, 40th U.S. President (1911–2004).

One of the most divisive issues has been over teaching evolution in public school science classes. For most scientists, it's "a fact" that human beings (and other creatures) evolved from earlier animals through a process known as "natural selection" over several million years (National Science Board, 2014). In contrast, *creationism*, based on a fundamentalist interpretation of the Bible, argues that God created humans in their present form about 10,000 years ago. Forty percent of Americans espouse creationism, and another 33 percent believe in evolution, but under God's guidance. Only 22 percent believe God has nothing to do with evolution (Gallup, 2021).

Despite the constitutional separation of church and state, since 2006 local governments have passed 67 anti-evolution education bills that teach creationism rather than evolution in science classes. In early 2017, eight states proposed laws to protect teachers who present creationism as a scientific theory (Matzke, 2016; Embury-Dennis, 2017).

The government has blurred the distinction between church and state in other ways. In 2014, the U.S. Supreme Court ruled that for-profit companies don't have to pay for their employees' contraceptives—as the Affordable Care Act (Obamacare) mandated—if doing so conflicted with the owners' religious beliefs. Betsy DeVos, the U.S. Secretary of Education (2017 to 2021), was an ardent supporter of using taxpayer-funded vouchers for private religious schools and has a lifelong dedication to "building God's kingdom through education" (Rizga, 2017).

The Johnson Amendment, enacted in 1954, prohibits tax-exempt institutions like churches from endorsing or opposing political candidates. Even though 90 percent of the nation's Evangelical board members and nearly 100 religious organizations have urged Congress to maintain the Johnson Amendment, President Trump signed the Executive Order "Presidential Executive Order Promoting Free Speech and Religious Liberty," which halted consequences for violations of the Johnson Amendment (Beckwith, 2017; Goodstein and Shear, 2017; Executive Office of the President, 2017).

Another indicator that religion is booming in the United States and elsewhere is the prevalence of **civil religion** (sometimes called *secular religion*), integrating religious beliefs into secular life. The Pledge of Allegiance was written in 1892; the phrase "under God" was added in 1954. Other examples of civil religion include the phrase "In God We Trust" on U.S. currency, prayer in public schools, a legally mandated National Day of Prayer held on the first Thursday of May, and many local jurisdictions' closing public schools, public libraries, and government offices on Good Friday, a Christian holy day.

civil religion (sometimes called *secular religion*) integrating religious beliefs into secular life.

13-7 Sociological Perspectives on Religion

You've learned that religion plays an important role in many people's lives. Why? How does society affect religion? And how does religion affect society? **Table 13.4** summarizes the major sociological perspectives on religion.

Table 13.4 Sociological Explanations of Religion

Theoretical Perspective	Level of Analysis	Key Points
Functionalist	Macro	Religion benefits society by providing a sense of belonging, identity, meaning, emotional comfort, and social control over deviant behavior.
Conflict	Macro	Religion promotes and legitimates social inequality, condones strife and violence between groups, and justifies oppression of poor people.
Feminist	Macro and micro	Religion subordinates women, excludes them from decision-making positions, and legitimizes patriarchal control of society.
Symbolic Interactionist	Micro	Religion provides meaning and sustenance in everyday life through symbols, rituals, and beliefs, and binds people together in a physical and spiritual community.

13-7a Functionalism: Religion Benefits Society

Durkheim (1961, p. 13, 43) described religion as a "permanent aspect of humanity" whose "essential task is to maintain, in a positive manner, the normal course of life." Religion can also be dysfunctional, but first, let's consider its benefits.

Religion Is a Social Glue

Religion fulfills a number of important functions at individual, community, and societal levels. All of the following, according to functionalists, contribute to a society's survival, stability, and solidarity:

- *Belonging and identity.* Communal worship and rituals increase social contacts, develop a sense of acceptance and identity, and reinforce people's feelings of belonging to a group. Eighty-five percent of Americans who say religion is not important to them still celebrate Christmas. Attending holiday activities with family or friends can boost a sense of bonding (Saad, 2019).
- *Purpose and emotional comfort.* Religion provides meaning in life and offers hope for the future. Religiosity is highest among the world's poorest countries because religion helps people cope with daily struggles to survive (Crabtree, 2010; WIN-Gallup International, 2013).
- *Well-being.* Around the world, actively religious people tend to report being happier. They also report drinking less alcohol and smoking less, both of which can be better for one's health (Marshall, 2019)
- *Social service.* Religious groups raise millions of dollars for the victims of natural disasters, distribute food and clothing, find shelter for displaced families, and help the poor and sick.
- *Social control.* Because many religious people fear punishment in the afterlife, they try to follow societal rules. Doing so suppresses deviant and antisocial behavior and promotes societal cohesion (Durkheim, 1961; Shariff and Aknin, 2014).

Religion Promotes Social Change

Religion can also spearhead social change. Mohandas Gandhi (1869–1948), a spiritual leader in India, worked for his country's independence from Great Britain through nonviolence and peaceful negotiations. In the United States, religious leaders, especially the Reverend Martin Luther King, Jr., were at the forefront of the civil rights movement during the late 1960s.

In 2000, Reform Judaism, which is the largest American Jewish denomination, made a step forward for marriage inclusivity when "the Central Conference of American Rabbis gave its full support to Reform rabbis who choose to officiate same-sex marriages" (Human Rights Campaign, 2023 "Stances of Faiths on LGBTQ Issues"). In 2011, after decades of debate, the Presbyterian Church, one of the nation's largest Protestant denominations, approved ordaining people with same-gender partners as ministers, elders, and deacons. Numerous Lutheran churches have also welcomed gay pastors. In 2014, and despite many of its congregations' threats of leaving, the Presbyterian Church voted to allow same-gender marriages. Other religious groups (e.g., United Church of Christ, Episcopal Church, Quakers, Reform and Conservative Judaism) have followed (Goodstein, 2014). Pope Francis created a tribunal to prosecute bishops who protect priests suspected of child sexual abuse. He has made it faster, easier, and cheaper for Roman Catholics to obtain marriage annulments and insists that priests welcome (rather than reject) divorced Catholics and their children.

Max Weber (1920/1958) asserted that religion sparks economic development. His study of Calvinism, a Christian sect that arose in Europe during the sixteenth century, led to his coining the term *Protestant*

John Gress/REUTERS

Megachurches are Christian congregations that have a regular weekly attendance of more than 2,000 and tend to be evangelical. They represent only 0.5 percent of all U.S. churches, but their number is growing and now at 1,750 (Bird and Thumma, 2011, 2020). From a functionalist perspective, why do you think megachurches are so popular?

ethic, a belief that hard work, diligence, self-denial, frugality, and economic success were signs of God's favor and that prosperity indicated divine blessing. According to Weber, the harder the early Calvinists worked and saved, the more likely they were to accumulate money, to become successful, and to drive the growth of capitalism.

Was Weber right about the relationship between the Protestant ethic and the rise of capitalism? The data are mixed. Some studies show that people are diligent not because of religious beliefs but simply because amassing savings provides resources when disaster strikes (as when droughts wipe out crops). On the other hand, a study of 59 high- and low-income countries found that religious beliefs that encourage hard work and thrift spur economic growth (Cohen, 2002; Barro and McCleary, 2003).

Is Religion Dysfunctional?

Functionalists emphasize its benefits but also recognize that religion can be dysfunctional when it harms individuals, communities, and societies. Religious intolerance can spark conflict between groups (e.g., vandalizing churches, mosques, and synagogues) and attacks on religious minorities. The United States has a long history, beginning with the Pilgrims in 1620, of people discriminating against other religious groups even though they themselves sought religious freedom (Davis, 2010).

The Roman Catholic Church and the Vatican protected predator priests for at least five decades (United Nations Committee on the Rights of the Child, 2014). Overall, religion meets important psychological needs, but strong group identity may breed narrow-mindedness, ethnocentrism, and bigotry toward nonmembers.

Critical Evaluation

One weakness is that functionalists, by emphasizing religion's benefits, imply that religion is indispensable to leading a good life. People who aren't religious do many good things for society, and some religious people commit heinous acts. In part because of priests' sexual abuse of young children, Americans' rating of the clergy's honesty and ethics fell from 61 percent in the late 1970s to 36 percent in 2019 (Saad, 2022 "Military Brass, Judges Among Professions at New Image Lows").

A second limitation is functionalists' exaggerating religion's positive effect on physical and emotional health. From survey data drawn from religiously active and non-active people, it is hard to say that people who are actively religious are healthier. Jewish people, who are one of the least religious U.S. groups, and religiously unaffiliated, including atheists, report higher well-being rates than do religious groups (Marshall, 2019; Newport et al., 2012; Leurent et al., 2014).

A third, and major, criticism is that functionalists gloss over numerous dysfunctional aspects of religion that generate wars, terrorism, and genocide. Conflict theory addresses these issues.

13-7b Conflict Theory: Religion Promotes Social Inequality

From a conflict perspective, religion promotes and reinforces social inequality. Throughout history and currently, religion has created discord and divisiveness within groups, between groups, and in the larger society.

"The Opium of the People"

Much conflict theory reflects the work of Karl Marx (1845/1972), who described religion as "the sigh of the oppressed creature" and "the opium of the people" because it encouraged passivity and acceptance of social and economic inequality. Marx viewed religion as a form of **false consciousness**, an acceptance of a system of beliefs that prevents people from protesting oppression. Contemporary conflict theorists don't view religion as an opiate, but they agree with Marx that religion is often used to justify violence and promotes social inequality.

Religion Justifies Intolerance and Violence

For thousands of years, many governments and religious leaders have condoned or perpetrated widespread violence in the name of religion. Religion promotes conflict when religious groups differentiate between "we" and "they" ("We're right and they're wrong," "Our God is 'the *real* God'"). Such self-righteousness condones aggression, oppression, and brutality within and across societies. For example:

- The "eternal war" between Sunnis and Shiites, two major Islamic sects, dates back more than 1,300 years, and there are long-standing conflicts and wars between Muslims and Jewish in Israel and Palestine (Crowley, 2014; "Religion: The New Strife," 2016).

---

false consciousness an acceptance of a system of beliefs that prevents people from protesting oppression.

- In Myanmar, Buddhist monks have incited bigotry and violence, mostly against Muslims, but also against women (Beech, 2013; BBC, 2015).
- In 2014 alone, religion-related terrorism and wars displaced almost 34 million people from their homes (Henne and Kishi, 2016).
- "Harassment of religious groups continues to be reported in more than 90% of countries" (Pew Research Center, 2020).

Religion Promotes Social Inequality

Conflict theorists view religion as a tool that dominant groups use to control society, protect their own interests, and derail social change. Roman Catholic bishops, for example, have opposed comprehensive health care because it might fund abortions, have publicly chastised pro-choice Catholic politicians, and have campaigned against same-gender marriage—all of which fuel social inequality and conflict (Doyle, 2011).

Private evangelical schools, colleges, and universities receive massive state and federal financial aid that diverts tax revenues from public schools and colleges. Many of the schools' textbooks describe LGBTQIA+ individuals and abortion rights supporters as "evil," and are hostile toward other religions, including nonevangelical Protestants, Jewish, and Catholics (Berkowitz, 2011; Tabachnick, 2011).

In some southern states, conservatives maintain that "religious freedom" should allow businesses to refuse services to LGBTQIA+ people (Bruinius, 2016). Doing so denies people basic civil liberties, seals them off from society, and creates resentment.

Critical Evaluation

Functionalists may overemphasize consensus and harmony, but conflict theorists, according to some critics, often ignore religion's role in challenging exclusionary policies. For example, more than 800 churches and synagogues, while also fighting deportation battles in courts, have mobilized to provide sanctuary for law-abiding undocumented immigrants and their U.S.-born children (Rodgers, 2017).

A second criticism is that conflict theorists attach too much credence to Marx's concept of false consciousness. Instead of passively submitting to religious oppression, many people revolt, complain, or resist. Many have sued employers, sometimes successfully, who denied requests to not work on holy days, and whose secular dress codes don't allow head scarves, turbans, or long beards (Trottman, 2013). Whether religious discrimination is real or imagined, many Americans protest, demonstrate, and seek legal remedies.

13-7c Feminist Theories: Religion Subordinates and Excludes Women

Feminist scholars agree with conflict theorists that religion can foster violence and inequality. They go further, however, by criticizing organized religions as sexist, patriarchal, and shutting women out of leadership positions.

Sexism, Patriarchy, and Subordination

"Religion is probably the most important force in continuing the oppression of women worldwide," Amanda Marcotte, a feminist blogger, writes, and "religiosity and sexism go hand in hand" (Marcotte, 2014). From a feminist perspective, most religions are patriarchal and support heteronormativity: They emphasize men's experiences and a male point of view, view women as subordinate to men, and prioritize opposite-sex relationships as the norm. According to the apostle Paul, "Wives should submit to their husbands in everything" (*Ephesians* 5: 24). The idea that Eve was created out of Adam's rib is often used to justify men's domination of women. Many religious teachings and institutions continue to propagate such beliefs (refer to Paludi and Ellens, 2016).

In Orthodox Judaism, a man's daily prayers include this line: "Blessed art thou, O Lord, our God, King of the Universe, that I was not born a woman." The Qur'an tells Muslims that men are in charge of women and that women should obey men. Almost all contemporary religions worship a male deity, and none of the major world religions treat people of all genders equally (Gross, 1996; Jeffreys, 2011).

Alessandro Albert/Getty Images

Dr. Amina Wadud is an American Muslim scholar who has led prayer as an imam.

Feminist scholars offer alternatives to patriarchal interpretations of Scripture. Some contend that Jesus was a feminist. For example, he defended women against men, encouraged women's intellectual pursuits when it wasn't the norm, and there are numerous passages in the Bible about women spreading Jesus' teachings (Gross, 1996; Ferguson, 2014).

Muslim feminists, similarly, note that men have interpreted sacred Islamic texts to ensure dominance and control by men. Women were among some of Muhammad's earliest converts, and the Qur'an has numerous passages that establish women's equal rights in inheritance and family roles (Menissi, 1991, 1996). Thus, women's subordination in many Islamic societies isn't due to religious tenets but to men's interpretations of Scripture to maintain their power and privilege (Smith, 1994).

In the Middle East and North Africa, 35 percent of the countries have religious police who, among other things, enforce strict segregation of the sexes and punish women who are perceived as behaving or dressing improperly (Theodorou and Henne, 2014). Thus, it's women's fault if men have lustful thoughts or assault women.

Women and Religious Leadership

In 2014, women earned 33 percent of theology degrees, a dramatic increase from only 2 percent in 1970. Still, women make up only 21 percent of the nation's clergy, lead only 11 percent of congregations, and earn 27 percent less than men (Chaves and Eagle, 2015; Emmert, 2015; Snyder et al., 2016).

Many women have active roles in their religious organizations, but they may face gender-based limitations in their participation. For instance, Pope Francis, leader of the Catholic Church, recently validated women serving as acolytes and lectors but affirmed that only men can serve as ordained priests (Whitehead, 2022).

In Islam, there are different perspectives on women serving in leadership positions as imams. Some followers of Islam contend that the Qur'an excludes women from leading prayers as imams. Others feel women can lead prayers but only for other women, not for men.

Critical Evaluation

Feminist perspectives on religion are limited in several ways. First, some feminist Muslim scholars have criticized Western feminists for misreading Islamic and other sacred scriptures and reducing practically all discussions of gender to the *hijab* (a veil or scarf that Muslim women wear) instead of focusing on justice for both women and men in marriage, employment, and other areas (Fakhraie, 2009; Daneshpour, 2015).

Second, religious women aren't as oppressed as many feminist scholars claim. For instance, the Qur'an does not specifically exclude women from being imams, but instead, fewer women may be imams because of gender-biased social influences and male-biased interpretation of the Qur'an. And an increasing number of women are imams in various countries such as India, the United States, France, and China (Borges, 2021; News Desk, 2021; Steinkopf-Frank, 2020). Muslim women do provide religious leadership in many ways "teaching in Islamic schools and universities, leading Quran study circles, preaching and otherwise providing religious guidance to the faithful" (Fam and Khan, 2021). U.S. Catholic women contribute billions each year during Sunday Masses, serve as nuns, and are directors of church initiatives for programs including young adult ministry, finances, and music (Miller, 2010; Mastrangelo, 2022). Even in the most conservative religions, women don't blindly submit to religious dogma. Instead, they may challenge existing doctrines, stop volunteering at church, don't attend weekly services, and establish their own parishes (Padgett, 2010; Miller, 2012).

13-7d Symbolic Interaction: Religion Is Socially Constructed

For symbolic interactionists, religion isn't innate but socially constructed. As a result, people can learn and interpret the same religion differently across cultures and over time. Symbols, rituals, and beliefs are three of the most common means of learning and internalizing religion.

Symbols

A *symbol* is anything that stands for or represents something else to which people attach meaning (refer to Chapter 3). Many religious symbols are objects (a cross, a steeple, a Bible), but also include behaviors (kneeling or bowing one's head), words ("Holy Father," "Allah,"

Marcel Mettelsiefen/Picture Alliance/Berlin Berlin Germany

"the Prophet"), and physical appearance (e.g., wearing head scarves, skull caps, turbans, clerical collars).

Religious symbols, like all symbols, are shorthand communication tools. Some interactionists define a religion as "a system of symbols" because it's a community that's unified by its symbols (Berger and Luckmann, 1966; Geertz, 1966). Religious symbols can also be divisive (e.g., a Ten Commandments monument on government property that alienates non-Christians and nonbelievers).

Rituals

A *ritual* (sometimes called a *rite*) is a formal and repeated behavior that unites people (see Chapter 3). Religious rituals, like secular ones, strengthen a participant's self-identity (Reiss, 2004). Religious rites of passage—*bat mitzvah* for girls and *bar mitzvah* for boys in the Jewish community, and first communion and confirmation for Catholic children—reinforce the individual's sense of belonging to a particular religious group. A group's rituals symbolize its spiritual beliefs and include a wide range of practices—praying, chanting, fasting, singing, dancing, and offering sacrifices.

All religions have rituals that mark significant life events like birth and marriage (refer to Chapters 3, 5, and 12). Death rituals are probably the most elaborate and sacred worldwide. They vary across religious groups and societies, but all of them comfort the living and show respect for the dead.

Beliefs

Rituals and symbols come from *beliefs*, convictions about what people think is true. Religious beliefs can be passive (believing in God but never attending services) or active (participating in rituals and ceremonies). Beliefs bind people together into a spiritual community.

One of the strongest beliefs worldwide is that prayer is important. Islam requires prayer five times a day. In the United States, 55 percent of U.S. adults say that they pray every day, and for a variety of reasons, including feeling close to God, as well as requesting better health, more money, and cures for sick pets. Even 20 percent of the "nones" pray every day (Wicker, 2009; Lipka, 2016). Prayer offers psychological and spiritual benefits such as comfort and a sense of unity among those who pray together, but depending on God (or a universal spirit) can also diminish people's motivation to actively shape their own lives (Schieman, 2010).

Critical Evaluation

A common criticism is that interactionists' focus on micro-level behavior ignores the ways that religion promotes social inequality at the macro level. Conflict theorists and feminist scholars, especially, maintain that people often use religion to justify violence and women's subordination.

Some critics also wonder if interactionists paint too rosy a picture of religion, even on a micro level, because people's beliefs can wreak considerable havoc. The Islamic zealots who targeted the World Trade Center on 9/11 didn't see themselves as "terrorists" but as "true believers": They were carrying out "God's will" by imposing their religious beliefs on others or destroying their "religious enemies" (Juergensmeyer, 2003).

Chapter Review

Key Takeaways

13-1 How have education and schooling changed in the United States? U.S. *education* and *schooling* have changed in four important ways: Universal education has expanded, community colleges have flourished, public higher education has burgeoned, and student diversity has greatly increased.

13-2 What are the key components of the different sociological perspectives on education? Refer to Table 13.1 on page 277.

13-3 What are some current issues in U.S. education? Compared with their counterparts in a number of other countries, many U.S. students are performing poorly in elementary and high schools—especially in mathematics and the sciences. U.S. teachers' salaries are low, many are poorly prepared, out of field, and have less control over curricula than ever before. A growing number of critics contend that standardized tests (such as SATs and ACTs) are little more than gatekeeping tools that exclude lower socioeconomic students from access to higher education, and that the tests scores don't predict college success. Despite grade inflation, high school and college dropout rates are high, and cheating is widespread.

13-4 **What is the difference between religion, religiosity, and spirituality?** Some form of religion exists in all societies and cultures. *Religion* unites believers into a community. Every known society distinguishes between *sacred* and *profane (secular)* activities. Religion, *religiosity*, and spirituality differ; for example, people who describe themselves as religious may not attend services.

Example: Sacred vs. Secular

The corporate mission of Chick-fil-A (a franchise that prepares sandwiches), as stated on a plaque at company headquarters, is "to glorify God." Chick-fil-A is the only national fast-food chain that closes on Sunday so employees can go to church, and prospective employees are asked about their religious activities. Some franchise operators are delighted with the religious emphasis; others believe that a business should stay out of its workers' personal lives (Schmall, 2007). What do *you* think?

13-5 **What are the different types of religious organizations? What are some of the major world religions?** People express their religious beliefs most commonly through organized groups, including *cults* (also called *new religious movements [NRMs]*), *sects*, *denominations*, and *churches*. Some NRMs, which usually are organized around a *charismatic leader*, have been short-lived, whereas others have become established religions with highly organized bureaucracies. Worldwide, the largest religious group is Christians, followed by Muslims, but no religious group comes close to being a global majority. The third largest group is people who don't have a religious affiliation.

13-6 **How is religion in the United States changing? How is it staying the same?** Religion in the United States is complex and diverse. About half of U.S. adults have changed their religion since childhood, many opting for no religion at all. Mainline Protestant groups have declined, whereas evangelicals have surged. Religious participation varies by gender, age, race, ethnicity, and social class. Some sociologists maintain that *secularization* is increasing rapidly in the United States; others contend that this claim is greatly exaggerated, especially as witnessed by the growth of *fundamentalism* and the prevalence of *civil religion*.

13-7 **What are the key components of the different sociological perspectives on religion?** Refer to Table 13.4 on page 296.

Key Terms

Education (p. 276)
Schooling (p. 276)
Achievement gap (p. 278)
Hidden curriculum (p. 279)
Credentialism (p. 279)
Tracking (p. 282)
Implicit bias (p. 283)
Meritocracy (p. 287)
Religion (p. 290)
Sacred (p. 290)
Profane (p. 290)
Religiosity (p. 290)
Cult (p. 290)
New religious movement (NRM) (p. 290)
Charismatic leader (p. 290)
Sect (p. 291)
Denomination (p. 291)
Church (p. 291)
Ecclesia (also called a *state religion*) (p. 291)
Secularization (p. 294)
Fundamentalism (p. 295)
Civil religion (sometimes called *secular religion*) (p. 296)
False consciousness (p. 298)

Discussion and Critical Thinking Questions

1. In a classic experiment, two researchers randomly selected 20 percent of the students at a grade school and told teachers that these children would probably be high achievers. Because the teachers saw these students as smarter than others, they treated them differently, such as giving them more chances to participate in class, praising them more often, and criticizing them less frequently. At the end of the school year, the students who had been singled out as potential achievers scored higher on IQ tests. The researchers concluded that the higher scores were due largely to teachers' labeling and differential treatment. More recent studies have found similar results of this self-fulfilling prophecy (Rosenthal and Jacobson, 1968; "Study Shows Positive Results . . . ," 2002).
   - **a.** Why did some scholars criticize this study as unethical?
   - **b.** How can self-fulfilling prophecies have positive or negative effects in high school and college?
2. How are the social institutions of education and religion similar? How do they differ?
3. Should colleges use only GPAs in admission decisions? If so, are GPAs less biased than standardized tests like SATs and ACTs in measuring a student's academic abilities?
4. In at least most states, the highest-paid public employees are college football and basketball coaches, not governors, college presidents, or faculty members. Most of these highly paid coaches make over $1 million per year (Stebbins, 2020). Should colleges slash the coaches' salaries and use the money to decrease student loan debts? Why or why not?
5. If religious organizations are involved in politics, should they pay state and federal taxes?

True or False?

How Much Do You Know about Education and Religion?

The answers to 1 and 4 are true. The answers to 2, 3, 5, and 6 are false.

1. Women now earn 58 percent of bachelor's degrees and a greater percentage of master's and doctorate's as well.
2. More teachers are leaving the profession. There are many reasons for this, including low pay and teachers' decreasing job satisfaction.
3. From 1995 to 2019, the average student loan debt at graduation increased by 70 percent.
4. Levels of religiosity are decreasing in the United States. Fewer people indicate that religion is very important in their lives, and fewer people are attending religious services.
5. Islam is the only major world religion that is expected to increase. By 2050, Islam is expected to have a greater number of followers than Christianity.
6. Within the U.S., religious influences are present in public educational settings. This includes support for teaching religiously based creationism versus evolution in some science classes.

14 Health and Medicine

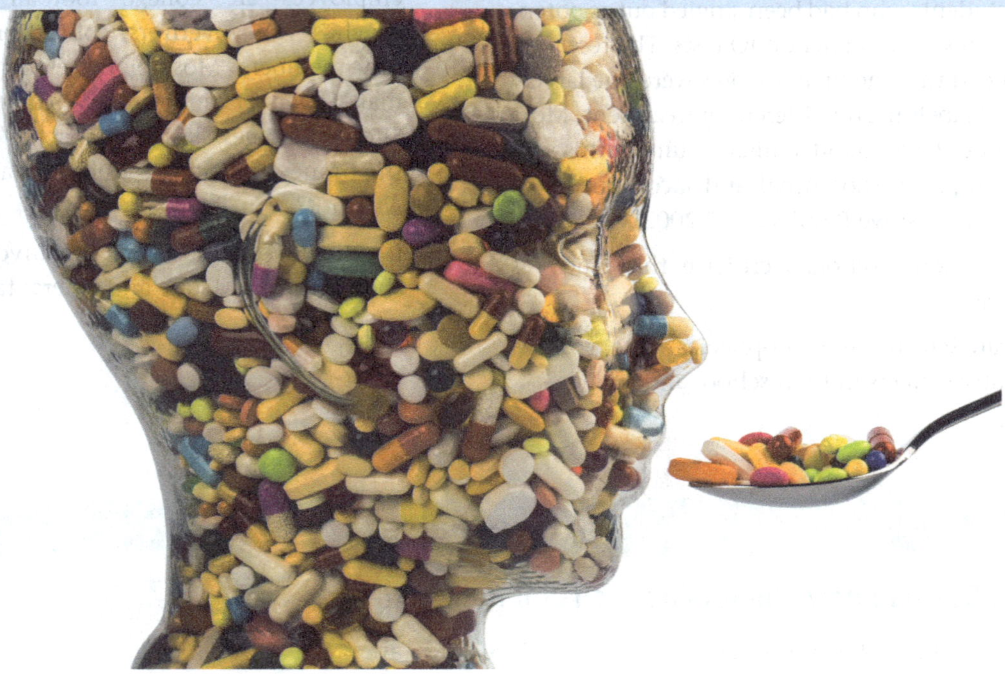

Lisa-S/Shutterstock

Learning Objectives

After studying this chapter, you will be able to…

14-1 Describe social epidemiology and related global health gaps.

14-2 Explain how and why environmental, demographic, and lifestyle factors affect U.S. health and illness.

14-3 Compare U.S. with other countries' health care systems in terms of access, costs, and outcomes.

14-4 Evaluate the theoretical explanations of health and medicine.

True or False?

How Much Do You Know about Health and Medicine?

1. The rate of uninsured people in the United States is increasing.
2. Lower income nations have lower life expectancy rates than higher income nations.
3. The percentage of Americans taking three or more prescriptions has increased.
4. Hereditary (genetic) factors are the sole determinant of whether a person will develop a chronic disease.
5. The U.S. spends more per person on healthcare compared with other high-income nations, so it has higher quality healthcare.
6. Race affects people's experiences with health, but social class does not.

The answers are at the end of the chapter.

Most of us would probably agree with American poet Ralph Waldo Emerson, who wrote, in 1860, "Health is the first wealth." **Health** is the state of physical, mental, and social well-being. We can usually determine physical health by using objective measures like cholesterol levels and blood pressure. Mental and social well-being are more difficult to gauge because they rely on subjective definitions that change over time.

Be sure to take the True or False quiz in the chapter opener to better understand how much you know about these topics.

What do you think?

The government should be responsible for making sure that all Americans have health care.

1	2	3	4	5	6	7
	strongly agree				strongly disagree	

14-1 Global Health and Illness

How does health differ around the world? And why? Social epidemiology addresses both questions.

14-1a Social Epidemiology

All people experience *disease,* a disorder that impairs a person's normal physical and/or mental condition. **Social epidemiology** examines how societal factors affect the distribution of disease within a population. Why, for example, do people live much longer in some countries than in others?

In answering this and other questions, epidemiologists look at two factors. One is *incidence,* the number of new cases of a health problem that occur in a given population during a given time period (e.g., in 2019, there were 1.4 million new diabetes cases among Americans aged 18 and older). The other measure is *prevalence,* the total number of cases (extent) of an illness or health problem within a population or at a particular point in time (e.g., in 2019, 11.3 percent of the U.S. population had diabetes) (Centers for Disease Control and Prevention, 2022). Epidemiological studies show large health differences across countries.

14-1b Global Health Disparities

The World Bank classifies countries into four broad groups based on gross national income (GNI) per capita. **Figure 14.1** shows the number and location of these countries and compares them on life expectancy, infant mortality, and per capita health expenditures. These three variables are among the most frequently used to measure health status. They indicate a population's living standards, people's average socioeconomic status, and the quality and financial resources that a country devotes to **health care**, the prevention, treatment, and management of illness.

The high-income countries make up only 16 percent of the world's population, but have the greatest access to clean water, sanitation, food, and health services. In contrast: more than 1.7 billion people worldwide lack basic sanitation; it's estimated that 10 percent of the world's people consume food irrigated by wastewater; and almost half are at risk of dying of malaria (World Health Organization, 2022 "Sanitation"; World Health Organization, 2023 "Malaria").

High income populations live longer, have lower infant mortality rates, and spend more on health (**Figure 14.1**), but experience "diseases of wealth" (e.g., diabetes, heart disease, various cancers). Compared with 10 other high-income countries, the United States has a large and growing "health disadvantage" and comes in last overall on rankings that use "71 performance measures across five domains — access to care, care process, administrative efficiency, equity, and health care outcomes" (Woolf and Aron, 2013; Luhby, 2021; Schneider et al., 2021).

14-2 Health and Illness in the United States

Why do Americans have a health disadvantage, including high disability rates? Compared with other high-income nations, non-medical social determinants contribute to large health disparities in the United States.

health the state of physical, mental, and social well-being.

social epidemiology examines how societal factors affect the distribution of disease within a population.

health care the prevention, management, and treatment of illness.

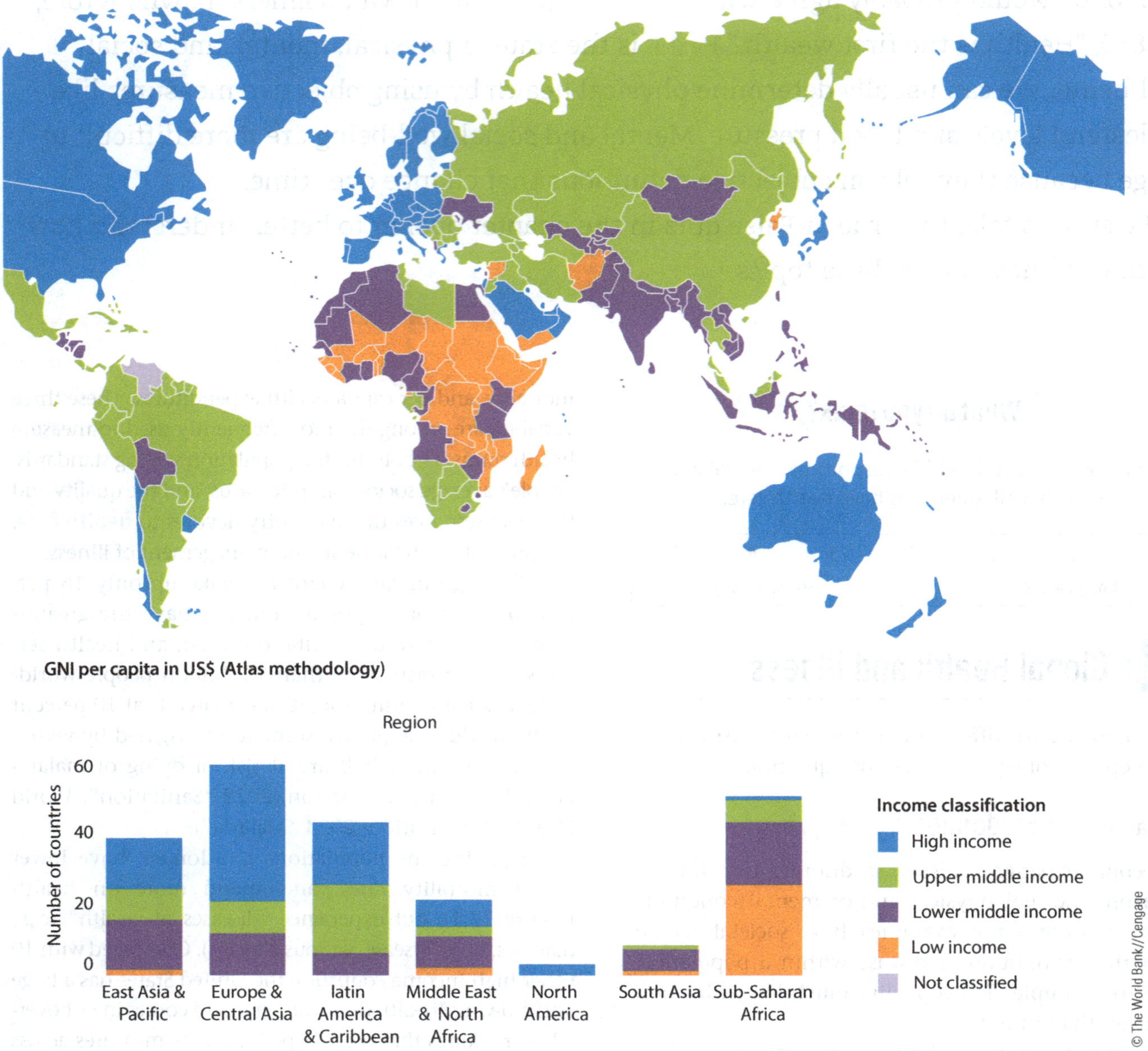

	GNI (2021)	Life Expectancy at Birth (2021)	Infant Deaths per 1,000 Live Births (2021)	Per Person Total Expenditure on Health (2020)
Low	<$1,085	62 years	47	$37
Lower middle	$1,086–4,255	67 years	33	$99
Upper middle	$4,256–13,205	75 years	10	$570
High	>$13,205	80 years	4	$6,180

Figure 14.1 Worldwide Health Gaps

Sources: Hamadeh et al., 2022; The World Bank "Life Expectancy at Birth," 2023; The World Bank, "Mortality Rate, Infant (per 1,000 live births," 2023; The World Bank, "Current Health Expenditure per Capita (Current U.S Dollars)," 2023.

14-2a Understanding Disability

A **disability** is any physical or mental impairment, temporary or permanent, that limits a person's ability to perform a basic life activity. These activities include walking; doing errands alone; bathing, dressing, or feeding oneself; and working.

Almost a third of Americans age 18 and over have a disability. Age is a major predictor of disability, affecting 58 percent of adults age 65 and older compared with 25 percent of those ages 18 to 64. Disability also increases with age—from 26 percent for people age 65 to 74 to 73 percent for those age 85 and older. Besides age, disability rates are higher for women (36 percent) than men (27 percent), for people living below the poverty line (41 percent versus 23 percent in other socioeconomic categories), and are highest among Native Americans (41 percent) (National Center for Health Statistics, 2017).

Four percent of American children have a disability (Young, 2021). Yet, one in four Americans have a disability that impacts "major life activities" which represents 61 million adults (Centers for Disease Control and Prevention, 2018). Why do so many Americans have disabilities? First, as you'll learn shortly, behavior previously considered normal has been medicalized. Second, medical advances help many people survive diseases, car accidents, and wars, but they may have lifelong health problems. Third, as the proportion of the population age 65 and older increases, so does the proportion living with disabilities.

14-2b Social Determinants of Health and Illness

No single variable explains well-being. Hereditary factors affect the likelihood of developing specific types of cancers. For example, a biological parent can pass a mutated BRCA1 or BRCA2 gene down to their children. By the age of 70, a person with a mutated BRCA2 gene has a 45 percent chance of developing breast cancer, a person with a mutated BRCA1 has a 55 to 65 percent chance, and person with neither has only a 12 percent chance (National Cancer Institute, 2020). Environmental factors play a role in cancer as well. Avoiding risk factors and using evidence-based prevention strategies can reduce cancer numbers by 30 to 50 percent. Yet, people may not always have control over their exposure to risk factors. For instance, in low- and middle-income countries, "Cancer-causing infections, … are responsible for approximately 30% of cancer cases," and comprehensive cancer treatment is available in 90 percent of high-income countries, but in less than 15 percent of low income (World Health Organization, 2022 "Cancer"). Whether it's cancer or other diseases, environmental, demographic, and lifestyle factors—all of which are interrelated—have a significant impact on health and illness.

Christophe Calais/Corbis Historical/Getty Images

Advances in battlefield medicine and body armor mean that many soldiers are surviving injuries, but they have shattered bones, brain damage, and amputated limbs, and may suffer from posttraumatic stress disorder (PTSD) or other mental health problems. Among post-9/11 veterans, 34 percent have a service-connected disability (U.S. Department of Veterans Affairs, 2017).

Environmental Factors

Many environmental hazards affect our bodies. For example:

- There's a strong association between early-life exposure to air pollution and autism, schizophrenia, and learning disabilities (Roberts et al., 2013; Allen et al., 2014; Imbriani et al., 2021).
- Thousands of children in Flint, Michigan—a predominantly low-income, Black city—are expected to suffer irreversible developmental problems because of lead contamination in the water supply (Milman and Glenza, 2016).
- Many toxic chemicals in food and everyday products (e.g., detergents, shampoos, shaving products, and makeup) have been linked to asthma, some types of cancer, birth defects, and children's impaired brain development (Blake, 2014; Grandjean and Landrigan, 2014; Attina et al., 2016).
- Norovirus—a very contagious virus—is the leading cause of disease outbreaks from contaminated food. About 20 million Americans get sick from norovirus each year, 110,000 are hospitalized, and about 900 die. Outbreaks on cruise ships account for only 1 percent of all norovirus cases; 70 percent of cases are due to food

---

disability any physical or mental impairment that limits a person's ability to perform a basic life activity.

workers who are infected, don't wash their hands, or go to work when sick (Hall et al., 2014; Varinsky and Ede-Osifo, 2021).

Access to health care can prolong life, but an estimated 250,000 Americans die each year because of preventable medical errors (e.g., doctors operating on the wrong patient or body part, patient infections). Medication errors—misdiagnosing an illness, prescribing the wrong drugs, or giving patients drugs that interact dangerously—injure about 1.3 million Americans and kill about 7,000 every year. Moreover, each year, thousands of patients leave operating rooms with surgical sponges in their bodies that cause infections, lifetime digestive problems, removal of intestines, and even death (Eisler, 2013; Laliberte, 2016; Makary and Daniel, 2016).

Antibiotic resistance is a growing problem. It causes over 2.8 million infections and an estimated 35,000 deaths in the United States each year. Bacteria have become more resistant because of the overuse of antibiotics: People take them for illnesses for which antibiotics aren't effective, and farmers feed them to animals to promote growth. Less attention has been placed on developing new antibiotics, largely because pharmaceutical companies reap much higher profits by developing very expensive and specialized medicines for small populations with chronic conditions, including multiple sclerosis, cancer, HIV, and hepatitis C. Recently, a new antibiotic was created, using artificial intelligence, that is showing promising signs that it can ward off drug-resistance bacteria ("The Drugs Don't Work," 2014, p. 54; Tozzi, 2014; Centers for Disease Control and Prevention, 2021 "National Infection and Death Estimates"; Goodman, 2023).

Demographic Factors

Religion, family size, marital status, urban-rural residence, and other variables affect health. Four of the most important demographic factors are age, gender, social class, and race and ethnicity.

Age. Not surprisingly, age is the single best predictor of illness and death. Death rates drop sharply shortly after birth, begin to rise at about age 45, and escalate, particularly after age 78.

From birth through elementary school, children have health problems over which they have no control. One concern is *Fetal alcohol spectrum disorders (FASD)* which includes a range of permanent birth defects caused by a pregnant person drinking alcohol during pregnancy. It is difficult to know the full impact of FADS on the population. It is estimated that 1 out of 1,000 babies born in the United States have FADs, yet other studies indicate that 1 to 5 percent of school age children show signs of these conditions. The disorders include intellectual infirmities, speech and language delays, poor social skills, mental disabilities, and physical abnormalities like congenital heart defects (National Organization on Fetal Alcohol Syndrome, 2012; Weinhold, 2012; Centers for Disease Control and Prevention, 2023, "Data & Statistics on FASDs").

Many adolescents have health problems because of their own choices which may be socially influenced. This includes smoking, vaping, misusing alcohol and other substances, and texting while driving. Such risky behavior may be due to peer pressure, to the ongoing development of some parts of the teen brain, and to increasing independence as adolescents break away from parental supervision (refer to Chapter 4).

Health problems emerge and increase during one's late thirties because of genes, lifestyle choices (which we'll address shortly), and because physical decline is normal and inevitable as we age. No matter how well tuned we keep our bodies, the parts start wearing down: Reflexes slow, hearing and eyesight dim, and stamina and muscle strength decrease.

After age 65, chronic rather than acute diseases comprise the majority of health problems, including disability. **Chronic diseases** (e.g., asthma and high blood pressure) are long-term or lifelong illnesses that develop gradually or are present from birth. In contrast, **acute diseases** (e.g., chicken pox) are illnesses that strike suddenly and often disappear rapidly but can cause incapacitation and sometimes death. Chronic diseases increase as people age (**Figure 14.2**). *Dementia*, a loss of mental abilities, most commonly occurs during one's 70s. The most debilitating form of dementia is **Alzheimer's disease**, a progressive, degenerative disorder that attacks the brain and impairs memory, thinking, and behavior.

Gender. The gender gap in life expectancy has decreased since 1975, but women, on average, still live longer than men. Gender roles affect overall life expectancy, for instance the expectation that boys and men should be strong (refer to Chapter 9). Men of all racial-ethnic groups are two to three times more likely than women to die in motor vehicle crashes, to be victims of homicide, to smoke and drink alcohol, to misuse drugs, and to work in dangerous occupations (e.g., construction, law enforcement). Women are also more likely than men to have a regular physician, to consult a doctor when something's

---

chronic diseases long-term or lifelong illnesses that develop gradually or are present from birth.

acute diseases illnesses that strike suddenly and often disappear rapidly.

Alzheimer's disease a progressive, degenerative disorder that attacks the brain and impairs memory, thinking, and behavior.

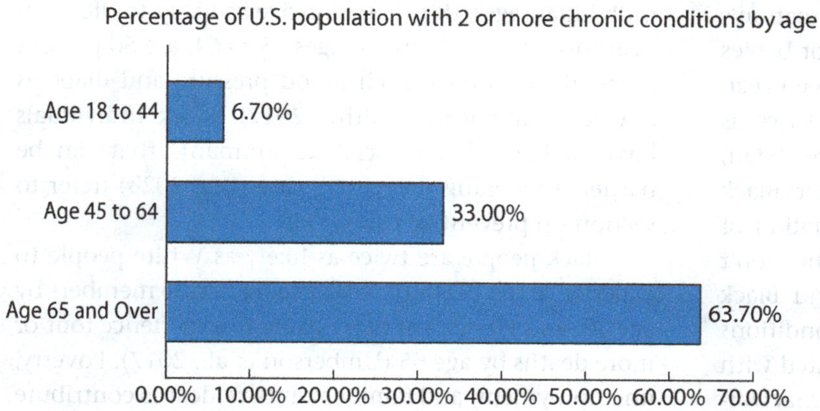

Figure 14.2 Chronic Illness Increases with Age

Source: Based on Boersma et al.. 2020. "Prevalence of Multiple Chronic Conditions Among US Adults, 2018"

wrong, and to keep medical appointments (Levine, 2014; National Center for Health Statistics, 2017).

Depression is the single most common mental health problem among adolescents, begins at about age 12, and affects 30 percent of U.S. high school students. Girls (20 percent) are more likely than boys (9 percent) to experience depression, but the reasons are similar: substance misuse, bullying (either as a victim or perpetrator), sexual or physical abuse, parental divorce, and a family history of mental disorders. In adulthood, and across all ages, women are more likely than men to be taking antidepressant medications (Murphey et al., 2013; Federal Interagency Forum on Child and Family Statistics, 2017).

Depression may lead to suicide. Among high school students, adolescent girls are twice as likely as adolescent boys to attempt suicide. Across all ages, however, boys and men are four times more likely than girls and women to actually die by suicide, and the highest rates are among White men age 75 and older (Curtin et al., 2016).

Social Class. Social class has a strong effect on health. The wealthiest Americans live 10 to 15 years longer than those with the lowest incomes. As early as age 45, the lower the family income, the greater the likelihood that adults will experience two or more chronic diseases. Living in a low-income neighborhood increases the likelihood of stress (which affects diet, smoking, and alcohol/drug usage), a sedentary lifestyle because of limited recreational facilities, and less access to health care (Beckles and Truman, 2013; Hummer and Hernandez, 2013; Dickman and Woolhandler, 2017).

Policymakers have tried to eliminate *food deserts*, geographic areas where access to fresh, healthy, and affordable food is limited or nonexistent because grocery stores are too far away. Social class disparities persist even when consumers have access to the same food, however. Supermarkets, compared with convenience stores, offer more healthy foods. Because low-income workers might make long commutes or have several jobs to support their families, they are more likely to require foods that are based on convenience than those that require preparation ("Dietary Inequality," 2016; Zagorsky and Smith, 2017).

People in lower social classes are also more likely to have dangerous jobs. In 2021, there were 5,190 job-related deaths. Among driving-based jobs (such as sales and trucking) there were 1,032 deaths and among construction and extraction jobs there were 951. Almost 2.6 million people experienced non-fatal workplace injuries or illnesses caused by factors such as falls, overexertion, and accidents using equipment (Bureau of Labor Statistics, 2022 "National Census of Fatal Occupational Injuries in 2021"; Bureau of Labor Statistics, 2022 "Employer-Reported Workplace Injuries and Illnesses, 2021").

Gender roles in many societies encourage boys and men to engage in risk-taking activities and behaviors which can affect their life expectancy.

Race and Ethnicity. A good measure of a population's health is its *infant mortality rate*, the number of babies under age 1 who die per 1,000 live births in a given year. Infants born to Black women are more than twice as likely to die before age 1 as those born to White, Asian, and Latina women (**Figure 14.3**). The higher Black infant mortality rate may be due to a combination of factors: More babies born before 24 weeks who don't survive, less access to quality health care, and Black mothers' greater likelihood of having some conditions (high blood pressure, diabetes) that are associated with negative birth outcomes (Chen et al., 2014; MacDorman et al., 2014). Black women also have higher maternal mortality rates: 69 per every 100,000 pregnancies compared to White women with 26. This is affected by Black women's experiences with health care including reports of providers who "did not or were slow to listen to patients," and much higher rates of maltreatment during pregnancy (including verbal mistreatment or denying help) (Hill et al., 2022).

Even when infants survive and thrive, race and ethnicity affect lifespan. You learned in Chapter 12 that Black individuals have lower life expectancy rates than Asian, Latina/o/x, and White individuals. The death rate for Black people has declined about 25 percent since 1999. Compared with White people, however, Black individuals ages 18 to 49 are twice as likely to die from heart disease and, between ages 35 to 64, are 50 percent more likely to have high blood pressure and diabetes ("African American Health," 2017). Black individuals have higher risk for social determinants that can be barriers to seeking preventive care (Lea, 2023) (refer to section on preventive care).

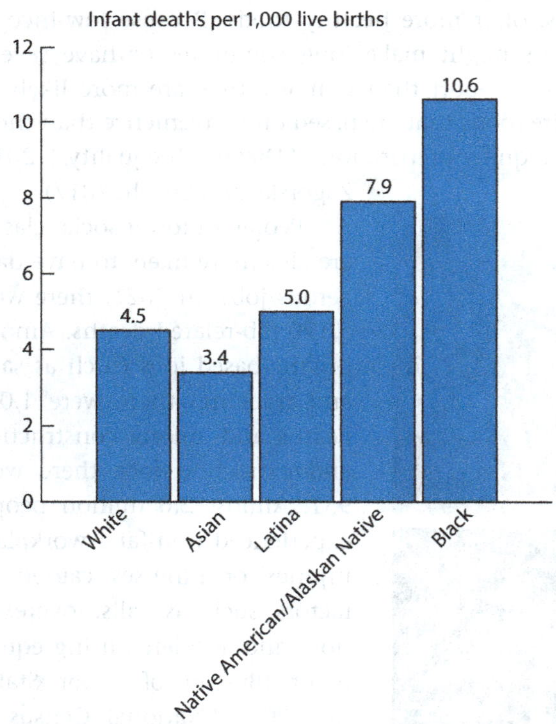

Figure 14.3 U.S. Infant Mortality Rates by Mother's Race and Ethnicity, 2019

Source: Based on Center for Disease Control and Prevention, 2019, "User Guide to the 2019 Period/2018 Cohort Linked Birth/Infant Death Public Use File."

Black people are twice as likely as White people to experience the death of two or more family members by age 30, and 90 percent more likely to experience four or more deaths by age 65 (Umberson et al., 2017). Poverty, unemployment, and family members' deaths contribute to lower quality health and lower life expectancy.

Despite generally lower education attainment and income levels, Latina/o/xs have higher life expectancy rates than Black or White individuals. One reason may be that only the healthiest individuals migrate to the United States. Another may be that, compared with other racial and ethnic groups, Latina/o/xs are the least likely to engage in harmful behavior or to die from drug misuse. The longer Latina/o/xs live in the United States, however, the higher their rates of heart disease, high blood pressure, some types of cancer, and diabetes. Acculturation may result in improved access to health care, but also to adopting harmful behaviors (e.g., smoking, excessive alcohol consumption), less physical activity, and less healthy diets (Dominguez et al., 2015; Riosmena et al., 2015).

Lifestyle Factors

Lifestyle choices can impair our health or improve it by reducing likelihood of death from illness and disease. Among the top ten causes of death are heart disease, COVID-19, unintentional injury, and diabetes which we'll cover in this section (Centers for Disease Control and Prevention, 2023 "Deaths and Mortality"). What lifestyle choices can reduce the likelihood of developing these health conditions? Activities that can prevent chronic diseases include: quit smoking, eat healthy and get regular exercise, stay up-to-date on preventive care, and curb substance misuse (Centers for Disease Control and Prevention, 2023 "How You Can Prevent Chronic Diseases").

Smoking. Worldwide and in the United States, tobacco use, primarily cigarette smoking, is the leading cause of preventable disease, disability, and death. U.S. smoking has declined from 43 percent of adults in 1965 to 14 percent in 2018 (American Lung Association, 2023). Who is more likely to use a tobacco product (including cigarettes, cigars, e-cigarettes, and chewing tobacco)? Tobacco use is higher among men than women and decreases as education levels increase (**Figure 14.4**). Smoking rates among people living below the poverty line are nearly twice as high as among those above

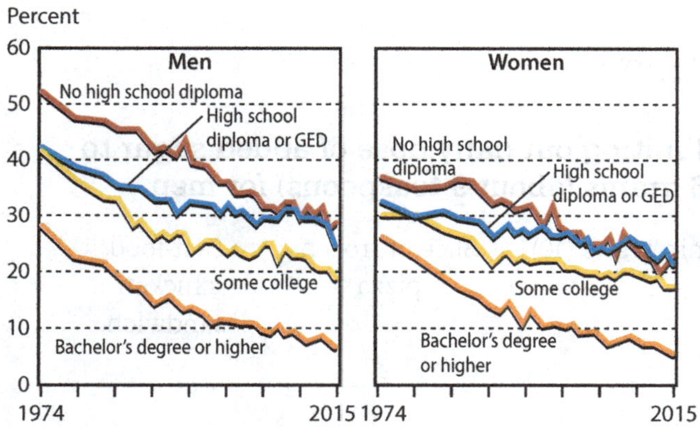

Note: The data are based on adults aged 25 years and over.

Figure 14.4 Cigarette Smoking Has Decreased, but Varies by Gender and Education Level

Source: National Center for Health Statistics, 2017, Figure 10.

the poverty line (Centers for Disease Control and Prevention, 2019).

There are now fewer people who smoke compared to in the past. Still, tobacco use is responsible for about 1 in 5 deaths annually, and, on average, people who smoke die about 10 years earlier than those who do not smoke. Because tobacco harms nearly every human organ, smoking is linked to cancer, heart disease, stroke, Type 2 diabetes, macular degeneration, erectile dysfunction, lung diseases (including emphysema and bronchitis), and birth defects (U.S. Department of Health and Human Services, 2014).

Prevention is difficult because the tobacco industry spends almost $10 billion a year to market its products; the number of smoking scenes among the top-grossing children's movies rose from 564 in 2010 to 809 in 2016, a 43 percent increase; and 69 percent of middle and high school students are exposed to electronic cigarette ads in retail stores, on the Internet, in magazines or newspapers, on TV, or in the movies (Singh et al., 2016; Tynan et al., 2017). During 2016, states collected almost $26 billion from tobacco taxes and legal settlements but spent less than 2 percent of the money on tobacco control programs (Jamal et al., 2016).

Diego Cervo/Shutterstock.com

From 2013 to 2016, the number of middle and high school students using electronic cigarettes almost tripled, rising from 780,000 to 2.2 million students (Jamal et al., 2017). To attract adolescents, the e-cigarette industry uses celebrities, cartoon characters, and "kid-friendly" flavors like chocolate, gummy bears, and mint candy (Zimmerman, 2014).

Public health experts are debating whether electronic cigarettes, or e-cigarettes, are safer than tobacco cigarettes. E-cigarettes deliver nicotine in a liquid smoke-free vapor (often called "vaping"), but don't contain the same carcinogens and other toxic chemicals as regular cigarettes. According to the U.S. Surgeon General, because e-cigarettes contain nicotine, they can harm the developing adolescent brain and lead to addiction. Among adolescents who have never smoked, for example, those who try e-cigarettes are likely to start smoking conventional cigarettes within a year. Some researchers maintain, however, that it's too soon to know if vaping leads to smoking traditional cigarettes, because both usages have been declining among middle and high school students (U.S. Department of Health and Human Services, 2016; Hines et al., 2017; Kozlowski and Warner, 2017).

Healthy Eating and Exercise. Healthy eating practices and regular exercise both help prevent, delay, and manage chronic diseases including heart disease and Type 2 diabetes as well as maintaining a healthy weight (Centers for Disease Control and Prevention, 2023 "How You Can Prevent Chronic Diseases"; Reinberg, 2023; Mayo Clinic Staff, 2021).

Chronic health issues such as heart disease and diabetes impact both individuals and wider society. In 2021, 1 in 5 deaths were related to heart disease—which accounts for nearly 700,000 people. It's estimated that heart disease costs the United States almost $240 billion per year in health care, medication, and productivity loss. The effects of diabetes are also wide reaching. Approximately 11 percent of the adult population in the United States has diabetes (Type 1 or 2) and around one-third of adults have prediabetes. People with diabetes have $9,601 more in medical costs each year. The United States spends more on diabetes than any other chronic condition—$327 billion annually and these costs are increasing (Centers for Disease Control and Prevention, 2023 "Heart Disease"; Centers for Disease Control and Prevention, 2022 "Health and Economic Benefits of Diabetes Interventions"; American Diabetes Association, 2018).

In the United States, changes are happening with healthy eating guidelines, attitudes, and practices. Current U.S. dietary guidelines use the MyPlate visual

How Much Sugar Are You Really Eating?

The American Heart Association recommends limiting our daily dose of added sugar to 24 grams (about 6 teaspoons) for women and 36 grams (about 9 teaspoons) for men.

Vitamin-enhanced water (20 ounces)

Whitebox Media/Alamy Stock Photo

32 g of sugar

Blueberry muffin

Foodcollection.com/Alamy Stock Photo

22 g of sugar

Prepared coleslaw (1 cup)

Ingram Publishing/Alamy Stock Photo

23 g of sugar

Thin Mints (10)

ignatius 63/Shutterstock.com

26 g of sugar

Slice of frozen pizza

Markus Mainka/Shutterstock

6 g of sugar

Fast-food chicken sandwich

Giancarlo Polacchini/Shutterstock

16 g of sugar

format and encourage eating a variety of healthy food (**Figure 14.5**). Americans are paying more attention to healthy eating, but actual eating practices may not be aligned. Around 72 percent of Americans think healthy eating is very important for "increasing chances for a long and healthy life". Around 50 percent of Americans aim to eat a healthy diet with this being slightly more common among people born between 1946 to 1964 (58 percent). Also, Americans think people today pay more attention to eating healthy (54 percent) compared with 20 years ago. Still, when asked about actual eating habits, only 29 percent of Americans think people are eating healthier now than in the past. Barriers to healthy eating include high cost (46 percent), lack of time (23 percent), and lack of knowledge about food preparation (20 percent) (Buchholz, 2023; Pew Research Center, 2016, "Public Views About Americans' Eating Habits").

PeterBajohr/Getty Images

Figure 14.5 MyPlate Visual for Healthy Eating

Source: U.S. Department of Agriculture, 2023

Regular exercise can help address chronic illness and help with mood, energy levels, and sleep. It is recommended that adults get regular aerobic activity (150 minutes per week) and muscle strengthening activity (twice per week). Seventy-one percent of American adults think getting physical exercise is important, but only 28 percent meet the exercise guidelines. Common challenges that deter people from exercising are lack of time, being part of social groups that does not share fitness interest, no motivation or energy, lacking resources or workout equipment, family responsibilities, and work (Mayo Clinic Staff, 2021; U.S. Department of Health and Human Services, 2018; Pew Research Center, 2016; Meghjani, 2023; American Heart Association, 2018).

Preventive Care. *Preventive care* involves regular health care aimed at the prevention of health concerns, including disease or illness. This kind of care includes screenings, check-ups, and providing guidance to patients regarding health habits and health care.

Although preventive care is important, only 8 percent of Americans get regular health screenings. Across the United States, only 15 out of the 50 states have better than average preventive health care scores (Batarseh et al., 2020). Barriers to keeping up with preventive care include the availability of reliable transportation, availability of child care, having health insurance, health literacy, the ability to take time off from work, and available support systems (family or friends) (Lea, 2023). Other challenges to keeping up with preventive care include cost and frustration with scheduling systems and long appointment wait times (Businesswire, 2022; Brenan, 2023 "Record High in U.S. Put Off Medical Care").

Vaccinations, which are a form of preventive care, have been a hot topic in recent years prompted by the COVID-19 pandemic. In the United States, the pandemic led to over 6 million hospitalizations and 1.1 million deaths. There are several recommended preventive actions to suppress the spread of the virus including handwashing, testing for COVID-19 when needed, staying home if symptoms emerge, and getting vaccinated. Among Americans, 81 percent have had one COVID-19 vaccination dose and 70 percent have completed a vaccination series. Overall, the rates of COVID-19 related hospitalizations and deaths are decreasing (Centers for Disease Control and Prevention, 2023 "About COVID-19").

Substance Misuse. Unintentional injury is the nation's third leading lifestyle-related and preventable cause of death, any many of these deaths can be attributed to substance misuse and overdose. **Substance misuse**, a harmful overindulgence in or dependence on a drug or other chemical. Nearly 85,000 Americans die each year because of excessive drinking, and more than 4 million visit emergency rooms for alcohol-related problems ("Alcohol and Public Health," 2014; Ahmad et al., 2023).

Excessive alcohol use includes *binge drinking*, having 4 or more drinks on a single occasion for women or 5 or more drinks on a single occasion for men, generally within about 2 hours. *Heavy drinking* involves consuming 15 or more drinks per week for men, and 8 or more drinks for women. Some 25 percent of Americans binge drink and 7 percent engage in heavy alcohol consumption (National Center for Health Statistics, 2017).

Table 14.1 Some Consequences of Alcohol Misuse in College

Researchers estimate that, each year, among college students age 18 to 24…	
1,519	Die from alcohol-related unintentional injuries, including motor vehicle crashes
25%	Have academic difficulties because of alcohol consumption, for instance, missing class or falling behind in schoolwork
696,000	Are assaulted by another student who's been drinking
13%	Meet the criterion for an alcohol use disorder (for those age 18 to 22).

Sources: Based on National Institute on Alcohol Abuse and Alcoholism, 2023.

Excessive alcohol use results in myriad short- and long-term health risks that include injuries, violence, risky sexual behaviors, miscarriage and stillbirth, physical and mental birth defects, alcohol poisoning, unemployment, psychiatric problems, heart disease, several types of cancer, and liver disease. **Table 14.1** summarizes some of the consequences of excessive drinking for college students, their families, and the larger community.

Illicit (illegal) *drugs* include marijuana/hashish, cocaine and crack, heroin, hallucinogens (e.g., LSD, PCP), inhalants, and any prescription-type psychotherapeutic drug, including stimulants and sedatives, used nonmedically. More than 10 percent of Americans (about 34 million people) age 12 and older use illegal drugs (**Figure 14.6**). Illicit drug use is more common

substance misuse a harmful overindulgence in or dependence on a drug or other chemical.

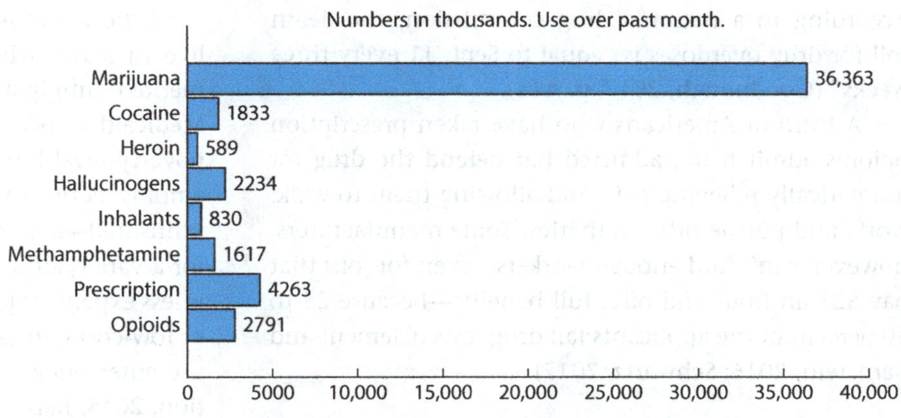

Figure 14.6 Illicit Drug Use Among Americans Age 12 and Older, 2021

Source: Based on Substance Abuse and Mental Health Services Administration, 2022 "Section 1: Illicit Drug Use/Misuse Tables".

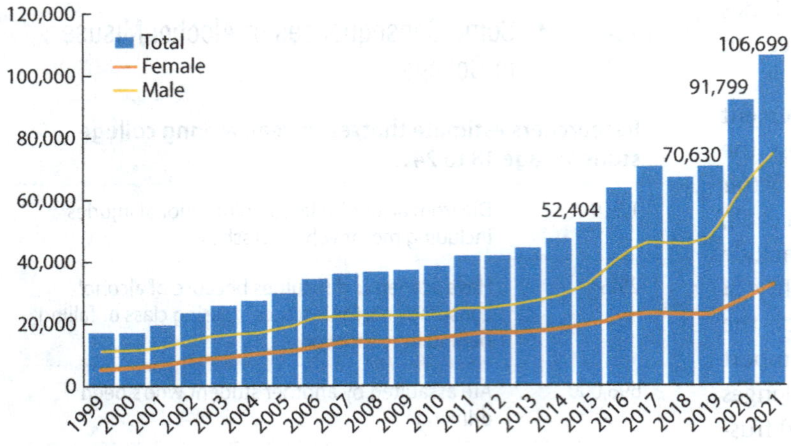

*includes deaths with underlying causes of unintentional drug poisoning (X40–X44), suicide drug poisoning (X60–X64), homicide drug poisoning (X85), or drug poisoning of undetermined intent (Y10–Y14), as coded in the International Classification of Diseases, 10th Revision.

Figure 14.7 National Drug-Involved Overdose Deaths* Numbers Among All Ages, by Gender, 1999-2021

Source: National Institute on Drug Abuse, 2023

among men than women and among people age 16 to 25, and least common among Asian Americans (National Center for Health Statistics, 2017).

Over 20 states have fully legalized marijuana, many others have mixed legalization (e.g., some allowing medical uses), and only 6 states still fully criminalize it (DISA, 2023), marijuana remains the most commonly used illicit drug (**Figure 14.6**).

Opioids include illegal drugs (e.g., heroin), synthetic drugs (e.g., fentanyl), and legally prescribed pain relievers (e.g., oxycodone, morphine). In the late 1990s, doctors started prescribing more and stronger opioids to treat arthritis, back problems, diabetes, and other chronic pain. As opioid-driven drug overdose deaths surged (**Figure 14.7**), health practitioners blamed opioids for "our nation's deadliest drug epidemic ever." According to a national drug commission, the death toll for drug overdoses is "equal to Sept. 11 every three weeks" (Goodnough, 2017, p. A14).

A third of Americans who have taken prescription opioids admit being addicted but defend the drug for dramatically relieving pain and allowing them to walk, work, and pursue other activities. Some manufacturers, however, can't find enough workers—even for jobs that pay $25 an hour and offer full benefits—because 25 to 50 percent of the applicants fail drug tests (Clement and Bernstein, 2016; Schwartz, 2017).

We've explored some of the reasons why some people are healthier than others. Who gets health care? And who pays for it?

14-3 Health Care: United States and Global

Health care encompasses a number of components. One of the most important is **medicine**, a social institution that deals with illness, injury, and other health problems (refer also to Chapter 6 on institutions).

14-3a U.S. Health Care Coverage

In 2010, Congress passed the Patient Protection and Affordable Care Act, also called the Affordable Care Act (ACA) or "Obamacare." The ACA's goal was to give more Americans under age 65 affordable, quality health insurance, and to reduce the growth in health care spending.

Has the ACA worked? The uninsured decreased from 18 percent (42 million people) in 2013, before the ACA took effect, to 10 percent (28 million) in 2017. In 2022, the uninsured rate reached an all-time low of 3.7 percent. The ACA generated the greatest gains for people living below or just above the poverty level, Latina/o/xs, and residents born in another country—the groups least likely to have health insurance (Barnett and Vornovitsky, 2016; "Health Insurance Coverage," 2017; National Center for Health Statistics, 2017; U.S. Department of Health and Human Services, 2022).

Who are most likely to be uninsured? People who live in states where Medicaid doesn't cover undocumented immigrants, those who haven't enrolled in Medicaid, or people with incomes just above the federal poverty level but can't afford the programs offered by employers or private insurers. The largest group of the uninsured—about 46 percent—doesn't have coverage for a variety of reasons: Paying out-of-pocket as needed is less expensive than health insurance, they rely on free or low-cost clinics, go without care or medicine, and/or use emergency rooms (Robert Wood Johnson Foundation, 2015; Kaiser Family Foundation, 2016).

How do politics affect health policies? Repealing the Affordable Care Act has been the rallying cry of the Republican Party ever since the law passed.

medicine a social institution that deals with illness, injury, and other health problems.

Table 14.2 Some Affordable Care Act Pros and Cons

Pros	Cons
• Millions of uninsured Americans now have health insurance. Young adults can stay under their parents' health plans until age 26.	• People who don't have health insurance (through an employer, Medicare, Medicaid, the military, or a private company) risk paying a penalty.
• Covers pre-existing conditions (e.g., diabetes, heart disease, cancer, HIV/AIDS), and there are no time limits on care.	• Healthy people pay for patients with pre-existing conditions. The costliest five percent of these patients account for nearly half of all health care spending.
• Lowers overall health care costs in the long run, because preventive services (e.g., screening for cancer or other diseases) have no out-of-pocket payments.	• Tens of millions of low- and middle-income people with private insurers earn too much to qualify for ACA or Medicaid but struggle to pay medical bills.
• Offers access to and comparisons of health insurance options through exchanges.	• Some major insurers have dropped out of exchanges, especially in rural counties, leaving consumers with few choices and rising premium costs.

Sources: Kim and Roland, 2015; Cox and Levitt, 2017; Infinit Healthcare, 2017; Tozzi, 2017; Tracer and Recht, 2017.

During the 2016 presidential election, Donald Trump described the Affordable Care Act as "disastrous" and vowed—repeatedly and to thunderous applause—to "immediately repeal and replace Obamacare." The ACA is still being challenged. In 2021, the Supreme Court dismissed a case that aimed to repeal the Affordable Care Act. Two years later, a federal judge in Texas ruled that some parts of the Affordable Care Act can't be enforced such as requiring insurers to cover preventive care services, cost-free to patients. These services included "some cancer, heart and STD screenings, and tobacco programs" (Luhby and Sneed, 2023; de Vogue and Duster, 2021).

Nationally, however, 35 percent of Americans, many of whom voted for Trump and were recipients of the ACA, didn't know that the ACA and Obamacare are the same (Dropp and Nyhan, 2017). **Table 14.2** presents some of the pros and cons of the ACA.

14-3b Who Pays for Health Care?

About 66 percent of Americans have private health insurance, through an employer or union, that they buy directly from an insurance company or through someone outside the household. Of the insured, employer-based insurance covers 54 percent of the population; public health insurance funded by taxpayers (Medicare, Medicaid, military) covers another 37 percent (Keisler-Starkey and Bunch, 2022; refer also to Chapter 12 on Medicare and Medicaid).

Since 1999, workers at both large and small firms have been paying higher premiums, deductibles, copayments, and other out-of-pocket costs, whereas wages, including those of middle-class families, have stagnated.

Most people pay a variety of state and federal taxes to support programs like Medicare and Medicaid. Medicare pays almost 60 percent of all health care costs for people age 65 and over, regardless of income (Federal Interagency Forum on Aging-Related Statistics, 2016). Thus, even billionaires are eligible for Medicare. Medicaid, another government insurance program, provides medical care, regardless of age, for people living below the poverty level (refer to Chapter 12).

14-3c Why Is U.S. Health Care So Expensive?

Compared with other nations, the United States spends a lot on health care — per person, nationally, and as a percentage of the gross domestic product. The United States actually spends more on health care than 34 other high-income countries, but it still ranks last among the top 11 in quality of care, access to care, efficiency, and health outcomes (Schneider, 2021).

Recently, health care expenses consumed nearly $1 of every $5 in the economy (**Table 14.3**). Why, then, are many Americans worse off than their counterparts in other high-income countries? One reason is that around 25 percent of U.S. health care spending is wasteful, and this accounts for $760 billion to $935 billion annually (Centers for Medicare & Medicaid Services, 2022; Japsen, 2019).

In Canada, Germany, Great Britain, France, Japan, Sweden, and other high-income countries, the government picks up most of the health care bill. Patients and health care workers don't have to submit bills to several

Table 14.3 The Increasing Cost of U.S. Health Care, 1980–2021

	1980	1990	2000	2010	2021
Average cost per person	$1,100	$2,864	$4,878	$8,402	$12,914
National health expenditure	$256 billion	$724 billion	$1.4 trillion	$2.6 trillion	$4.3 trillion
Percent of gross domestic product (GDP)	9%	13%	14%	18%	18%

Notes: The "average cost per person" includes medical care, supplies, drugs, and health insurance.

Sources: Based on Centers for Medicare & Medicaid Services, 2010, Table 1; 2014, Table 1; Centers for Medicare & Medicaid Services, 2022.

insurance providers, resulting in considerable administrative savings. The government, nonprofit organizations, or large groups (cities and industries) have considerable power in keeping down drug costs and setting prices for health care providers and services. There are also private hospitals, but most are nonprofit (OECD, 2014).

In the United States, "hospitals are the most expensive part of the world's most expensive health system" where 1 out of every 7 healthcare dollars are spent. Even nonprofit hospitals are businesses: The top paid executives at some of these hospitals were paid between $10 and $22 million. In total, the salaries for the top executive at 82 hospitals equaled $298 million ("Prescription for Change," 2013; Andrzejewski, 2019).

Unlike other countries where the government sets a national price for each drug, the United States lets pharmaceutical companies charge whatever they want. Congressional lawmakers—heavily lobbied by the pharmaceutical industry, the American Hospital Association, and other groups—have forbidden Medicare, the nation's largest medical insurer and the world's largest buyer of prescription drugs, to negotiate drug prices or the cost of health products. Thus, for example, the price of EpiPen, a lifesaving injection for severe allergic reactions, jumped more than 400 percent—from $94 to $609—after a pharmaceutical company acquired the product. A drug that cures Hepatitis C, a liver disease, costs just $4 a pill in India, compared with $1,000 in the United States (Agorist, 2016; Atwater, 2016).

Medical technology also increases health care costs. Doctors in the United States perform 71 percent more CT scans than do doctors in Germany. Technology can save lives, but American doctors, compared with those in many other countries, are more likely to adopt new and expensive machines that are highly profitable for both physicians and hospitals. Technology helps us live longer, but also increases a patient's, insurer's, and nation's medical costs (Brill, 2015; Frakt, 2017). **Figure 14.8** provides more examples of why U.S. health care costs are so high. Because of the high costs, each year up to 750,000 Americans engage in *medical tourism*, traveling to another country for medical care ("Medical Tourism," 2015). A growing number of people are also opting to not get medical care. When asked "have you or a member of your family put off any sort of medical treatment because of the cost you would have to pay?" 38 percent of Americans said yes compared with 19 percent who said the same in 2001. Those who delayed care for a serious condition were more likely to be women or younger adults (age 18 to 49) or to have lower incomes (Brenan, 2023).

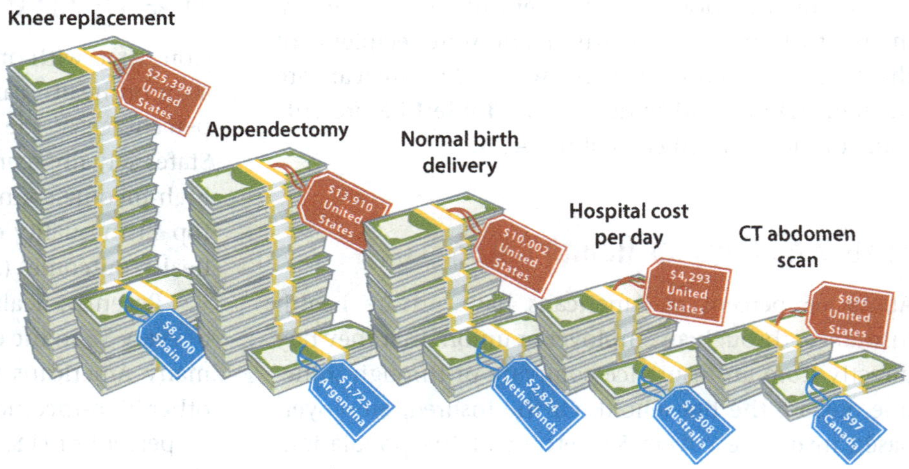

Figure 14.8 Should You Travel Abroad for Medical Care?

Would you be better off living outside the United States when it comes to health care costs? Except for Argentina, all of these countries have higher life expectancy rates than the United States.

Source: Based on International Federation of Health Plans, 2014. Refer also to Howard, 2014, for a comparison of surgery costs in the United States and other countries.

14-4 Sociological Perspectives on Health and Medicine

Sociologists focus on different aspects of health, health care, and medicine. **Table 14.4** summarizes these perspectives.

14-4a Functionalism: Good Health and Medicine Benefit Society

For functionalists, good health and medicine are critical for a society's survival and stability. Thus, countries try to develop a medical care system that, ideally, benefits the entire population regardless of age, sex, race, ethnicity, social class, or other characteristics.

Health Care and Other Institutions

Health care systems are connected to many institutions. Despite ongoing debates about the ACA and the government's requiring people to buy health insurance, millions of individuals and families are now eligible for government assistance to pay for coverage, and can get free preventive health services (e.g., diabetes tests, colonoscopies, routine vaccinations).

The government is deeply involved in health care in other ways. It funds much of the scientific research at universities as well as federal agencies like the Centers for Disease Control and Prevention and the National Institutes of Health, both of which deal with prevention, treatment, and health care policy. At the national and state levels, numerous government agencies are responsible for passing and enforcing regulations regarding new drugs, medical procedures, and access to medical care.

Health care affects the economy because it's the nation's largest employer. "Demand for health services will rise as Obamacare expands insurance and Americans grow older, fatter, and sicker" ("The Health Paradox," 2013, p. 27). Our huge health care spending threatens funding for other public programs, including education and transportation. Functionalists would point out, however, that the trillions spent on health care have generated millions of jobs at hospitals, drug companies, insurance companies, nursing homes, and technology firms (refer to Chapter 11).

Besides generating jobs, health care systems shape corporate policies which, in turn, benefit families. For instance, the pharmacy chain CVS banned sales of all tobacco products, Target is promoting employees' wellness and adding more organic and natural food to grocery aisles, and some restaurants (e.g., McDonald's, Wendy's, Red Lobster) have taken sodas off their children's menus and reduced sodium and fat in kids' meals (D'Innocenzio, 2015; Martin, 2015).

The Sick Role

Talcott Parsons (1951), an influential sociologist, introduced the concept of the **sick role**, a social role that

sick role a social role that excuses people from normal obligations because of illness.

Table 14.4 Sociological Perspectives on Health and Medicine

Perspective	Level of Analysis	Key Points
Functionalism	Macro	• Health and medicine are critical in ensuring a society's survival and are closely linked to other institutions. • Illness is dysfunctional because it prevents people from performing expected roles. • Sick people are expected to seek professional help and get well.
Conflict	Macro	• There are gross inequities in the health care system. • The medical establishment is a powerful social control agent. • A drive for profit ignores people's health needs.
Feminist	Macro and micro	• Women are less likely than men to receive high-quality health care. • Gender stratification in medicine and the health care industry reduces women's earnings. • Men control women's health.
Symbolic Interaction	Micro	• Illness and disease are socially constructed. • Labeling people as ill increases their likelihood of being stigmatized. • Medicalization has increased the power of medical associations, parents, and mental health advocates and the profits of pharmaceutical companies.

How sick is too sick to miss work or class? Are menstrual cramps enough? What about a severe cold? A hangover? And when does being sick become dysfunctional—after a day, a week, a month?

excuses people from normal obligations because of illness. In Parsons' model, sick people aren't responsible for their condition and, therefore, have legitimate reasons for not performing their usual social roles ("I missed the exam because I had COVID").

Parsons emphasized that the sick role is temporary, and that individuals must seek medical help to hasten their recovery. Otherwise, people will view them as hypochondriacs, slouchers, and malingerers who aren't living up to their responsibilities. Thus, from a functionalist perspective, the sick role is legitimate if it's short-lived, but dysfunctional if people feign illness to shirk their duties long term in the family, the workplace, or other groups.

The Healthcare Provider as Gatekeeper

From a functionalist perspective, physicians (medical doctors and doctors of osteopathy), physicians' assistants (PA's), and nurse practitioners (NP's) play a key gatekeeping role in limiting the sick role. They verify a person's condition as sick and provide an excuse for temporarily not performing necessary roles. They also designate a patient as "recovered" and ready to meet societal role expectations again. These professionals' specialized knowledge gives them considerable authority in defining health and illness.

Critical Evaluation

Functionalist theories are limited for several reasons. First, health care policies can be divisive. For example, 57 percent of Americans say that the government should pay for health care, but 40 percent disagree (Brenan, 2023 "Majority in U.S. Still Say Government Should Ensure Healthcare"). Moreover, many of the job growth areas in the health care sector are occupations with low annual salaries such as home health and personal care aids ($27,000) and physical therapist aids ($28,000) (Smith, 2021). Thus, it cannot be broadly said that all workers are benefiting from the booming health care industry.

Second, and despite the Affordable Care Act, 25 percent of Americans say that either they or members of their family don't receive needed medical care because they can't afford the out-of-pocket costs. About 80 percent of low-wage workers don't receive any paid sick leave. When workers have paid sick days, 69 percent don't take them, even when they're ill, because they worry about falling behind or having coworkers take on extra work (Henderson, 2016; Frankel, 2017) (refer to Chapter 11). Thus, many people don't or can't assume even a short-lived sick role.

Third, because physicians are powerful gatekeepers, many prosper by prolonging sick roles. The more times that doctors have appointments with patients or visit them at the hospital, and the more tests that doctors order, the higher their salaries. Conflict theorists, especially, maintain that much of the health care industry is dysfunctional because access to health care varies considerably, and only the wealthy don't have to worry about receiving and paying for the best available medical care.

14-4b Conflict Theory: Health Care and Medicine Don't Benefit Everyone

For conflict theorists, medicine and the health care industry benefit some groups much more than others. In contrast to functionalists, conflict theorists argue that the medical system reinforces social inequality, exerts social control to maintain the status quo, and is often driven by a profit motive rather than a concern for people's well-being.

Social Control of Medicine and Health Care

From a conflict perspective, those at the top of the medical hierarchy have considerable control. Physicians, for instance, have almost absolute power in diagnosing an illness, providing treatment, and deciding on medical procedures.

The health care industry can maintain the status quo because of the **medical-industrial complex**, a network of business enterprises that influences medicine and health care. The medical-industrial complex includes many groups—doctors, nurses, hospitals, lawyers, nursing homes and hospices, insurance companies,

medical-industrial complex a network of business enterprises that influences medicine and health care.

drug manufacturers, accountants, banks, and real estate and construction businesses.

Who benefits from the medical-industrial complex? All of those involved, especially pharma (also called Big Pharma), the pharmaceutical drug industry, and hospitals. The annual profit margins of some prominent drug manufacturers, like Amgen, are 43 percent, compared with 3 to 6 percent for Ford and General Motors (Love, 2017).

Big hospitals have been buying up previously independent cancer clinics and doctors' offices. Doing so gives them the power to raise prices for private insurers and government programs, particularly Medicare. As a result, they can charge higher rates, sometimes three times as much, for cancer drugs and outpatient visits and services (Bach, 2015; Bai and Anderson, 2015).

Use of additional prescription drugs is also adding to healthcare profits. The percentage of Americans taking three or more prescription drugs increased from 11 percent in 1998 to more than one-third in 2021. With people taking more prescriptions, this has increased the overall costs they must pay. In addition, prices for dozens of drugs that treat everything from blood pressure to multiple sclerosis have doubled since 2007, and the prices of some generic drugs have soared by 1,000 percent or more. Drug prices keep rising, despite widespread complaints, because Big Pharma can charge whatever it wants. "Prescription drug prices in the U.S. more than 2.5 times as high as those in other similar high-income nations" (McPhillips, 2023; Jaret, 2015; Love, 2017; National Center for Health Statistics, 2017; Parasrampuria and Murphy, 2022).

The United States and New Zealand are the only high-income countries that allow direct-to-consumer prescription drug marketing ("Ask your doctor about..."). The ads, which began in the 1980s, increase a drug's sales. People who are aware of drug advertising are more likely to ask their doctors for a particular brand-name medication. Medical providers may feel pressure to accommodate their patients' requests, which, in turn, increases a drug's price and sales (Timmermans and Oh, 2010; Niederdeppe et al., 2013).

Health, Profit, and Waste

From a conflict perspective, health care is big business. The median household income in 2021 was $70,784—while the average healthcare executive made $2.8 million. Much of an executive's earnings are based on increasing stock prices rather than improving patient care. That means selling more drugs, raising prices above inflation, performing more tests and procedures, and getting more people into the hospital (U.S. Census Bureau, 2022 "Median Household Income"; Herman, 2017).

Stockbyte/Thinkstock

End-of-life care is a billion-dollar industry. Compared with their nonprofit counterparts, for-profit hospices spend less on nursing per patient, provide fewer services, and push out patients whose care becomes expensive. Medicare provides about 90 percent of funding for hospices and nursing homes, but rarely punishes either for violating health and safety rules (Hallman and Shifflett, 2014; Whoriskey and Keating, 2014).

Large hospitals are crowded with "senior VPs, VPs of this, that, and the other," whose salaries are two to three times higher than those of surgeons. A Wisconsin surgeon discovered that a brief outpatient appendectomy he had performed for a fee of $1,700 generated over $12,000 in hospital bills. Thus, "the biggest bucks are earned not through the delivery of care, but from overseeing the business of medicine" (Healthcare-NOW! 2014; Rosenthal, 2014: SR4).

The medical-industrial complex profits in other ways. Some popular diet and other supplements contain dangerous chemicals that pose health risks. Several of the Food Drug Administration's (FDA) top officials are former lobbyists or industry leaders. In other cases, sugar and food industries fund doctors who publish articles in prominent medical journals that play down the link between sugar and heart disease (O'Connor, 2015; Kearns et al., 2016).

At least 30 percent of federally funded health care is wasteful. "Improper" payments (e.g., to ineligible beneficiaries, to dead people, and for unallowable services) totaled $144 billion in 2016, a 250 percent increase since 2003. Between 10 and 20 percent of all surgeries are unnecessary. Doctors often use aggressive and expensive treatment in the last month of life for three-quarters of patients younger than 65 with terminal cancer. Patients and families may insist on such treatment, not realizing that it'll bring "nothing but emotional and physical misery," because physicians, nurses, and other professionals aren't trained in end-of-life conversations (Chen et al., 2016; "Uncle Sam Overpays," 2017).

Other wasteful spending includes the expensive services that have the same health outcomes as lower-priced alternatives, preventable hospital injuries and readmissions, unnecessary tests and treatments to guard against liability in malpractice lawsuits, high prices for drugs, and prescriptions for drugs that work no better than a placebo. Huge agencies like Medicare and Medicaid are particularly vulnerable to waste. The private contractors employed to pay the government's millions of health care bills are under pressure to process claims as quickly and inexpensively as possible. "The cheapest way to process a claim is to pay it without question" (Rosenberg, 2016; Office of Inspector General, 2017; "Uncle Sam Overpays," 2017).

Critical Evaluation

The most common criticism is that conflict theorists often overlook the contributions of health care systems. Without them, people would suffer more, die at a younger age, and have a lower quality of life.

Second, doctors, medical scientists, and some medical associations (e.g., American Academy of Family Physicians) have been among the most vocal critics of unneeded surgery, tests, and procedures. Examples include cardiac stress tests and recommending vitamins and other supplements that don't benefit health. Others have urged ending expensive end-of-life care that simply prolongs the dying process. Nearly 120 of the nation's leading oncologists have publicly protested drug companies' "out of control greed," which is bankrupting cancer patients (Guallar et al., 2013; Agnvall, 2014; Tefferi et al., 2015).

Is Being Fashionable Hazardous to Women's Health?

Squeezing into skinny jeans and tight pants can cause nerve compression, numbness, and digestive problems. Narrow-toed shoes and stilettos wreak havoc: blisters, bunions, corns, calluses, hammertoes, nerve damage, back and neck pain, painful inflammation, stress fractures, and ankle sprains. Still, according to a podiatrist, "Women will wear their high-heeled shoes until their feet are bloody stumps" (Ianzito, 2013, p. 3; Gleiber, 2014).

Many cosmetics contain chemicals that can trigger skin problems (e.g., itching, rashes, acne). Others—contaminated with bacteria, yeasts, or molds—can lead to a range of problems from simple rashes to serious infections that can cause swelling and breathing difficulties. A study that tested 32 commonly sold lipsticks and lip glosses found that they contained lead, cadmium, chromium, aluminum, and five other metals—some at potentially toxic levels that are linked to stomach tumors. Some hair products contain hazardous ingredients like formaldehyde and mercury that can result in losing large clumps of hair (Liu et al., 2013; Lipton and Abrams, 2016).

Except for some hair color additives, the FDA doesn't require companies to test products for safety or list toxic ingredients. The U.S. Food and Drug Administration limits just nine cosmetic ingredients for safety reasons—placing the United States behind 80 other nations in their cosmetic regulations (Inouye, 2021).

maron/AGE Fotostock

Finally, most people aren't simply victims of a malicious medical-industrial complex. Many make unhealthy lifestyle choices (unprotected sex, drug misuse), and complain rather than change their behavior when employers impose health insurance penalties for smoking. Others join forces with drug companies to lobby the FDA to decrease current restrictions on potent opioids (Meier and Lipton, 2013; Brody, 2015).

14-4c Feminist Theories: Health Care and Medicine Benefit Men More Than Women

Both feminist and conflict theorists emphasize the connection between health and inequality. Feminist scholars go further by addressing the health costs of being a woman, gender stratification in medicine and health care, and men's control over women's choices.

The Health Costs of Being a Woman

Women and men face different health care issues. For example, women are more likely than men to

- get inaccurate results from treadmill stress tests to detect heart problems because the scoring system is based on "only middle-age men" (Cremer et al., 2017);
- have chronic pain, be prescribed pain relievers, receive higher doses, use opioids for longer periods, become addicted, and, early in a pregnancy, risk birth defects and preterm births (Ailes et al., 2015; American Society of Addiction Medicine, 2016);
- be unpaid family caregivers, even if they're ill, disabled, or older (Hooyman, 2016); and
- suffer a fatal heart attack because they often receive less aggressive treatment than men (Sagon, 2017).

For decades, heart disease "has been studied in men, for men, and by men." Even though heart disease is the leading cause of women's death and kills more women than men, only 35 percent of participants in heart-related studies are women. Such research gaps "leave women's health to chance" (Johnson et al., 2014; Merz, 2015: A16).

During their reproductive years, 32 percent of U.S. women have a cesarean section (C-section), up from only 5 percent in 1970. The main reason for the increase is hospital administrators' and maternity clinicians' fear of lawsuits if a vaginal birth has a negative outcome (Morris, 2014; Centers for Disease Control, 2023 "Births-Delivery Method"). Because C-sections are major surgical procedures, the risks include infection, blood clots, and injury to other organs, particularly the bladder.

U.S. maternal mortality rates have increased. American women are now three to four times as likely to die from pregnancy-related complications as their counterparts in Britain, the Czech Republic, Germany, Greece, Finland, Italy, or Japan, and Black women's maternal mortality rates are more than three times higher than those of White women. The upturn is probably due to increasingly common health conditions (e.g., hypertension and diabetes) that make delivery more dangerous ("Maternal Mortality," 2015; "Pregnancy Mortality Surveillance System," 2017).

Gender Stratification in Health Care Occupations

In health care jobs, men consistently earn more than women (**Table 14.5**). Among physical therapists, for example, men earn almost $71,000 a year compared with $68,000 for women even when both have the same level of education, years of experience, and work responsibilities (refer also to Chapters 8 and 11).

Feminist scholars attribute much of the gender wage gap to men's gatekeeping. Because many registered nurses, pharmacists, and physical therapists (occupations that have more women than men) now receive doctoral degrees, they want to use the honorific title of "doctor." Doing so would win more respect from patients, help women land top administrative jobs that pay more, provide more autonomy in treating patients, and bring higher fees from health insurers. Physicians oppose the idea for several reasons: They want to maintain their prestige in the health care industry; they treat patients first, whereas nurses and others play only secondary roles, and they have considerably more education and training to diagnose and treat illness and disease (Harris, 2011).

Men's Control of Women's Health

Feminist scholars maintain that men control many aspects of women's health. Catholic bishops, all of whom are men, condemn contraceptives as sinful, even though 98 percent of U.S. Catholic women have used birth control (Jones and Dreweke, 2011).

In 2014, the Supreme Court ruled, 5-4 (*Burwell v. Hobby Lobby Stores*), that privately held for-profit corporations don't have to pay for their employees' contraceptive coverage, as the ACA requires, if doing so conflicts with the owners' religious beliefs. In effect, a religious objection can trump a federal law, and businesses can shift insurance costs to taxpayers or private insurers. According to some Johns Hopkins doctors, "The Supreme Court decision has started us down a dangerous path on which the religious beliefs of a third party enter the examination room and interfere with the

Table 14.5 Women Earn Less Than Men in Health Care Occupations, 2016

	Men	Women	Percentage of Women in This Occupation
All health care practitioners	**$69,264**	**$53,872**	**75**
Pharmacists	$108,992	$95,628	60
Physicians and surgeons	$121,836	$76,752	38
Physical therapists	$70,096	$67,912	65
Registered nurses	$65,572	$59,436	89
Clinical laboratory technologists and technicians	$55,692	$43,368	67
Emergency medical technicians and paramedics	$42,692	$33,800	34

Note: The data are median annual earnings of full-time workers.

Source: Based on Bureau of Labor Statistics, 2017, Table 39.

doctor-patient relationship" (Singal et al., 2014: 17; refer also to Rosenfeld, 2014). All of the justices who ruled against women's contraceptive coverage were men.

In 2022, the Supreme Court, consisting of mostly men, voted to overturn *Roe v. Wade* therefore eliminating federal protection of abortion care throughout the United States. In many states, lawmakers have passed laws to restrict or block girls' and women's ability to get abortion care (refer to Chapter 9). They aren't passing similar laws to limit the availability of Viagra and other erectile dysfunction medications because "Viagra is a wonderful drug" (Beadle, 2012). The Hobby Lobby chain doesn't have to pay for IUDs and emergency contraceptive pills but covers Viagra and other erectile dysfunction drugs.

Critical Evaluation

Feminist theories are limited in several ways. First, they sometimes gloss over the fact that social class, rather than gender, has a big effect on people's health and receiving health care services (Montez and Zajacova, 2013).

Second, men also experience gender bias in health care. They suffer one-third of all hip fractures but, because osteoporosis is considered a "woman's disease," there's little research on men, and guidelines about testing older men's bone density are vague (Sagon, 2017).

Third, the medical profession has expanded professional opportunities increasing the roles and statuses of positions beyond medical doctor. For instance, nurse practitioners (NP) can prescribe medication in all states, and 85 percent of NPs are women. In Florida, NPs with 3,000 hours of experience, supervised by a physician, can operate their own practices for primary care. The profession of medical doctors has also become more inclusive: 37 percent of active physicians are women (Wiesen, 2023; Mensik, 2023; Daily Nurse, 2020).

Similar to conflict theorists, feminist scholars rarely address lifestyle choices that affect people's health. Among adult women, for example, 14 percent smoke, and at least 8 percent use illicit drugs (National Center for Health Statistics, 2017). Also, women aren't always passive but stand up to men's domination of health laws. For example, a Republican-controlled Senate failed to repeal the Affordable Care Act because two Republican women voted "no," and despite "intense pressure" from Republican male leaders (Leonhardt, 2017: A23). Finally, women may support policies that limit women's access to medical services such as abortion care. Supreme Court Justice Amy Comey Barret voted to overturn *Roe v. Wade*, and 8 Congresswomen voted against building abortion rights into federal law (Carbonaro, 2022).

14-4d Symbolic Interaction: The Social Construction of Health and Illness

Symbolic interactionists focus on how we define and construct views about health, illness, and medicine and then implement these definitions in everyday life. Social constructions include labeling, stigmatizing behavior, and medicalizing attitudes and behaviors as normal or sick.

The Social Construction of Illness

Medical models assume that illness is anything that deviates from normal biological functioning and that an illness has specific features that a doctor can recognize. In contrast, interactionists view illness and medicine as social constructions that can change over time. In 1956, the American Medical Association declared that alcohol misuse is a treatable disease that's due to a genetic predisposition rather than a person's lacking moral character or self-discipline. Medical designation opens the door for

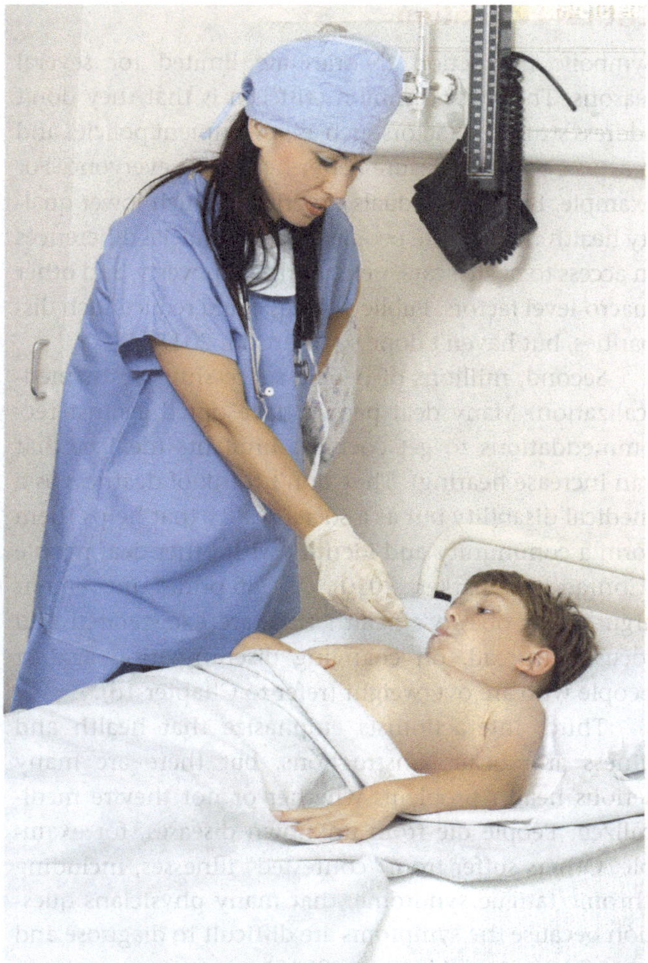

Leah-Anne Thompson/Shutterstock.com

Nurse practitioners (NPs), most of whom are women, complete a master's or doctoral program that includes diagnosing and treating patients, managing acute and chronic illnesses, and prescribing medications. Because of a projected shortage of doctors, NPs are pressing 34 states to relax laws requiring doctors to oversee their work. Many doctors' groups, led by the AMA, are fighting the NPs, arguing that less supervision would jeopardize patient safety. Several national studies have found no evidence that NPs provide lower-quality care than physicians (Beck, 2013; Pettypiece, 2013). How might you explain the AMA's resistance from a feminist perspective?

research funding and medical intervention, including therapy with a licensed therapist, medications, and inpatient or outpatient care (Newberry, 2013; Pollack, 2013).

With or without medical intervention, individuals construct and manage their illnesses differently. Some people's social worlds shrink when they become immersed in the day-to-day aspects of managing a chronic illness like rheumatoid arthritis: They become increasingly cut off from everyday routines if they're unable to work, spend less time with family and friends, or can't move about freely. Others create new self-identities by describing themselves as survivors (of breast cancer, for instance), exchanging information about treatment options, and participating in local or national fundraising events (Conrad and Barker, 2010).

Social class also affects social constructions of illness. As educational level rises, people report poorer health, presumably because they're more aware of medical knowledge and evaluate their own physical well-being more critically (Schnittker, 2009).

The Medicalization of Health

In 1952, the American Psychiatric Association (APA) published the first edition of the *Diagnostic and Statistical Manual of Mental Disorders* (DSM), which has become the preeminent global guide for mental illness. The DSM relies on subjective definitions and criteria but is influential in labeling (or unlabeling) mental illness. The number of disorders increased from 106 in 1952 to nearly 300 in the most recent version of the DSM-5. Previous editions added "binge eating" (frequent overeating), "bereavement" (mourning the loss of a loved one), "caffeine intoxication" (drinking too much coffee), "disruptive mood dysregulation disorder" (children's temper tantrums), and "hoarding" (accumulating too much stuff) (American Psychiatric Association, 2013).

The DSM's ever-changing diagnoses and labels are an example of **medicalization**, a process that defines a nonmedical condition or behavior as an illness, disorder, or disease that requires medical treatment. The APA once classified identifying gay as a psychological disorder but dropped sexual orientation from its roster of mental illnesses in 1973. The DSM affects patients, doctors, insurers, pharmaceutical companies, and taxpayers because psychiatrists and other physicians use it to bill insurance companies. DSM-based diagnoses also determine whether people get special services at school, qualify for disability benefits, are stigmatized, and even whether they can adopt children.

Medicalization is a lucrative business. Pharmaceutical corporations have reaped enormous profits because everyday normal anxieties, discomforts, and stresses (frustration with traffic, boredom with routine housekeeping chores, feeling sad or personally insecure) can be "fixed" by popping a Prozac or other pill (Herzberg, 2009). The more behaviors that the DSM defines as mental illness, the more likely psychiatrists are to increase their number of patients. Mental rights advocates and parents also benefit from medicalization, as when children who are diagnosed with attention-deficit hyperactivity disorder

medicalization a process that defines a nonmedical condition or behavior as an illness, disorder, or disease that requires medical treatment.

(ADHD) get insurance coverage or special treatment, including more attention from teachers and health specialists (Conrad and Barker, 2010; Rochman, 2012).

According to critics, the DSM encourages misdiagnosis, overdiagnosis, the medicalization of normal behavior, and the prescription of a large number of unnecessary drugs, especially in the case of ADHD. In 2014, more than 10,000 2- to 3-year-olds (that's right, 2- to 3-year-olds!), particularly those on Medicaid, were being medicated for ADHD. Between ages 2 and 5, 44 percent of children diagnosed with ADHD take medications (Schwarz, 2014; Danielson et al., 2017).

Attitudes vary by country, but many European parents, teachers, and doctors are reluctant to use ADHD medication for what they consider routine childhood behavioral problems. Even though some Europeans believe that ADHD is a disease, "parents are loath to get their child labeled" (Kelley, 2013, p. 26). Thus, America's high ADHD rates and drug sales illustrate cultural differences in the medicalization of health and illness.

Labeling and Stigma

From an interactionist perspective, conditions and behaviors that are diagnosed and labeled as illness or disease change over time, differ among groups, and vary across countries. Medicalization and labeling stigmatize some illnesses and diseases more than others. Impotence in men, which commonly increases as men age and was stigmatized, is now called "erectile dysfunction" and treated with drugs, regardless of age, to enhance a man's sexual experience. Ads and commercials ("Cialis is ready when you are") have flooded magazines, newspapers, and online sites.

Women's sexuality is also being medicalized. The FDA approved the sale of flibanserin to increase postmenopausal women's sex drive. In clinical trials, only 10 percent of women said that the pill increased their sexual desire. Flibanserin has dangerous side effects, like fainting, because it interacts with other medications and alcohol. Women's sex drives fluctuate throughout life, diminish due to bad relationships or stress, and there's nothing wrong with low sexual desire (Nagoski, 2015; Weisman, 2015; refer also to Chapter 9). Pharmaceutical companies, however, can reap huge profits by convincing women that their sexuality is abnormal.

Labeling and stigma can also prevent seeking medical help. Because of fatigue, feeling overwhelmed, and sleep deprivation, almost 30 percent of resident physicians (doctors still being trained) have either severe symptoms or a diagnosis of depression. Few seek medical help, however, because admitting such problems carries a stigma that might affect obtaining a medical license (Mata et al., 2015; Hill, 2017).

Critical Evaluation

Symbolic interaction theories are limited for several reasons. The most common criticism is that they don't address structural factors such as government policies and practices that can ensure better health for everyone. For example, Black individuals experience much lower quality health than White people because of racial differences in access to health care, neighborhood poverty, and other macro-level factors. Public policies could reduce such disparities, but haven't done so (Johnson, 2017).

Second, millions of people successfully resist medicalization. Many deaf people have opted against recommendations to get cochlear implants (devices that can increase hearing). They don't think of deafness as a medical disability but as a social reality that helps them form a community and identify with other deaf people (Conrad and Barker, 2010). A "fat pride" movement argues that obesity's health risks are exaggerated and focuses, instead, on changing discrimination against people who are overweight (refer to Chapter 16).

Third, interactionists emphasize that health and illness are social constructions, but there are many serious health problems whether or not they're medicalized. People die from unknown diseases, for example. Others suffer from "contested" illnesses, including chronic fatigue syndrome, that many physicians question because the symptoms are difficult to diagnose and treat (Institute of Medicine, 2015).

Tom Williams/CQ-Roll Call, Inc./Getty Images

How do authority figures affect people's health perceptions and behavior? Dr. Mehmet Oz is a highly respected and published cardiothoracic surgeon, and host of the popular *The Dr. Oz Show*. Products he endorses on his show "are almost guaranteed to fly off the shelves" (Weathers, 2014). A study of 40 randomly selected episodes of *The Dr. Oz Show* found that scientific data supported only 46 percent of Dr. Oz's recommendations. The researchers cautioned consumers to be skeptical about any information on television medical talk shows because only "up to one half of recommendations are based on believable or somewhat believable evidence" (Korowynk et al., 2014, p. 4).

Chapter Review

Key Takeaways

14-1 What are the key parts for understanding global health and illness? *Health* varies among individuals and societies, but all people experience *disease*. In addressing the question of why some people are healthier than others, health and medical practitioners, researchers, and sociologists examine *social epidemiology*, which includes both the incidence and prevalence of illness or health problems within a population during a specific time period. Worldwide, high-income populations live longer, have lower infant mortality rates, and spend more on health but experience "diseases of wealth" such as diabetes, heart disease, and various cancers.

14-2 What are the main factors affecting health and illness in the United States? Many people live with a *disability*; others die at an earlier age than expected for a variety of macro- and micro-level reasons. The three most important reasons for illness and early death are environmental, demographic, and lifestyle choices.

After age 65, people are more likely to experience *chronic diseases* rather than *acute diseases*. Women tend to live longer than men but have more chronic diseases, such as arthritis, asthma, cancer, and mental illness, and experience higher depression rates than men.

Among U.S. racial-ethnic groups, black babies have the *highest infant mortality rate*, and Latina/o/xs are the least likely to die from illicit drugs and prescription drug misuse. Our lifestyle choices also improve or impair health. Lifestyle factors that can improve health include not smoking, eating healthy and getting regular physical activity, keeping up with preventive care, and curbing substance misuse.

14-3 How does health care in the United States compare with health care globally? *Medicine* is a vital part of *health care* in diagnosing, treating, and preventing illness, injury, and other health impairments. The United States is one the richest nations in the world, but only the very wealthy don't have to worry about receiving and paying for the best medical care available.

Since 2000, employer-based health insurance coverage has deteriorated. Consequently, workers at both large and small firms have been making higher contributions to premiums and paying higher deductibles or copayments. *Medicare* pays many of the medical costs of Americans age 65 and over, regardless of income. *Medicaid*, another government program, provides medical care for those living below the poverty level.

The United States spends more on health care than any other nation in the world, but compared with other high-income countries, covers a smaller percentage of the total population, and has lower quality health outcomes. Unlike people in other high-income countries, Americans have fewer and shorter doctor visits, fewer days of inpatient hospital care, and pay considerably more for prescription drugs and routine medical services (such as doctor's office visits and normal birth deliveries).

In 2010, Congress passed the Patient Protection and Affordable Care Act, also called the Affordable Care Act (ACA) or "Obamacare." The ACA's goal is to give more Americans ages 64 and under access to affordable, quality health insurance and to reduce the growth in U.S. health care spending.

The uninsured include people who live in states where Medicaid doesn't cover those undocumented immigrants, those who haven't enrolled in Medicaid, and people with incomes just above the federal poverty level who can't afford the programs offered by employers or private insurers, and those who forgo coverage by choice.

14-4 How do sociological perspectives on health and medicine compare? Refer to Table 14.4 on page 317.

Key Terms

Health (p. 305)
Social epidemiology (p. 305)
Health care (p. 305)
Disability (p. 307)
Chronic diseases (p. 308)
Acute diseases (p. 308)
Alzheimer's disease (p. 308)
Substance misuse (p. 313)
Medicine (p. 314)
Sick role (p. 317)
Medical-industrial complex (p. 318)
Medicalization (p. 323)

Discussion and Critical Thinking Questions

1. According to a survey, 33 percent of U.S. patients who spent at least one night in a hospital during 2011 reported that nurses weren't available when needed or didn't respond quickly to requests for help. Many nurses, who are as annoyed about this situation as are patients, maintain that the problem isn't a nurse shortage but an inadequate registered nurse-to-patient staffing ratio with 57 percent saying that staffing has gotten worse. Under California's landmark law, which went into effect in 2004, nurses may care for no more than five patients on a medical-surgical unit, four pediatric patients, three patients in labor and delivery, or two patients in intensive care. These minimum nurse staffing requirements have resulted in fewer patient deaths, higher quality patient care, higher nurse morale, and lower burnout rates. In 2023, a bill was introduced to Congress with the outcome still to be determined (Robert Wood Johnson Foundation, 2010; Chen, 2012; Mensik, 2023).
    a. This chapter shows that many U.S. hospitals are run by giant corporations that focus on profits rather than patient care and safety. Should the U.S. health care system eliminate all for-profit hospitals?
    b. Whether a hospital is nonprofit or for-profit, should there be a federal law like California's that establishes nurse–patient ratios? If so, who will pay for the higher health care costs?
2. Is government-paid health care a basic human right? Or should it be limited to taxpayers?
3. Is it wasteful to spend resources on people aged 80 and older who have surgery in the last week or month of life?
4. Are federal, state, and county governments infringing on individual rights by passing (or trying to pass) laws on smoking, sugary drinks, texting while driving, and other individual choices that can affect health? Would people be better off if the government banned corporations from inundating people with ads for alcohol, tobacco, prescription drugs, and harmful food?
5. Employers have the right to not hire people who use illegal drugs or smoke. Explain whether and why you agree or disagree with such rights.

True or False?

How Much Do You Know about Health and Medicine?

The answers to 2 and 3 are true. The answers to 1, 4, 5, and 6 are false.

1. In 2022, the uninsured rate reached an all-time low of 3.7 percent. This is in part due to the Affordable Care Act which gave more Americans under age 65 affordable, quality health insurance.
2. In 2021, lower-income nations had average life expectancy of 62 years compared with higher income nations at 80 years.
3. The percentage of Americans taking three or more prescription drugs increased from 11 percent in 1998 to more than one-third in 2021.
4. No single variable explains well-being. Hereditary (genetic) factors, environmental factors, and other conditions (such as socioeconomic considerations) need to be considered.
5. Although the U.S. spend more (nearly $13,000 per person) than other high-income nations, it ranked last in an overall assessment of access to care, care process, administrative efficiency, equity, and health care outcomes.
6. Social class also affects health. For instance, the wealthiest Americans live 10 to 15 years longer than those with the lowest incomes and those with lower incomes have greater likelihood of experiencing two or more chronic diseases as adults.

15 Population, Urbanization, and the Environment

US Army Photo/Alamy Stock Photo

Learning Objectives

After studying this chapter, you will be able to…

15-1 Explain key aspects of the study of populations and population change.

15-2 Evaluate the main population growth theories.

15-3 Explain why cities in the U.S. and around the world are changing including the consequences of urbanization.

15-4 Explain the major environmental issues affecting people and communities.

15-5 Discuss different perspectives on whether sustainable development is achievable.

True or False?

How Much Do You Know about Population, Urbanization, and the Environment?

1. The total fertility rate (TFR) in the United States has been declining.
2. The pace of worldwide population growth is speeding up.
3. Over recent decades, suburban communities in the United States have become less racially and ethnically diverse.
4. Around the world, an estimated 2 billion people consume contaminated drinking water.
5. Worldwide there is not enough food produced to feed the entire population.
6. The United States ranks last among all nations for its environmental practices.

The answers are at the end of the chapter.

In November 2022, the world hit a population milestone of nearly 8 billion people, and population is projected to increase to 9.7 billion by 2050. Our planet has a record number of inhabitants, and many are living longer than ever before. This chapter examines how such changes affect the world's population, urbanization, and environment. Let's begin with population.

Be sure to take the True or False quiz in the chapter opener to better understand how much you know about these topics.

What do you think?

Individuals can do little to prevent climate change.

1	2	3	4	5	6	7
	strongly agree				strongly disagree	

15-1 Population Changes

Global population has grown rapidly since 1800 and more than tripled since 1900 (**Figure 15.1**). By 2100, the world population will climb to 10.4 billion. "More than half of the projected increase in global population up to 2050 will be concentrated in just eight countries: the Democratic Republic of the Congo, Egypt, Ethiopia, India, Nigeria, Pakistan, the Philippines and the United Republic of Tanzania" (United Nations, 2022, p. i).

The world adds about 400,000 people each day, but population changes vary across countries. By 2023, India is projected to surpass China as the world's largest country (Associated Press, 2022). By 2050, Nigeria will replace the United States as the third largest country. It's projected that Pakistan, Mexico, and Russia will drop off the list while Congo, Ethiopia, and the Philippines will join the top ten (**Table 15.1**).

demography the scientific study of human populations.

population people who share a geographic territory.

fertility the number of babies born during a specified period in a particular society.

crude birth rate (also called the *birth rate*) the number of live births per 1,000 people in a population in a given year.

total fertility rate (TFR) the average number of children born to a woman during her lifetime.

Information about these and other changes comes from **demography**, the scientific study of human populations. A **population** is a group of people who share a geographic territory. A territory can be as small as a town or as vast as the planet, depending on a researcher's focus. Demographers analyze populations in terms of size, composition, distribution, and change. They also study personally relevant topics, including your probability of getting married or divorced, the kind of job you'll probably have, how many times you'll move, and how long you'll probably live (McFalls, 2007).

15-1a Why Populations Change

Populations are never static. Their growth or decline involves three key factors: fertility (how many people are born), mortality (how many die), and migration (how many move from one area to another).

Fertility: Adding New People

The study of population changes begins with **fertility**, the number of babies born during a specified period in a particular society. Demographers use several fertility measures depending on the level of specificity needed. One of the most commonly used measures is the **crude birth rate** (also called the *birth rate*), the number of live births for every 1,000 people in a population in a given year. "Crude" implies that the rate is a general measure of a society's childbearing. It's based on the total population rather than more specific measures (e.g., a woman's age or marital status). In 2022, the crude birth rate was 18 worldwide, which is a decrease from 20 in 2015. The 2022 crude birth rate for the United States was 12. This rate was highest for Niger (47) and lowest for both Saint Pierre and Miquelon (6.5) (U.S. Census Bureau, 2023).

Another standard measure, the **total fertility rate (TFR)**, is the average number of children born to a woman during her lifetime. In the early 1900s, women in the U.S. had an average TFR of 3.5, compared with

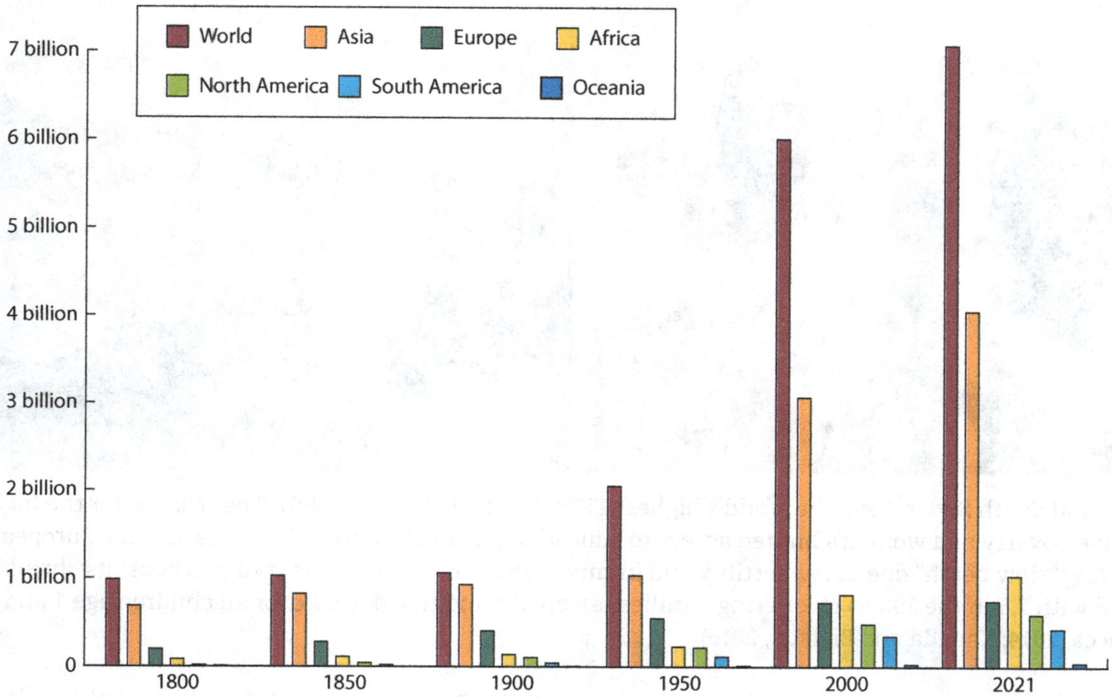

Figure 15.1 World Population Growth Through History

Source: Based on Hannah Ritchie et al. 2023. "Population Growth". Published online at OurWorldInData.org. Retrieved from: 'https://ourworldindata.org/population-growth'

only 1.7 in 2021. Worldwide, TFRs are much higher in lower-income nations (4.6) than in high-income nations (1.5), ranging from a low of 0.8 in Hong Kong, Republic of Korea (South Korea), and Palau to a high of 6.8 in Niger (The World Bank, 2023 "Fertility rate,

Table 15.1 World's Ten Most Populous Countries, 2022 and 2050

2022		2050	
Country	**Population (In Millions)**	**Country**	**Population (In Millions)**
1. China	1,410	1. India	1,622
2. India	1,390	2. China	1,362
3. United States	337	3. Nigeria	428
4. Indonesia	277	4. United States	389
5. Pakistan	2438	5. Indonesia	318
6. Nigeria	225	6. Congo	241
7. Brazil	217	7. Brazil	238
8. Bangladesh	166	8. Ethiopia	196
9. Russia	142	9. Bangladesh	194
10. Mexico	129	10. Philippines	158

Source: Based on U.S. Census Bureau, 2023 "International Database (IDB)"

total (births per woman)"). Fertility rates vary for many reasons, including contraceptive access, education, and use. Fertility rates within a country vary too. In the U.S., there have been higher TFRs among women from other nations of origin (2.02) compared with women born in the United States (1.69). Yet, the gap between the two is narrowing, with the TFR for women from other nations of origin declining at a faster rate (Bialik, 2017; Camarota and Zeigler, 2021).

To maintain a stable population, each woman must have an average of two children, the *replacement rate* for herself and her partner. TFRs above 2.1 indicate that a country's population is increasing and getting younger. Rates below 2.1 mean that a country's population is decreasing and getting older. The older a country's population, the more difficult it is to support people who are living into their eighties and longer and to provide resources for the young (United Nations Population Division, 2017; refer also to Chapter 12 on the old-age dependency ratio).

In nearly all countries, fertility tends to fall as people earn higher incomes and women earn more education. Some nations, particularly those in East Asia, worry that their "ultra low" TFRs of 1.3 or less aren't at replacement levels. For instance, in Tokyo, Seoul, and Bangkok college graduates who are women outnumber those who are men and women's labor force participation is high. According to

AP Images/John Heilprin

Agencja Fotograficzna Caro/Alamy Stock Photo

Africa's Niger and South Sudan have the world's highest TFRs, 7.6 and 6.7, respectively. The reasons for the high fertility include poverty and women's limited access to education and birth control. In contrast, some European towns are facing "slow death" due to low fertility and immigration rates. Germany is trying to boost its shrinking TFR (1.5 compared with 2.5 in the 1960s) by offering families tax breaks and free day care for all children age 1 and older (Rossi and Jucca, 2014; Kaneda and Bietsch, 2016).

cultural norms, however, women should give up work after the birth of a child. And, employed or not, women do at least three more hours of housework a day than men. Consequently, many women aren't marrying or are postponing marriage, reducing their likelihood of having children. Because nonmarital births are taboo and rare in Asia, TFRs have plummeted ("Asia's New Family Values," 2015).

Italy and Japan have among the lowest fertility rates in the world (1.2 and 1.4, respectively). Italy's government has promoted a "Fertility Day" campaign to encourage couples to have more babies. Many working women would like to have more children, but don't do so because the government offers limited child care services. Because of low birth rates and restrictive immigration policies, Japan has one of the most rapidly aging societies worldwide. One of the results is 10 million abandoned properties or "ghost houses." The vacant houses have been inherited by people who don't want or need them, but they can't sell the abandoned properties because of a shortage of buyers (Soble, 2015; Pianigiani, 2016; Hornyak, 2023).

Mortality: Subtracting People

The second factor in population change is **mortality**, the number of deaths in a population during a specified period. Demographers typically measure mortality using the **crude death rate** (also called the *death rate*), the number of deaths per 1,000 people in a population in a given year. In 2022, the death rate was 8.4 for the United States. Worldwide, the lowest death rate was 1.4 in Qatar and the highest was 16.4 in Serbia (U.S. Census Bureau, 2023).

Higher-income countries, compared with most lower-income countries, have lower death rates, but also large proportions of people age 65 and older who are no longer employed but require costly medical services. Thus, a better measure of a population's health is the *infant mortality rate*, the number of deaths of infants younger than 1 year per 1,000 live births. Generally, as the standard of living improves—access to clean water, adequate sanitation, and medical care—the infant mortality rate decreases. In 2021, the average infant mortality rate was 4 in higher-income countries and 47 in the lower-income countries (The World Bank, 2023 "Mortality Rate, Infant (per 1,000 live births)") (refer also to Chapter 14).

Life expectancy is the average number of years that people who were born at about the same time can expect to live. Globally, life expectancy is 73 years for all people, 71 for men and 76 for women. Again, however, there are considerable variations across countries—from a high of 85 in Monaco to a low of 51 in Lesotho. The United States, with a life expectancy of 79, ranks below many other high-income countries (The World Bank, 2023 "Life Expectancy at Birth (years)"). In almost every society, people with a higher socioeconomic status live longer and healthier lives because often their occupations are physically safer, and their resources include clean drinking water, sanitation, and medical services.

mortality the number of deaths in a population during a specified period.

crude death rate (also called the *death rate*) the number of deaths per 1,000 people in a population in a given year.

Migration: Adding and Subtracting People

The third demographic factor in population change is **migration**, the movement of people into or out of a specific geographic area. There are two types of migration: international and internal. *International migration*, movement to another country, includes *emigrants* (people who move out of a country) and *immigrants* (people who move into a country).

Migrants are people who choose to leave their country to seek a better life elsewhere. *Refugees* (sometimes called *displaced persons*) are people who are forced to flee their country—often with no warning—to escape war, persecution, torture, or death. An international migrant can have several of these characteristics. For example, my family members and I were refugees who fled Lithuania (by horse and wagon) from Communism during World War II, emigrated to displaced persons camps in Germany, and then immigrated to the United States.

International Migration. The number of international migrants grew from about 79 million in 1960 to nearly 280 million in 2020 (a 350 percent increase), and migrants make up 3.6 percent of the world's population today. The greatest shifts of people have been from one low-income country to another (e.g., from Indonesia to Saudi Arabia) or from a lower-income to higher-income country (e.g., from Mexico to the United States). Also, most emigrants move to a neighboring country (e.g., Mexico to the United States, for example, rather than from Mexico to Canada). The United States has the largest number of the world's migrants (nearly 51 million). Proportionately, however, only 15 percent of the U.S. population is foreign born, compared with 88 percent in the United Arab Emirates, 77 percent in Qatar, and 73 percent in the British Virgin Islands (Connor, 2016; International Organization for Migration, 2022).

There are currently more refugees worldwide than at any time in recorded history. In 2022, nearly 103 million people had been forcibly displaced from their homes because of conflict, violence, or human rights violations. Of that group, 32.5 percent are refugees. More than 70 percent of all forcibly displaced people come from just five countries—Syrian Arab Republic, Venezuela, Ukraine, Afghanistan, and South Sudan (United Nations High Commissioner for Refugees, 2021). People also migrate to another country for economic reasons or leave bad schools, high crime areas, and because of natural disasters. Thus, the reasons for international migration today are similar to those of European immigrants who came to the United States during the twentieth century.

Internal Migration. International migration has increased, but 92 percent of adults worldwide who change residences do so because of *internal migration*, movement within a country. Over a five-year period, 24 percent of American adults move within the country, making the United States "one of the most geographically mobile countries in the world" (Chandler, 2016). In parts of rural Spain, people have migrated to cities for jobs, education, and access to public transit and health care. As a result, abandoned villages are for sale. One village, where the asking price is about $230,000, includes 100 acres with half a dozen houses and two sprawling farms with room for 70 cattle. The biggest houses "have hardwood floors and five bedrooms overlooking an orchard with peaches, figs, walnuts, apples, and pears . . . and a little river full of trout" (Frayer, 2015).

Why do people move? The most common reasons (42 percent) are housing-related (e.g., "Wanted new or better home/apartment," eviction), 28 percent are family-related (e.g., change in marital status), and 20 percent are job-related (e.g., new job, job transfer, to look for work) ("Americans Moving . . .," 2016).

Socioeconomic status, the economy, and natural disasters also affect migration. From the late 1940s through the mid 1980s, around 20 percent of people moved within the county each year. That rate began declining for many reasons. With higher levels of home

Dona_Bozzi/Shutterstock.com

Over 12 million Syrians (more than half of the country's population) have been forcibly displaced from the civil war in the Syrian Arab Republic (Syria) that began in 2011. Of this total, nearly 7 million people are displaced within the country, and another 5.4 million are refugees living in countries nearby (United Nations High Commissioner for Refugees, 2021). "Many are…moving from village to village, up to as many as 20 times, before they finally made it across an international border" (Bengali, 2014, p. 10).

migration the movement of people into or out of a specific geographic area.

ownership and two-income earner families, people became less likely to move. Recently, those numbers decreased even more because of economic challenges including those related to the COVID-19 pandemic. From 2021 to 2022, only 8.7 percent of people in the U.S. migrated internally. Natural disasters can also affect internal migration. By 2015, there were more Puerto Ricans living in the United States than on the island of Puerto Rico, primarily because many low-income residents sought better jobs and schools in the United States. After Hurricane Maria devastated much of the island in 2017, between 10 and 14 percent of Puerto Rico's 3.4 million people, including many young professionals, moved to the mainland or planned to do so (Dorell, 2017; Luscombe, 2017; Frey, 2023).

15-1b Population Composition and Structure

Demographers study age and gender to understand a population's composition and structure. Two of the most common measures are *sex ratios* and *population pyramids*.

Sex Ratios

A **sex ratio** is the proportion of men to women in a population. A sex ratio of 100 means that there are equal numbers of men and women; a ratio of 95 means that there are 95 men for every 100 women (fewer men than women). Worldwide, without human intervention, 103 to 107 boys are born for every 100 girls. In the United States, the sex ratio is 105 at birth. In China and San Marino, it is higher than the average at 109 (CIA, 2023).

These sex ratio imbalances are attributed primarily to *female infanticide* (sometimes called *gendercide*)—the intentional killing of baby girls. In some countries, including China and India, there's a preference for boys because they are expected to carry on the family name, care for elderly parents, inherit property, and play a central role in family rituals. Consequently, and particularly in rural areas, hundreds of thousands of infant girls die every year because of neglect, abandonment, and starvation. Others are aborted after ultrasound scanners reveal the child's sex ("Sex Selection," 2017).

China is a good example of some of the unintended negative consequences of lopsided sex ratios. In 1979, China passed a one-child policy to combat poverty and overpopulation. The policy succeeded in limiting population growth because of at least 600 million abortions and sterilizations. The country's TFR fell from 7.5 in the 1960s to 1.5 by 2010, but well below the 2.1 replacement rate to maintain a constant population (Eberstadt, 2013; "Family Planning . . . ," 2014).

Because of the unbalanced sex ratio for the last 30 years, China now has approximately 35 million more young men than women. That means a large number of men won't be able to find women to marry—a problem that has already increased the illegal trafficking of women from lower-income neighboring countries, including Cambodia, Myanmar, and Vietnam—and there'll be millions of aging bachelors (Tsai, 2012; "The Marriage Squeeze . . . ," 2015; Feng, 2021).

China's working-age population also started declining in 2015, resulting in fewer workers to support a growing older population. To reduce the load of elder care, the government eased the one-child policy in 2013 and in 2015 by allowing married couples to have two children. China's leaders want to encourage increasing family size among those living in urban areas, with middle-incomes, and higher levels of educational attainment. This intended group hasn't shown much interest in doing so because of high costs of living and expenses of raising a second child. As of 2021, China's has a low TFR (1.2) well below the replacement rate (Buckley, 2015; Walsh, 2015; The World Bank, 2021).

Population Pyramids

A **population pyramid** is a graphic depiction of a population's age and sex distribution at a given point in time. **Figure 15.2** shows three different population pyramids from 2022. Niger is a young country with around 60 percent its population under age 20. In contrast, the Republic of Korea (South Korea) is an old country, and the United States is somewhere in the middle.

The shape of the pyramid (a triangle for Niger, a rectangle for the United States, and a diamond for Republic of Korea) has future implications for young and old countries. The Republic of Korea has a relatively small number of women ages 15 to 44 (in their reproductive years) and more people ages 45 to 79. This suggests that there'll be a scarcity of workers to support an aging population and a greater need for social services for older people than for children and adolescents (refer to Chapter 12). Thus, population pyramids give us a snapshot of a country's demographic profile and indicate some of the problems that countries are likely to face in the future.

sex ratio the proportion of men to women in a population.

population pyramid a graphic depiction of a population's age and sex distribution at a given point in time.

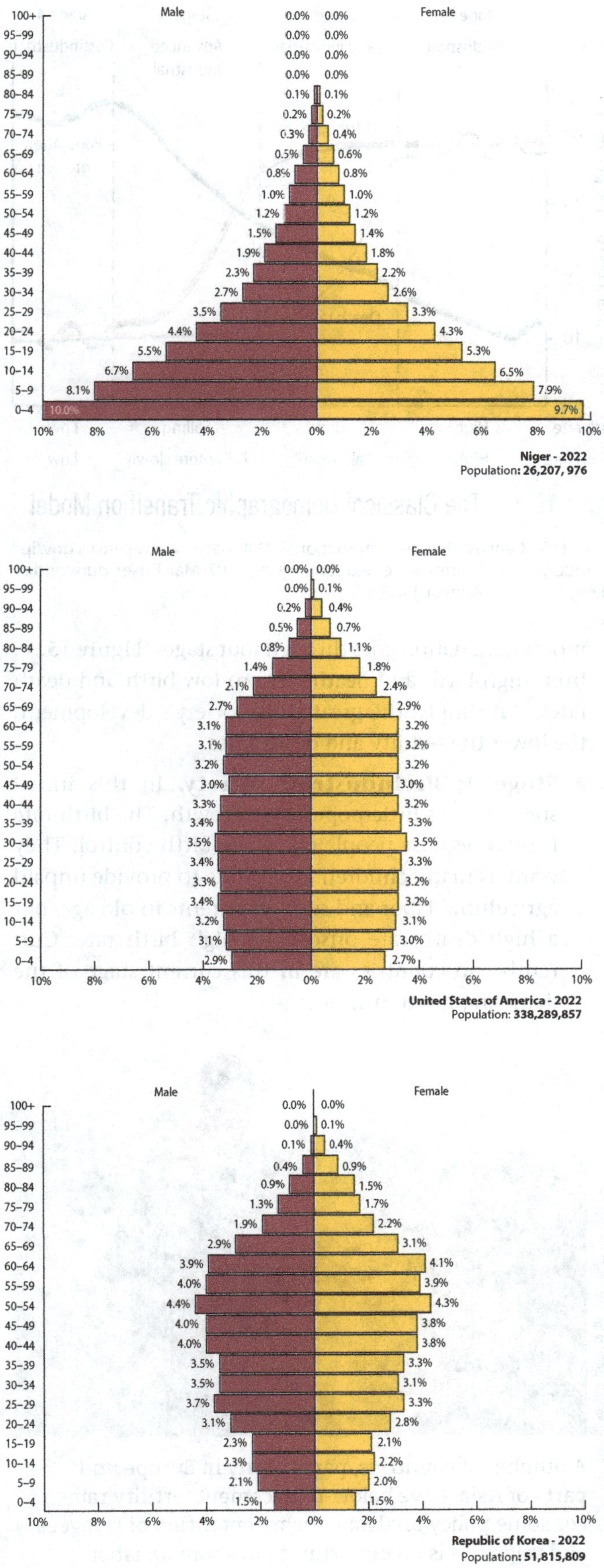

Figure 15.2 Population Pyramids, 2022

Source: PopulationPyramid.net, 2023 (accessed June 13, 2023).

15-2 Population Growth: A Ticking Bomb?

Some of the world's largest countries will grow even more by 2050 (**Table 15.1**). So, has population growth gotten out of hand? There are many views on this question, but two of the most influential have been Malthusian theory (which argues that the world can't sustain its unprecedented population surge), and demographic transition theory (which maintains that population growth is slowing).

Malthusian Theory

For many demographers, population growth is a ticking bomb. They subscribe to **Malthusian theory**, which maintains that the population is growing faster than the food supply needed to sustain it. This theory is named after Thomas Malthus (1766–1834), an English economist, clergyman, and college professor who argued that humans are being added to the earth more quickly than the earth's ability to produce sufficient food.

According to Malthus (1798/1965), population grows at a *geometric rate* (2, 4, 8, and so on), whereas the food supply grows at an *arithmetic rate* (1, 2, 3, 4, and so on). That is, two parents can have 4 children and 16 grandchildren within 50 years. The available number of acres of land, farm animals, and other sources of food can

Hulton Archive/Stringer/Getty images

Thomas Malthus

Malthusian theory the population is growing faster than the food supply needed to sustain it.

increase in that time period, but certainly not quadruple. In effect, then, the food supply will not keep up with population growth. Because there are millions (now billions) of people who are able to have children, the results could be catastrophic, with masses of people living in poverty or dying of starvation.

Malthus posited that two types of checks affect population size. *Positive checks* (famine, disease, war) limit reproduction, raise the death rate, and lower the overall population. *Preventive checks* (contraception, postponing marriage, abortion, and extramarital sex) also limit reproduction by reducing birth rates and, consequently, ensure a higher standard of living for all (Malthus, 1872/1991).

Malthusian theory has had a lasting influence. *Neo-Malthusians* (or New Malthusians) agree that the population is exploding beyond food supplies. The world's population reached its first billion in 1800 (**Figure 15.1** on page 329). As a result of this growth, according to some influential neo-Malthusians, the earth has become a "dying planet"—a world with insufficient food and a rapidly expanding population that pollutes the environment (Ehrlich, 1971; Ehrlich and Ehrlich, 2008).

Although there was progress in reducing hunger worldwide, more recently there has been a shift backwards. The COVID-19 pandemic moved the number of people experiencing hunger up 150 million to a total 828 million in 2021 (World Health Organization, 2022). Chronic hunger is connected with stunting (both physical and cognitive) as well as chronic diseases. It can also limit the ability to participate in activities such as work and school. Enough food is produced to feed more than the world's population, yet "one-third of food produced for human consumption is lost or wasted." This food would be enough to feed all of those who experience hunger—times two! (World Food Programme, 2020). Also, internal wars restrict distribution, fueling crises in the countries that need food the most (Druker, 2015).

Demographic Transition Theory

Some demographers are more optimistic than neo-Malthusians. **Demographic transition theory** maintains that population growth is kept in check and stabilizes as countries experience economic and technological development, which, in turn, affects birth and death rates. According to this theory, population growth changes as societies undergo industrialization, modernization, technological progress, and urbanization. During these

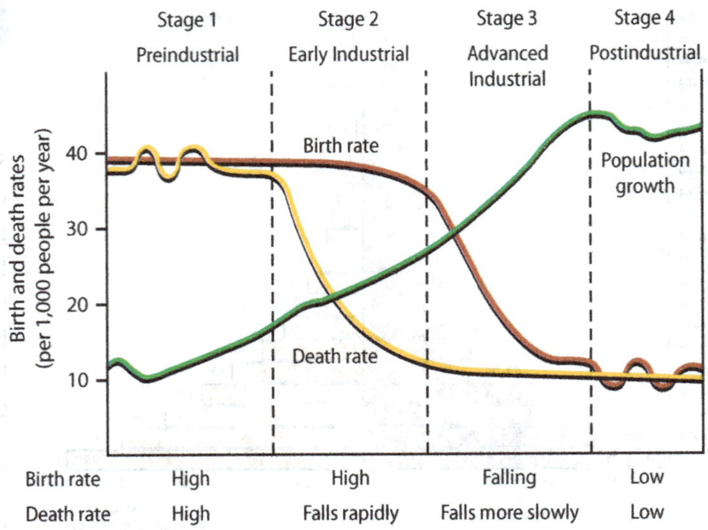

Figure 15.3 The Classical Demographic Transition Model

Source: U.S. Census Bureau, International Database, www.census.gov/ipc/www/idb/pyramids.html. Accessed January 20, 2007; Max Roser, ourworldindata.org (accessed August 19, 2017).

processes, a nation goes through four stages (**Figure 15.3**), from high birth and death rates to low birth and death rates. Put simply, the greater the society's development, the lower the fertility and death rates.

- **Stage 1: Preindustrial society.** In this initial stage, there's little population growth. The birth rate is high because people rarely use birth control: They want as many children as possible to provide unpaid agricultural labor and support parents in old age, but a high death rate offsets the high birth rate. Currently, no countries are in this current stage of the demographic transition.

Sean Gallup/Getty Images

A number of countries, particularly in Europe and parts of Asia, have below-replacement fertility rates. For some policy analysts, the recent influx of refugees and migrants is an opportunity to shore up labor forces with young workers whose taxes will support graying and dwindling native-born populations (OECD/European Union, 2015; Portes, 2015).

demographic transition theory population growth is kept in check and stabilizes as countries experience economic and technological development.

- **Stage 2: Early industrial society.** There's significant population growth because the birth rate is higher than the death rate. The birth rate may even increase over what it was in Stage 1 because mothers and their children benefit from improved health care. Parents may still have large numbers of children because they fear that many of them will die, but the death rate declines because of better sanitation, better nutrition, and medical advances (e.g., immunizations and antibiotics). Most of the world's lowest income countries—including Uganda, Afghanistan, and Yemen—are currently in Stage 2.
- **Stage 3: Advanced industrial society.** As the infant mortality rate declines, parents have fewer children. Effective birth control reduces family size. The decrease in childcare responsibilities, in turn, enables women to work outside the home. India, Mexico, and South Africa are currently in Stage 3.
- **Stage 4: Postindustrial society.** In this stage, the demographic transition is complete, and the society has low birth and death rates. Women tend to be well educated and to have full-time jobs or careers. If there's little immigration, the population may even decrease because the birth rate is low. This is the case today in Canada, Japan, Singapore, Hong Kong, Australia, New Zealand, the United States, and Germany.

Critical Evaluation

The dire predictions of Malthus and his successors that global population growth would lead to worldwide famine, disease, and poverty haven't come true. There has been a decrease in the number of people who live in extreme poverty (subsisting on less than $2.15 a day). In 1990, 1.8 billion people (35 percent of the world's population). Still today, there are many people living in extreme poverty—648 million people (8 percent of the world's population) (refer to Chapter 8).

Despite neo-Malthusians' fears, global fertility is half of what it was in 1972. The population of some industrialized countries is declining because people aren't having enough babies to replace themselves. These countries are experiencing **zero population growth (ZPG)**, a stable population level that occurs when each woman has no more than two children.

Fearing that there won't be enough young workers to pay for social security and the rising cost of health care for aging populations, some low-birth nations with fertility rates below ZPG are paying women to have more children. Russia has revived a Soviet-era title of "Mother Heroine" with $16,000 monetary compensation for women who birth 10 children; Japan has expanded its day care facilities and offers families a monthly allowance of $75 per child younger than 15; Germany and France have liberal parental leave; the Republic of Korea is increasing the stipend to parents of children under age one from about $230 to $765 by 2024; China is encouraging married couples to have two children; and Singapore, besides giving couples almost $5,000 for having a child, lets families with babies go to the front of the queue for high quality, affordable public housing (Hales, 2014; "Pro-natalism," 2015; Reid, 2022; Ahn, 2023).

North Wind Picture Archives/Alamy Stock Photo

During the Industrial Revolution, many cities, like this one in Hamburg, Germany, constructed canals that provided an inexpensive means of transportation and distributing goods.

Some neo-Malthusians maintain, however, that it's irresponsible for *any* country to encourage higher fertility rates. They worry about the consequences of adding 3 billion more inhabitants to the planet in less than 50 years, especially for many lower-income countries "with desperate economic outlooks" (Sachs, 2005; Shorto, 2008).

One result of population growth is urban growth. Cities attract people because of jobs and cultural activities, but urbanization has also created numerous problems.

15-3 Urbanization

If you've flown over the United States, you've probably noticed that people tend to cluster in and around cities. After sunset, some areas glow with lights, whereas others are engulfed in darkness. The average person, in the United States and worldwide, is more likely to live in a city than a rural area, and this trend is rising.

zero population growth (ZPG) a stable population level when each woman has no more than two children.

A *city* is a geographic area where a large number of people live relatively permanently and make a living primarily through nonagricultural activities. **Urbanization**, which increases the size of cities, is the movement of people from rural to urban areas. Most of this discussion focuses on U.S. cities, but let's begin with a brief look at global urbanization.

15-3a Urbanization: A Global View

In 2008, for the first time in history, a majority of the world's population lived in urban areas. Fifty-six percent of people worldwide live in urban areas, and by 2050, nearly 70 percent of the world's population is projected to be urban (The World Bank, 2023 "Urban Development").

Origin and Growth of Cities

Cities are one of the most striking features of modern life, but they've existed for centuries. About 7,000 years ago, for example, people built small cities in the Middle East and Latin America to protect themselves from attackers and to increase trade. By 1800, 56 cities in Western Europe had a population of 40,000 or more (Chandler and Fox, 1974; De Long and Shleifer, 1992).

Before the Industrial Revolution, which began in the late eighteenth century, urban settlements in Europe, India, and China developed largely because people figured out how to use natural resources (e.g., mining coal and transporting water for irrigation and consumption). The Industrial Revolution spurred ever-increasing numbers of people to move to cities in search of jobs, schooling, and improved living conditions. As a result, the urban population surged—from 3 percent of the world's population in 1800 to 14 percent in 1900 (Sjoberg, 1960; Mumford, 1961). As industrialization advanced, urbanization increased.

World Urbanization Trends

Africa and Asia are home to nearly 90 percent of the world's rural population, but both are urbanizing faster than other regions (**Table 15.2**). Between 2014 and 2050, just three countries—India, China, and Nigeria—are expected to account for 37 percent of the world's urban population growth (United Nations, 2014).

Many of the world's largest cities are **megacities**, metropolitan areas with at least 10 million inhabitants. In 1950, the world had only two megacities: Tokyo (11.3 million) and New York-Newark (12.3 million). By 2030, there'll be 41 megacities. Only two of them will be in the United States—New York-Newark (19.9 million), and Los Angeles-Long Beach-Santa Ana (13.3 million)—and both will be much smaller than other cities including Tokyo (37.2 million), Delhi (36.1 million), and Shanghai (30.8 million). Thus, the number of megacities is growing, and they're much bigger than in the past (United Nations, 2014).

Should the explosive growth of cities and megacities concern us? Generally, cities provide jobs, offer better health services, and have more educational opportunities, but not everyone benefits from such advantages.

Table 15.2 Urbanization Around the World

Percentage of People Living in Urban Areas	1950	2014	2050 (Projected)
World	**29**	**54**	**66**
Africa	14	40	56
Asia	18	48	64
Latin America and the Caribbean	41	80	86
North America	64	81	87
Europe	51	73	82
Oceania	62	71	74

Source: Based on United Nations Department of Economic and Social Affairs, 2012, Table 2, and United Nations, 2014, Table 1.

Qilai Shen/Bloomberg/Getty Images

In Beijing, China, with nearly 22 million people, rentable bikes are popular, but bike lanes are often blocked by parked cars, can end without warning, and have a "menagerie of moving objects," including "delivery guys on motorcycles, tourists meandering in rickshaws, and construction crews hauling bricks in rusted carts" (Yeung, 2023; Larson, 2017, p. 50).

urbanization people's movement from rural to urban areas.

megacities metropolitan areas with at least 10 million inhabitants.

People with lower incomes, who live in cities, are often crowded into blighted areas where children are less likely to be enrolled in school, sanitation is inadequate, and the economic gap between the haves and have-nots is widening (Laneri, 2011).

15-3b Urbanization in the United States

Like many other countries, the United States is becoming more urban. How, specifically, has the urban landscape changed? And what are some of the consequences?

How Urban America Is Changing

During the Industrial Revolution, millions of Americans in agricultural areas migrated to cities to find jobs. As a result, between 1900 and 2000, the urban population surged from 39 to 79 percent. Rural areas cover 97 percent of the nation's land but contain only 14 percent of the population (about 46 million people), down from 55 percent in 1910 (Riche, 2000; Dobis et al., 2021).

Suburbanization. As cities grew more crowded, dirtier, and noisier, urban growth sparked *suburbanization*, people moving to communities just outside a city. In 1920, about 15 percent of Americans lived in the suburbs compared with 45 percent today (Palen, 2014; Frey, 2022 "Today's Suburbs Are Symbolic of America's Rising Diversity").

During the 1950s, two-thirds of urban dwellers moved to suburbs. The federal government, fearing a return to the economic depression of the 1930s, underwrote the construction of much new housing in the suburbs. The general public obtained low-interest mortgages, veterans were offered the added incentive of being able to purchase a home with a $1 down payment, and massive highway construction programs enabled commuting by car. As a result, suburbs mushroomed (Rothman, 1978).

Suburban life is appealing. Some of the benefits include more privacy and space, one's own yard and garages, better schools, safer streets, and, if there are jobs nearby, shorter commutes. Suburban elites have sprawling "McMansions," considerable acreage, and gated communities.

Especially since 1980, however, much of the social and physical separation between cities and suburbs has blurred. The suburbs are becoming more racially and ethnically diverse shifting from 79 percent of White, non-Latina/o/x (non-Hispanic) residents in 1990 down to 54 percent in 2020. The suburbs are also experiencing more growth in numbers of people living in poverty compared with urban areas (Fry, 2018; Frey, 2022 "Today's Suburbs Are Symbolic of America's Rising Diversity").

Patrick Lienin/AGE Fotostock

Is this your dream house? Compared with the average American home of around 2,500 square feet, the typical "tiny house" is between 100 and 400 square feet. Living smaller is becoming popular because of environmental concerns and much lower financial costs. Prices range from $4,000 to $200,000 depending, like a traditional house, on location, cost of the land, building materials, and so on. Some residents and local governments refuse to allow tiny houses, fearing they'll drive down property values (Willett, 2015; Beitsch, 2016; Sullivan, 2017).

Edge Cities and Exurbs. Originally, most suburbs were bedroom communities for commuters with jobs in the city. Over the last few decades, suburbanization has generated **edge cities**, business centers that are within or close to suburban residential areas and include offices, schools, shopping, entertainment, malls, hotels, and medical facilities.

People have also created **exurbs**, small, usually prosperous communities beyond a city's suburbs. About 18 percent of Americans live in exurbs, and counties containing far-flung exurbs are growing faster than many urban counties. The average exurbanite is White, middle or upper-middle income, married with children, and a "super commuter" (who travels two or more hours a day for work) (Berube et al., 2006; Badger, 2015; Frey, 2015).

Metropolitan Statistical Areas. Together, suburbs, edge cities, and exurbs form **metropolitan statistical areas** (MSAs), also called *metro areas*, that consist of a central city of at least 50,000 people and the urban areas linked to it. The vast majority of Americans (84 percent) live in 366 MSAs.

---

edge cities business centers that are within or close to suburban residential areas.

exurbs small, usually prosperous communities beyond a city's suburbs.

metropolitan statistical area (MSA, also called *metro area*) a central city of at least 50,000 people and urban areas linked to it.

Since 2015, 10 of the 15 fastest-growing metro areas have been in the South, with four of the top five in Texas (e.g., Frisco and McKinney in the Dallas–Fort Worth–Arlington MSA) (U.S. Census Bureau Newsroom, 2017).

Recently, there has been fluctuation in movement toward urban or suburban areas. From 2011 to 2014, for the first time in nearly a hundred years, metro areas grew faster than suburbs (Frey, 2014). Two groups—young professionals and baby boomers—drove the trend in metro living. Because of the shift to young professionals postponing marriage and parenthood (refer to Chapter 12), they were less concerned about the quality of schools. Moreover, working in cities reduced commute times that would be required for living in suburbs and commuting to urban centers. Baby boomers, particularly those who are retiring, moved to metro areas for a variety of reasons (e.g., better public transportation, greater access to cultural events, social services for older people, no longer being healthy enough to maintain a suburban house) (Kneebone and Berube, 2013; Westcott, 2014). Beginning in 2015, higher growth in suburban areas again overtook that of urban areas. From 2020 to 2021, 51 urban areas experienced population losses. The COVID-19 pandemic and economy both prompted movement from city to suburbs. People again sought more affordable housing, and the shift to remote work options added a choice to live in suburban areas away from urban jobs (Frey, 2022 "Big Cities Saw Historic Population Losses").

Some Consequences of Urbanization

Cities offer many benefits, including a vast array of culturally diverse restaurants, shops, and activities, but urbanization also creates problems. Some of the drawbacks include urban sprawl, increased traffic congestion, and a scarcity of affordable housing.

Urban Sprawl. **Urban sprawl**—the rapid, unplanned, and uncontrolled spread of development into regions adjacent to cities—is widespread. According to some estimates, urbanization in the Southeast will increase by up to 190 percent by 2060. Development on that scale will result in losing 15 percent of agricultural land, 12 percent of grasslands, and 10 percent of forests (Terando et al., 2014).

Urban sprawl has created rapid *job sprawl*, which occurs when companies move jobs from metropolitan areas to suburbs. The more distant the suburb, the less likely low-income people are to hear about employment opportunities through informal networks, to afford houses in these areas, and to have transportation to the jobs (Raphael and Stoll, 2010).

Some large metro areas (e.g., Washington, D.C., New York, Boston, Atlanta, Miami, and Denver) are implementing "walkable urbanism" to curb urban sprawl. That is, the developments are neighborhoods where everyday destinations (e.g., work, school, restaurants) are concentrated and within walking distances. For the most part, however, most urban development is built around driving rather than walking (Leinberger and Lynch, 2014).

Traffic Congestion. Generally, the only way to get around in urban sprawl areas is by car. This means that most suburban households face the costs of buying, fueling, insuring, and maintaining multiple cars. The U.S. population has increased by 23 percent over the last 25 years, but total highway miles have increased by only 5 percent, resulting in greater traffic congestion within and outside cities. More than 10 million Americans (8 percent of all people who don't work at home) now travel 60 minutes or longer one way, a proportion that has increased by 95 percent since 1990. Traffic snarls and long commutes increase air pollution and stress, waste fuel and time, and decrease the time that people have for family and leisure activities (McKenzie, 2013; Jones, 2017).

Lack of Affordable Housing. In some cases, the people with lower-incomes are pushed out by **gentrification**, a process in which upper-middle-income and affluent people buy and renovate houses and stores in downtown urban neighborhoods. Governments in many older cities encourage gentrification to increase dwindling populations, to revitalize urban areas, and to augment tax revenues. Gentrification can benefit an entire community because it breaks up concentrated poverty pockets, the new residents may demand improvements in schools and crime control, and an influx of retail stores can generate new jobs. Rent increases, however, have displaced many—mainly financially disadvantaged people of color and small businesses. Also, some longtime Black residents have pointed out that aggressive policing tactics, like stopping and frisking people on the street, increase when wealthier people move into a neighborhood (Buntin, 2015; Mendoza, 2016; Feder, 2020).

Residential Segregation. Since 1970, racial residential segregation has decreased. Metropolitan areas, especially, are becoming racially more diverse (**Figure 15.4**). Among metropolitan areas with a population of 500,000 or more, the least segregated are in the South and West,

---

urban sprawl rapid, unplanned, and uncontrolled spread of development into regions adjacent to cities.

gentrification upper-middle-income and affluent people buy and renovate houses and stores in downtown urban neighborhoods.

Erik McGregor/Getty Images

Gentrification improves old city neighborhoods and increases property values, but also displaces residents—who are often financially disadvantaged people of color.

and the most segregated are mainly in the Northeast and Midwest. There may be a "continued easing of the color line" because racial segregation has waned even in long-standing "hypersegregated" cities like Chicago and New York. The nation's 20 most multiethnic metropolitan regions are now "global neighborhoods" that have substantial numbers of White, Black, Latina/o/x, and Asian individuals (Iceland et al., 2013: 119; Frey, 2015).

Whereas racial residential segregation has decreased, socioeconomic residential segregation is more pronounced than in the past, and the most residentially segregated are low-income Black and Latina/o/x individuals. Since 2000, for example, the number of Black and Latina/o/x people living in high-poverty areas has nearly doubled, rising from 7 to almost 14 million people (Bischoff and Reardon, 2013; Jargowsky, 2015).

Residential segregation along income lines has increased primarily because of the long-term rise in economic inequality (refer to Chapters 8 and 11). Some wealthy neighborhoods have even fought off state laws that mandate constructing affordable housing for low- and middle-income families. According to the high-income homeowners, building less expensive houses would raise crime rates, lower property values, and "There's plenty of affordable housing *in neighboring communities*" (McCabe, 2014: 39, emphasis added).

15-3c Sociological Explanations of Urbanization

How and why do cities change? In answering this and other questions, functionalists underscore urban development, conflict theorists emphasize the impact of

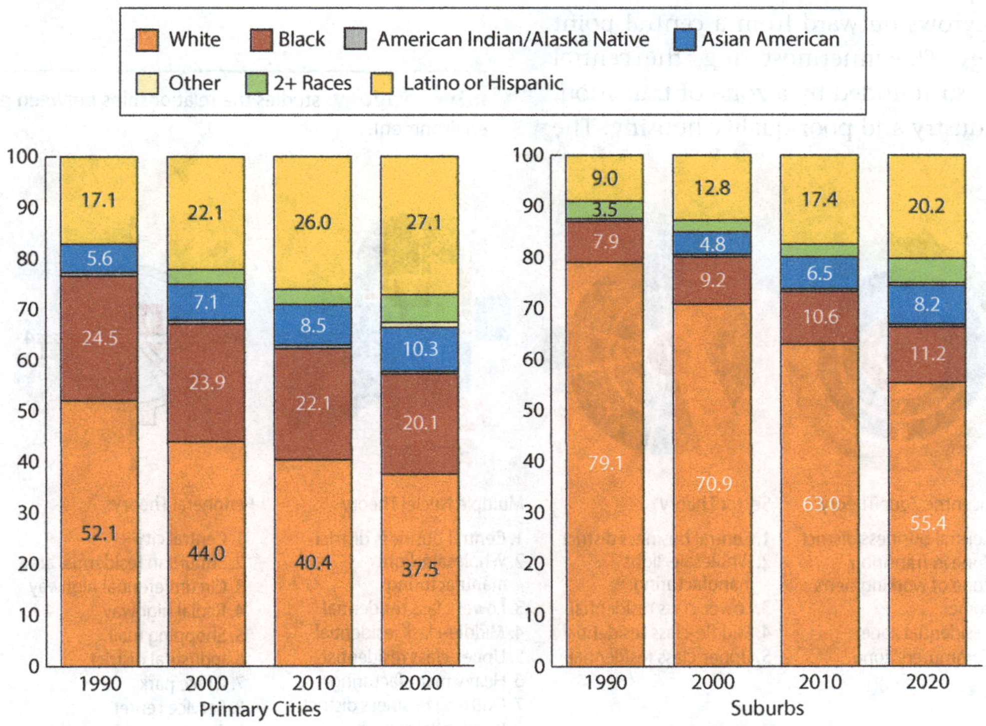

Note: Primary cities are defined by Brookings Metro and include up to three cities in metropolitan area name with populations exceeding 100,000. Suburbs are defined as the metropolitan area lying outside of primary cities

Figure 15.4 U.S. Metropolitan Areas and Suburbs Are Getting Racially More Diverse

Source: Based on Frey, 2022, "Today's Suburbs Are Symbolic of America's Rising Diversity" Figure 2.

Table 15.3 Sociological Explanations of Urbanization

Perspective	Level of Analysis	Key Points
Functionalist	Macro	Cities serve many important social and economic functions, but urbanization can also be dysfunctional.
Conflict	Macro	Driven by greed and profit, large corporations, banks, developers, and other capitalist groups shape cities' growth or decline.
Feminist	Macro and micro	Whether they live in cities or suburbs, women generally experience fewer choices and more constraints than men.
Symbolic Interactionist	Micro	City residents differ in their types of interaction, lifestyles, and perceptions of urban life.

capitalism and big business, feminist scholars focus on women's safety and space, and symbolic interactionists examine the quality of city life. **Table 15.3** summarizes these perspectives.

Functionalism: How and Why Cities Change

Urban ecology studies the relationships between people and urban environments. Over the years, social scientists have revised urban ecology theories (**Figure 15.5**).

Sociologists Robert Park and Ernest Burgess (1921) proposed *concentric zone theory* to explain the distribution of social groups within urban areas. According to this model, a city grows outward from a central point in a series of rings. The innermost ring, the central business district, is surrounded by a zone of transition, which contains industry and poor-quality housing. The third and fourth rings have housing for the working and middle-income individuals. The outermost ring is occupied by people who live in the suburbs and commute daily to work in the central business district.

Economist Homer Hoyt's (1939) *sector theory* refined concentric zone theory. He proposed that cities develop in sectors (instead of rings) that radiate from the central business district depending on various economic and social activities. Some sectors are predominantly industrial, some contain stores and offices, and others, generally farther away from the central business district, are middle- and upper-income residential areas.

urban ecology studies the relationships between people and urban environments.

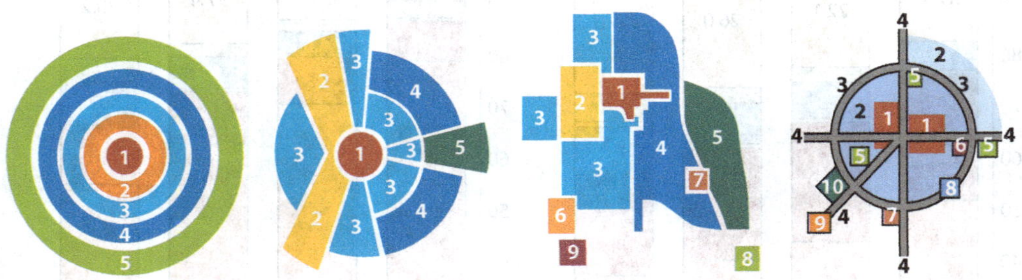

Concentric Zone Theory
1. Central business district
2. Zone in transition
3. Zone of workingmen's homes
4. Residential zone
5. Commuters' zone

Sector Theory
1. Central business district
2. Wholesale, light manufacturing
3. Lower-class residential
4. Middle-class residential
5. Upper-class residential

Multiple Nuclei Theory
1. Central business district
2. Wholesale, light manufacturing
3. Lower-class residential
4. Middle-class residential
5. Upper-class residential
6. Heavy manufacturing
7. Outlying business district
8. Residential suburb
9. Industrial suburb

Peripheral Theory
1. Central city
2. Suburban residential area
3. Circumferential highway
4. Radial highway
5. Shopping mall
6. Industrial district
7. Office park
8. Service center
9. Airport complex
10. Combined employment and shopping center

Figure 15.5 Four Models of City Growth and Change

Source: Based on Park and Burgress, 1921; Hoyt, 1939; Harris and Ullman, 1945; and Harris, 1997.

Geographers Chauncey Harris and Edward Ullman (1945) developed another influential model, *multiple nuclei theory*, which proposed that a city contains more than one center around which activities revolve. For example, a "minicenter" often includes an outlying business district with stores and offices that are accessible to middle- and upper-income residential neighborhoods, whereas airports typically attract hotels and warehouses.

As cities grew after World War II, these models no longer described urban spaces. Thus, Harris (1997) proposed a *peripheral* theory of urban growth, which emphasized the development of suburbs around a city but away from its center. As suburbs and edge cities burgeon, highways that link the city's central business district to outlying areas and beltways that loop around the city provide relatively easy access to airports, the downtown, and surrounding areas.

Functionalists also examine urban dysfunctions, including overcrowding, poverty, deviant behavior, and environmental destruction. For instance, many older industrial cities, such as Detroit, have high housing vacancy rates. Although Detroit's rate has declined from 29 percent in 2010 to 18 percent in 2020, that still means nearly 1 in 5 homes remain empty. Abandoned buildings can lead to property deterioration and increase crime rates and a neighborhood's social disorganization (Williams, 2013; Citizens Research Council of Michigan, 2022).

audioscience/Shutterstock.com

Very concentrated, high poverty areas are by-products of urbanization. Megacities—particularly those in India, Egypt, Pakistan, Kenya, and Mexico—have several.

Conflict Theory: The Impact of Capitalism and Big Business

For functionalists, people's choices shape urban changes. In contrast, **new urban sociology**, a perspective heavily influenced by conflict theory, views urban changes as largely the result of decisions made by powerful capitalists and high-income groups. That is, economic and political factors favorable to the few who are rich determine urban growth or decline. When a local government wants to rejuvenate parts of the inner city, it typically offers tax breaks, changes zoning laws, and allows real estate, construction, and banking industries to seek profits with little regard for the needs of low-income households or the homeless (Macionis and Parrillo, 2007).

Baltimore, like many other cities, subsidizes private development. Taxpayers have financed billions of dollars in city projects for multinational real estate and banking corporations. The profits are channeled to CEOs and stockholders, while city residents continue paying for (inadequate) schools, parks, libraries, and other public buildings and spaces (Kreitner, 2016).

For conflict theorists, urban space is a commodity that's bought and sold for profit. It's not the average American, they argue, but bankers, developers, politicians, and influential businesspeople who determine how to use urban space. Because increasing a property's value is a higher priority than community needs, lower-income people are crowded into dilapidated neighborhoods (Logan and Molotch, 1987; Gottdiener and Hutchison, 2000).

Some analysts are especially critical of the tech sector for being "bad urbanists." To illustrate, in 2011 Twitter received generous tax incentives to move its new headquarters into one of San Francisco's lowest income neighborhoods. City officials expected that the company's presence would revitalize the community. Instead, Twitter employees ate all their meals in the dining area, rarely leaving the building to shop because small businesses couldn't afford the area's skyrocketing rents. The neighborhood's evictions increased 38 percent by 2013 because longtime residents couldn't pay the soaring rent increases (Arieff, 2013; Butler, 2014; Steinmetz, 2014).

Feminist Theories: Gender, Poverty, and Safety

Feminist theories emphasize gender-related urban constraints. Women's exclusion from urban planning results in "a male perspective and is in men's interest" (WUNRN, 2017). For example, violence against women and girls occurs in public urban areas around the world. Harassment and sexual violence, occurring in parks, streets, shopping areas, and on public transit, contributes to

new urban sociology views urban changes as largely the result of decisions made by powerful capitalists and high-income groups.

"gender exclusion in metropolitan public space." In response, a few cities provide public transportation to prevent crimes against women. One study of world-wide metropolitan areas found 13 with women-only transit options (Metropolis and Women in Cities International, 2018).

Urban areas have a poverty rate of 12 percent and, as you learned in chapter 8, women and single parents are more likely to have lower incomes (Economic Research Service, 2019). Because of a lack of affordable housing in urban areas, many women and their children may be forced to live under the same roof with an aggressor. If public transportation isn't available or adequate, low-income women are cut off from job opportunities. Low-income mothers are expected to work. However, low-wage jobs, coupled with a lack of affordable childcare, create role conflicts, undermine mental health, and increase the likelihood of intimate partner violence (Jacobs et al., 2016; WUNRN, 2017).

Symbolic Interaction Theory: How People Experience City Life

Symbolic interactionists are most interested in the impact of urban life on city residents. In a classic essay, sociologist Louis Wirth (1938, p. 14) described the city as a place where "our physical contacts are close, but our social contacts are distant."

Wirth defined the city as a large, dense, and socially and culturally diverse area. These characteristics produce *urbanism*, a way of life that differs from that of rural dwellers. Wirth saw urbanites as more tolerant than residents of small towns or rural areas of a variety of lifestyles, religious practices, and attitudes.

He also believed that urbanism has negative consequences, including alienation, friction because of physical congestion, impersonal relationships, and a disintegration of kinship and friendship ties. Some studies have supported Wirth's theory of urbanism (refer to Guterman, 1969).

Critical Evaluation

Functionalists tend to overlook urbanization's negative political and economic impact, especially when profit and greed guide urban planning. Conflict theory seems to assume that residents are helpless victims as developers and corporations raze low-income houses. In fact, environmental groups have had considerable success in pushing through legislation to maintain and even increase open public spaces and build energy-saving homes in low-income neighborhoods (Moore, 2008).

Feminist sociologists have made important contributions through studies of the everyday lives of low-income women, especially in central cities (refer to Chapter 11), but urban areas actually have lower poverty rates (11 percent) compared with rural areas (15 percent). Also urbanization has received much less attention. Urbanites are more diverse than some symbolic interactionists claim. People living in cities aren't necessarily more self-centered or isolated than those in small towns or rural areas. Instead, many have close family bonds, friends, and satisfying relationships with coworkers (Wilson, 1993; Kotkin, 2016). Also, compared with people living in rural areas, those in urban areas are more optimistic about their financial futures and have less risk of death from heart disease, cancer, and unintentional injury among others (Parker et al., 2018; Centers for Disease Control and Prevention, 2023 "About Rural Health").

You've learned that the world's population is growing rapidly and becoming more urbanized. Both population growth and urbanization are taxing the planet's limited resources.

15-4 The Environment

Consider the following:

- Each year, Americans throw away 40 percent of available food—equaling 119 billion pounds. That could provide 130 billion meals (Feeding America, 2023 "Food Waste and Rescue").
- The average American throws away 5 pounds of solid waste every day. Fifty percent of this waste goes into landfills. As it decomposes, it emits gases, including methane, which contributes to global warming (Powell et al., 2015; U.S. Environmental Protection Agency, 2022 "National Overview: Facts and Figures on Materials, Wastes and Recycling").
- It's estimated that "recreational water activities are related to 90 million illnesses each year in the U.S." Illnesses range from gastrointestinal issues to less common ones such as flesh-eating bacteria (Ducharme, 2019).
- The United States uses 1.2 billion pounds of pesticides a year, but only 0.01 percent reaches the intended target. The other 99.99 percent contaminates the food, air, and water (Hsu et al., 2014).

Such environmental problems threaten our **ecosystem**, a community of living and non-living organisms that share a physical environment. Plants, animals, and humans depend on each other for survival.

ecosystem a community of living and non-living organisms that share a physical environment.

Danshutter/Shutterstock.com

Among lower-income countries, 25 percent of girls aren't in school compared with 17 percent of boys. One reason for this difference is that girls are more likely than boys to be responsible for collecting the family's water, making it difficult for them to attend school ("Global WASH Fast Facts," 2017).

Because the ecosystem is interconnected worldwide, what happens in one country affects others. Water, air pollution, and climate change are three interrelated factors that are endangering the global ecosystem.

15-4a Water

An expanding world population, extreme weather patterns, and industrial pollution are jeopardizing already limited water supplies. Because of a lack of access to clean water, 2 billion people around the world drink contaminated water. It is estimated that 829,000 people die each year from diarrhea resulting from water-related issues. In some places, getting water can require long trips, which can pose safety concerns and musculoskeletal disorders (World Health Organization, 2022 "Drinking-Water").

Availability and Consumption

Worldwide, water scarcity is our biggest problem. Just 3 percent of the earth's water is fresh, and two-thirds of it is locked up in the ground, glaciers, and ice caps. That leaves about 1 percent of the earth's water for the world's more than 8 billion people. The world's demand for water has tripled over the last half century.

How Much Water It Takes to Make...

In 1950, Americans used 150 billion gallons of water every day compared with 400 billion gallons today. It takes water to make everything. For example, it takes:

Jacob Kearns/Shutterstock.com

2,600 gallons to make a pair of blue jeans

Stocksnapper/Shutterstock.com

713 gallons to make a cotton shirt

Christopher Elwell/Shutterstock.com

634 gallons to make an average hamburger

magicoven/Shutterstock

53 gallons to produce a cup of to-go latte coffee

Evgeny Karandaev/Shutterstock.com

1.5 gallons to produce an average 18 ounces of bottled water

Sources: Based on Connell, 2011; Postel, 2012.

Water scarcity, which affects around 50 percent of people around the world, is projected to increase. More than half of Earth's 37 largest aquifers (underground layers of rock that store water) are being depleted "at alarming rates" to keep pace with demands from agriculture, growing populations, and industries like mining (WWAP, 2014; Richey et al., 2015; United Nations, 2015; UNICEF, 2023 "Water Scarcity").

Some refer to water as "blue gold" because it's becoming one of the earth's most precious and scarce commodities. Water shortages—rather than oil or diamonds—are behind conflicts and even wars in a number of countries. According to a past Secretary-General of the United Nations, "Too often, where we need water, we find guns..." (World Water Assessment Program, 2009, p. 20; Reynolds, 2012).

Higher-income nations have greater access to clean water than lower-income nations and they use more and pay less for it. The average person in the United States uses about 163 gallons of water per day compared with 40 gallons in Germany, 30 in Denmark, 23 in China, and less than 3 in Mozambique. On average, Americans pay only $.48 per gallon compared with $1.65 a gallon in Denmark. In the developing world, people typically pay five times as much as Europeans (Maxwell, 2012). Thus, in many countries, clean water is a luxury rather than a basic human right.

Threats to Water Supplies

Precipitation (in the form of rain, snow, sleet, or hail) is the ecosystem's main source of water. Clean water has been depleted for many reasons, including pollution, privatization, and waste.

Pollution and environmental racism. Freshwater sources are polluted by pesticides and fertilizers coming off agricultural land in addition to pollution from sewage, industrial waste, and oil. As a result, around half the water in rivers, lakes, and streams in the United States is "too polluted for swimming, fishing, or drinking" (Ali, 2022).

Environmental racism, also called *environmental injustice*, refers to the exposure of lower-income people, especially people of color, to environmental hazards. Because of environmental policies and practices, neighborhoods that are home to higher numbers of people of color are routinely the site of toxic waste, sewerage treatment plants, landfills, illegal dumps, incinerators, polluting power plants, and air and water pollution. These and other hazards increase the likelihood of illnesses and diseases, including cancer, asthma, and children's stunted brain development (Krajicek, 2016).

For almost two years, the residents of Flint, Michigan—a city with a predominantly low-income, Black population—complained of burning skin, hair loss, seizures, strange red splotches on their hands and faces, and anemic children. The mayor and city officials repeatedly told residents the water was fine, until an environmental engineering professor confirmed that Flint's water supply was contaminated with lead. In mid-2016, an independent panel concluded that "the Flint water crisis is a clear case of environmental injustice" (Flint Water Advisory Task Force, 2016, p. 55).

Corporate polluters and fracking. Few corporate polluters are fined, and many are exempt from major environmental laws, including the 1974 Safe Water Drinking Act. Consider hydraulic fracturing, or *fracking*, a process that injects water, sand, and chemicals at high pressure to extract gas and oil from rock that lies deep underground. Many U.S. lawmakers—some of them with deep ties to the industry that finances political campaigns—endorse fracking to increase energy supplies (Blake, 2014).

Fracking has risks because gases can escape into drinking water. Some of the 750 chemicals injected into the ground contaminate water, increasing the risks of cancer, infertility, and other health problems (Kassotis et al., 2014; EPA, 2015).

Fracking has also increased the number and magnitude of earth tremors and earthquakes from Colorado to the Atlantic coast. In Oklahoma, the number of earthquakes measuring 3 or higher on the Richter scale jumped from 1 in 2007 to 907 in 2015. No place in the world has ever experienced earthquakes at such a rate in such a short time. Cities and counties nationwide have passed 430 measures to control or ban fracking, but companies typically prevail despite opposition (Shauk and Olson, 2014; Skoumal et al., 2015; Philips, 2016).

Privatization. Water is a big business because of *privatization*, transferring some or all of the assets or operations of public systems into private hands. Perrier, Evian, Coca-Cola, PepsiCo—and particularly the French giants Vivendi and Suez—have been buying the rights to extract water in the United States and other countries at will from aquifers, then bottling and selling it around the world.

In 2019, bottled water companies spent more than $203 million on advertising to compete with each other and to convince consumers that bottled water is healthier and safer than soda and tap water. In fact, tap water is safer. FDA regulations allow bottles to be contaminated

---

environmental racism (also called *environmental injustice*) exposure of lower-income people, especially people of color, to environmental hazards.

with small amounts of bacteria, including *E. coli*. And, unlike tap water, which is always moving, bacteria increase after bottling because microbes "can attach to the plastic, munch on organic material in the water, and multiply" (Castella-McDonald, 2021; Lohan, 2015; Matanoski, 2017, p. 21).

Bottling water is lucrative for corporations, but there are many environmental and health drawbacks. Around the world, over "1 million bottles of water are sold every minute"(Ramirez, 2023). This water comes from groundwater sources needed for drinking water. Plastic from these roughly 600 billion bottles adds to plastic pollution. Bottled water can also contain chemicals that can settle into the body (blood, kidney, and liver), leading to serious health problems (e.g., cancer, thyroid disease, and liver damage). Recently, the EPA proposed the first national standard for drinking water to combat these "forever chemicals" (Christensen, 2023).

For-profit companies own approximately 10 percent of U.S. community water systems. Proponents maintain that, compared with government, private businesses improve water infrastructure, decrease local mismanagement (as in Flint), operate more efficiently, and spend billions of dollars fixing problems. Yet, one study of water bills for the 500 biggest community water systems in the United States found that privatized water is connected to higher prices which can be less practical for low-income households. Opponents to privatized water also suggest that a business is more likely than the government to place a lien on consumers' property if they fall behind in paying water and sewer bills (Food & Water Watch, 2016; Ivory et al., 2016; Millsap, 2016; Zhang et al., 2022).

Waste. Most water problems are due to human mismanagement and waste, not nature. Of all available water worldwide, agriculture consumes about 70 percent, industry uses 20 percent, and 10 percent is residential (WWAP, 2014). In agriculture, many irrigation systems are inefficient, farmers and agribusiness (large agriculture companies) often grow water-hungry crops like cotton and sugarcane in arid areas, and pesticide and chemical fertilizer runoff from fields pollutes streams, rivers, and lakes.

There are issues with systems that transport water and water waste. Most of the nation's pipes, especially in cities along the eastern seaboard, are nearly 200 years old, and some are made of wood. Each day in the United States, there are around 850 water-main breaks causing billions of gallons of water to be lost. A water pipe that leaks or bursts wastes huge amounts of drinking water, damages streets and homes, and seeps dangerous pollutants into drinking water. A contributing factor is that there is no nationwide agency for central planning of water unlike energy which has oversight by Energy Information Administration and Department of Energy (Lloyd, 2022; Duhigg, 2010; Pearlstine, 2013).

15-4b Air Pollution, Global Warming, and Climate Change

Global warming is a serious environmental problem. Let's begin with air pollution, the major cause of climate change and global warming.

Air Pollution: Some Sources and Causes

Worldwide, 92 percent of people breathe unhealthy outdoor air. About three million die each year from cardiovascular and respiratory diseases linked to outdoor air pollution (WHO, 2016).

There are many reasons for air pollution, but four are the most common. First, a major source of air pollution is the burning of *fossil fuels*, substances obtained from the earth, including coal, petroleum, and natural gas. The exhaust gases of cars, trucks, and buses contain poisons—sulfur dioxide, nitrogen oxide, carbon dioxide (CO_2), and carbon monoxide. Power plants that produce electricity by burning coal or oil account for nearly 40 percent of U.S. carbon dioxide emissions.

Second, manufacturing plants that produce consumer goods pour pollutants into the air. Formaldehyde-based vapors that can lead to cancer and respiratory problems are emitted by many household and personal care products: pressed wood (often used in furniture), plastic grocery bags, waxed paper, latex paints, detergents, nail polish, cosmetics, and shampoos ("Formaldehyde," 2012; refer also to Chapter 14).

Delhi, India, is the world's most polluted city, but air quality in Pakistan's urban areas is, on average, worse (Smith, 2017).

Third, winds blow contaminants in the air across borders and oceans. Air pollution originating in Europe has been tracked to Asia, the Arctic, and even rural areas in the western United States, where there's little industry or automobile traffic. Recently, smoke from Canadian wildfires blanketed the northeastern United States, lowering visibility and making breathing difficult. Some areas had air quality index scores of 400—exceeding the scores for both "unhealthy" (100) and "dangerous" (300) (Cooper et al., 2010; Zhang et al., 2017; Clifford, 2023).

A fourth reason for air pollution is that government policies and enforcement vary from one administration to another. President Nixon signed the Clean Air Act (1970) into law and President Carter added amendments to strengthen it. It's estimated that in 20 years, the Clean Air Act prevented 200,000 early deaths and 700,000 chronic bronchitis cases (U.S. Environmental Protection Agency, 2022 "Impacts of Climate Change"). Other administrations have not been as supportive. For instance, the Bush administration blocked the efforts of 18 states to cut emissions from cars and trucks because it believed that tougher regulations would hurt the U.S. economy ("No Action on Greenhouse Gases," 2008).

President Obama promised to decrease air pollution, but environmental groups accused his administration of being too soft on the fossil-fuel industry by not limiting their carbon dioxide emissions. The rules also excluded Native American reservations—which have some of the most polluting coal-fired power plants in the country—from meeting state goals to reduce carbon dioxide emissions by 2030 (Eilperin and Bernstein, 2014; Magill, 2014).

U.S. presidents of both political parties have attended U.N.–sponsored global climate change conventions since 1995, supporting programs to reduce global pollution. In 2017, however, President Trump pulled out of the Paris talks to reduce planet-warming emissions. He has also removed dozens of Obama-era environmental rules that would have closed hundreds of coal-fired power plants and replaced them with wind and solar farms and would have limited oil and gas drilling in America's Arctic and Atlantic waters (Davenport and Rubin, 2017; Rushe et al., 2017).

global warming increase in the average temperature of earth's atmosphere.

greenhouse effect heating of earth's atmosphere because of the presence of certain atmospheric gases.

climate change a change in overall temperatures and weather conditions over time.

Global Warming and the Greenhouse Effect

Global warming is an increase in the average temperature of earth's atmosphere. The warming is due to several factors. Some are natural, such as changes in solar radiation, the earth's orbit, and the frequency or intensity of volcanic activity. At least 95 percent of global warming is due to human activities, however, and not natural climate swings (Intergovernmental Panel on Climate Change, 2014).

Global warming begins with the **greenhouse effect**, the heating of earth's atmosphere because of the presence of certain atmospheric gases. When the sun's heat enters the atmosphere, some of it is absorbed by earth's surface, and some of it is reflected back to space. Greenhouse gases in the atmosphere trap some of this heat. Heat is necessary to support life, but when greenhouse gases increase, earth becomes warmer than it would be otherwise, endangering public health and the welfare of current and future generations (EPA, 2009).

Air pollutants, primarily CO_2, ignite the greenhouse effect. Between 2000 and 2010, greenhouse-gas emissions grew at 2.2 percent a year—almost twice as fast as in the previous 30 years. The 20 largest global economies (e.g., China, United States, European Union, India) account for 82 percent of global CO_2 emissions that come from burning fossil fuels, primarily from coal-fired power plants. Some of the countries are shifting to natural gas and other energy sources, but coal-fired plants produce 46 percent of their CO_2 emissions (Intergovernmental Panel on Climate Change, 2014; Olivier et al., 2016).

Some Effects of Climate Change

Climate change is a change in overall temperatures and weather conditions over time. Global warming probably began thousands of years ago when humans started changing the planet by cutting down and burning forests to grow food (Fischman, 2009). Such *deforestation*, clearing massive amounts of trees, affects global climate changes because forests recycle carbon dioxide into oxygen.

Regardless of when it began, climate change is "unequivocal" and is having a profound impact on every ecosystem worldwide. Since record-keeping began in 1880, 2015 through 2022 have been the warmest recorded. Between 1750 and 2021, five countries produced around half of all CO_2 emissions. The United States, accounting for 20 percent of all climate-changing pollution, was the leading offender, followed by China (11 percent) and Russia (7 percent). Scientists predict that extreme weather, especially heat waves, will produce heavier rainfall in some regions; more floods, landslides, and uncontained wildfires; stronger hurricanes and tornadoes; and more intense droughts around the world (O'Grady, 2021; U.S. Environmental Protection

Agency, 2022 "Impacts of Climate Change"; World Meteorological Organization, 2023; Evans, 2021).

Many changes have already occurred because of climate change. For example:

- Higher CO_2 levels are turning oceans more acidic, resulting in some shellfish (e.g., clams and crabs) dying or not developing (Spotts, 2009).
- Global weather- and climate-related disasters rose from 743 between 1971 and 1980 to 3,496 between 2001 and 2010. Climate change will lead to an increase in big storms (like Hurricanes Harvey and Irma) that cause floods, landslides, and other natural disasters (World Meteorological Organization, 2014; Prein et al., 2017).
- Excessive heat in the United States claims more lives each year (158) than floods (88), lightning (37), tornadoes (71), and hurricanes (45) (Erdman, 2023).
- The world's largest ice sheets in Antarctica and western Greenland are melting by an average of 3 feet a year. The Arctic Ocean could be largely free of sea ice by the late 2030s, only two decades from now. The melting glaciers could contribute to sea levels rising 3 to 6 feet by 2100, flooding up to 70 percent of the populations in Florida, Georgia, South Carolina, and Louisiana (Hauer et al., 2016; AMAP, 2017; Blunden and Arndt, 2017).
- Some Alaskan native communities are facing disaster. Climate change has thinned the Arctic ice so much that it's become too dangerous to hunt whales, a traditional means of sustenance, or even to live in the state's coastal communities. As the ice melts, the towns are in "imminent danger" from erosion and being washed to sea during storms (Mooney, 2015).

Such data show that our planet is ailing. Can people slow some of earth's devastation through better environmental policies and practices, particularly by endorsing sustainable development?

15-5 Is Sustainable Development Possible?

Sustainable development refers to economic activities that don't threaten the environment. There are reasons to be both pessimistic and optimistic about achieving sustainable development.

Reasons to Be Pessimistic

Which country is the greenest? Not the United States. The Environmental Performance Index, which ranks 180 nations on factors like water quality, air pollution, and protecting the ecosystem, ranks the United States (Wolf et al., 2022):

Agata/Alamy Stock Photo

Fluffy and ultra-plush toilet paper is made by chopping down and grinding up the pulp of trees that are decades or even a century old. U.S. corporations have introduced "earth-friendly" toilet paper, but "customers are unwavering in their desire for the softest paper possible" (Fahrenthold, 2009: A1).

- 43rd overall and 20th among 22 higher-income democracies. Denmark, the United Kingdom, and Finland were in the top three.
- 57th in the specific category of ecosystem vitality. This includes an assessment of ecosystem services, agriculture, and water resources.
- 22nd in health factors such as air quality, sanitation of drinking water, and waste management.
- 101st in overall climate policy.

Why does the United States rank so low in environmental performance? There are many reasons, but three important and interrelated factors are profit, politics, and personal choices.

Profit. Does profit—capitalism's basic drive—override environmental hazards? Exxon (now ExxonMobil), the

sustainable development economic activities that don't threaten the environment.

world's largest oil company, started conducting cutting-edge climate research in the mid-1950s. When the internal studies showed that rising CO_2 levels were likely to cause global warming, the top executives cut much of the climate research budget, began to organize against air pollution regulations, and funded climate denial research and activities (Banerjee et al., 2015).

Even well-intentioned businesses dismantle eco-friendly ventures if they don't bring quick profits. For example, both Chevron Corporation and BP (British Petroleum) pulled back from clean energy projects because the profits from non-renewable oil and gas sales were much higher (Upton, 2014).

Some environmentalists are especially dismayed by *greenwashers*, companies and other organizations that pollute the planet while presenting an environmentally responsible public image. In 2000, for instance, BP rebranded itself "Beyond Petroleum" and changed its logo to a green-and-yellow sunburst, but the majority of its business produces fossil fuels (Elgin, 2014). Governments in France, Norway, and the United Kingdom are beginning to take action to protect consumers (Sengupta, 2022).

Politics. You learned earlier that environmental policies reflect a president's views. During the Trump presidency, many environmental protections were rolled back, and he appointed a chief of the EPA who contended that nature, not human activities (e.g., burning fossil fuels), causes global warming and who dismissed at least five academic scientists from the EPA that will be replaced by chemical and fossil fuel company applicants. Upon President Biden taking office, he vowed to reinstate over 100 environmental protections that were eliminated or weakened. Many of these policies affecting water, air, and climate change are still under review (Davenport, 2021; Goodell, 2017).

These and other political pressures, policies, and practices help explain America's low environmental rankings globally. The United States is "one of the very few large energy-consuming economies that doesn't have national energy reduction targets" across all sectors to lower greenhouse gas emissions (Kallakuri et al., 2016).

A recent national survey of 1,500 middle and high school science teachers found that 30 percent—because of personal beliefs or pressure from parents or school administrators—taught that global warming is due to natural, not human, causes. Also, 62 percent didn't know that more than 95 percent of climate scientists attribute recent global warming to human activities. A conservative think tank known for attacking climate science has sent 200,000 copies of a glossy book to public school teachers (and some college professors) that describes global warming as "another fake crisis" created by Democrats (Plutzer et al., 2016; Stager, 2017). Thus, politics also fuels ignorance of and misinformation about environmental issues.

Fotos593/Shutterstock.com

Almost 15 million tons of plastics enter oceans each year (Geyer et al., 2017). Because plastic trash doesn't degrade, it kills oceanic marine life and endangers ecosystems.

Personal Choices. About 75 percent of Americans say that "the country should do whatever it takes to protect the environment." However, we recycle or compost only 35 percent of municipal waste, even though 73 percent of the population has curbside recycling. Also, people want cleaner energy and endorse wind farms, but "not in my backyard" (NIMBY) (EPA, 2016; Sustainable Packaging Coalition, 2016; Anderson, 2017; Hemphill and Perry, 2017). Thus, are Americans eco-friendly only when it's convenient?

> Are Americans eco-friendly only when it's convenient?

Reasons to Be Optimistic

There's been considerable progress since 1970, when the United States celebrated its first Earth Day. Let's look briefly at government, business, and individual contributions to the environment.

Government. Primarily because of federal rules and regulations, the United States ranks second among 23 of the world's top energy-consuming countries in how efficiently buildings use energy. It has the most mandatory appliances and equipment standards, and stringent energy codes for new residential and commercial buildings (Kallakuri et al., 2016).

Elena Eliisseeva/Shutterstock.com

Imagewise Ltd/Shutterstock

Many individuals and businesses are contributing to sustainable development. Hawaii has some of the nation's greenest cities, admired for their excellent air quality and extensive use of rooftop solar panels. KFC, a restaurant chain, introduced their "Scoff-ee" edible coffee cups at Seattle's Best Coffee in Britain, to decrease the environmental impact of packaging (Strom, 2015).

Emissions from the six major pollutants declined by 74 percent, all while the U.S. economy, population, and energy increased (US. Environmental Protection Agency, 2019). Most cars no longer burn leaded gasoline, and ozone-destroying chlorofluorocarbons (CFCs) have been generally phased out. The recycling rate in the United States increased from around 7 percent in 1960 to 32 percent in 2019 (U.S. Census Bureau, 2021 "America Recycles Day").

Current efficiency standards for appliances, lighting, and other equipment that were implemented during the early 1980s will save the United States the equivalent of two years of energy use ($1.1 trillion by 2035) and slash greenhouse gas emissions. Also, because of serious droughts, local governments in California and some Texas cities are recycling wastewater, including "toilet-to-tap" drinking water (Lowenberger et al., 2012; Erbentraut, 2016).

More states have embraced clean energy sources, which include wind, solar, and hydroelectric among others. The three states that produce the most renewable forms of energy are Texas, Washington, and California. Renewable energy in Texas lowered costs by $423 per customer using the state's power grid. Much of Texas' renewable energy is wind, and President Biden recently announced an expansion of offshore wind infrastructure (Gilligan, 2022; Gold, 2023)

Some government-level progress has been mixed. For example, the EPA has imposed stiffer penalties on automakers that overstate their fuel efficiency but allows the use of 82 chemicals, including pesticides that Europe restricts or bans (Hakim, 2015; Kessler, 2015).

Businesses. Many companies have found that being green is good for their profits, image, and environment, and are now more eco-friendly. For example:

- Each pair of Levi's new Waste<Less jeans is composed of eight recycled plastic bottles (Berfield, 2012).
- Ford, Honda, and Toyota use recycled materials—jeans, sweaters, plastic bottles, and sugar cane by-products—for dashboards, tires, seat covers, and cushions (Fleming, 2014).
- Several companies, including Best Buy and Sprint, have already met their 2020 goal of cutting energy consumption by 20 percent (Martin, 2016).
- Perdue Farms, the country's fourth-largest poultry producer, no longer uses growth-producing antibiotics that pose health risks to people. The chickens grow just as fast and efficiently as in the past, and getting rid of the drugs hasn't affected the company's profitability (Philpott, 2016).
- Among others, Ikea, Nike, and H&M (a clothing retail company) use sustainable cotton grown by farmers who use less water and fewer pesticides (Kessenides, 2017).

Economic self-interest rather than a concern for the environment sometimes drives such eco-friendly practices. Some corporations are eager to reduce their reliance on water-intensive products like cotton and nuts. Regardless of motives, environmentally friendly policies promote sustainability.

Individuals. In general, Americans still slightly prioritize protecting the environment (50 percent) over

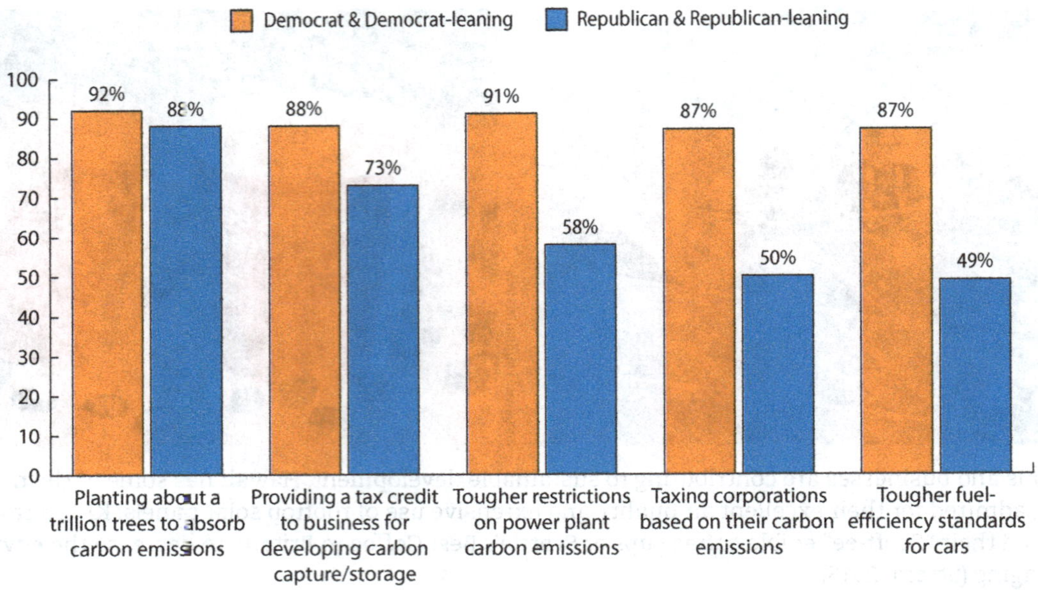

Figure 15.6 Percent of U.S. Adults Who Favor Proposals to Reduce Climate Change by Political Party

Source: Based on Funk, 2021

economic growth (42 percent). Most Americans (71 percent) say the U.S. should focus on developing alternative energy sources over expanding the production of fossil fuels (oil, coal, and natural gas). Yet, Americans are also divided, mostly along party lines, regarding climate change proposals (Funk, 2021; Saad, 2021) (**Figure 15.6**).

People's environmental beliefs and behavior aren't always consistent. About 75 percent of Americans say they're concerned with helping the environment as they go about their lives, but only 15 percent always bring their own shopping bags, and only 12 percent always choose cleaning products based on whether the ingredients would help or hurt the environment (Funk and Kennedy, 2016).

How can individual behavior change? A 2020 study found that climate change education may be the single most useful solution. Specifically, if 16 percent of middle and high school students in middle- and high-income countries studied climate change, it could reduce CO_2 by almost 19 gigatons by 2050. Why? These youth would gain personal connections to climate change and shift their behaviors throughout their lives (Cho, 2023).

Chapter Review

Key Takeaways

15-1 What are some of the key considerations for the study of population and population change? *Demography* examines the interplay among *fertility, mortality,* and *migration.* The *crude birth rate, total fertility rate, crude death rate,* and infant mortality rate measure a population's life expectancy and health.

Push and pull factors affect international and internal migration. Demographers also use *sex ratios* and *population pyramids* to understand a population's composition and structure.

15-2 How do different population growth theories compare? Demographers who believe that population growth is a ticking bomb subscribe to *Malthusian theory,* which argues that the world's food supply won't keep up with population growth. *Demographic transition theory,* in contrast, maintains that population growth is kept in check and stabilizes as countries experience greater economic and technological development.

15-3 **What are the main effects of urbanization both in the United States and around the world?** Globally and in the United States, cities and *urbanization* mushroomed during the twentieth century and are expected to increase. As more people move from rural to urban areas, many of the world's largest cities are becoming *megacities*. In the United States, urban growth has led to suburbanization, *edge cities, exurbs, gentrification*, and *urban sprawl*.

Sociologists offer several perspectives on how and why cities change, and the consequences. Refer to **Table 15.3** on page 340.

15-4 **What are the major environmental issues?** Population growth and urbanization are changing the planet's *ecosystem* and, many argue, endangering plants, animals, and humans. Water and air pollution and global warming are good examples of threats to the ecosystem in the United States and globally. Clean water has been depleted for many reasons, including pollution, privatization, waste, and mismanagement.

Four of the most common sources and causes of air pollution are burning fossil fuels, manufacturing plants that spew pollutants into the air, winds that carry contaminants across borders and oceans, and lax governmental policies. Air pollution, which can lead to the *greenhouse effect*, is a major cause of *climate change* and *global warming*.

15-5 **What are different perspectives on whether sustainable development is achievable?** The rise of environmental problems has sparked discussions about *sustainable development*. Those who are pessimistic about achieving sustainable development show that, worldwide, the United States has one of the worst records on environmental performance, largely because of the close ties between government officials and corporations.

Others are optimistic about achieving sustainable development and point to examples such as decreases in the emission of major air pollutants and some large U.S. corporations' switching to practices that decrease pollution and energy consumption.

Key Terms

Demography (p. 328)
Population (p. 328)
Fertility (p. 328)
Crude birth rate (also called the *birth rate*) (p. 328)
Total fertility rate (TFR) (p. 328)
Mortality (p. 330)
Crude death rate (*also called the death rate*) (p. 330)
Migration (p. 331)
Sex ratio (p. 332)
Population pyramid (p. 332)
Malthusian theory (p. 333)
Demographic transition theory (p. 334)
Zero population growth (ZPG) (p. 335)
Urbanization (p. 336)
Megacities (p. 336)
Edge cities (p. 337)
Exurbs (p. 337)
Metropolitan statistical area (MSA, also called *metro area*) (p. 337)
Urban sprawl (p. 338)
Gentrification (p. 338)
Urban ecology (p. 340)
New urban sociology (p. 341)
Ecosystem (p. 342)
Environmental racism (also called *environmental injustice*) (p. 344)
Global warming (p. 346)
Greenhouse effect (p. 346)
Climate change (p. 346)
Sustainable development (p. 347)

Discussion and Critical Thinking Questions

1. Are population, urbanization, and the environment related? Or are they separate issues?
2. Has immigration or emigration affected your life, your community, or a nearby community? What have been some of the positive and negative outcomes?
3. Does Malthusian theory or demographic transition theory come closer to your perspective on population growth? Why?
4. Should federal policies punish groups that degrade the environment and its resources? Would such penalties endanger small businesses? What about individuals? Would punitive government measures encourage people to be more environmentally conscious? Or spark protests because the enforcement would be seen as too authoritarian in a democratic country where people have free will?
5. Who should be responsible for e-waste—the companies that make the products, or the consumers who buy them? Would your buying habits change if you had to pay to recycle old electronics?

True or False?

How Much Do You Know about Population, Urbanization, and the Environment?

The answers to 1, 2 and 4 are true. The answers to 3, 5, and 6 are false.

1. In the early 1900s, women in the United States had a TFR of 3.5 which declined to 1.7 in 2021.
2. The pace of population growth worldwide is speeding up. It took until 1800 for the world to reach a population of one billion. By 1927 the population reached 2 billion, and less than one hundred years later, the population skyrocketed to 8 billion in 2022.
3. Over recent decades, both suburban and urban communities in the United States have become more racially and ethnically diverse.
4. Worldwide, an estimated 2 billion people consume contaminated drinking water and 829,000 people die each year from diarrhea resulting from water related issues.
5. Enough food is produced to feed 14 billion people which is more than the world's population of 8 billion, but one-third of food is lost or wasted.
6. The U.S. ranks 43rd among 180 nations for its environmental practices.

16 Social Change: Collective Behavior, Social Movements, and Technology

FG Trade/Getty Images

Learning Objectives

After studying this chapter, you will be able to…

16-1 Compare the major types of collective behavior, their theories, and functions.

16-2 Compare major types of social movements, their theories, and functions.

16-3 Describe recent technological advances, their benefits, costs, and ethical controversies.

True or False?

How Much Do You Know about Social Change?

1. Crowds can consist of like-minded people who assemble in a location to pursue a common goal.
2. Rumors don't cause actual harm to people.
3. Disasters (such as hurricanes, earthquakes, or fires) often prompt chaos.
4. Social movements may aim to resist a particular social change.
5. Most Americans are comfortable with how their data is being handled by companies and government.

The answers are at the end of the chapter.

This chapter examines **social change**, the transformations of societies and social institutions over time. Some collective behavior is short-lived with few long-term societal changes; others, particularly social movements, can have lasting effects. Technology has also played a critical role in sparking social change. Let's begin with collective behavior.

Be sure to take the True or False quiz in the chapter opener to better understand how much you know about these topics.

What do you think?

I sometimes feel overwhelmed by keeping up with texting and social media.

1	2	3	4	5	6	7
	strongly agree				strongly disagree	

16-1 Collective Behavior

Do you consume energy drinks? Have a tattoo? Search for the label "natural" on food packages? Have you ever joined a club or sports team? Texted? Taken a selfie? If so, you've engaged in collective behavior.

16-1a What Is Collective Behavior?

Collective behavior is the spontaneous and unstructured behavior of a large number of people. Collective behavior encompasses a wide range of actions, including riots, fads, fashion, panic, rumors, and responses to disasters.

Collective behavior has two important characteristics. First, it's an act rather than a state of mind. You may *feel* panic when a tornado threatens your town, but you don't engage in collective behavior until you actually *leave* your home and head for a safer location.

Second, collective behavior varies in its degree of spontaneity and structure. Panic, the least structured form of collective behavior, is typically short-lived. Fads (e.g., diets, "ugly" holiday sweaters) are more structured. They may last several years, are planned, and may be expensive. Other forms of collective behavior, including pro- and antiabortion groups, become highly institutionalized social movements that include a staff, budget, and lobbying.

16-1b Why Collective Behavior Occurs

Sociologists have proposed many explanations of collective behavior. Four of the most influential are contagion, convergence, emergent norm, and structural strain theory. Each has strengths and weaknesses.

Contagion Theory

Contagion theory proposes that individuals act emotionally and irrationally due to a crowd's almost hypnotic influence. This perspective is rooted in the work of French scholar Gustave Le Bon (1841–1931), who asserted that people in crowds, particularly mobs and riots, change radically. The anonymity of the crowd, a feeling of power, and an infectious "mob mind" embolden people to abandon personal responsibility and to engage in antisocial, and often violent, behavior (Le Bon, 1896/1968).

American sociologists later refined Le Bon's theory. They emphasized "milling," a process of people moving about, talking to one another, and becoming increasingly excited. Milling produces a "circular reaction" in which the participants become more intense and more unified in attacking a target (Blumer, 1969). Because people feel anonymous in crowds, they abandon self-control and adopt behavior that they'd reject in other settings (Brown and Goldin, 1973).

Convergence Theory

Contagion theory asserts that crowds are highly suggestible and out of control. In contrast, *convergence theory* proposes that crowds consist of like-minded people who deliberately assemble in a place to pursue a common goal. Thus, collective behavior occurs when people who share similar values, beliefs, attitudes, emotions, and goals come together in, or converge on, a certain location.

social change transformations of societies and social institutions over time.

collective behavior spontaneous and unstructured behavior of a large number of people.

Thearon W. Henderson/Getty Images Sport/Getty Images

In 2016, NFL quarterback Colin Kaepernick (front) and some of his teammates started "taking a knee" instead of standing during the national anthem, to protest racism and police violence. Players across the NFL and at some high school and college football games did the same to show solidarity. Which collective behavior theory, if any, explains these protests? Why?

Instead of the crowd affecting individuals, as contagion theory claims, convergence theory argues that individuals influence a crowd. Examples of convergence theory include marches, rallies, and acts of civil disobedience that oppose a government's domestic or foreign policies.

Emergent Norm Theory

Emergent norm theory posits that crowd members establish rational behavioral norms in response to a precipitating crisis. When Hurricane Harvey deluged Houston and surrounding towns, for example, people used their own boats, jet skis, and even air mattresses to rescue strangers from roofs and partly submerged cars and volunteered at shelters. In effect, then, people aren't spontaneously "infected" with the emotions of others, as contagion theory postulates. Instead, crowd members develop norms as a situation unfolds (Turner and Killian, 1987).

Structural Strain Theory

Structural strain theory (also called *value-added theory*) proposes that collective behavior occurs if six conditions are present (Smelser, 1962). These conditions are "value-added" in the sense that each condition leads to the next one, ending in an episode of collective behavior:

1. *Structural conduciveness* refers to the social conditions that foster collective behavior. When channels for expressing a grievance either aren't available or fail, like-minded people may resort to protests.
2. *Structural strain* occurs when societal problems (e.g., discrimination, war) make people angry, frustrated, or interfere with their everyday lives.
3. In the *growth and spread of a generalized belief* stage, people begin to view a situation as a widespread problem instead of their own fault, attribute the problem to a person or group, and believe that something should be done.
4. *Precipitating factors* are incidents or events that trigger action.
5. During the *mobilization for action* stage, leaders emerge.
6. In the *social control* stage, opposing groups prevent, interrupt, or repress those advocating social change. Government officials, the police, community and business leaders, courts, the mass media, and other social control agents—all of whom benefit from the status quo—may ridicule or quash the emerging collective behavior.

Critical Evaluation

Over time, scholars found little evidence for the contagion theory's claim that crowds are typically irrational or out of control. Crowds can sway emotions but can also be effective in meeting their objectives, and even mobs can be focused and rational (Couch, 1968; McPhail, 1991).

Convergence theories are useful in describing crowds that are deliberate and planned but have several limitations. They don't explain why some people who share the same attitudes join a protest, rally, or demonstration while others don't, and they don't account for changes in crowd behavior, no matter how well planned (Berk, 1974; McPhail, 1991).

Emergent norm theory helps explain why many crowds are orderly and focused rather than mob-like. Nonetheless, the theory doesn't explain which norms emerge, why, and how they differ in crowds. In the case of large crowds—those numbering in the thousands, for instance—how are emergent norms disseminated so quickly and accepted by the participants? It's also not clear why some participants conform to emergent norms while others ignore them and act irrationally (e.g., looting stores and vandalizing property) (Brown and Goldin, 1973; Berk, 1974).

Structural strain theory helps to predict when and where collective behavior might occur and offers insights on why, at every stage, collective behavior may either fade or escalate. The model doesn't explain all forms of collective behavior, however. In the case of fads and rumors, as you'll learn shortly, all six stages

Center for Disease Control

A day after the CDC confirmed the first U.S. Ebola case in Texas and emphasized that the risk of infection was low, there were hundreds of social media rumors that the disease had spread to Iowa and other states. Why did the rumors proliferate? Almost 67 percent of Americans believed that Ebola spreads easily (it doesn't), 40 percent expected a major outbreak in the United States (highly unlikely), and 26 percent were personally terrified that they or someone in their immediate family would get the Ebola virus (also highly unlikely) (Harvard School of Public Health, 2014).

don't necessarily occur. A second limitation is that the sequence of stages isn't necessarily the same as Smelser outlined. The theory also doesn't say how much structural strain must exist to spark collective behavior (Rule, 1988; Locher, 2002).

16-1c Types of Collective Behavior

There are many forms of collective behavior; some are more fleeting or harmful than others (Turner and Killian, 1987). Let's begin with rumors, one of the most common types of collective behavior.

Rumors

There were widespread rumors that on January 1, 2000, a glitch (Y2K) in operating systems would cause computers around the world to crash, leading to global power outages, banks losing all of their customers' accounts, and even airplanes falling from the skies. None of this occurred.

A **rumor** is unfounded information that people spread quickly. Through modern communication technology, a rumor can spread to millions of people within seconds and incite riots, panic, or widespread anxiety. Because of the Y2K rumor, thousands of people built underground shelters, and millions stocked up on bottled water, canned food, batteries, and medical supplies.

Most rumors (that rock stars Elvis Presley and John Lennon are alive, for example) are harmless. Others can be damaging. After the 2020 presidential election, rumors of voter fraud began to spread. It caused election administrators and workers to face threats from people who felt cheated (Cox, 2022). Dominion Voting Systems, the company that produced voting machines used in many states, was targeted by Fox News (and others), which damaged its reputation. Although Dominion sued Fox, and they eventually reached a $787 million settlement, it's not clear whether the company will be able to bounce back (Bauder et al., 2023). The 2020 election rumors also cost American taxpayers over $500 million in extra expenditures for greater security measures for future elections and chipped away at Americans' faith in the electoral system (Olorunnipa and Lee, 2021).

Rumors are typically false, so why do so many people believe them? First, rumors often deal with an important subject about which—especially during uncertain times related to politics, economics, public health, or natural disasters—people are anxious, insecure, or stressed. This makes people especially suggestible. In Hurricane Katrina's aftermath, the media reported numerous rumors of carjacking, murders, thefts, and rapes, the overwhelming majority of which subsequently proved to be false.

Second, there may be little factual information to counter a rumor, too much information available to reasonably sift through, or people distrust the sources with accurate information. For example, there were many rumors about COVID-19 vaccines. One study found approximately 600 rumors about COVID-19 vaccines that were spread via Google, Twitter, Facebook, and other online platforms. Rumors ranged from the World Health Organization saying the vaccines don't work to "Bill Gates explains that the COVID vaccine will use experimental technology and permanently alter your DNA." These rumors contributed to mistrust in political and medical systems (including pharmaceutical companies), which added to some people's hesitancy to get vaccinated (Islam et al., 2021, p. 9).

Third, rumors offer entertainment, diversion, and drama in our otherwise mundane daily lives. In the early years of the COVID-19 pandemic, typical social activities were limited, and people got bored. This opened the door for alternate forms of entertainment, including the consumption of rumors, and social media sources that share rumors became more popular (Campion-Vincent,

rumor unfounded information that people spread quickly.

2005; Heath, 2005; Dockterman, 2022). Gossip and urban legends are two of the most common types of rumor.

Gossip. **Gossip** is the act of spreading rumors, often negative, about other people's personal lives. Someone once said that "no one gossips about other people's virtues." Because of its tendency to be derogatory, gossip makes us feel superior. Gossip is also interesting, entertaining, and exposes hypocrisy (Epstein, 2011).

Sometimes, individuals gossip to control other people's behavior and to reinforce a community's moral standards. Comments about someone's drug misuse or relationship infidelity strengthen norms of what's deviant or unacceptable and bonds us with like-minded people (refer to Chapter 4). Having knowledge about rumors and participating in gossip (even about people we don't personally know, like celebrities) can connect us with others by providing a common topic for discussion. In other cases, people gossip because they resent or envy someone's physical appearance or accomplishments. Besides hurting personal and workplace relationships, gossip often creates fear, resentment, and stress—all of which can increase the likelihood of anxiety, depression and even escalate to physical violence (Sunstein, 2009; Drexler, 2014; Dockterman, 2022; Anander, 2022).

Urban Legends. Another form of rumor is **urban legends** (also called *contemporary legends* and *modern legends*), stories—funny, horrifying, or just odd—that supposedly happened somewhere. Some of the most common and enduring urban legends, but with updated variations, deal with food contamination or health concerns. For instance, the claim that part of a human finger was found in chili from Wendy's, that eating Mentos and diet soda will cause one's stomach to explode, or that the tryptophan in Thanksgiving turkey will put everyone to sleep. There are also health-focused urban legends that are not connected to food, for example, that a person can get pregnant by swallowing a fertilized octopus egg or that flushing an airplane toilet while still seated can result in having one's intestines sucked out.

Why do urban legends persist much longer than gossip? First, they reflect contemporary anxieties and fears—about contaminated food, unscrupulous companies, and corrupt and unresponsive governments. Second, urban legends are cautionary tales that warn us to watch out in a dangerous world. Third, we tend to believe urban legends because we learn them from people we trust, especially family members, coworkers, and friends. Fourth, rejecting misinformation and outright lies is hard work because doing so requires rethinking already-held beliefs. Finally, urban legends—like

Elliott Cowand Jr/Shutterstock

When Popeye's debuted its chicken sandwich in 2019, it created a craze. People waited in long lines. The sandwich garnered attention from both social media and traditional news outlets- some of whom ran taste test competitions and added to people's concerns that the sandwich would sell out. It did sell out- in just two weeks (Maynard, 2019).

the one about alligators living in New York City's sewer system or that AriZona tea contains human urine—are fun to tell and "too beguiling to fade away" (Brunvand, 2001; Ellis, 2005; Emery, 2015).

Panic and Mass Hysteria

At an Indonesian soccer (football) match in 2022, police fired tear gas to stop fans fighting in the stands and running on to the field. The tear gas triggered fans to flee for exits. People were trampled and suffocated. One hundred and twenty-five people died and around 300 were injured. It was one of the deadliest sporting disasters in history (Associated Press, 2022 "At Least 125 Dead at Indonesian Soccer Match"; De la Fuente et al., 2022).

The injuries and deaths from this incident were due to **panic**—a collective flight, often irrational, from a real or perceived danger. The danger seems so overwhelming that people desperately jam an escape route, jump from high buildings, leap from a sinking ship, or sell off their stock. Fear drives panic: "Each person's concern is with his [or their] own safety and personal security, whether the danger is physical, psychological, social, or financial" (Lang and Lang, 1961, p. 83).

---

gossip rumors, often negative, about other people's personal lives.

urban legends (also called *contemporary legends* and *modern legends*) rumors about stories that supposedly happened somewhere.

panic collective flight, often irrational, from a real or perceived danger.

Panic is similar to **mass hysteria**—an intense, fearful, and anxious reaction to a real or imagined threat by large numbers of people. At the outbreak of the COVID-19 pandemic, concerned people bought all available masks, hand sanitizer, and (most surprising) toilet paper. Some stores had to place limits on these purchases. "An Australian newspaper went so far as printing eight extra pages in a recent edition—emergency toilet paper, the newspaper said, should Aussies run out" (Andrew, 2020). Unlike panic, which usually subsides quickly, mass hysteria may last longer because warnings—especially about health and food—reinforce our general fears and anxieties about life's dangers.

BETTY CROCKER MAKEOVER

In 1921, General Mills created Betty Crocker, a fictitious woman, to answer thousands of questions about baking that came in from consumers every year. As fashions and hairstyles changed, so did Betty Crocker's image. The original image of a stern, gray-haired older woman has changed over the years so that, by 1996, she had a darker complexion and wore casual attire. Can you think of other brands that have changed their image over the years to keep up with changing trends?

Fashion

As the "Betty Crocker Makeover" on the next page shows, General Mills has been modernizing "Betty" since the 1930s. Doing so may reinforce people's cynical views about corporate manipulation. On the other hand, updating Betty's image and fashion may help explain why General Mills has successfully attracted new generations of consumers over the decades, and why Betty Crocker products are now sold worldwide.

Fashion is a popular way of dressing during a particular time or among a particular group of people. Over the years, Black women's hairstyles have changed—from Afros in the late 1960s, to straightened hair during the 1980s, to braids, cornrows, dreadlocks, hair extensions, and coloring more recently. All reflect gender politics, racial solidarity, generational differences, identity, and changing images of beauty (Banks, 2000; Desmond-Harris, 2009).

People who want to be fashionable buy clothes with prominent labels, but what's fashionable changes from year to year. Thus, now many clothing items "are only worn seven to ten times." To the benefit of clothing makers, the longevity of clothing ownership is changing too as the "average person keeps clothing for half as long as they did 15 years ago" (Greenwood, 2016). Fashion also includes periodic changes in the popularity of furniture, music, language usage, books, cars and trucks, sports, recreational activities, the names parents give their children, and even the breeds of dogs that people own.

Why do fashions, particularly clothes, change fairly quickly? One reason is that designers, manufacturers, and retailers must continuously generate demand for new products and services to maintain a profit. Retailers have created a number of "shopping holidays" to move merchandise, including clothes (e.g., Amazon's "Prime Day," Kohl's senior discount Wednesdays, and many large department stores' "Black Friday" sales the day after Thanksgiving). Second, many people keep up with fashion because they don't want to seem different, out-of-date, or dowdy. Third, shopping for new clothes and other products decreases the boredom of everyday routine. Also, clothes and other merchandise are status symbols that signal being an insider: "Others...will admire me for...making stylish choices" (Best, 2006, p. 85–86; for classic analyses of fashion and collective behavior, refer to Veblen, 1899/1953; Packard, 1959; and Bourdieu, 1984).

---

mass hysteria an intense, fearful, and anxious reaction to a real or imagined threat by large numbers of people.

fashion a popular way of dressing during a particular time or among a particular group of people.

Fads and Crazes

A **fad** is a trend that's popular for a short time. Fads include *products* (e.g., bean bag chairs, hoverboards, pet rocks), *activities* (e.g., twerking, roller skating, paleo and keto diets), and *popular personalities and television characters* (e.g., the Lone Ranger during the 1950s, Guy Fieri or the Kardashians more recently).

A *craze*, a type of fad, is a temporary, widespread, and obsessive enthusiasm for a particular activity or object. Recent examples include fidget spinners, Hatchimals, and Powerball lotteries that pay millions of dollars. Crazes, like fads, are usually harmless, but there are exceptions. For instance, the blackout challenge on Tik Tock where people hold their breath until they black out (Felbin et al., 2023).

Why do fads and crazes pop up? A major reason is profit. Because children and adolescents are especially likely to adopt fads, manufacturers create numerous products and activities that they hope will catch on. The products include toys, sportswear, and new cereals. The hottest fads are usually the must-have holiday toys that children plead for every year. A few months later, the toy may be thrown away or stuffed in the back of a closet.

Some people dismiss fads as "ridiculous" or "silly," but they serve several functions. In a mass society, where people often feel anonymous, a fad can develop strong in-group feelings and a sense of belonging, especially among people who have similar interests and attitudes. Fads can also be fun, promise to resolve a nagging problem (e.g., concerns over physical appearance), and help us keep up with technological changes (Marx and McAdam, 1994; Best, 2006).

Most fads are soon forgotten, but some become established. Pez candy dispensers, which originated in 1952 and cost 49 cents, are still inexpensive (around $2.00 each) and are sold in more than 60 countries. The products have been continuously updated to connect with what's popular; offering dispensers with characters like Yoda, sayings such as "Let's Taco Bout It" (taco dispenser), and holiday themes. To continue people's interest in this fad, Pez even opened a visitor center (Paul, 2002; Pez.com, 2023). Other fads, like streaking (running around nude in public places), reemerge from time to time ("Streaking," 2005).

Disasters

A **disaster** is an unexpected event that causes widespread damage, destruction, distress, and loss. Some disasters are due to *social causes* like war, genocide, terrorist attacks, and civil strife. Some are due to *technological causes*, including oil spills, nuclear accidents, burst dams, building collapses. Others—like floods, landslides, earthquakes, hurricanes, and tsunamis—have environmental origins (Marx and McAdam, 1994).

Disasters often inspire organized behavior rather than chaos. Instead of panicking, most people are rational, cooperative, and altruistic. They often opt for prosocial behaviors and care for family and community members instead of fleeing, and thousands of volunteers offer financial, medical, and other help (Clarke, 2002; Rose et al., 2022).

Publics, Public Opinion, and Propaganda

A **public** is a group of people, not necessarily in direct contact with each other, who are interested in a particular issue. A public is different than the general public, which consists of everyone in a society.

There are as many publics as there are issues—abortion, gun control, pollution, health care, and same-sex marriage, to name just a few. Even within one organization or institution, there may be several publics that are concerned about entirely different issues. At a college, students may be most concerned about the cost of tuition, faculty may spend much time discussing instructional technology, and administrators may be more focused on the financial well-being of the institution.

The interaction within a public is often indirect rather than face-to-face—as through the mass media,

Rachel Weill/FoodPix/Jupiter Images

fad a trend that's popular for a short time.

disaster an unexpected event that causes widespread damage, destruction, distress, and loss.

public a group of people, not necessarily in direct contact with each other, who are interested in a particular issue.

Library of Congress Prints and Photographs Division [LC-USZC4-3859]

One of the best-known examples of propaganda in the United States is this Uncle Sam poster, designed in 1917, and used to recruit soldiers for World Wars I and II. Opponents of the Vietnam War used it as an antiwar poster during the 1960s and 1970s.

social media, blogs, or professional journals. Because publics aren't organized groups with memberships, they're often transitory. Publics expand or contract as people lose or develop interest in an issue. A public may surge during a highly publicized and controversial incident, like removing a patient's life support, but then evaporate quickly. In other cases, publics organize and become enduring social movements (a topic we'll examine shortly).

Some publics express themselves through **public opinion**, widespread attitudes on a particular issue. Public opinion (1) is a verbalization rather than an action; (2) is about a matter that concerns many people; and (3) involves a controversial issue (Turner and Killian, 1987). Like publics, public opinions wax and wane over time. People's interest in crime and education drops, for example, when they're more concerned about other issues such as jobs, income, or health.

Public opinion can be swayed by **propaganda**, spreading information (or misinformation) to influence people's attitudes or behavior. Propaganda isn't a type of collective behavior but affects it in several important ways. First, it can create attitudes that will arouse collective outbursts (e.g., strikes or riots). Second, propaganda can attempt or succeed in preventing collective outbursts, as when corporations convince employees that job losses are due to a weak economy rather than to offshoring (refer to Chapter 11). Third, propaganda tries to gain followers for a cause, whatever it might be (Smelser, 1962). Propaganda is institutionalized in advertising, political campaign literature, and government policies. It's conveyed in many ways: the mass media, social media, political speeches, religious groups, rumor, and symbols (e.g., flags, bumper stickers).

Crowds

Much collective behavior is scattered geographically, but crowds are concentrated in a limited physical space. A **crowd** is a temporary gathering of people who share a common interest or participate in a particular event. Whether it's a few dozen people or millions, crowds come together for a specific reason, such as a religious leader's death, a concert, or a riot.

Crowds differ in their motives, interests, and emotional level:

- A *casual crowd* is a loose collection of people who have little in common except for being in the same place at the same time and participating in a common activity or event. There's little if any interaction, the gathering is temporary, and there's little emotion. Examples include people watching a street

iStock.com/Tanuki Photography

"Public mourning" includes strangers coming together to show their respect to the dead. Less than a day after Apple's co-founder Steve Jobs' death in 2011, fans across the globe created memorials in Japan and other countries. Is this kind of collective behavior an example of a conventional, expressive, or acting crowd? Why?

public opinion widespread attitudes on a particular issue.

propaganda spreading information (or misinformation) to influence people's attitudes or behavior.

crowd a temporary gathering of people who share a common interest or participate in a particular event.

performer, spectators at the scene of a fire, and shoppers at a busy mall.

- A *conventional crowd* is a group of people that assembles for a specific purpose and follows established norms. Unlike casual crowds, conventional crowds are structured, their members may interact, and they conform to rules that are appropriate for the situation. Examples include people attending religious services, funerals, graduation ceremonies, and parades.
- An *expressive crowd* is a group of people who show strong emotions toward some object or event. The feelings—which can range from joy to grief—pour out freely as the crowd reacts to a stimulus. Examples include attendees at religious revivals, revelers during Mardi Gras, and enthusiastic fans at a football game.
- An *acting crowd* is a group of people who have intense emotions and a single-minded purpose. The event may be planned, but acting crowds can also be spontaneous. Examples include people fleeing a burning building, soccer fans storming a field, and college students having a water balloon fight.
- A *protest crowd* is a group of people who assemble to achieve a specific goal. Protest crowds demonstrate their support of or opposition to an idea or event. Most demonstrations—like prolife or abortion rights protests, boycotts, and labor strikes—are usually peaceful. Peaceful protesters can become aggressive, however, resulting in destruction and violence (Blumer, 1946; McPhail and Wohlstein, 1983).

One type of crowd can easily change into another. A conventional crowd at a nightclub can turn into an acting crowd if a fire erupts; people panic and flee for safety. Any of the five types of crowds—from casual to protest—can become a mob or a riot.

Mobs

A **mob** is a highly emotional and disorderly crowd that uses force, the threat of force, or violence against a specific target. The target can be a person, group, or property. Mobs often arise when people are demanding radical societal changes, like removing a corrupt government official. In other cases, especially when authority breaks down, people who take advantage of a situation may engage in mob behavior.

In January 2021, U.S. Congress was set to certify the 2020 presidential election votes from the Electoral College when a mob entered the Capitol building. Approximately 2,000 people vandalized and looted the Capitol. Five people were killed (just before, during, or after the event) and many others were injured.

Riots

Compared with mobs, riots usually last longer. A **riot** is a violent crowd that directs its hostility at a wide and shifting range of targets. Unlike mobs, which usually have a specific target, rioters unpredictably attack whomever or whatever gets in their way. Most riots arise out of long-standing anger, frustration, or dissatisfaction that may have smoldered for years or even decades. Some of these long-term tensions are due to discrimination, poverty, unemployment, economic deprivation, or other unaddressed grievances.

There are numerous protests in the United States every year, but race riots have been the most violent and destructive. More than 150 U.S. cities experienced race riots after the assassination of Martin Luther King Jr. in 1968. In 1992, riots broke out in 11 cities after four White police officers were acquitted of beating Rodney King, a Black motorist, in Los Angeles. The violence resulted in deaths and considerable property damage because of fires and looting.

In 2014 and 2015, riots erupted in Ferguson, Missouri, and Baltimore, Maryland, after the controversial deaths of Black men at the hands of the police. As in the past, looting and/or burning local businesses accompanied both of the riots. A few years later, white supremacists marched into Charlottesville, Virginia, to protest the planned removal of a statue of Robert E. Lee, a Confederate general, from a city park. Many carried Confederate flags, gave Nazi salutes, and displayed swastikas and KKK symbols. Counter protesters, including some with Black Lives Matter signs, confronted the marchers, violence ensued, and a car that barreled into a group of anti-supremacist demonstrators killed a young woman.

Riots are usually expressions of deep-seated hostility, but this isn't always the case. After the Philadelphia Eagles won the 2018 Superbowl Championship, riots broke out across the city. People lit fires, damaged cars, and some "kicked in a Macy's window to take selfies with mannequins" (Lardieri, 2018). Such "celebration riots" are due to extreme enthusiasm and excitement rather than anger or frustration and can lead to "an orgy of gleeful destruction" (Locher, 2002, p. 95).

Much collective behavior, like a mob or riot, is spontaneous and short-lived. In contrast, because social movements are usually structured and enduring, they create or suppress social changes.

---

mob a highly emotional and disorderly crowd that uses force, the threat of force, or violence against a specific target.

riot a violent crowd that directs its hostility at a wide and shifting range of targets.

16-2 Social Movements

There are hundreds of social movements in the United States alone. Why are they so prevalent? And why do they matter?

16-2a What Is a Social Movement?

A **social movement** is a large and organized group of people who want to promote or resist a particular social change. Many important social movements have occurred around the world and in the United States: the abolition of slavery, women's right to vote, unions, the movement against apartheid in South Africa, the #MeToo movement, Black Lives Matter, and marriage equality (Ewen, 1998, pp. 81–82; Amnesty International, 2020).

Unlike other forms of collective behavior, social movements are goal-oriented, deliberate, structured, and can have a lasting impact on a society. And in contrast to many other forms of collective behavior (like crowds, mobs, and riots), the people who make up a social movement are dispersed over time and space, and usually have little face-to-face interaction (Turner and Killian, 1987; Lofland, 1996).

Some social movements are relatively small such as the movement in the U.S. to decriminalize the use of psychedelic drug psilocybin for medical purposes. Others social movements are large and have subgroups that appeal to different segments of the population. The environmental movement has many smaller groups, including Earth First!, Greenpeace, the National Audubon Society, the Union of Concerned Scientists, the Wilderness Society, and plogging (jogging while picking up trash).

16-2b Types of Social Movements

Sociologists generally classify social movements according to their goals (changing some aspect of society or resisting change) and the amount of change they seek (limited or widespread). Some social movements are perceived as more threatening than others because they challenge the existing social order (**Table 16.1**).

Alternative social movements focus on changing some people's attitudes or behavior in a specific way. They typically emphasize self-improvement which may center on spirituality or physical well-being. These movements are the least threatening to the status quo because they seek limited change and only for some people (Cadge and Bender, 2004).

Redemptive social movements (also called *religious* or *expressive movements*) propose a dramatic change, but only for some people. They're usually based on spiritual or supernatural beliefs, promising some form of salvation or rebirth. Examples include any religious movements that actively seek converts (e.g., Jehovah's Witnesses, some Christian evangelical groups) (refer to Chapter 13).

Reformative social movements want to change everyone, but only in a particular way. These movements, the most common type in U.S. society, don't want to replace the existing economic, political, or social class arrangements. Examples include groups that champion the rights of people with disabilities, LGBTQIA+ individuals, crime victims, and animals.

social movement a large and organized group of people who want to promote or resist a particular social change.

iStock.com/Coast-to-Coast

Black Lives Matter, a civil rights movement, emerged after George Zimmerman, a White man, was acquitted of killing Trayvon Martin, an unarmed Black teenager, in a Florida suburb in 2013. The group's chapters regularly protest racial profiling, police brutality, and racial inequality in the criminal justice system.

Table 16.1 Five Types of Social Movements

Movement	Goal	Examples
Alternative	Change some people in a specific way	Alcoholics Anonymous, transcendental meditation
Redemptive	Change some people, but completely	Jehovah's Witnesses, born-again Christians
Reformative	Change everyone, but in specific ways	Gay rights advocates, Mothers Against Drunk Driving (MADD)
Resistance	Preserve status quo by blocking or undoing change	Antiabortion groups, white nationalists
Revolutionary	Change everyone completely	Right-wing militia groups, Communism, ISIS in the Middle East

Resistance social movements (also called *reactionary movements*) try to preserve the status quo by blocking change or undoing change that has already occurred. Resistance movements are often called *countermovements* because they usually form immediately after an earlier movement has created change. For example, shortly after the Supreme Court decision in *Roe v. Wade* (1973), which legalized abortion in the United States, antiabortion groups arose and sought to reverse that decision. Following the Supreme Court's decision in *Dobbs v. Jackson* (2022), which overturned *Roe v. Wade*, a new countermovement is growing in response.

Revolutionary social movements want to completely destroy the existing social order and replace it with a new one. These movements range from utopian groups that withdraw from society and try to create their own to terrorists who use violence and intimidation such as ISIL (Islamic State of Iraq and the Levant).

16-2c Why Social Movements Emerge

A social movement is "an answer either to a threat or a hope" (Touraine, 2002, p. 89), but not everyone who feels threatened or hopeful joins a social movement. Why not? Let's explore four explanations, beginning with the oldest.

Mass Society Theory

Early on, sociologists believed that the people who formed social movements felt powerless, insignificant, and isolated in modern mass societies, which are impersonal, industrialized, and highly bureaucratized. Thus, according to *mass society theory*, social movements offer a sense of belonging to people who feel alienated and disconnected from others (Kornhauser, 1959).

Critical Evaluation

Mass society theory may explain why some people form extreme political movements like Fascism and Nazism, but subsequent research has shown that movement organizers are typically not isolated but well-integrated into their families and communities. Also, historically, many political activists in the United States, including those behind the civil rights and women's rights movements during the late 1960s, weren't powerless but came from relatively privileged backgrounds (McAdam and Paulsen, 1994).

Relative Deprivation Theory

Relative deprivation theory is broader than mass society theory. **Relative deprivation** is a gap between what people have and what they think they should have compared with others in a society. What people think they should have includes certain amounts of money, social status, power, or privilege.

Relative deprivation theorists note two other elements. First, people often feel that they *deserve* better than they have ("I've worked hard all my life"). Second, they believe that they *can't attain their goals through conventional channels* ("I've contacted my senators, but they ignore me"). Thus, shared beliefs combined with unfulfilled expectations can trigger change-oriented social movements (Davies, 1979; Morrison, 1971).

Relative deprivation also includes people who want to protect their declining power and privilege. For example, the white nationalist/neo-Nazi movement attracts people, particularly lower-income White people, who believe that people of color and immigrants threaten the economic and political status that benefitted White people in the past.

Critical Evaluation

Relative deprivation theory helps explain why some social movements emerge. Critics point out, however, that there's a certain degree of relative deprivation in all societies, but people don't always react by forming or joining social movements. Relative deprivation theory also doesn't explain why some people join movements even though they don't view themselves as deprived, and don't expect to gain anything personally if the movement succeeds (Gurney and Tierney, 1982; Johnson and Klandermans, 1995).

Resource Mobilization Theory

It takes more than feeling alienated (mass society theory) or disadvantaged (relative deprivation theory) to sustain a social movement. Instead, according to *resource mobilization theory*, a social movement will succeed if it can put together (or mobilize) an organization and leadership dedicated to advancing its cause (Oberschall, 1995; Gamson, 1990). Other important resources include money, devoted volunteers, paid staff, access to the media, effective communication systems, special technical or legal knowledge and skills, equipment, physical space, alliances with like-minded groups, and lobbyists who influence legislation (refer to Chapter 11).

Critical Evaluation

A key contribution of resource mobilization theory is its emphasis on structural factors (like organization and leadership) in explaining why some social

---

relative deprivation a gap between what people have and what they think they should have compared with others in a society.

movements thrive whereas others shrivel. A major criticism, however, is that resource mobilization theory largely ignores the role of relative deprivation in a social movement's formation. If there aren't large numbers of dissatisfied people to initiate a movement, even plentiful resources won't be able to sustain it (Klandermans, 1984; Buechler, 2000).

New Social Movements Theory

New social movements theory, which became prominent during the 1970s, emphasizes the linkages between culture, politics, and ideology. Unlike the earlier perspectives, new social movements theory proposes that many recent movements (like those that work for peace, LGBTQIA+ rights, and environmental protection) promote the rights and welfare of *all* people rather than specific groups in particular countries (Laraña et al., 1994; Melucci, 1995). Thus, the new social movements theory is especially interested in "the struggle to liberate the voices of the dispossessed" (Schehr, 1997, p. 6).

If there aren't large numbers of dissatisfied people to initiate a movement, even plentiful resources won't be able to sustain it.

For these theorists, recent social movements differ from older ones in several ways. They attract a disproportionate number of people who are well-educated and relatively affluent, who represent a wide array of professions (e.g., educators, scientists, actors, businesspeople, political leaders), and who share a broad goal—improving the quality of life for all people around the world. Moreover, new social movements pursue goals or advance values that may not personally benefit its members, such as eradicating diseases in developing countries (Obach, 2004).

Critical Evaluation

Unlike earlier perspectives, new social movements theory helps us understand collective behavior that crosses international boundaries. According to some critics, however, neither this perspective nor the groups that it examines are novel. Some social movements (like feminism and environmentalism) have been around for a long time and still focus on the same basic issues, like women's unequal positions and environmental destruction. In addition, some scholars point out that educated middle-class or wealthy activists were as common in the old social movements as in more recent ones (Rose, 1997; Buechler, 2000; Sutton, 2000).

Critics also contend that new social movements theory often overstates people's altruistic motivations. Many people who join environmental groups do so for reasons referred to as NIMBY (not in my backyard). That is, they're concerned about their own community's environment, but show little interest in ecological threats to people elsewhere.

These four theories, despite their limitations, offer useful insights into social movements because "no single theory is sufficient to explain the complexities of any social movement" (Blanchard, 1994, p. 8). The next question is why some social movements flourish and others collapse.

16-2d The Stages of Social Movements

Most social movements are short-lived. Some never really get off the ground; others meet their goals and disband. Social movements generally go through four stages: emergence, organization, institutionalization, and decline (**Figure 16.1**) (Tilly, 1978).

Emergence

During *emergence*, the first stage of a social movement, a number of people are distressed about some condition and want to change it. One or more individuals, serving as agitators or prophets, emerge as leaders. They verbalize the feelings of the discontented, crystallize the issues, and push for action. If leaders don't get much support, the movement may die. If, on the other hand, the discontent resonates among growing numbers of

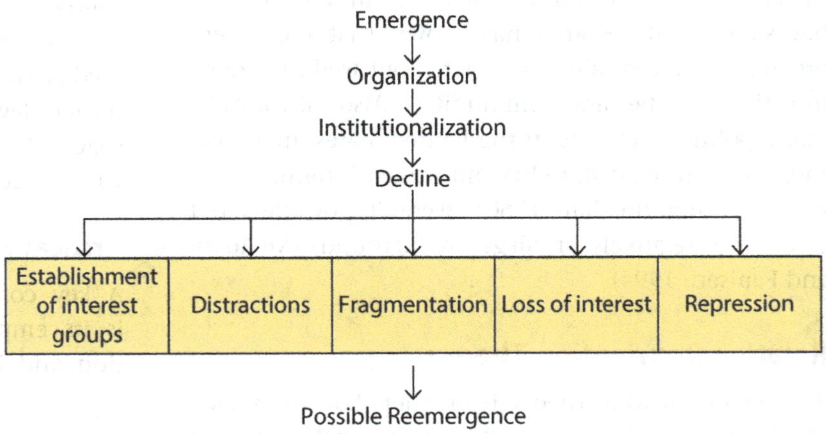

Figure 16.1 Typical Stages of a Social Movement

people, public awareness increases, and the movement attracts like-minded people.

Organization

Once people's consciousness has been raised, the second stage is *organization*. The most active members form alliances, seek media coverage, develop strategies and tactics, recruit members, and acquire the necessary resources. The organization establishes a division of labor in which leaders make policy decisions and followers perform necessary tasks like preparing mass mailings, developing websites, and responding to phone calls and emails. At this stage, the movement may develop chapters at local, regional, national, and international levels (refer to Obach, 2015).

Institutionalization

As a movement grows, it becomes *institutionalized* and more bureaucratic: The number of staff positions increases, members draw up bylaws, the organization may hire outsiders (writers, attorneys, and lobbyists) to handle some of the necessary tasks, and the leaders may spend more of their time on speaking tours, in media interviews, and at national or international meetings. The movement becomes more stable and self-sufficient, and the original leaders may move on to better paying and more influential positions in government or the private sector.

Decline

Almost all social movements end sooner or later. ("Why Is the NRA a Successful Social Movement?" describes one exception.) This *decline*, the last stage, may take several forms:

AP Images/Damian Dovarganes

Food movements have burgeoned worldwide, especially in urban centers. Many U.S. cities have institutionalized food carts and trucks by passing or revising regulations that support food vending. Doing so encourages neighborhood entrepreneurs to use available city spaces, attracts "foodies," and offers residents and tourists foods from a variety of cultures, organic, and gourmet "culinary experiences" (Hanser and Hyde, 2014, p. 47).

- If a social movement is successful, it can become an *interest group* and a part of society's fabric. A small antismoking movement that began in the mid-1970s now has the enthusiastic support of numerous prestigious organizations, including the American Cancer Society, the American Heart Association, the American Medical Association, the World Health Organization, and governing bodies within and outside the United States (Wolfson, 2001).
- Those involved in a social movement may become *distracted* because the group loses sight of its original goals, their enthusiasm diminishes, or both. In the mid-1960s, when Ralph Nader initially condemned the auto industry for car safety defects, he gained a large following. As Nader and his consumer rights groups expanded their focus to include environmental issues and corporate crimes, many of the initial followers lost interest.
- A social movement may experience *fragmentation* because the participants disagree about goals, strategies, or tactics. The environmental movement encompasses numerous groups that focus on different issues—air quality, marine life, land use, and global warming. Participants may also drift away because of time constraints, strategy disagreements, health problems, or similar reasons (refer to Blee, 2015).
- Social movements may decline because of *repression*. Many autocratic governments quash dissent. With current technology, repression can be carried out through "digital repression" which may involve use of spyware, mass surveillance, or censorship. A government can crush an emerging social movement by arresting protestors and imprisoning or even executing leaders (refer to Chapter 11) (Earl et al., 2022; National Intelligence Council, 2022).

Social movements can wane for a combination of reasons. In 2011, the Occupy Wall Street (OWS) movement protested socioeconomic inequality, corporate greed, and corporate power over the U.S. government. The group's slogan, "We are the 99%," referred to the income inequality between the top 1 percent and the rest of the population. Despite similar demonstrations in 70 major U.S. cities and over 600 communities, and 900 cities worldwide, the movement faded away. Why? OWS raised awareness of the injustice of this nation's inequality, but ultimately failed because it lacked leadership, clear objectives, solutions, organization, and a long-term commitment (Madrick, 2013; Sandbu, 2013).

Why Is the NRA a Successful Social Movement?

A majority of Americans say gun violence is a "very big problem" in the country (60 percent), and most support preventing people with mental illness from purchasing firearms (88 percent). Also, roughly half of Americans favor stricter gun laws (58 percent) (Schaeffer, 2023).

A brief exploration of the stages of the National Rifle Association (NRA) helps explain why, despite public consensus to the contrary, this social movement has successfully increased gun rights. For example, half the states in the U.S. allow open carrying of firearms without a permit, and 26 states allow concealed carrying of weapons without a permit (Buchholz, 2023).

1. *Emergence* The NRA was founded in 1871 by two former Union veterans to improve their troops' shooting skills. The NRA built a rifle range on Long Island, New York, and organized rifle competitions and clubs. The leaders attracted people—soldiers, hunters, and sportsmen—who shared the common goal of becoming better marksmen.
2. *Organization* In 1871, NRA leaders formed a board of directors to elect a president, elected other officers (vice president, treasurer, and so on), moved the headquarters to Washington, D.C., established a committee to lobby for laws that befitted the NRA, and created a legislative affairs division to update members on upcoming bills. In effect, then, the NRA set up a division of labor, developed strategies, and formed alliances with political decision makers.
3. *Institutionalization* After the passage of the Gun Control Act of 1968, the NRA hired well-paid lobbyists, funded political action committees (refer to Chapter 11), supported gun-rights candidates, and opposed gun-control candidates. It developed multilayered executive staff positions, elected or paid celebrity spokespersons (e.g., Chuck Norris, Charlton Heston), and formed official chapters in every state.

The NRA also implemented a "savvy business plan" that offers members—for only $45 for a year—an official membership ID card and decal, discounts on everything from travel to wine clubs, free items (hunting knives, tactical backpacks, heavy-duty duffel bags), a free subscription to one of their many magazines, and insurance for them

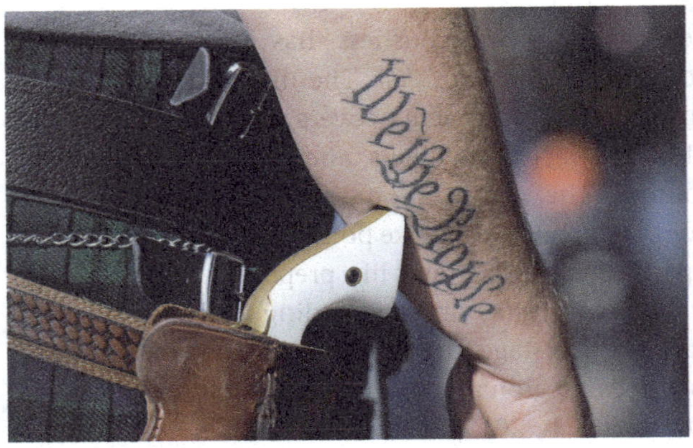

AP Images/Eric Gay

and their guns. The NRA has embedded its fundamental values and objectives into culturally acceptable activities, which helps maintain membership numbers. The NRA is estimated to have between three and five million members, and in 2022, the organization brought in $97 million in membership dues ("Gun Laws: A Shot and a Beer," 2014; National Rifle Association, 2023; BBC News, 2023).

Decline The NRA isn't facing imminent decline. It offers shooting, training, and gun safety programs for many diverse groups (e.g., women, children, hunters, competitive shooters, law enforcement), which broadens the NRA's member base and loyalty. It has a network of over 15,000 affiliated businesses, associations, and clubs that generate significant revenue and help lobby against gun control. In addition, the number of gun owners (potential NRA supporters) continues to be large in the U.S. Three out of ten American adults own a firearm themselves, and four out of ten live in a home with one. There is also more diversity among new gun owners. For two years in a row, around 20 percent of new firearm owners were Black, and another 20 percent were Latina/o/x—higher than their representation in the overall population. Also, half of new gun owners are now women, which is another group that has a growing interest in firearm ownership (Elinson, 2021; Schaeffer, 2023).

Perhaps most importantly, to many Americans, guns represent freedom and independence. As a result, 21 percent of gun owners, compared with only 12 percent of non-gun owners, contact public officials at least once a year, urging more lenient gun laws.

Sources: Davidson, 1998; Rodengen, 2002; Melzer, 2009; Winkler, 2011; Bellini, 2012; BBC News, 2016; Law Center to Prevent Gun Violence, 2017; Parker et al., 2017; and undated NRA online sites.

A social movement that declines can sometimes experience a resurgence. For example, there have been several waves of the women's rights movement in the United States since the mid-nineteenth century. The first wave ensured women's right to vote in 1920; the second wave expanded employment and educational rights during the 1960s and 1970s; the third wave, during the 1990s, focused on economic and other inequalities experienced by women of different social classes, sexual orientations, and nationalities; and a fourth wave has been growing with the #MeToo movement and aims to counter the backslide of rights for women and LGBTQIA+ individuals. Despite considerable backlash over the past hundred years, American feminism has appealed to diverse groups, agitated in a multitude of spheres ranging from athletics to religion, and has used a variety of strategies to press for social change (Cobble et al., 2014).

16-2e Why Social Movements Matter

On an *individual level*, many people enjoy a variety of rights as workers, consumers, voters, and victims—rights that are attributed to highly dedicated people who were determined to change inequitable laws and customs.

On an *institutional level*, social movements can change general practices. Shopping for healthy food is much easier today than it was before the 1990s, when veggie burgers, tofu, and nutrition labeling were practically nonexistent. Now, mainstream grocery stores have large organic food sections, a variety of fruits and vegetables, and breads and cereals made with whole grains, nuts, and less salt, sugar, and chemical additives as well. The "farm-to-table" movement, with its emphasis on locally grown produce and meat, has benefited people's health and regional economies. In effect, then, vegetarian and consumer groups have changed the way many farms operate and have loosened corporate control of food products (O'Connor, 2015; Kowitt, 2015).

On a *societal level*, social movements have had a major impact in the United States and globally. Most of the world's great religions began as protest movements (refer to Chapter 13). Also, democratic governments in the United States and other countries grew out of the activities of revolutionary groups that sought greater political and economic freedom (Giugni et al., 1999; della Porta and Diani, 1999).

You've read that social movements can generate or resist change. Technology is another important source of societal change.

16-3 Technology and Social Change

This brief "history" of how to cure an earache has circulated on the Internet:

- 2000 B.C.—Here, eat this root.
- 1000 A.D.—That root is heathen, say this prayer.
- 1850 A.D.—That prayer is superstition, drink this potion.
- 1940 A.D.—That potion is snake oil, swallow this pill.
- 1985 A.D.—That pill is ineffective, take this antibiotic.
- 2000 A.D.—That antibiotic is artificial. Here, eat this root.

Teens Organize for Social Change

After a gunman took the lives of 17 students and teachers at Marjorie Stoneman Douglas High School, teenage survivors responded by building the March For Our Lives Movement. Youth leaders, including Emma Gonzalez and David Hogg, encouraged hundreds of thousands of people to gather for the "largest single day of protest against gun violence in history" (March for Our Lives, 2023). In the years since, this organization has continued its efforts by meeting lawmakers and prompting public conversation about gun policies in the United States. They are bringing attention to passing laws to ban assault weapons, halting high-capacity magazine sales, and putting requirements in place for background checks on all firearm purchases (Facing History and Ourselves, 2021)

Tom Williams/Getty Images

As this anecdote suggests, despite technological progress over the centuries, some of the old remedies are enjoying renewed popularity (peoplespharmacy.com, for instance, offers home treatments for everything from arthritis to whooping cough). Still, technological advances have brought enormous social changes, including benefits, costs, and ethical concerns.

16-3a Some Recent Technological Advances

Technology, the application of scientific knowledge for practical purposes, is vital to human life: "For good or ill, [technologies] are woven inextricably into the fabric of our lives, from birth to death, at home, in school, in paid work" (MacKenzie and Wajcman, 1999, p. 3; refer to Chapter 3 for early technological inventions).

In the coming decades, technology is likely to change our lives more dramatically than ever before. Millions of Americans' homes already use smart technology, such as controlling lights or appliances via the Internet. Homes of the future may include robots to detect harmful emissions or intruders as well as help with meal preparation or pets. There may be sensor systems to determine if people are taking the medicines and to make recommendations for healthier food choices when people open the fridge (Austin, 2019). Transportation technology may change, too. Several companies now have autonomous cars that are almost driverless, with abilities such as assisting the driver with staying in the lane and automatically stopping at stoplights (Threewitt and Clarke, 2023). Let's briefly explore a few of the most influential technological advances that are changing our lives.

Computer Technology

In 1887, English mathematician Charles Babbage designed the first programmable computer. Since then, computers have gone through seven generations of evolution. *Artificial intelligence (AI)*, a branch of computer science, is developing theories and computer systems to perform tasks that normally require human intelligence (e.g., speech recognition, decision making, and understanding). AI advances have spurred the development of *robots*, machines that are programmed to perform humanlike functions. Industrial robots, known as "collaborative robots" or "cobots," already work alongside people in doing tasks like stacking heavy packages. For $2,000, you can now buy a laundry-folding robot. By 2025, according to some experts, AI and robots will affect nearly every aspect of daily life (Smith et al., 2014; Salkin, 2017).

STEPHANE DE SAKUTIN/Getty Images

Some nursing homes are using robots to socialize with residents, teach them exercises, and chat with children in their classrooms. One woman, who hadn't spoken in four months, surprised everyone when she blurted out, "I'm well," after a robot asked how she was doing and then carried on a brief conversation (McFarland, 2015; Leland, 2022).

Robots have performed repetitive, dirty, or dangerous human tasks, including toxic waste cleanup, mining, minefield sweeping, underwater and space exploration, and locating bombs. Software is also being used in new ways. For good or bad, ChatGPT can help write essays or emails, search for jobs, review computer code, create art, and many other things. More than 70 percent of Americans think ChatGPT is somewhat, very, or extremely useful. There are still uncertainties about this AI. Out of those who have heard of ChatGPT, almost half are "concerned," half think it should slow down, and half think there should be federal agency regulation agency for AI ("Artificial Intelligence," 2015; Kharif, 2015; Marr, 2023; Vogels, 2023; Kacholia, 2023).

Biotechnology

Biotechnology is a broad term that applies to all practical uses of living organisms. It covers anything from the use of microorganisms (e.g., yeast) to ferment beer to *genetic engineering*, sophisticated techniques that can change the makeup of cells and move genes across species to produce new organisms.

In medicine, biotechnology includes genetic testing/screening that can determine a child's biological parents, a person's ancestry, inherited disorders, and the chance of developing or passing on a genetic disease. Pig heart valves are already transplanted into patients. It may soon be possible to transplant pig livers, hearts, and other organs, saving some of the 22 people a day who die waiting for an organ transplant. There has also been progress with technology to assist amputees in

technology the application of scientific knowledge for practical purposes.

controlling prosthetics using their minds which could help approximately 2.1 million people with limb loss in the U.S. ("Biomedical Engineering," 2017; Kolata, 2017; Kosowatz, 2020).

Nanotechnology

Another promising innovation is *nanotechnology*, the ability to build objects one atom or molecule at a time. The key characteristic of these objects is their tiny size. Nanotechnology is based on structures measured in nanometers, a unit of measurement equal to 1 billionth of a meter, or 1/80,000th the width of a human hair.

Scientists are also working on nanoparticles that hunt down and kill tumor cells, and nanosensors—microscopic robots with legs, propellers, and cameras—that would live in the bloodstream and send messages to smartphones about signs of infection, an impending heart attack, or other issues that are early warnings for disease or death. Swallowing little pills that contain nanosensors may become a routine part of doctors' office visits in the future (Cha, 2014; "Hunting as a Pack," 2014).

These and other technological innovations promise longer and healthier lives in the future, but what about the present? What are some of the current benefits and drawbacks of technological changes?

16-3b Some Benefits and Drawbacks of Recent Technologies

The ever-accelerating pace of change means that we're using technologies that didn't exist just a decade ago. The average American embraces modern technology but is also wary of some advancements. For example, a majority doesn't want some medical enhancements that could improve people's cognitive or physical capacities (78 percent) and around half would not want to use a robotic exoskeleton to enhance their own strength (54 percent) **(Table 16.2)** (Rainie et al., 2022).

Table 16.2 Americans Are Divided About Some Medical Advances

Percentage who say they wouldn't want these enhancements for themselves or their children	
Gene editing to greatly reduce a baby's risk of disease	49
Robotic exoskeleton to enhance strength	54
Brain chip implant for much improved cognitive abilities	78

Source: Based on Rainie et al., 2022

A computer scientist has noted that every technology has a "dark side" because practically every benefit also has a cost (Lohr, 2011, p. A1). Here are a few contemporary examples.

Collaborative Consumption

- **Benefits:** Technology has spurred "collaborative consumption" (also called a "sharing economy" and "access economy") that enables people to earn and save money by sharing goods and services. Airbnb matches budget travelers and people with a spare room or other lodging for rent, RelayRides allows people to rent your car, and taxi-like services (e.g., Lyft, Uber, Sidecar) bring drivers and passengers together. Crowdfunding, such as Kickstarter, helps entrepreneurs connect with people who can provide crucial financial backing for their startups.
- **Drawbacks:** The sharing economy has created conflict: Airbnb hosts have sometimes ignored zoning laws, don't always pay city or state taxes, neighbors have complained about noise and a steady stream of strangers, and their added costs (e.g. cleaning fees) are increasing. Taxicab companies have protested that ridesharing (Uber and Lyft) has decreased their business because the drivers have avoided all regulations, inspections, fees, and insurance requirements (Goodale, 2015; Stein, 2015). Crowdfunding sites can have high fees, require a lot of time, and put the entrepreneur at risk of having someone else steal their idea (Vukelich, 2023).

3-D Printing

- **Benefits:** 3-D printing—a process for making a physical object from a three-dimensional digital

SS MIRZA/AFP/Getty Images

In restaurants, there is increasing automation and even utilization of robots. For example, robot waiters can help deliver food to customers. Are restaurant jobs for people at risk?

model—is revolutionizing manufacturing. Because the process can reduce manufacturing costs by 25 percent or more, the CEO of an aluminum corporation describes 3-D printing as "the beginning of... a second Industrial Revolution" (Geier, 2014, p. 76). Among other things, 3-D printing produces medical implants (such as knee replacements), synthetic heart valves, prosthetics that are easy to use, models of buildings, artificial human bones, and even cars that are safe and fuel efficient. Desktop 3-D printers produce a wide range of objects like toys, clocks, ceramics, and wrenches.

- **Drawbacks:** As with any new technology, 3-D printers have negative impacts: They consume 50 to 100 times more electrical energy than traditional molds and increase usage of plastics that aren't recycled. They can also pose health risks by emitting toxins into the home or workplace. Moreover, people can print weapons, including plastic guns. Currently, some components of weapons still need to be made out of traditional metals (and can likely be detected by metal detectors or scanners) but these weapons are more difficult for law enforcement to trace, and the tech keeps getting better (Gilpin, 2014; Vallance, 2022).

Health and Medicine

- **Benefits:** People who use health-related mobile gadgets and apps can monitor their blood pressure, calories, and physical fitness. The number of robotic procedures increases by 30 percent each year. Robotic surgery reduces physician fatigue because they don't have to stand over the patient for hours, and a robot's "hands" can reach into tight spots and move in ways that human hands can't. In addition, minimally invasive robotic surgeries usually result in less blood loss and faster recoveries because there's a smaller incision to heal.
- **Drawbacks:** Health-related apps typically collect less user information than other types of apps, but most mobile health apps (88 percent) can access and possibly share data from users. In addition, there are "inconsistent privacy practices" in these medical apps (Pifer, 2021). There are drawbacks of robotic procedures, too. At times, robot surgeries don't work, which requires a shift to a traditional surgical method, and the robotic approaches often take longer. It's also more expensive, with setup costs around $2 million, along with costs for servicing of equipment and additional work hours for the extra time (Bakalar, 2021).

Monitoring Systems

- **Benefits:** New technology allows people who need assistance to live at home rather than in an institutional setting. Sensors can track people's medications and falls. Family or friends can monitor their friend or relative remotely via cellular connections, checking if they go outside or return home safely. These devices can also provide ways to know if abuse by caretakers is occurring.
- **Drawbacks:** Medicaid pays for some aging-at-home technology, but Medicare doesn't. Even if families can afford the technology, some people may feel intimidated by even one-button devices and don't use them. The remote options may also limit in-person visits by family or friends, therefore reducing important social interactions (Abrahms, 2014; Tsukayama, 2014; Kelly, 2021).

Jobs and Privacy Issues

Some policy analysts fear that technology will produce more jobs for robots than for humans, even in occupations like teaching, sales, nursing, and stock trading. Driverless cars could displace millions of truck drivers, bus drivers, and others who drive for a living. Others contend that computers will never replace jobs that require "people skills" like teamwork, creativity, managing diverse employees, leadership, decision-making, and face-to-face interaction with clients or customers (Colvin, 2014; Arntz et al., 2016).

There's considerably more consensus that almost every technological advance reduces privacy and increases data breaches. The scale and impact of data breaches can be huge. In 2022 alone, there were 1800 breaches affecting over 400 million people. During the 2016 presidential election, the Russian government infiltrated the Democratic National Committee's computer network, hoping to undermine Hillary Clinton's campaign. The hackers also targeted voting equipment in at least 21 states and created several hundred fake Facebook and Twitter accounts to spread anti-Clinton messages during the election. Unfortunately, warnings about hacking and data breaches are sometimes ignored, and protecting data continues to be an issue (Davis, 2015; Yeung, 2016; Perlroth et al., 2017; Griffith, 2023) (refer also to Chapter 7).

A lot of software, including smartphone apps and games like Candy Crush Saga and Pokémon Go, have lengthy, legalese-filled "terms of service" agreements. Users rarely read these documents, but clicking on "I agree" gives a company "perpetual, irrevocable, transferable, worldwide, royalty-free license to use, copy,

"Hey! Elbows off the table."

modify ... publicly display, publicly perform, and distribute your User Content" (Ziccarelli, 2016, p. 15).

Many "data mining" companies routinely collect information as people click from site to site. Much of this web tracking is done anonymously, but a new crop of "snooper" sites makes it easier than ever before for anyone with Internet access to assemble and sell personal information, including your name, Social Security number, address, what you buy, whom you love, and which sites you've visited.

Google compiles enough data to build comprehensive portfolios of most users—who they are, where they go, and what they do. To create a new stream of profit, Facebook sells marketers detailed information about its users. Marketers, in turn, can flood millions of websites and mobile apps with targeted ads. At many department stores, cameras have become so sophisticated that companies can analyze what shoppers are looking at and even their mood (Efrati, 2013; "Marketing in the Digital Age," 2015; Komando, 2016).

Americans are uneasy about data collection and say they have little control over it in present-day society. Six in ten Americans say it's "not possible to go through daily life without having their data collected," and most are concerned about data use by companies (79 percent) and government (64 percent). And many don't understand how companies (59 percent) or government (78 percent) are using the data they collect (Auxier et al., 2019).

16-3c Some Ethical Issues

DNA testing has given millions of people information about their genetic predispositions for diseases like cystic fibrosis, cancer, and Huntington's disease (an incurable neurological disorder). Having such information helps doctors and patients make better health care decisions, but the expanding use of genetic testing has an unforeseen consequence: More people who don't show any symptoms are being told they have genes for potentially fatal diseases that may develop "next month, next year, when you are 60 years old, or never" (Marcus, 2013, p. D1).

Once people have the information, they may become "patients-in-waiting" who undergo continuous screening and worry about their condition. Asymptomatic patients with an inherited risk for a genetic heart muscle disease may have a defibrillator implanted, which involves regular maintenance, possible equipment failures, infections, and other complications. Some doctors believe that such preventive measures for patients-in-waiting are unethical because they're unnecessary, risky, and people live in limbo not knowing whether or when a disease will develop (Marcus, 2013).

Editing of DNA in human embryos is now possible, and the technique could potentially prevent a long list of inheritable diseases, such as cystic fibrosis, heart disorders, Huntington's disease, and breast and ovarian cancer. However, 84 percent of people worry that gene editing could be used in morally unacceptable ways. The capabilities of this DNA science are progressing. In 2018, a biophysicist announced that he created the first gene-edited babies. He spent three years in prison in China as a result. Others fear that such research could lead to wealthy parents requesting "designer babies" that are smarter, taller, healthier, and stronger—giving them a competitive edge. Fifty-five percent of Americans are concerned that widespread use of gene-editing could lead to more income inequality. Designer babies aren't technically possible, but some geneticists are denouncing any experiments that involve genetically modified human embryos as unethical and irresponsible (Ma et al., 2017; Stein, 2017; Rainie et al., 2022; Cohen, 2022). Thus, technological advances are fraught with both promise and pitfalls.

Chapter Review

Key Takeaways

16-1 **What are key components of collective behavior?** Social scientists offer several explanations of *collective behavior.* Contagion theory proposes that individuals act emotionally and irrationally due to a crowd's almost hypnotic influence. According to convergence theory, crowds consist of like-minded people who deliberately assemble in a place to pursue a common goal. Emergent norm theory emphasizes the importance of social norms in shaping crowd behavior. Structural strain theory proposes that collective behavior occurs only if six conditions are present.

There are many types of collective behavior, some more short-lived or harmful than others. *Rumors, gossip,* and *urban legends* are typically untrue, but many people believe and pass them on for a number of reasons, such as anxiety or to reinforce a community's moral standards. In contrast, *panic* and *mass hysteria* can have dire consequences, including death. *Fashions* and *fads* are harmless because they usually last only a short time and change over time. In contrast, a *disaster* is an unexpected event due to social, technological, or natural causes that result in widespread damage and destruction.

Publics, public opinion, and *propaganda* also affect large numbers of people. *Crowds* vary in their motives, interests, and emotional level. A casual crowd, for example, has little, if any, interaction, the gathering is temporary, and there is little emotion. On the other hand, protest crowds, especially in *mobs* and *riots,* can wreak considerable havoc on property and result in death.

16-2 **What are the main types of social movements? What are the stages of social movements?** Unlike collective behavior, *social movements* are typically organized and have long-lasting effects. Some of the most common social movements are alternative, redemptive, reformative, resistance, and revolutionary (refer to **Table 16.1** on page 362). Sociologists have offered several explanations for the emergence of social movements, including mass society theory, relative deprivation theory, resource mobilization theory, and new social movements theory. Each theory has strengths and weaknesses in explaining social movements.

Social movements generally go through four stages: emergence, organization, institutionalization, and decline. Decline occurs when a social movement is successful and becomes a part of society's fabric; when the members become distracted because the group loses sight of its original goals and/or their enthusiasm wanes; when the membership fragments because the participants disagree about goals, strategies, or tactics; or when a government quashes dissent. Social movements are important because they can create or resist change at the individual, institutional, and societal levels.

16-3 **How do technology and social change affect society?** *Technology* also generates changes. Some of the most important technological advances have included computer technology, biotechnology, and nanotechnology—all of which have changed our lives. Technology has both benefits and costs, however. For example, the Internet and other forms of telecommunication technology can bring people together but can also intrude on our privacy. In addition, technological advances raise numerous ethical questions, such as their greater availability to higher-income people.

Key Terms

Social change (p. 354)
Collective behavior (p. 354)
Rumor (p. 356)
Gossip (p. 357)
Urban legends (also called *contemporary legends* and *modern legends*) (p. 357)
Panic (p. 357)
Mass hysteria (p. 358)
Fashion (p. 358)
Fad (p. 359)
Disaster (p. 359)
Public (p. 359)
Public opinion (p. 360)
Propaganda (p. 360)
Crowd (p. 360)
Mob (p. 361)
Riot (p. 361)
Social movement (p. 362)
Relative deprivation (p. 363)
Technology (p. 368)

Discussion and Critical Thinking Questions

1. A classic example of mass hysteria is Orson Welles's "Men from Mars" radio broadcast, based on the H. G. Wells novel The War of the Worlds (1898), on Halloween eve, 1938. Welles interrupted a dance program with a "news bulletin" that there was an invasion from Mars at a New Jersey farm and that the aliens were about to conquer Earth. Welles announced several times that the broadcast was fictional, but many listeners missed the announcements. Thousands fled their homes to escape the monsters. According to some scholars, the mass hysteria was triggered by many Americans' widespread anxieties about getting involved in World War II and a general apprehension about world conditions (Cantril et al., 1952).
   a. Many people, including students, think that the "Men from Mars" example is dated because it wouldn't happen today. Why, then, did so many Americans believe the rumors about Y2K in 2000, with millions close to panicking?
   b. In mid-2008, as the stock market plunged, millions of Americans sold their stock or transferred their 401(k) income to savings accounts. Is structural strain theory still useful in explaining such collective behavior?
2. Do you listen to or pass along gossip, rumors, and/or urban legends? Why or why not?
3. Have you ever participated in any of the five types of crowds that this chapter describes? If so, which one(s), and why?
4. Have you ever been involved in a social movement? If yes, which one(s), and why? If no, why not?
5. What, personally, have been some of the benefits and costs of technology in your life?
6. If you could change one thing in this world, what would it be? And how would you accomplish it?

True or False?

How Much Do You Know about Social Change?

The answers to 1 and 4 are true. The answers to 2, 3, and 5 are false.

1. Convergence theory points out that crowds can consist of like-minded people who deliberately assemble in a place to pursue a common goal, such as marches, rallies, and acts of civil disobedience that oppose a government's domestic or foreign policies.
2. Rumors can cause harm. If believed, they can break down people's trust in important systems (e.g., the electoral system), can lead to threats, and can cause psychological strain.
3. Disasters often inspire organized behavior rather than chaos. Most people are rational, cooperative, and altruistic, opting to care for family and community members, and volunteers offer financial, medical, and other forms of help.
4. A social movement can either promote social change or aim to resist it- such as the white nationalist movement.
5. Seventy-nine percent of Americans are concerned about how their personal data is being used by companies, and 64 percent have concerns about the government's use of it.

References

A

"2012 Survey of Business Owners." 2015. American Factfinder, Table SB1200CSA01. Accessed October 3, 2015 (factfinder.census.gov).

"2013 Indian Gaming Industry Report." 2015. 500 Nations. Accessed October 4, 2015 (500nations.com).

"75%—A Nation of Flag Wavers." 2011. Pew Research Center. Accessed July 15, 2011 (www.pewresearch.org).

AAA Foundation for Traffic Safety. 2015. "2014 Traffic Safety Culture Index." January. Accessed July 21, 2015 (aaafoundation.org).

Aaron, Daniel G., and Michael B. Siegel. 2017. "Sponsorship of National Health Organizations by Two Major Soda Companies." *American Journal of Preventive Medicine* 52 (January): 20–30.

AARP Magazine. 2014. "You're Old, I'm Not." February 3, 40–42.

AAUW. 2013. *The Simple Truth About the Gender Pay Gap, Fall 2013 Edition*. Washington, DC: American Association of University Women.

AAUW. 2017. "The Simple Truth About the Gender Pay Gap." Spring. Accessed May 20, 2017 (aauw.org).

Abbey-Lambertz, Kate. 2017. "How Activists Destroyed Bill O'Reilly's Reputation with Advertisers." *Huffington Post*, March 11. Accessed March 13, 2017 (huffingtonpost.com).

Abelson, Reed. 2016. "Survey Finds Workers Pay a Bigger Share for Health Care." *New York Times*, September 15, B2.

Abrahams, Jessica. 2015. "Are Men Natural Born Criminals? The Prison Numbers Don't Lie." *The Telegraph*, January 13. Accessed April 22, 2017 (telegraph.co.uk).

Abramitzky, Ran, and Leah Boustan. 2022. "Why the Children of Immigrants Are the Ones Getting Ahead in America." *Time Magazine*, June 1. Accessed April 13, 2023 (https://time.com/6182715/immigrants-children-us-mobility/).

Abrams, Samuel J. 2021. "The U.S. Still Believes in Hard Work and Upward Mobility." *Washington Examiner*, February 26. Accessed April 13, 2023 (https://www.aei.org/op-eds/the-us-still-believes-in-hard-work-and-upward-mobility/).

Abrahms, Sally. 2014. "Is This the End of the Nursing Home?" *AARP Bulletin*, March, 20, 22.

Abrams, Rachel. 2016. "Barbie Now Comes in Tall, Short and Curvy." *New York Times*, January 29, B2, B9.

Abramson, Ashley, 2021. "The Impact of Parental Burnout: What Psychological Research Suggests About How to Recognize and Overcome It." American Psychological Association, October 1. Accessed March 11, 2023 (https://www.apa.org/monitor/2021/10/cover-parental-b¥urnout).

Abril, Danielle, and Drew Harwell. 2021." Keystroke Tracking, Screenshots, and Facial Recognition: The boss may be watching long after the pandemic ends." *The Washington* Post, September 24. Accessed September 7, 2023 (https://www.washingtonpost.com/technology/2021/09/24/remote-work-from-home-surveillance/).

Abudabbeh, Nuha. 1996. "Arab Families." Pp. 333–346 in *Ethnicity and Family Therapy*, 2nd edition, edited by Monica McGoldrick, Joe Giordano, and John K. Pearce. New York: Guilford Press.

Academy of Medical Royal Colleges. 2011. "Induced Abortion and Mental Health: A Systematic Review of the Mental Health Outcomes of Induced Abortion, Including Their Prevalence and Associated Factors." National Collaborating Centre for Mental Health, December. Accessed February 9, 2013 (www.aomrc.org.uk).

Acciai, Francesco, Aggie J. Noah, and Glenn Firebaugh. 2015. "Pinpointing the Sources of the Asian Mortality Advantage in the USA." *Journal of Epidemiology and Community Health* 69 (October): 1006–1011.

Acierno, Ron, Melba Hernandez, Amanda B. Amstadter, Heidi S. Resnick, Kenneth Steve, Wendy Muzzy, and Dean G. Kilpatrick. 2010. "Prevalence and Correlates of Emotional, Physical, Sexual, and Financial Abuse and Potential Neglect in the United States: The National Elder Mistreatment Study." *American Journal of Public Health* 100 (February): 292–297.

ACT, Inc. 2014. "Improving College and Career Readiness and Success for Everyone." Accessed June 21, 2014 (www.act.org).

Adams, Bert N., and R. A. Sydie. 2001. *Sociological Theory*. Thousand Oaks, CA: Pine Forge Press.

Adamson, David M. 2010. "The Influence of Personal, Family, and School Factors on Early Adolescent Substance Use." Rand Health. Accessed August 20, 2011 (www.rand.org).

Adkins, Amy. 2015. "Majority of U.S. Employees Not Engaged Despite Gains in 2014." Gallup, January 28. Accessed October 7, 2015 (gallup.com).

Adkins, Amy. 2015. "Only 35% of U.S. Managers Are Engaged in Their Jobs." Gallup, April 2. Accessed August 22, 2015 (gallup.com).

Administration for Children and Families. 2021. "National Data Shows Number of Children in Foster Care Decreases for the Third Consecutive Year." November 19. Accessed February 18, 2023 (https://www.acf.hhs.gov/media/press/2021/national-data-shows-number-children-foster-care-decreases-third-consecutive-year).

"African American Health." 2017. Centers for Disease Control and Prevention, May. Accessed August 15, 2017 (cdc.gov).

Agence France-Presse. 2013. "Bill Gates' 'Hand in Pocket' Draws Criticism in South Korea." April 23. Accessed November 10, 2013 (www.rawstory.com).

Agnvall, Elizabeth. 2014. "10 Tests to Avoid." *AARP Bulletin*, March, 12, 14, 16.

Agorist, Matt. 2016. "A Pill That Cures Hepatitis Costs Just $4, but If You Live in America It's $1,000." AlterNet, February 22. Accessed August 15, 2017 (alternet.org).

Agress, Lynne. 2017. "Immigrants: Our Best and Brightest." *Baltimore Sun*, May 2, 11.

Aguila, Raul. 2014. "What We Hate About Holiday Travel." *Consumer Reports*, November, 7.

Aguilar, Leslie, and Linda Stokes. 1995. *Multicultural Customer Service: Providing Outstanding Service Across Cultures*. New York: McGraw Hill.

Ahmad, Farida B., Jodi A. Cisewski, Jiaquan Xu, and Robert N. Anderson. 2023. "Provisional Mortality Data — United States, 2022." Centers for Disease Control and Prevention, May 5. Accessed June 11, 2023 (https://www.cdc.gov

/mmwr/volumes/72/wr/mm7218a3.htm).

Ahmed, Yasmin. 2020. "Developers in U.S. Earn Almost 3 Times More Than Indian Coders, Developers in Pakistan Too, Earn More." *India Today*, Feb 21. Accessed May 10, 2023 (https://www.indiatoday.in/technology/news/story/developers-in-us-earn-almost-3-times-more-than-indian-coders-developers-in-pakistan-too-earn-more-1648667-2020-02-21).

Ahmetov, Ildus I., Emiliya S Egorova, Leysan J Gabdrakhmanova, and Olga N Fedotovskaya. 2016. "Genes and Athletic Performance: An Update." *Medicine and Sport Science* 61 (June): 41–54.

Ahn, Ashley. 2023. "South Korea Has the World's Lowest Fertility Rate, A Struggle with Lessons for Us All." NPR, March 19. Accessed June 27, 2023 (https://www.npr.org/2023/03/19/1163341684/south-korea-fertility-rate).

Ahn, Lauren, and Michelle Ruiz. 2015. "Survey: 1 in 3 Women Has Been Sexually Harassed at Work." Cosmopolitan, February 16. Accessed May 22, 2017 (cosmopolitan.com).

Ahrens, Jacqueline, Fiona Brennan, Sarah Eaglesham, Audrey Buelo, Yvonne Laird, Jillian Manner, Emily Newman, and Helen Sharpe. 2022. "A Longitudinal and Comparative Content Analysis of Instagram Fitness Posts." *International Journal of Environmental Research and Public Health*, 19(6845), 6845. Accessed July 30, 2023 (https://doi-org.ju.idm.oclc.org/10.3390/ijerph19116845)

Ailes, Elizabeth C., et al. 2015. "Opioid Prescription Claims among Women of Reproductive Age—United States, 2008–2012." *MMWR* 64 (2): 37–41.

Akbarian, Firouzeh, and Sofia Bettiza. 2022. "Iranian Women Who Need Certificates to Prove They Are Virgins." BBC, August 11. Accessed April 23, 2023 (https://www.bbc.com/news/world-middle-east-62423983).

Aizer, Anna, and Joseph Doyle. 2013. " Juvenile Incarceration, Human Capital, and Future Crime: Evidence from Randomly-Assigned Judges." Access November 22, 2023 (https://www.nber.org/system/files/working_papers/w19102/w19102.pdf).

Akbulut-Yuksel, Mevlude, and Adriana D. Kugler. 2016. "Intergenerational Persistence of Health in the U.S.: Do Immigrants Get Healthier as They Assimilate?" National Bureau of Economic Research, February. Accessed June 20, 2017 (nber.org).

Akcigit, Ufuk, John Grigsby, and Tom Nicholas. 2017. "Immigration and the Rise of American Ingenuity." National Bureau of Economic Research, February. Accessed June 20, 2017 (nber.org).

Akechi, Hironori, Atsushi Senju, Helen Uibo, Yukiko Kikuchi, Toshikazu Hasegawa, and Jari K. Hietanen. 2013. "Attention to Eye Contact in the West and East: Autonomic Responses and Evaluative Ratings." *PLoS One* 8 (3): e59312. Accessed December 5, 2013 (www.plosone.org).

Akers, Ronald L. 1997. *Criminological Theories: Introduction and Evaluation*, 2nd edition. Los Angeles: Roxbury.

Alas, Horus. 2020. "Will Other Countries Continue to Outpace the U.S. in Voter Turnout?" *Newsweek*, October 30. Accessed May 12, 2023 (https://www.usnews.com/news/best-countries/articles/2020-10-30/these-countries-have-had-the-highest-voter-turnout).

"Alcohol and Public Health." 2014. Centers for Disease Control and Prevention, March 14. Accessed July 30, 2014 (www.cdc.gov).

Alexander, Dan, and Michela Tindera. 2021. "Joe's Regular Joes." *Forbes*, June/July. Accessed May 10, 23 (forbes.com).

Ali, Shirin. 2022. "About Half of US Water 'Too Polluted' For Swimming, Fishing or Drinking, Report Finds." The Hill, March 28. Accessed June 28, 2023 (https://thehill.com/changing-america/sustainability/environment/600070-about-half-of-us-water-too-polluted-for-swimming/).

Al-Jassem, Diana. 2011. "Women in Polygamous Marriages Suffering Psychological Torture." ArabNews, March 8. Accessed September 12, 2012 (www.arabnews.com).

Allen, Jonathan. 2022. "New Study Estimates 1.6 million in U.S. Identify as Transgender." Reuters, June 10. Accessed April 14, 2023 (https://www.reuters.com/world/us/new-study-estimates-16-million-us-identify-transgender-2022-06-10/).

Allen, Joshua, et al. 2014. "Early Postnatal Exposure to Ultrafine Particulate Matter Air Pollution: Persistent Ventriculomegaly, Neurochemical Disruption, and Glial Activation Preferentially in Male Mice." *Environmental Health Perspectives*, June 5, advance publication. Accessed July 30, 2014 (www.ehponline.org).

Allen, Kathy Grannis. 2015. "Retailers Estimate Holiday Return Fraud Will Cost $2.2 Billion in 2015." National Retail Federation, December 17. Accessed April 14, 2017 (nrf.com).

Allport, Gordon W. 1954. *The Nature of Prejudice*. Reading, MA: Addison-Wesley.

Al-Mahmood, Syed Z. 2013. "Bangladesh to Raise Workers' Pay." *Wall Street Journal*, May 13, B4.

Almeida and Salas-Betsch. 2023 "Fact Sheet: The State of Women in the Labor Market in 2023." The Center for American Progress, February 6. Accessed May 12, 2023 (https://www.americanprogress.org/article/fact-sheet-the-state-of-women-in-the-labor-market-in-2023/).

Alper, Becca A. 2022. "4. Religious Groups' Views on Climate Change." Pew Research Center, November 17. Accessed January 30, 2023 (https://www.pewresearch.org/religion/2022/11/17/religious-groups-views-on-climate-change/).

Alsever, Jennifer. 2014. "Prison Labor's New Frontier." *Fortune*, June 16, 60.

Altintas, Evrim. 2016. "The Widening Education Gap in Developmental Child Care Activities in the United States, 1965–2013." *Journal of Marriage and Family* 78 (February): 26–42.

Alter, Charlotte. 2016. "Seeing Sexism from Both Sides: What Trans Men Experience." *Time*, June 27, 24.

Altonji, Joseph G., Sarah Cattan, and Iain Ware. 2010. "Identifying Sibling Influence on Teenage Substance Use." National Bureau of Economic Research, October. Accessed July 24, 2011 (www.nber.org).

Alvaredo, Facundo, Anthony B. Atkinson, Thomas Piketty, and Emmanuel Saez. 2013. "The Top 1 Percent in International and Historical Perspective." *Journal of Economic Perspectives* 27 (Summer): 3–20.

Alvaredo, Facundo, Bertrand Garbinti, and Thomas Piketty. 2017. "On the Share of Inheritance in Aggregate Wealth: Europe and the USA, 1900–2010." *Economica* 84 (April): 239–260.

Alves, Joice. 2023. "A Man's World: Gender Parity Will Take More than Five Generations." *Reuters*, March 8. Accessed April 19, 2023 (https://www.reuters.com/world/mans-world-gender-parity-will-take-more-than-five-generations-2023-03-08/#:~:text=Several%20investors%20and%20regulators%20are,projected%20136%20years%20in%202021).

AMAP. 2017. "Snow, Water, Ice and Permafrost in the Arctic: Summary for Policy-makers." Accessed August 20, 2017 (amap.no).

American Academy of Pediatrics. 2016. "American Academy of Pediatrics Announces New Recommendations for Children's Media Use." October 21. Accessed February 28, 2017 (aap.org).

American Addiction Centers. 2022. "How Does Drug & Alcohol Addiction Affect Families?" August 18. Accessed March 25, 2023 (https://americanaddictioncenters.org/rehab-guide/guide-for-families-i).

American Council on Education. 2023. "ACE'S American College President Study Finds Higher Ed Leadership Continues Slow March to Parity." April 14. Accessed, April 21, 2023 (https://www.acenet.edu/News-Room/Pages/2023-ACPS.aspx).

American Association of Community Colleges. 2021. "DataPoints: Education,

Income and Poverty." October 26. Accessed April 13, 2023 (https://www.aacc.nche.edu/2021/10/26/datapoints-education-income-and-poverty/).

American Association of Suicidology. 2009. "Elderly Suicide Fact Sheet." June 23. Accessed December 17, 2009 (www.suicidology. org).

American Association of University Professors. 2022. "The Annual Report on the Economic Status of the Profession, 2021–22." June. Accessed April 15, 2023 (https://www.aaup.org/file/AAUP_ARES_2021%E2%80%932022.pdf).

American Bar Association. 2021. "Five Facts About the Troubled Teen Industry," Accessed February 18, 2023 (https://www.americanbar.org/groups/litigation/committees/childrens-rights/practice/2021/5-facts-about-the-troubled-teen-industry/)

American Bar Association. 2017. "A Current Glance at Women in the Law: January 2017." Accessed May 25, 2017 (americanbar.org/women).

American Diabetes Association. 2018. "Economic Costs of Diabetes in the U.S. in 2017." March 22. Accessed June 11, 2023 (https://diabetesjournals.org/care/article/41/5/917/36518/Economic-Costs-of-Diabetes-in-the-U-S-in-2017).

American Heart Association. 2018. "Breaking Down Barriers to Fitness." Accessed June 11, 2023 (https://www.heart.org/en/healthy-living/fitness/getting-active/breaking-down-barriers-to-fitness).

"American Indian and Alaska Native Heritage Month: November 2016." 2016. U.S. Census Bureau News, November 2. Accessed November 22, 2016 (census.gov).

American Lung Association. 2023. Overall Tobacco Trends. Accessed July 29, 2023 (https://www.lung.org/research/trends-in-lung-disease/tobacco-trends-brief/overall-tobacco-trends).

The American National Red Cross. 2020. "Red Cross Emblem Symbolizes Neutrality, Impartiality." June 4. Accessed February 11, 2023 (https://www.redcross.org/about-us/news-and-events/news/2020/red-cross-emblem-symbolizes-neutrality-impartiality.html).

American Psychiatric Association. 2023. "About DSM-5-TR." Accessed June 19, 2023 (https://www.psychiatry.org/psychiatrists/practice/dsm/about-dsm).

American Psychological Association. 2014. "About Transgender People, Gender Identity, and Gender Expression." Accessed March 2, 2015 (apa.org).

American Psychological Association. 2014. "Are Teens Adopting Adults' Stress Habits?" February 11. Accessed April 12, 2014 (www.stressinamerica.org).

American Psychological Association, 2019. "Stress in America." Accessed May 11, 2023 (https://www.apa.org/news/press/releases/stress/2019/stress-america-2019.pdf).

American Society of Addiction Medicine. 2016. "Opioid Addiction: 2016 Facts & Figures." Accessed August 15, 2017 (asam.org).

American Society of Plastic Surgeons. 2019. "More Men Undergo Plastic Surgery as the Daddy-Do-Over Trend Rises in Popularity." June 12. Accessed April 16, 2023 (https://www.plasticsurgery.org/news/press-releases/more-men-undergo-plastic-surgery-as-the-daddy-do-over-trend-rises-in-popularity#:~:text=A%20new%20report%20from%20the,29%20percent%20increase%20since%202000).

American Sociological Association. 1999. *Code of Ethics and Policies and Procedures of the ASA Committee on Professional Ethics*. Accessed January 10, 2010 (www.asanet.org).

"Americans Moving at Historically Low Rates, Census Bureau Reports." 2016. Census Bureau News, November 16. Accessed August 20, 2017 (census.gov).

AmeriCorps. 2021. "Volunteering and Civic Life in America Research Summary." Accessed March 11, 2023 (https://americorps.gov/sites/default/files/document/volunteering-civic-life-america-research-summary.pdf).

Amnesty International USA. 2017. "Amnesty International Report 2016/17: The State of the World's Human Rights." February 22. (https://www.amnesty.org/en/documents/pol10/4800/2017/en/)

Amnesty International. 2020. "9 Powerful Social Change Movements You Need to Know About." Accessed June 25, 2023 (https://www.amnesty.org.au/9-powerful-social-change-movements-you-need-to-know-about/).

Amnesty International. 2022. "Death Penalty 2021: Facts and Figures." May 24. Accessed March 25, 2023 (https://www.amnesty.org/en/latest/news/2022/05/death-penalty-2021-facts-and-figures/).

Amrein-Beardsley, Audrey. 2014. "Recommended Readings on VAMs." VAMboozled. Accessed June 21, 2014 (www.vamboozled.com).

Amusa, Malena. 2010. "'Precious' Pushes Past Controversy to Oscar Night." Women's eNews, March 5. Accessed March 8, 2010 (www.womensenew.org).

Anander, Lori. 2022. "Rumors, Gossip, and Your Health." WebMd, February 13. Accessed June 24, 2023 (https://www.webmd.com/balance/health-rumors-gossip).

Andersen, Margaret L., and Patricia Hill Collins. 2010. "Why Race, Class, and Gender Still Matter." Pp. 1–16 in *Race, Class, and Gender: An Anthology,"* 7th edition, edited by Margaret L. Andersen and Patricia Hill Collins. Belmont, CA: Wadsworth.

Anderson, James F., and Laronistine Dyson. 2002. *Criminological Theories: Understanding Crime in America*. Lanham, MD: University Press of America.

Anderson, Monica. 2015. "6 Facts About Americans and Their Smartphones." Pew Research Center, April 1. Accessed August 14, 2015 (pewresearch.org).

Anderson, Monica. 2015. "A Rising Share of the U.S. Black Population Is Foreign Born." Pew Research Center, April 9. Accessed October 4, 2015 (pewresearch.org).

Anderson, Monica. 2016. "Parents, Teens and Digital Monitoring." Pew Research Center, January 7. Accessed March 14, 2017 (pewresearch.org).

Anderson, Monica. 2017. "For Earth Day, Here's How Americans View Environmental Issues." Pew Research Center, April 20. Accessed August 25, 2017 (pewresearch.org).

Anderson, Monica, and Andrew Perrin. 2017. "Tech Adoption Climbs Among Older Adults." Pew Research Center, May 17. Accessed June 14, 2017 (pewresearch.org).

Anderson, Monica, Emily A. Vogels, Andrew Perrin, and Lee Rainie. 2022. "Connection, Creativity and Drama: Teen Life on Social Media in 2022." Pew Research Center, November 16. Accessed September 7, 2023 (https://www.pewresearch.org/internet/2022/11/16/connection-creativity-and-drama-teen-life-on-social-media-in-2022/).

Anderson, Monica, Emily A. Vogels, and Erica Turner. 2020. "The Virtues and Downsides of Online Dating. Pew Research Center, February 6. Accessed September 7, 2023 (https://www.pewresearch.org/internet/wp-content/uploads/sites/9/2020/02/PI_2020.02.06_online-dating_REPORT.pdf).

Andrew, Scottie. 2020. "The Psychology Behind Why Toilet Paper, of All Things, is the Latest Coronavirus Panic Buy." *CNN*, March 9. Accessed August 2, 2023 (https://www.cnn.com/2020/03/09/health/toilet-paper-shortages-novel-coronavirus-trnd/index.html).

Andrews, Edmund L. 2015. "Stanford Study Finds Blacks and Hispanics Typically Need Higher Incomes Than Whites to Live in Affluent Neighborhoods." Stanford Report, June 25. Accessed October 4, 2015 (news.stanford.edu).

Andrzejewski, Adam. 2014. "The Federal Transfer Report." Open the Books, March. Accessed April 12, 2014 (www.openthebooks.com).

Andrzejewski, Adam. 2019. "Top U.S. "Non-Profit" Hospitals & CEOs Are Racking Up Huge Profits." Forbes, June 26. (https://www.forbes.com/sites/adamandrzejewski/2019/06/26/top-u

-s-non-profit-hospitals-ceos-are-racking-up-huge-profits/?sh=7a7dec4319df).

Anestis, Michael D., and Joye C. Anestis. 2015. "Suicide Rates and State Laws Regulating Access and Exposure to Handguns." *American Journal of Public Health* 105 (October): 2049–2058.

Angwin, Julia, Terry Parris Jr., and Surya Mattu. 2016. "Facebook Doesn't Tell Users Everything It Really Knows About Them." AlterNet, March 4. Accessed December 29, 2016 (alternet.org).

Annie E. Casey Foundation. 2014. "Race for Results: Building a Path to Opportunity for All Children." Accessed June 21, 2014 (www.aecf.org).

Anteby, Michael, and Rakesh Khurana. n.d. "A New Vision." Historical Collections, Baker Library Harvard Business School. Accessed March 11, 2023 (https://www.library.hbs.edu/hc/hawthorne/).

Anwar, Yasmin. 2021. "Are Facial Expressions the Same Around the World?" *Greater Good* Magazine, January 8. Accessed September 7, 2023 (https://greatergood.berkeley.edu/article/item/are_facial_expressions_the_same_around_the_world).

Aptekar, Sofya. 2016. "Celebrating New Citizens, Defining the Nation." *Contexts*, 15 (Spring): 46–51.

Arab American Institute. 2021. "National Arab American Demographics." Accessed May 3, 2023 (https://www.aaiusa.org/demographics).

Arab American Institute Foundation. 2012. "Quick Facts About Arab Americans." Accessed March 22, 2014 (www.aaiusa.org).

Aragão, Carolina. 2023. "Gender Pay Gap in U.S. Hasn't Changed Much in Two Decades" Pew Research Center, March 1. Accessed April 16, 2023 (https://www.pewresearch.org/fact-tank/2023/03/01/gender-pay-gap-facts/).

Arendt, Hannah. 2004. *The Origins of Totalitarianism.* New York: Schocken.

Arias, Elizabeth, Betzaida Tejada-Vera, Farida Ahmad, and Kenneth D. Kochanek. 2021. "Provisional Life Expectancy Estimates for 2020. Vital Statistics Rapid Release; no 15." National Center for Health Statistics, July. Accessed May 22, 2023 (https://www.cdc.gov/nchs/data/vsrr/vsrr015-508.pdf).

Arieff, Allison. 2013. "What Tech Hasn't Learned from Urban Planning." *New York Times*, December 13. Accessed September 19, 2014 (www.nytimes.com).

Ariès, Phillippe. 1962. *Centuries of Childhood.* New York: Vintage.

Arntz, Melanie, Terry Gregory, and Ulrich Zierahn. 2016. "The Risk of Automation for Jobs in OECD Countries: A Comparative Analysis." OECD Publishing, Paris. Accessed September 2, 2017 (oecd-library.org).

"Artificial Intelligence." 2015. *The Economist*, October 3, 81–82.

ASA Research Department. 2013. "Recruitment and Retention of Sociology Majors." *ASA Footnotes* 41 (January): 1, 4.

Asch, Solomon. 1952. *Social Psychology.* Englewood Cliffs, NJ: Prentice-Hall.

"The Ashley Madison Hack." 2015. *The Economist*, August 22, 14.

Asi, Maryam, and Daniel Beaulieu. 2013. "Arab Households in the United States: 2006–2010." U.S. Census Bureau, May. Accessed March 22, 2014 (www.census.gov).

"Asia's New Family Values." 2015. *The Economist*, August 22, 36.

"Asian-American and Pacific Islander Heritage Month: May 2017." 2017. U.S. Census Bureau, March 14. Accessed June 20, 2017 (census.gov).

"Asians Fastest-Growing Race or Ethnic Group in 2012, Census Bureau Reports." 2013. Newsroom, U.S. Census Bureau, June 13. Accessed March 22, 2014 (www.census.gov).

Associated Press. 2019. "Still Some Extra Hurdles for LGBTQ Candidates, Poll Shows." NBC News, June 26. Accessed May 12, 2023 (https://www.nbcnews.com/feature/nbc-out/still-some-extra-hurdles-lgbtq-candidates-poll-shows-n1021951).

Associated Press. 2022. "French Court Rules Against 'Burkini' Swimwear for Religious Reasons." NBC News., June 22. Accessed February 12, 2023 (https://www.nbcnews.com/news/world/france-court-rules-burkini-swimwear-religious-reasons-rcna34833).

Associated Press. 2022. "At Least 125 Dead at Indonesian Soccer Match, Tear Gas Triggers Panic." *Sports Illustrated*, October 1. Accessed June 24, 2023 (https://www.si.com/soccer/2022/10/02/more-than-120-dead-after-fans-fled-indonesian-soccer-match-in-panic).

Associated Press. 2022. "World Population Will Reach 8 Billion This Year, U.N. Projects." *NBC News*, July 12. Accessed June 27, 2023 (https://www.nbcnews.com/news/world/world-population-will-reach-8-billion-year-un-projects-rcna37763).

Associated Press and NORC. 2016. "Hillary Clinton's Candidacy and the State of Gender Discrimination in the United States." August. Accessed June 26, 2017 (apnorc.org).

Association of Certified Fraud Examiners. 2016. "Report to the Nations on Occupational Fraud and Abuse: 2016 Global Fraud Study." Accessed April 2, 2016 (acfe.com).

AT&T Newsroom. 2015. "Smartphone Use While Driving Grows Beyond Texting to Social Media, Web Surfing, Selfies, Video Chatting." May 19. Accessed July 21, 2015 (att.com).

Atchley, Robert C., and Amanda S. Barusch. 2004. *Social Forces and Aging: An Introduction to Social Gerontology,* 10th edition. Belmont, CA: Wadsworth.

Attina, Teresa M., et al. 2016. "Exposure to Endocrine-Disrupting Chemicals in the USA: A Population-Based Disease Burden and Cost Analysis." *The Lancet: Diabetes & Endocrinology* 4 (December): 996–1003.

Atwater, Peter. 2016. "The Wild EpiPen Price Hike Points to a Looming Pharmaceutical Crisis." *Time*, September 12–19, 32.

Aubrey, Allison. 2022. "The U.S. Diet Is Deadly. Here Are 7 Ideas to Get Americans Eating Healthier." NPR, August 31. Accessed November 17, 2023 https://www.npr.org/sections/health-shots/2022/08/31/1120004717/the-u-s-diet-is-deadly-here-are-7-ideas-to-get-americans-eating-healthier).

Aughinbaugh, Alison, Omar Robles, and Hugette Sun. 2013. "Marriage and Divorce: Patterns by Gender, Race, and Educational Attainment." *Monthly Labor Review* (October): 1–19.

Auguste, Byron, Paul Kihn, and Matt Miller. 2010. "Closing the Talent Gap: Attracting and Retaining Top-Third Graduates to Careers in Teaching." McKinsey & Company, September. Accessed March 20, 2012 (www.mckinsey.com).

Aunola, Kaisa, and Jari-Erik Nurmi. 2005. "The Role of Parenting Styles in Children's Problem Behavior." *Child Development* 76 (November/December): 1144–1159.

Auster, Ellen R., and Ajnesh Prasad. 2016. "Why Do Women Still Not Make It to the Top? Dominant Organizational Ideologies and Biases by Promotion Committees Limit Opportunities to Destination Positions." *Sex Roles* 75 (5/6): 177–196.

Austin, Algernon. 2013. "Native Americans and Jobs." Economic Policy Institute, December 17. Accessed March 22, 2014 (www.epi.org).

Austin, Patrick Lucas. 2019. "What Will Smart Homes Look Like 10 Years from Now?" *Time Magazine*, July 25. Accessed June 25, 2023 (https://time.com/5634791/smart-homes-future/).

Autor, David H. 2014. "Skills, Education, and the Rise of Earnings Inequality Among the 'Other 99 Percent.'" *Science* 344 (May): 843–851.

Auxier, Brooke, and Monica Anderson. 2021. "Social Media Use in 2021." Pew Research Center, April 2. Accessed September 7, 2023 (https://www.pewresearch.org/internet/2021/04/07/social-media-use-in-2021/).

Auxier, Brooke, Monica Anderson, Andrew Perrin, and Erica Turner. 2020. "Parenting Children in the Age of Screens." Pew Research Center, July 28. Accessed February 18, 2023 (https://www.pewresearch.org/internet/2020/07/28

/parenting-children-in-the-age-of-screens/).

Auxier, Brooke. Lee Rainie, Monica Anderson, Andrew Perrin, Madhu Kumar, and Erica Turner. 2019. "Americans and Privacy: Concerned, Confused and Feeling Lack of Control Over Their Personal Information." Pew Research Center, November 15. Accessed June 27, 2023 (https://www.pewresearch.org/internet/2019/11/15/americans-and-privacy-concerned-confused-and-feeling-lack-of-control-over-their-personal-information/).

Azofeifa, Alejandro, Margaret E. Mattson, and Rob Lyerla. 2015. "Driving Under the Influence of Alcohol, Marijuana, and Alcohol and Marijuana Combined Among Persons Aged 16–25 Years—United States, 2002–2014." *Morbidity and Mortality Weekly Report* 64 (48): 1325–1329.

B

Babbie, Earl. 2021. *The Practice of Social Research, 15th edition.* Boston, MA: Cengage.

Babcock, Philip, and Mindy Marks. 2011. "The Falling Time Cost of College: Evidence from Half a Century of Time Use Data." *The Review of Economics and Statistics* 83 (May): 468–478.

Bach, Peter B. 2015. "Seeking a Cure for Drug-Price Insanity." *Fortune*, September 15, 63–64.

Bachrach, Deborah, and Jonah Frohlich. 2016. "Retail Clinics Drive New Health Care Utilization and That Is a Good Thing." Health Affairs Blog, May 20. Accessed April 3, 2017 (healthaffairs.org).

Bakalar, Nicholas. 2021. "Are Robotic Surgeries Really Better?" *The New York Times,* August 16, Accessed June 24, 2023 (https://www.nytimes.com/2021/08/16/well/live/robotic-surgery-benefits.html).

Badger, Emily. 2015. "New Census Data: Americans Are Returning to the Far-Flung Suburbs." *Washington Post,* March 26. Accessed August 20, 2017 (washingtonpost.com).

Bae, Jin Yung, Anderson, Evan, Silver, Diana, & Macinko, James. 2014. "Child Passenger Safety Laws in the United States, 1978-2010: Policy Diffusion in the Absence of Strong Federal Intervention. *Social Science & Medicine (1982), 100,* 30–37. Accessed November 22, 23 (https://doi.org/10.1016/j.socscimed.2013.10.035).

Bagri, Neha Thirani, and Nida Najar. 2015. "Chicken: It's What's for Dinner in an Indian State, Even Among Caged Carnivores." *New York Times,* March 30, A4.

Bahadur, Nina. 2014. "'Misconception' Reveals the Dark, Misleading World of Crisis Pregnancy Centers." *Huffington Post,* September 18. Accessed September 29, 2015 (huffingtonpost.com).

Bai, Ge, and Gerard F. Anderson. 2015. "Extreme Markup: The Fifty US Hospitals with the Highest Charge-to-Cost Ratios." *Health Affairs* 34 (6): 922–928.

Bainbridge, William S. 1997. *The Sociology of Religious Movements.* New York: Routledge.

Baldor, Lolita C., and Zeke Miller. 2021. "Biden Reverses Trump Ban on Transgender People in Military." *AP News,* January 25. Accessed April 19, 2023 (https://apnews.com/article/biden-reverse-ban-transgender-military-f0ace4f9866e0ca0df021eba75b3af20).

Bales, Robert F. 1950. *Interaction Process Analysis.* Reading, MA: Addison-Wesley.

Baltimore Sun. 2012. "Elderly Man Dies in Fire at Single-Family Home." December 9, 5.

Bandura, Albert, and Richard H. Walters. 1963. *Social Learning and Personality Development.* New York: Holt, Rinehart & Winston.

Bandy, Tawana. 2012. "What Works for Male Children and Adolescents: Lessons from Experimental Evaluations of Programs and Interventions." Child Trends Research Brief, August. Accessed September 22, 2012 (www.childtrends.org).

Banerjee, Neela, Lisa Song, and David Hasemyer. 2015. "Exxon: The Road Not Taken." Inside Climate News, September 16. Accessed August 20, 2017 (insideclimatenews.org).

Banfield, Edward C. 1974. *The Unheavenly City Revisited.* Boston: Little, Brown.

Banks, Ingrid. 2000. *Hair Matters: Beauty, Power, and Black Women's Consciousness.* New York: New York University Press.

Barash, David P. 2012. "The Evolutionary Mystery of Homosexuality." *Chronicle Review,* November 21, B4–B5.

Barbara Lee Foundation. 2017. "Modern Family: How Women Candidates Can Talk About Politics, Parenting, and Their Personal Lives." Accessed June 26, 2017 (barbaraleefoundation.org).

Barnes, Robert. 2011. "Limits on Video Games Rejected." *Washington Post,* June 28, A1.

Barreto, Michelle, Micheal K. Ryan, and Manuela T. Schmitt, eds. 2009. *The Glass Ceiling in the 21st Century: Understanding Barriers to Gender Equality.* Washington, DC: American Psychological Association.

Barro, Josh. 2015. "How Unpopular Is Trump's Muslim Ban? Depends on How You Ask." *New York Times,* December 16, A21.

Barro, Robert J., and Rachel M. McCleary. 2003. "Religion and Economic Growth Across Countries." *American Sociological Review* 68 (October): 760–781.

Barroso, Amanda. 2020. "More Than Half of Americans Say Marriage Is Important but Not Essential to Leading a Fulfilling Life." Pew Research, February 14. Accessed May 19, 2023 (https://www.pewresearch.org/short-reads/2020/02/14/more-than-half-of-americans-say-marriage-is-important-but-not-essential-to-leading-a-fulfilling-life/).

Barroso, Amanda, Kim Parker, and Richard Fry. 2019. "Majority of Americans Say Parents Are Doing Too Much for Their Young Adult Children." Pew Research Center, October 23. Accessed April 13, 2023 (https://www.pewresearch.org/social-trends/2019/10/23/majority-of-americans-say-parents-are-doing-too-much-for-their-young-adult-children/).

Barstow, David, and Suhasini Raj. 2015. "Caste Quotas in India Come Under Attack." *New York Times,* August 31, A1, A3.

Batarseh, Feras A., Iya Ghassib, Deri (Sondor) Chong, and Po-Hsuan Su. 2020. "Preventive Healthcare Policies in the US: Solutions for Disease Management Using Big Data Analytics." *Journal of Big Data.* 2020; 7(1): 38. Accessed June 11, 2023 doi: 10.1186/s40537-020-00315-8

Barthel, Michael, Amy Mitchell, and Jesse Holcomb. 2016. "Many Americans Believe Fake News Is Sowing Confusion." Pew Research Center, December 15. Accessed December 17, 2016 (pewresearch.org).

Bartlett, Thomas, and Paula Wasley. 2008. "Just Say 'A': Grade Inflation Undergoes Reality Check." *Chronicle of Higher Education,* September 5. Accessed June 21, 2014 (www.chronicle.com).

Bartlett, Tom. 2011. "Caffeine Is Definitely Good/Bad for You." *Chronicle of Higher Education,* January 20. Accessed January 25, 2011 (www.chronicle.com/blogs).

Barton, Allen H. 1980. "A Diagnosis of Bureaucratic Maladies." Pp. 27–36 in *Making Bureaucracies Work,* edited by Carol H. Weiss and Allen H. Barton. Beverly Hills, CA: Sage.

Basken, Paul. 2016. "Data Could Help Scholars Persuade, If Only They Were Willing to Use It." *Chronicle of Higher Education,* January 29, A23–A24.

Basken, Paul. 2016. "To Curtail Violence, Researchers Say, Reduce Economic Inequality." *Chronicle of Higher Education,* September 2, A34.

Bass, Frank, and Dakin Campbell. 2013. "Poor Neighborhoods See Branches Disappear." *Bloomberg Businessweek,* May 13–19, 48–49.

Bates, Nancy, and Theresa J. DeMaio. 2013. "Measuring Same-Sex Relationships." *Contexts* 12 (Winter): 66–69.

Bauder, David, Randall Chase, and Geoff Mulvihill. 2023. "Fox, Dominion Voting Systems Reach $787 Million Settlement Over False Election

Claims." PBS, Associated Press, April 18. Accessed June 24, 2023 (https://www.pbs.org/newshour/politics/fox-dominion-voting-systems-reach-settlement-over-false-election-claims).

Baum, Sandy, Jennifer Ma, and Kathleen Payea. 2013. "Education Pays 2013: The Benefits of Higher Education for Individuals and Society." College Board. Accessed June 21, 2014 (www.collegeboard.org).

Baumrind, Diana. 1968. "Authoritarian versus Authoritative Parental Control." *Adolescence* 3 (11): 255–272.

Baumrind, Diana. 1989. "Rearing Competent Children." Pp. 349–378 in *Child Development Today and Tomorrow*, edited by William Damon. San Francisco: Jossey-Bass.

Bax, Pauline. 2013. "In Ghana, Death Has Become Big Business." *Bloomburg Businessweek*, August 29–September 1, 23–26.

Bazelon, Emily. 2009. "2 Kids + 0 Husbands = Family." *New York Times Magazine*, February 1, 30.

BBC News. 2016. "How Has the US Gun Lobby Been So Successful?" January 27. Accessed September 2, 2017 (bbc.com).

BBC News. 2022. "How Many US Mass Shootings Have There Been in 2023?" August 27. Accessed September 16, 2023 (https://www.bbc.com/news/world-us-canada-41488081).

BBC News. 2023. "U.S. Gun Control: What Is the NRA and Why Is It So Powerful?" April 13. Accessed June 27, 2023 (https://www.bbc.com/news/world-us-canada-35261394).

Beadle, Amanda Peterson. 2012. "Minnesota Senator Thinks Abortion Pill Is Wrong, While Viagra Is a 'Wonderful Drug.'" Alternet, May 3. Accessed May 24 (www.alternet.org).

Beard, Henry, and Christopher Cerf. 2015. *Spinglish: The Definitive Dictionary of Deliberately Deceptive Language*. New York: Blue Rider Press.

Beck, Melinda. 2013. "Nurse Practitioners Seek Right to Treat Patients on Their Own." *Wall Street Journal*, August 15, A3.

Becker, Amanda. 2020. "White Women Had Doubts. They Voted for Trump Anyway." *The 19th News*. November 7. Accessed January 29, 2023. (https://19thnews.org/2020/11/white-women-had-doubts-they-voted-for-trump-anyway/).

Becker, Howard S. 1963. *Outsiders: Studies in the Sociology of Deviance*. New York: Free Press.

Becker, Howard, and Ruth H. Useem. 1942. "Sociological Analysis of the Dyad." *American Sociological Review* 7 (February): 3–26.

Becker, Sam. 2015. "Places Where People Tip (and Don't Tip) Around the World." Money & Career Cheat Sheet, September 4. Accessed February 14, 2017 (cheatsheet.com).

Beckles, Gloria L., and Benedict I. Truman. 2013. "Education and Income—United States, 2009 and 2011." *MMWR* 62 (November 22): 9–19.

Beckwith, Ryan Teague. 2017. "Read President Trump's Remarks at the National Prayer Breakfast." *Time*, February 2. Accessed July 31, 2017 (time.com).

Beech, Hannah. 2013. "The Face of Buddhist Terror." *Time*, July 1, 42–50.

Begley, Sharon. 2010. "Sins of the Grandfathers." *Newsweek*, November 8, 48–50.

Beil, Laura. 2012. "How Much Would You Pay for Three More Months of Life?" *Newsweek*, September 3, 40–44.

Beitsch, Rebecca. 2016. "Tiny Houses Are Trendy, Minimalist and Often Illegal." PBS News Hour, July 6. Accessed August 20, 2017 (pbs.org).

Belkin, Douglas, and Caroline Porter. 2012. "Web Profiles Haunt Students." *Wall Street Journal*, October 4, A3.

Belknap, Joanne. 2007. *The Invisible Woman: Gender, Crime, and Justice*, 3rd edition. Belmont, CA: Wadsworth Press.

Bell, Claes. 2015. "Budgets Can Crumble in Times of Trouble." Bankrate, January 7. Accessed August 26, 2015 (bankrate.com).

Bellini, Jason. 2012. "A Brief History of the NRA." *Wall Street Journal*, December 20. Accessed September 2, 2017 (wsj.com).

Bendavid, Naftali. 2013. "Countries Expand Recognition for Alternative 'Intersex' Gender." *Wall Street Journal*, October 31, A9.

Benecchi, Liz. 2021. "Recidivism Imprisons American Progress." Harvard Political Review, August 8. Accessed March 25, 2023 (https://harvardpolitics.com/recidivism-american-progress/).

Bengali, Shashank. 2014. "Toll of Refugees Fleeing Syrian Strife Tops 3 Million, U.N. Says." *Baltimore Sun*, August 30, 10.

Bennett, Jessica, and Jesse Ellison. 2010. "'I Don't': The Case Against Marriage." *Newsweek*, June 21, 42–45.

Bennhold, Katrin, and Rick Gladstone. 2016. "Women Have Led over 70 Nations, but Experts See Pipeline Problem in U.S." *New York Times*, November 4, P11.

Benokraitis, Nijole V. 1997. *Subtle Sexism: Current Practices and Prospects for Change, 2nd edition*. Thousand Oaks, CA: Sage.

Benokraitis, Nijole V. 2015. *Marriages & Families: Changes, Choices, and Constraints*, 8th edition. Upper Saddle River, NJ: Prentice Hall.

Benson, Michael L. 2002. *Crime and the Life Course*. Los Angeles: Roxbury.

Benton, Thomas H. 2011. "A Perfect Storm in Undergraduate Education." *Chronicle of Higher Education*, February 25, A43, A45. Part 2 in April 8, A45–A46.

Berenson, Tessa. 2016. "How Latinos Drive America's Economic Growth." *Time*, August 26, 32–35.

Beres, Damon. 2016. "Here's How Much People Trust Facebook." *Huffington Post*, April 27. Accessed May 1, 2016 (huffingtonpost.com).

Berfield, Susan. 2012. "Levi's Has a New Color for Blue Jeans: Green." *Bloomberg Businessweek*, October 22–28, 26–28.

Berfield, Susan. 2013. "Fast-Food Workers of the World, Unite!" *Bloomberg Businessweek*, December 9–15, 20–22.

Berger, Peter L., and Thomas Luckmann. 1966. *The Social Construction of Reality: A Treatise in the Sociology of Knowledge*. New York: Doubleday.

Berk, Richard A. 1974. *Collective Behavior*. Dubuque, IA: Wm. C. Brown Company.

Berkos, Kristen M., Terre H. Allen, Patricia Kearney, and Timothy G. Plax. 2001. "When Norms Are Violated: Imagined Interactions as Processing and Coping Mechanisms." *Communication Monographs* 68 (September): 289–300.

Berkowitz, Bill. 2011. "Why Is Jerry Falwell's Evangelical University Getting Filthy Rich Off Your Tax Money?" AlterNet, June 29. Accessed June 29, 2011 (www.alternet.org).

Berlin, Overton B., and Paul Kay. 1969. *Basic Color Terms*. Berkeley: University of California Press.

Berrett, Dan. 2014. "Some Elite Colleges Reject AP Credits in Favor of New Core Classes." *Chronicle of Higher Education*, February 21, A13.

Bersani, Bianca E. 2014. "A Game of Catch-Up? The Offending Experience of Second-Generation Immigrants." *Crime & Delinquency* 60 (February): 60–84.

Bertrand, Marianne, and Jessica Pan. 2011. "The Trouble with Boys: Social Influences and the Gender Gap in Disruptive Behavior." National Bureau of Economic Research, October. Accessed April 14, 2012 (www.nber.org).

Berube, Alan, and Natalie Holmes. 2015. "Some Cities Are Still More Unequal Than Others—An Update." Brookings Institution, March 17. Accessed September 31, 2015 (brookings.edu).

Berube, Alan, Audrey Sinter, Jill H. Wilson, and William H. Frey. 2006. "Finding Exurbia: America's Fast-Growing Communities at the Metropolitan Fringe." Brookings Institution. Accessed February 2, 2008 (www.brookings.edu).

Best, Joel. 2006. *Flavor of the Month: Why Smart People Fall for Fads*. Berkeley: University of California Press.

Betcher, R. William, and William S. Pollack. 1993. *In a Time of Fallen Heroes: The Re-Creation of Masculinity*. New York: Atheneum.

Bettinger, Eric P., Bridget T. Long, Philip Oreopoulos, and Lisa Sanbonmatsu. 2012. "The Role of Application Assistance and Information in College Decisions: Results from the H&R Block FAFSA Experiment." *Quarterly Journal of Economics* 127 (August): 1205–1242.

Bharadwaj, Prashant, Mallesh M. Pai, and Agne Suziedelyte. 2015. "Mental Health Stigma." National Bureau of Economic Research, June. Accessed July 22, 2015 (nber.org).

Bhardwaj, Deepika. 2014. "Delhi Takes Forward 'Kiss of Love' Protest Amid Police Barricades and Assaults." *Times of India*, November 10. Accessed September 27, 2015 (timesofindia.indiatimes.com).

Bhat, Rajesh. 2022. "Women and The Struggles of Networking." LinkedIn, April 2. Accessed March 11, 2023 (https://www.linkedin.com/pulse/women-struggles-networking-rajesh-bhat/?trk=articles_directory).

Bialik, Carl. 2011. "Irreconcilable Claim: Facebook Causes 1 in 5 Divorces." *Wall Street Journal*, March 12. Accessed November 8, 2011 (www.online.wsj.com).

Bialik, Kristen. 2016. "Most Americans Favor Stricter Environmental Laws and Regulations." Pew Research Center, December 14. Accessed August 25, 2017 (pewresearch.org).

Bialik, Kristen. 2017. "6 Facts About U.S. Mothers." Pew Research Center, May 11. Accessed August 25, 2017 (pewresearch.org).

Bialik, Kristen, and Anthony Cilluffo. 2017. "6 Facts About Black Americans for Black History Month." Pew Research Center, February 22. Accessed June 20, 2017 (pewresearch.org).

Bick, Alexander, Bettina Brüggemann, and Nicola Fuchs-Schundeln. 2017. "Hours Worked in Europe and the US: New Data, New Answers." Institute for the Study of Labor, April 12. Accessed May 25, 2017 (wiwi.uni-frankfurt.de).

Bienasz, Gabrielle. 2023. "Uber, Lyft Drivers Are Getting Kicked Off Their Driver Apps Spontaneously." Entrepreneur, March 7. Accessed April 28, 2023 (https://www.entrepreneur.com/business-news/report-uber-lyft-drivers-face-suspensions-discrimination/447115).

Biggs, Antonia M., Ushma D. Upadhyay, Charles E. McCulloch, and Diana G. Foster. 2017. "Women's Mental Health and Well-being 5 Years After Receiving or Being Denied an Abortion: A Prospective, Longitudinal Cohort Study." *JAMA Psychiatry* 74 (2): 169–178.

Bilefsky, Dan. 2017. "E.U. Court, in 'Bold Step,' Allows Banning Head Scarves at Work." *New York Times*, March 15, A1, A7.

Billingsley, Andrew. 1992. *Climbing Jacob's Ladder: The Enduring Legacy of African-American Families*. New York: Simon & Schuster.

"Biomedical Engineering." 2017. *The Economist*, April 1, 72–73.

Bird, Warren, and Scott Thumma. 2020. "Megachurch 2020: The Changing Reality in America's Largest Churches." Hartford Insistute for Religious Research. Accessed July 30, 2023 (http://hirr.hartsem.edu/megachurch/2020_Megachurch_Report.pdf).

Bird, Warren, and Scott Thumma. 2011. "A New Decade of Megachurches: 2011 Profile of Large Attendance Churches in the United States." Leadership and Network. Accessed June 21, 2014 (www.leadnet.org).

Birnbaum. Emily. 2022. "Tech spent big on lobbying last year." *Politico*, January 24. Accessed May 12, 2023 (https://www.politico.com/newsletters/morning-tech/2022/01/24/tech-spent-big-on-lobbying-last-year-00001144).

Bischoff, Kendra, and Sean F. Reardon. 2013. "Residential Segregation by Income, 1970–2009." Russell Sage Foundation, October 16. Accessed September 19, 2014 (www.russellsage.org).

Biunno, J. B. 2016. "Inmates Reveal How Smartphones Are Smuggled into Prison." WKRG News, March 16. Accessed March 16, 2017 (wkrg.com).

Bivens, Josh, and Jori Kandra. 2021. "CEO Pay Has Skyrocketed 1,460% Since 1978." Economic Policy Institute, October 4. Accessed May 12, 2023 (https://www.epi.org/publication/ceo-pay-in-2021/).

Black, Lindsey I., and Dzifa Adjaye-Gbewonyo. 2021. "Urgent Care Center and Retail Health Clinic Utilization Among Adults: United States, 2019." National Center for Health Statistics Data Brief, June. Accessed March 11, 2023 (https://www.cdc.gov/nchs/products/databriefs/db409.htm#:~:text=In%202019%2C%2029.2%25%20of%20adults,in%20the%20past%2012%20months).

Black, M. C., et al. 2011. "National Intimate Partner and Sexual Violence Survey: 2010 Summary Report." Centers for Disease Control and Prevention. Accessed November 2013 (cdc.gov).

Black, Sandra E., Jane A. Lincove, Jenna Cullinane, and Rachel Veron. 2014. "Can You Leave High School Behind?" National Bureau of Economic Research, January. Accessed June 21, 2014 (www.nber.org).

Black, Sandra E., Paul J. Devereux, Petter Lundborg, and Kaveh Majlesi. 2015. "Poor Little Rich Kids? The Determinants of the Intergenerational Transmission of Wealth." National Bureau of Economic Research, December. Accessed March 1, 2016 (nber.org).

Blake, Mariah. 2014. "Are Any Plastics Safe?" *Mother Jones*, March/April, 19–25, 60–61.

Blanchard, Dallas A. 1994. *The Anti-Abortion Movement and the Rise of the Religious Right: From Polite to Fiery Protest.* New York: Twayne Publishers.

Blau, Francine D., and Christopher Mackie, eds. 2016. *The Economic and Fiscal Consequences of Immigration*. Washington, DC: The National Academies Press.

Blau, Francine D., and Lawrence M. Kahn. 2016. "The Gender Wage Gap: Extent, Trends, and Explanations." National Bureau of Economic Research, March. Accessed March 13, 2017 (nber.org).

Blau, Peter M. 1986. *Exchange and Power in Social Life*, revised edition. New Brunswick, NJ: Transaction.

Blau, Peter M., and Marshall W. Meyer. 1987. *Bureaucracy in Modern Society*, 3rd edition. New York: Random House.

Blay, Zeba. 2015. "How Feminist TV Became the New Normal." *Huffington Post*, June 18. Accessed September 29, 2015 (huffingtonpost.com).

Blee, Kathleen. 2015. "How Grassroots Groups Lose Political Imagination." *Contexts* 14 (1): 32–37.

BLS News Release. 2016. "Volunteering in the United States—2015." U.S. Department of Labor, February 25. Accessed March 30, 2017 (bls.gov).

BLS News Release. 2017. "The Employment Situation—May 2017." U.S. Bureau of Labor Statistics, June 2. Accessed June 5, 2017 (bls.gov).

BLS Reports. 2017. "A Profile of the Working Poor, 2015." U.S. Bureau of Labor Statistics, April. Accessed May 5, 2017 (bls.gov).

BLS Reports. 2017. "Women in the Labor Force: A Databook." U.S. Bureau of Labor Statistics, April. Accessed May 10, 2017 (bls.gov).

Blumenstyk, Goldie. 2015. "When a Degree Is Just the Beginning." *Chronicle of Higher Education*, September 18, B4–B7.

Blumer, Herbert. 1946. "Collective Behavior." Pp. 65–121 in *New Outline of the Principles of Sociology*, edited by Alfred M. Lee. New York: Barnes & Noble.

Blumer, Herbert. 1969. *Symbolic Interactionism: Perspective and Method*. Englewood Cliffs, NJ: Prentice Hall.

Board of Governors of the Federal Reserve System. 2023. "Distribution of Household Wealth in the U.S. since 1989." March 24. Accessed April 13, 2023 (https://www.federalreserve.gov/releases/z1/dataviz/dfa/distribute/chart/#quarter:133;series:Net%20worth;demographic:income;population:all;units:levels;range:2007.4,2022.4).

Board of Trustees, The. 2013. *The 2013 Annual Report of the Board of Trustees of the Federal Old-Age and Survivors Insurance and Federal Disability Insurance Trust Funds*. Accessed July 28, 2013 (www.socialsecurity.gov).

Boersma, Peter, Lindsey I. Black, and Brian W. Ward. 2020. "Prevalence of Multiple Chronic Conditions Among US Adults, 2018." Centers for Disease Control and Prevention, September 17. Accessed June 11, 2023 (https://www.cdc.gov/pcd/issues/2020/20_0130.htm#T1_down).

Bogage, Joseph. 2015. "Why Companies Are Speaking Up About Gay Marriage." *Washington Post,* June 26. Accessed September 30, 2015 (washingtonpost.com).

Bogle, Kathleen A. 2008. *Hooking Up: Sex, Dating, and Relationships on Campus.* New York: New York University Press.

Bohannon, Paul, ed. 1971. *Divorce and After.* New York: Doubleday.

Bollag, Burton. 2007. "Credential Creep." *Chronicle of Higher Education,* June 22, A10–A12.

Bolter, Jessica. 2022. "Immigration Has Been a Defining, Often Contentious, Element Throughout U.S. History." Migration Policy Institute, January 6. Accessed September 26, 2023 (https://www.migrationpolicy.org/article/immigration-shaped-united-states-history).

Bonacich, Edna. 1972. "A Theory of Ethnic Antagonism: The Split Labor Market." *American Sociological Review* 37 (October): 547–559.

Booker, Brakkton. 2021. " Former Michigan Gov. Rick Snyder Charged In Flint Water Crisis." NPR, January 13. Accessed March 11, 2023 (https://www.npr.org/2021/01/13/956592508/new-charges-in-flint-water-crisis-including-former-michigan-gov-rick-snyder).

Bookman, Todd, and Brett Neeley. "Labor Unions Appear Set for More State-Level Defeats in 2017." NPR, January 30. Accessed June 19, 2017 (npr.org).

Borelli, Gabriel. 2022. "About Six-in-Ten Americans Say Legalization of Same-Sex Marriage Is Good for Society." Pew Research Center, November 15. Accessed April 21, 2023 (https://www.pewresearch.org/fact-tank/2022/11/15/about-six-in-ten-americans-say-legalization-of-same-sex-marriage-is-good-for-society/).

Borges, Anelise. 2021. "Meet France's First Female Imam Who Is on a Mission to Modernise Islam." Euronews, April 22. Accessed June 14, 2023 (https://www.euronews.com/my-europe/2021/04/21/meet-france-s-first-female-imam-who-is-on-a-mission-to-modernise-islam).

Borgerhoff Mulder, Monique. 2009. "Serial Monogamy as Polygyny or Polyandry?" *Human Nature* 20 (Summer): 130–150.

Borjas, George J. 2017. "The Earnings of Undocumented Immigrants." National Bureau of Economic Research, March. Accessed June 20, 2017 (nber.org).

Bosk, Charles. 1979. *Forgive and Remember: Managing Medical Failure.* Chicago: University of Chicago Press.

Boslaugh, Sarah. 2007. *Secondary Data Sources for Public Health: A Practical Guide.* New York: Cambridge University Press.

Botstein, Leon. 2014. "The SAT Is Part Hoax, Part Fraud." *Time,* March 24, 17.

Bourdieu, Pierre. 1984. *Distinction: A Social Critique of the Judgement of Taste.* Translated by Richard Nice. Cambridge, MA: Harvard University Press.

Bourdieu, Pierre. 1986. "The Forms of Capital." Pp. 241–258 in *Handbook for Theory and Research for the Sociology of Education,* edited by J. G. Richardson. Westport, CT: Greenwood.

Bouvard, Véronique, et al. 2015. "Carcinogenicity of Consumption of Red and Processed Meat." *Lance Oncology,* October 26. Accessed December 15, 2016 (iarc.fr).

Bowie, Liz. 2016. "Schools Teaching Etiquette as an Antidote to Lack of Respect." *Baltimore Sun,* May 16, 1, 11.

Bowles, Samuel, and Herbert Gintis. 1977. *Schooling in Capitalist America: Educational Reform and the Contradictions of Economic Life.* New York: Basic Books.

Boyle, Matthew. 2013. "Yes, Real Men Drink Beer and Use Skin Moisturizer." *Bloomberg Businessweek,* October 7–13, 32–33.

Bradshaw, York W., and Michael Wallace. 1996. *Global Inequalities.* Thousand Oaks, CA: Pine Forge Press.

Brady, Erin. 2021. "Americans Who Identify with No Religion Now Account for 29 Percent of Population: Poll." *Newsweek,* December 14. Accessed June 14, 2023 (https://www.newsweek.com/americans-who-identify-no-religion-now-account-29-percent-population-poll-1659344).

Braff, Danielle. 2016. "Single Mothers by Choice." *Baltimore Sun,* April 3, 7.

Braff, Danielle. 2020. "The New Helicopter Parents Are on Zoom." *The New York Times,* September 28. Accessed May 19, 2023 (https://www.nytimes.com/2020/09/28/parenting/helicopter-parent-remote-learning.html).

Braga, Anthony A. 2003. "Systematic Review of the Effects of Hot Spots Policing on Crime." Unpublished paper. Accessed May 14, 2005 (www.campbellcollaboration.org).

Brainard, Jeffrey, and Jia You. 2018. "What a Massive Database of Retracted Papers Reveals about Science Publishing's 'Death Penalty'" *Science,* October 25. Accessed January 29, 2023 (https://www.science.org/content/article/what-massive-database-retracted-papers-reveals-about-science-publishing-s-death-penalty).

Braitman, Keli A., Neil K. Chaudhary, and Anne T. McCart. 2011. "Effect of Passenger Presence on Older Drivers' Risk of Fatal Crash Involvement." Insurance Institute for Highway Safety, March. Accessed July 4, 2011 (www.iihs.org).

Brandon, Emily. 2022. "The Social Security Retirement Age Increases in 2022." *U.S. News,* January 24. Accessed May 19, 2023 (https://money.usnews.com/money/retirement/social-security/articles/the-social-security-retirement-age-increases).

Breining, Sanni N., Joseph J. Doyle, Jr., David N. Figlio, Krzysztof Karbownik, and Jeffrey Roth. 2017. "Birth Order and Delinquency: Evidence from Denmark and Florida." National Bureau of Economic Research, January. Accessed March 16, 2017 (nber.org).

Brenan, Megan. 2019. "40% of Americans Believe in Creationism." Gallup, July 26. (https://news.gallup.com/poll/261680/americans-believe-creationism.aspx).

Brenan, Megan. 2021. "Gender Disparities in Views of Women's Equality Persist." Gallup, October 15. Accessed April 16, 2023 (https://news.gallup.com/poll/355958/gender-disparities-views-women-equality-persist.aspx).

Brenan, Megan. 2021. "Americans' Trust in Government Remains Low." Gallup, September 30. Access May 12, 2023 (https://news.gallup.com/poll/355124/americans-trust-government-remains-low.aspx).

Brenan, Megan. 2021. "Record-High Worry in U.S. About Hunger, Race Relations." Pew Research Institute, March 26. Accessed May 3, 2023 (https://news.gallup.com/poll/341954/record-high-worry-hunger-race-relations.aspx?utm_source=alert&utm_medium=email&utm_content=morelink&utm_campaign=syndication).

Brenan, Megan. 2022. "Worry About Crime in U.S. at Highest Level Since 2016." Gallup, April 7. Accessed March 25, 2023 (https://news.gallup.com/poll/391610/worry-crime-highest-level-2016.aspx).

Brenan, Megan. 2022. "Americans Say Birth Control, Divorce Most 'Morally Acceptable'." Gallup, June 9. Accessed February 11, 2023 (https://news.gallup.com/poll/393515/americans-say-birth-control-divorce-morally-acceptable.aspx).

Brenan, Megan. 2022. "Record-Low 38% Extremely Proud to Be American." Gallup, June 29. Accessed February 12, 2023 (https://news.gallup.com/poll/394202/record-low-extremely-proud-american.aspx).

Brenan, Megan. 2023 "Record High in U.S. Put Off Medical Care Due to Cost In 2022." Gallup, January 17. Accessed June 11, 2023 (https://news.gallup.com/poll/468053/record-high-put-off-medical-care-due-cost-2022.aspx).

Brenan, Megan. 2023. "Majority in U.S. Still Say Government Should Ensure Healthcare." Gallup, January 26. Accessed June 11, 2023 (https://news.gallup.com/poll/468401/majority-say-gov-ensure-healthcare.aspx#:~:text=--%20A%2057%25%20majority%20of%20U.S.%20adults%20believe,private%20insurance%20rather%20than%20run%20by%20the%20government.).

Brennan, Dan. 2021. "Benefits of Extracurricular Activities for Kids." WebMD, March 2. Accessed February 18, 2023 (https://www.webmd.com/parenting/benefits-extracurricular-activities-kids).

Brescoll, Victoria L., and Eric L. Uhlmann. 2008. "Can an Angry Woman Get Ahead? Status Conferral, Gender, and Expression of Emotion in the Workplace." *Psychological Science* 19 (March): 268–275.

Brill, Steven. 2015. "What I Learned from My $190,000 Surgery." *Time*, January 19, 34–43.

Brody, Jane E. 2015. "Limit Children's Screen Time, and Your Own" *New York Times*, July 14, D7.

Brody, Jane E. 2015. "Screens Separate the Obsessed from Life." *New York Times*, July 7, D7.

Brody, Jane E. 2015. "The More We Learn, the More We Ignore." *New York Times*, October 13, D7.

Brooks, Arthur C. 2007. "I Love My Work." *The American* (online), September–October. Accessed September 30, 2007 (www.theamericanmag.com).

Brown, Anna. 2016. "The Challenges of Translating the U.S. Census Questionnaire into Arabic." Pew Research Center, June 3. Accessed June 20, 2017 (pewresearch.org).

Brown, Anna. 2016. "What Americans Say It Takes to Be Middle Class." Pew Research Center, February 4. Accessed July 26, 2016 (pewresearch.org).

Brown, Anna. 2020. "54% in U.S. Say Marriage Is Important to a Fulfilling Life, Not Essential." Pew Research Center, August 20. Accessed May 19, 20230 (https://www.pewresearch.org/short-reads/2020/02/14/more-than-half-of-americans-say-marriage-is-important-but-not-essential-to-leading-a-fulfilling-life/).

Brown, Anna. 2021. "Growing Share of Childless Adults in U.S. Don't Expect to Ever Have Children." Pew Research Center, November 19. Accessed May 19, 2023 (https://www.pewresearch.org/short-reads/2021/11/19/growing-share-of-childless-adults-in-u-s-dont-expect-to-ever-have-children/).

Brown, Archie. 2010. "Signposts: Why Did Communism End When It Did?" *History Today*, March. Accessed April 12, 2014 (www.historytoday.com).

Brown, DeNeen L. 2009. "The High Cost of Poverty: Why the Poor Pay More." *Washington Post*, May 18, C1.

Brown, Kelly. 2022. "Intensive Outpatient Programs." Accessed March 25, 2023 (https://www.addictiongroup.org/treatment/therapies/iop/#:~:text=Costs%20range%20from%20%243000%20to%20%2410%2C000%20for%2030-days.,typically%20cost%20about%20%24250%20to%20%24350%20per%20day).

Brown, Michael, and Amy Goldin. 1973. *Collective Behavior: A Review and Reinterpretation of the Literature*. Pacific Palisades, CA: Goodyear.

Brown, Susan I. 2005. "How Cohabitation Is Reshaping American Families." *Contexts* 4 (Summer): 33–37.

Bruinius, Harry. 2013. "Whose Holidays Are They?" *Christian Science Monitor*, December 16, 21–23.

Bruinius, Harry. 2016. "A New Twist in Religious Liberty Debate." *Christian Science Monitor Weekly*, April 25 & May 2, 21–23.

Brummelman, Eddie, Sander Thomaes, Stefanie A. Nelemans, Bram Orobio de Castro, Geertjan Overbeek, and Brad J. Bushman. 2015. "Origins of Narcissism in Children." *Proceedings of the National Academy of Sciences* 112 (March 25): 3659–3662.

Brunvand, Jan H. 2001. *The Truth Never Stands in the Way of a Good Story*. Urbana and Chicago: University of Illinois Press.

Bryant, Karl. 2014. "Teaching the Nature-Nurture Debate." *Contexts* 13 (Fall): 22–23.

Buchanan, Larry, Andrew W. Lehren, Jugal K. Patel, and Adam Pearce. 2017. "How Much People in the Trump Administration Are Worth." *New York Times*, April 3. Accessed May 5, 2017 (nytimes.com).

Buchheit, Paul. 2015. "The Corporate Debt to Society: $10,000 Per Household, Per Year." AlterNet, February 22. Accessed August 26, 2015 (alternet.org).

Buchheit, Paul. 2015. "The Number 1 Thing Rich People Get Dead Wrong About Poor People." AlterNet, November 20. Accessed June 4, 2017 (alternet.org).

Buchheit, Paul. 2017. "The Rich Pay Fewer Taxes Than the Poor, and Get More Services." AlterNet, March 20. Accessed June 4, 2017 (alternet.org).

Buchholz, Katharina. 2023. "Charted: Here's How Americans Are Approaching Healthy Eating." World Economic Forum, March 1. Accessed Jun 11, 2023 (https://www.weforum.org/agenda/2023/03/health-food-us/).

Buckley, Chris. 2015. "China Approves a Two-Child P olicy to Help Economy." *New York Times*, October 30, A1, A6.

Budiman, Abby, and Neil G. Ruiz. 2021. "Asian Americans and Their Origins: Key facts." Pew Research Center, April 29. Accessed June 13, 2023 (https://www.pewresearch.org/short-reads/2021/04/29/key-facts-about-asian-origin-groups-in-the-u-s/#:~:text=Those%20of%20Indian%20%2875%25%29%2C%20Malaysian%20%2865%25%29%2C%20Mongolian%20%2860%25%29,Bhutanese%20%2815%25%29%20have%20at%20least%20a%20bachelor%E2%80%99s%20degree).

Buechler, Steven M. 2000. *Social Movements in Advanced Capitalism: The Political Economy and Cultural Construction of Social Activism*. New York: Oxford University Press.

Buffington, Catherine, Benjamin Cerf Harris, Christina Jones, and Bruce A. Weinberg. 2016. "STEM Training and Early Career Outcomes of Female and Male Graduate Students: Evidence from UMETRICS Data Linked to the 2010 Census." *American Economic Review* 106 (May): 333–338.

Buggey, Hope, and Shannon Kelley. 2023. "Americans Cite Cost of Heathy Food as Biggest Barrier to a Heart-Healthy Diet, According to Cleveland Clinic Survey." Cleveland Clinic, February 1. Accessed June 11, 2023 (https://newsroom.clevelandclinic.org/2023/02/01/americans-cite-cost-of-heathy-food-as-biggest-barrier-to-a-heart-healthy-diet-according-to-cleveland-clinic-survey/).

Bullock, Karen. 2005. "Grandfathers and the Impact of Raising Grandchildren." *Journal of Sociology and Social Welfare* 32 (March): 43–59.

Buntin, John. 2015. "The Myth of Gentrification." *Slate*, January 14. Accessed October 23, 2015 (slate.com).

Bureau of Indian Affairs. 2023. "Tribal Consultation Written Comments Received American Rescue Plan – Indian Affairs Appropriations." Accessed May 3, 2023 (https://www.bia.gov/sites/default/files/dup/assets/as-ia/pdf/Comment_Compilation_ARP_FINAL_508_Compliant.pdf).

Bureau of Labor Statistics. 2017. "Demographics." Accessed July 20, 2017 (bls.gov).

Bureau of Labor Statistics. 2017. "Household Data Annual Averages, Table 39." Accessed April 10, 2017 (bls.gov).

Bureau of Labor Statistics. 2021. "Occupational and Labor Statistics." May. Accessed April 10, 2023 (https://www.bls.gov/oes/current/oes372012.htm).

Bureau of Labor Statistics. 2022. "Employer-Reported Workplace Injuries and Illnesses, 2021." November 9. Accessed June 11, 2023 (https://www.bls.gov/news.release/osh.nr0.htm).

Bureau of Labor Statistics. 2022. "National Census of Fatal Occupational Injuries in 2021" December 16. Accessed June 11, 2023 (https://www.bls.gov/news.release/pdf/cfoi.pdf).

Bureau of Labor Statistics, 2023. "Household Data, Annual Averages. 11. Employed Persons by Detailed Occupation, Sex, Race, and Hispanic or Latino Ethnicity." January 25. Accessed September 17, 2023 (https://www.bls.gov/cps/cpsaat11.htm).

Bureau of Labor Statistics, 2023. "Economic News Release: Employment Situation." May 5. Accessed May 11, 2023 (https://www.bls.gov/news.release/empsit.toc.htm).

Bureau of Labor Statistics. 2023. "Economic News Release: Union Members Summary." January 19. Accessed May 12, 2023 (https://www.bls.gov/news.release/union2.nr0.htm).

Bureau of Prisons. 2021. "Annual Determination of Average Cost of Incarceration Fee (COIF)" National Archives, Federal Register, September 1. Accessed March 25, 2023 (https://www.federalregister.gov/documents/2021/09/01/2021-18800/annual-determination-of-average-cost-of-incarceration-fee-coif).

Businesswire. 2021. "COVID Cyber Crime: 74% of Financial Institutions Experience Significant Spike in Threats Linked To COVID-19." April 28. Accessed March 25, 2023 (https://www.businesswire.com/news/home/20210428005365/en/COVID-Cyber-Crime-74-of-Financial-Institutions-Experience-Significant-Spike-in-Threats-Linked-To-COVID-19).

Businesswire. 2022. "Notable Survey: 61% of Patients Skip Medical Appointments Due to Scheduling Hassles." November 14. Accessed June 11, 2023 (https://www.businesswire.com/news/home/20221114005097/en/Notable-Survey-61-of-Patients-Skip-Medical-Appointments-Due-to-Scheduling-Hassles).

Busteed, Brandon. 2015. "Is College Worth It? That Depends." Gallup, April 8. Accessed July 30, 2017 (gallup.com).

Butler, Kiera. 2014. "Google's Magic Bus." *Mother Jones*, May–June, 70–71.

Butler, Kiera. 2015. "A Tale of Two Diseases." *Mother Jones*, May/June, 68.

C

Cadge, Wendy, and Courtney Bender. 2004. "Yoga and Rebirth in America: Asian Religions Are Here to Stay." *Contexts* 3 (Winter): 45–51.

Cahalan, M., L. W. Perna, M. Yamashita, R. Ruiz, and K. Franklin. 2017. *Indicators of Higher Education Equity in the United States: 2017 Trend Report.* Pell Institute for the Study of Higher Education, Council for Education Opportunity, and Alliance for Higher Education and Democracy. Accessed July 31, 2017 (pellinstitute.org).

Calacal, Celisa. 2017. "Not a Single Black Woman Heads a Top Fortune 500 Company." AlterNet, June 8. Accessed June 20, 2017 (alternet.org).

Calderon, Valerie J., and Preety Sidhu. 2014. "Business Leaders Say Knowledge Trumps College Pedigree." Gallup, February 25. Accessed June 21, 2014 (www.gallup.com).

Calderon, Valerie J., and Susan Sorenson. 2014. "Americans Say College Degree Leads to a Better Life." Gallup, April 7. Accessed June 21, 2014 (www.gallup.com).

Callahan, David. 2004. *The Cheating Culture: Why More Americans Are Doing Wrong to Get Ahead.* Boston: Houghton Mifflin Harcourt.

Camarota, Steven A., and Karen Zeigler. 2021. "Fertility Among Immigrants and Native-Born Americans: Difference Between the Foreign-Born and the Native-Born Continues to Narrow." Center for Immigration Studies, February 16. Accessed June 27, 2023 (https://cis.org/Report/Fertility-Among-Immigrants-and-NativeBorn-Americans).

Cameron, Deborah. 2007. *The Myth of Mars and Venus.* New York: Oxford University Press.

Campion-Vincent, Véronique. 2005. "From Evil Others to Evil Elites: A Dominant Pattern in Conspiracy Theories Today." Pp. 103–122 in *Rumor Mills: The Social Impact of Rumor and Legend,* edited by Gary Alan Fine, Veronique Campion-Vincent, and Chip Heath. New Brunswick, NJ: Transaction Publishers.

Campo-Flores, Arian. 2013. "Religious Dorms Sprout Up." *Wall Street Journal*, September 3, A3.

Cano, Regina Garcia. 2015. "S.D. Tribe to Open Nation's 1st Pot Resort." *Baltimore Sun*, September 30, 10.

Cantor, David, et al. 2015. "Report on the AAU Campus Climate Survey on Sexual Assault and Sexual Misconduct." Westat, September 21. Accessed July 31, 2017 (westat.com).

Cantrell, Steven, and Thomas J. Kane. 2013. "Ensuring Fair and Reliable Measures of Effective Teaching." MET Project, Policy and Practice Brief. Accessed June 21, 2014 (www.metproject.org).

Cantril, Hadley, Hazel Gaudet, and Herta Herzog. 1952. *The Invasion from Mars.* Princeton, NJ: Princeton University Press.

Carbonaro, Giulia. 2022. "These 8 Republican Women Voted Against Codifying Roe v. Wade." Newsweek, May 12. Accessed May 12, 2023 (https://www.*newsweek*.com/these-8-republican-women-voted-against-codifying-roe-v-wade-1705915).

Card, David, and Laura Giuliano. 2016. "Can Tracking Raise the Test Scores of High-Ability Minority Students?" National Bureau of Economic Research, March. Accessed July 31, 2017 (nber.org).

Carey, Benedict. 2010. "Revising Book on Disorders of the Mind." *New York Times,* February 10, 1.

Carl, Traci. 2002. "Amid Latte, Mocha Craze, Coffee Growers Go Hungry in Paradise." Accessed September 19, 2002 (www.yahoo.com).

Carlisle, Madeleine. 2022. "Here's How Much the Supreme Court Justices Made Last Year." Time Magazine, June 9. Accessed May 12, 2023 (https://time.com/6186294/supreme-court-salary-book-deals/).

Carnagey, Nicholas L., and Craig A. Anderson. 2005. "The Effects of Reward and Punishment in Violent Video Games on Aggressive Affect, Cognition, and Behavior." *Psychological Science* 16 (November): 882–889.

Carnevale, Anthony P., and Ban Cheah. 2013. "Hard Times: College Majors, Unemployment, and Earnings." Georgetown Center on Education and the Workforce, May. Accessed June 24, 2014 (www.cew.georgetown.edu).

Carney, Ginny. 1997. "Native American Loanwords in American English." *Wicazo SA Review,* 12 (Spring): 189–203

Carnoy, Martin, and Emma García. 2017. "Five Key Trends in U.S. Student Performance." Economic Policy Institute, January 12. Accessed July 31, 2017 (epi.org).

Carnoy, Martin, and Richard Rothstein. 2013. "What Do International Tests Really Show About U.S. Student Performance?" Economic Policy Institute, January 28. Accessed June 21, 2014 (www.epi.org).

Carpenter, Christopher S., Chandler B. McClellan, and Daniel I. Rees. 2016. "Economic Conditions, Illicit Drug Use, and Substance Use Disorders in the United States." National Bureau of Economic Research, February. Accessed March 16, 2017 (nber.org).

Carrell, Scott E., Mark Hoekstra, and James E. West. 2015. "The Impact of Intergroup Contact on Racial Attitudes and Revealed Preferences." National Bureau of Economic Research, February. Accessed May 1, 2015 (nber.org).

Carson, E. Ann. 2021. "Prisoners in 2020 – Statistical Tables." U.S. Department of Justice, December. Accessed March 25, 2023 (https://bjs.ojp.gov/content/pub/pdf/p20st.pdf).

Carson, E. Ann, and Anderson, Elizabeth. 2016. "Prisoners in 2015." U.S. Department of Justice, December. (https://bjs.ojp.gov/content/pub/pdf/p15.pdf).

Case, Anne, and Angus Deaton. 2017. "Mortality and Morbidity in the 21st Century."

Brookings, March 23. Accessed May 22, 2017 (brookings.edu).

Cason, Mike. 2017. "Roy Moore Running for Senate, Resigns from Supreme Court to Challenge Luther Strange." *Montgomery News,* April 26. Accessed May 25, 2017 (al.com/news/Montgomery).

Castella-McDonald, Issac. 2021. "Bottled Water 3500 Times Worse Than Tap Water, New Study Shows." Impakter, in Business, Climate Change, Environment, Health, Society, August 9. Accessed June 28, 2023 (https://impakter.com/bottled-water-worse-than-tap-water/#:~:text=According%20to%20the%20Beverage%20Marketing,normalised%20is%20misleading%20and%20dangerous).

Catalano, Shannan. 2013. "Intimate Partner Violence: Attributes of Victimization, 1993–2011." Bureau of Justice Statistics, November. Accessed November 25, 2014 (www.bjs.gov).

Catalyst. 2017. "Pyramid: Women in S&P 500 Companies." March 1. Accessed April 10, 2017 (catalyst.org).

Catalyst. 2017. "Women CEOs of the S&P 500." March 14. Accessed April 2, 2017 (catalyst.org).

"Catching Up." 2013. *The Economist,* January 12, 67–68.

CDC/NCHS National Vital Statistics System. 2017. "National Marriage and Divorce Rate Trends." Centers for Disease Control and Prevention, March 17. Accessed July 8, 2017 (cdc.gov/nchs).

Cellini, Stephanie R., Signe-Mary McKernan, and Caroline Ratcliffe. 2008. "The Dynamics of Poverty in the United States: A Review of Data, Methods, and Findings." *Journal of Policy Analysis and Management* 27 (Summer): 577–605.

Census Bureau News. 2016. "The Majority of Children Live with Two Parents, Census Bureau Reports." News release, November 17. Accessed January 3, 2017 (census.gov).

Center for American Women and Politics (CAWP). 2023. "Women in Elective Office 2023." New Brunswick, NJ: Center for American Women and Politics, Eagleton Institute of Politics, Rutgers University-New Brunswick. Accessed April 16, 2023 (https://cawp.rutgers.edu/facts/current-numbers/women-elective-office-2023).

Center for Behavioral Health Statistics and Quality. 2016. "*Key Substance Use and Mental Health Indicators in the United States: Results from the 2015 National Survey on Drug Use and Health.*" Accessed August 15, 2017 (samhsa.gov).

Center for Economic and Policy Research. 2015. "The Rise of Discouraged Workers." March 25. Accessed May 23, 2017 (cepr.net).

Center for Responsive Politics. 2017. "2016 Outside Spending, by Super Pac." Open Secrets, June 25. Accessed June 26, 2017 (opensecrets.org).

Center for the Study of Global Christianity. 2007. "Global Table 5: Status of Global Mission, Presence and Activities, AD 1800–2005." Gordon-Conwell Theological Seminary. Accessed August 24, 2007 (www.gcts.edu).

Center on Budget and Policy Priorities. 2016. "Policy Basics: Where Do Our Federal Dollars Go?" March 4. Accessed May 5, 2017 (cbpp.org).

Centers for Disease Control and Prevention. 2018. "CDC: 1 in 4 US Adults Live with a Disability." August 16. Accessed June 10, 2023 (https://www.cdc.gov/media/releases/2018/p0816-disability.html#:~:text=One%20in%204%20U.S.%20adults%20%E2%80%93%2061%20million,2%20in%205%20adults%20age%2065%20and%20older).

Centers for Disease Control and Prevention. 2019. "Tobacco Product Use and Cessation Indicators Among Adults — United States, 2018." November 15. Accessed June 11, 2023 (https://www.cdc.gov/tobacco/data_statistics/mmwrs/byyear/2019/mm6845a2/index.html).

Centers for Disease Control and Prevention. 2019. "User Guide to the 2019 Period/2018 Cohort Linked Birth/Infant Death Public Use File." Accessed October 6, 2023 (https://ftp.cdc.gov/pub/Health_Statistics/NCHS/Dataset_Documentation/DVS/period-cohort-linked/19PE18CO_linkedUG.pdf).

Centers for Disease Control and Prevention. 2020. "Youth Risk Behavior Surveillance — United States, 2019." Morbidity and Mortality Weekly Report. Accessed April 19. 2023 (https://www.cdc.gov/healthyyouth/data/yrbs/pdf/2019/su6901-H.pdf).

Centers for Disease Control and Prevention. 2021. "National Infection and Death Estimates for Antimicrobial Resistance." Accessed June 11, 2023 (https://www.cdc.gov/drugresistance/national-estimates.html).

Centers for Disease Control and Prevention. 2022. "Health and Economic Benefits of Diabetes Interventions." Accessed June 11, 2023 (https://www.cdc.gov/chronicdisease/programs-impact/pop/diabetes.htm).

Centers for Disease Control and Prevention. 2022. "CDC 2022 National Diabetes Statistics Report." January 26. Accessed June 10, 2023 (https://nationaldppcsc.cdc.gov/s/article/CDC-2022-National-Diabetes-Statistics-Report#:~:text=The%20Centers%20for%20Disease%20Control,prediabetes%20in%20the%20United%20States).

Centers for Disease Control and Prevention. 2022. "Fast Facts: Preventing Child Abuse and Neglect." April 6. Accessed May 22, 2023 (https://www.cdc.gov/violenceprevention/childabuseandneglect/fastfact.html).

Centers for Disease Control and Prevention. 2023. "About Rural Health." May 9. Accessed June 27, 2023 (https://www.cdc.gov/ruralhealth/about.html).

Centers for Disease Control and Prevention. 2023. "Covid-19." Accessed June 11, 2023 (https://www.cdc.gov/coronavirus/2019-nCoV/index.html).

Centers for Disease Control and Prevention. 2023. "Deaths and Mortality." January 17. Accessed July 29, 2023 (https://www.cdc.gov/nchs/fastats/deaths.htm).

Centers for Disease Control and Prevention. 2023. "Births – Delivery Method." Accessed June 11, 2023 (https://www.cdc.gov/nchs/fastats/delivery.htm).

Centers for Disease Control and Prevention. 2023. "About COVID-19." Accessed July 29, 2023 (https://www.cdc.gov/coronavirus/2019-ncov/your-health/about-covid-19.html).

Centers for Disease Control and Prevention. 2023. "Data & Statistics on FASDs." Accessed June 11, 2023 (https://www.cdc.gov/ncbddd/fasd/data.html#:~:text=Using%20medical%20and%20other%20records,areas%20of%20the%20United%20States.&text=The%20most%20recent%20CDC%20study,to%209%20years%20of%20age).

Centers for Disease Control and Prevention. 2023. "Disparities in Suicide." Accessed August 30, 2023 (https://www.cdc.gov/suicide/facts/disparities-in-suicide.html).

Centers for Disease Control and Prevention. 2023. "Heart Disease Facts." Accessed July 29, 2023 (https://www.cdc.gov/heartdisease/facts.htm).

Centers for Disease Control and Prevention. 2023. "How You Can Prevent Chronic Diseases." Accessed June 11, 2023 (https://www.cdc.gov/chronicdisease/about/prevent/index.htm).

Centers for Medicare & Medicaid Services. 2010. "NHE Tables for Selected Calendar 1960–2010." April 11. Accessed April 24, 2010 (www.cms.gov).

Centers for Medicare & Medicaid Services. 2014. "National Health Expenditure Projections 2012–2022." May 6. Accessed July 26, 2014 (www.cms.gov).

Centers for Medicare & Medicaid Services. 2022. "CMS Office of the Actuary Releases 2021-2030 Projections of National Health Expenditures." March 28. Accessed June 10, 2023 (https://www.cms.gov/newsroom/press-releases/cms-office-actuary-releases-2021-2030-projections-national-health-expenditures).

Ceron, Ella. 2022. "One in Three U.S. Workers Are Earning Less Than $15 an Hour." *Bloomberg,* March 22. Accessed May 10, 2023 (https://www.bloomberg.com/news/articles/2022-03-22/federal

-minimum-wage-1-in-3-us-workers-make-less-than-15-an-hour#:~:text=Share%20this%20article&text=Nearly%2052%20million%20U.S.%20workers,published%20Tuesday%20by%20Oxfam%20America).

Ceron, Ella. 2022. "The Proportion of Elected City Offices Held by Women Rose by Just 3% in 2022." *Bloomberg,* April 6. Accessed April 16, 2023 (https://www.bloomberg.com/news/articles/2022-04-06/women-make-up-less-than-a-third-of-mayors-city-councils-other-elected-offices#xj4y7vzkg).

Cha, Ariana Eunjung. 2014. "'Smart Pills' with Chips, Cameras and Robotic Parts Raise Legal, Ethical Questions." *Washington Post,* May 24. Accessed October 8, 2014 (www.washingtonpost.com).

Cha, Ariana Eunjung. 2016. "Bad Eating Habits Start Early." *Baltimore Sun,* April 14, 8.

Chambliss, William J. 1969. *Crime and the Legal Process.* New York: McGraw-Hill.

Chambliss, William J., and Robert B. Seidman. 1982. *Law, Order, and Power,* 2nd edition. Reading, MA: Addison-Wesley.

Chamie, Joseph. 2017. "Out-of-Wedlock Births Rise Worldwide." Yale Global Online, March 16. Accessed July 17, 2017 (yaleglobal.yale.edu).

Chancel, Lucas, Thomas Piketty, Emmanuel Saez, and Gabriel Zucman 2022. "World Inequality Report." World Inequality Lab. Accessed April 13, 2023 (https://wir2022.wid.world/www-site/uploads/2023/03/D_FINAL_WIL_RIM_RAPPORT_2303.pdf).

Chandler, Adam. 2016. "Why Do Americans Move So Much More Than Europeans?" *The Atlantic,* October 21. Accessed June 27, 2023 (https://www.theatlantic.com/business/archive/2016/10/us-geographic-mobility/504968/).

Chandler, Tertius, and Gerald Fox. 1974. *3000 Years of Urban Growth.* New York: Academic Press.

Chandra, Anjani, William D. Mosher, and Casey Copen. 2011. "Sexual Behavior, Sexual Attraction, and Sexual Identity in the United States: Data from the 2006–2008 National Survey of Family Growth." National Health Statistics Reports, No. 36, March 3. Accessed March 2, 2014 (www.cdc.gov/nchs).

Chapman, Isabell. 2018. "This photo shows how diaper-changing stations in men's restrooms are still too rare." CNN, October 2. Accessed April 21, 2023 (https://www.cnn.com/2018/10/02/health/diaper-change-mens-restrooms-trnd/index.html).

"Characteristics of Same-Sex Households." 2017. U.S. Census Bureau, Table 1, April 7. Accessed July 20, 2017 (census.gov).

Chasin, Barbara H. 2004. *Inequality & Violence in the United States: Casualties of Capitalism,* 2nd edition. Amherst, NY: Humanity Books.

Chaves, Mark, and Alison Eagle. 2015. "National Congregations Study." Accessed July 31, 2017 (soc.duke.edu/natcong).

Chen, Alice, Emily Oster, and Heidi Williams. 2014. "Why Is Infant Mortality Higher in the US Than in Europe?" National Bureau of Economic Research, September. Accessed December 5, 2014 (www.nber.org).

Chen, Frances S., Julia A. Minson, Maren Schöne1, and Markus Heinrichs. 2013. "In the Eye of the Beholder: Eye Contact Increases Resistance to Persuasion." *Psychological Science* 24 (November): 2254–2261.

Chen, Juan, and David T. Takeuchi. 2011. "Intermarriage, Ethnic Identity, and Perceived Social Standing among Asian Women in the United States." *Journal of Marriage and Family* 73 (August): 876–888.

Chen, Michelle. 2012. "Health Care System Leaves Patients Frustrated—Nurses Work for a Solution." AlterNet, June 4. Accessed June 5, 2012 (www.alternet.org).

Chen, Ronald C., et al. 2016. "Aggressive Care at the End-of-Life for Younger Patients with Cancer: Impact of ASCO's Choosing Wisely Campaign." Paper presented at the American Society of Clinical Oncology meeting, Chicago, Illinois, June 6.

Cherokee Nation. 2023. "Osiyo." Accessed May 3, 2023 (https://www.cherokee.org/#:~:text=Today%2C%20the%20Cherokee%20Nation%20is,reservation%20boundaries%20in%20northeastern%20Oklahoma).

Chesler, Phyllis. 2006. "The Failure of Feminism." *Chronicle of Higher Education,* February 26, B12.

Chesney-Lind, Meda, and Lisa Pasko. 2004. *The Female Offender: Girls, Women, and Crime,* 2nd edition. Thousand Oaks, CA: Sage.

Chetty, Raj, David Grusky, Maximilian Hell, Nathaniel Hendren, Robert Manduca, and Jimmy Narang. 2016. "The Fading American Dream: Trends in Absolute Income Mobility Since 1940." National Bureau of Economic Research, December. Accessed March 15, 2017 (nber.org).

Chetty, Raj, John N. Friedman, Emmanuel Saez, Nicholas Turner, and Danny Yagan. 2017. "Mobility Report Cards: The Role of Colleges in Intergenerational Mobility." Federal Reserve Bank of Boston, January. Accessed May 5, 2017 (bostonfed.org).

Chetty, Raj, and Nathaniel Hendren. 2016. "The Impacts of Neighborhoods on Intergenerational Mobility II: County-Level Estimates." National Bureau of Economic Research, December. Accessed March 15, 2017 (nber.org).

Chetty, Raj, Nathaniel Hendren, Patrick Kline, Emmanuel Saez, and Nicholas Turner. 2014. "Is the United States Still a Land of Opportunity? Recent Trends in Intergenerational Mobility." National Bureau of Economic Research, January. Accessed February 6, 2014 (www.nber.org).

"Child Brides in West Africa: Girls Fight Back." 2014. *The Economist,* August 23, 42.

Child Trends Data Bank. 2013. "Youth Voting." June. Accessed April 20, 2014 (www.childtrendsdatabank.org).

Child Welfare Information Gateway. 2016. "Foster Care Statistics 2014." Children's Bureau, March. Accessed February 28, 2017 (childwelfare.gov).

Children's Defense Fund. 2014. "The State of America's Children 2014." Accessed February 14, 2014 (www.childrensdefense.org).

Chin, Margaret M. 2016. "Asian Americans, Bamboo Ceilings, and Affirmative Action." *Contexts* 15 (Winter): 70–73.

Chira, Susan. 2017. "New Era After O'Reilly? Women Aren't So Sure." *New York Times,* April 21, B7.

Chmielewski, Anna K., and Sean F. Reardon. 2016. "Education." *Pathways,* 45–50.

Cho, Renee. 2023. "Climate Education in the U.S.: Where It Stands, and Why It Matters." Columbia Climate School, States of the Planet, February 9. Accessed Jun 28, 2023 (https://news.climate.columbia.edu/2023/02/09/climate-education-in-the-u-s-where-it-stands-and-why-it-matters/).

Choi, Jung Hyun. 2023. "Unmasking the Real Gender Homeownership Gap." The Urban Institute, March 28. Accessed May 22, 2023 (https://www.urban.org/urban-wire/unmasking-real-gender-homeownership-gap#:~:text=In%201990%2C%2033%20percent%20of,only%208%20percent%20in%201990).

Choma, Russ. 2015. "One Member of Congress = 18 American Households: Lawmakers' Personal Finances Far from Average." Open Secrets, January 12. Accessed October 7, 2015 (opensecrets.org).

Christensen, Jen. 2023. "EPA Proposes First Standards to Make Drinking Water Safer From 'Forever Chemicals'." *CNN,* March 15. Accessed June 27, 2023 (https://amp.cnn.com/cnn/2023/03/14/health/epa-pfas-standards-wellness/index.html).

Christina, Greta. 2011. "Wealthy, Handsome, Strong, Packing Endless Hard-Ons: The Impossible Ideals Men Are Expected to Meet." Independent Media Institute, June 20. Accessed June 21, 2011 (www.alternet.org).

Christina, Greta. 2016. "7 of the Less-Noted but Still Very Sexist Attacks on Hillary Clinton." AlterNet, August 19. Accessed May 15, 2016 (alternet.org).

Chronicle of Higher Education. 2014. "Data Point." April 18, A23.

Churchill, Ward. 1997. *A Little Matter of Genocide: Holocaust and Denial in the Americas, 1492 to the Present*. San Francisco: City Lights Books.

CIA. 2023. "The World Factbook." Accessed June 14, 2023 (https://www.cia.gov/the-world-factbook/).

Ciabattari, Teresa, Kathleen S. Lowney, Renee A. Monson, Mary Scheuer Senter, and Jeffrey Chin. 2018. "Linking Sociology Majors to Labor Market Success." *Teaching Sociology*, 46(3): 191-207. Accessed January 29, 2023 (https://www.jstor.org/stable/26589028).

Cillizza, Chris. 2014. "Our C-Minus Government (and Why We Deserve It)." *Washington Post*, October 15. Accessed November 23, 2014 (www.brookings.edu).

Citizens Against Government Waste. 2023. "2023 Congressional Pig Book Summary." Accessed May 10, 2023 (https://www.cagw.org/reporting/pig-book).

Citizens Research Council of Michigan. 2022. "Examining Detroit's Vacancy Rate Drop." March. Accessed June 27 (https://news.stanford.edu/2020/12/01/gentrification-disproportionately-affects-minorities/).

Clancy, Laura, and Sarah Austin. 2023. "Fewer Than a Third of UN Member States Have Ever Had a Woman Leader." Pew Research Center, March 28. Accessed April 21, 2023 (https://www.pewresearch.org/fact-tank/2023/03/28/women-leaders-around-the-world/#:~:text=Women%20currently%20serve%20as%20the,a%20Pew%20Research%20Center%20analysis).

Clarke, Lee. 2002. "Panic: Myth or Reality?" *Contexts*. (Fall) 1: 21-26. (https://doi.org/10.1525/ctx.2002.1.3.21).

Claxton, Gary, et al. 2016. "Employer Health Benefits: 2016 Annual Survey." Kaiser Family Foundation, September 14. Accessed August 15, 2017 (kff.org).

Claybourn, Cole. 2022. "ACT vs. SAT: How to Decide Which Test to Take." U.S. News and World Report, December 1. Accessed June 10, 2023 (https://www.usnews.com/education/best-colleges/articles/act-vs-sat-how-to-decide-which-test-to-take).

Clayson, Dennis E., and Debra A. Haley. 2011. "Are Students Telling Us the Truth? A Critical Look at the Student Evaluation of Teaching." *Marketing Education Review* 21 (Summer): 101–112.

Clement, Scott, and Lenny Bernstein. 2016. "One-Third of Long-Term Users Say They're Hooked on Prescription Opioids." *Washington Post*, December 9. Accessed August 15, 2017 (washingtonpost.com).

Clements, David D. 2012. *Corporations Are Not People: Why They Have More Rights Than You Do and What You Can Do About It*. San Francisco: Berrett-Koehler.

Clemmons, Nakia S., Paul A. Gastanaduy, Amy P. Fiebelkorn, Susan B. Redd, and Gregory S. Wallace. 2015. "Measles—United States, January 4 -April 2, 2015." *MMWR* 64 (April 17): 373–6.

Clifford, Stephanie, and Jessica Silver-Greenberg. 2016. "Bank Tellers, with Low Pay and High Access, Pose a Rising Security Risk." *New York Times*, February 2, A22–A23.

Clifford, Stephanie, and Quentin Hardy. 2013. "Attention, Shoppers: Store Is Tracking Your Cell." *New York Times*, July 15, A1.

Clifford, Tyler. 2023. "US East Coast Blanketed in Veil of Smoke from Canadian Fires." Reuters, June 8. Accessed June 14, 2023 (https://www.reuters.com/business/environment/us-states-under-air-quality-alerts-canadian-smoke-drifts-south-2023-06-07/).

Clifton, Jim. 2016. "Explaining Trump: Widespread Government Corruption." Gallup, January 6. Accessed June 27, 2017 (gallup.com).

Cloke, Kenneth, and Joan Goldsmith. 2002. *End of Management and the Rise of Organizational Democracy*. San Francisco: Jossey-Bass.

Cloud, John. 2010. "How to Recruit Better Teachers." *Time*, September 20, 47–52.

Cobble, Dorothy Sue, Linda Gordon, and Astrid Henry. 2014. "What 'Lean In' Leaves Out." *Chronicle Review*, September 16, B4–B5.

Cohen, Jere. 2002. *Protestantism and Capitalism: The Mechanisms of Influence*. New York: Aldine de Gruyter.

Cohen, Jon. 2022. "As Creator of 'CRISPR Babies' Nears Release from Prison, Where Does Embryo Editing Stand?" *Science*, March 21. Accessed June 27, 2023 (https://www.science.org/content/article/creator-crispr-babies-nears-release-prison-where-does-embryo-editing-stand).

Cohen, Lawrence J., and Anthony T. DeBenedet. 2012. "Penn State Cover-Up: Groupthink in Action." *Time*, July 17. Accessed April 3, 2017 (ideas.time.com).

Cohen, Patricia. 2015. "The New Shop Class." *New York Times*, March 11, Bi, B4.

Cohen, Patricia. 2015. "When Company Is Fined, Taxpayers Often Share Bill." *New York Times*, February 4, B1, B9.

Cohen, Patricia. 2016. "With Pay Rising, Millions Climb Out of Poverty." *New York Times*, September 26, A1, A3.

Cohen, Patricia. 2017. "In Rural Iowa, a Future Rests on Immigrants." *New York Times*, May 30, A1, A12.

Cohn, D'Vera. 2012. "Divorce and the Great Recession." Pew Research Center, May 2. Accessed May 22, 2014 (www.pewsocialtrends.org).

Cohn, D'Vera. 2013. "Love and Marriage." Pew Research Social & Demographic Trends, February 13. Accessed May 16, 2014 (www.pewsocialtrends.org).

Cohn, D'Vera, Eileen Patten, and Mark Hugo. 2014. "Puerto Rican Population Declines on Island, Grows on U.S. Mainland." Pew Research Center, August 11. Accessed September 19, 2014 (www.pewresearch.org).

Cohn, D'Vera, Gretchen Livingston, and Wendy Wang. 2014. "After Decades of Decline, a Rise in Stay-at-Home Mothers." Pew Research Center, April 8. Accessed April 12, 2014 (www.pewresearch.org).

Cohn, D'Vera, Paul Taylor, Mark H. Lopez, Catherine A. Gallagher, Kim Parker, and Kevin T. Maass. 2013. "Gun Homicide Rate Down 49% Since 1993 Peak; Public Unaware." Pew Research Center, May 7. Accessed October 12, 2013 (www.pewresearch.org).

Cohn, Meredith. 2014. "Kids Are Outfitted with New Hands Made on 3-D Printers." *Baltimore Sun*, September 29, 1, 11.

Cohn, D'Vera, Juliana Menasce Horowitz, Rachel Minkin, Richard Fry, and Kiley Hurst. 2022. "The Demographics of Multigenerational Households." Pew Research Center, March 24. Accessed May 19, 2023 (https://www.pewresearch.org/social-trends/2022/03/24/the-demographics-of-multigenerational-households/#:~:text=From%201971%20to%202021%2C%20the,1971%20to%2018%25%20in%202021).

Colapinto, John. 1997. "The True Story of John/ Joan." *Rolling Stone* (December 11): 54–73, 92–97.

Colapinto, John. 2001. *As Nature Made Him: The Boy Who Was Raised as a Girl*. New York: Harper Perennial.

Colapinto, John. 2004. "What Were the Real Reasons Behind David Reimer's Suicide?" *Slate*, June 3. Accessed April 24, 2008 (www.slate.com).

Colarossi. Natalie 2020. "The Highest Voter Turnouts in History of U.S. Elections." *Newsweek*, November 5. Accessed May 12, 2023 (https://www.newsweek.com/highest-voter-turnouts-history-us-elections-1545235).

Colby, Sandra L., and Jennifer M. Ortman. 2015. "Projections of the Size and Composition of the U.S. Population: 2014 to 2060." U.S. Census Bureau, Current Population Reports, March. Accessed July 20, 2017 (census.gov).

"The Collaboration Curse." 2016. *The Economist*, January 23, 63.

College Board. 2013. "2013 College-Bound Seniors: Total Group Profile Report." Accessed June 21, 2014 (www.collegeboard.com).

College Board. 2014. "The 10th Annual AP Report to the Nation." February 11. Accessed June 21, 2014 (apreport.collegeboard.com).

College Board. 2022. "AP Program Results: Class of 2022." Accessed June 13, 2023 (https://reports.collegeboard.org/ap-program-results/class-of-2022).

Collins, William J., and Marianne H. Wanamaker. 2017. "Up from Slavery? African American Intergenerational Economic Mobility Since 1880." National Bureau of Economic Research, May. Accessed June 10, 2017 (nber.org).

Colvin, Geoff. 2012. "The Art of the Self-Managing Team." *Fortune*, December 3, 22–23.

Colvin, George. 2014. "In the Future, Will There Be Any Work Left for People to Do?" *Fortune*, June 16, 193–202.

Colvin, Mark, and John Pauly. 1983. "A Critique of Criminology: Toward an Integrated Structural-Marxist Theory of Delinquency Production." *American Journal of Sociology* 89 (November): 513–551.

Commisceo Global. 2016. "Global Guide to Culture, Customs, and Etiquette." Accessed March 13, 2017 (commisceo-global.com).

Conley, Dalton. 1999. *Being Black, Living in the Red: Race, Wealth, and Social Policy in America.* Berkeley: University of California Press.

Conley, Dalton. 2004. *The Pecking Order: Which Siblings Succeed and Why.* New York: Pantheon.

Connell, Dave. 2011. "30 Days of H20: Final Days." The Nature Conservancy, February 22. Accessed June 6, 2012 (blog.nature.org).

Connor, Phillip. 2016. "International Migration: Key Findings from the U.S., Europe and the World." Pew Research Center, December 15. Accessed August 25, 2017 (pewresearch.org).

Connor, Phillip. 2016. "Nearly Half of Refugees Entering the U.S. This Year Are Muslim." Pew Research Center, August 16. Accessed June 20, 2017 (pewresearch.org).

Conrad, Peter, and Kristin K. Barker. 2010. "The Social Construction of Illness: Key Insights and Policy Implications." *Journal of Health and Social Behavior* 51 (Suppl.): S67–S79.

"Consanguineous Marriage." 2016. *Economist*, April 16, 39–40.

Considine, Austin. 2011. "For Asian-American Stars, Many Web Fans." *New York Times*, July 31, ST6.

ConsumerReports.org. 2009. "6 Top Reasons for Not Having Sex." February. Accessed March 4, 2014 (www.consumerreports.org).

Cook, Thomas D., and Donald T. Campbell. 1979. *Quasi-Experimentation: Design and Analysis Issues for Field Settings.* Chicago: Rand McNally.

Cooky, Cheryl, Michael A. Messner, and Michela Musto. 2015. "'It's Dude Time!': A Quarter Century of Excluding Women's Sports in Televised News and Highlight Shows." *Communication & Sport* 3 (September): 261–287.

Cooley, Charles Horton. 1909/1983. *Social Organization: A Study of the Larger Mind.* New Brunswick, NJ: Transaction Books.

Cooper, David. 2015. "Given the Economy's Growth, the Federal Minimum Wage Could Be Significantly Higher." Economic Policy Institute, April 14. Accessed October 7, 2015 (epi.org).

Cooper, O. R., et al. 2010. "Increasing Springtime Ozone Mixing Ratios in the Free Troposphere over Western North America." *Nature* 463 (January 21): 344–348.

Copen, Casey E., Anjani Chandra, and Isaedmarie Febo-Vazquez. 2016. "Sexual Behavior Sexual Attraction, and Sexual Orientation Among Adults Ages 18–44 in the United States: Data from the 2011–13 National Survey of Family Growth." *National Health Statistics Reports* 88 (January 7). Accessed May 25, 2017 (cdc.gov/nchs).

Copen, Casey E., Kimberly Daniels, and William D. Mosher. 2013. "First Premarital Cohabitation in the United States: 2006–2010 National Survey of Family Growth." *National Health Statistics Reports*, April 4. Accessed May 16, 2014 (www.cdc.gov/nchs).

Copen, Casey E., Kimberly Daniels, Jonathan Vespa, and William D. Mosher. 2012. "First Marriages in the United States: Data from the 2006–2010 National Survey of Family Growth." *National Health Statistics Reports*, March 22. Accessed May 16, 2014 (www.cdc.gov/nchs).

Corak, Miles. 2013. "Income Inequality, Equality of Opportunity, and Intergenerational Mobility." *Journal of Economic Perspectives* 27 (Summer): 79–102.

Corak, Miles. 2016. "Economic Mobility." *Pathways*, Special Issue, 51–57.

Coren, Arthur. 2011. "Turning a Blind Eye: Faculty Who Ignore Student Cheating." *Journal of Academic Ethics* 9 (4): 291–305.

Corporation for National & Community Service. 2016. "Volunteering and Civic Life in America." Accessed April 2, 2017 (nationalservice.gov).

Couch, Carl J. 1968. "Collective Behavior: An Examination of Some Stereotypes." *Social Problems* 15 (Winter): 310–322.

Chelsea, Cox. 2022. " 'We're Going to Hang You': DOJ Cracks Down on Threats to Poll Workers." CNBC, October 27. Accessed June 24, 2023 (https://www.cnbc.com/2022/10/27/were-going-to-hang-you-doj-cracks-down-on-threats-to-election-workers-ahead-of-high-stakes-midterms.html).

Cox, Chelsey. 2022. "'We're Going to Hang You': DOJ Cracks Down on Threats to Election Workers Ahead of High-Stakes Midterms." *CNBC*, October 22. Accessed August 2, 2023 (https://www.cnbc.com/2022/10/27/were-going-to-hang-you-doj-cracks-down-on-threats-to-election-workers-ahead-of-high-stakes-midterms.html).

Cox, Cynthia, and Larry Levitt. 2017. "What's the Near-Term Outlook for the Affordable Care Act?" Kaiser Family Foundation, August 4. Accessed August 15, 2017 (kff.org).

Coy, Peter. 2016. "Rising Profits Don't Lift Workers' Boats." *Bloomberg Businessweek*, May 9–15, 19–20.

Crabtree, Steve. 2010. "Religiosity Highest in World's Poorest Nations." Gallup, June 3. Accessed August 31, 2012 (gallup.com).

Craig, Lyn. 2015. "How Mothers and Fathers Allocate Time to Children: Trends, Resources, and Policy Context." *Family Focus*, Spring, F9–F10.

Cramer, Rachel. 2022. "Iowa State Researcher Evaluates the Uptick in Women's Alcohol Consumption: Stress, Stereotypes, Marketing." Iowa State University News Service, August 30. Accessed July 23, 2023 (https://www.research.iastate.edu/news/iowa-state-researcher-evaluates-the-uptick-in-womens-alcohol-consumption-stress-stereotypes-marketing/).

Creamer, John, Emily A. Shrider, Kalee Burns, and Frances Chen. 2022. "Poverty in the United States, 2021. U.S. Census Bureau, September 13. Accessed April 13, 2023 (https://www.census.gov/library/publications/2022/demo/p60-277.html).

Cremer, Paul C., Yuping Wu, and Haitham M. Ahmed. 2017. "Use of Sex-Specific Clinical and Exercise Risk Scores to Identify Patients at Increased Risk for All-Cause Mortality." *JAMA Cardiology* 2 (1): 15–22.

Cressey, Donald R. 1953. *Other People's Money: A Study in the Social Psychology of Embezzlement.* Glencoe, IL: Free Press.

Cross, Rob, Reb Rebele, and Adam Grant. 2016. "Collaborative Overload." *Harvard Business Review* 94 (January/February): 74–79.

Crowley, Michael. 2014. "Iraq's Eternal War." *Time*, June 30, 28–34.

Cruikshank, Margaret. 2009. *Learning to Be Old: Gender, Culture, and Aging.* Lanham, MD: Rowman & Littlefield.

Cuadra, Deanna. 2022. "Should the U.S. Be Taking Notes from Europe when It Comes to PTO?" *Employee Benefit News*,

January 4. Accessed May 12, 2023 (https://www.benefitnews.com/news/should-the-u-s-have-the-same-pto-as-the-u-k-and-the-eu).

Cubanski, Juliette, Tricia Newman, Shannon Griffin, and Anthony Damico. 2016. "Medicare Spending at the End of Life: A Snapshot of Beneficiaries Who Died in 2014 and the Cost of Their Care." Kaiser Family Foundation, July 14. Accessed September 4, 2017 (kff.org).

Cudd, Ann E., and Nancy Holstrom. 2010. *Capitalism, For and Against: A Feminist Debate*. New York: Cambridge University Press.

Culbert, Samuel A. 2011. "Why Your Boss Is Wrong About You." *New York Times*, March 3, A25.

Cullen, Lisa T., and Coco Masters. 2008. "We Just Clicked." *Time*, January 28, 86–89.

Cumberworth, Erin. 2010. "Homeboy Industries." *Pathways*, Summer, 22–23.

Cummings, William. 2017. "Millennials Differ from Other Generations in Almost Every Regard. Here's the Data." *USA Today*, April 19. Accessed April 20, 2017 (usatoday.com).

Cunningham, Scott, Jason M. Lindo, Caitlin Myers, and Andrea Schlosser. 2017. "How Far Is Too Far? New Evidence on Abortion Clinic Closures, Access, and Abortions." National Bureau of Economic Research, April. Accessed July 19, 2017 (nber.org).

Currie, E. 1985. *Confronting Crime: An American Challenge*. New York: Pantheon.

Curtin, Sally C., Margaret Warner, and Holly Hedegaard. 2016. "Increase in Suicide in the United States, 1999–2014." Centers for Disease Control and Prevention, April. Accessed December 21, 2016 (cdc.gov).

Curtin, Sally C., Stephanie J. Ventura, and Gladys M. Martinez. 2014. "Recent Declines in Nonmarital Childbearing in the United States." National Center for Health Statistics, August. Accessed November 30, 2014 (www.nchs.gov).

Curtiss, Susan. 1977. *Genie*. New York: Academic Press.

Cyberbullying Research Center. 2016. "2016 Cyberbullying Data." November 26. Accessed March 13, 2017 (cyberbullying.org).

D

D'Innocenzio, Anne. 2015. "Target Stores Aim to Instill Healthier Habits." *Baltimore Sun*, September 17, 10.

Dahl, Robert A. 1961. *Who Governs? Democracy and Power in an American City*. New Haven, CT: Yale University Press.

Dahlin, E. 2011. "There's No 'I' in Innovation." *Contexts* 10 (Fall): 22–27.

Daily Nurse. 2020. "Florida Nurse Practitioners Gain Full Practice Authority." March 24. Accessed June 11, 2023 (https://dailynurse.com/florida-grants-advanced-nps-full-practice-authority/).

Daly, Mary C., and Leila Bengali. 2014. "Is It Still Worth Going to College?" Federal Reserve Bank of San Francisco, May 5. Accessed June 21, 2014 (www.frbsf.org).

Daneshpour, Manijeh. 2015. "A Mini-Narrative About My Praxis as a Muslim Feminist." *Family Focus* 57.4 (Winter): F14–F16.

Daniel, Mary-Alice. 2014. "What White People Need to Learn." AlterNet, February 7. Accessed March 22, 2014 (www.alternet.org).

Danielson, Melissa L., Susanna N. Visser, Mary Margaret Gleason, Georgina Peacock, Angelika H. Claussen, and Stephen J. Blumberg. 2017. "A National Profile of Attention-Deficit Hyperactivity Disorder Diagnosis and Treatment Among US Children Aged 2 to 5." *Journal of Developmental & Behavioral Pediatrics*, July 14. Accessed August 15, 2017 (journals.lww.com/jrnldbp).

Dank, Meredith, et al. 2014. "Estimating the Size and Structure of the Underground Commercial Sex Economy in Eight Major US Cities." Urban Institute, March. Accessed November 20, 2014 (www.urban.org).

Dao, James. 2013. "In Debate over Military Sexual Assault, Men Are Overlooked Victims." *New York Times*, June 24, A12.

Daskal, Lolly. 2016. "Tangled in Rigid Rules." *Baltimore Sun*, September 25, 19.

Dastagir. Alia E. 2020. "What Kamala Harris put up with." *USA Today*, October 9. Accessed May 12, 2023 (https://www.usatoday.com/story/news/nation/2020/10/08/vp-debate-kamala-harris-and-sexism-racism-black-women-face/5924748002/).

Davenport, Coral. 2017. "Denialists in Charge." *New York Times*, March 29, A19.

Davenport, Coral, and Alissa J. Rubin. 2017. "Trump Signs Rule to Block Efforts on Aiding Climate." *New York Times*, March 29, A1, A19.

Davenport, Coral. 2021 "Restoring Environmental Rules Rolled Back by Trump Could Take Years." *New York Times*, January 22. Accessed June 14, 2023 (https://www.nytimes.com/2021/01/22/climate/biden-environment.html).

Davey, Monica. 2014. "Immigrants Seen a Way to Refill Detroit Ranks." *New York Times*, January 24, A12.

Davidai, Shai, and Thomas Gilovich. 2015. "Building a More Mobile America—One Income Quintile at a Time." *Perspectives on Psychological Science* 10 (1): 60–71.

Davidson, Osha Gray. 1998. *Under Fire: The NRA and the Battle for Gun Control*. Iowa City: University of Iowa Press

Davies, James C. 1979. "The J-Curve of Rising and Declining Satisfaction as a Cause of Revolution and Rebellion." Pp. 413–436 in *Violence in America: Historical and Comparative Perspectives*, edited by Hugh D. Graham and Ted R. Gurr. Beverly Hills, CA: Sage.

Davis, Alyssa. 2016. "In U.S., Concern About Crime Climbs to 15-Year High." Gallup, April 6. Accessed April 15, 2017 (gallup.com).

Davis, Julie Hirschfeld. 2015. "Hacking Exposed 21 Million in U.S., Government Says." *New York Times*, July 10, A1, A3.

Davis, Kenneth C. 2010. "America's True History of Religious Tolerance." *Smithsonian*, October. Accessed September 28, 2010 (www.smithsonian.com).

Davis, Kingsley, and Wilbert E. Moore. 1945. "Some Principles of Stratification." *American Sociological Review* 10 (April): 242–249.

Dawson, Steve. 2017. "What Happened to Otto Warmbier—News & Updates." *Gazette Review*, January 16. Accessed February 14, 2017 (gazettereview.com).

De la Fuente, Homero, Masrur Jamaluddin, and Issy Ronald. 2022. "Kanjuruhan Stadium: Football World Mourns the Victims of the Indonesia Stadium Crush." *CNN*, October 2. Accessed June 24, 2023 (https://www.cnn.com/2022/10/02/football/indonesia-stadium-tragedy-football-reaction-spt-intl).

de las Casas, Bartolome. 1992. *The Devastation of the Indies: A Brief Account*. Baltimore, MD: Johns Hopkins University Press.

De Long, J. Bradford, and Andrei Shleifer. 1992. "Princes and Merchants: European City Growth Before the Industrial Revolution." National Bureau of Economic Research, December. Accessed January 10, 2008 (www.nper.org).

de Vogue, Ariane, and Jeremy Diamond. 2015. "Supreme Court Rules in Favor of Same-Sex Marriage Nationwide." CNN, June 27. Accessed September 29, 2015 (cnn.com).

de Vogue, Ariane, and Chandelis Duster. 2021. "Supreme Court Dismisses Challenge to Affordable Care Act, Leaving It in Place." *CNN*, June 26. Accessed June 11, 2023 (https://www.cnn.com/2021/06/17/politics/supreme-court-affordable-care-act-obamacare/index.html).

de Vries, Gijs. 2022. "The SDGs Require a Stronger Role for Culture in Development." OECD Development Matters, January 21. Accessed February 12, 2023 (https://oecd-development-matters.org/2022/01/21/the-sdgs-require-a-stronger-role-for-culture-in-development/).

Death Penalty Information Center. 2017. "Facts About the Death Penalty."

April 28. The Sentencing Project. Accessed April 29, 2017 (deathpenalty-info.org).

Deaton, Angus, and Arthur A. Stone. 2013. "Grandpa and the Snapper: The Wellbeing of the Elderly Who Live with Children." National Bureau of Economic Research, June. Accessed July 10, 2013 (www.nber.org).

Deegan, Mary Jo. 1986. *Jane Addams and the Men of the Chicago School, 1892–1918*. New Brunswick, NJ: Transaction Books.

Deerwater. Eric W. 2018. "Study Provides New Insights on Why College Students Engage in Hookups." PsyPost, November 8. Accessed April 21, 2023 (https://www.psypost.org/2018/11/study-provides-new-insights-on-why-college-students-engage-in-hookups-52541).

Deerwater, Raina. 2022. "GLAAD'S 2021-2022 Where We Are on TV Report: LGBTQ Representation Reaches New Record Highs." GLAAD, February 16. Accessed June 19, 2023 (https://glaad.org/glaads-2021-2022-where-we-are-tv-report-lgbtq-representation-reaches-new-record-highs/).

DeGraw, David. 2014. "How the Ultra-Rich .01% Have Sucked Up Even More of America's Wealth Than You Think." AlterNet, August 22. Accessed November 10, 2014 (www.alternet.org).

DeKeseredy, Walter S. 2011. *Violence Against Women: Myths, Facts, Controversies.* Toronto: University of Toronto Press.

della Porta, Donatella, and Mario Diani. 1999. *Social Movements: An Introduction.* Malden, MA: Blackwell.

Deloitte and Alliance for Board Diversity. 2023. "Missing Pieces Report: The Board Diversity Census of Women and Minorities on *Fortune 500* Boards, 6th edition" Accessed March 11, 2023 (https://www2.deloitte.com/us/en/pages/center-for-board-effectiveness/articles/missing-pieces-report-board-diversity.html).

Denny, C. H., J. Tsai, R. L. Floyd, and P. P. Green. 2009. "Alcohol Use Among Pregnant and Nonpregnant Women of Childbearing Age—United States, 1991–2005." *MMWR* 58 (May 22): 529–532.

DeParle, Jason. 2012. "Harder for Americans to Rise from Lower Rungs." *New York Times*, January 5, A1.

Deprez, Esmé E., and William Selway. 2014. "Guns Allowed All Over—Except Near Politicians." *Bloomberg Businessweek*, May 19–25, 27–28.

DeSilver, 2014. "Facing Challenges, Pollsters Broaden Experiments with New Methodologies." Pew Research Center, November 24. Accessed November 17, 2023 (https://www.pewresearch.org/short-reads/2014/11/24/facing-challenges-pollsters-broaden-experiments-with-new-methodologies-2/).

DeSilver, Drew. 2017. "5 Facts About the Minimum Wage." Pew Research Center, January 4. Accessed June 16, 2017 (pewresearch.org).

DeSilver, Drew. 2017. "U.S. Students' Academic Achievement Still Lags that of Their Peers in Many Other Countries." Pew Research Center, February 15. Accessed July 31, 2017 (pewresearch.org).

DeSilver, Drew. 2022. "Some Members of Congress Have Been There for Decades, but Seats Typically Change Hands More Frequently." Pew Research Center, June 9. Accessed May 12, 2023 (https://www.pewresearch.org/short-reads/2022/06/09/some-members-of-congress-have-been-there-for-decades-but-seats-typically-change-hands-more-frequently/).

DeSilver, Drew. 2022. "Turnout in U.S. Has Soared in Recent Elections but By Some Measures Still Trails That Of Many Other Countries." Pew Research Center, November 1. Accessed May 12, 2023 (https://www.pewresearch.org/short-reads/2022/11/01/turnout-in-u-s-has-soared-in-recent-elections-but-by-some-measures-still-trails-that-of-many-other-countries/).

Desmond-Harris, Jenee. 2009. "Why Michelle's Hair Matters." *Time*, September 7, 55–57.

Deveny, Kathleen. 2008. "They're No Baby Einsteins." *Newsweek*, January 14, 61.

Dewan, Angela. 2022. "Scientists Slam Joe Rogan's Podcast Episode with Jordan Peterson as 'Absurd' and 'Dangerous'." *CNN*, January 28. Accessed January 29, 2023 (https://www.cnn.com/2022/01/27/us/joe-rogan-jordan-peterson-climate-science-intl/index.html).

Dewan, Shaila, and Robert Gebeloff. 2012. "Among the Wealthiest One Percent, Many Variations." *New York Times*, January 15, A1.

Diamant, Jeff. 2023. "The Religious Composition of the 118th Congress." Pew Research Center, January 3. Accessed June 14, 2023 (https://www.pewresearch.org/religion/2023/01/03/faith-on-the-hill-2023/).

Diamant, Jeremy, and Besheer Mohamed. 2023. "What the Data Says about Abortion in the U.S." Pew Research Center, January 11. Accessed April 21, 2023 (https://www.pewresearch.org/fact-tank/2023/01/11/what-the-data-says-about-abortion-in-the-u-s-2/).

Diamond, Jeremy. 2017. "Trump Falsely Claims US Murder Rate Is 'Highest' in 47 Years." CNN, February 7. Accessed April 15, 2017 (cnn.com).

Dias, Elizabeth, and Eliana Dockterman. 2016. "The Fine Print." *Time*, October 31, 32–35.

Dickey, Jack. 2015. "Save Our Vacation." *Time*, June 1, 44–49.

Dickler, Jessica. 2022. "67% of Pandemic 'Boomerang Kids' Are Still Living with Mom and Dad." CNBC News, September 6. Accessed May 19, 2023 (https://www.cnbc.com/2022/09/06/many-pandemic-boomerang-kids-still-live-with-mom-and-dad.html).

Dickman, Samuel L., David U. Himmelstein, and Steffie Woolhandler. 2017. "Inequality and the Health-Care System in the USA." *The Lancet* 389 (April 8): 1431–1441.

"Dietary Inequality." 2016. *The Economist*, August 13, 19

Dietrich, Sandy, and Erik Hernandez. 2022. "Language Use in the United States: 2019." U.S. Census Bureau, September 1. Accessed May 3, 2023 (https://www.census.gov/library/publications/2022/acs/acs-50.html).

Dilberti, Melissa Kay, and Health L. Schwartz, 2023. "Educator Turnover Has Markedly Increased, but Districts Have Taken Actions to Boost Teacher Ranks: Selected Findings from the Sixth American School District Panel Survey." Rand. Accessed June 14, 2023 (https://www.rand.org/content/dam/rand/pubs/research_reports/RRA900/RRA956-14/RAND_RRA956-14.pdf).

Dilworth-Anderson, Peggye, Linda M. Burton, and William L. Turner. 1993. "The Importance of Values in the Study of Culturally Diverse Families." *Family Relations* 42 (July): 238–242.

Dimmick, Brian, and Aaron Madrid Aksoz. 2022. "Voting with a Disability: Breaking Down Barriers to the Ballot." American Civil Liberties Union, October 26. Accessed May 12, 2023 (https://www.aclu.org/news/voting-rights/voting-with-a-disability).

Ding, Ming, et al. 2015. "Association of Coffee Consumption with Total and Cause-Specific Mortality in 3 Large Prospective Cohorts." *Circulation* 132 (November): 2305–2315.

Dhingra, Pahwan. 2022. "How Indian American spelling bee dominance may fuel educational inequities." The Conversation, June 3. Accessed May 3, 2023 (https://theconversation.com/how-indian-american-spelling-bee-dominance-may-fuel-educational-inequities-184376).

Dirks, Sandhya. 2022. "Stories About Crime Are Rife with Misinformation and Racism, Critics Say." National Public Radio, August 8. Accessed January 29, 2023 (https://www.npr.org/2022/11/08/1134550280/stories-about-crime-are-rife-with-misinformation-and-racism-critics-say).

DISA. 2023. "Marijuana Legality by State." Global Solutions, Inc. Accessed February 11, 2023 (https://disa.com/maps/marijuana-legality-by-state).

DiSalvo, David. 2010. "Are Social Networks Messing with Your Head?" *Scientific American Mind* 20 (January/February): 48–55.

Ditmar, Kelly. 2017. "Finding Gender in Election 2016: Lessons from Presidential Gender Watch." Barbara Lee Foundation. Accessed June 26, 2017 (barbaraleefoundation.org).

"DNA Reveals White Supremacist's Black Heritage." 2013. *Baltimore Sun*, November 14, 8.

Dobis, Elizabeth A, Thomas Krumel, John Cromartie, Kelsey L. Conley, Austin Sanders, and Ruben Ortiz. 2021. "Rural America at a Glance: 2021 Edition." Economic Information Bulletin No. (EIB-230) November. Accessed June 27, 2023 (https://www.ers.usda.gov/publications/pub-details/?pubid=102575).

Dockterman, Eliana. 2013. "The Digital Parent Trap. Should Your Kids Avoid Tech—or Embrace It?" *Time*, August 19, 54.

Dockterman, Eliana. 2016. "A Barbie for Every Body." *Time*, February 8, 43–51.

Dockterman, Eliana. 2022. "How 2022 Become the Year of Gossip." *Time Magazine*, December 13. Accessed June 18, 2023 (https://time.com/6240467/year-of-gossip-2022/).

Doeringer, Peter B., and Michael J. Piore. 1971. *Internal Labor Markets and Manpower Analysis*. Lexington, MA: Heath-Lexington Books.

"Dog-lovers v. Dog-eaters: Pet Food." 2015. *The Economist*, June 20, 46.

Doherty, Carroll. 2015. "Continue Bipartisan Support for Expanded Background Checks on Gun Sales." Pew Research Center, August 13. Accessed August 23, 2015 (pewresearch.org).

Doherty, Carroll, and Vianny Gomez. 2022 "Inflation Is the Top Problem Facing the Country, Americans Say." Pew Research Center, May 12. Accessed June 13, 2023 (https://www.pewresearch.org/short-reads/2022/05/12/by-a-wide-margin-americans-view-inflation-as-the-top-problem-facing-the-country-today/).

Doherty, Carroll, and Alec Tyson. 2015. "Most Say Government Policies Since Recession Have Done Little to Help Middle Class, Poor." Pew Research Center, March 4. Accessed August 26, 2015 (pewresearch.org).

Doherty, Carroll, Juliana M. Horowitz, and Michael Dimock. 2014. "Most See Inequality Growing, but Partisans Differ over Solutions." Pew Research Center, January 23. Accessed April 12, 2014 (www.pewresearch.org).

Dolan, Eric W. 2018. "Study Provides New Insights on Why College Students Engage in Hookups." PsyPost, November 8. Accessed April 21, 2023 (https://www.psypost.org/2018/11/study-provides-new-insights-on-why-college-students-engage-in-hookups-52541).

Dolnick, Sam. 2011. "Dance, Laugh, Drink, Save the Date: It's a Ghanaian Funeral." *New York Times*, April 12, A1.

Domhoff, G. William. 2013. "Interlocking Directorates in the Corporate Community." Who Rules America, October. Accessed April 12, 2014 (whorulesamerica.net).

Dominguez, Kenneth, et al. 2015. "Vital Signs: Leading Causes of Death, Prevalence of Diseases and Risk Factors, and Use of Health Services Among Hispanics in the United States—2009–2013." *MMWR* 64 (17): 469–478.

Donadio, Rachel. 2013. "When Italians Chat, Hands and Fingers Do the Talking." *New York Times*, July 1, A6.

Donegan, Moira. 2023. "American Teenage Girls Are Experiencing High Levels of Emotional Distress. Why?" *The Guardian*, Feb 16. Accessed February 17, 2023 (https://www.theguardian.com/commentisfree/2023/feb/16/teenage-girls-mental-health-survey).

Dorell, Oren. 2017. "Who Will Rebuild Puerto Rico as Young Professionals Leave Island After Hurricane Maria?" *USA Today*, October 12. Accessed October 31, 2017 (usatoday.com).

Douglas, Susan J., and Meredith W. Michaels. 2004. *The Mommy Myth: The Idealization of Motherhood and How It Has Undermined Women*. New York: Free Press.

Dovidio, John F. 2009. "Racial Bias, Unspoken but Heard." *Science* 326 (December 18): 1641–1642.

Dow, Dawn Marie. 2016. "The Deadly Challenges of Raising African American Boys: Navigating the Controlling Image of the 'Thug'." *Gender & Society* 30 (April): 161–188.

Downs, Edward, and Stacy L. Smith. 2010. "Keeping Abreast of Hypersexuality: A Video Game Character Content Analysis." *Sex Roles* 62 (June): 721–733.

Doyle, Francis X. 2011. "Bishops' Missed Opportunity." *Baltimore Sun*, November 14, 15.

Dreid, Nadia. 2016. "Meet the Professor Who's Calling Out Clickbait." *Chronicle of Higher Education*, December 2, A24.

Dreweke, Joerg. 2017. "U.S. Abortion Rate Reaches Record Low Amidst Looming Onslaught Against Reproductive Health and Rights." *Guttmacher Policy Review* 20 (January): 15–19.

Drexler, Peggy. 2012. "The New Face of Infidelity." *Wall Street Journal*, October 21, C3.

Drexler, Peggy. 2013. "The Delicate Protocol of Hugging." *Wall Street Journal*, September 14–15, C3.

Drexler, Peggy. 2014. "Why We Love to Gossip." *Huffington Post*, November 14. Accessed October 25, 2015 (huffingtonpost.com).

"Driven to Distraction." 2017. *The Economist*, February 11, 11.

Dropp, Kyle, and Brendan Nyhan. 2017. "Many Don't Know Obamacare and Affordable Care Act Are the Same." *New York Times*, February 8, A10.

Drucker, Peter F. 2001. *Management Challenges for the 21st Century*. New York: HarperBusiness.

"The Drugs Don't Work." 2014. *The Economist*, May 3, 54.

Druker, Steven M. 2015. *Altered Genes, Twisted Truth*. Salt Lake City, UT: Clear River Press.

Drutman, Lee, and Alexander Furnas. 2014. "K Street Pays Top Dollar for Revolving Door Talent." Sunlight Foundation, January 21. Accessed April 20, 2014 (sunlightfoundation.com).

Du Bois, W. E. B. 1986. *The Souls of Black Folk*. New York: Library of America.

Ducharme, Jamie. 2019. "Your Summer Beach Trip Could Put You at Risk of Getting Sick. Here's What You Need to Know." *Time Magazine*, July 24. Accessed August 2, 2023 (https://time.com/5631608/swimming-illness-risks/).

Due, Linnea. 1995. *Joining the Tribe: Growing Up Gay & Lesbian in the '90s*. New York: Doubleday.

Dugan, Andrew. 2017. "US Divorce Rate Dips, but Moral Acceptability Hits New High." Gallup, July 7. Accessed July 20, 2017 (gallup.com).

Dugan, Andrew, and Brad Hoffman. 2014. "Record Low Say Own Representative Deserves Re-Election." Gallup, January 24. Accessed April 20, 2014 (www.gallup.com).

Duhigg, Charles. 2010. "Saving U.S. Water and Sewer Systems Would Be Costly." *New York Times*, March 14, A1.

Dumaine, Brian. 2015. "Can a [Billionaire] Hedge Fund Manager Fix Income Inequality?" *Fortune*, September 1, 87–91.

Dunifon, Rachel, and Lori Kowaleski-Jones. 2007. "The Influence of Grandparents in Single-Mother Families." *Journal of Marriage and Family* 69 (May): 465–481.

Durden, Tyler. 2011. "Here Are the 29 Public Companies with More Cash Than the US Treasury." Zero Hedge, July 15. Accessed January 9, 2012 (www.zerohedge.com).

Durkheim, Émile. 1893/1964. *The Division of Labor in Society*. New York: Free Press.

Durkheim, Émile. 1897/1951. *Suicide: A Study in Sociology*, translated by John A. Spaulding and George Simpson, edited by George Simpson. New York: Free Press.

Durkheim, Émile. 1898/1956. *Education and Sociology*, translated by Sherwood D. Fox. Glencoe, IL: Free Press.

Durkheim, Émile. 1961. *The Elementary Forms of the Religious Life*. New York: Collier Books.

Dunleavy. Brian P. 2020. "CDC: 40% of U.S. Teens Are Sexually Active." United Press International, May 6. Accessed April 21, 2023 (https://www.upi.com

/Health_News/2020/05/06/CDC-40-of-US-teens-are-sexually-active/8811588709258/).

Dye, Thomas R., and Harmon Ziegler. 2003. *The Irony of Democracy: An Uncommon Introduction to American Politics*, 12th edition. Belmont, CA: Wadsworth.

E

Eagan, Kevin, Jennifer B. Lozano, Sylvia Hurtado, and Matthew H. Case. 2014. *The American Freshman: National Norms Fall 2013*. Los Angeles: Higher Education Research Institute, UCLA.

Earl, Jennifer, Thomas V. Maher, and Jennifer Pan. 2022. "The Digital Repression of Social Movements, Protest, and Activism: A Synthetic Review." Science Advances, March 9. Accessed Jaune 25, 2023 (https://www.science.org/doi/10.1126/sciadv.abl8198).

Eavis, Peter. 2015. "Judge's Ruling Against 2 Banks Points to Wide Fraud in '08 Crash." *New York Times*, May 12, A1, B2.

Eberstadt, Nicholas. 2009. "Poor Statistics." *Forbes*, March 2, 26.

Eberstadt, Nicholas. 2013. "China's Coming One-Child Crisis." *Wall Street Journal*, November 27, A15.

Eberstadt, Nicholas. 2016. *Men Without Work: America's Invisible Crisis*. West Conshocken, PA: Templeton Press.

Eckenrode, John, Elliott G. Smith, Margaret E. McCarthy, and Michael Dineen. 2014. "Income Inequality and Child Maltreatment in the United States." *Pediatrics* 33 (March): 454–461.

Eckhoff, Vickery. 2015. "Exposing America's Billionaire Welfare Ranchers." AlterNet, March 24. Accessed August 26, 2015 (alternet.org).

Eckholm, Erik. 2014. "Aid in Dying' Movement Takes Hold in Some States." *New York Times*, February 8, A1.

Eckstein, Rick, Rebecca Schoenike, and Kevin Delaney. 1995. "The Voice of Sociology: Obstacles to Teaching and Learning the Sociological Imagination." *Teaching Sociology* 23 (October): 353–363.

Economic Commission for Latin America and the Caribbean, 2022. "ECLAC: At Least 4,473 Women Were Victims of Femicide in Latin America and the Caribbean in 2021." November 24. Accessed February 12, 2023 (https://www.cepal.org/en/pressreleases/eclac-least-4473-women-were-victims-femicide-latin-america-and-caribbean-2021#:~:text=The%20highest%20femicide%20rates%20in,cases%20for%20every%20100%2C000%20women).

"Economic News Release." 2017. U.S. Bureau of Labor Statistics, June 2. Accessed June 5, 2017 (bls.gov).

Economic Research Service, 2019. "Data Show U.S. Poverty Rates in 2019 Higher in Rural Areas Than In Urban For Racial/Ethnic Groups." Accessed June 27, 2023 (https://www.ers.usda.gov/data-products/chart-gallery/gallery/chart-detail/?chartId=101903#:~:text=Across%20all%20races%20and%20ethnicities,urban)%20areas%20at%2011.9%20percent).

The Economist. 2019. "American Millennials Think They Will Be Rich." April 22. Accessed April 13, 2023 (https://www.economist.com/graphic-detail/2019/04/22/american-millennials-think-they-will-be-rich).

Edin, Kathryn, and Timothy J. Nelson. 2013. *Doing the Best I Can: Fatherhood in the Inner City*. Berkeley and Los Angeles: University of California Press.

eDiplomat. 2016. "Cultural Etiquette Around the World." Accessed February 14, 2017 (ediplomat.com).

"Education." 2016. *The Economist*, December 10, 59–61.

Education Trust. 2023. "Completion and Success." Accessed June 13, 2023 (https://edtrust.org/issue/completion-success/).

EEOC. 2017. "Age Discrimination and Outdated Views of Older Worker Persist, Experts Tell Commission." U.S. Equal Employment Opportunity Commission, June 14. Accessed July 10, 2017 (eeoc.gov).

Efrati, Amir. 2013. "Google's Data-Trove Dance." *Wall Street Journal*, July 31, B1, B4.

Egan, Mark L., Gregor Matvos, and Amit Seru. 2017. "When Harry Fired Sally: The Double Standard in Punishing Misconduct." National Bureau of Economic Research, March. Accessed March 13, 2017 (nber.org).

Eggebeen, David J. 2005. "Cohabitation and Exchanges of Support." *Social Forces* 83 (May): 1097–1110.

Egley, Arlen, and James C. Howell. 2013. "Highlights of the 2011 National Youth Gang Survey." Office of Juvenile Justice and Delinquency Prevention, September. Accessed January 12, 2014 (www.ojjdp.gov).

Ehrlich, Paul. 1971. *The Population Bomb*, 2nd edition. San Francisco: Freeman.

Ehrlich, Paul R., and Anne H. Ehrlich. 2008. *The Dominant Animal: Human Evolution and the Environment*. Washington, DC: Island Press.

Eickmeyer, Kasey J. Eickmeyer, Krista K. Payne, Susan L. Brown, and Wendy D. Manning. 2017. "Crossover in the Median Age at First Marriage and First Birth: Thirty-Five Years of Change." National Center for Family and Marriage Research (NCFMR), Bowling Green State University, November 22. Accessed February 11, 2023 (https://www.bgsu.edu/ncfmr/resources/data/family-profiles/eickmeyer-payne-brown-manning-crossover-age-first-marriage-birth-fp-17-22.html).

Eilperin, Juliet, and Lenny Bernstein. 2014. "Environmental Groups Say Obama Needs to Address Climate Change More Aggressively." *Washington Post*, January 16. Accessed September 19, 2014 (www.washingtonpost.com).

Einhorn, Erin. 2022. "Hundreds of Colleges Dropped the SAT and ACT. Here's What Happened." NBC News, April 10. Accessed June 13, 2023 (https://www.nbcnews.com/news/us-news/college-admissions-test-sat-act-rcna23574).

Eisenberg, Abne M., and Ralph R. Smith, Jr. 1971. *Nonverbal Communication*. Indianapolis: Bobbs-Merrill.

Eisler, Peter. 2013. "What Surgeons Leave Behind Costs Some Patients Dearly." *USA Today*, March 8. Accessed July 26, 2014 (www.usatoday.com).

Ekman, Paul, and Wallace V. Friesen. 1984. *Unmasking the Face: A Guide to Recognizing Emotions from Facial Clues*. Palo Alto, CA: Consulting Psychologists Press.

Ekins, Emily, and David Kemp. 2021. "Poll: 72% of Americans Say Immigrants Come to the United States for Jobs and to Improve Their Lives." CATO Institute, April 27. Accessed April 26, 2023 (https://www.cato.org/blog/poll-72-americans-say-immigrants-come-us-jobs-improve-their-lives-53-say-ability-immigrate).

Elgin, Ben. 2014. "An Oil Giant Dims the Lights on Green Power." *Bloomberg Businessweek*, June 2–8, 21–22.

Eligon, John, and Robert Gebeloff. 2016. "Segregation, the Neighbor That Won't Leave." *New York Times*, August 21, A1.

Elinson, Zusha. 2021. "Women Are Nearly Half of New Gun Buyers, Study Finds." *The Wall Street Journal*, September 16. Accessed June 28, 2023 (https://www.wsj.com/articles/women-are-nearly-half-of-new-gun-buyers-study-finds-11631792761).

Eliot, Lise. 2012. *Pink Brain, Blue Brain: How Small Differences Grow into Troublesome Gaps—And What We Can Do About It*. Oxford, England: Oneworld Publications.

Ellis, Bill. 2005. "Legend/AntiLegend: Humor as an Integral Part of the Contemporary Legend Process." Pp. 123–140 in *Rumor Mills: The Social Impact of Rumor and Legend*, edited by Gary Alan Fine, Veronique Campion-Vincent, and Chip Heath. New Brunswick, NJ: Transaction Publishers.

Ellis, Renee. 2013. "Changes in Coresidence of Grandparents and Grandchildren." *Family Focus* (Summer): F13, F15, F17.

Ellis, Renee R., and Tavia Simmons. 2014. "Coresident Grandparents and Their Grandchildren: 2012." *U.S. Census Bureau*, October. Accessed October 25, 2014 (www.census.gov).

Embury-Dennis, Tom. 2017. "US States Consider Laws Allowing Creationism to be Taught by Science Teachers." *Independent*, March 16. Accessed July 31, 2017 (independent.co.uk).

Emery, David. 2015. "Does AriZona Tea Contain Human Urine?" Urban Legends, July 25. Accessed October 25, 2015 (urbanlegends.about.com).

Emmert, Ashley. 2015. "The State of Female Pastors." *Christianity Today*, October 15. Accessed July 31, 2017 (christianitytoday.com).

Emmons, William R., and Bryan J. Noeth. 2015. "Race, Ethnicity, and Wealth." Federal Reserve Bank of St. Louis, February. Accessed September 16, 2015 (stlouisfed.org).

"Employer-Reported Workplace Injuries and Illnesses—2015." 2016. Bureau of Labor Statistics News Release, October 27. Accessed August 15, 2017 (bls.gov).

"Employment Projections." 2017. U.S. Bureau of Labor Statistics, April 14. Accessed June 5, 2017 (bls.gov).

"Ending the Shame." 2013. *The Economist*, September 14, 16, 18.

England, Paula. 2005. "Emerging Theories of Care Work." *Annual Review of Sociology* 31 (August): 381–399.

Entmacher, Joan, Lauren Frohlich, Katherine G. Robbins, Emily Martin, and Liz Watson. 2014. "Underpaid & Overloaded: Women in Low-Wage Jobs." National Women's Law Center. Accessed November 23, 2014 (www.nwlc.org).

EPA. 2009. "EPA's Endangerment Findings." December 7. Accessed May 1, 2010 (www.epa.gov).

EPA. 2015. "Assessment of the Potential Impacts of Hydraulic Fracturing for Oil and Gas on Drinking Water Resources. Executive Summary." June. Accessed October 23, 2015 (epa.gov).

Epstein, Joseph. 2011. *Gossip: The Untrivial Pursuit*. New York: Houghton Mifflin Harcourt.

Erbentraut, Joseph. 2016. "If Farm-to-Table Eating Is Cool, What About Toilet-to-Tap Drinking?" *Huffington Post*, March 16. Accessed August 20, 2017 (huffingtonpost.com).

Erdman, Jonathan. 2023. "America's Top Weather Killer Is Not Tornadoes, Flooding, Lightning or Hurricanes – It's Heat." The Weather Channel, June 21. Accessed June 28. 2023 (https://weather.com/safety/heat/news/2021-06-03-heat-america-fatalities).

Erikson, Kai T. 1966. *Wayward Puritans: A Study in the Sociology of Deviance*. New York: John Wiley & Sons.

Ernst & Young. 2013. "Navigating Today's Complex Business Risks: Europe, Middle East, India and Africa Fraud Survey 2013." Accessed January 12, 2014 (www.ey.com).

Esipova, Neil, Anita Pagliese, and Julie Ray. 2013. "381 Million Adults Worldwide Migrate Within Countries." Gallup, May 15. Accessed September 19, 2014 (www.gallup.com).

Espelage, Dorothy L., Jun Sung Hong, Sarah Rinehart, and Namrata Doshi. 2016. "Understanding Types, Locations, & Perpetrators of Peer-to-Peer Sexual Harassment in U.S. Middle Schools: A Focus on Sex, Racial, and Grade Differences." *Children and Youth Services Review* 71 (December): 174–183.

Esping-Andersen, Gosta. 1990. *The Three Worlds of Welfare Capitalism*. Cambridge, United Kingdom: Polity Press.

ESPN. 2022. "Washington Finally Shed the Name, but 37 High Schools Haven't." November 2. Accessed May 3, 2022 (https://www.espn.com/nfl/story/_/id/34874030/washington-changed-37-high-schools-use-word).

Esterl, Mike. 2013. "The Natural Evolution of Food Labels." *Wall Street Journal*, November 6, B1–B2.

Etter, Lauren, and Shruti Singh. 2017. "Big Meat Braces for a Labor Shortage." *Bloomberg Businessweek*, February 13–19, 19–20.

Etzioni, Amitai. 1975. *A Comparative Analysis of Complex Organizations*. New York: Free Press.

European Parliament. 2023. "Understanding the Gender Pay Gap: Definition and Causes." *EU Monitor*, April 12. Accessed April 21, 2023 (https://www.eumonitor.eu/9353000/1/j9vvik7m1c3gyxp/vl59mpupm0vq?ctx=vk4jic6t1dxz).

Evans, Becky. 2013. "Bill Gates 'Disrespects' South Korea's Female President by Shaking Hands While Keeping Other in His Pocket." MailOnline, April 23. Accessed November 10, 2013 (dailymail.co.uk).

Evans, Harold, Gail Buckland, and David Lefer. 2006. *They Made America: From the Steam Engine to the Search Engine: Two Centuries of Innovators*. New York: Little, Brown.

Evans, Simon. 2021. "Analysis: Which Countries Are Historically Responsible for Climate Change? - Carbon Brief." Carbon Brief, May 10. Accessed June 28, 2023 (https://www.carbonbrief.org/analysis-which-countries-are-historically-responsible-for-climate-change/).

Eversley, Melanie. 2017. "Report: Anti-Muslim Groups Triple in U.S. Amid Trump Hate Rhetoric." *USA Today*, February 15. Accessed February 20, 2017 (usatoday.com).

Evich, Helena Bottemiller. 2016. The Right Takes on School Lunch." Politico, May 18. Accessed November 17, 2023 (https://www.npr.org/sections/health-shots/2022/08/31/1120004717/the-u-s-diet-is-deadly-here-are-7-ideas-to-get-americans-eating-healthier).

Ewen, Lynda Ann. 1998. *Social Stratification and Power in America: A View from Below*. Six Hills, NY: General Hall.

Ewert, Stephanie, and Robert Kominski. 2014. "Measuring Alternative Educational Credentials: 2012." U.S. Census Bureau, January. Accessed June 21, 2014 (www.census.gov).

Ewing, Walter A., Daniel E. Martínez, and Rubén G. Rumbaut. 2015. "The Criminalization of Immigration in the United States." American Immigration Council, July. Accessed June 10, 2017 (americanimmigrationcouncil.org).

Executive Office of the President. 2017. "Promoting Free Speech and Religious Liberty." Federal Register, May 9. Accessed June 13, 2023 (https://www.federalregister.gov/documents/2017/05/09/2017-09574/promoting-free-speech-and-religious-liberty).

Expedia. 2017. "In-Flight Turbulence: Expedia.com Airplane Etiquette Study Shows Seat-Kicking Edges Bad Parenting as Most Aggravating Behavior." January 17. Accessed March 13, 2017 (viewfinder.expedia.com).

Expedia. 2023. "Vacation Deprivation: Report 2023." April 27. Accessed May 12, 2023 (https://www.expedia.com/stories/wp-content/uploads/2023/04/Expedia-Vacation-Deprivation-Global-Report-2023_Final-US-Small-compressed-1.pdf).

F

Facing History and Ourselves. 2021. "How the Parkland Students Pulled off a Massive National Protest in Only 5 Weeks." January 11. Accessed June 27, 2023 (https://www.facinghistory.org/resource-library/how-parkland-students-pulled-massive-national-protest-only-5-weeks).

Fahrenthold, David A. 2009. "Environmentalists Seek to Wipe Out Plush Toilet Paper." *Washington Post*, September 24, A1.

Fakhraie, Fatemeh. 2009. "Feminists Don't Understand Muslim Women." Double X, May 20. Accessed April 26, 2010 (www.doublex.com).

Falk, Gene, and Patrick A. Landers 2023. "The Temporary Assistance for Needy Families (TANF) Block Grant: Responses to Frequently Asked Questions."

Congressional Research Service, February 27. Accessed April 13, 2023 (https://sgp.fas.org/crs/misc/RL32760.pdf).

Fam, Mariam, and Aysha Khan. 2021. "Women Seek Diverse Paths to Leadership in Islamic Spaces." *AP News*, December 8. Accessed June 13, 2023 (https://apnews.com/article/africa-religion-islam-14cfe5f40ffa234a7e783dc9a53b98ea).

"Families and Living Arrangements." 2017. U.S. Census Bureau. Accessed July 4, 2017 (census.gov).

"Family Planning: One-Child Proclivity." 2014. *The Economist*, July 19, 40.

Famous Black Inventors. 2019. "Dr. Shirley Ann Jackson." Accessed February 12, 2023 (https://www.black-inventor.com/dr-shirley-jackson#:~:text=at%20Bell%20Laboratories.-,Dr.,caller%20ID%20and%20call%20waiting).

Fang, Ferric C., R. Grant Steen, and Arturo Casadevall. 2012. "Misconduct Accounts for the Majority of Retracted Scientific Publications." *Proceedings of the National Academy of Sciences* 109 (42): 17028–17033.

Fantz, Ashley, Steve Almasy, and AnneClaire Stapleton. 2015. "Muslim Teen Ahmed Mohamed Creates Clock, Shows Teachers, Gets Arrested." CNN, September 16. Accessed April 28, 2023 (https://www.cnn.com/2015/09/16/us/texas-student-ahmed-muslim-clock-bomb/index.html).

FAO, IFAD and WFP. 2015. "The State of Food Insecurity in the World. Meeting the 2015 International Hunger Targets: Taking Stock of Uneven Progress." Accessed May 13, 2017 (fao.org).

Farley, Sally D., Amie M. Ashcraft, Mark F. Stasson, and Rebecca L. Nusbaum. 2010. "Nonverbal Reactions to Conversational Interruption: A Test of Complementarity Theory and the Status/Gender Parallel." *Journal of Nonverbal Behavior* 34 (December): 193–206.

"Father's Day: June 18, 2017." 2017. U.S. Census Bureau Newsroom, June 8. Accessed June 20, 2017 (census.gov).

Feagin, Joe R. 2001. "Social Justice and Sociology: Agendas for the Twenty-First Century." *American Sociological Review* 66 (February): 1–20.

Feagin, Joe R., and Clairece Booher Feagin. 2008. *Racial and Ethnic Relations*, 8th ed. Upper Saddle River, NJ: Prentice Hall.

Featherstone, Liza. 2016. "Feminists Misunderstood the Presidential Election from Day One." *The Guardian*, November 15. Accessed January 5, 2017 (theguardian.com).

Feder, Sandra. 2020. "Stanford Professor's Study Finds Gentrification Disproportionately Affects Minorities." Stanford News, December 1. Accessed June 27, 2023 (https://news.stanford.edu/2020/12/01/gentrification-disproportionately-affects-minorities/).

Federal Bureau of Investigation. 2014. "*Crime in the United States, 2013*". Fall. Accessed September 2, 2015 (fbi.gov).

Federal Bureau of Investigation. 2016. "2015 Internet Crime Report." Accessed March 13, 2017 (ic3.gov).

Federal Bureau of Investigation. 2017. "Transnational Organized Crime." Accessed April 2, 2017 (fbi.gov).

Federal Bureau of Investigation. 2020. Crime in the United States, 2019." Accessed March 25, 2023 (https://ucr.fbi.gov/crime-in-the-u.s/2019/crime-in-the-u.s.-2019/topic-pages/persons-arrested).

Federal Bureau of Investigation. 2021. "Internet Crime Report: 2020." March 17. Accessed March 25, 2023 (https://www.ic3.gov/Media/PDF/AnnualReport/2020_IC3Report.pdf).

Federal Election Commission. 2021. "Statistical Summary of 24-Month Campaign Activity of the 2019-2020 Election Cycle." April 2. Accessed May 12, 2023 (https://www.fec.gov/updates/statistical-summary-24-month-campaign-activity-2019-2020-election-cycle/).

Federal Election Commission. 2023. "Statistical Summary of 21-Month Campaign Activity of the 2021-2022 Election Cycle." January 5. Accessed May 12, 2023 (https://www.fec.gov/updates/statistical-summary-of-21-month-campaign-activity-of-the-2021-2022-election-cycle/#:~:text=Based%20on%20reports%20filed%20with,on%2Dhand%20of%20%-241.8%20billion).

Federal Interagency Forum on Aging-Related Statistics. 2016. "Older Americans: 2016 Key Indicators of Well-Being." Accessed July 20, 2017 (agingstats.gov).

Federal Interagency Forum on Child and Family Statistics. 2017. "America's Children: Key National Indicators of Well-Being, 2017." Accessed August 15, 2017 (childstats.gov).

The Feed. 2021. "AP Courses Falling Short as a Tool for Narrowing Equity Gaps." Georgetown University, September 10. Accessed June 13, 2023 (https://feed.georgetown.edu/access-affordability/ap-courses-falling-short-as-a-tool-for-narrowing-equity-gaps/#:~:text=Around%2070%20percent%20of%20the,from%20just%2055%2C442%20in%201970).

Feeding America. 2023. "Food Waste and Food Rescue." Accessed June 28, 2023 (https://www.feedingamerica.org/our-work/reduce-food-waste).

Feeding America. 2023. "How We Fight Food Waste in the U.S." Accessed February 11, 2023 (https://www.feedingamerica.org/our-work/reduce-food-waste#:~:text=Each%20year%2C%20119%20billion%20pounds,food%20in%20America%20is%20wasted).

Feingold, Spencer. 2022. "The Four Key Ways Disinformation is Spread Online." World Economic Forum, Aug 9. Accessed January 29, 2023 (https://www.weforum.org/agenda/2022/08/four-ways-disinformation-campaigns-are-propagated-online/).

Felbin, Sarah, Sabrina Talbert, and Addison Aloian. 2023. "The 'Blackout Challenge' Has Resurfaced on TikTok, And It's Still Just as Dangerous as It Was 17 Years Ago." *Women's Health*, January 17. Accessed June 25, 2023 (https://www.womenshealthmag.com/health/a38603617/blackout-challenge-tik-tok-2021/).

Feldman, Jamie. 2017. "Michelle Obama Reveals Barack 'Wore That Same Tux' for Years." *Huffington Post*, June 7. Accessed June 7, 2017 (huffingtonpost.com).

Fell, James. 2017. "'Get the Hell Out of My Race': The First Woman To Run The Boston Marathon On How She's Silenced Critics Over The Years." *Women's Health*, April 17. Accessed April 19, 2023 (https://www.womenshealthmag.com/fitness/a19972882/kathrine-switzer-boston-marathon/).

"Female Genital Cutting." 2016. *The Economist*, June 18, 63–64.

Feng, John. 2021. "China Has Nearly 35 million More Single Men Than Women." *Newsweek*, May 18. Accessed June 27, 2023 (https://www.newsweek.com/china-has-nearly-35-million-more-single-men-women-1592486).

Ferdman, Roberto A. 2015. "France Is Making It Illegal for Supermarkets to Throw Away Edible Food." *Washington Post*, May 22. Accessed July 26, 2015 (washingtonpost.com).

Ferguson, Christopher J., and Cheryl K. Olson. 2013. "Video Game Violence Use Among 'Vulnerable' Populations: The Impact of Violent Games on Delinquency and Bullying Among Children with Clinically Elevated Depression or Attention Deficit Symptoms." *Journal of Youth and Adolescence* 42 (August), published online.

Ferguson, Emily. 2014. "11 Reasons Christianity Needs Feminism." *Huffington Post*, May 21. Accessed June 21, 2014 (www.huffingtonpost.com).

Fernald, Anne, Virginia A. Marchman, and Adriana Weisleder. 2013. "SES Differences in Language Processing Skill and Vocabulary are Evident at 18 Months." *Developmental Science* 16 (March): 234–248.

Ferree, Myra M. 2005. "It's Time to Mainstream Research on Gender." *Chronicle Review*, August 12, B10.

File, Thom, and Camille Ryan. 2014. "Computer and Internet Use in the United States: 2013." U.S. Census Bureau, American Community Survey Reports, November. Accessed August 22, 2015 (census.gov).

File, Thom, and Sarah Crissey. 2010. "Voting and Registration in the Election of November 2008: Population Characteristics." U.S. Census Bureau, Current Population Reports. Accessed May 15, 2010 (www.census.gov).

Filipovic, Jill. 2016. "Barbie's Problem Is Far Beyond Skin-Deep." *Time*, February 8, 51.

Filipovic, Jill. 2017. "What an All-Male Photo Op Really Says." *New York Times*, March 29, D2.

"Financial Crime." 2015. *The Economist*, May 23, 61.

Finelli, Renata, Filomena Mottola, and Ashok Agarwal. 2022. "Impact of Alcohol Consumption on Male Fertility Potential: A Narrative Review." *International Journal of Environmental Research on Public Health*. 2022 Jan; 19(1): 328. doi: 10.3390/ijerph19010328.

Finer, Lawrence B., and Mia R. Zolna. 2016. "Declines in Unintended Pregnancy in the United States, 2008–2011." *New England Journal of Medicine* 374 (March 3): 843–852.

Fingerhut, Hannah. 2017. "About Seven-in-Ten Americans Oppose Overturning *Roe v. Wade*." Pew Research Center, January 3. Accessed May 19, 2017 (pewresearch.org).

Finkelhor, David, Heather Turner, Sherry Hamby, and Richard Ormrod. 2011. "Polyvictimization: Children's Exposure to Multiple Types of Violence, Crime, and Abuse." Office of Justice Programs, October. Accessed March 7, 2012 (www.ojp.usdoj.gov).

Fischman, Josh. 2009. "Global Warming Before Smokestacks." *Chronicle Review*, November 6, B11–B12.

Fitch, Kimberly, and Sangeeta Agrawal. 2015. "Female Bosses Are More Engaging Than Male Bosses." Gallup, May 7. Accessed April 4, 2017 (gallup.com).

Flabbi, Luca, Mario Macis, Andrea Moro, and Fabiano Schivardi. 2016. "Do Female Executives Make a Difference? The Impact of Female Leadership on Gender Gaps and Firm Performance." National Bureau of Economic Research, December. Accessed April 10, 2017 (nber.org).

Flaherty, Colleen. 2020. "Faculty Members Struggle with Burnout." Inside Higher Ed, September 14. Accessed November 17, 2023 (https://www.insidehighered.com/news/2020/09/14/faculty-members-struggle-burnout).

Flasphaler, Julia. 2017. "Watch: Patagonia Ad Calls Out Corporate Takeover of Public Land." AlterNet, August 25. Accessed August 27, 2017 (alternet.org).

Flavin, Jeanne. 2001. "Feminism for the Mainstream Criminologist: An Invitation." *Journal of Criminal Justice* 29 (July/August): 271–285.

Fleegler, Eric W., et al. 2013. "Firearm Legislation and Firearm-Related Fatalities in the United State." *JAMA Internal Medicine*, March 6. Accessed January 15, 2014 (www.archinte.jamanetwork.com).

Fleming, Charles. 2014. "Ford Greenlights Recycling." *Baltimore Sun*, July 5 8.

Fleming, John. 2015. "Millennials Most Trusting on Safety of Personal Information." Gallup, May 11. Accessed August 14, 2015 (gallup.com).

Flesher, John, and Michael Biesecker. 2017. "EPA Moves to Withdraw Obama Clean-Water Rule." *Baltimore Sun*, June 28, 6.

Flint Water Advisory Task Force. 2016. "Final Report." Office of Governor Rick Snyder, State of Michigan, March. Accessed August 20, 2017 (michigan.gov).

Florida, Richard. 2013. "The Striking Relationship Between Gun Safety Laws and Firearm Deaths." City Lab, March 7. Accessed April 13, 2017 (citylab.com).

Fogg, Piper. 2005. "Don't Stand So Close to Me." *Chronicle of Higher Education*, April 29, A10–A12.

Fomby, Paula, and Angela Estacion. 2011. "Cohabitation and Children's Externalizing Behavior in Low-Income Latino Families." *Journal of Marriage and Family* 73 (February): 46–66.

Fomby, Paula, and Cynthia Osborne. 2017. "Family Instability, Multipartner Fertility, and Behavior in Middle Childhood." *Journal of Marriage and Family* 79 (February): 75–93.

Food and Water Watch. 2016. "The State of Public Water in the United States." February 16. Accessed August 20, 2017 (foodandwaterwatch.org).

"Food Companies Propose Cutting Back on Junk Food Marketing Aimed at Children." 2011. *Washington Post*, July 14. Accessed July 20, (www.washingtonpost.com).

Food Marketing Workgroup. 2007. "Food Marketing in Other Countries." Center for Science in the Public Interest, February 16. Accessed February 12, 2023 (https://www.foodmarketing.org/resources/food-marketing-101/food-marketing-in-other-countries/#:~:text=Marketing%20in%20schools,prohibit%20all%20in%2Dschool%20marketing).

Ford, Peter. 2011. "Why Japan Will Rebound." *Christian Science Monitor*, March 28, 16–29.

"Foreign Born." 2016. U.S. Census Bureau, July 6. Accessed June 20, 2017 (census.gov).

"Formaldehyde." 2012. Environmental Protection Agency. Accessed September 20, 2014 (www.epa.gov).

Foroohar, Rana. 2011. "Whatever Happened to Upward Mobility?" *Time*, November 14, 27–34.

Fraire, John. 2014. "Why Your College Should Dump the SAT." *Chronicle of Higher Education*, May 2, A44.

Frakt, Austin. 2017. "A Big Culprit of Increased Health Spending? Technology." *New York Times*, January 26, A3.

Frank, Robert. 2017. "Plenty of Billionaires, but Few Are Women." *New York Times*, January 1, BU3.

Frayer, Lauren. 2015. "In Spain, Entire Villages Ae Up for Sale—and They're Going Cheap." NPR, August 23. Accessed August 24, 2017 (npr.org).

Freedom House. 2023. "Global Freedom Status: 2023." Accessed May 12, 2023 (https://freedomhouse.org/explore-the-map?type=fiw&year=2023).

Freese, Jeremy. 2008. "Genetics and the Social Science Explanation of Individual Outcomes." *American Journal of Sociology* 114 (Suppl.): S1–S35.

Fremstad, Shawn. 2012. "The Poverty Rate Is Higher Than the Federal Government Says It Is." Center for Economic and Policy Research, September 21. Accessed February 6, 2014 (www.cepr.net).

Frey, William H. 2014. "New Projections Point to a Majority Nation in 2044." Brookings, December 12. Accessed October 4, 2015 (brookings.edu).

Frey, William H. 2014. "The Suburbs: Not Just for White People Anymore." *New Republic*, November 14. Accessed October 23, 2015 (newrepublic.com).

Frey, William H. 2014. "Will This Be the Decade of Big City Growth?" Brookings Institution, May 23. Accessed September 19, 2014 (brookings.edu).

Frey, William H. 2015. *Diversity Explosion: How New Racial Demographics Are Remaking America*. Washington, DC: Brookings Institution.

Frey, William H. 2016. "White Neighborhoods Get Modestly More Diverse, New Census Data Show." Brookings, December 13. Accessed August 20, 2017 (brookings.edu).

Frey, William H. 2017. "Census Shows Pervasive Decline in 2016 Minority Voter Turnout." Brookings, May 18. Accessed May 22, 2017 (brookings.edu).

Frey, William H. 2021. "Turnout in 2020 Election Spiked Among Both Democratic and Republican Voting Groups, New Census Data Show." Brookings, May 5. Accessed May 12, 2023 (https://www.brookings.edu/research/turnout-in-2020-spiked-among-both-democratic-and-republican-voting-groups-new-census-data-shows/0).

Frey, William H. 2022. "Today's Suburbs Are Symbolic of America's Rising Diversity: A 2020 Census Portrait." Brookings, June 15. Accessed June 27 (https://www.brookings.edu/articles/todays-suburbs-are-symbolic-of-americas-rising-diversity-a-2020-census-portrait/#:~:text=This%20rose%20to%2030%25%20in,(55.4%25%20versus%2057.8%25).

Frey, William H. 2022. "Big Cities Saw Historic Population Losses While Suburban Growth Declined During the Pandemic." Brookings, July 11. Accessed June 27, 2023 (https://www.brookings.edu/articles/big-cities-saw-historic-population-losses-while-suburban-growth-declined-during-the-pandemic/#:~:text=Among%20the%20nation's%2056%20major,(see%20-downloadable%20Table%20C).

Frey, William H. 2023 "American's Local Migration Reached A Historic Low in 2022, But Long-Distance Moves Picked Up," Brookings, February 2. Accessed June 27, 2023 (https://www.brookings.edu/articles/americans-local-migration-reached-a-historic-low-in-2022-but-long-distance-moves-picked-up/#:~:text=In%20 2021%2D22%2C%20the%20overall,20%25%20of%20Americans%20 moved%20annually).

Friedman, Howard S. 2012. "American Voter Turnout Lower Than Other Wealthy Countries." *Huffington Post*, July 10. Accessed April 20, 2014 (www.huffingtonpost.com).

Friedman, Jaclyn. 2011. *What You Really Really Want: The Smart Girl's Shame-Free Guide to Sex and Safety.* Berkeley, CA: Seal Press.

Friedrichs, David O. 2009. *Trusted Criminals: White Collar Crime in Contemporary Society,* 4th edition. Belmont, CA: Wadsworth Publishing.

Frohwirth, Lon, Ann M. Moore, and Renata Maniaci. 2013. "Perceptions of Susceptibility to Pregnancy Among Women Obtaining Abortions." *Social Science & Medicine* 99 (December): 18–26.

"From Alpha to Omega." 2015. *The Economist*, August 15, 61.

Fry, Richard. 2016. "For First Time in Modern Era, Living with Parents Edges Out Other Living Arrangements for 18- to 34-Year-Olds." Pew Research Center, May 24. Accessed March 1, 2017 (pewresearch.org).

Fry, Richard. 2017. "Women May Never Make up Half of the U.S. Workforce." Pew Research Center, September 12. Accessed January 31, 2017 (pewresearch.org).

Fry, Richard. 2020. "2023 The Pace of Boomer Retirements Has Accelerated in the Past Year." Pew Research Center, November 9. Accessed May 19, 2023 (https://www.pewresearch.org/short-reads/2020/11/09/the-pace-of-boomer-retirements-has-accelerated-in-the-past-year/).

Fry, Richard. 2022. "College Grads in U.S. Tend to Partner With Each Other – Especially If Their Parents Also Graduated from College." Pew Research Center, September 7. Accessed May 19, 2023 (https://www.pewresearch.org/short-reads/2022/09/07/college-grads-in-u-s-tend-to-partner-with-each-other-especially-if-their-parents-also-graduated-from-college/).

Fry, Richard, and Kim Parker. 2021. "Rising Share of U.S. Adults Are Living Without a Spouse or Partner." Pew Research Center, October 5. Accessed February 11, 2023 https://www.pewresearch.org/social-trends/2021/10/05/rising-share-of-u-s-adults-are-living-without-a-spouse-or-partner/).

Fry, Richard, Carolina Aragáo, Kiley Hurst, and Kim Parker. 2023. "In a Growing Share of U.S. Marriages, Husbands and Wives Earn About the Same." Pew Research Center, April 13. April 22, 2023 (https://www.pewresearch.org/social-trends/2023/04/13/in-a-growing-share-of-u-s-marriages-husbands-and-wives-earn-about-the-same/#:~:text=In%2029%25%20 of%20marriages%20today,16%25%20 have%20a%20breadwinner%20wife).

Fry, Richard, Brian Kennedy, and Cary Funk. 2021. "STEM Jobs See Uneven Progress in Increasing Gender, Racial and Ethnic Diversity." Pew Research Center, April 1. Accessed June 13, 2023 (https://www.pewresearch.org/science/2021/04/01/stem-jobs-see-uneven-progress-in-increasing-gender-racial-and-ethnic-diversity/).

Fry, Richard, and Jeffrey S. Passel. 2014. "In Post-Recession Era, Young Adults Drive Continuing Rise in Multi-Generational Living." Pew Research Center, July 17. Accessed July 25, 2014 (www.pewresearch.org).

Fry et al., 2023. Chapter 11, page 231.

Fryar, Cheryl D., Margaret D. Carroll, and Cynthia L. Ogden. 2016. "Prevalence of Overweight and Obesity Among Children and Adolescents Aged 2–19 Years: United States, 1963–1965 Through 2013–2014." National Center for Health Statistics, July. Accessed March 1, 2017 (cdc.gov/nchs).

Fryer, Roland G., Jr., Devah Pager, and Jorg L. Spenkuch. 2011. "Racial Disparities in Job Finding and Offered Wages." National Bureau of Economic Research, September. Accessed December 7, 2011 (www.nber.org).

Funk, Cary. 2021. "Key findings: How Americans' Attitudes about Climate Change Differ by Generation, Party and Other Factors." Pew Research Center, May 26. Accessed June 28, 2023 (https://www.pewresearch.org/short-reads/2021/05/26/key-findings-how-americans-attitudes-about-climate-change-differ-by-generation-party-and-other-factors/).

Funk, Cary, and Brian Kennedy. 2016. "The Politics of Climate." Pew Research Center, October 4. Accessed August 25, 2017 (pewresearch.org).

Funk, Cary, Brian Kennedy, and Elizabeth Sciupac. 2016. "U.S. Public Wary of Biomedical Technologies to 'Enhance' Human Abilities." Pew Research Center, July 26. Accessed September 2, 2017 (pewresearch.org).

Funk, Cary, and Mark Hugo Lopez. 2022. "A Brief Statistical Portrait of U.S. Hispanics." Pew Research Center, June 14. Accessed February 11, 2023 (https://www.pewresearch.org/science/2022/06/14/a-brief-statistical-portrait-of-u-s-hispanics/).

G

Gallup. 2021. "Evolution, Creationism, Intelligent Design." Accessed June 13, 2023 (https://news.gallup.com/poll/21814/evolution-creationism-intelligent-design.aspx). |

Gallup. 2023. "Immigration." Accessed April 26, 2023 (https://news.gallup.com/poll/1660/immigration.aspx#:~:text=Trend%20on%20Americans'%20views%20of,believe%20 they%20should%20be%20decreased).

Gallup. 2023. "Party Affiliation." Accessed May 12, 2023 (https:/news.gallup.com/poll/15370/party-affiliation.aspx).

Gallup. 2023 "Marriage." Accessed May 19, 2023 (https:/news.gallup.com/poll/117328/marriage.aspx).

Galston, William A. 2013. "The Eroding American Middle Class." *Wall Street Journal*, November 13, A15.

Gamson, William. 1999. *Popular Culture & High Culture: An Analysis and Evaluation of Taste.* Revised and updated edition. New York: Basic Books.

Ganem, Joseph. 2016. "Science in Self-Interest." *Baltimore Sun*, August 21, 25.

Gans, Herbert J. 1999. *Popular Culture and High Culture: An Analysis and Evaluation of Taste.* Revised and updated edition. New York: Basic Books.

Gans, Herbert J. 2005. "Wishes for the Discipline's Future." *Chronicle Review*, August 12, B9.

Garcia-Hallett, Janet, Kanika Samuels-Wortley, Tri Keah Henry, and Andrea S. Boyles. 2022, "Introduction to Special Issue: Reclaiming Our Stories: Centering the Voices, Experiences, and Expertise of Black, Indigenous, and Women of Color (BIWOC) on the Carceral State." Journal of Qualitative Criminal Justice and Criminology, 11 (3). Accessed March 25, 2023 (https://www.qualitativecriminology.com/pub/ebz3rnro/release/1#special-issue-content).

Gardner, Marilyn. 2008. "Happiness Is a Warm 'Thank You.'" *Christian Science Monitor,* January 28, 13, 16.

Gardner, Matthew, and Steve Wamhoff. 2021. "55 Corporations Paid $0 in Federal Taxes on 2020 Profits." Institute on Taxation and Economic Policy, April 2. Accessed April 13, 2023 (https://itep.org/55-profitable-corporations-zero-corporate-tax/).

Garfinkel, Harold. 1967. *Studies in Ethnomethodology.* Englewood Cliffs, NJ: Prentice-Hall.

Garza, Lisa Maria, and Timothy Williams. 2016. "Teenager in 'Affluenza' Case Is Sentenced to Jail." *New York Times*, April 4, A11.

Gates, Gary J. 2014. "LGB Families and Relationships: Analyses of the 2013 National Health Interview Survey."

Williams Institute, October. Accessed July 20, 2017 (law.ucla.edu/williamsinstitute).

Gates, Gary J. 2017. "In US, More Adults Identifying as LGBT." Gallup, January 11. Accessed May 20, 2017 (gallup.com).

Gavin, Lorrie, et al. 2013. "Vital Signs: Repeat Births among Teens—United States, 2007–2010." *MMWR* 62 (April 5): 249–255.

Gayer, Ted, Austin J. Drukker, and Alexander K. Gold. 2016. "Tax-Exempt Municipal Bonds and the Financial of Professional Sports Stadiums." Brookings, September. Accessed April 10, 2017 (brookings.edu).

Ge, Yanbo, Christopher R. Knittel, Don MacKenzie, and Stephen Zoepf. 2016. "Racial and Gender Discrimination in Transportation Network Companies." National Bureau of Economic Research, October. Accessed June 20, 2017 (nber.org).

Gecewicz, Claire. 2018. "'New Age' Beliefs Common Among Both Religious and Nonreligious Americans." Pew Research Center, October 1. Access July 30, 2023 (https://www.pewresearch.org/short-reads/2018/10/01/new-age-beliefs-common-among-both-religious-and-nonreligious-americans/#:~:text=Overall%2C%20roughly%20six%2Din%2D,%25)%20and%20astrology%20(29%25).

Gee, Lisa Christensen, Matthew Gardner, Misha E. Hill, and Meg Wiehe. 2017. "Undocumented Immigrants' State & Local Tax Contributions." Institute on Taxation and Economic Policy, March. Accessed June 10, 2017 (itep.org).

Geertz, Clifford. 1966. "Religion as a Cultural System." Pp. 1–46 in *Anthropological Approaches to the Study of Religion*, edited by Michael Banton. London: Tavistock.

Geier, Ben. 2014. "Using 3-D Printing to Make Jet Engines." *Fortune*, December 1, 76.

Geiger, Abigail. 2016. "16 Striking Findings from 2016." Pew Research Center, December 21. Accessed June 20, 2017 (pewresearch.org).

Geiger, Abigail. 2016. "Sharing Chores a Key to Good Marriage, Say Majority of Married Adults." Pew Research Center, November 30. Accessed May 19, 2017 (pewresearch.org).

Geiger, Abigail. 2016. "Support for Marijuana Legalization Continues to Rise." Pew Research Center, October 12. Accessed March 1, 2017 (pewresearch.org).

Geiger, Abigail, and Lauren Kent. 2017. "Number of Women Leaders Around the World Has Grown, but They're Still a Small Group." Pew Research Center, March 8. Accessed May 19, 2017 (pewresearch.org).

"Gender, Education and Work." 2015. *The Economist*, March 7, 61–62.

"Gender Equality Universally Embraced, but Inequalities Acknowledged." 2010. Pew Research Center, Global Attitudes Project, July 1. Accessed November 1, 2011 (www.pewglobal.org).

"The Gender Gap in Religion Around the World." 2016. Pew Research Center, March 22. Accessed July 31, 2017 (pewresearch.org).

Genocide Education Project. 2014. "Modern Era Genocides." Accessed April 28, 2023 (https://genocideeducation.org/resources/modern-era-genocides/).

Gershenson, Seth. 2015. "The Alarming Effect of Racial Mismatch on Teacher Expectations." Brookings, August 18. Accessed March 1, 2017 (brookings.edu).

Gerth, H. H., and C. Wright Mills, eds. 1946. *Max Weber: Essays in Sociology*. New York: Oxford University Press.

Geyer, Roland, Jenna R. Jambeck, and Kara Lavender Law. 2017. "Production, Use, and Fate of All Plastics Ever Made." *Science Advances* 3 (July 19): e1700782.

Gilbert, Dennis. 2011. *The American Class Structure in an Age of Growing Inequality*, 8th edition. Thousand Oaks, CA: Pine Forge Press.

Gillespie, Lane, and Karen Bennett. 2023. "Average Wedding Cost: Wedding Planning Tips For 2023." Bankrate, March 8. Accessed May 19, 2023 (https://www.bankrate.com/personal-finance/average-wedding-cost/).

Gillespie, Michael Allen. 2014. "Grade Degradation." *Chronicle Review*, January 24, B2.

Gillett, Francesca. 2021. "Covid-19: Five Ways to Make Hugging Safer, from the Experts." *BBC*, May 14. Accessed September 7, 2023 (https://www.bbc.com/news/uk-57083571).

Gilliam, Walter S., Angela N. Maupin, Chin R. Reyes, Maria Accavitti, and Frederick Shic. 2016. "Do Early Educators' Implicit Biases Regarding Sex and Race Relate to Behavior Expectations and Recommendations of Preschool Expulsions and Suspensions?" Yale Child Study Center, September 28. Accessed July 31, 2017 (ziglercenter.yale.edu).

Gilligan, Chris. 2022. "10 States That Produce the Most Renewable Energy." U.S. News & World Report, July 27. Accessed June 28, 2023 (https://www.usnews.com/news/best-states/slideshows/these-states-use-the-most-renewable-energy).

Gilpin, Lindsey. 2014. "The Dark Side of 3D Printing: 10 Things to Watch." TechRepublic, March 5. Accessed October 8, 2014 (www.techrepublic.com).

Gilson, Dave. 2014. "Don't Tread on Me." *Mother Jones*, January/February, 28–33.

Gimpelson, Vladimir, and Daniel Treisman. 2015. "Misperceiving Inequality." National Bureau of Economic Research, May. Accessed September 16, 2015 (nber.org).

Giugni, Marco, Doug McAdam, and Charles Tilley, eds. 1999. *How Social Movements Matter*. Minneapolis: University of Minnesota Press.

Giving USA Foundation. 2017. "Giving USA 2017." June 12. Accessed July 3, 2017 (givingusa.org).

Gladstone, Rick. 2015. "Toll from Hajj Stampede Reaches 2,411 in New Estimate." *New York Times*, December 11, A4.

Glauber, Bill. 2011. "Do Monarchies Still Matter?" *Christian Science Monitor*, February 21, 27–31.

Gleiber, Michael A. 2014. "The High Price Our Bodies Pay for High Fashion." *Huffington Post*, October 26. Accessed December 5, 2014 (www.huffingtonpost.com).

Glenn, David. 2011. "For Business Majors, Easy Does It." *Chronicle of Higher Education*, April 22, A1, A3–A5.

Glionna, John M. 2009. "South Korean Kids Get a Taste of Boot Camp." *Los Angeles Times*, August 22, A1.

"Global Christianity: A Report on the Size and Distribution of the World's Christian Population." 2011. Pew Research Center, December 1. Accessed April 8, 2012 (www.pewresearch.org).

Global Fund Community Foundations. 2022 "The Women Who Stood Up for the "Witches" of Northern Ghana: Community Philanthropy's Role in Challenging Stigma and Discrimination." June 14. Accessed March 12, 2023 (https://globalfundcommunityfoundations.org/blog/the-women-who-stood-up-for-the-witches-of-northern-ghana-community-philanthropys-role-in-challenging-stigma-and-discrimination/).

Global WASH Fast Facts. 2013. "Information on Water, Sanitation, and Hygiene." Centers for Disease Control and Prevention, November 8. Accessed September 19, 2014 (www.cdc.gov).

"Global WASH Fast Facts." 2017. Centers for Disease Control and Prevention, April 11. Accessed August 20, 2017 (cdc.gov).

Goffman, Alice. 2014. *On the Run: Fugitive Life in an American City*. Chicago: University of Chicago Press.

Goffman, Erving. 1959. *The Presentation of Self in Everyday Life*. New York: Doubleday Anchor Books.

Goffman, Erving. 1961. *Asylums: Essays on the Social Situation of Mental Patients and Other Inmates*. Garden City, NY: Anchor Books.

Goffman, Erving. 1963. *Stigma: Notes on the Management of Spoiled Identity*. Englewood Cliffs, NJ: Prentice-Hall.

Goffman, Erving. 1967. *Interaction Ritual: Essays on Face-to-Face Behavior*. New York: Anchor Books.

Goffman, Erving. 1969. *Strategic Interaction.* Philadelphia: University of Pennsylvania Press.

Gold, Russell. 2023. "The Texas GOP's War on Renewable Energy." *Texas Monthly*, June. Accessed June 28, 2023 (https://www.texasmonthly.com/news-politics/texas-republican-war-on-renewable-energy/).

Goldberg, Michelle. 2015. "Feminist Writers Are So Besieged by Online Abuse That Some Have Begun to Retire." *Washington Post*, February 20. Accessed September 30, 2015 (washingtonpost.com).

Golden, Daniel. 2006. *The Price of Admission: How America's Ruling Class Buys Its Way into Elite Colleges—and Who Gets Left Outside the Gates.* New York: Crown.

Goldin, Claudia, and Lawrence F. Katz. 2016. "Women Working Longer: Facts and Some Explanations." National Bureau of Economic Research, September. Accessed July 19, 2017 (nber.org).

Goldin, Claudia, Sari Pekkala Kerr, Claudia Olivetti, and Erling Barth. 2017. "The Expanding Gender Earnings Gap: Evidence from the LEHD-2000 Census." *American Economic Review: Papers & Proceedings* 107 (5): 110–114.

Golightley, Sarah. 2020. "Troubling the 'Troubled Teen' Industry: Adult Reflections on Youth Experiences of Therapeutic Boarding Schools." *Global Studies of Childhood,* 10(1), 53–63. (https://doi.org/10.1177/2043610619900514).

Goodale, Gloria. 2015. "'Sharing Economy': Helpful or Selfish?" *Christian Science Monitor Weekly*, March 30, 21–23.

Goode, Erica. 2013. "U.S. Prison Populations Decline, Reflecting New Approach to Crime." *New York Times*, July 26, A11.

Goodell, Jeff. 2017. "Scott Pruitt's Crimes Against Nature." *Rolling Stone*, July 27. Accessed August 20, 2017 (rollingstone.com).

Goodman, Brenda. 2023. "A New Antibiotic, Discovered with Artificial Intelligence, May Defeat a Dangerous Superbug." *CNN News*, May 26. Accessed June 11, 2023 (https://www.cnn.com/2023/05/25/health/antibiotic-artificial-intelligence-superbug/index.html).

Goodnough, Abby. 2010. "Making It Clear That a Clear Parking Space Isn't." *New York Times*, December 28, A10.

Goodnough, Abby. 2017. "Panel Wants Opioid Crisis Declared Emergency." *New York Times*, August 1, A14.

Goodstein, Laurie. 2014. "Presbyterians Vote to Allow Same-Sex Marriages." *New York Times*, June 20, A11.

Goodstein, Laurie, and Michael D. Shear. 2017. "Trump's Order on Religious Liberty Pleases Some, but Lets Down Conservatives." *New York Times*, May 5, A20.

Gopnik, Alison, Andrew N. Meltzoff, and Patricia K. Kuhl. 2001. *The Scientist in the Crib: What Early Learning Tells Us About the Mind.* New York: Perennial.

Gottdiener, Mark, and Ray Hutchison. 2000. *The New Urban Sociology,* 2nd edition. New York: McGraw-Hill.

Gottlieb, Mark. 2011. "Attention to Duty." *Christian Science Monitor,* June 6, 44.

Gottman, John M. 1994. *What Predicts Divorce? The Relationships Between Marital Processes and Marital Outcome.* Hillsdale, NJ: Lawrence Erlbaum Associates.

Goudreau, Jenna. 2012. "A New Obstacle for Professional Women: The Glass Escalator." *Forbes*, May 21. Accessed December 18, 2013 (www.forbes.com).

Gould, Skye, and Jenny Cheng. 2018. "How to Properly Shake Hands in 14 Different Countries." *Business Insider,* March 23. Accessed September 7, 2023 (https://www.businessinsider.com/handshakes-around-the-world-2-2017).

Gouldner, Alvin W. 1962. "Anti-Minotaur: The Myth of a Value-Free Sociology." *Social Problems* 9 (Winter): 199–212.

Govender, Serusha. 2014. "American Foods the Rest of the World Thinks Are Strange." Fox News, May 13. Accessed February 14, 2017 (foxnews.com).

"Government Resolves to Start Making Sense Under New Law Forbidding Federal Gibberish." 2011. *Washington Post,* May 20. Accessed May 21, 2011 (www.washingtonpost.com).

Governors Highway Safety Association. 2014. "Distracted Driving Laws." Accessed January 12, 2014 (www.ghsa.org).

Goyette, Braden. 2013. "Oprah Talks to Swiss Newspaper About Alleged Racist Encounter." *Huffington Post*, August 14. Accessed March 22, 2014 (www.huffingtonpost.com).

Graham, Carol. 2015. "The High Costs of Being Poor in America: Stress, Pain, and Worry." Brookings Institution, February 19. Accessed September 31, 2015 (brookings.edu).

Graham, Carol. 2016. "The Rich Even Have a Better Kind of Stress Than the Poor." Brookings, February 10. Accessed December 23, 2016 (brookings.edu).

Graif, Corina, and Robert J. Sampson. 2009. "Spatial Heterogeneity in the Effects of Immigration and Diversity on Neighborhood Homicide Rates." *Homicide Studies* 13 (August): 242–260.

Grall, Timothy. 2013. "Custodial Mothers and Fathers and Their Child Support: 2011." U.S. Census Bureau, October. Accessed May 16, 2014 (www.census.gov).

Gramlich, John. 2016. "Voters' Perceptions of Crime Continue to Conflict with Reality." Pew Research Center, November 16. Accessed December 17, 2016 (pewresearch.org).

Gramlich, John. 2017. "Most Violent and Property Crimes in the U.S. Go Unsolved." Pew Research Center, March 1. Accessed March 2, 2017 (pewresearch.org).

Gramlich, John. 2017. "Nearly Half of Americans Have a Family Member or Close Friend Who's Been Addicted to Drugs." Pew Research Center. October 26. Accessed March 25, 2023 (https://www.pewresearch.org/fact-tank/2017/10/26/nearly-half-of-americans-have-a-family-member-or-close-friend-whos-been-addicted-to-drugs/).

Gramlich, John. 2020. "What the Data Says (and Doesn't Say) About Crime in the United States." Pew Research Center, November 20. Accessed March 12, 2023 (https://www.pewresearch.org/fact-tank/2020/11/20/facts-about-crime-in-the-u-s/).

Gramlich, John. 2021. "America's Incarceration Rate Falls to Lowest Level Since 1995." Pew Research Center, August 16. Accessed March 25, 2023 (https://www.pewresearch.org/fact-tank/2021/08/16/americas-incarceration-rate-lowest-since-1995/#:~:text=At%20the%20end%20of%202019%2C%20there%20were%20just,every%20100%2C000%20adult%20residents%20ages%2018%20and%20older).

Grandjean, Philippe, and Philip J. Landrigan. 2014. "Neurobehavioral Effects of Developmental Toxicity." *The Lancet Neurology* 13 (March): 330–338.

Grauerholz, Liz, and Sharon Bouma-Holtrop. 2003. "Exploring Critical Sociological Thinking." *Teaching Sociology* 31 (October): 485–496.

Graves, Joseph L., Jr. 2001. *The Emperor's New Clothes: Biological Theories of Race at the Millennium.* New Brunswick, NJ: Rutgers University Press.

Gray, Alex. 2017. "These Are the World's Biggest Corporate Giants." World Economic Forum, January 16. Accessed May 15, 2017 (weforum.org).

Gray, Paul S., John B. Williamson, David R. Karp, and John R. Dalphin. 2007. *The Research Imagination: An Introduction to Qualitative and Quantitative Methods.* New York: Cambridge University Press.

Greeley, Brendan. 2013. "How Inequality Became a Household Word." *Bloomberg Businessweek*, December 16–22, 16–17.

Green, Jeff. 2016. "At Work and Out of the Closet in the Heartland." *Bloomberg Businessweek*, March 3–April 3, 22–23.

Green, Jeff. 2016. "Detroit's Comeback Has an Arabic Accent." *Bloomberg Businessweek*, February 22–26, 18–19.

Green, Jeff, and Tim Higgins. 2016. "LGBT Inc." *Bloomberg Businessweek*, May 2–8, 26–28.

Greenberg, Julie, Arthur McKee, and Kate Walsh. 2013. "Teacher Prep Review: A Review of the Nation's Teacher

Preparation Programs 2013." National Council on Teacher Quality. Accessed June 21, 2014 (www.nctq.org).

Greenfield, Rebecca. 2015. "You Take the Vacation. Your Boss Pays the Bill." *Bloomberg Businessweek*, May 18–24, 22–23.

Greenwood, Shannon, Andrew Perrin, and Maeve Duggan. 2016. "Social Media Update 2016." Pew Research Center, November 11. Accessed March 14, 2017 (pewresearch.org).

Greenwood, Veronqiue. 2016. "Keeping Clothes Out of the Garbage." *Anthropecene Magazine*. Accessed June 25, 2023 (https://www.anthropocenemagazine.org/sustainablefashion/).

Gregory, Sean. 2017. "The Jobs That Weren't Saved." *Time*, May 29, 35–41.

Griffith, Eric. 2023. "The Worst Data Breaches in 2022: Were You a Victim?" *PCMag*, January 30. Accessed June 27, 2023 (https://www.pcmag.com/news/cybercrime-in-2022-fewer-data-breaches-but-more-victims).

Grim, Brian J., and Alan Cooperman. 2014. "Religious Hostilities Reach Six-Year High." Pew Research Center, January 14. Accessed June 21, 2014 (www.pewresearch.org).

Grobart, Sam. 2016. "People Think They Are Middle-Class, Whether They Make $22K or $200K." *Bloomberg Businessweek*, September 19–25, 26.

Gross, Rita M. 1996. *Feminism and Religion: An Introduction*. Boston: Beacon Press.

Gross, Terry. 2022. "College Is Increasingly Out of Reach for Many Students. What Went Wrong?." NPR, August 3. Accessed June 13, 2023 (https://www.npr.org/2022/08/03/1115380528/college-is-increasingly-out-of-reach-for-many-students-what-went-wrong).

Guallar, Eliseo, Saverio Stranges, Cynthia Mulrow, Lawrence J. Appel, and Edgar R. Miller III. 2013. "Enough Is Enough: Stop Wasting Money on Vitamin and Mineral Supplements." *Annals of Internal Medicine* 159 (12): 850–851.

"Gun Laws: A Shot and a Beer." 2014. *The Economist*, May 3, 28.

Gurkoff, Joe, and Anna Ranieri. 2012. *How Can I Help You? What You Can (and Can't) Do to Counsel a Friend, Colleague or Family Member with a Problem*. North Charleston, SC: CreateSpace Independent Publishing Platform.

Gurney, Joan M., and Kathleen T. Tierney. 1982. "Relative Deprivation and Social Movements: A Critical Look at Twenty Years of Theory and Research." *Sociological Quarterly* 23 (Winter): 33–47.

Gurrentz, Benjamin, and Yeris Mayol-Garcia. 2021. "Marriage, Divorce Widowhood Remain Prevalent Among Older Populations." U.S. Census Bureau, April 22. Accessed May 19, 2023 (https://www.census.gov/library/stories/2021/04/love-and-loss-among-older-adults.html#:~:text=At%20least%209%20in%2010,%25)%20ages%2015%20or%20older).

Guterman, Stanley S. 1969. "In Defense of Wirth's 'Urbanism as a Way of Life.'" *American Journal of Sociology* 74 (March): 492–499.

Guttmacher Institute. 2013. "U.S. Women Who Have Abortions." Infographics. Accessed February 11, 2012 (www.guttmacher.org).

Guttmacher Institute. 2014. "Facts on Induced Abortion in the United States." February. Accessed March 2, 2014 (guttmacher.org).

Guttmacher Institute. 2017. "Policy Trends in the States: 2016." January 3. Accessed May 5, 2017 (guttmacher.org).

Guy, Gery P., Jr., et al. 2017. "Vital Signs: Changes in Opioid Prescribing in the United States, 2006–2015." *MMWR* 66 (26): 697–704.

H

Haas, Steven A., and David R. Schaefer. 2014. "With a Little Help from My Friends? Asymmetrical Social Influence on Adolescent Smoking Initiation and Cessation." *Journal of Health and Social Behavior* 55 (June): 126–143.

Hacker, Jacob S. 2002. *The Divided Welfare State: The Battle over Public and Private Social Benefits in the United States*. New York: Cambridge University Press.

Hacker, Jacob S., and Paul Pierson. 2010. *Winner-Take-All Politics: How Washington Made the Rich Richer and Turned Its Back on the Middle Class*. New York: Simon and Schuster.

Hagan, Frank E. 2008. *Introduction to Criminology: Theories, Methods, and Criminal Behavior*, 6th edition. Thousand Oaks, CA: Sage.

Haigney, Sophie. 2022. "Anatomy of a Product Placement." *The New York Times*, June 24. Accessed February 12, 2023 (https://www.nytimes.com/interactive/2022/06/23/arts/product-placement.html).

Haines, Julia, Kaia Hubbard, and Christopher Wolf. 2023. "Where State Abortion Laws Stand Without Roe." *U.S. News & World Report*, July 14. Accessed July 31, 2023 (https://www.usnews.com/news/best-states/articles/a-guide-to-abortion-laws-by-state).

Hahn, Heather, Julia Isaacs, Sara Edelstein, Ellen Steele, and C. Eugene Steuerle. 2014. "Kids' Share 2014: Report on Federal Expenditures on Children Through 2013." Urban Institute. Accessed July 1, 2017 (urban.org).

Hakim, Danny. 2015. "Banned Abroad: Differences in Rules for Pesticides Point Up U.S.-Europe Trade Bumps." *New York Times*, February 24, B1, B4.

Hales, Emily. 2014. "Incentives for Having Kids: Solution for Falling Birth Rates?" *Deseret News*, June 26. Accessed September 21 (www.newsok.com).

Halikias, Dimitrios, and Richard V. Reeves. 2017. "Ladders, Labs, or Laggards? Which Public Universities Contribute Most." Brookings, July 11. Accessed July 31, 2017 (brookings.edu).

Hall, Aron J., Mary E. Wikswo, Kimberly Pringle, L. Hannah Gould, and Umesh D. Parashar. 2014. "Vital Signs: Foodborne Norovirus Outbreaks—United States, 2009–2012." *MMWR* 63 (22): 491–495.

Hall, Edward T. 1959. *The Silent Language*. New York: Doubleday.

Hall, Edward T. 1966. *The Hidden Dimension*. Garden City, NY: Doubleday.

Hallman, Ben, and Shane Shifflett. 2014. "When Hospices Mistreat the Dying, They Almost Never Get Punished." *Huffington Post*, December 30. Accessed October 16, 2015 (projects.huffingtonpost.com).

Halperin, David M. 2012. "How to Be Gay." *Chronicle Review*, September 7, B13–B17.

Halsey, Ashley, III. 2010. "Study: Older People Are Driving More, Having Fewer Accidents." *Washington Post*, June 22, A10.

Hamadeh, Nada, Catherine Von Rompaey, Eric Metreau, and Shweta Grace Eapen. 2022. "New World Bank Country Classifications by Income Level: 2022-2023." World Bank Blogs, July 1. Accessed June 10, 2023 (https://blogs.worldbank.org/opendata/new-world-bank-country-classifications-income-level-2022-2023).

Hamby, Sherry, David Finkelhor, Heather Turner, and Richard Ormrod. 2011. "Children's Exposure to Intimate Partner Violence and Other Family Violence." Office of Justice Programs, October. Accessed May 20, 2014 (www.ojp.usdoj.gov).

Hamilton, Malcolm B. 2001. *The Sociology of Religion*, 2nd edition. New York: Routledge.

Hammadi, Saad. 2013. "Making Asia's Factories Safer." *Christian Science Monitor Weekly*, November 11, 16.

Haney, Craig, Curtis Banks, and Philip Zimbardo. 1973. "Interpersonal Dynamics in a Simulated Prison." *International Journal of Criminology and Psychology* 1: 69–97.

Hansen, Claire. Horus Alas, and Elliott Davis Jr. 2023. "Where Is Marijuana Legal? A Guide to Marijuana Legalization." U.S. News and World Report, January 20. Accessed February 11, 2023 (https://www.usnews.com/news/best-states/articles/where-is-marijuana-legal-a-guide-to-marijuana-legalization).

Hansen, Lawrence A. 2010. "Noxious Groupthink." *Chronicle of Higher Education,* November 12, B6–B8.

Hanson, Melanie. 2023. "Average Student Loan Debt." Education Data Initiative. Accessed June 13, 2023 (https://educationdata.org/average-student-loan-debt#:~:text=The%20average%20federal%20student%20loan,them%20have%20federal%20loan%20debt).

Hanser, Amy, and Zachary Hyde. 2014. "Foodies Remaking Cities." *Contexts* 13 (Summer): 44–49.

Hanushek, Eric A., and Ludger Woessman. 2005. "Does Educational Tracking Affect Performance and Inequality? Differences-in-Differences Evidence Across Countries." National Bureau of Economic Research, February. Accessed July 25, 2007 (www.nber.org).

Hardoon, Deborah. 2017. "An Economy for the 99%." Oxfam, January. Accessed November 22, 2023 (https://oi-files-d8-prod.s3.eu-west-2.amazonaws.com/s3fs-public/file_attachments/bp-economy-for-99-percent-160117-summ-en.pdf).

Hardy, Adam. 2022. "The Wage Gap Between College and High School Grads Just Hit a Record High." Money, February 14. Accessed June 13, 2023 (https://money.com/wage-gap-college-high-school-grads/).

Harlow, Harry E., and Margaret K. Harlow. 1962. "Social Deprivation in Monkeys." *Scientific American* 206 (November): 137–146.

Harnish, Verne. 2014. "5 Ways to Tone Your Operations." *Fortune,* October 27, 46.

Harrell, Erika, and Lynn Langton. 2013. "Victims of Identity Theft, 2012." U.S. Department of Justice, Office of Justice Programs, December. Accessed January 12, 2014 (www.ojp.usdoj.gov).

Harrell, Erika, Lynn Langton, Marcus Berzofsky, Lance Couzens, and Hope Smiley-McDonald. 2014. "Household Poverty and Nonfatal Violent Victimization, 2008–2012." Bureau of Justice Statistics, November. Accessed September 11, 2015 (bjs.gov).

Harring, Alex. 2020. "More Than Half of U.S. States Have Statewide Mask Mandates." CNBC, July 20. Accessed February 11, 2023 (https://www.cnbc.com/2020/07/20/more-than-half-of-us-states-have-statewide-mask-mandates.html).

Harris, Benjamin, and Sydney Schreiner Wertz. 2022. "Racial Differences in Economic Security: The Racial Wealth Gap." U.S. Department of the Treasury, September 15. Accessed April 13, 2023 (https://home.treasury.gov/news/featured-stories/racial-differences-economic-security-racial-wealth-gap).

Harris, Chauncey D. 1997. "'The Nature of Cities' and Urban Geography in the Last Century." *Urban Geography* 18 (1): 15–35.

Harris, Chauncey D., and Edward L. Ullman. 1945. "The Nature of Cities." *Annals* 242: 7–17.

Harris, Gardiner. 2011. "When the Nurse Wants to Be Called 'Doctor.'" *New York Times,* June 6, A23.

Harris, Gardiner. 2013. "India's New Focus on Rape Shows Only the Surface of Women's Perils." *New York Times,* January 12. Accessed March 4, 2014 (www.nytimes.com).

Harris, Jennifer L., et al. 2013. "Fast Food FACTS 2013: Measuring Progress in Nutrition and Marketing to Children and Teens." Yale Rudd Center for Food Policy & Obesity. Accessed November 10, 2013 (www.fastfoodmarketing.org).

Harter, Jim, and Amy Adkins, 2017. "Are Your Star Employees Slipping Away?." Gallup, February 24. Accessed November 17, 2023 (https://www.gallup.com/workplace/236351/star-employees-slipping-away.aspx).

Harter, Jim. 2022. "Employee Engagement vs. Employee Satisfaction and Organizational Culture." Gallup, August 13. Accessed March 11, 2023 (https://www.gallup.com/workplace/236366/right-culture-not-employee-satisfaction.aspx).

Harter, Jim. 2023 "U.S. Employee Engagement Needs a Rebound in 2023." Gallup, January 25. Accessed May 11, 2023 (https://www.gallup.com/workplace/468233/employee-engagement-needs-rebound-2023.aspx).

Harter, Jim, and Sangeeta Agrawal. 2014. "Many Baby Boomers Reluctant to Retire." Gallup, January 20. Accessed May 16, 2014 (www.gallup.com).

Hartig, Hannah, John Lapinski, and Stephanie Psyllos. 2016. "Poll: More Voters Say Trump Doesn't Respect Women After Lewd Tape Surfaces." NBC News, October 10. Accessed January 5, 2017 (nbcnews.com).

Harvard School of Public Health. 2014. "Poll Finds Many in U.S. Lack Knowledge About Ebola and Its Transmission." August 21. Accessed October 8, 2014 (www.hsph.harvard.edu).

Haskins, Ron. 2008. "Immigration: Wages, Education, and Mobility." Pp. 81–90 in *Getting Ahead or Losing Ground: Economic Mobility in America,* edited by Julia B. Isaacs, Isabel V. Sawhill, and Ron Haskins. Washington, DC: Brookings Institution.

Haskins, Ron. 2015. "$2 a Day: A More Complete Picture." Brookings Institution, September 14. Accessed September 31, 2015 (brookings.edu).

Haqqi, Ty. 2022. "5 Biggest Multinational Companies in the World." Insider Monkey, November 19. Accessed May 12, 2023 (https://www.insidermonkey.com/blog/5-biggest-multinational-companies-in-the-world-1087487/5/).

Harris, Katelyn. 2020. "Forty Years of Falling Manufacturing Employment." U.S. Bureau of Labor Statistics, November. Accessed May 12, 2023 (https://www.bls.gov/opub/btn/volume-9/forty-years-of-falling-manufacturing-employment.htm).

Hartmans, Avery, Sarah Jackson, and Azmi Haroun. 2023. "The Rise and Fall of Elizabeth Holmes, the Former Theranos CEO Found Guilty of Wire Fraud and Conspiracy Who is Now Trying to Delay Her Prison Sentence After Having Her Second Child." Business Insider. January 20. Accessed March 12, 2023 (https://www.businessinsider.com/theranos-founder-ceo-elizabeth-holmes-life-story-bio-2018-4).

Hattry, Harry. 2020. "Evaluation Guide for Public Service Program Managers." Urban Institute, December 16. Accessed January 29, 2023 (https://www.urban.org/research/publication/evaluation-guide-public-service-program-managers).

Hauer, Mathew E., Jason M. Evens, and Deepak R. Mishra. 2016. "Millions Projected to Be at Risk from Sea-Level Rise in the Continental United States." *Nature Climate Change* 6 (April): 691–695.

Hawai'i Free Press. 2012. "Census Bureau: Hawaii 7th Highest Poverty Rate in US." November 15. Accessed February 14, 2014 (www.hawaiifreepress.com).

Hayes, Danny, and Jennifer L. Lawless. 2016. *Women on the Run: Gender, Media, and Political Campaigns in a Polarized Era*. New York: Cambridge University Press.

Hayford, Sarah R., and Karen Benjamin Guzzo. 2015. "The Single Mother by Choice Myth." *Contexts* 14 (Fall): 70–72.

"Health Care in America." 2015. *The Economist,* March 7, 63–64.

"The Health Paradox." 2013. *The Economist,* May 11, 27–28.

Health Resources and Services Administration. 2010. "Assuring Access to Essential Health Care." Accessed July 15, 2011 (www.plainlanguage.gov).

Healthcare-NOW! 2014. "Health Insurance CEO Pay Skyrockets in 2013." May 5. Accessed July 26, 2014 (www.healthcare-now.org).

"Health Insurance Coverage." 2017. Centers for Disease Control and Prevention, March 31. Accessed August 15, 2017 (cdc.gov).

Heath, Chip. 2005. "Introduction." Pp. 81–85 in *Rumor Mills: The Social Impact of Rumor and Legend,* edited by Gary Alan Fine, Veronique Campion-Vincent, and Chip Heath. New Brunswick, NJ: Transaction Publishers.

Heath, Jennifer. 2008. *The Veil: Women Writers on Its History, Lore, and Politics.* Berkeley: University of California Press.

Hechter, Michael, and Karl-Dieter Opp. 2001. "Introduction." Pp. xi–xx in *Social Norms,* edited by Michael Hechter and Karl-Dieter Opp. New York: Russell Sage Foundation.

Hedges, Chris. 2011. "Our Public Schools Are Churning Out Drones for the Corporate State." AlterNet, April 11. Accessed April 14, 2011 (www.alternet.org).

Hee Lee, Ye Michelle. 2017. "Fact-Checking Trump's Rhetoric on Crime and the 'American Carnage'." *Washington Post,* January 30. Accessed April 15, 2017 (washingtonpost.com).

Heffer, Simon. 2010. "America Is the Acceptable Face of Cultural Imperialism." *Telegraph Media Group,* July 24. Accessed November 10, 2013 (telegraph.co.uk).

Hegewisch, Ariane, and Heidi Hartmann. 2015. "The Gender Wage Gap: 2014." Institute for Women's Policy Research, September. Accessed September 29, 2015 (iwpr.org).

Heider, Eleanor R., and Donald C. Olivier. 1972. "The Structure of the Color Space in Naming and Memory for Two Languages." *Cognitive Psychology* 3: 337–354.

Heilbroner, R. L., and L. C. Thurow. 1998. *Economics Explained: Everything You Need to Know About How the Economy Works and Where It's Going.* New York: Touchstone.

Heim, Joe. 2015. "Indian Americans Dominate the National Spelling Bee. Why Should They Take Abuse on Social Media for It?" *Washington Post,* May 25. Accessed October 4, 2015 (washingtonpost.com).

Heisler, Candace. 2012. "Elder Abuse." Office for Victims of Crime. Accessed May 16, 2014 (www.ovc.ncjrs.gov).

Helhoski, Anna. 2020. "Why Rich Students Get More Financial Aid Than Poor Ones." *AP News,* April 2. Accessed June 13, 2023 (https://apnews.com/article/270e2f52ea8b4acfac923d5f6631008f).

Helliwell, John, Richard Layard, and Jeffrey Sachs (eds). 2013. *World Happiness Report 2013.* United Nations. Accessed February 6, 2014 (www.unsdsn.org).

Hemez, Paul, and Chanell Washington. 2021. "Percentage and Number of Children Living with Two Parents Has Dropped Since 1968." U.S. Census Bureau, April 12. Accessed May 20, 2023 (https://www.census.gov/library/stories/2021/04/number-of-children-living-only-with-their-mothers-has-doubled-in-past-50-years.html).

Hemphill, Thomas A., and Mark J. Perry. 2017. "Wind Power's Future in U.S. Could Be Thwarted by Grassroots Opposition." Real Clear Energy, June 25. Accessed August 20, 2017 (realclearenergy.org).

Henderson, Alex. 2016. "Sick at Work: Staggering Number of Americans Go to Work with Bad Colds or Flu." AlterNet, May 30. Accessed August 15, 2017 (alternet.org).

Hendricks, Sara. 2018. "The 15 Biggest Differences Between French and American Parenting." *Business Insider,* March 1. Accessed February 12, 2023 (https://www.businessinsider.com/differences-french-and-american-parenting-2018-3).

Henne, Peter. 2015. "Latest Trends in Religious Restrictions and Hostilities." Pew Research Center, February 26. Accessed July 31, 2017 (pewresearch.org).

Henne, Peter, and Katayoun Kishi. 2016. "5 Key Findings About Global Restriction on Religion." Pew Research Center, May 4. Accessed July 31, 2017 (pewresearch.org).

Henry, Patrick. 2022. "Economic Inequality Has Deepened During the Pandemic. That Doesn't Mean It Can't Be Fixed." World Economic Forum, April 7. Accessed April 13, 2023 (https://www.weforum.org/agenda/2022/04/economic-inequality-wealth-gap-pandemic/).

Herman, Bob. 2017. "The Sky-High Pay of Health Care CEOs." Axios, July 24. Accessed August 15, 2017 (axios.com).

Hernandez, Joe. 2021. "Why There Are Few Openly Gay Athletes In Men's Professional Sports." NPR, July 21. Accessed April 29, 2023 (https://www.npr.org/2021/07/21/1018404859/openly-gay-athletes-in-mens-pro-sports-few).

Herzberg, David. 2009. *Happy Pills in America: From Miltown to Prozac.* Baltimore, MD: Johns Hopkins University Press.

Hibbard, Roberta, Jane Barlow, and Harriet MacMillan. 2012. "Psychological Maltreatment." American Academy of Pediatrics. Accessed May 4, 2013 (www.pediatrics.aappublications.org).

Hicks, Maureen Soyars. 2017. "Flexible Jobs Give Workers Choices." U.S. Bureau of Labor Statistics, Monthly Labor Review, May. Accessed June 5, 2017 (bls.gov).

Higgins, Tim. 2015. "Public Opinion: The Polling Industry Cuts the Cord." *Bloomberg Businessweek,* November 23–29, 30.

Hightower, Jim. 2015. "CEOs Call for Wage Increases for Workers! What's the Catch?" AlterNet, August 26. Accessed August 26, 2015 (alternet.org).

Hilbert, Richard A. 1992. *The Classical Roots of Ethnomethodology: Durkheim, Weber, and Garfinkel.* Chapel Hill: University of North Carolina Press.

Hilger, Nathaniel. 2016. "Upward Mobility and Discrimination: The Case of Asian Americans." National Bureau of Economic Research, October. Accessed June 20, 2017 (nber.org).

Hill, Adam. 2017. "Breaking the Stigma—A Physician's Perspective on Self-Care and Recovery." *New England Journal of Medicine* 376 (March 23): 1103–1105.

Hill, Catey. 2015. "5 Times It's More Expensive to Be a Woman." Market Watch, April 14. Accessed September 29, 2015 (marketwatch.com).

Hill, Catherine, Christianne Corbett, and Andresse St. Rose. 2010. "Why So Few? Women in Science, Technology, Engineering, and Mathematics." American Association of University Women. Accessed October 25, 2011 (www.aauw.org).

Hill, Jason, Christina Stearns, and Chelsea Owens. 2015. "Education and Certification Qualifications of Departmentalized Public High-School-Level Teachers of Selected Subjects: Evidence from the 2011–12 Schools and Staffing Survey." National Center for Education Statistics, June. Accessed July 30, 2017 (nces.ed.gov).

Hill, Latoya, Samantha Artiga, and Usha Rnaji. 2022. "Racial Disparities in Maternal and Infant Health: Current Status and Efforts to Address Them." Kaiser Family Foundation, November 1. Accessed Jun 11, 2023 (https://www.kff.org/racial-equity-and-health-policy/issue-brief/racial-disparities-in-maternal-and-infant-health-current-status-and-efforts-to-address-them/).

Hillaker, B. D., H. E. Brophy-Herb, F. A. Villarruel, and B. E. Hass. 2008. "The Contributions of Parenting to Social Competencies and Positive Values in Middle School Youth: Positive Family Communication, Maintaining Standards, and Supportive Family Relationships." *Family Relations* 57 (December): 591–601.

Hillin, Taryn. 2014. "The Divorce Mistakes You Don't Even Know You're Making." *Huffington Post,* March 18. Accessed March 18, 2014 (www.huffingtonpost.com).

Hinduja. Sameer. 2021. "Cyberbullying Statistics 2021 | Age, Gender, Sexual Orientation, and Race." Cyberbullying Research Center, October 21. Accessed September 7, 2023 (https://cyberbullying.org/cyberbullying-statistics-age-gender-sexual-orientation-race).

Hines, Jonas Z., Steven C. Fiala, and Katrina Hedberg. 2017. "Electronic Cigarettes as an Introductory Tobacco Product Among Eighth and 11th Grade Tobacco Users—Oregon, 2015." *MMWR* 66 (23): 604–606.

Hinkle, Stephen, and John Schopler. 1986. "Bias in the Evaluation of In-Group and Out-Group Performance." Pp. 196 -212 in *Psychology of Everyday Intergroup Relations,* 2nd edition, edited by Stephen Worchel and William G. Austin. Chicago: Nelson-Hall

Hirschi, Travis. 1969. *Causes of Delinquency.* Berkeley: University of California Press.

Hiss, William C., and Valerie W. Franks. 2014. "Defining Promise: Optional Standardized Testing Policies in American College and University Admissions." National Association for College Admission Counseling, February 5. Accessed June 21, 2014 (www.nacacnet.org).

Hlavka, Heather R. 2014. "Normalizing Sexual Violence: Young Women Account for Harassment and Abuse." *Gender & Society* 20 (June): 1–22.

Hoecker-Drysdale, Susan. 1992. *Harriet Martineau: First Woman Sociologist.* Providence, RI: Berg.

Hoffman, Ari, and Jeanne Batalova. 2023. "Venezuelan Immigrants in the United States." Migration Policy Institute, February 15. Accessed April 28, 2023 (https://www.migrationpolicy.org/article/venezuelan-immigrants-united-states).

Hokayem, Charles, and Misty L. Heggeness. 2014. "Living in Near Poverty in the United States: 1966–2012." U.S. Census Bureau, May. Accessed May 16, 2014 (www.census.gov).

Hollander, Matthew. 2015. "How to Be a Hero: Insight from the Milgram Experiment." Huffpost, February 27. Accessed November 17, 2023 (https://www.npr.org/sections/health-shots/2022/08/31/1120004717/the-u-s-diet-is-deadly-here-are-7-ideas-to-get-americans-eating-healthier).

Holloway, Kali. 2015. "8 Myths That Fuel the Assault on Abortion Rights." AlterNet, July 25. Accessed September 29, 2015 (alternet.org).

Holloway, Kali. 2016. "African-American Women Now Top the List of Most-Educated Group in the Country." AlterNet, June 2. Accessed June 20, 2017 (alternet.org).

Holloway, Kim. 2016. "Yes, Hillary Still Has to Suffer the Sexism of the Dude TV Pundits." AlterNet, February 4. Accessed May 15, 2017 (alternet.org).

Holmes, Elizabeth. 2013. "Grooming Tips from Giant Men." *Wall Street Journal*, November 27, D1.

Holt, Steve. 2018. "The Psychology of Boston's Snow Parking Wars." *Bloomberg*, January 22. Accessed September 7, 2023 (https://www.bloomberg.com/news/articles/2018-01-22/the-psychology-of-boston-s-snow-parking-space-savers).

Homans, George. 1974. *Social Behavior: Its Elementary Forms,* revised edition. New York: Harcourt Brace Jovanovich.

Hondagneu-Sotelo, Pierrette. 2001. *Domestica: Immigrant Workers Cleaning and Caring in the Shadows of Affluence.* Berkeley: University of California Press.

Hong, Gihoon, and John McLaren. 2015. "Are Immigrants a Shot in the Arm for the Local Economy?" National Bureau of Economic Research, February. Accessed June 20, 2017 (nber.org).

Hooke, Alexander E. 2014. "Speak Not, Lest Ye Be Judged." *Baltimore Sun*, May 27, 13.

Hooyman, Nancy R. 2016. "Social and Health Disparities in Aging: Gender Inequities in Long-Term Care." American Society on Aging, August 8. Accessed August 15, 2017 (asaging.org).

Hornyak, Tim. 2023. "Japan Has Millions of Empty Houses. Want to Buy One for $25,000?" *The New York Times*, April 22. Accessed June 27, 2023 (https://www.nytimes.com/2023/04/17/realestate/japan-empty-houses.html).

Horowitz, Juliana Menasce. 2022. "More Than Half of Americans in Their 40s Are 'Sandwiched' Between an Aging Parent and Their Own Children." Pew Research Center, April 8. Accessed February 18, 2023 (https://www.pewresearch.org/fact-tank/2022/04/08/more-than-half-of-americans-in-their-40s-are-sandwiched-between-an-aging-parent-and-their-own-children/).

Horowitz, Juliana Menasce, Nikki Graf, and Gretchen Livingston. 2019. "Why People Get Married or Move in with a Partner." Pew Research Center, November 6. Accessed May 19, 2023 (https://www.pewresearch.org/social-trends/2019/11/06/why-people-get-married-or-move-in-with-a-partner/#:~:text=The%20decision%20to%20get%20married,children%20someday%2C%20convenience%20or%20finances).

Horowitz, Juliana Menasce, Ruth Igielnik, and Rakesh Kochar. 2020. "Most Americans Say There Is Too Much Economic Inequality in the U.S., but Fewer Than Half Call It a Top Priority." Pew Research Center, January 9. Accessed April 13, 2023 (https://www.pewresearch.org/social-trends/2020/01/09/most-americans-say-there-is-too-much-economic-inequality-in-the-u-s-but-fewer-than-half-call-it-a-top-priority/).

Horowitz, Juliana Menasce, and Kim Parker. 2023. "How Americans View Their Jobs." Pew Research Center, March 30. Accessed May 12, 2023 (https://www.pewresearch.org/social-trends/2023/03/30/how-americans-view-their-jobs/).

Horowitz, Juliana, Kim Parker, Nikki Graf, and Gretchen Livingston. 2017. "Americans Widely Support Paid Family and Medical Leave, but Differ over Specific Policies." Pew Research Center, March 23. Accessed July 20, 2017 (pewresearch.org).

Hotz, V. Joseph, and Juan Pantano. 2013. "Strategic Parenting, Birth Order and School Performance." National Bureau of Economic Research, October. Accessed October 26, 2014 (www.nber.org).

Howard, Beth. 2014. "Should You Have Surgery Abroad?" *AARP Magazine*, October/November, 26–30.

Howard, Jeff. 2003. "Still at Risk: The Causes and Costs of Failure to Educate Poor and Minority Children for the Twenty-First Century." Pp. 81–97 in *A Nation Reformed? American Education 20 Years After A Nation at Risk,* edited by David T. Gordon and Patricia A. Graham. Cambridge, MA: Harvard Education Press.

Hoxie, Josh. 2017. "The Myths Behind Inequality in Our Country." Institute for Policy Studies, May 3. Accessed May 4, 2017 (ips-dc.org).

Hoyt, Homer. 1939. *The Structure and Growth of Residential Neighborhoods in American Cities.* Washington, DC: Federal Housing Administration.

Hsin, Amy, and Yu Xie. 2014. "Explaining Asian Americans' Academic Advantage over Whites." *Proceedings of the National Academy of Sciences*, Early Edition, May 5. Accessed June 21, 2014 (www.pnas.org).

Hsu, Angel, et al. 2014. "The 2014 Environmental Performance Index: Full Report and Analysis." Yale Center for Environmental Law & Policy. Accessed September 19, 2014 (www.epi.yale.edu).

Huang, Jon, Samuel Jacoby, Michael Strickland, and K. K. Rebecca Lai. 2016. "Election 2016: Exit." *New York Times*, November 8. Accessed January 5, 2017 (nytimes.com).

Hubbard, Kaia. 2021. "The 10 States with the Most Racial Diversity." *U.S. News and World Report*, August 19. Accessed April 28, 2023 (https://www.usnews.com/news/best-states/slideshows/the-10-most-racially-diverse-states-in-the-us).

Hubbard, Ruth. 1990. *The Politics of Women's Biology.* New Brunswick, NJ: Rutgers University Press.

Huffington Post. 2011. "New American Bible Changes Some Words, Including 'Holocaust.'" March 3. Accessed July 16, 2011 (www.huffingtonpost.com).

Huffman, Matt L. 2013. "Organizations, Managers, and Wage Inequality." *Sex Roles* 68 (February): 216–222.

Hughes, Everett C. 1945. "Dilemmas and Contradictions of Status." *American Journal of Sociology* 50: 353–359.

Huis, Arnold van, et al. 2013. "Edible Insects: Future Prospects for Food and Feed Security." Food and Agriculture Organization of the United Nations. Accessed February 14, 2017 (fao.org).

Huizinga, David, Shari Miller, and the Conduct Problems Prevention Research Group. 2013. "Developmental Sequences of Girls' Delinquent Behavior." Office of Juvenile Justice and Delinquency Prevention, December.

Accessed January 12, 2014 (www.ojp.usdoj.gov).

Human Rights Campaign. 2023. "Marriage Equality Around the World." Accessed April 23, 2023 (https://www.hrc.org/resources/marriage-equality-around-the-world).

Human Rights Campaign, 2023. "Stances of Faiths on LGBTQ Issues: Reform Judaism - Human Rights Campaign." Accessed June 13, 2023 (https://www.hrc.org/resources/stances-of-faiths-on-lgbt-issues-reform-judaism).

"Human Rights Violations." 2015. Unite to End Violence Against Women. Accessed September 26, 2015 (un.org).

Human Rights Watch. 2021. ""Everything I Have to Do is Tied to a Man." March 29. Accessed April 21, 2023 (https://www.hrw.org/report/2021/03/29/everything-i-have-do-tied-man/women-and-qatars-male-guardianship-rules).

Hummer, Robert A., and Elaine M. Hernandez. 2013. "The Effect of Educational Attainment on Adult Mortality in the United States." *Population Bulletin* 68 (June): 1–16.

Hunnicutt, Gwen. 2009. "Varieties of Patriarchy and Violence Against Women: Resurrecting 'Patriarchy' as a Theoretical Tool." *Violence Against Women* 15 (May): 553–573.

Hunt, Melissa G., Rachel Marx, Courtney Lipson, and Jordyn Young. 2018. "No More FOMO: Limiting Social Media Decreases Loneliness and Depression." *Journal of Social and Clinical Psychology*, 37(10), 751–768. Accessed September 7, 2023 (https://doi.org/10.1521/jscp.2018.37.10.751).

Hunter, Lori M. 2014. "Unmarried Baby Boomers Face Disadvantages as They Grow Older." Population Reference Bureau, February. Accessed May 16, 2014 (www.prb.org).

"Hunting as a Pack." 2014. *Economist Technology Quarterly*, December 6, 6.

Hurst, Kiley. 2022. "Rising Share of Americans See Women Raising Children on Their Own, Cohabitation as Bad for Society." Pew Research Center, March 11. Accessed May 20, 2023 (https://www.pewresearch.org/short-reads/2022/03/11/rising-share-of-americans-see-women-raising-children-on-their-own-cohabitation-as-bad-for-society/).

Hurwitz, Michael. 2011. "The Impact of Legacy Status on Undergraduate Admissions at Elite Colleges and Universities." *Economics of Education Review* 30 (June): 480–492.

Hymowitz, Kay. 2020. "The Gender Wage Gap Shrinks, While the Parenthood Gap Grows." Institute for Family Studies, October 27. Accessed May 12, 2023 (https://ifstudies.org/blog/the-gender-wage-gap-shrinks-while-the-parenthood-gap-grows).

I

Ianzito, Christina. 2013. "Killer Heels." *Baltimore Sun*, August 1, 3.

IBM Journal Staff. 2016. "Cyber Security & Data Protection: A Glimpse into the Dark World of Organized Cybercrime." *IBM Journal*, May 11. Accessed April 15, 2017 (ibmjournal.com).

Iceland, John, Gregory Sharp, and Jeffrey M. Timberlake. 2013. "Sun Belt Rising: Regional Population Change and the Decline in Black Residential Segregation, 1970–2009." *Demography* 50 (February): 97–123.

Identity Theft Resource Center. 2017. "Data Breach Reports: 2016 End of Year Report." January 18. Accessed April 2, 2017 (idtheftcenter.org).

Igielnik, Ruth, Scott Keeter, and Hannah Hartig. 2021. "Behind Biden's 2020 Victory: An Examination of the 2020 Electorate, Based on Validated Voters." Pew Research Center, June 30. Accessed January 29, 2023. (https://www.pewresearch.org/politics/2021/06/30/behind-bidens-2020-victory/).

Imbriani, Giovanni, Alessandra Panico, Tiziana Grassi, Adele Idolo, et al. 2021. "Early-Life Exposure to Environmental Air Pollution and Autism Spectrum Disorder: A Review of Available Evidence." International Journal of Environmental Research and Public Health 18(3):1204. DOI:10.3390/ijerph18031204

"In First Month, Views of Trump Are Already Strongly Felt, Deeply Polarized." 2017. Pew Research Center, February 16. Accessed June 20, 2017 (pewresearch.org).

Independent Television Service. 2003. "Race the Power of an Illusion: What Is This Thing Called Race?" Accessed March 22, 2014 (www.itvs.org).

"Inequality." 2017. *The Economist*, January 17, 21–22.

Infinit Healthcare. 2017. "The Pros and Cons of the Affordable Care Act." March 2. Accessed August 15, 2017 (infinithealthcare.com).

Inouye, Brandin. 2021. "On Protecting Consumers from Toxics in Cosmetics, U.S. Lags at Least 80 Countries." EWG, August 10. Accessed June 11, 2023 (https://www.ewg.org/news-insights/news/2021/08/protecting-consumers-toxics-cosmetics-us-lags-least-80-countries).

Institute of Medicine. 2015. Beyond Myalgic Encephalomyelitis/Chronic Fatigue Syndrome: Redefining an Illness. Washington, DC: National Academies Press.

Insurance Institute for Highway Safety. 2020. "Fatality Facts 2020: Males and Females." May. Accessed January 29, 2023. (https://www.iihs.org/topics/fatality-statistics/detail/males-and-females#age-differences).

Insurance Institute for Highway Safety. 2022. "Older Drivers." July. Accessed January 29, 2023 (https://www.iihs.org/topics/older-drivers#:~:text=Based%20on%20data%20reported%20by%20states%20to%20the,National%20Household%20Travel%20Survey%2C%20they%20drive%20fewer%20miles).

Interact. 2022. "How Common Is Intersex." Interact Intersex Society of North America Accessed April 14, 2023 (https://isna.org/faq/frequency/).

Intergovernmental Panel on Climate Change. 2014. "Climate Change 2014: Impacts, Adaptation and Vulnerability; Summary to Policymakers." Accessed September 19, 2014 (www.ipcc.ch).

International Federation of Health Plans. 2014. "2013 Comparative Price Report: Variation in Medical and Hospice Prices by Country." March 14. Accessed July 26, 2014 (www.ifhp.com).

International Institute for Democracy and Electoral Assistance. 2014. "VAP Turnout." Accessed April 20, 2014 (www.idea.int).

International Institute for Democracy and Electoral Assistance. 2022. "Global Democracy Weakens in 2022." November 30. Accessed May 12, 2023 (https://www.idea.int/news-media/news/global-democracy-weakens-2022).

International Labour Organization, Walk Free Foundation, and International Organization for Migration. 2022. "Global Estimate of Modern Slavery, Forced Labour and Forced Marriage." September. Accessed April 13, 2023 (https://cdn.walkfree.org/content/uploads/2022/09/12142341/GEMS-2022_Report_EN_V8.pdf).

International Organization for Migration. 2022. "World Migration Report 2022." UN Migration. Accessed June 27, 2023 (https://publications.iom.int/books/world-migration-report-2022).

Inter-Parliamentary Union. 2017. "Women in Parliaments: World Classification." May 1. Accessed May 25, 2017 (ipu.org).

Ioannidis, John. 2005. "Contradicted and Initially Stronger Effects in Highly Cited Clinical Research." *JAMA* 294 (July 13): 218–228.

Ipsos. 2021. "LGBT+ Pride 2021 Global Survey." Accessed April 21, 2023 (https://www.ipsos.com/sites/default/files/ct/news/documents/2021-06/LGBT%20Pride%202021%20Global%20Survey%20Report_3.pdf).

Iqbal, Sarosh, Rubina Zakar, Florian Fischer, and Muhammed Sakria Sakar. 2022. "Consanguineous Marriages and Their Association with Women's Reproductive Health and Fertility Behavior in Pakistan: Secondary Data Analysis from

Demographic and Health Surveys, 1990–2018." BMC Women's Health 22, 118. Accessed May 13, 2023 (https://doi.org/10.1186/s12905-022-01704-2).

Irving, Shelley K., and Tracy A. Loveless. 2015. "Dynamics of Economic Well-Being: Participation in Government Programs, 2009–2012: Who Gets Assistance?" U.S. Census Bureau, May. Accessed September 11, 2015 (census.gov).

Islam, Saiful, Abu-Hena Mostofa Kamal, Alamgir Kabir, Dorothy L. Southern, Sazzad Hossain Khan, S. M. Murshid Hasan, Tonmoy Sarkar, Shayla Sharmin, Shiuli Das, Tuhin Roy, Golam Dostogir Harun, Abrar Ahmad Chughtai, Nusrat Homaira, and Holly Seale. 2021. "COVID-19 Vaccine Rumors and Conspiracy Theories: The Need for Cognitive Inoculation Against Misinformation to Improve Vaccine Adherence." *PloS One*. May 12;16(5):e0251605. Doi: 10.1371/journal.pone.0251605.

Ivory, Danielle, Ben Protess, and Griff Palmer. 2016. "In American Towns, Pumping Private Profit from Public Works." *New York Times*, December 25, A1.

Ivry, Bob, Bradley Keoun, and Phil Kuntz. 2011. "Secret Fed Loans Gave Banks $13 Billion Undisclosed to Congress." *Bloomberg Businessweek*, November 27. Accessed August 26, 2015 (loomberg.com).

J

Jackson, D. D. 1998. "'This Hole in Our Heart': Urban Indian Identity and the Power of Silence." *American Indian Culture and Research Journal* 22 (4): 227–254.

Jackson, Janine. 2017. "'It Is Not at All Typical to Stifle Basic Scientific Information.'" FAIR, February 13. Accessed March 3, 2017 (fair.org).

Jackson, Shelly L., and Thomas L. Hafemeister. 2011. "Financial Abuse of Elderly People vs. Other Forms of Elder Abuse: Assessing Their Dynamics, Risk Factors, and Society's Response." U.S. Department of Justice, February. Accessed March 5, 2012 (www.ncjrs.gov).

Jackson, Shelly L., and Thomas L. Hafemeister. 2013. "Understanding Elder Abuse." National Institute of Justice, June. Accessed May 16, 2014 (www.nij.gov).

Jacobe, Dennis. 2013. "U.S. Small Businesses Struggle to Find Qualified Employees." Gallup, February 15. Accessed April 12, 2014 (www.gallup.com).

Jacobs, Anna W., Terrence D. Hill, Daniel Tope, and Laureen K. O'Brien. 2016. "Employment Transitions, Child Care Conflict, and the Mental Health of Low-Income Urban Women with Children." *Women's Health Issues* 26 (4): 366–376.

Jacobs, Jerry A. 2005. "Multiple Methods in *ASR*." *Footnotes* 33 (December): 1, 4.

Jagacinski, Carolyn M. 2013. "Women Engineering Students: Competence Perceptions and Achievement Goals in the Freshman Engineering Course." *Sex Roles* 69 (December): 644–657.

Jaimungal, Candice. 2022. "Half of Americans Have an American Flag at Home." YouGovAmerica, July 02, Accessed February 11, 2023. (https://today.yougov.com/topics/politics/articles-reports/2020/07/02/half-americans-have-american-flag-home).

Jamal, Ahmed, et al. 2017. "Tobacco Use Among Middle and High School Students—United States, 2011–2016." *MMWR* 66 (23): 597–603.

"Jail Bait." 2016. *The Economist*, December 10, 33.

Jandt, Fred E. 2001. *Intercultural Communication: An Introduction*. Thousand Oaks, CA: Sage.

Janis, Irving L. 1972. *Victims of Groupthink: A Psychological Study of Foreign-Policy Decisions and Fiascoes*. Boston: Houghton Mifflin.

Janis, Irving L. 1982. *Groupthink: A Psychological Study of Policy Decisions and Fiascoes*. Boston: Houghton Mifflin Company.

Janofsky, Michael. 2003. "Young Brides Stir New Outcry on Utah Polygamy." *New York Times*, February 28, 1.

Japsen, Bruce. 2019. "U.S. Health System Waste Hits $935 Billion A Year." *Forbes*, October 7. Accessed June 11, 2023 (https://www.forbes.com/sites/brucejapsen/2019/10/07/us-health-system-waste-hits-935-billion-a-year/?sh=5ed877ac2f40).

Jaret, Peter. 2015. "Prices Spike for Some Generics." *AARP Bulletin*, July–August, 8, 10.

Jargon, Julie. 2013. "Not-So-Fast Food." *Wall Street Journal*, November 15, B2.

Jargowsky, Paul A. 2015. "The Architecture of Segregation: Civil Unrest, the Concentration of Poverty, and Public Policy." The Century Foundation, August 9. Accessed August 20, 2017 (tcf.org).

Jaspers, Loes, Frederik Feys, Wichor M. Bramer, Oscar H. Franco, Peter Leusink, and Ellen T. M. Laan. 2016. "Efficacy and Safety of Flibanserin for the Treatment of Hypoactive Sexual Desire Disorder in Women: A Systematic Review and Meta-analysis." *JAMA Internal Medicine* 176 (4): 453–462.

Jeffreys, Sheila. 2011. *Man's Dominion: The Rise of Religion and the Eclipse of Women's Rights*. New York: Routledge.

Jeglic, Elizabeth L. 2021. "Understanding the Role of Women in Sex Trafficking." Psychology Today, December 30. Accessed April 21, 2023 (https://www.psychologytoday.com/us/blog/protecting-children-sexual-abuse/202112/understanding-the-role-women-in-sex-trafficking).

Jenkin, G., N. Madhvani, L. Signal, and S. Bowers. 2014. "A Systematic Review of Persuasive Marketing Techniques to Promote Food to Children on Television." *Obesity Reviews* 15 (April): 281–293.

Jewkes, Rachel, Emma Fulu, Tim Roselli, and Claudia Garcia-Moreno. 2013. "Prevalence of and Factors Associated with Non-Partner Rape Perpetration: Findings from the UN Multi-Country Cross-Sectional Study on Men and Violence in Asia and the Pacific." *Lancet Global Health* 1 (October): e208–218.

Johnson, Dave. 2014. "8 Ways Being Poor Is Wildly Expensive in America." AlterNet, January 29. Accessed February 14, 2014 (www.alternet.org).

Johnson, Hank, and Bert Klandermans, eds. 1995. *Social Movements and Culture*. Minneapolis: University of Minnesota Press.

Johnson, Heather L. 2016. "Pipelines, Pathways, and Institutional Leadership." American Council on Education. Accessed May 25, 2017 (acenet.edu).

Johnson, Ileana. 2010. "Why Has Communism Failed?" *Orthodoxy Today*, October 30. Accessed April 12, 2014 (www.orthodoxytoday.org).

Johnson, Jenna. 2016. "Trump's Rhetoric on Muslims Plays Well with Fans, but Horrifies Others." *Washington Post*, February 29. Accessed March 1, 2017 (washingtonpost.com).

Johnson, Kevin, and H. Darr Beiser. 2013. "Aging Prisoners' Costs Put Systems Nationwide in a Bind." *USA Today*, July 11. Accessed January 15, 2014 (www.usatoday.com).

Johnson, Kirk. 2011. "Between Young and Old, a Political Collision." *New York Times*, June 4, A10.

Johnson, Paula A., Therese Fitzgerald, Alina Salganicoff, Susan F. Wood, and Jill M. Goldstein. 2014. "Sex-Specific Medical Research: Why Women's Health Can't Wait." Brigham and Women's Hospital. Accessed October 17, 2015 (brighamandwomens.org).

Johnson, Rucker C. 2017. "Health." *Pathways*, Special Issue, 27–31.

Johnston, Pamela. 2005. "Dressing the Part." *Chronicle of Higher Education*, August 10. Accessed August 11, 2005 (www.chronicle.com).

Jones, Ann. 2013. *How the Wounded Return from America's Wars—The Untold Story*. Chicago, IL: Haymarket Books.

Jones, Charise. 2017. "Los Angeles, New York and San Francisco Are Most Congested U.S. Cities." *USA Today*, February 20. Accessed August 20, 2017 (usatoday.com).

Jones, Jeffrey M. 2012. "Gender Gap in 2012 Vote Is Largest in Gallup's History." Gallup, November 9. Accessed April 20, 2014 (www.gallup.com).

Jones, Jeffrey M. 2013. "U.S. Blacks, Hispanics Have No Preferences on Group Labels." Gallup, July 26. Accessed March 22, 2014 (www.gallup.com).

Jones, Jeffrey M. 2015. "Majority in U.S. Now Say Gays and Lesbians Born, Not Made." Gallup, May 20. Accessed September 29, 2015 (gallup.com).

Jones, Jeffrey M. 2016. "New Low of 52% 'Extremely Proud' to Be Americans." Gallup, July 1. Accessed July 1, 2016 (gallup.com).

Jones, Jeffrey M. 2017. "Independent Political ID in U.S. Lowest in Six Years." Gallup, January 6. Accessed June 27, 2017 (gallup.com).

Jeffrey, Jones. 2019. "Majority Rejects Idea of Haves, Have-Nots Divide in U.S." Gallup, April 3. Accessed July 23, 2023 (https://news.gallup.com/poll/248216/majority-rejects-idea-haves-nots-divide.aspx).

Jones, Jeffrey M. 2020. "Prevalence of Living Wills in U.S. Up Slightly." Gallup, June 22. Accessed May 19, 2023 (https://news.gallup.com/poll/312209/prevalence-living-wills-slightly.aspx).

Jones, Jeffrey M. 2020. "Is Marriage Becoming Irrelevant?" Gallup, December 28. Accessed May 19, 2023 (https://news.gallup.com/poll/316223/fewer-say-important-parents-married.asp).

Jones, Jeffrey M. 2021. "One in 10 LGBT Americans Married to Same-Sex Spouse." Gallup, February 24. Accessed May 19, 2023 (https://news.gallup.com/poll/329975/one-lgbt-americans-married-sex-spouse.aspx).

Jones, Jeffrey M. 2021. "In U.S., Black Confidence in Police Recovers From 2020 Low." Gallup, July 14. Accessed March 25, 2023 (https://news.gallup.com/poll/352304/black-confidence-police-recovers-2020-low.aspx).

Jones, Jeffrey M. 2023. "Americans Remain Divided on Preferred Immigration Levels." Gallup, July 23. Accessed April 26, 2023 (https://news.gallup.com/poll/352664/americans-remain-divided-preferred-immigration-levels.aspx).

Jones, Jeffrey M. 2022. "Americans Offer Gloomy State of the Nation Report." Gallup, February 2. Accessed April 14, 2023 (https://news.gallup.com/poll/389309/americans-offer-gloomy-state-nation-report.aspx).

Jones, Jeffrey. 2022. "Trust in Federal Government Branches Continues to Falter." Gallup, October 11. Accessed May 12, 2023 (https://news.gallup.com/poll/402737/trust-federal-government-branches-continues-falter.aspx).

Jones, Jeffrey M. 2022. "Belief in God in U.S. Dips to 81%, a New Low." Gallup, June 17. Accessed June 14, 2023 (https://news.gallup.com/poll/393737/belief-god-dips-new-low.aspx).

Jones, Jeffrey M. 2023 "U.S. LGBT Identification Steady at 7.2%" Gallup. Accessed April 14, 2023 (https://news.gallup.com/poll/470708/lgbt-identification-steady.aspx).

Jones, Jo, and William D. Mosher. 2013. "Fathers' Involvement with Their Children: United States, 2006–2010." *National Health Statistics Reports*, December 20. Accessed May 16, 2014 (www.cdc.gov/nchs).

Jones, Nicholas A., and Jungmiwha Bullock. 2012. "The Two or More Races Population: 2012." U.S. Census Bureau, September. Accessed March 22, 2014 (www.census.gov).

Jones, R. K., L. Frohwirth, and A. M. Moore. 2013. "More Than Poverty: Disruptive Events Among Women Having Abortions in the USA." *Journal of Family Planning and Reproductive Health Care* 39 (January): 36–43.

Jones, Rachel K., and Jenna Jerman. 2017. "Abortion Incidence and Service Availability in the United States, 2014." *Perspectives on Sexual and Reproductive Health* 49 (1): 17–27.

Jones, Rachel K., and Joerg Dreweke. 2011. "Countering Conventional Wisdom: New Evidence on Religion and Contraceptive Use." Guttmacher Institute, April. Accessed April 24, 2012 (www.guttmacher.org).

Jones, Rachel K., Lori F. Frohwirth, and Nakeisha M. Blades. 2016. "'If I Know I Am on the Pill and I Get Pregnant, It's an Act of God': Women's Views on Fatalism, Agency and Pregnancy." *Contraception* 93 (6): 551–555.

Jones, Robert P., Daniel Cox, E. J. Dionne, Jr., William A. Galston, Betsy Cooper, and Rachel Lienesch. 2016. "How Immigration and Concerns About Cultural Changes Are Shaping the 2016 Election." Public Religion Research Institute, June 23. Accessed June 20, 2017 (publicreligion.org).

Jones, Sophia. 2014. "Turkish Activists Say Their Country Is Sliding Backward on Women's Rights." *Huffington Post*, April 18. Accessed April 20, 2014 (www.huffingtonpost.com).

Jonsson, Jan O., David B. Grusky, Reinhard Pollak, and Matthew Di Carlo. 2009. "Recent Trends in Social Mobility in the United States: A New Approach to Modeling Trend in Big Class, Gradational, and Microclass Reproduction." Stanford Center for the Study of Poverty and Inequality, September. Accessed September 22, 2011 (www.inequality.com).

Jonsson, Patrik. 2007. "In US Justice, How Much Bias?" *Christian Science Monitor*, September 21, 1, 10.

Jordan, William. 2013. "Americans Divided over Blackface Halloween Make Up." YouGov, November 1. Accessed March 18, 2014 (www.today.yougov.com).

Joshi, Suchi P., Jochen Peter, and Patti M. Valkenburg. 2011. "Scripts of Sexual Desire and Danger in US and Dutch Teen Girl Magazines: A Cross-National Content Analysis." *Sex Roles* 64 (April): 463–474.

Juergensmeyer, Mark. 2003. "Thinking Globally About Religion." Pp. 3–13 in *Global Religions: An Introduction*, edited by Mark Juergensmeyer. New York: Oxford University Press.

Julian, Tiffany. 2012. "Work-Life Earnings by Field of Degree and Occupation for People with a Bachelor's Degree: 2011." U.S. Census Bureau, October. Accessed June 21, 2014 (www.census.gov).

K

Kabali, Hilda K., et al. 2015. "Exposure and Use of Mobile Media Devices by Young Children." *Pediatrics* 136 (December): 1044–1050.

Kachola, Suhan. 2023 "Voters Are Concerned About ChatGPT and Support More Regulation of AI." Data for Progress, May 10. Accessed June 27, 2023 (https://www.dataforprogress.org/blog/2023/5/10/voters-are-concerned-about-chatgpt-and-support-more-regulation-of-ai).

Kaeble, Danielle, and Lauren Glaze. 2016. "Correctional Populations in the United States, 2015." Bureau of Justice Statistics, December. Accessed March 2, 2017 (bjs.gov).

Kaeble, Danielle, and Thomas P. Bonczar. 2017. "Probation and Parole in the United States, 2015." Bureau of Justice Statistics, February 2. Accessed March 2, 2017 (bjs.gov).

Kahan, Dan M., Hank Jenkins-Smith, and Donald Braman. 2011. "Cultural Cognition of Scientific Consensus." *Journal of Risk Research* 14 (2): 147–174.

Kahn, Shulamit, and Donna Ginther. 2017. "Women and STEM." National Bureau of Economic Research, June. Accessed July 31, 2017 (nber.org).

Kaiser Family Foundation. 2016. "Key Facts About the Uninsured Population." Fact Sheet, September. Accessed August 15, 2017 (kff.org).

Kalev, Alexandra, Frank Dobbin, and Erin Kelly. 2006. "Best Practices or Best Guesses? Assessing the Efficacy of Corporate Affirmative Action and Diversity Policies." *American Sociological Review* 71 (August): 589–617.

Kallakuri, Chetana, Shruti Vaidyanathan, Meegan Kelly, and Rachel Cluett. 2016. "The 2016 International Energy Efficiency Scorecard." American Council for and Energy-Efficient Economy, July Accessed August 20, 2017 (aceee.org).

Kalogrides, Demetra, Susanna Loeb, and Tara Beteille. 2013. "Systematic Sorting: Teacher Characteristics and Class Assignments." *Sociology of Education* 86 (April): 103–123.

Kamp, Karin. 2016. "Missing from the Debates: Climate Change, Poverty, Campaign Finance and More." Moyers & Company, October 19. Accessed May 13, 2017 (billmoyers.com).

Kane, Nazneen. 2015. "Stay-at-Home Fatherhoods." *Contexts* 14 (Spring): 74–76.

Kaneda, Toshiko, and Kristin Bietsch. 2016. "2015 World Population Data Sheet." Washington, DC: Population Reference Bureau. Accessed September 13, 2014 (prb.org).

Kang, Cecilia. 2015. "Facebook Doubles Black Hires, but They Only Make Up About 1.5% of Staff." *Washington Post*, July 1. Accessed October 4, 2015 (washingtonpost.com).

Kann, Laura, Tim McManus et al. 2016. "Youth Risk Behavior Surveillance—United States, 2015." *MMWR* 65 (6): 1–180.

Kantor, Jodi, and David Streitfeld. 2015. "Amazon's Bruising, Thrilling Workplace." *New York Times*, August 16, A1.

Kaplan, David A. 2010. "The Best Company to Work For." *Fortune*, February 8, 57–72.

Kaplan, Juliana. 2022. "Biden Says Nearly 1.2 Million Women Haven't Returned to Work Because There's No Affordable Childcare." *Business Insider*, April 7. Accessed May 12, 2023 (https://www.businessinsider.com/biden-million-women-not-working-no-affordable-childcare-labor-shortage-2022-4).

Kassotis, Christopher D., Donald E. Tillitt, J. Wade Davis, Annette M. Hormann, and Susan C. Nagel. 2014. "Estrogen and Androgen Receptor Activities of Hydraulic Fracturing Chemicals and Surface and Ground Water in a Drilling-Dense Region." *Endocrinology* 155 (March): 897–907.

Kasumovic, Michael M., and Jeffrey H. Kuznekoff. 2015. "Insights into Sexism: Male Status and Performance Moderates Female-Directed Hostile and Amicable Behavior." *PloS One* 10 (7): e0131613. Doi:10.1371/journal.pone.0131613.

Katz, Jackson. 2006. *The Macho Paradox: Why Some Men Hurt Women and How All Men Can Help*. Naperville, IL: Sourcebooks.

Katz, Lawrence F., and Alan B. Krueger. 2016. "The Rise and Nature of Alternative Work Arrangements in the United States, 1995–2015." National Bureau of Economic Research, September. Accessed December 6, 2016 (nber.org).

Katzenbach, Jon. 2003. *Why Pride Matters More Than Money: The Power of the World's Greatest Motivational Force.* New York: Crown Business.

Kaufman, Alexander C. 2015. "Why Marriage Equality Is Great for the Economy." *Huffington Post*, June 26. Accessed September 29, 2015 (huffingtonpost.com).

Kay, Aaron C., Martin V. Day, Mark P. Zanna, and A. David Nussbaum. 2013. "The Insidious (and Ironic) Effects of Positive Stereotypes." *Journal of Experimental Social Psychology* 49 (March): 287–291.

Kearney, Melissa S., and Phillip B. Levine. 2014. "Teen Births Are Falling: What's Going On?" Brookings Institution, March. Accessed May 16, 2014 (www.brookings.edu).

Kearney, Melissa S., Benjamin H. Harris, Elisa Jácome, and Lucie Parker. 2014. "Ten Economic Facts About Crime and Incarceration in the United States." The Hamilton Project, May. Accessed November 4, 2014 (www.hamiltonproject.org).

Kearns, Cristin E., Laura A. Schmidt, and Stanton A. Glantz. 2016. "Sugar Industry and Coronary Heart Disease Research: A Historical Analysis of Internal Industry Documents." *JAMA Internal Medicine* 176 (November): 1680–1685.

Keating, Elizabeth, and Cynthia English. 2015. "U.S. Girls Less Confident They Can Learn Computer Science." Gallup, November 17. Accessed July 30, 2017 (gallup.com).

Keeter, Scott. 2010. "Ask the Expert." Pew Research Center, December 29. Accessed December 30, 2010 (pewresearch.org).

Keeter, Scott, Kyley McGeeney, Ruth Igielnik, Andrew Mercer, and Nancy Mathiowetz. 2015. "From Telephone to the Web: The Challenge of Mode of Interview Effects in Public Opinion Polls." Pew Research Center, May 13. Accessed July 22, 2015 (pewresearch.org).

Keisler-Starkey, Katherine, and Lisa N. Bunch. 2022. "Health Insurance Coverage in the United States: 2021." U.S. Census Bureau, September 13. Accessed June 11, 2023 (https://www.census.gov/library/publications/2022/demo/p60-278.html).

Kelderman, Eric. 2016. "Chill on Funding Still Puts Limits on Gun-Violence Research." *Chronicle of Higher Education*, January 22, A10–A11.

Kelly, Jack. 2023. "The 'Sandwich Generation' Is Financially Taking Care of Their Parents, Kids and Themselves." *Forbes,* February 24. Accessed May 19, 2023 (https://www.forbes.com/sites/jackkelly/2023/02/24/the-sandwich-generation-is-financially-taking-care-of-their-parents-kids-and-themselves/?sh=68ee996b2af4).

Kelley, Trista. 2013. "The Big Bucks in Keeping Kids Focused." *Bloomberg Businessweek, October 14–20, 25–27.*

Kelly, Heather. 2021. "For Seniors Using Tech to Age in Place, Surveillance Can Be the Price of Independence." *The Washington Post*, June 19. Accessed June 27, 2023 (https://www.washingtonpost.com/technology/2021/11/19/seniors-smart-home-privacy/).

Kelly, Heather, and Amily Guskin. 2021. "Americans Widely Distrust Facebook, TikTok and Instagram with Their Data, Poll Finds." *The Washington Post,* December 22. Accessed September 7, 2023 (https://www.washingtonpost.com/technology/2021/12/22/tech-trust-survey/).

Kelly, William R. 2016. *The Future of Crime and Punishment: Smart Policies for Reducing Crime and Saving Money*. Lanham, MD: Rowman & Littlefield.

Kemp, Alice Abel. 1994. *Women's Work: Degraded and Devalued*. Englewood Cliffs, NJ: Prentice Hall.

Kempner, Joanna, Clifford S. Perlis, and John F. Merz. 2005. "Ethics: Forbidden Knowledge." *Science* 307 (February 11): 854.

Kena, Grace, Susan Aud, Frank Johnson, Xiaolei Wang, Jijun Zhang, Amy Rathbun, Sidney Wilkinson-Flicker, and Paul Kristapovich. 2014. "The Condition of Education 2014." National Center for Education Statistics, May. Accessed June 21, 2014 (www.nces.ed.gov).

Kendall, Diana. 2002. *The Power of Good Deeds: Privileged Women and the Social Reproduction of the Upper Class*. Lanham, MD: Rowman & Littlefield.

Kendrick, Sam, and Nancy Jo Keppel. 2022. "Scripting Sex in Courtship: Predicting Genital Contact in Date Outcomes." *Sexuality & Culture* 26: 1190–121. Accessed April 23, 2023. Accessed July 31, 2023 (https://doi.org/10.1007/s12119-021-09938-2).

Kennedy, Courtney, et al. 2016. "Evaluating Online Nonprobability Surveys." Pew Research Center, May 2. Accessed May 8, 2016 (pewresearch.org).

Kerr, Emma, and Ilana. Kowarski. 2022. "25 Highest-Paying Associate Degree Jobs." *U.S. News and World Report*, February 23 Accessed June 13, 2023 (https://www.usnews.com/education/best-colleges/paying-for-college/slideshows/10-community-college-degree-jobs-that-pay-50k?slide=15).

Kerr, Emma, and Sarah Wood. 2022. "See 20 Years of Tuition Growth at National Universities." *US News and World Report*, September 13. Accessed June 13, 2023 (https://www.usnews.com/education/best-colleges/paying-for-college/articles/see-20-years-of-tuition-growth-at-national-universities).

Kessem, Limor. 2016. "2016 Cybercrime Reloaded: Our Predictions for the Year Ahead." Security Intelligence, January 15. Accessed April 18, 2017 (securityintelligence.com).

Kessenides, Dimitra. 2017. "Sustainable Cotton." *Bloomberg Businessweek*, January 6, 62.

Kessler, Aaron M. 2015. "E.P.A. Issues Stiffer Rules on Vehicle Fuel Ratings." *New York Times*, February 24, B1, B8.

Kessler, Glenn. 2016. "Fact Check: How Much Help Did Trump's Father Give His Son?" *Washington Post*, September 26. Accessed May 10, 2017 (washingtonpost.com).

Ketcheson, Felicia, Lisa King, and J. Don Richardson. 2018. "Association Between Social Support and Mental Health Conditions in Treatment-Seeking Veterans and Canadian Armed Forces Personnel." *Journal of Military, Veteran and Family Health*. 4 (January): 20-32. Doi: 10.3138/jmvfh.2017-0001

Kettner, Peter M., Robert M. Moroney, and Lawrence L. Martin. 1999. *Designing and Managing Programs: An Effectiveness-Based Approach*, 2nd edition. Thousand Oaks, CA: Sage.

Khan, Shamus R. 2012. *Privilege: The Making of an Adolescent Elite at St. Paul's School*. Princeton, NJ: Princeton University Press.

Kharif, Olga. 2015. "A Technology that Reveals Your Feelings." *Bloomberg Businessweek*, August 10–23, 38–39.

Khaw, Lyndal B. L., and Jennifer L. Hardesty. 2009. "Leaving an Abusive Partner: Exploring Boundary Ambiguity Using the Stages of Change Model." *Journal of Family Theory & Review* 1 (March): 38–53.

Kidder, Katherine, Amy Schafer, Phillip Carter, and Andrew Swick. 2017. "From College to Cabinet: Women in National Security." Center for a New American Security. Accessed May 25, 2017 (cnas.org).

Kiefer, Francine. 2015. "Behind Landmark Sex-Trafficking Bills, 20 Women Senators." *Christian Science Monitor Weekly*, March 30, 14.

Kiel, Paul, and Dan Nguyen. 2015. "Bailout Tracker." ProPublica, September 14. Accessed September 15, 2015 (projects.propublica.org).

Kilgore, James. 2015. "What I Learned in Prison." *Chronicle Review*, August 7, B20.

Killoren, Sarah E., and Andrea L. Roach. 2014. "Sibling Conversations About Dating and Sexuality: Sisters as Confidants, Sources of Support, and Mentors." *Family Relations* 63 (April): 232–243.

Killoren, Sarah E., Edna C. Alfaro, Anna K. Lindell, and Cara Streit. 2014. "Mexican American College Students' Communication with Their Siblings." *Family Relations* 63 (October): 513–525.

Kilmann, R. H. 2011. *Quantum Organizations: A New Paradigm for Achieving Organizational Success and Personal Meaning*. Newport Coast, CA: Kilmann Diagnostics.

Kim, Steven, and James Roland. 2015. "The Affordable Care Act." Healthline, June 15. Accessed August 15, 2017 (healthline.com).

Kindy, Kimberly. 2013. "Tribes' Fight with Government Shakes Contractors." *Baltimore Sun*, December 29, 8.

Kinsey Institute. 2011. "Continuum of Human Sexuality." Accessed November 24, 2012 (www.indiana.edu/~kinsey).

Kinsey, Alfred C., Wardell B. Pomeroy, and Clyde E. Martin. 1948. *Sexual Behavior in the Human Male*. Philadelphia: Saunders.

Kirkegaard, Jacob Funk. 2015. "The True Levels of Government and Social Expenditures in Advanced Economies." Peterson Institute for International Economics, March. Accessed September 11, 2015 (piie.org).

Kirstein, Marielle, Joerg Dreweke, Rachel K. Jones, and Jesse Philbin. 2022. "100 Days Post-Roe: At Least 66 Clinics Across 15 US States Have Stopped Offering Abortion Care." Guttmacher Institute, October 2. Accessed May 22, 2023 (https://www.guttmacher.org/node/300216/printable/print).

Kivisto, Peter, and Dan Pittman. 2001. "Goffman's Dramaturgical Sociology: Personal Sales and Service in a Commodified World." Pp. 311–334 in *Illuminating Social Life: Classical and Contemporary Theory Revisited*, 2nd edition. Thousand Oaks, CA: Pine Forge Press.

Klandermans, Bert. 1984. "Mobilization and Participation: Social Psychological Explanations of Resource Mobilization Theory." *American Sociological Review* 49 (October): 583–600.

Kneebone, Elizabeth. 2016. "Suburban Poverty Is Missing from the Conversation About America's Future." Brookings, September 15. Accessed December 23, 2016 (brookings.edu).

Kneebone, Elizabeth, and Alan Berube. 2013. *Confronting Suburban Poverty in America*. Washington, DC: Brookings Institution.

Kneebone, Elizabeth, and Alan Berube. 2014. "Does the Suburbanization of Poverty Mean the War on Poverty Failed?" Brookings, January 8. Accessed February 6, 2014 (www.brookings.edu).

"Knife-edge Lives." 2016. *The Economist*, September 17, 41.

Koba, Mark. 2013. "Workers Unite! (So You Can Become Capitalists)." CNBC, December 13. Accessed December 19, 2013 (www.nbcnews.com).

Koch, Angela. 2023. "Here's What Employees Actually Want from a Leader." Business.com, March 6. Accessed March 11, 2023 (https://www.business.com/articles/what-employees-want-from-leaders/).

Kochhar, Rakesh, and Richard Fry. 2015. "5 Takeaways About the American Middle Class." Pew Research Center, December 10. Accessed March 1, 2017 (pewresearch.org).

Kochhar, Rakesh, and Russ Oates. 2014. "Attitudes About Aging: A Global Perspective." Pew Research Center, January 30. Accessed September 19, 2014 (www.pewresearch.org).

Kochhar, Rakesh, and Stella Sechopoulos. 2022. "How the American Middle Class Has Changed in the Past Five Decades." Pew Research Center, April 20. Accessed April 13, 2023 (https://www.pewresearch.org/fact-tank/2022/04/20/how-the-american-middle-class-has-changed-in-the-past-five-decades/).

Kochhar, Rakesh. 2023. "The Enduring Grip of the Gender Pay Gap." Pew Research Center, March 1. Accessed April 16, 2023 (https://www.pewresearch.org/social-trends/2023/03/01/the-enduring-grip-of-the-gender-pay-gap/).

Koenig, David, and Scott Mayerowitz. 2015. "1 or 2 Airlines Controlling Many Airports." *Baltimore Sun*, July 15, 14.

Kolata, Gina. 2017. "Gain May Usher in Pig-to-Human Transplants." *New York Times*, August 11, 3A.

Kolesnikov, Andrei. 2022. "Putin's War Has Moved Russia from Authoritarianism to Hybrid Totalitarianism." Carnegie Endowment for International Peace. April 19. Accessed May 12, 2023 (https://carnegieendowment.org/2022/04/19/putin-s-war-has-moved-russia-from-authoritarianism-to-hybrid-totalitarianism-pub-86921#:~:text=This%20is%20a%20sign%20of,a%20fully%20mature%20authoritarian%20regime).

Komando, Kim. 2016. "Facebook Is Watching and Tracking You More Than You Probably Realize." *USA Today*, March 18. Accessed September 2, 2017 (usatoday.com).

Komyati, Ariane. 2018. "How France's Food Waste Law Helps Those in Need." The Borgen Project, August 24. Accessed February 12, 2023 (https://borgenproject.org/frances-food-waste-law/#:~:text=In%20July%202016%2C%20the%20French,is%20usable%20as%20animal%20feed)

Kornhauser, William. 1959. *The Politics of Mass Society*. New York: Free Press.

Korownyk, Christina, et al. 2014. "Televised Medical Talk Shows—What They Recommend and the Evidence to Support Their Recommendations: A Prospective Observational Study." *BMJ Open*, 14 (December): 1–9.

Kosowatz, John. 2020. "Three Advances in Prosthetics." The American Society of Mechanical Engineers, October 14. Accessed June 27, 2023 (https://www.asme.org/topics-resources/content/three-advances-in-prosthetics).

Kotkin, Joel. 2016. *The Human City: Urbanism for the Rest of Us*. Evanston, IL: Agate B2.

Kosova, Weston, and Pat Wingert. 2009. "Crazy Talk." *Newsweek*, June 8, 54–62.

Kowarski, Ilana, and Cole Claybourn. 2023. "25 Highest-Paying Associate Degree Jobs." *U.S. News*, June 26. Accessed July 19, 2023 (https://www.usnews.com/education/best-colleges/paying-for-college/slideshows/10-community-college-degree-jobs-that-pay-50k?slide=15).

Kowitt, Beth. 2015. "The War on Big Food." *Fortune*, June 1, 61–70.

Kozlowski, Lynn T., and Kenneth E. Warner. 2017. "Adolescents and E-Cigarettes: Objects of Concern May Appear Larger Than They Are." *Drug and Alcohol Dependence* 174 (May): 209–214.

Kozol, Jonathan. 2005. *The Shame of the Nation: The Restoration of Apartheid Schooling in America*. New York: Crown.

Krache, D. 2008. "How to Ground a 'Helicopter Parent.'" CNN, August 19. Accessed July 10, 2009 (www.cnn.com).

Krajicek, David J. 2016. "7 Toxic Assaults on Communities of Color Besides Flint: The Dirty Racial Politics of Pollution." AlterNet, January 23. Accessed March 25, 2016 (alternet.org).

Kramer, Laura, and Ann Beutel. 2015. *The Sociology of Gender: A Brief Introduction*, 4th edition. New York: Oxford University Press.

Kramer, Stephanie. 2020. "Polygamy Is Rare Around the World and Mostly Confined to a Few Regions." Pew Research Center, December 7. Accessed May 19, 2023 (https://www.pewresearch.org/short-reads/2020/12/07/polygamy-is-rare-around-the-world-and-mostly-confined-to-a-few-regions/).

Kramer, Stephanie. 2021. "White Evangelicals More Likely Than Other Christians to Say People Should Prioritize Marriage, Procreation." Pew Research Center, December 16. Accessed February 11, 2023 (https://www.pewresearch.org/fact-tank/2021/12/16/white-evangelicals-more-likely-than-other-christians-to-say-people-should-prioritize-marriage-procreation/).

Krantz, Matt. 2015. "9 CEOs Paid 800 Times More Than Their Workers." *USA Today*, August 5. Accessed October 7, 2015 (usatoday.com).

Kraska, Peter B. 2004. *Theorizing Criminal Justice: Eight Essential Orientations*. Long Grove, IL: Waveland Press.

Krasnova, Hanna, Helena Wenninger, Thomas Widjaja, and Peter Buxmann. 2013. "Envy on Facebook: A Hidden Threat to Users' Life Satisfaction?" Unpublished paper presented at the 11th International Conference on Wirtschaftinformatik, February 27–March 1, Leipzig, Germany. Accessed December 5, 2013 (www.aisel.aisnet.org).

Kraus, Michael W., and Jacinth J. X. Tan. 2015. "Americans Overestimate Social Class Mobility." *Journal of Experimental Social Psychology* 58 (January): 101–111.

Kreager, Derek A., and Jeremy Staff. 2009. "The Sexual Double Standard and Adolescent Peer Acceptance." *Social Psychology Quarterly* 72 (June): 143.

Krebs, Cathy. 2021. "Five Facts About the Troubled Teen Industry." American Bar Association. October 22. Accessed July 23, 2023 (https://www.americanbar.org/groups/litigation/committees/childrens-rights/practice/2021/5-facts-about-the-troubled-teen-industry/).

Kreeger, Karen Y. 2002. "Sex-Based Differences Continue to Mount." *The Scientist* 16 (February 18). Accessed November 20, 2013 (www.the-scientist.com).

Kreider, Rose M., and Renee Ellis. 2011. "Number, Timing, and Duration of Marriages and Divorces: 2009." U.S. Census Bureau, Current Population Reports, May. Accessed March 10, 2012 (www.census.gov).

Kreitner, Ronald. 2016. "Baltimore Subsidies Line Developers' Pockets at the City's Expense." *Baltimore Sun*, March 1, 13.

Krisher, Tom, and Michael Biesecker. 2016. "VW to Spend $10 Billion to Settle Claims." *Baltimore Sun*, June 24, 12.

Krishnamurthy, Prasad, and Aaron Edlin. 2014. "Affirmative Action and Stereotypes in Higher Education Admissions." National Bureau of Economic Research, October. Accessed December 5, 2014 (www.nber.org).

Kristof, Nicholas. 2014. "A Nation of Takers?" *New York Times*, March 26, A31.

Kristof, Nicholas. 2016. "Growing up Poor in America." *New York Times*, October 28, SR1.

Krogstad, Jens Manuel. 2014. "Asian American Voter Turnout Lags Behind Other Groups; Some Non-Voters Say They're 'Too Busy.'" Pew Research Center, April 9. Accessed April 20, 2014 (www.pewresearch.org).

Krogstad, Jens Manuel, and Jynnah Radford. 2017. "Key Facts About Refugees to the U.S." Pew Research Center, January 30. Accessed June 20, 2017 (pewresearch.org).

Krogstad, Jens Manuel, Jeffrey S. Passel, Mohamed Moslimani, and Luis Not-Bustamante. 2023. "Key facts about U.S. Latinos for National Hispanic Heritage Month." Pew Research Center, November 22. Accessed May 3, 2023 (https://www.pewresearch.org/short-reads/2023/09/22/key-facts-about-us-latinos-for-national-hispanic-heritage-month/).

Kroll, Luisa, and Kerry A. Dolan, 2013. "Forbes Billionaires 2013: The Top 20." Forbes, April 1. Accessed November 22, 2023 (https://www.forbesindia.com/article/worlds-billionaires-2013/forbes-billionaires-2013-the-top-20/34945/1). Chapter 8, page 152.

Kromer, Braedyn, and David Howard. 2013. "Labor Force Participation and Work Status of People 65 Years and Older." U.S. Census Bureau, January. Accessed May 16, 2014 (www.census.gov).

Kross, Ethan, et al. 2013. "Facebook Use Predicts Declines in Subjective Well-Being in Young Adults." *PloS One* 8 (8): e69841. Accessed December 5, 2013 (www.plosone.org).

Krueger, Alan B. 2016. "Where Have All the Workers Gone?" Prepared for Federal Reserve Bank of Boston, 60th Economic Conference, October 14. Accessed May 15, 2017 (bostonfed.org).

Krumwiede, Kip. 2017. "IMA Global Salary Survey for 2016." Association of Accountants and Financial Professionals in Business. Accessed June 21, 2017 (imanet.org).

Kueppers, Courtney. 2016. "Should Colleges Ever Disinvite a Controversial Speaker?" *Chronicle of Higher Education*, March 25, A6.

Kuhn, Manford H., and Thomas S. McPartland. 1954. "An Empirical Investigation of Self-Attitudes." *American Sociological Review* 19 (1): 68–76.

Kulczycki, Andrei, and Arun P. Lobo. 2002. "Patterns, Determinants, and Implications of Intermarriage Among Arab Americans." *Journal of Marriage and the Family* 64 (February): 202–210.

Kunkel Dale L., Jessica S. Castonguay, and Christine R. Filer. 2015. "Evaluating Industry Self-Regulation of Food Marketing to Children." *American Journal of Preventive Medicine* 49 (August): 181–187.

Küntzle, Julia, and Paul Blondé. 2013. "The Unhappy Fate of Ghanaian Witches." VICE United States, October 22. Accessed February 14, 2014 (www.vice.com).

Kurtzleben, Daniel. 2021. "We Just Got Our Clearest Picture Yet of How Biden Won In 2020." National Public Radio, July 1. Accessed January 29, 2023 (https://www.npr.org/2021/06/30/1011781661/we-just-got-our-clearest-picture-yet-of-how-biden-won-in-2020).

Kurzman, Charles, and David Schanzer. 2015. "The Other Terror Threat." *New York Times*, June 16, A27.

Kwok, Alvin C., et al. 2011. "The Intensity and Variation of Surgical Care at the End of Life: A Retrospective Cohort

Study." *The Lancet*, October 6. Accessed April 15, 2012 (www.lancet.com).

Kymlicka, Will. 2012. *Multiculturalism: Success, Failure, and the Future*. Washington, DC: Migration Policy Institute.

L

La France, Betty H., David D. Henningsen, Aubrey Oates, and Christina M. Shaw. 2009. "Social-Sexual Interactions? Meta-Analyses of Sex Differences in Perceptions of Flirtatiousness, Seductiveness, and Promiscuousness." *Communication Monographs* 76 (September): 263–285.

La Puma, John. 2014. "Don't Ask Your Doctor About 'Low T.'" *New York Times*, February 4, A21.

Lardieri, Alexa. 2018. "Celebrations and Riots Break Out in Philly Following Super Bowl Win." *U.S. News & World Report*, February 5. Accessed June 25, 2013 (https://www.usnews.com/news/national-news/articles/2018-02-05/celebrations-and-riots-break-out-in-philly-following-super-bowl-win).

Lakoff, Robin. T. 1990. *Talking Power: The Politics of Language*. New York: Basic Books.

Laliberte, Richard. 2016. "12 Ways the Health Care System May Be Harming You." *AARP Bulletin*, September, 18–24.

Lamb, Sharon, and Lyn Mikel Brown. 2007. *Packaging Girlhood: Rescuing Our Daughters from Marketers' Schemes*. New York: St. Martin's Press.

Lamble, Lucy. 2019. "Only Six Countries in the World Give Women and Men Equal Legal Work Rights." *The Guardian*, March 1. Accessed February 11, 2023 (https://www.theguardian.com/global-development/2019/mar/01/only-six-countries-in-the-world-give-women-and-men-equal-legal-rights).

Landgrave, Michelangelo, and Alex Nowrasteh. 2017. "Criminal Immigrants: Their Numbers, Demographics, and Countries of Origin." Cato Institute, March 15. Accessed May 20, 2017 (cato.org).

Landivar, Liana C. 2013. "Disparities in STEM Employment by Sex, Race, and Hispanic Origin." U.S. Census Bureau, September. Accessed June 21, 2014 (www.census.gov).

Landivar, Christin. 2023. "New Childcare Data Shows Prices Are Untenable for Families." U.S. Department of Labor Blog, January 24. Accessed September 18, 2023 (https://blog.dol.gov/2023/01/24/new-childcare-data-shows-prices-are-untenable-for-families).

Landsberger, Henry A. 1958. *Hawthorne Revisited*. Ithaca, NY: Cornell University Press.

Laneri, Raquel. 2011. "Slumdog Millions." *Forbes*, May 9, 100–101.

Lang, James M. 2013. "How Orwell and Twitter Revitalized My Course." *Chronicle of Higher Education*, November 1, A35–A37.

Lang, Kurt, and Gladys Engel Lang. 1961. *Collective Dynamics*. New York: Thomas Y. Crowell.

Langreth, Robert. 2013. "Do Robot Surgeons Do No Harm?" *Bloomberg Businessweek*, March 14, 17–18.

Lankford, Adam. 2015. "Mass Shooters, Firearms, and Social Strains: A Global Analysis of an Exceptionally American Problem." Paper presented at the American Sociological Association, 110th Annual Meeting, August 23, Chicago, Illinois.

Lannutti, Pamela J., Melanie Laliker, and Jerold L. Hale. 2001. "Violations of Expectations and Social-Sexual Communication in Student/ Professor Interactions." *Communication Education* 50 (January): 69–82.

Laraña, Enrique, Hank Johnston, and Joseph R. Gusfield, eds. 1994. *New Social Movements: From Ideology to Identity*. Philadelphia: Temple University Press.

Lareau, Annette. 2011. *Unequal Childhoods: Class, Race, and Family Life*, 2nd edition. Berkeley: University of California Press.

LaRossa, Ralph, and Donald C. Reitzes. 1993. "Symbolic Interactionism and Family Studies." Pp. 135–163 in *Sourcebook of Family Theories and Methods: A Contextual Approach*, edited by Pauline G. Boss, William J. Doherty, Ralph LaRossa, Walter R. Schumm, and Suzanne K. Steinmetz. New York: Plenum Press.

Larrimore, Jeff, Sam Dodini, and Logan Thomas. 2016. "Report on the Economic Well-Being of U.S. Households in 2015." Board of Governors of the Federal Reserve System, May. Accessed May 5, 2017 (federalreserve.gov).

Larson, Christina. 2017. "Over the Handlebars." *Bloomberg Businessweek*, August 14, 50.

Laughlin, Lynda. 2014. "A Child's Day: Living Arrangements, Nativity, and Family Transitions: 2011 (Selected Indicators of Child Well-Being)." U.S. Census Bureau, December. Accessed August 6, 2015 (census.gov).

Lavy, Victor, and Edith Sand. 2015. "On the Origins of Gender Human Capital Gaps: Short and Long Term Consequences of Teachers' Stereotypical Biases." National Bureau of Economic Research, January. Accessed September 16, 2015 (nber.org).

Law Center to Prevent Gun Violence. 2017. "Open Carry." Accessed September 2, 2017 (smartgunlaws.org).

Lawless, Jennifer L., and Richard L. Fox. 2013. "Girls Just Wanna Not Run: The Gender Gap in Young Americans' Political Ambition." Women & Politics Institute, March. Accessed April 21, 2014 (www.american.edu).

Lawless, Jennifer L. and Richard L. Fox. 2022. "Running for Office is Still for Men—Some Data on the "Ambition Gap"." Brookings, February 8. Accessed May 12, 2023 (https://www.brookings.edu/blog/fixgov/2022/02/08/running-for-office-is-still-for-men-some-data-on-the-ambition-gap/).

Lawn, Richard. 2013. *Experiencing Jazz*, 2nd edition. New York: Routledge.

Lawrence, Lee. 2013. "Was God Expelled?" *Christian Science Monitor*, June 17, 26–32.

Lawrence, Lee. 2015. "The Overbooked Generation." *Christian Science Monitor Weekly*, May 11, 26–32.

Lazear, Edward P., Kathryn L. Shaw, and Christopher T. Stanton. 2013. "The Value of Bosses." National Bureau of Economic Research, June. Accessed December 10, 2013 (www.nber.org).

Lea, Dan. 2023 "Health Disparities in Preventive Screenings for African Americans – Mayo Clinic." South Florida Hospital News, February 4. Accessed June 11, 2023 (https://southfloridahospitalnews.com/health-disparities-in-preventive-screenings-for-african-americans-mayo-clinic/).

LeBlanc, Paul. 2017. "The Countries That Allow Transgender Troops to Serve in Their Armed Forces." CNN, July 27. Accessed August 12, 2017 (cnn.com).

Le Bon, Gustave. 1896/1968. *The Crowd: A Study of the Popular Mind*, 2nd edition. Dunwoody, GA: Norman S. Berg.

Lee, David M., James Nazroo, Daryl B. O'Connor, Margaret Blake, and Neil Pendleton. 2015. "Sexual Health and Well-being Among Older Men and Women in England: Findings from the English Longitudinal Study of Ageing." *Archives of Sexual Behavior*, January 27, 1–12. Accessed September 29, 2015 (link.springer.com).

Lee, Jennifer, and Min Zhou. 2015. *The Asian American Achievement Paradox*. New York, NY: Russell Sage Foundation.

Lee, Lynda. 2023. "Minority Business Ownership Differs by Sector." U.S. Census Bureau, January 4. Accessed September 26, 2023 (https://www.census.gov/library/stories/2023/01/who-owns-americas-businesses.html).

Lee, Marlene. 2009. "Aging, Family Structure, and Health." Population Reference Bureau, October. Accessed April 10, 2010 (www.prb.org).

Lee, Yoojung. 2022. "World's Average Adult Will Be Worth $100,000 by 2024, Says Credit Suisse." Bloomberg, September 20. Accessed September 17, 2023. (https://www.bloomberg.com/news/articles/2022-09-20/world-will-get-more-millionaires-after-2022-s-wealth-destruction).

Legatum Prosperity Index. 2013. "The Prosperity Index 2013." Accessed February 6, 2014 (www.prosperity.com).

Leinberger, Christopher B., and Patrick Lynch. 2014. "Foot Traffic Ahead: Ranking Walkable Urbanism in America's Largest Metros." Smart Growth America. Accessed September 19, 2014 (www.smartgrowthAmerica.org).

Leland. John. 2022. "Can Robots Save Nursing Homes?" *The New York Times,* April 21. Accessed June 27, 2023 (https://www.nytimes.com/2022/04/21/realestate/nursing-home-robots.html).

Lemert, Edwin M. 1951. *Social Pathology: A Systematic Approach to the Theory of Sociopathic Behavior.* New York: McGraw-Hill.

Lemert, Edwin M. 1967. *Human Deviance, Social Problems and Social Control.* Englewood Cliffs, NJ: Prentice Hall.

Lemieux, Frederick. 2016. "6 Things Americans Should Know About Mass Shootings." Scientific American, June 13. Accessed December 3, 2016 (scientificamerican.com).

Lencioni, Patrick. 2002. *The Five Dysfunctions of a Team: A Leadership Fable.* San Francisco: Jossey-Bass.

Lengermann, Patricia Madoo, and Jill Niebrugge-Brantley. 1992. "Contemporary Feminist Theory." Pp. 308–357 in *Contemporary Sociological Theory,* 3rd edition, edited by George Ritzer. New York: McGraw-Hill.

Lenhart, Amanda, Aaron Smith, Monica Anderson, Maeve Duggan, and Andrew Perrin. 2015. "Teens, Technology & Friendships." Pew Research Center, August 6. Accessed August 14, 2015 (pewresearch.org).

Lenhart, Amanda, and Maeve Duggan. 2014. "Couples, the Internet, and Social Media." Pew Research Center, February 11. Accessed May 25, 2017 (pewresearch.org).

Lenhart, Amanda, Michele Ybarra, Kathryn Zickuhr, and Myeshia Price-Feeney. 2016. "Online Harassment, Digital Abuse, and Cyberstalking in America." Data & Society Research Institute and Center for Innovative Public Health Research, November 21. Accessed March 13, 2017 (datasociety.net).

Leonardsen, Dag. 2004. *Japan as a Low-Crime Nation.* New York: Palgrave Macmillan.

Leonhardt, David. 2017. "The Americans Who Saved Health Insurance." *New York Times,* August 1, A23.

Leppert, Rebecca. 2023. "A Look at Black-Owned Businesses in the U.S." Pew Research Center, February 21. Accessed May 3, 2023 (https://www.pewresearch.org/short-reads/2018/09/11/latinos-are-more-likely-to-believe-in-the-american-dream-but-most-say-it-is-hard-to-achieve/).

Leppert, Rebecca, and Drew DeSilver. 2023. "118th Congress Has a Record Number of Women." Pew Research Center, January 3. Accessed April 16, 2023 (https://www.pewresearch.org/fact-tank/2023/01/03/118th-congress-has-a-record-number-of-women/).

Leung, Maxwell. 2013. "Jeremy Lin's Model Minority Problem." *Contexts* 12 (Summer): 53–56.

Leurent, B., et al. 2013. "Spiritual and Religious Beliefs as Risk Factors for the Onset of Major Depression: An International Cohort Study." *Psychological Medicine* 43 (10): 2109–2120.

LeVay, Simon. 2011. *Gay, Straight, and the Reason Why: The Science of Sexual Orientation.* New York: Oxford University Press.

Levin, Diane E., and Jean Kilbourne. 2009. *So Sexy So Soon: The New Sexualized Childhood and What Parents Can Do to Protect Their Kids.* New York: Ballantine Books.

The Levin Institute. 2013. "Pop Culture." State University of New York. Accessed November 10, 2013 (www.globalization101.org).

Levin, Jack, and Jim Nolan. 2017. *The Violence of Hate: Understanding Harmful Forms of Bias and Bigotry,* 4th edition. Lanham, MD: Rowman & Littlefield.

Levine, Carrie, Matt Vasilogambros, Tim Henderson, and Pratheek Rebala. 2020. "How One Rural County Struggles to Find Polling Places." The Center for Public Integrity, October 30. Accessed May 12, 2023 (https://publicintegrity.org/politics/elections/ballotboxbarriers/rural-county-struggles-to-find-polling-places/).

Levine, Hallie. 2014. "5 Reasons Women Live Longer Than Men." *Health,* October 13. Accessed October 16, 2015 (news.health.com).

Levine, Lawrence W. 1998. *Highbrow/Lowbrow: The Emergence of Cultural Hierarchy in America.* Cambridge, MA: Harvard University Press.

Levinson, Lauren. 2016. "What Percentage of White Women Voted for Trump?" Pop Sugar, November 30. Accessed January 5, 2017 (popsugar.com).

Levinthal, Dave, Lateshia Beachum, and Carrie Levine. 2017. "Supreme Court a Millionaire's Club." Public Integrity, June 22. Accessed June 26, 2017 (publicintegrity.org).

Levintova, Hannah. 2016. "Minor Threats." *Mother Jones,* September/October, 36–41, 64.

Lewin, Kurt, Ronald R. Lippit, and Ralph K. White. 1939. "Patterns of Aggressive Behavior in Experimentally Created 'Social Climates.'" *Journal of Social Psychology* 10 (2): 271–301.

Lewin, Tamar. 2013. "New Milestone Emerges: Baby's First iPhone App." *New York Times,* October 26, A17.

Lewis, David Levering. 1993. *W. E. B. Du Bois: Biography of a Race, 1868–1919.* New York: Henry Holt.

Lewis, Oscar. 1966. "The Culture of Poverty." *Scientific American* 115 (October): 19–25.

Lewis, Ricki. 2014. "Noninvasive Prenatal DNA Test Okay for Low-Risk Pregnancies." Medscape, February 27. Accessed October 8, 2014 (www.medscape.com).

Lewis, Tanya. 2014. "Twins Separated at Birth Reveal Staggering Influence of Genetics." *Live Science,* August 11. Accessed July 23, 2023 (https://www.livescience.com/47288-twin-study-importance-of-genetics.html).

Li, Shan. 2015. "Toy-makers Building a New Model for Girls." *Baltimore Sun,* March 1, 6.

"Life Size Barbie Shows Young Girls the Dangers of Unrealistic Body Expectations." 2011. Diets in Review, May 3. Accessed October 25, 2011 (www.dietsinreview.com).

Light, Michael T., and Jeffrey T. Ulmer. 2016. "Explaining the Gaps in White, Black, and Hispanic Violence Since 1990: Accounting for Immigration, Incarceration, and Inequality." *American Sociological Review* 81 (April): 290–315.

Lilly, J. Robert, Francis T. Cullen, and Richard A. Ball. 1995. *Criminological Theory: Context and Consequences,* 2nd edition. Thousand Oaks, CA: Sage.

Lindberg, Laura D., Isaac Maddow-Zimet, and Heather Boonstra. 2016. "Changes in Adolescents' Receipt of Sex Education, 2006–2013." *Journal of Adolescent Health* 58 (6): 621–627.

Lindberg, Laura D., and Leslie M. Kantor. 2022. "Adolescents' Receipt of Sex Education in a Nationally Representative Sample, 2011–2019." *Journal of Adolescent Health* 70 (February): 290-297.

Lindberg, Laura D., Lauren Firestein, and Cynthia Beavin. 2021. "Trends in U.S. Adolescent Sexual Behavior and Contraceptive Use, 2006-2019." *Contraception X* (3). Accessed April 21, 2023 (https://doi.org/10.1016/j.conx.2021.100064).

Lindner, Andrew M. 2012. "An Old Tool with New Promise." *Contexts* 11 (Winter): 70–72.

Lindsey, Linda L. 2005. *Gender Roles: A Sociological Perspective,* 4th edition. Upper Saddle River, NJ: Prentice Hall.

Lino, Mark, Kevin Kuczynski, Nestor Rodriguez, and TusaRebecca Schap. 2017. "Expenditures on Children by Families, 2015." U.S. Department of Agriculture, Center for Nutrition Policy and Promotion. Accessed February 28, 2017 (cnpp.usda.gov).

Linton, Ralph. 1936. *The Study of Man.* New York: Appleton-Century-Crofts.

Linton, Ralph. 1964. *The Study of Man: An Introduction.* New York: Appleton-Century-Crofts.

Lipari, Rachel N., and Struther L. Van Horn. 2017. "Children Living with Parents Who Have a Substance Use Disorder." The Substance Abuse and Mental Health Services Administration (SAMHSA), August 24. Accessed February 18, 2023 (https://www.samhsa.gov/data/sites/default/files/report_3223/ShortReport-3223.html).

Lipka, Michael. 2014. "Few in LGBT Community See Pro Sports as Friendly." Pew Research Center, February 10. Accessed May 19, 2017 (pewresearch.org).

Lipka, Michael. 2016. "10 Facts About Atheists." Pew Research Center, May 4. Accessed June 1, 2017 (pewresearch.org).

Lipka, Michael. 2016. "5 Facts About Prayer." Pew Research Center, May 4. Accessed July 31, 2017 (pewresearch.org).

Lipka, Michael. 2017. "Muslim and Islam: Key Findings in the U.S. and Around the World." Pew Research Center, February 27. Accessed July 31, 2017 (pewresearch.org).

Lipka, Michael, and Sandra Stencel. 2013. "Racial and Ethnic Groups View 'Radical Life Extension' Differently." Pew Research Center, August 6. Accessed May 16, 2014 (www.pewresearch.org).

Lipman, Joanne. 2015. "Let's Expose the Gender Pay Gap." *New York Times,* August 13, A19.

Lipton, Eric, and Julie Creswell. 2016. "Documents Show How Wealthy Hid Millions Abroad." *New York Times,* May 6, A1, A14.

Lipton, Eric, and Rachel Abrams. 2016. "Their Hair Fell Out, and F.D.A. Can't Do a Thing." *New York Times,* August 16, A1, B2.

Liu, Sa, S. Katherine Hammond, and Ann Rojas-Cheatham. 2013. "Concentrations and Potential Health Risks of Metals in Lip Products." *Environmental Health Perspectives* 121 (June): 705–710.

Livingston, Gretchen. 2015. "It's No Longer a 'Leave It to Beaver' World for American Families—but It Wasn't Back Then, Either." Pew Research Center, December 30. Accessed July 20, 2017 (pewresearch.org).

Livingston, Gretchen. 2016. "5 Facts About Immigrant Mothers and U.S. Fertility Trends." Pew Research Center, October 26. Accessed June 20, 2017 (pewresearch.org).

Livingston, Gretchen. 2017. "Among U.S. Cohabiters, 18 Percent Have a Partner of a Different Race or Ethnicity." Pew Research Center, June 8. Accessed July 20, 2017 (pewresearch.org).

Livingston, Gretchen. 2018. "Stay-at-Home Moms and Dads Account for about One-in-Five U.S. Parents. "Pew Research Center, November 24. Accessed May 12, 2023 (https://www.pewresearch.org/short-reads/2018/09/24/stay-at-home-moms-and-dads-account-for-about-one-in-five-u-s-parents).

Livingston, Gretchen, and Anna Brown. 2014. "Birth Rate for Unmarried Women Declining for First Time in Decades." Pew Research Center, August 11. Accessed August 14, 2014 (www.pewresearch.org).

Livingston, Gretchen, and Anna Brown. 2017. "Intermarriage in the U.S. 50 Years After *Loving v. Virginia.*" Pew Research Center, May 18. Accessed June 20, 2017 (pewresearch.org).

Livingston, Gretchen, and D'Vera Cohn. 2013. "Long-Term Trend Accelerates Since Recession: Record Share of New Mothers Are College Educated." Pew Research Center, May 10. Brookings Institution, March. Accessed May 16, 2014 (pewresearch.org).

Llana, Sara M., and Sibylla Brodzinsky. 2012. "Latin America's Silent Scourge." *Christian Science Monitor,* November 26, 18–20.

Lloyd, Robin. 2022. "A Growing Drinking Water Crisis Threatens American Cities and Town. Scientific American." September 9. Accessed June 28, 2023 (https://www.scientificamerican.com/article/a-growing-drinking-water-crisis-threatens-american-cities-and-towns/?amp=true).

"Lobbying Database." 2017. Center for Responsive Politics, Open Secrets. Accessed June 26, 2017 (opensecrets.org).

Locher, David A. 2002. *Collective Behavior.* Upper Saddle River, NJ: Prentice Hall.

Lofland, John. 1996. *Social Movement Organizations: Guide to Research on Insurgent Realities.* New York: Aldine De Gruyter.

Logan, John, and Harvey Molotch. 1987. *Urban Fortunes: The Political Economy of Place.* Berkeley: University of California Press.

Lohan, Tara. 2015. "Why America's Deadly Love Affair with Bottles Water Has to Stop." AlterNet, September 17. Accessed October 23, 2015 (alternet.org).

Lohr, Steve. 2011. "Computers That See You and Keep Watch over You." *New York Times,* January 1, A1.

Lonas, Lexi. 2021. "Navajo Nation Becomes Largest Tribe in US After Pandemic Saw Climb in Enrollment." The Hill, May. Accessed September 26, 2023 (https://thehill.com/blogs/blog-briefing-room/news/554910-navajo-nation-becomes-largest-tribe-in-us-after-pandemic-saw/).

López, Gustavo, and Kristen Bialik, 2017. "Key Findings about U.S. Immigrants." Pew Research Center, May 25. Reprinted in The California-Mexico Studies Center. Accessed November 20, 2023 (https://www.california-mexicocenter.org/key-findings-about-u-s-immigrants/).

López, Gustavo, and Jynnah Radford. 2017. "Statistical Portrait of the Foreign-Born Population in the United States." Pew Research Center, May 3. Accessed June 20, 2017 (pewhispanic.org).

Lopez, Mark H., and Ana Gonzalez-Barrera. 2013. "What Is the Future of Spanish in the United States?" Pew Research Center, September 5, 2013. Accessed November 10, 2013 (www.pewresearch.org).

Lopez, Mark Hugo, Ana Gonzalex-Barrera, and Jens Manuel Krogstad. 2018. "Latinos Are More Likely to Believe in the American Dream, But Most Say It Is Hard to Achieve." Pew Research Center, September 11. Accessed May 3, 2023 (https://www.pewresearch.org/short-reads/2018/09/11/latinos-are-more-likely-to-believe-in-the-american-dream-but-most-say-it-is-hard-to-achieve/).

Louie, J. D. 2015. "10 Examples of Confusing Etiquette in Other Countries." Listverse, February 3. Accessed February 14, 2017 (listverse.com).

Louis, Catherine Saint. 2015. "Many Children Under 5 Are Left to Their Mobile Devices, Survey Finds." *New York Times,* November 2, A16.

Loumarr, Ess. 2015. "How Much Weight Can You Lose by Not Drinking Soda?" Live Strong, January 28. Accessed August 23, 2015 (livestrong.com).

Love, Robert. 2017. "Why Drugs Cost So Much." *AARP Bulletin,* May, 16–23.

Loveless, Tom. 2014. "The 2014 Brown Center Report on American Education: How Well Are American Students Learning?" Brown Center on Education Policy, March. Accessed June 21, 2014 (www.brookings.edu).

Lowenberger, Amanda, Joanna Mauer, Andrew deLaski, Marianne DiMascio, Jennifer Amann, and Steven Nadel. 2012. "The Efficiency Boom: Cashing in on the Savings from Appliance Standards." American Council for an Energy-Efficient Economy, and Appliance Standards Awareness Project, March. Accessed June 22, 2012 (www.aceee.org).

Lowery, Wesley. 2014. "Senate Republicans Unanimously Reject Equal Pay Bill." *Washington Post,* April 9. Accessed April 21, 2014 (www.washingtonpost.com).

Lublin, Joann S. 2013. "Tyco's Former CEO Set for 2014 Parole." *Wall Street Journal,* December 4, B2.

Luckey, John R. 2008. "CRS Report for Congress: The United States Flag: Federal

Law Relating to Display and Associated Questions." Congressional Research Service, April 14. Accessed July 20, 2011 (www.senate.gov).

Lugaila, Terry. A. 1998. "Marital Status and Living Arrangements: March 1998 (Update)." U.S. Census Bureau, Current Population Reports. Accessed August 8, 2000 (www.census.gov).

Lugo, Luis, Alan Cooperman, and Cary Funk. 2013. "Living to 120 and Beyond: Americans' Views on Aging, Medical Advances, and Radical Life Extension." Pew Research Center, August 6. Accessed May 16, 2014 (www.pewresearch.org).

Luhby, Tami. 2021. "US Comes in Last in Health Care Rankings of High-Income Countries." *CNN*, August 4. Accessed June 10, 2023 (https://www.cnn.com/2021/08/04/health/us-health-care-rankings/index.html).

Luhby, Tami, and Tierney Sneed. 2023. "Federal Judge Says Insurers No Longer Have to Provide Some Preventive Care Services, Including Cancer and Heart Screenings, at No Cost." *CNN*, March 31. Accessed June 11, 2023 (https://www.cnn.com/2023/03/30/politics/affordable-care-act-preventive-care-reed-oconnor/index.html).

Lundberg, Shelly, and Robert A. Pollak. 2007. "The American Family and Family Economics." *Journal of Economic Perspectives* 21 (Spring): 3–26.

Lundberg, Shelly, Robert A. Pollak, and Jenna E. Stearns. 2016. "Family Inequality: Diverging Patterns in Marriage, Cohabitation, and Childbearing." National Bureau of Economic Research, March. Accessed July 19, 2017 (nber.org).

Luscombe, Richard. 2017. "Arrival of Puerto Ricans Post-Hurricane Maria Could Have Big Impact on Florida." *The Guardian*, October 12. Accessed November 1, 2017 (theguardian.com).

Lutz, William. 1989. *Doublespeak*. New York: Harper & Row.

Lynas, Mark. 2015. "Even in 2015, the Public Doesn't Trust Scientists." *Washington Post*, January 30. Accessed July 22, 2015 (washingtonpost.com).

Lynch, Eleanor W., and Marci J. Hanson, eds. 1999. *Developing Cross-Cultural Competence: A Guide for Working with Children and Their Families*, 2nd edition. Baltimore, MD: Paul H. Brookes.

Lynn, David B. 1969. *Parental and Sex Role Identification: A Theoretical Formulation.* Berkeley, CA: McCutchen.

Lynn, M., and M. Todoroff. 1995. "Women's Work and Family Lives." Pp. 244–271 in *Feminist Issues: Race, Class, and Sexuality*, edited by Nancy Mandell. Scarborough, Ontario: Prentice Hall Canada.

M

Ma, Hong, et al. 2017. "Correction of a Pathogenic Gene Mutation in Human Embryos." *Science* 548 (August): 413–419.

Maccoby, Eleanor E., and John A. Martin. 1983. "Socialization in the Context of the Family: Parent-Child Interaction." Pp. 1–101 in *Socialization, Personality, and Social Development: Vol. 4. Handbook of Child Psychology*, edited by E. Mavis Hetherington. New York: Wiley.

MacDorman, Marian F., T. J. Mathews, Ashna D. Mohangoo, and Jennifer Zeitlin. 2014. "International Comparisons of Infant Mortality and Related Factors: United States and Europe, 2010." *National Vital Statistics Reports* 63 (5): 1–7.

Mach, Annie L., and Ada S. Cornell. 2013. "Federal Employees Health Benefits Program (FEHBP): Available Health Insurance Options." Congressional Research Service, November, 13. Accessed November 17, 2014 (www.crs.gov).

Macionis, John J., and Vincent N. Parrillo. 2007. *Cities and Urban Life*, 4th edition. Upper Saddle River, NJ: Prentice Hall.

MacKenzie, Donald, and Judy Wajcman, eds. 1999. *The Social Shaping of Technology*, 2nd edition. Philadelphia: Open University Press.

Madden, Mary, Amanda Lenhart, Sandra Cortesi, Urs Gasser, Maeve Duggan, and Aaron Smith. 2013. "Teens, Social Media, and Privacy." Pew Research Center, May 21. Accessed December 4, 2013 (www.pewinternet.org).

Madrick, Jeff. 2013. "The Anti-Economist: The Fall and Rise of Occupy Wall Street." *Harper's Magazine*, March, 9–11.

Mahler, Daniel Gerszon, Nishant Yonzan, Christoph Lakner, R. Andres Castaneda Aguilar, and Haoya Wu. 2021. "Updated Estimates of the Impact of COVID-19 on Global Poverty: Turning the Corner on the Pandemic in 2021?" World Bank Blogs, June 24. Accessed July 23, 2023 (https://blogs.worldbank.org/opendata/updated-estimates-impact-covid-19-global-poverty-turning-corner-pandemic-2021).

Magill, Bobby. 2014. "Clean Power Plan Exempts Major CO_2 Emitters." Climate Central, June 17. Accessed September 19, 2014 (www.climatecontrol.org).

Makary, Martin A., and Michael Daniel. 2016. "Medical Error—The Third Leading Cause of Death in the US." *British Medical Journal* 353 (5): i2139.

Makinen, Julie. 2013. "Law on Elder Visits in Effect." *Baltimore Sun*, August 6, 6.

Makoni, Abbianca. 2022. The Fortune 500 List Has A 'Record Number' Of Black CEOs — But There Are Still Only 6 Of Them." People of Color in Tech, May 31. Accessed September 26, 2023 (https://peopleofcolorintech.com/front/the-fortune-500-list-has-a-record-number-of-black-ceos-but-theres-still-only-6-of-them/).

Malone, Clare. 2016. "Clinton Couldn't Win over White Women." FiveThirtyEight, November 9. Accessed January 5, 2017 (fivethirtyeight.com).

Malthus, Thomas Robert. 1798/1965. *An Essay on Population*. New York: Augustus Kelley.

Malthus, Thomas Robert. 1872/1991. *An Essay on the Principle of Population*, 7th edition. London: Reeves & Turner.

Mangan, Katherine. 2013. "Comanche Nation College Tries to Rescue a Lost Tribal Language." *Chronicle of Higher Education*, June 14, A18–A19.

Mangan, Katherine. 2013. "Life After Steel." *Chronicle of Higher Education*, April 26, A33–A35.

Mangan, Katherine. 2015. "Stack Those Credentials." *Chronicle of Higher Education*, September 18, B14–B17.

Mann, Michael E. 2017. "No Apologies, No Regrets." *Chronicle of Higher Education*, February 10, A6.

Manning, Wendy D., and Jessica A. Cohen. 2012. "Premarital Cohabitation and Marital Dissolution: An Examination of Recent Marriages." *Journal of Marriage and Family* 74 (April): 377–387.

Manning, Wendy D., and Pamela J. Smock. 2005. "Measuring and Modeling Cohabitation: New Perspectives from Qualitative Data." *Journal of Marriage and Family* 67 (November): 989–1002.

Manyika, James, Michael Chui, Mehdi Miremadi, Jacques Bughin, Katy George, Paul Willmott, and Martin Dewhurst. 2017. "A Future That Works: Automation, Employment, and Productivity." McKinsey Global Institute, January. Accessed May 25, 2017 (mckinsey.com/mgi).

Manza, Jeff, and Clem Brooks. 2016. "Why Aren't Americans Angrier About Rising Inequality?" *Pathways*, Winter, 22–26.

March for Our Lives. 2023. "Mission & Story – March for Our Lives." Accessed June 27, 2023 (https://marchforourlives.com/mission-story/).

Marcec, Dan. 2017. "Equilar/Associated Press CEO Pay Study 2017." Equilar, May 23. Accessed June 23, 2017 (equilar.com).

Marcotte, Amanda. 2014. "Is Religion Inherently Oppressive?" AlterNet, April 9. Accessed June 21, 2014 (www.alternet.org).

Marcus, Amy Dockser. 2013. "Genetic Testing Leaves More Patients Living in Limbo." *Wall Street Journal*, November 19, D1–D2.

Margolis, Hillary. 2021. "European Union Court Oks Bans on Religious Dress at Work." Human Rights Watch, June 19. Accessed June 13, 2023 (https://www

.hrw.org/news/2021/07/19/european-union-court-oks-bans-religious-dress-work#:~:text=Protection%20of%20religious%20freedom%20%E2%80%93%20for,people%20who%20wear%20religious%20dress).

Marikar, Sheila. 2015. "The Company Boosts Growth by Treating Its Employees Like Royalty." *Baltimore Sun*, June 7, 4.

"Marital Status." 2017. U.S. Census Bureau, Figure MS-2, April 4. Accessed July 20, 2017 (census.gov).

"Marketing in the Digital Age." 2015. *The Economist*, August 29, 51–52.

Markey, Richard M., Charlotte N. Markey, and Juliana E. French. 2014. "Violent Video Games and Real-World Violence: Rhetoric Versus Data." *Psychology of Popular Media Culture*, August 18. APA PsycNet. Accessed August 6, 2015 (psycnet.apa.org).

Marken, Stephanie. 2019. "Half in U.S. Now Consider College Education Very Important." Gallup, December 30. Accessed July 29, 2023 (https://www.gallup.com/education/272228/half-consider-college-education-important.aspx).

Markowitz, David. 2021. "Who Lied More During Their First 100 Days: Biden, Trump, Or Obama?" *Forbes*, August 30. Accessed January 29, 2023 (https://www.forbes.com/sites/davidmarkowitz/2021/04/30/who-lied-more-during-their-first-100-days-biden-trump-or-obama/?sh=7689544a1a89).

Marks, Rachel. 2021. "The U.S. Census Allows People to Identity as One or More Races and Select Their Ethnicity." U.S. Census Bureau, August 4. Accessed September 26, 2023 (https://www.census.gov/newsroom/blogs/random-samplings/2021/08/improvements-to-2020-census-race-hispanic-origin-question-designs.html).

Marquand, Robert. 2011. "Europe Rejects Multiculturalism." *Christian Science Monitor*, February 21, 14.

Marr, Bernard. 2023. "The Best Examples of What you Can do with ChatGPT." *Forbes*, March 1. Accessed June 27, 2023 (https://www.forbes.com/sites/bernardmarr/2023/03/01/the-best-examples-of-what-you-can-do-with-chatgpt/?sh=2ee14f52df11).

"Marriage in Japan." 2016. *The Economist*, April 16, 35–36.

"The Marriage Squeeze in India and China." 2015. *The Economist*, April 18, 36–37.

Marshall, Joey. 2019. "Are Religious People Happier, Healthier?" Pew Research Center, January 31. Accessed June 13, 2023. (https://www.pewresearch.org/short-reads/2019/01/31/are-religious-people-happier-healthier-our-new-global-study-explores-this-question/#:~:text=2%20There%20is%20not%20a,%E2%80%9Cvery%20good%E2%80%9D%20overall%20health).

Martin, Andrew. 2015. "Making Washington Fall in Love with Pizza Again." *Bloomberg Businessweek*, March 9–15, 35–36.

Martin, Chris. 2016. "The Greening of Adidas." *Bloomberg Businessweek*, May 9–15, 46–47.

Martin, Joyce A., Brady E. Hamilton, Michelle J. K. Osterman, Anne K. Driscoll, and T. J. Matthews. 2017. "Births: Final Data for 2015." National Center for Health Statistics, National Vital Statistics System, January 5. Accessed July 1, 2017 (cdc.gov/nchs).

Martin, Michael. 2021. "Computer and Internet Use in the United States: 2018." U.S. Census Bureau, April 2021. Accessed June 3, 2023 (https://www.census.gov/content/dam/Census/library/publications/2021/acs/acs-49.pdf).

Martin, Molly A. 2008. "The Intergenerational Correlation in Weight: How Genetic Resemblance Reveals the Social Role of Families." *American Journal of Sociology* 114 (Suppl.): S67–S105.

Martin, Sophie. 2013. "29 Non-Americans on the Food They Find Disgusting." Thought Catalog, October 3. Accessed February 14, 2017 (thoughtcatalog.com).

Martin, Steven, Nan Astone, and Elizabeth Peters. 2014. "Fewer Marriages, More Divergence: Marriage Projections for Millennials to Age 40." Urban Institute, April 29. Accessed July 25, 2014 (www.urban.org).

Martinez, Anthony, and Cheridan Christnacht. 2021. "Women Are Nearly Half of U.S. Workforce but Only 27% of STEM Workers." U.S. Census Bureau, January 26. Accessed September 7, 2023 (https://wellness.osu.edu/sites/default/files/documents/2022/05/OCWO_ParentalBurnout_3674200_Report_FINAL.pdf).

Martino, Steven C., Rebecca L. Collins, Stephanie Ann Kovalchik, Claude Messan Setodji, Elizabeth J. D'Amico, Kirsten Becker, William G. Shadel, and Anagha Alka Tolpadi. 2018. "Drinking It In: The Impact of Youth Exposure to Alcohol Advertising." Rand Corporation. Accessed February 17, 2023 (https://www.rand.org/pubs/research_briefs/RB10015.html#:~:text=Exposure%20to%20alcohol%20ads%20led,lasted%20up%20to%20two%20days).

Martins, Vinicius, J. B., et al. 2011. "Long-Lasting Effects of Undernutrition." *International Journal of Environmental Research and Public Health* 8 (June): 1817–1846.

Maruschak, Laura M., Jennifer Bronson, and Mariel Alper. 2021. "Survey of Prison Inmates, 2016: Parents in Prison and Their Minor Children." U.S. Department of Justice, March. Accessed February 18, 2023 (https://bjs.ojp.gov/content/pub/pdf/pptmcspi16st.pdf).

Marx, Gary T., and Douglas McAdam. 1994. *Collective Behavior and Social Movements: Process and Structure*. Upper Saddle River, NJ: Prentice Hall.

Marx, Karl. 1844/1964. *Economic and Philosophic Manuscripts of 1844*. New York: International Publishers.

Marx, Karl. 1845/1972. "The German Ideology." Pp. 110–164 in *The Marx-Engels Reader*, edited by Robert C. Tucker. New York: W. W. Norton.

Marx, Karl. 1867/1967. *Capital*, edited by Friedrich Engels. New York: International Publishers.

Marx, Karl. 1934. *The Class Struggles in France*. New York: International Publishers.

Marx, Karl. 1964. *Karl Marx: Selected Writings in Sociology and Social Philosophy*, translated by T. B. Bottomore. New York: McGraw-Hill.

Masci, David. 2016. "Key Findings About Americans' Views on Religious Liberty and Nondiscrimination." Pew Research Center, September 28. Accessed May 19, 2017 (pewresearch.org).

Masci, David. 2016. "Q&A: Why Millennials Are Less Religious Than Older Americans." Pew Research Center, January 8. Accessed July 31, 2017 (pewresearch.org).

Massey, Douglas S. 2007. *Categorically Unequal: The American Stratification System*. New York: Russell Sage Foundation.

Mast, Nina and Daniel Perez. 2023. "Historic Job Growth in 2022 Reflects Strong but Uneven Economic Recovery." Economic Policy Institute, January 25. Accessed May 10, 2023 (https://www.epi.org/blog/historic-job-growth-in-2022-reflects-strong-but-uneven-economic-recovery-state-and-local-lawmakers-should-prioritize-rebuilding-the-public-sector-in-2023/).

Mastrangelo, Victoria. 2022. "100 Women in Leadership in the Catholic Church." Fem Catholic, March 7. Accessed June 13, 2023 (https://www.femcatholic.com/post/100-women-in-leadership-in-the-catholic-church).

Mata, Douglas A., et al. 2015. "Prevalence of Depression and Depressive Symptoms Among Resident Physicians: A Systematic Review and Meta-Analysis." *JAMA* 314 (22): 2373–2383.

Matanoski, Joseph. 2017. "The Germy Case Against Bottled Water." *Baltimore Sun*, January 8, 21.

"Maternal Mortality." 2015. *The Economist*, July 18, 28.

Mather, Mark, Linda A. Jacobsen, and Kelvin M. Pollard. 2015. "Aging in the United States." Population Reference Bureau, December. Accessed July 20, 2017 (prb.org).

Matjasko, Jennifer L., Phyllis H. Niolon, and Linda A. Valle. 2013. "The Role of Economic Factors and Economic Support in Preventing and Escaping from Intimate Partner Violence." *Journal of Policy Analysis and Management* 32 (October): 122–141.

Mattel Inc. 2022. "Barbie: Careers." Accessed February 18, 2023 (http://www.barbie-media.com/images/careers.html).

Matthew, Dayna Bowen, and Richard V. Reeves. 2016. "6 Charts Showing Race Gaps Within the American Middle Class." Brookings, October 21. Accessed June 20, 2017 (brookings.edu).

Matthew, Dayna Bowen, and Richard V. Reeves. 2017. "Trump Won White Voters, but Serious Inequities Remain for Black Americans." Brookings, January 13. Accessed June 28, 2017 (brookings.edu).

Matzke, Nicholas J. 2016. "The Evolution of Antievolution Policies After *Kitzmiller v. Dover*." *Science* 351 (6268): 28–30.

Maxwell, Steve. 2012. "Water Is *Still* Cheap: Demonstrating the True Value of Water." *American Water Works Association* 104 (May): 31–36.

Maynard, Micheline. 2019. "How Popeyes' Fried Chicken Sandwich Became the Fast Food Phenomenon Of 2019." *Forbes*, December 22. Accessed June 29, 2023 (https://www.forbes.com/sites/michelinemaynard/2019/12/22/how-popeyes-fried-chicken-sandwich-became-the-fast-food-phenomenon-of-2019/?sh=6f816c492c43).

Mayo, Elton. 1945. *The Problems of an Industrial Civilization.* Cambridge, MA: Harvard University Press.

Mayo, Michael. 2012. "Home for the Holidays: Convicted Killers Goodman, LeVin." *Sun Sentinel*, December 19. Accessed January 12, 2014 (www.articles.sun-sentinel.com).

Mayo Clinic Staff. 2021 "Exercise: 7 Benefits of Regular Physical Activity." Mayo Clinic, October 7. Accessed June 11, 2023 (https://www.mayoclinic.org/healthy-lifestyle/fitness/in-depth/exercise/art-20048389).

Mayo Clinic Staff. 2022. "Hormone Therapy: Is It Right for You?" Mayo Clinic. December 6. Accessed January 29, 2023 (https://www.mayoclinic.org/diseases-conditions/menopause/in-depth/hormone-therapy/ART-20046372).

Mayol-Garcia, Yeris. 2022. "Girls Take Lessons, Join Clubs More Often Than Boys lo Boys Play More Sports." U.S. Census Bureau, July 26. Accessed February 18, 2023 (https://www.census.gov/library/stories/2022/07/children-continue-to-be-involved-in-extracurricular-activities.html).

McAdam, Doug, and Ronnelle Paulsen. 1994. "Specifying the Relationship Between Social Ties and Activism." *American Journal of Sociology* 99 (November): 640–667.

McCabe, Brian J. 2014. "When Property Values Rule." *Contexts* 13 (Winter): 38–43.

McCabe, Janice. 2016. "Friends with Academic Benefits." *Contexts* 15 (Summer): 22–29.

McCabe, Janice, Emily Fairchild, Liz Grauerholz, Bernice A. Pescosolido, and Daniel Tope. 2011. "Gender in Twentieth-Century Children's Books." *Gender & Society* 25 (April): 197–226.

McCarthy, Julie. 2015. "An Angry Young Man Leads Protests Against India's Affirmative Action." NPR, August 31. Accessed September 11, 2015 (npr.org).

McCarthy, Justin. 2015. "Majority in U.S. Maintain Need for Third Major Party." Gallup, September 25. Accessed October 7, 2015 (gallup.com).

McCarthy, Justin. 2015. "Majority in U.S. Still Say Moral Values Getting Worse." Gallup, June 2. Accessed July 1, 2016 (gallup.com).

McCarthy, Justin. 2015. "More Americans Say Crime Is Rising in U.S." Gallup, October 22. Accessed April 15, 2017 (gallup.com).

McCarthy, Justin. 2016. "Americans' Views Shift on Toughness of Justice System." Gallup, October 20. Accessed April 15, 2017 (gallup.com).

McCarthy, Justin. 2016. "Satisfaction with Acceptance of Gays in U.S. at New High." Gallup, January 18. Accessed May 20, 2017 (gallup.com).

McCarthy, Justin. 2017. "Americans More Positive About Effects of Immigration." Gallup, June 28. Accessed June 27, 2017 (gallup.com).

McCarthy, Justin. 2017. "Americans Split over New LGBT Protections, Restroom Policies." Gallup, May 18. Accessed May 20, 2017 (gallup.com).

McCarthy, Justin. 2017. "US Support for Gay Marriage Edges to New High." Gallup, May 15. Accessed May 20, 2017 (gallup.com).

McCarthy, Justin. 2018. "Over Four in 10 in U.S. Now Say Teen Sex Morally Acceptable." Gallup, June 7. Accessed April 21, 2023 (https://news.gallup.com/poll/235415/four-say-teen-sex-morally-acceptable.aspx).

McCarthy, Justin. 2021. "U.S. Approval of Interracial Marriage at New High of 94%." Gallup, September 10. Accessed September 26, 2023 (https://news.gallup.com/poll/354638/approval-interracial-marriage-new-high.aspx).

McCarthy, Justin. 2022. "Same-Sex Marriage Support Inches Up to New High of 71%." Gallup, June 1. Accessed September 17, 2023 (https://news.gallup.com/poll/393197/same-sex-marriage-support-inches-new-high.aspx).

McClendon, David, and Aleksandra Sandstrom. 2016. "Child Marriage Is Rare in the U.S., Though This Varies by State." Pew Research Center, November 1. Accessed July 20, 2017 (pewresearch.org).

McCoy, Charles, and Roscoe Scarborough. 2015. "Why Do We Watch Trashy TV?" *Washington Post*, May 27. Accessed July 26, 2015 (washingtonpost.com).

McDermott, Rose, James H. Fowler, and Nicholas A. Christakis. 2013. "Breaking Up Is Hard to Do, Unless Everyone Else Is Doing It Too: Social Network Effects on Divorce in a Longitudinal Sample." *Social Forces* 92 (2): 491–519.

McDevitt, Jack, Jack Levin, and Susan Bennett. 2002. "Hate Crime Offenders: An Expanded Typology." *Journal of Social Issues* 58 (Summer): 303–318.

McDonald, Kim A. 1999. "Studies of Women's Health Produce a Wealth of Knowledge on the Biology of Gender Differences." *Chronicle of Higher Education*, June 25, A19, A22.

McDonough, Katie. 2014. "The GOP's Other War on Women: 5 Gender Battlegrounds Beyond Abortion and Contraception." AlterNet, March 10. Accessed April 21, 2014 (www.alternet.org).

McDuff, Daniel, Jeffrey M. Girard, and Rana el Kaliouby. 2017. "Large-Scale Observational Evidence of Cross-Cultural Differences in Facial Behavior." *Journal of Nonverbal Behavior* 41 (March): 1–19.

McEwan, Melissa. 2012. "Unbelievable: Man Beats Wife, Judge Orders Him to Take Her Out to Red Lobster and the Bowling Alley." AlterNet, February 9. Accessed February 10, 2012 (www.alternet.org).

McFadden, Robert D., and Angela Macropoulos. 2008. "Wal-Mart Employee Trampled to Death." *New York Times*, November 29, A16.

McFalls, Joseph A., Jr. 2007. "Population: A Lively Introduction, 5th edition." *Population Bulletin* 62 (March): 1–31.

McGirt, Ellen. 2017. "Google Searches Its Soul." *Fortune*, February 1, 50–56.

McHale, Susan M. 2001. "Free-Time Activities in Middle Childhood: Links with Adjustment in Early Adolescence." *Child Development* 76 (November/December): 1764–1778.

McHale, Susan M., Kimberly A. Updegraff, and Shawn D. Whiteman. 2012. "Sibling Relationships and Influences in Childhood and Adolescence." *Journal of Marriage and Family* 74 (October): 913–930.

McIntosh, Peggy. 1995. "White Privilege and Male Privilege: A Personal Account of Coming to See Correspondences Through Work in Women's Studies." Pp. 76–87 in *Race, Class, and Gender: An Anthology*, 2nd edition, edited by Margaret L. Andersen and Patricia Hill Collins. Belmont, CA: Wadsworth.

McKenna, Phil. 2017. "China's Carbon Emissions Falling, While Trump Points U.S. in Opposite Direction." Inside Climate News, March 1. Accessed August 20, 2017 (insideclimatenews.org).

McKenzie, Brian. 2013. "Out-of-State and Long Commutes: 2011." U.S. Census Bureau, February. Accessed September 19, 2014 (www.census.gov).

McKernan, Signe-Mary, Caroline Ratcliffe, Eugene Steuerle, and Sisi Zhang. 2013. "Less Than Equal: Racial Disparities in Wealth Accumulation." Urban Institute, April. Accessed February 6, 2014 (www.urban.org).

McKinnon, Mark. 2011. "Do We Still Need Unions? No: Let's End a Privileged Class." *Newsweek,* March 7, 19.

McNeill, David. 2012. "Japanese Fraud Case Highlights Weaknesses in Scientific Publishing." *Chronicle of Higher Education,* October 12, A16–A17.

McPhail, Clark. 1991. *The Myth of the Madding Crowd.* New York: deGruyter.

McPhail, Clark, and Ronald T. Wohlstein. 1983. "Individual and Collective Behavior Within Gatherings, Demonstrations, and Riots." *Annual Review of Sociology* 9 (August): 579–600.

McPhate, Mike. 2016. "Record Number of False Convictions Were Overturned in 2015, Study Finds." *New York Times,* February 4, A13.

McPheat, Sean. 2019 "The Pros and Cons of Self-Managed Teams: Are They the Panacea of Development?" LinkedIn, January 9. Accessed March 11, 2023 (https://www.linkedin.com/pulse/pros-cons-self-managed-teams-panacea-development-sean-mcpheat/).

McPhillips, Deidre. 2023. "Drug Costs Lead Millions in the US to Not Take Medications as Prescribed, according to CDC." *CNN,* June 2 Accessed June 11, 2023 (https://www.cnn.com/2023/06/02/health/prescription-drug-costs-rationing/index.html).

McRae, Susan. 1999. "Cohabitation or Marriage?" Pp. 172–190 in *The Sociology of the Family,* edited by Graham Allan. Malden, MA: Blackwell Publishers.

Mead, George Herbert. 1934. *Mind, Self, and Society.* Chicago: University of Chicago Press.

Mead, George Herbert. 1964. *On Social Psychology.* Chicago: University of Chicago Press.

Mears, Ashley. 2020. *Very Important People: Status and Beauty in the Global Party Circuit.* Princeton, NJ: Princeton University Press.

Medical News Today. 2023. "How Diabetes Affects Men Vs. Women." January 30. Accessed February 17, 2023 (https://www.medicalnewstoday.com/articles/diabetes-affects-men-women#sexual-health).

"Medical Tourism." 2015. Centers for Disease Control and Prevention, February 23. Accessed October 17, 2015 (cdc.gov).

Mehta, Vinita. 2013. "Does Grief Counseling Cause More Harm Than Good?" Psychology Today, May 22. Accessed January 29, 2023. (https://www.psychologytoday.com/us/blog/head-games/201305/does-grief-counseling-cause-more-harm-good).

Meier, Barry, and Eric Lipton, 2013. "F.D.A. Shift on Painkillers Was Years in the Making." *New York Times,* October 28, A1.

Gawlik, Kate, and Bernadette Mazurek Melnyk. 2021. "Pandemic Parenting: Examining the Epidemic of Working Parental Burnout and Strategies to Help." The Ohio State University: Office of the Chief Wellness Officer and College of Nursing, May. Accessed March 11, 2023. (https://wellness.osu.edu/sites/default/files/documents/2022/05/OCWO_ParentalBurnout_3674200_Report_FINAL.pdf).

Meghjani, Tanaz. 2023. "Most Americans Aren't Getting Enough Exercise, Study Finds." *Time Magazine, Bloomberg,* January 27. Accessed June 11, 2023 (https://time.com/6250870/americans-exercise-habits-fitness/).

Melton, Glennon. 2013. "5 Reasons Social Media Is Dangerous for Me." *Huffington Post,* October 1. Accessed December 4, 2013 (www.huffingtonpost.com).

Melucci, Alberto. 1995. "The New Social Movements Revisited: Reflections on a Sociological Misunderstanding." Pp. 107–119 in *Social Movements and Social Classes: The Future of Collective Action,* edited by Louis Maheu. Thousand Oaks, CA: Sage.

Melzer, Scott. 2009. *Gun Crusaders: The NRA's Culture War.* New York: New York University Press.

Mencher, Steve. 2013. "Visit Your Parents or Be Sued: Is That Un-American?" AARP, July 8. Accessed November 2, 2013 (www.aarp.org).

Mendelsohn, Oliver, and Maria Vicziany. 1998. *The Untouchables, Subordination, Poverty and the State in Modern India.* New York: Cambridge University Press.

Mendoza, Jessica. 2016. "The Price Isn't Right." *Christian Science Monitor Weekly,* April 25 & May 2, 26–31.

Mertens, Richard. 2015. "In Small-Town America, Open Arms." *Christian Science Monitor Weekly,* March 23, 21–23.

Menissi, Fatima. 1991. *The Veil and the Male Elite: A Feminist Interpretation of Women's Rights in Islam.* Reading, MA: Addison-Wesley.

Menissi, Fatima. 1996. *Women's Rebellion and Islamic Memory.* Atlantic Highlands, NJ: Zed Books.

Mensik, Hailey. 2023. "Women Making Up More of Physician Workforce." Heathcare Dive, January 28. Accessed June 11, 2023 (https://www.healthcaredive.com/news/AAMC-us-physician-workforce-women-specialties/640621/#:~:text=Some%2037%25%20of%20active%20physicians,according%20to%20the%20AAMC%20report).

Mercer, Andrew, Claudia Deane, and Kyley McGeeney. 2016. "Why 2016 Election Polls Missed Their Mark." Pew Research Center, November 9. Accessed November 11, 2016 (pewresearch.org).

Merelli, Annalisa. 2017. "Only 4.2% of Fortune 500 Companies Are Run by Women." Quartz, March 7. Accessed April 10, 2017 (qz.com).

Merriam-Webster. 2022. "We Added 370 New Words to the Dictionary for September 2022." Accessed February 11, 2023 (https://www.merriam-webster.com/words-at-play/new-words-in-the-dictionary).

Merton, Robert K. 1938. "Social Structure and Anomie." *American Sociological Review* 3 (December): 672–682.

Merton, Robert K. 1948/1996. "The Self-Fulfilling Prophecy." Pp. 183–201 in *Robert K. Merton: On Social Structure and Science,* edited by Piotr Sztompka. Chicago: University of Chicago Press.

Merton, Robert K. 1949. "Discrimination and the American Creed." Pp. 99–126 in *Discrimination and National Welfare,* edited by Robert M. MacIver. New York: Harper.

Merton, Robert K. 1968. *Social Theory and Social Structure.* New York: Free Press.

Merton, Robert K., and Alice K. Rossi. 1950. "Contributions to the Theory of Reference Group Behavior." Pp. 40–105 in *Continuities in Social Research,* edited by Robert K. Merton and Paul L. Lazarsfeld. New York: Free Press.

Merz, C. Noel Bairey. 2015. "Improving Gender Equity in Medical Research (Letter to the Editor)." *New York Times,* July 27, A16.

Messerli, Joe. 2012. "Should Affirmative Action Policies, Which Give Preferential Treatment Based on Minority Status, Be Eliminated?" BalancedPolitics.org, January 7. Accessed June 21, 2014 (www.balancedpolitics.org).

Messing, Jill T., Jacquelyn Campbell, Janet Sullivan Wilson, Sheryll Brown, and Beverly Patchell. 2014. "Police Departments' Use of the Lethality Assessment Program: A Quasi-Experimental Evaluation." National Institute of Justice, March 31. Accessed September 2, 2015 (www.nij.gov).

Meston, Cindy M., and David M. Buss. 2007. "Why Humans Have Sex." *Archives of Sexual Behavior* 36 (August): 477–507.

Metcalf, Tom. 2022. "Global Wealth Hit Record High of $530 Trillion in 2021." *Bloomberg,* June 22. Accessed April 13,

2023 (https://www.bloomberg.com/news/articles/2022-06-09/global-wealth-hit-record-high-of-530-trillion-in-2021-bgc-says#xj4y7vzkg).

Meteyer, Karen B., and Maureen Perry-Jenkins. 2009. "Dyadic Parenting and Children's Externalizing Symptoms." *Family Relations* 58 (July): 289–302.

Metropolis and Women in Cities International. 2018. "Mapping Metropolitan Gender Policies." Accessed June 27, 2023 (https://www.metropolis.org/sites/default/files/resources/Mapping_metropolitan_gender_policies_0.pdf).

"M-Health: Health and Appiness." 2014. *The Economist*, February 1, 56–57.

Migdon, Brooke. 2022. "Here Are the States Where You Can (and Cannot) Change Your Gender Designation on Official Documents." *The Hill*, Changing America, May 31. Accessed April 16. 2023 (https://thehill.com/changing-america/respect/diversity-inclusion/3507206-here-are-the-states-where-you-can-and-cannot-change-your-gender-designation-on-official-documents/).

Michels, Robert. 1911/1949. *Political Parties.* Glencoe, IL: Free Press.

McKinsey and Company. 2022. "Freelance, Side Hustles, and Gigs: Many More Americans Have Become Independent Workers." August 23. Accessed May 10, 2023 (https://www.mckinsey.com/featured-insights/sustainable-inclusive-growth/future-of-america/freelance-side-hustles-and-gigs-many-more-americans-have-become-independent-workers).

Migration Policy Institute (MPI). 2021. "Regions of Birth for Immigrants in the United States, 1960-Present." Accessed September 26, 2023 (https://www.migrationpolicy.org/programs/data-hub/charts/regions-immigrant-birth-1960-present).

Milgram, Stanley. 1963. "Behavioral Study of Obedience." *Journal of Abnormal and Social Psychology* 67 (4): 371–378.

Milgram, Stanley. 1965. "Some Conditions of Obedience and Disobedience to Authority." *Human Relations* 18 (February): 57–76.

"Millennials Outnumber Baby Boomers and Are Far More Diverse, Census Bureau Reports." 2015. U.S. Census Bureau, June 25. Accessed October 4, 2015 (census.gov).

Miller, Claire C. 2013. "Angry over U.S. Surveillance, Tech Giants Bolster Defenses." *New York Times*, November 1, A1.

Miller, D. W. 2001. "DARE Reinvents Itself—With Help from Its Social-Scientist Critics." *Chronicle of Higher Education*, October 16, A12–A14.

Miller, Elizabeth, et al. 2015. "A School Health Center Intervention for Abusive Adolescent Relationships: A Cluster RCT." *Pediatrics* 135 (January): 76–85.

Miller, Lisa. 2012. "Feminism's Final Frontier? Religion." *Washington Post*, March 8. Accessed March 9, 2012 (www.washingtonpost.com).

Mills, C. Wright. 1956. *The Power Elite.* New York: Oxford University Press.

Mills, C. Wright. 1959. *The Sociological Imagination.* New York: Oxford University Press.

Mills, Theodore M. 1958. "Some Hypotheses on Small Groups from Simmel." *American Journal of Sociology* 63 (May): 642–650.

Millsap, Adam. 2016. "Privatizing Water Facilities Can Help Cash-Strapped Municipalities." *Forbes*, October 5. Accessed August 20, 2017 (forbes.com).

Milman, Oliver, and Jessica Glenza. 2016. "At Least 33 US Cities Used Water Testing 'Cheats' over Lead Concerns." *The Guardian*, June 2. Accessed August 15, 2017 (theguardian.com).

Ming, Liu. 2011. "Chinese College Drops Plan to Discourage Kissing and Other 'Uncivilized Behavior' on Campus." *Global Times*, April 22. Accessed October 25, 2011 (www.china.globaltimes.cn).

Mischel, Walter. 1966. "A Social Learning View of Sex Differences." Pp. 57–81 in *The Development of Sex Differences*, edited by Eleanor E. Maccoby. Stanford, CA: Stanford University Press.

Mitnik, Pablo A., and David B. Grusky. 2015. "Economic Mobility in the United States." The Pew Charitable Trusts, July. Accessed September 11, 2015 (pewtrusts.org).

Mohdin, Aamna. 2016. "American Women Voted Overwhelmingly for Clinton, Except the White Ones." Quartz, November 9. Accessed January 5, 2017 (qz.com).

Monk, Ellis P., Jr. 2015. "The Cost of Color: Skin Color, Discrimination, and Health among African-Americans." *American Journal of Sociology* 121 (September): 396–444.

Montanaro, Domenico, Rachel Wellford, and Simone Pathe. 2014. "Money Is Pretty Good Predictor of Who Will Win Elections." Public Broadcasting Service, November 11. Accessed November 23, 2014 (www.pbs.org).

Montez, Jennifer Karas, and Anna Zajacova. 2013. "Explaining the Widening Education Gap in Mortality Among U.S. White Women." *Journal of Health and Social Behavior* 54 (June): 165–181.

Montlake, Simon. 2016. "Maine's Minimum Wage Experiment." *Christian Science Monitor Weekly*, May 16, 20–23.

Mooney, Chris. 2015. "The Remote Alaskan Village That Needs to Be Relocated Due to Climate Change." *Washington Post*, February 24. Accessed October 23, 2015 (washingtonpost.com).

Moore, Elizabeth S. 2006. "It's Child's Play: Advergaming and the Online Marketing of Food to Children." Henry J. Kaiser Family Foundation, July. Accessed April 12, 2008 (www.kff.org).

Moore, Kathleen. 2008. "Low-Income Homes Green—and Affordable." *Daily Gazette*, July 1. Accessed September 21, 2008 (www.dailygazette.com).

Moore, Peter. 2015. "A Third of Young Americans Say They Aren't 100% Heterosexual." YouGov, August 20. Accessed September 29, 2015 (today.yougov.com).

Morello, Carol. 2014. "Census to Change the Way It Counts Gay Married Couples." *Washington Post*, May 26. Accessed October 23, 2014 (www.washingtonpost.com).

Morgan, Jacob. 2017. "The Gig Economy Explained." INC, April 14. Accessed May 15, 2017 (inc.com).

Morgan, Rachel E., and Alexandra Thompson. 2022. "Criminal Victimization, 2020 – Supplemental Statistical Tables." U.S. Department of Justice, February. Accessed March 12, 2023 (https://bjs.ojp.gov/content/pub/pdf/cv20sst.pdf).

Morgenson, Gretchen. 2004. "No Wonder C.E.O.s Love Those Mergers." *New York Times*, July 18, C1.

Morin, Rich. 2016. "Behind Trump's Win in Rural White America: Women Joined Men in Backing Him." Pew Research Center, November 17. Accessed January 5, 2017 (pewresearch.org).

Morin, Rich, and Seth Motel. 2012. "A Third of Americans Now Say They Are in the Lower Class." Pew Social & Demographic Trends, September 10. Accessed February 6, 2014 (www.pewsocialtrends.org).

Morin, Rich, Kim Parker, Renee Stepler, and Andrew Mercer. 2017. "Behind the Badge." Pew Research Center, January 11. Accessed April 10, 2017 (pewresearch.org).

Morris, Desmond. 1994. *Bodytalk: The Meaning of Human Gestures.* New York: Crown Trade Paperbacks.

Morris, Edward W., and Brea L. Perry. 2017. "Girls Behaving Badly? Race, Gender, and Subjective Evaluation in the Discipline of African American Girls." *Sociology of Education* 90 (2): 127–148.

Morris, Theresa. 2014. "C-Section Epidemic." *Contexts* 13 (Winter): 70–72.

Morris, Kevin, and Coryn Grange. 2021. "Large Racial Turnout Gap Persisted in 2020 Election." Brennan Center for Justice, August 6. Accessed May 12, 2023 (https://www.brennancenter.org/our-work/analysis-opinion/large-racial-turnout-gap-persisted-2020-election).

Morrison, Aaron. 2022. "Number of Hate Groups Declined in 2021, But Proud Boys Chapters Surging, Says SPLC." Associated Press. March 9. Accessed February 12,

2023 (https://www.pbs.org/newshour/nation/number-of-hate-groups-declined-in-2021-but-proud-boys-chapters-surging-says-splc).

Morrison, Denton E. 1971. "Some Notes Toward Theory on Relative Deprivation, Social Movements, and Social Change." *American Behavioral Scientist* 14 (May–June): 675–690.

Moslimani, Mohamad. 2022. "Key Facts about Black Eligible Voters in 2022." Pew Research Center, October 12. Accessed May 12, 2023 (https://www.pewresearch.org/short-reads/2022/10/12/key-facts-about-black-eligible-voters-in-2022/).

Moss-Racusin, Corrine A., John F. Dovidio, Victoria L. Brescoli, Mark J. Graham, and Jo Handelsman. 2012. "Science Faculty's Subtle Gender Biases Favor Male Students." *Proceedings of the National Academy of Sciences* 109 (October 9): 16474–16479.

Motel, Seth, and Eileen Patten. 2013. "Statistical Portrait of the Foreign-Born Population in the United States, 2011." Pew Research Hispanic Trends Project, January 29. Accessed March 22, 2014 (www.pewhispanic.org).

Motivans, Mark. 2013. "Federal Justice Statistics 2010—Statistical Tables." Bureau of Justice Statistics, December. Accessed January 12, 2014 (www.bjs.gov).

Moyer, Imogene L. 2001. *Criminological Theories: Traditional and Nontraditional Voices and Themes*. Thousand Oaks, CA: Sage.

Moyer, Imogene L. 2003. "Jane Addams: Pioneer in Criminology." *Women & Criminal Justice* 14 (3/4): 1–14.

Mozzafarian, Dariush, and David S. Ludwig. 2015. "Stop Fearing Fat." *New York Times*, July 9, A25.

Mudallal, Zainab. 2015. "Nearly Half of Americans Didn't Take a Vacation Day in 2014." Quartz, January 5. Accessed October 7, 2015 (qz.com).

Mui, Ylan Q. 2015. "Rather Than Raises, Firms Raise Perks." *Baltimore Sun*, July 31, 12.

Mullen, Ann. 2012. "The Not-So-Pink Ivory Tower." *Contexts* 4 (Fall): 34–38.

Mumford, Lewis. 1961. *The City in History: Its Origins, Transformations, and Its Prospects*. New York: Harcourt, Brace.

Munoz-Boudet, Ana Maria, and Ana Revenga. 2017. "Breaking the STEM Ceiling for Girls." Brookings, March 7. Accessed July 31, 2017 (brookings.edu).

Munsell, Christina R., Jennifer L. Harris, Vishnudas Sarda, and Marlene B. Schwartz. 2015. "Parents' Beliefs About the Healthfulness of Sugary Drink Options: Opportunities to Address Misperceptions." *Public Health Nutrition*, March 11, 1–9 published online ahead of print. Accessed October 25, 2015 (http://dx.doi.org/10.1017/S1368980015000397).

Murdock, George P. 1940. "The Cross-Cultural Survey." *American Sociological Review*, 5: 361–370.

Murphey, David, and P. Mae Cooper. 2015. "Parents Behind Bars: What Happens to Their Children?" Child Trends, October. Accessed February 28, 2017 (childtrends.org).

Murphey, David, Megan Barry, and Brigitte Vaughn. 2013. "Mental Health Disorders." Child Trends, January. Accessed July 26, 2014 (www.childtrends.org).

Murphy, John. 2004. "S. Africa's New Goal: Economic Equality." *Baltimore Sun*, April 27, 1A, 4A.

Murphy, Kevin. 2013. "Hallmark Responds to 'Gay' Ornament Controversy with Regret." *Huffington Post*, November 1. Accessed November 1, 2013 (www.huffingtonpost.com).

Murray, Charles. 1984. *Losing Ground: American Social Policy, 1950–1980*. New York: Basic Books.

Murray, Charles. 2012. *Coming Apart: The State of White America, 1960–2010*. New York: Crown Forum.

Musick, Kelly, Ann Meier, and Sarah Flood. 2016. "How Parents Fare: Mothers' and Fathers' Subjective Well-Being in Time with Children." *American Sociological Review* 81 (5): 1069–1095.

Myers, Joe. 2016. "This Is What Gun Ownership Looks Like Around the World." World Economic Forum, January 6. Accessed May 1, 2017 (weforum.org).

Myers, John P. 2007. *Dominant–Minority Relations in America: Convergence in the New World*, 2nd edition. Boston: Allyn & Bacon.

N

Nace, Trevor. 2018. "Only Two-Thirds of American Millennials Believe The Earth Is Round." *Forbes*, April 4. Accessed January 30, 2023 (https://www.forbes.com/sites/trevornace/2018/04/04/only-two-thirds-of-american-millennials-believe-the-earth-is-round/?sh=4e2f84c07ec6).

Nagoski, Emily. 2015. "Nothing Is with Your Sex Drive." *New York Times*, January 27, A21.

Naili, Hajer. 2011. "Study Details Sex-Traffic in Post-Saddam Iraq." Women's e-News, November 9. Accessed February 25, 2012 (www.womensenews.org).

Najam, Adil. 2016. "America's Nobel Success Is the Story of Wrong Immigrants." The Conversation, October 18. Accessed June 8, 2017 (theconversation.com).

"National African-American History Month: February 2017." 2017. U.S. Census Bureau. Accessed June 20, 2017 (census.gov).

National Association of Hispanic Real Estate Professionals. 2022. "State of Hispanic Homeownership Report." Accessed April 28, 2023 (https://nahrep.org/shhr/).

National Cancer Institute. 2020. "BRCA Gene Mutations: Cancer Risk and Genetic Testing." National Institute of Health. Accessed June 10, 2023 (https://www.cancer.gov/about-cancer/causes-prevention/genetics/brca-fact-sheet).

National Cancer Institute. 2023. "Cancer Stat Facts: Prostate Cancer." National Institute of Health. Accessed May 20, 2023 (https://seer.cancer.gov/statfacts/html/prost.html).

National Foundation for American Policy. 2021. "NFAO Policy Brief, October 2021, Immigrant and Nobel Prizes: 1901-2021." Accesses April 28, 2023 (https://nfap.com/wp-content/uploads/2021/10/Immigrants-and-Nobel-Prizes-1901-to-2021.NFAP-Policy-Brief.October-2021.pdf).

National Center for Education Statistics. 2016. "Career and Technical Education Statistics." Accessed June 13, 2023 (https://nces.ed.gov/surveys/ctes/figures/fig_2018057_A9.asp).

National Center for Education Statistics. 2019. "Education at a Glance 2019: Putting U.S. Data in a Global Context." NCES Blog, September 10. (https://nces.ed.gov/blogs/nces/post/education-at-a-glance-2019-putting-u-s-data-in-a-global-context).

National Center for Education Statistics. 2022. "Degrees Conferred by Race, Ethnicity, and Sex." Accessed May 12, 2023 (https://nces.ed.gov/fastfacts/display.asp?id=72

National Center for Education Statistics. 2022. "Fast Facts: Expenditures." Accessed March 25, 2023 (https://nces.ed.gov/fastfacts/display.asp?id=66).

National Center for Education Statistics. 2023. "Enrollment Rates of Young Children." Accessed June 13, 2023 (https://nces.ed.gov/programs/coe/pdf/2023/cfa_508.pdf)

National Center for Education Statistics. 2023. "NAEP Report Card: Reading." Accessed June 14, 2023 (https://nces.ed.gov/nationsreportcard/reading/).

National Center for Education Statistics. 2023 "NAEP Report Card: Math." Accessed June 14, 2023 (https://nces.ed.gov/nationsreportcard/mathematics/).

National Center for Employee Ownership. 2022. "The Employee Ownership 100: America's Largest Majority Employee-Owned Companies." October. Accessed March 11, 2023 (https://www.nceo.org/articles/employee-ownership-100).

National Center for Health Statistics. 2015. "Health, United States, 2014." May. Accessed October 8, 2015 (cdc.gov/nchs).

National Center for Statistics and Analysis. 2016. "Distracted Driving 2014." National Highway Traffic Safety Administration, April. Accessed December 25, 2016 (nrd.nhtsa.dot.gov).

National Center on Elder Abuse. 2012. "Statistics/Data." Accessed May 16, 2014 (www.ncea.aoa.gov).

National Conference of State Legislatures. 2014. "State Laws Regarding Marriages Between First Cousins." Accessed May 16, 2014 (www.ncsl.org).

National Council on Aging. 2021. "Get the Facts on Elder Abuse." February 23. Accessed May 22, 2023 (https://www.ncoa.org/article/get-the-facts-on-elder-abuse).

"National Grandparents Day 2016: Sept. 11." 2016. Census Bureau, Facts for Features, July 29. Accessed February 28, 2017 (census.gov).

"National Grandparents Day 2017: Sept 10." 2017. U.S. Census Bureau, July 10. Accessed July 20, 2017 (census.gov).

National Institute of Mental Health. 2011. "The Teen Brain: Still Under Construction." Accessed November 20, 2013 (www.nimh.nih.gov).

National Institute of Mental Health. 2020. "Suicide." Accessed January 30, 2023. (https://www.nimh.nih.gov/health/statistics/suicide)

National Institute on Alcohol Abuse and Alcoholism. 2022. "Underage Drinking." National Institutes of Health, October. Accessed February 22, 2023 (https://www.niaaa.nih.gov/publications/brochures-and-fact-sheets/underage-drinking).

National Institute on Alcohol Abuse and Alcoholism. 2022. "Women and Alcohol." National Institutes of Health, September. Accessed February 22, 2023 (https://www.niaaa.nih.gov/publications/brochures-and-fact-sheets/underage-drinking).

National Institute on Alcohol Abuse and Alcoholism. 2023. "College Drinking," March. Accessed June 11, 2023 (https://www.niaaa.nih.gov/publications/brochures-and-fact-sheets/college-drinking).

National Institute on Drug Abuse. 2017. "Overdose Death Rates." January. Accessed August 15, 2017 (drugabuse.gov).

National Institute on Drug Use, 2023. "Drug Overdose Death Rates." Accessed June 11, 2023 (https://nida.nih.gov/research-topics/trends-statistics/overdose-death-rates).

National Intelligence Council. 2022. "Digital Repression Growing Globally, Threatening Freedoms." October 31. Accessed June 25, 2023 (https://www.dni.gov/files/ODNI/documents/assessments/NIC-Declassified-Assessment-Digital-Repression-Growing-April2023.pdf).

National Oceanic and Atmospheric Administration. 2014. "Heat Wave: A Major Summer Killer." NOAA Watch. Accessed September 19, 2014 (www.noaawatch.gov).

National Organization on Fetal Alcohol Syndrome. 2012. "FASD: What Everyone Should Know." Accessed July 25, 2014 (www.nofas.org).

National Public Radio. 2010. "Black Male Privilege?" Interview transcript, March 4. Accessed March 10, 2010 (www.npr.org).

National Retail Federation. 2016. "Retailers See Increase in Organized Retail Crime." October 18. Accessed April 19, 2017 (nrf.com).

National Rifle Association. 2023. "Home of the NRA." Accessed June 28, 2023 (https://home.nra.org/).

National Science Board. 2014. *Science and Engineering Indicators 2014*. Arlington, VA: National Science Foundation.

National Survey of Student Engagement. 2016. "Engagement Insights: Survey Findings on the Quality of Undergraduate Education." Accessed July 31, 2017 (nsse.indiana.edu).

National Survey of Student Engagement. 2021. "Introducing NSSE Pulse: A Special Fall 2020 Survey Option; Findings Now Available." Accessed July 29, 2023 (https://nsse.indiana.edu/nsse/survey-instruments/pulse/index.html).

National Survey of Student Engagement. 2022. "Introducing NSSE Pulse: A Special Fall 2020 Survey Option; Findings Now Available." Accessed June 13, 2023 (https://nsse.indiana.edu/nsse/survey-instruments/pulse/index.html).

National Women's Law Center. 2014. "Women's Unemployment Rises Despite Job Gains, Growth Concentrated in Low-Wage Sectors, NWLC Analysis Shows." April 4. Accessed April 12, 2014 (www.nwlc.org).

"The Nation's Older Population Is Still Growing, Census Bureau Reports." 2017. June 22. Accessed July 20, 2017 (census.gov).

Navarro, Andrea 2022. "Landmark Deal with Union Agreement Also Includes Negotiations for 'Salary." *Bloomberg*, May 12. Accessed May 12, 2023 (https://www.bloomberg.com/news/articles/2022-05-12/gm-hikes-mexico-plant-wages-by-8-5-in-landmark-deal-with-union#xj4y7vzkg).

Nechepurenko, Ivan. 2014. "Russia Reverses Birth Decline—But for How Long?" *The Moscow Times*, June 22. Accessed September 21 (www.themoscowtimes.com).

Neider, Linda L., and Chester A. Schriesheim, eds. 2005. *Understanding Teams*. Greenwich, CT: Information Age.

Nelson, Libby, and Emily Crockett. 2017. "Sexual Assault Allegations Against Donald Trump: 15 Women Say He Groped, Kissed, or Assaulted Them." Vox, January 19. Accessed March 14, 2017 (vox.com).

Neuhaus, Jessamyn. 2010. "Marge Simpson, Blue-Haired Housewife: Defining Domesticity on *The Simpsons." The Journal of Popular Culture* 43 (August): 761–781.

Neuman, Scott. 2021. "Under Half of U.S. Adults Belong to A Religious Congregation, Survey Says." NPR, March 30. Accessed June 13, 2023 (https://www.npr.org/2021/03/30/982671783/fewer-than-half-of-u-s-adults-belong-to-a-religious-congregation-new-poll-shows).

"The New American Father." 2013. Pew Research Center, June 14. Accessed May 16, 2014 (www.pewresearch.org).

New York Times. 2016. "Transcript: Donald Trump's Taped Comments About Women." October 8. Accessed March 14, 2017 (nytimes.com).

Newberry, Sydne. 2013. "What's in a Name? Calling Obesity a Disease Could Help Improve Chronic Disease Outcomes." The RAND Blog, October 16. Accessed July 26, 2014 (www.rand.org).

Newkirk, Margaret, and Gigi Douban. 2012. "Legal Immigrants Wanted for Dirty Jobs." *Bloomberg Businessweek*, October 8–14, 34, 36.

Newmark, David, Ian Burn, and Patrick Button. 2017. "Age Discrimination and Hiring of Older Workers." Federal Reserve Bank of San Francisco, February 27. Accessed July 20, 2017 (frbsf.org).

Newport, Frank. 2015. "Most U.S. Smartphone Owners Check Phone at Least Hourly." Gallup, July 9. Accessed August 14, 2015 (gallup.com).

Newport, Frank. 2016. "American Public Opinion, Terrorism and Guns." Gallup, June 13. Accessed April 15, 2017 (gallup.com).

Newport, Frank. 2017. "Americans' Confidence in Institutions Edges Up." Gallup, June 26. Accessed July 30, 2017 (gallup.com).

Newport, Frank. 2017. "Middle-Class Identification in US at Pre-Recession Levels." Gallup, June 21. Accessed June 23, 2017 (gallup.com).

Newport, Frank. 2017. "More U.S. Protestants Have No Specific Denominational Identity." Gallup, July 18. Accessed July 30, 2017 (gallup.com).

Newport, Frank, Jeffrey M. Jones, and Lydia Saad. 2014. "State of the Union: The Public Weighs In on 10 Key Issues." Gallup, January 31. Accessed March 22, 2014 (www.gallup.com).

News Desk. 2021. "Female Imam Leads Prayers in US Mosque." Global Village Space, November 23. Accessed June 13, 2023 (https://www.globalvillagespace.com/female-imam-leads-prayers-in-us-mosque/).

Newton, Derek. 2020. "Another Problem with Shifting Education Online: A Rise in Cheating." *The Washington Post*, August 7. Accessed September 30, 2023 (https://www.washingtonpost.com/local/education/another-problem-with-shifting-education-online-a-rise-in-cheating/2020/08/07/1284c9f6-d762-11ea-aff6-220dd3a14741_story.html).

Ngabirano, Anne-Marcelle. 2017. "'Pink Tax' Forces Women to Pay More Than Men." *USA Today*, March 27. Accessed March 28, 2017 (usatoday.com).

Nguyen, Frances. 2016. "Women Workers Fight Back Against Sexual Harassment and Assault." Women's Media Center, December 20. Accessed December 22, 2016 (womensmediacenter.com).

Nichols, Andrew Howard, and José Luis Santos. 2016. "A Glimpse Inside the Coffers: Endowment Spending at Wealthy Colleges and Universities." The Education Trust, August. Accessed July 31, 2017 (edtrust.org).

Nichols, James. 2014. "Cecil Chao, Hong Kong Billionaire, Doubles Reward for Any Man Who Can Make Lesbian Daughter Straight." *Huffington Post*, January 24. Accessed March 2, 2014 (www.huffingtonpost.com).

Niederdeppe, Jeff, Sahara Byrne, Rosemary J. Avery, and Jonathan Cantor. 2013. "Direct-to-Consumer Television Advertising Exposure, Diagnosis with High Cholesterol, and Statin Use." *Journal of General Internal Medicine 28* (July): 886–893.

Nielsen. 2016. "The Nielsen Total Audience Report: Q1 2016." Accessed March 13, 2017 (https://www.nielsen.com/insights/2016/the-total-audience-report-q1-2016/).

Nielsen, 2018. "Time Flies: U.S. Adults Now Spend Nearly Half a Day Interacting with Media." June. Accessed November 17, 2023 (https://www.nielsen.com/insights/2018/time-flies-us-adults-now-spend-nearly-half-a-day-interacting-with-media/#:~:text=According%20to%20the%20first%2Dquarter,or%20generally%20interacting%20with%20media).

Nielsen Company, The. 2008. "College Spring Break Study." February 27. Accessed April 10, 2008 (www.alcoholstats.com).

Nietzel, Michael T. 2022. "Legacy College Admissions Come Under Fire in New Report." *Forbes*, October 30. Accessed June 13, 2023 (https://www.forbes.com/sites/michaeltnietzel/2022/10/30/legacy-college-admissions-come-under-fire-in-new-report/?sh=4e220dad5f07).

Nietzel, Michael T. 2022. "More Than 80% Of Four-Year Colleges Won't Require Standardized Tests for Fall 2023 Admissions." *Forbes*, November 15. Accessed July 29, 2023 (https://www.forbes.com/sites/michaeltnietzel/2022/11/15/more-than-80-of-four-year-colleges-wont-require-standardized--tests-for-fall-2023-admissions/?sh=292f8c107fb9).

Niquette, Mark, and Richard Rubin. 2014. "States Target Corporate Cash Stashed Overseas." *Bloomberg Businessweek*, April 21–27, 27–28.

Nir, Sarah Maslin. 2015. "The Price of Nice Nails." *New York Times*, May 7. Accessed October 4, 2015 (nytimes.com).

"No Action on Greenhouse Gases." 2008. *Baltimore Sun*, July 12, 2A.

Noe-Bustamante, Lauren Mora, and Mark Hugo Lopez. 2020. "About One-in-Four U.S. Hispanics Have Heard of Latinx, but Just 3% Use It." Pew Research Center, August 11. Accessed September 26, 2023 (https://www.pewresearch.org/hispanic/2020/08/11/about-one-in-four-u-s-hispanics-have-heard-of-latinx-but-just-3-use-it/).

Nolette, James. 2015. "Using Research to Move Policing Forward." *NIJ Journal* 276 (December): 46–51.

Norman, Jim. 2016. "Americans' Confidence in Institutions Stays Low." Gallup, June 13. Accessed April 4, 2017 (gallup.com).

Norman, Jim. 2016. "Millennials Like Sanders, Dislike Election Process." Gallup, May 11. Accessed January 9, 2017 (gallup.com/poll).

Norman, Jim. 2016. "Most U.S. Retirees Living Well, but Nonretirees Unconvinced." Gallup, April 29. Accessed July 20, 2017 (gallup.com).

Nossiter, Adam. 2011. "Hinting at an End to a Curb on Polygamy, Interim Libyan Leader Stirs Anger." *New York Times*, October 30, A6.

"Number of Jobs Held, Labor Market Activity, and Earnings Growth Among the Youngest Baby Boomers: Results from a Longitudinal Survey." 2012. Bureau of Labor Statistics, News Release, July 25. Accessed November 18, 2013 (www.bls.gov).

Nyseth, Hollie, Sarah Shannon, Kia Heise, and Suzy Maves McElrath. 2011. "Embedded Sociologists." *Contexts* 10 (Spring): 44–50.

O

O'Brien, Jodi, and Peter Kollock. 2001. *The Production of Reality: Essays and Readings on Social Interaction*, 3rd edition. Thousand Oaks, CA: Pine Forge Press.

O'Connor, Anahad. 2014. "New Concern About Testosterone and Heart Risks." *New York Times*, January 30, A13.

O'Connor, Anahad. 2015. "Study Warns of Diet Supplement Dangers Kept Quiet by F.D.A." *New York Times*, April 8, A1, A12.

O'Connor, Siobhan. 2015. "Farm-to-Fridge Is Heating Up." *Time*, August 24, 28.

O'Donnell, James. 2023. "Three Numbers That Help Explain What Happened to Inequality During the Pandemic." PBS, Frontline, March 14. Accessed April 13, 2023 (https://www.pbs.org/wgbh/frontline/article/wealth-wages-inequality-pandemic/).

O'Donnell, Victoria. 2007. *Television Criticism*. Thousand Oaks, CA: Sage.

O'Grady. Cathleen. 2021. "Climate Change 'Unequivocal' and 'Unprecedented,' Says New U.N. Report." Science, August 9. Accessed June 28, 2023 (https://www.science.org/content/article/climate-change-unequivocal-and-unprecedented-says-new-un-report).

O'Keefe, Ed, Matea Gold, and David A. Fahrenthold. 2015. "Bush Tax Forms Show Income of $29 Million Since 2007." *Washington Post*, June 30. Accessed August 31, 2015 (washingtonpost.com).

O'Keefe, Kevin. 2014. "TV's Renaissance for Strong Women Is Happening in a Surprising Place." *The Atlantic*, October 9. Accessed September 29, 2015 (theatlantic.com).

Oakes, Jeannie. 1985. *Keeping Track: How Schools Structure Inequality*. New Haven, CT: Yale University Press.

Obach, Brian K. 2004. *Labor and the Environmental Movement: The Quest for Common Ground*. Cambridge, MA: MIT Press.

Obach, Brian K. 2015. "A Fracking Fracas Demonstrates Movement Potential." *Contexts* 14 (Fall): 72–75.

Oberschall, Anthony. 1995. *Social Movements: Ideologies, Interests, and Identities*. New Brunswick, NJ: Transaction.

Ochs, Susan M. 2016. "At Banks, the Buck Stops Short." *New York Times*, September 15, A25.

OECD. 2016. "Income Inequality Update." November. Accessed May 5, 2017 (oecd.org).

OECD. 2016. "PISA 2015 Results in Focus." Accessed July 30, 2017 (oecd.org).

OECD. 2017. "Continued Slowdown in Productivity Growth Weighs Down on Living Standards." May 18. Accessed June 22, 2017 (oecd.org).

OECD. 2022. "Education at a Glance 2022: OECD Indicators." Accessed June 13, 2023 (https://www.oecd-ilibrary.org/education/education-at-a-glance-2022_3197152b-en).

OECD. 2023. "GDP Per Hours Worked." Accessed May 12, 2023 (https://data.oecd.org/lprdty/gdp-per-hour-worked.htm#indicator-chart).

OECD/European Union. 2015. *Indicators of Immigrant Integration 2015: Settling In*. Paris: OECD Publishing.

Office for Human Research Protections. 2016. "Regulations." U.S. Department of Health and Human Services, February 12. Accessed March 24, 2017 (hhs.gov/ohrp).

Office of Inspector General. 2017. "Semiannual Report to Congress: October 1, 2016 to March 31, 2017." U.S. Department of Health and Human Services. Accessed August 15, 2017 (oig.hhs.gov).

Office of the Deputy Chief of Staff for Intelligence. 2006. "Arab Cultural Awareness: 58 Factsheets." U.S. Army Training and Doctrine Command, Ft. Leavenworth, Kansas, January. Accessed February 15, 2006 (www.fas.org).

Office of the Inspector General. 2015. "Qualifying for Disability Benefits in Puerto Rico Based on an Inability to Speak English." Social Security Administration, April. Accessed October 4, 2015 (oig.ssa.gov).

OfficeTime. 2015. "Top 10 Time Killers." Accessed August 23, 2015 (officetime.net).

Ogburn, William F. 1922. *Social Change with Respect to Culture and Original Nature.* New York: Dell.

Ogunro, Nola. 2012. "10 Great Paying Jobs You Can Get Without a 4 Year College Degree." Alternet, September 21. Accessed June 23, 2014 (www.alternet.org).

"Older Americans Month: May 2017." 2017. U.S. Census Bureau, March 27. Accessed July 20, 2017 (census.gov).

Oliphant, Baxter. 2016. "Support for Death Penalty Lowest in More Than Four Decades." Pew Research Center, September 29. Accessed March 1, 2017 (pewresearch.org).

Oliphant, J. Baxter, and Andy Cerda. 2022. "Republicans and Democrats have different top priorities for U.S. immigration policy." Pew Research Center, September 8. Accessed April 26, 2023 (https://www.pewresearch.org/short-reads/2022/09/08/republicans-and-democrats-have-different-top-priorities-for-u-s-immigration-policy/).

Oliver, Melvin L., and Thomas M. Shapiro. 1995. *Black Wealth/White Wealth: A New Perspective on Racial Inequality.* New York: Routledge.

Olivier, Jos G. J., Greet Janssens-Maenhout, Marilena Muntean, and Jeroen A. H. W. Peters. 2016. "Trends in Global CO_2 Emissions: 2016 Report." PBL Netherlands Environmental Assessment Agency. Accessed August 20, 2017 (edgar.jrc.europa.eu).

Olmstead, Kenneth, and Aaron Smith. 2017. "Americans and Cybersecurity." Pew Research Center, January 26. Accessed March 1, 2017 (pewresearch.org).

Olsen, Lauren D. 2016. "'It's on the MCAT for a Reason': Premedical Students and the Perceived Utility of Sociology." *Teaching Sociology* 44(2): 72–83.

Olson, Jonathan R. 2010. "Choosing Effective Youth-Focused Prevention Strategies: A Practical Guide for Applied Family Professionals." *Family Relations* 59 (April): 207–220.

Olson, Theodore B. 2010. "The Conservative Case for Gay Marriage." *Newsweek*, January 18, 48–53.

Olorunnipa, Toluse, and Michelle Ye Hee Lee. 2021 "Trump's Election Fraud Falsehoods Have Cost Taxpayers $519 million — and Counting." Washington Post, February 21. Accessed June 14, 2023 (https://www.washingtonpost.com/politics/interactive/2021/cost-trump-election-fraud/).

"On Views of Race and Inequality, Blacks and Whites Are Worlds Apart." 2016. Pew Research Center, June 27. Accessed June 20, 2017 (pewresearch.org).

Oreopoulos, Philip, and Uros Petronijevic. 2013. "Making College Worth It: A Review of Research on the Returns to Higher Education." Accessed June 21, 2014 (www.ticas.org).

Ortman, Jennifer M., Victoria A. Velkoff, and Howard Hogan. 2014. "An Aging Nation: The Older Population in the United States." U.S. Census Bureau, May. Accessed May 16, 2014 (www.census.gov).

Orth, Taylor. 2022. "Airport Security: Despite the Inconvenience, Most Americans Say It's Effective." YouGovAmerica., January 14. Accessed February 11, 2023 (https://today.yougov.com/topics/travel/articles-reports/2022/01/14/airport-security-despite-inconvenience-most-americ).

Osterman, Michelle J.K., Brady E. Hamilton, Joyce A. Martin, Anne K. Driscoll, and Claudia P. Valenzuela. 2023. "Births: Final Data for 2021." National Vital Statistics Reports, January 31. Accessed May 22, 2023 (https://www.cdc.gov/nchs/data/nvsr/nvsr72/nvsr72-01.pdf).

Oudekerk, Barbara, Dara Blachman-Demner, and Carrie Mulford. 2014. "Teen Dating Violence: How Peers Can Affect Risk & Protective Factors." National Institute of Justice, November. Accessed August 6, 2015 (nij.gov).

Owen, Daniela J., Amy M. S. Slep, and Richard E. Heyman. 2012. "The Effect of Praise, Positive Nonverbal Response, Reprimand, and Negative Nonverbal Response on Child Compliance: A Systematic Review." *Clinical Child and Family Psychology Review* 15 (December): 364–385.

Oxfam America. 2017. "Rigged Reform." April 12. Accessed April 15, 2017 (oxfamamerica.org).

P

Packard, Vance. 1959. *The Status Seekers.* New York: David McKay.

Padgett, Tim. 2010. "Robes for Women." *Time,* September 27, 53–55.

Palen, J. John. 2014. *The Urban World,* 10th edition. New York: Oxford University Press.

Paludi, Michele A., and J. Harold Ellens, eds. 2016. *Feminism and Religion: How Faiths View Women and Their Rights.* Santa Barbara, CA: Praeger Press.

Pandit, Sonia G. 2015. "Compete with the NRA?" *Baltimore Sun*, October 27, 13.

Pantoja, Melanie J. 2023. "Annual Update of the HHS Poverty Guidelines." Federal Register, Vol. 88, No. 12, January 19. Accessed April 13, 2023 (https://www.govinfo.gov/content/pkg/FR-2023-01-19/pdf/2023-00885.pdf).

Parasrampuria, Sonal, and Stephen Murphy. 2022. "Trends in Prescription Drug Spending, 2016-2021." U.S. Department of Health and Human Services, September 30. Accessed June 11, 2023 (https://aspe.hhs.gov/reports/trends-prescription-drug-spending).

Park, Robert, and Ernest Burgess. 1921. *Introduction to the Science of Sociology.* Chicago: University of Chicago Press.

Parker, Kim. 2012. "The Boomerang Generation: Feeling OK About Living with Mom and Dad." Pew Social & Demographic Trends, March 15. Accessed November 20, 2013 (pewsocialtrends.org).

Parker, Kim. 2021. "What's Behind the Growing Gap Between Men and Women in College Completion?" Pew Research Center, November 8. Accessed June 13, 2023 (https://www.pewresearch.org/short-reads/2021/11/08/whats-behind-the-growing-gap-between-men-and-women-in-college-completion/).

Parker, Kim, and Gretchen Livingston. 2016. "6 Facts About American Fathers." Pew Research Center, June 16. Accessed May 19, 2017 (pewresearch.org).

Parker, Kim, and Juliana Menasce Horowitz. 2015. "Women and Leadership." Pew Research Center, January 14. Accessed September 29, 2015 (pewresearch.org).

Parker, Kim, Horowitz, Juliana Menasce Horowitz, Anna Brown, Richard Fry, D'Vera Cohn, and Ruth Igielnik. 2018. "What Unites and Divides Urban, Suburban and Rural Communities." Pew Research Center, May 22. Accessed June 27, 2023 (https://www.pewresearch.org/social-trends/2018/05/22/what-unites-and-divides-urban-suburban-and-rural-communities/).

Parker, Kim, and Renee Stepler. 2017. "As U.S. Marriage Rate Hovers at 50%, Education Gap in Marital Status Widens." Pew Research Center, September 14. Accessed May 19, 2023 (https://www.pewresearch.org/short-reads/2017/09/14/as-u-s-marriage-rate-hovers-at-50-education-gap-in-marital-status-widens/).

Parker, Kim, Juliana Horowitz, Ruth Igielnik, Baxter Oliphant, and Anna Brown. 2017. "America's Complex Relationship with Guns." Pew Research Center, June 22. Accessed September 2, 2017 (pewresearch.org).

Parker, Kim, Rich Morin, Juliana Menasce Horowitz, and Mark Hugo Lopez. 2015. "Multiracial in America." Pew Research Center, June 11. Accessed June 13, 2015 (pewresearch.org).

Parker, Kim, Juliana Menasce Horowitx, and Anna Brown. 2022. "Americans' Complex Views on Gender Identity and Transgender Issues." Pew Research Center, June 28. Accessed April 21, 2023 (https://www.pewresearch.org/social-trends/2022/06/28/americans-complex-views-on-gender-identity-and-transgender-issues/).

Parker, Kim, Richard Fry, D'Vera Cohn, and Wendy Wang. 2011. "Is College Worth It? College Presidents, Public Assess Value, Quality and Mission of Higher Education." Pew Research Center, May 16. Accessed April 15, 2012 (www.pewsocialtrends.org).

Parker, Laura. 2016. "National Park Service Faces Sex Harassment Scandal." *National Geographic,* September 23. Accessed May 22, 2017 (nationalgeographic.com).

Parrott, Scott, and Caroline Titcomb Parrott. 2015. "U.S. Television's 'Mean World' for White Women: The Portrayal of Gender and Race on Fictional Crime Dramas." *Sex Roles* 73 (June): 70–82.

Parsons, Chelsea, and Eugenio Weigend. 2016. "America Under Fire: An Analysis of Gun Violence in the United States and the Link to Weak Gun Laws." Center for American Progress, October. Accessed March 1, 2017 (americanprogress.org).

Parsons, Talcott, and Robert F. Bales, eds. 1955. *Family, Socialization, and Interaction Process.* New York: Free Press.

Parsons, Talcott. 1951. *The Social System.* Glencoe, IL: Free Press.

Parsons, Talcott. 1954. *Essays in Sociological Theory,* revised edition. New York: Free Press.

Parsons, Talcott. 1959. "The School Class as a Social System: Some of Its Functions in American Society." *Harvard Educational Review* 29 (Fall): 297–313.

Parsons, Talcott. 1960. *Structure and Process in Modern Societies.* New York: Free Press.

Passel, Jeffrey S., and D'Vera Cohn. 2016. "Overall Number of U.S. Unauthorized Immigrants Holds Steady Since 2009." Pew Research Center, September 20. Accessed June 20, 2017 (pewresearch.org).

Passel, Jeffrey S., D'Vera Cohn, and Mark Hugo Lopez. 2011. "Hispanics Account for More Than Half of Nation's Growth in Past Decade." Pew Research Center, March 24. Accessed December 7, 2011 (www.pewhispanic.org).

Patel, Sujan. 2017. "7 Deadly Sins of Employee Communication." *Baltimore Sun,* January 5, 19.

Patten, Eileen. 2015. "Who Is Multiracial? Depends on How You Ask." Pew Research Center, November 6. Accessed June 20, 2017 (pewresearch.org).

Patten, Eileen. 2016. "The Nation's Latino Population Is Defined by Its Youth." Pew Research Center, April 20. Accessed June 20, 2017 (pewresearch.org).

Paul, Annie Murphy. 2010. *Origins: How the Nine Months Before Birth Shape the Rest of Our Lives.* New York: Free Press.

Paul, Noel. C. 2002. "The Birth of a Would-Be Fad." *Christian Science Monitor,* September 23, 11, 14–16.

Paul, Richard, and Linda Elder. 2007. *The Miniature Guide to Critical Thinking: Concepts and Tools.* Dillon Beach, CA: Foundation for Critical Thinking.

Paulson, Amanda. 2014. "Changes of Address." *Christian Science Monitor Weekly,* May 26, 12–13.

Payne, David, and Glenn Somerville. 2015. "Where the Jobs Will Be." *Baltimore Sun,* April 19, 3.

Payne, K. K., and J. Copp. 2013. "Young Adults in the Parental Home and the Great Recession." National Center for Family & Marriage Research. Accessed November 20, 2013 (www.hcfmr.bgsu.edu).

Pazol, Karen, Andreea A. Creanga, Kim D. Burley, and Denise J. Jamieson. 2014. "Abortion Surveillance—United States, 2011." *MMWR* 63 (11): 1–41.

PBS. 2021. "Exhaustive Fact Check Finds Little Evidence of Voter Fraud, but 2020's 'Big Lie' Lives On." December 17. Accessed May 12, 2023 (https://www.pbs.org/newshour/show/exhaustive-fact-check-finds-little-evidence-of-voter-fraud-but-2020s-big-lie-lives-on).

Pearlstine, Norman. 2013. "Fix This: Water." *Bloomberg Businessweek,* March 25–31, 46–51.

Peck, Ki-eun. 2017. "The Challenges to Muslim Integration in France." *McGill Journal of Politics,* January 31, Accessed February 12, 2023 (https://mjps.ssmu.ca/2017/01/31/the-challenges-to-muslim-integration-in-france/).

Pedersen, Paul. 1995. *The Five Stages of Culture Shock: Critical Incidents Around the World.* Westport, CT: Greenwood Press.

Peek, Sean. 2023. "How to Apply Rosabeth Moss Kanter's Management Theory." Business.com, March 1. Accessed March 11, 2023 (https://www.business.com/articles/management-theory-of-rosabeth-moss-kanter/).

Perlmutter, David D. 2001. "Students Are Blithely Ignorant; Professors Are Bitter." *Chronicle of Higher Education,* July 27, B20.

Perlroth, Nicole, Michael Wines, and Matthew Rosenberg. 2017. "Russian Election Hacking Efforts, Wider Than Previously Known, Draw Little Scrutiny." *New York Times,* September 1. Accessed September 2, 2017 (nytimes.com).

Perrin, Andrew. 2021. "Mobile Technology and Home Broadband 2021." Pew Research Center, June 3. Accessed September 7, 2023 (https://www.pewresearch.org/internet/wp-content/uploads/sites/9/2021/06/PI_2021.06.03_Mobile-Broadband_FINAL.pdf).

Perrin, Andrew, and Sara Atske, 2021. "About Three-in-Ten U.S. Adults Say They Are 'Almost Constantly' Online." Pew Research Center, March 21. Accessed November 17, 2023 (https://www.pewresearch.org/short-reads/2021/03/26/about-three-in-ten-u-s-adults-say-they-are-almost-constant-ly-online/).

Perry, Gina. 2013. *Behind the Shock Machine: The Untold Story of the Notorious Milgram Psychology Experiments.* New York: New Press.

Perry, Mark. J. 2021. "Women Earned the Majority of Doctoral Degrees in 2020 for the 12th Straight Year and Outnumber Men in Grad School 148 to 100." American Enterprise Institute, October 14. Accessed September 17, 2023 (https://www.aei.org/carpe-diem/women-earned-the-majority-of-doctoral-degrees-in-2020-for-the-12th-straight-year-and-outnumber-men-in-grad-school-148-to-100/).

PerryUndem. 2017. "The State of the Union on Gender Equality, Sexism, and Women's Rights." January 17. Accessed January 20, 2017 (perryundem.com).

Persell, Caroline Hodges, and Peter W. Cookson, Jr. 1985. "Chartering and Bartering: Elite Education and Social Reproduction." *Social Problems* 33 (December): 114–129.

Peter, Tom A. 2012. "Mistreatment of Afghan Women Caused by Far More Than Taliban." *Christian Science Monitor,* January 31. Accessed March 4, 2014 (www.csmonitor.com).

Petersen, Andrea. 2012. "Smarter Ways to Discipline Kids." *Wall Street Journal,* December 26, D1, D3.

Peterson, James L., Josefina J. Card, Marvin B. Eisen, and Bonnie Sherman-Williams. 1994. "Evaluating Teenage Pregnancy Prevention and Other Social Programs: Ten Stages of Program Assessment." *Family Planning Perspectives* 26 (May): 116–120, 131.

Peterson, Scott. 2008. "In Iran, Barbie Seen as Cultural Invader." *Christian Science Monitor,* September 15, 4.

Peterson-Withorn, Chase. 2023. "Forbes' 37th Annual World's Billionaires List:

Facts And Figures 2023." *Forbes*, Billionaires 2023, April 4. Accessed April 13, 2023 (https://www.forbes.com/sites/chasewithorn/2023/04/04/forbes-37th-annual-worlds-billionaires-list-facts-and-figures-2023/).

Petrosino, Anthony, Carolyn Turpin-Petrosino, Meghan E. Hollis-Peel, and Julia G. Lavenberg. 2013. "Scared Straight and Other Juvenile Awareness Programs for Preventing Juvenile Delinquency: A Systematic Review." Campbell Systematic Reviews. Accessed September 2, 2015 (campbellcollaboration.org).

Petrosyan, Ani. 2023. "Annual Number of Data Compromises and Individuals Impacted in the United States from 2005 to 2022." Statista, February 24. Accessed March 25, 2023 (https://www.statista.com/statistics/273550/data-breaches-recorded-in-the-united-states-by-number-of-breaches-and-records-exposed/#:~:text=In%202022%2C%20the%.20number%20of,have%20one%20thing%20in%20common)

Pettypiece, Shannon. 2013. "Anything You Can Do, I Can Do Better." *Bloomberg Businessweek*, March 11–17, 27–29.

Pettypiece, Shannon. 2016. "No Cheers When Walmart Packs Up." *Bloomberg Businessweek*, February 1–7, 18–19.

Pew Charitable Trusts. 2015. "The Precarious State of Family Balance Sheets." January. Accessed September 11, 2015 (pewtrusts.org).

Pew Hispanic Center. 2012. "When Labels Don't Fit: Hispanics and Their Views of Identity." Pew Research Center, April 4. Accessed March 22, 2014 (www.pewhispanic.org).

Pew-MacArthur Results First Initiative. 2018. "Targeted Evaluations Can Help Policymakers Set Priorities." Accessed January 29, 2023 (https://www.pewtrusts.org/-/media/assets/2018/03/rf_targeted_evaluation-brief.pdf).

Pew Research Center. 2014. "Most Say Religious Holiday Displays on Public Property Are OK." December 14. Accessed July 31, 2017 (pewresearch.org).

Pew Research Center. 2014. "Attendance at Religious Services – Religion in America: U.S. Religious Data, Demographics and Statistics." Accessed June 14, 2023 (https://www.pewresearch.org/religion/religious-landscape-study/attendance-at-religious-services/).

Pew Research Center. 2015. "The American Middle Class Is Losing Ground: No Longer the Majority and Falling Behind Financially." December 9. Accessed December 10, 2015 (pewsocialtrends.org).

Pew Research Center. 2016. "Choosing a New Church or House of Worship." August 23. Accessed July 31, 2017 (pewresearch.org).

Pew Research Center. 2016. "Public Views About Americans' Eating Habits." December 1. Accessed June 11, 2023 (https://www.pewresearch.org/science/2016/12/01/public-views-about-americans-eating-habits/).

Pew Research Center. 2017. "In America, Does More Education Equal Less Religion?" April 26. Accessed July 31, 2017 (pewresearch.org).

Pew Research Center. 2020. "Harassment of Religious Groups Continues in More than 90% of Countries." Pew Research Center, November 10. Accessed June 13, 2023 (https://www.pewresearch.org/religion/2020/11/10/harassment-of-religious-groups-continues-to-be-reported-in-more-than-90-of-countries/).

Pew Research Center. 2021. "Internet/Broadband Fact Sheet." April 7. Accessed September 7, 2023 (https://www.pewresearch.org/internet/fact-sheet/internet-broadband/).

Pew Research Center. 2021. "Measuring Religion in American Trends Panel." January 14. Accessed June 13, 2023 (https://www.pewresearch.org/religion/2021/01/14/measuring-religion-in-pew-research-centers-american-trends-panel/)

Pew Research Center. 2022. "America's Abortion Quandary." May 6. Accessed September 17, 2023 (https://www.pewresearch.org/religion/2022/05/06/americas-abortion-quandary/).

Pew Research Center. 2022. "Public Trust in Government: 1958-2022." June 6. Accessed May 12, 2023 (https://www.pewresearch.org/politics/2022/06/06/public-trust-in-government-1958-2022/).

Pew Research Center. 2022. "How U.S. Religious Composition Has Changed in Recent Decades." September 13. Accessed June 14, 2023 (https://www.pewresearch.org/religion/2022/09/13/how-u-s-religious-composition-has-changed-in-recent-decades/).

Pew Research Center. 2022. "1. Views of Election Administration and Confidence in Vote Counts." October 31. Accessed May 12, 2023 (https://www.pewresearch.org/politics/2022/10/31/views-of-election-administration-and-confidence-in-vote-counts/).

Pewewardy, Cornel. 1998. "Fluff and Feathers: Treatment of American Indians in the Literature and the Classroom." *Equity & Excellence in Education* 31 (April): 69–76.

Pez. 2023. "PEZ.com – Official Website for PEZ Candy USA – Home for All Things PEZ." Accessed June 25, 2023 (https://us.pez.com/).

Pflaumer, Alicia. 2011. "Texting Bride Video Goes Viral on the Web." *Christian Science Monitor,* November 1. Accessed November 3, 2011 (www.csmonitor.com).

Philips, Matthew. 2016. "Regulations Dry Up Wastewater Wells." *Bloomberg Businessweek*, March 21–27, 23–24.

Phillips, L. Taylor, and Brian S. Lowery. 2015. "The Hard-Knock Life? Whites Claim Hardships in Response to Racial Inequity." *Journal of Experimental Social Psychology* 61 (November): 12–18.

Phillips, Matthew. 2014. "Welders, America Needs You." *Bloomberg Businessweek*, March 24–April 6, 19–21.

Philpott, Tom. 2016. "Playing Chicken." *Mother Jones*, May/June, 41–47, 74.

Pianigiani, Gaia. 2016. "Italy's 'Fertility Day' Ads Anger Women Stymied by Lack of Support." *New York Times*, August 14, A4.

Pianta, Robert C., Jay Belsky, Renate Houts, Fred Morrison, and The National Institute of Child Health and Human Development (NICHD) Early Child Care Research Network. 2007. "Teaching: Opportunities to Learn in America's Elementary Classrooms." *Science* 315 (March 30): 1795–1796.

Piazza, Jo. 2016. "Women of Color Hit a 'Concrete Ceiling' in Business." *Wall Street Journal*, September 27. Accessed June 8, 2017 (wsj.com).

Picchi, Aimee. 2023. "Your College Major Can Influence Your Pay. Here Are the Best and Worst Majors." CBS News, Money Watch, March 2. Accessed April 13, 2023 (https://www.cbsnews.com/news/college-major-highest-lowest-incomes/).

Pickert, Reade. 2022. "U.S. Sees Record Job Growth in 2021 After Millions Lost in 2020." *Bloomberg*, January 7. Accessed May 10, 2023 (https://www.bloomberg.com/news/articles/2022-01-07/u-s-sees-record-job-growth-in-2021-after-millions-lost-in-2020#xj4y7vzkg).

Pierce, Lamar, Daniel Snow, and Andrew McAfee. 2013. "Cleaning House: The Impact of Information Technology Monitoring on Employee Theft and Productivity." Social Science Research Network, August 24. Accessed October 24, 2013 (www.ssrn.com).

Pifer, Rebecca. 2021. "Mobile Health Apps Have 'Serious Problems' with Patient Privacy, Study Warns." Healthcare Dive, June 16 Accessed June 27, 2023 (https://www.healthcaredive.com/news/mobile-health-apps-have-serious-problems-with-patient-privacy-study-warn/601943/).

Piketty, Thomas, Emmanuel Saez, and Gabriel Zucman. 2016. "Distributional National Accounts: Methods and Estimates for the United States." National Bureau of Economic Research, December. Accessed March 15, 2017 (nber.org).

Pipkin, Whitney. 2013. "Are Asian Americans Taking over the Internet?" *Asian*

Fortune, November 29. Accessed December 4, 2013 (www.asianfortune-news.com).

Pittz, Will. 2005. "Closing the Gap: Solutions to Race-Based Health Disparities." Applied Research Center & Northwest Federation of Community Organizations, June. Accessed April 20, 2007 (www.arc.org).

Planty, M., W. Hussar, T. Snyder, S. Provasnik, G. Kena, R. Dinkes, A. KewalRamani, and J. Kemp. 2008. *The Condition of Education 2008*. Washington, DC: National Center for Education Statistics, U.S. Department of Education.

Plateris, Alexander A. 1973. *100 Years of Marriage and Divorce Statistics: 1867–1967*. Rockville, MD: National Center for Health Statistics.

Plutzer, Eric, Mark McCaffrey, A. Lee Hannah, Joshua Rosenau, Minda Berbeco, and Ann H. Reid. 2016. "Climate Confusion Among U.S. Teachers." *Science* 351 (6274): 664–665.

Polderman, Tinca, J. C., et al. 2015. "Meta-Analysis of the Heritability of Human Traits Based on Fifty Years of Twin Studies." *Nature Genetics* 47 (May): 702–709.

Polgreen, Lydia. 2010. "One Bride for Two Brothers: A Custom Fades in India." *New York Times*, July 16, A4.

"Police Culture." 2015. *The Economist*, April 25, 28–29.

"Policing Philadelphia: Boots on the Street." 2013. *The Economist*, August 24, 33.

Polikoff, Morgan S., and Andrew C. Porter. 2014. "Instructional Alignment as a Measure of Teaching Quality." *Educational Evaluation and Policy Analysis* 36 (December): 399–416.

"The Politics of Financial Insecurity." 2015. Pew Research Center, January 8. Accessed October 7, 2015 (pewresearch.org).

PolitiFact. 2016. "Comparing Hillary Clinton, Donald Trump on the Truth-O-Meter." Accessed January 11, 2017 (politifact.com).

Pollack, Andrew. 2013. "A.M.A. Recognizes Obesity as a Disease." *New York Times*, June P19, B1.

Pollick, Michael. 2014. "Why Didn't Communism Work in Eastern Europe?" Conjecture Corporation, April 1. Accessed April 12, 2014 (www.wisegeek.org).

Polsby, Nelson W. 1959. "Three Problems in the Analysis of Community Power." *American Sociological Review* 24 (December): 796–803.

Pong, Suet-ling, Lingxin Hao, and Erica Gardner. 2005. "The Roles of Parenting Styles and Social Capital in the School Performance of Immigrant Asian and Hispanic Adolescents." *Social Science Quarterly* 86 (December): 928–950.

PopulationPyramid.net. 2023. "Population Pyramids of the World from 1950 to 2100." Accessed June 13 (https://www.populationpyramid.net/).

Porter, Eduardo. 2013. "America's Sinking Middle Class." *New York Times*, September 19, A1.

Porter, Eduardo. 2014. "A Global Boom, but Only for Some." *New York Times*, March 19, B1.

Porter, Nicole D. 2016. "The State of Sentencing 2015: Developments in Policy and Practice." The Sentencing Project. Accessed April 20, 2017 (sentencingproject.org).

Portes, Jonathan. 2015. "Immigration Is Good for Economic Growth. If Europe Gets It Right, Refugees Can Be Too." *Huffington Post*, September 15. Accessed October 23, 2015 (huffingtonpost.com).

Postel, Sandra. 2012. "Humanity's Growing Impact on the World's Freshwater." AlterNet, February 23. Accessed June 23, 2012 (alternet.org).

Potok, Mark. 2017. "The Year in Hate and Extremism." Southern Poverty Law Center, February 15. Accessed February 25, 2017 (splcenter.org).

Powell, Jon T., Timothy G. Townsend, and Julie B. Zimmerman. 2015. "Estimates of Solid Waste Disposal Rates and Reduction Targets for Landfill Gas Emissions." *Nature Climate Change*, September 21. Accessed October 23, 2015 (nature.com).

Powers, Charles H. 2004. *Making Sense of Social Theory: A Practical Introduction*. Lanham, MD: Rowman & Littlefield.

Pratt, Beverly M., Lindsay Hixson, and Nicholas A. Jones. 2015. "Measuring Race and Ethnicity Across the Decades: 1790–2010." U.S. Census Bureau, November 2. Accessed January 12, 2017 (census.gov).

"Pregnancy Mortality Surveillance System." 2017. Centers for Disease Control and Prevention, June 27. Accessed August 15, 2017 (cdc.gov).

Prein, Andreas, Roy M. Rasmussen, Kyoko Ikeda, Changhai Liu, Martyn P. Clark, and Greg J. Holland. 2017. "The Future Intensification of Hourly Precipitation Extremes." *Nature Climate Change* 7 (December): 48–52.

"Prescription for Change." 2013. *The Economist*, June 29, 61–62.

Preston, Caroline. 2016. "The 20 Most Generous Companies of the Fortune 500." *Fortune*, June 22. Accessed May 15, 2017 (fortune.com).

Preston, Julia. 2015. "States Are Divided by the Lines They Draw on Immigration." *New York Times*, March 30, A10, A12.

Price, Barbara Raffel, and Natalie J. Sokoloff, eds. 2004. *The Criminal Justice System and Women: Offenders, Prisoners, Victims, & Workers*, 3rd edition. New York: McGraw-Hill.

Princiotta, Daniel, Laura Lippman, Renee Ryberg, Hannah Schmitz, David Murphey, and Mae Cooper. 2014. "Social Indicators Predicting Postsecondary Success." Child Trends, April 1. Accessed June 23, 2014 (www.childtrends.org).

"Prisons." 2017. *The Economist*, March 18, 26.

Project on Government Oversight. 2010. "Letter to NIH on Ghostwriting Academics." November 29. Accessed July 3, 2011 (www.pogo.org).

Protect Our Defenders. 2016. "Facts on United States Military Sexual Violence." September. Accessed April 2, 2017 (protectourdefenders.com).

"Pro-natalism." 2015. *The Economist*, July 25, 47–48.

Pryor, John H. 2011. "The Changing First-Year Student: Challenges for 2011." Higher Education Research Institute at UCLA, January 27. Accessed April 10, 2012 (www.heri.ucla.edu).

Przybylski, Andrew K., and Allison F. Mishkin. 2015. "How the Quantity and Quality of Electronic Gaming Relates to Adolescents' Academic Engagement and Psychosocial Adjustment." *Psychology of Popular Media Culture*, March 2. APA PsycNet. Accessed August 6, 2015 (psycnet.apa.org).

Psych Central. 2021. "How Common Is Cheating and Infidelity Really?" October 19. Accessed February 11, 2023 (https://psychcentral.com/blog/how-common-is-cheating-infidelity-really#statistics).

Public Citizen. 2022. "Report: Corporate Prosecutions Reach Record Low Under Biden." October 25. Accessed March 25, 2023 (https://www.citizen.org/news/report-corporate-prosecutions-reach-record-low-under-biden/).

"Public Trust in Government Remains Near Historic Lows as Partisan Attitudes Shift." 2017 Pew Research Center, May 3. Accessed June 16, 2017 (pewresearch.org).

"Public Uncertain, Divided over America's Place in the World." 2016. Pew Research Center, May 5. Accessed February 5, 2017 (pewresearch.org).

Puddington, Arch, and Tyler Roylance. 2017. "Populists and Autocrats: The Dual Threat to Global Democracy." Freedom House. Accessed June 25, 2017 (freedomhouse.org).

Putnam, Hannah, Julie Greenberg, and Kate Walsh. 2014. "Easy A's and What's Behind Them." National Council on Teacher Quality, November. Accessed December 5, 2014 (www.nctq.org).

Putnam, Robert D. 2015. *Our Kids: The American Dream in Crisis*. New York: Simon & Schuster.

Pyne, Jaymes. 2020. "Boys Enjoy Educational Advantages Despite Being Less Engaged in School than Girls." Brookings, July 30. Accessed June 13, 2023 (https://www.brookings.edu/blog/brown-center-chalkboard/2020/07/30/boys-enjoy-educational-advantages-despite-being-less-engaged-in-school-than-girls/).

Q

Qin, Amy. 2015. "'Kingdom of Daughters' Draws Tourists to Its Matrilineal Society." *New York Times*, October 26, A1, A6.

Quinney, Richard. 1980. *Class, State, and Crime*. Boston: Little, Brown.

R

Rackin, Heather, and Christina M. Gibson-Davis. 2012. "The Role of Pre- and Postconception Relationships for First-Time Parents." *Journal of Marriage and Family* 74 (June): 389–398.

Radesky, Jenny, et al. 2014. "Patterns of Mobile Device Use by Caregivers and Children During Meals in Fast Food Restaurants." *Pediatrics* 133 (4): e843–e849.

Rainie, Lee, and Kathryn Zickuhr. 2015. "Americans' Views on Mobile Etiquette." Pew Research Center, August 26. Accessed August 26, 2015 (pewresearch.org).

Rainie, Lee, Cary Funk, Monica Anderson, and Alec Tyson. 2022. "7. Americans Are Closely Divided Over Editing a Baby's Genes to Reduce Serious Health Risk." Pew Research Center, March 17. Accessed June 27, 2023 (https://www.pewresearch.org/internet/2022/03/17/americans-are-closely-divided-over-editing-a-babys-genes-to-reduce-serious-health-risk/).

RAINN. 2023. "Victims of Sexual Violence: Statistics." Accessed March 25, 2023 (https://www.rainn.org/statistics/victims-sexual-violence).

Raj, Suhasini. 2015. "Goat Meat, Not Beef, Found in Home of Slain Indian." *New York Times*, December 30, A3.

Ramirez. Rachel. 2023. "The Plastic Water Bottle Industry Is Booming. Here's Why That's a Huge Problem." CNN, March 16. Accessed June 27, 2023 (https://amp.cnn.com/cnn/2023/03/16/world/plastic-water-bottles-un-report-climate/index.html).

Rampell, Catherine. 2014. "The Safety Net Catches the Middle Class More Than the Poor." *Washington Post*, April 7. Accessed March 8, 2014 (www.washingtonpost.com).

Ran, Florence Xiaotao, and Yuxin Lin. 2022. "Rethinking Remedial Programs to Promote College Student Success." Brookings Institute, February 15. Accessed June 13, 2023 (weldinghttps://www.brookings.edu/blog/brown-center-chalkboard/2022/02/15/rethinking-remedial-programs-to-promote-college-student-success/#:~:text=Up%20to%2065%25%20of%20community,six%20years%20of%20initial%20enrollment.).

Rank, Mark R., and Thomas A. Hirschl. 2015. "The Likelihood of Experiencing Relative Poverty over the Life Course." *PloS One* 10(7): e0133513. doi: 10.1371/journal.pone.0133513. Accessed September 16, 2015 (journals.plos.org).

Rano, Jason, and Jane Houlihan. 2012. "Myths on Cosmetic Safety." Skin Deep Cosmetics Database. Accessed May 10, 2012 (www.ewg.org).

Raphael, Steven, and Michael A. Stoll. 2010. "Job Sprawl and the Suburbanization of Poverty." Brookings Institute, Metropolitan Policy Program, March. Accessed May 1, 2010 (www.brookings.edu).

Raphelson, Samantha. 2018. "How the NRA Worked to Stifle Gun Violence Research." National Public Radio, April 5. Accessed January 29, 2023. (https://www.npr.org/2018/04/05/599773911/how-the-nra-worked-to-stifle-gun-violence-research).

Ray, Rebecca, Milla Sanes, and John Schmitt. 2013. "No-Vacation Nation Revisited." Center for Economic and Policy Research, May. Accessed April 12, 2014 (cepr.net).

Rayasam, Renuka. 2007. "Immigrants: The Unsung Heroes of the U.S. Economy." *U.S. News & World Report*, February 26, 58.

Reality Check Team. 2023. "Homosexuality: The Countries Where It Is Illegal to Be Gay." *BBC News*, March 31. Accessed April 21, 2023 (https://www.bbc.com/news/world-43822234).

Reaney, Patricia. 2012. "Average Cost of U.S. Weddings Hits $27,021." Reuters, May 23. Accessed June 2, 2012 (www.reuters.com).

Reardon, Sean F. 2011. "The Widening Academic Achievement Gap Between the Rich and the Poor: New Evidence and Possible Explanations." Pp. 91–116 in *Whither Opportunity? Rising Inequality and the Uncertain Life Chances of Low-Income Children*, edited by R. Murnane and G. Duncan. New York: Russell Sage Foundation Press.

Reardon, Sean F., Lindsay Fox, and Joseph Townsend. 2015. "Neighborhood Income Composition by Household Race and Income, 1990–2009." *Annals of the American Academy of Political and Social Science* 660 (July): 78–99.

Reaves, Brian A. 2017. "Police Response to Domestic Violence, 2006–2015." Bureau of Justice Statistics, May. Accessed July 15, 2017 (bjs.gov).

Redd, Zakia, Tahilin Sanchez Karver, David Murphey, Kristin Anderson Moore, and Dylan Knewstub. 2011. "Two Generations in Poverty: Status and Trends among Parents and Children in the United States, 2000–2010." Child Trends Research Brief, November. Accessed January 12, 2012 (www.childtrends.org).

Reddy, Sumathi. 2013. "'I Don't Smoke, Doc,' and Other Patient Lies." *Wall Street Journal*, February 19, D3.

Reeves, Jay. 2013. "Alabama Woman's Body Removed from Front Yard Grave." *U.S. News & World Report*, November 15. Accessed August 10, 2014 (www.us.news.com).

Reeves, Richard V. 2017. "Race Gaps in SAT Math Scores Are as Big as Ever." Brookings, July 11. Accessed February 1, 2017 (brookings.edu).

Reeves, Richard V., and Joanna Venator. 2015. "Sex, Contraception, or Abortion? Explaining Class Gaps in Unintended Childbearing." Brookings, February. Accessed September 29, 2015 (brookings.edu).

Reeves, Richard V., and Nathan Joo. 2016. "How Much Social Mobility Do People Really Want?" Brookings, January 12. Accessed December 23, 2016 (brookings.edu).

Reeves, Richard V., Isabel V. Sawhill, and Eleanor Krause. 2016. "The Most Educated Women Are the Most Likely to Be Married." Brookings, August 19. Accessed August 25, 2016 (brookings.edu).

Reeves, Richard V., Eliana Buckner, and Ember Smith. 2021. "The Unreported Gender Gap in High School Graduation Rates." Brookings, January 12. Accessed June 13, 2023 (https://www.brookings.edu/blog/up-front/2021/01/12/the-unreported-gender-gap-in-high-school-graduation-rates/).

Reger, Jo. 2012. *Everywhere and Nowhere: Contemporary Feminism in the United States*. New York: Oxford University Press.

Reich, Robert. 2014. "Why Widening Inequality Is Hobbling Equal Opportunity." AlterNet, February 6. Accessed February 10, 2014 (www.alternet.org).

Reid, Jenni. 2022. "Russia Is Offering a Hero's Medal and $16,000 to Women Who Have 10 Kids." *CNBC News*, August 18. Accessed June 28, 2023 (https://www.cnbc.com/2022/08/18/russia-offers-mother-heroine-medal-and-16800-for-having-10-children.html).

Reiman, Jeffrey, and Paul Leighton. 2010. *The Rich Get Richer and the Poor Get Prison: Ideology, Class, and Criminal Justice*, 9th edition. Upper Saddle River, NJ: Prentice Hall.

Rein, Lisa. 2011. "Federal Workers Tell Us What Should Be Cut from the Budget." *Washington Post,* April 14, B4.

Rein, Lisa, and Emily Wax-Thibodeaux. 2015. "Veterans Affairs Improperly Spent $6 Billion Annually, Senior Official Says." *Washington Post,* May 14. Accessed August 23, 2015 (washingtonpost.com).

Reinberg, Steven. 2023. "Why Is American Food So Unhealthy?" *US News & World Report,* Healthy Day, January 17. Accessed June 11, 2023 (https://www.usnews.com/news/health-news/articles/2023-01-17/why-is-american-food-so-unhealthy).

Reiner, Andrew. 2017. "The Power of Touch, Especially for Men. *The New York Times,* December 5. Accessed September 7, 2023 (https://www.nytimes.com/2017/12/05/well/family/gender-men-touch.html).

Reingold, Jennifer. 2016. "The Disappeared." *Fortune,* September 15, 100–108.

Reinhart, R. J. 2017. "More in US Say Government Is the Most Important Problem." Gallup, June 15. Accessed June 27, 2017 (gallup.com).

Reinhold, Steffen 2010. "Reassessing the Link Between Premarital Cohabitation and Marital Instability." *Demography* 47 (August): 719–733.

Reiss, Fraidy. 2015. "America's Child-Marriage Problem." *New York Times,* October 14, A23.

Reiss, Steven. 2004. "The Sixteen Strivings for God." *Zygon* 39 (June): 303–320.

"Religion: The New Strife." 2016. *The Economist,* May 14, 10–12.

"Religions of the World: Number of Adherents, Names of Houses of Worship. . . ." 2007. Religious Tolerance. Accessed August 24, 2007 (www.religioustolerance.org).

Reuben, Cynthia, and Patricia Pastor. 2015. "Percentage of Children and Adolescents Aged 4–17 Years with Serious Emotional or Behavioral Difficulties, by Poverty Status and Sex—National Health Interview Survey, 2011–2014." *Morbidity and Mortality Weekly Report* 64 (46): 1303.

Reynolds, Kelly A. 2012. "The Price of Drinking Water." *Water Conditioning & Purification* 54 (July): 50–52.

Rheault, Magali, and Kyley McGeeney. 2011. "Education Is a Key Predictor of Emotional Health After 65." Gallup, August 19. Accessed April 3, 2012 (www.gallup.com).

Ricciardelli, Rosemary, Kimberley A. Clow, and Philip White. 2010. "Investigating Hegemonic Masculinity: Portrayals of Masculinity in Men's Lifestyle Magazines." *Sex Roles* 63 (March): 64–78.

Rich, Motoko. 2013. "Subtract Teachers, Add Pupils: Math of Today's Jammed Schools." *New York Times,* December 22, A1.

Rich, Motoko. 2014. "Why Don't More Men Go into Teaching?" *New York Times,* September 6, SR3.

Richburg, Keith B. 2009. "States Seek Less Costly Substitutes for Prison." *Washington Post,* July 13, A1.

Riche, Martha Farnsworth. 2000. "America's Diversity and Growth: Signposts for the 21st Century." *Population Bulletin* 55 (June): 1–41.

Richey, Alexandra S., Brian F. Thomas, Min-Hui Lo, James S. Famiglietti, Sean Swenson, and Matthew Rodell. 2015. "Uncertainty in Global Groundwater Storage Estimates in a Total Groundwater Stress Framework." *Water Resources Research* 51 (July): 5198–5216.

Richman, Talia, and Trisha Powell Crain. 2022. "With Teachers in Short Supply, States Ease Job Requirements." AP News, October 13. Accessed June 13, 2023 (https://apnews.com/article/science-texas-alabama-race-and-ethnicity-0d83fcc256ec2efef37b0d5c24f418a5).

Richtel, Matt, and Alexei Barrionuevo. 2005. "Wendy's Restaurants." *New York Times,* April 22, A9.

Rideout, Victoria. 2013. "Zero to Eight: Children's Media Use in America 2013." Common Sense Media, Fall. Accessed November 18, 2013 (www.commonsensemedia.org).

Rideout, Victoria. 2014. "Children, Teens, and Reading: A Common Sense Media Research Brief." Accessed March 1, 2017 (commonsensemedia.org).

Rideout, Victoria. 2015. "The Common Sense Census: Media Use by Tweens and Teens." Accessed March 1, 2017 (commonsensemedia.org).

Riesman, David. 1953. *The Lonely Crowd.* New York: Doubleday.

Riffkin, Rebecca. 2016. "Majority of Americans Dissatisfied with Corporate Influence." Gallup, January 20. Accessed April 4, 2017 (gallup.com).

Rios, Edwin. 2016. "Block the Vote." *Mother Jones,* July/August, 7.

Riosmena, Fernando, Elisabeth Root, Jamie Humphrey, Emily Steiner, and Rebecca Stubbs. 2015. "The Waning Hispanic Health Paradox." *Pathways* (Spring): 25–29.

Ripley, Amanda. 2016. "What the U.S. Can Learn from Other Nations' Schools." *New York Times,* December 8, A3.

Risk Based Security. 2017. "Data Breach QuickView Report." January. Accessed April 2, 2017 (riskbasedsecurity.com).

Ritchie, Hannah, Lucas Rodés-Guirao, Edouard Mathieu, Marcel Gerber, Esteban Ortiz-Ospina, Joe Hasell, and Max Roser. 2023. "World Population Growth". OurWorldInData.org. Accessed June 15, 2023 (https://ourworldindata.org/world-population-growth).

Ritzer, George. 1992. *Contemporary Sociological Theory,* 3rd edition. New York: McGraw-Hill.

Ritzer, George. 1996. *The McDonaldization of Society: An Investigation into the Changing Character of Contemporary Social Life.* Thousand Oaks, CA: Pine Forge Press.

Ritzer, George. 2008. *The McDonaldization of Society,* 5th edition. Los Angeles: Pine Forge Press.

Rizga, Kristina. 2017. "Betsy DeVos Wants to Use America's Schools to Build 'God's Kingdom'." *Mother Jones* March/April. Accessed July 31, 2017 (motherjones.com).

Robbins, Katherine Gallagher, and Julie Vogtman. 2016. "Low-Wage Jobs Held Primarily by Women Will Grow the Most over the Next Decade." National Women's Law Center, April. Accessed June 22, 2017 (nwlc.org).

Robert Wood Johnson Foundation. 2010. "California Nurse Ratio Law Saves Lives, Improves Nurse Morale, Study Finds." May 16. Accessed June 6, 2012 (www.rwjf.org).

Robert Wood Johnson Foundation. 2015. "Understanding the Uninsured Now." June. Accessed October 16, 2015 (rwjf.org).

Roberts, Andrea L., et al. 2013. "Perinatal Air Pollutant Exposures and Autism Spectrum Disorder in the Children of Nurses' Health Study II Participants." *Environmental Health Perspectives* 121 (August): 978–984.

Roberts, James A., and Meredith E. David. 2016. "My Life Has Become a Major Distraction from My Cell Phone: Partner Phubbing and Relationship Satisfaction Among Romantic Partners." *Computers in Human Behavior* 54 (January): 134–141.

Roberts, Keith A. 2004. *Religion in Sociological Perspective,* 4th edition. Belmont, CA: Wadsworth.

Roberts, Laura M., and Robin J. Ely. 2016. "Why Did So Many White Women Vote for Donald Trump?" *Fortune,* November 18. Accessed January 5, 2017 (fortune.com).

Roberts, Yvonne. 2014. "Addicted to Email? The Germans Have an Answer." AlterNet, September 1, 2015. Accessed September 2, 2014.

Robertson, Campbell. 2016. "Chief Justice in Alabama Is Suspended a Second Time." *New York Times,* October 1, A9.

Robinson, Laurie O., and Jeff Slowikowski. 2011. "Scary—and Ineffective." *Baltimore Sun,* February 1, 11.

Rochman, Bonnie. 2012. "The End of an Epidemic?" *Time,* February 6, 16.

Rochman, Bonnie. 2013. "Hover No More: Helicopter Parents May Breed Depression and Incompetence in Children."

Time, February 22. Accessed November 20, 2013 (www.healthland.time.com).

Rockel, Nick. 2022. "Directors Overboard: How Many Seats Is Too Many?" *Fortune*, December 2. Accessed May 12, 2023 (https://fortune.com/2022/12/02/modern-board-seats-directors/).

Roden, Lee. 2017. "Why Sweden Is NOT the 'Rape Capital of the World'." The Local, February 21. Accessed April 24, 2017 (thelocal.se).

Rodengen, Jeffrey L. 2002. *NRA: An American Legend*. Fort Lauderdale, FL: Write Stuff Enterprises.

Rodgers, Bill. 2017. "Sanctuary Churches in US Mobilize to Help Undocumented." VOA News, March 21. Accessed July 31, 2017 (voanews.com).

Rodgers, Timothy. 2016. "Organized Crime." LinkedIn, December 5. Accessed March 1, 2017 (linkedin.com).

Rodrigue, Edward, and Richard V. Reeves. 2015. "Cutting Poverty by Increasing Program Participation." Brookings Institution, June 11. Accessed September 31, 2015 (brookings.edu).

Roethlisberger, F. J., and William J. Dickson. 1939/1942. *Management and the Worker: An Account of a Research Program Conducted at the Western Electric Company, Hawthorne Works, Chicago*. Cambridge, MA: Harvard University Press.

Rohrlich, Justin. 2010. "Why White-Collar Criminals Don't Fear Getting Caught." Minyanville Media, Inc., November 12. Accessed January 12, 2014 (www.minyanville.com).

Roman, Caterina Gouvis, et al. 2012. "Social Networks, Delinquency, and Gang Membership: Using a Neighborhood Framework to Examine the Influence of Network Composition and Structure in a Latino Community." The Urban Institute, February. Accessed January 12, 2014 (www.urban.org).

Romer, Dan. 2011. "After 11 Years of Setting the Record Straight, Stories About Holiday Suicides Still Outnumber Those Debunking the Myth." Annenberg Public Policy Center, December 13. Accessed December 14, 2011 (www.annenbergpublicpolicycenter.org).

Ronfeldt, Matthew, Hamilton Lankford, Susanna Loeb, and James Wyckoff. 2011. "How Teacher Turnover Harms Student Achievement." National Bureau of Economic Research, June. Accessed April 14, 2012 (nber.org).

Roscigno, Vincent J. 2010. "Ageism in the American Workplace." *Contexts* 9 (Winter): 16–21.

Rose, Fred. 1997. "Toward a Class-Cultural Theory of Social Movements: Reinterpreting New Social Movements." *Sociological Forum* 12 (September): 461–494.

Rose, Lucie, Guillaume Dezecache, Tara Powell, Sylvie Chokron, and Klara Kovarski. 2022. "The Emergency of Prosociality: A Developmental Perspective on Altruism and Other Prosocial Behavior in the Face of Disaster." Current Directions in Psychological Science, 31(6), 486–492. Accessed June 25, 2023 (https://doi.org/10.1177/09637214221114090).

Rose, Peter I. 1997. *They and We: Racial and Ethnic Relations in the United States*, 5th edition. New York: McGraw-Hill.

Rosen, Ruth. 2016. "US Presidential Race: The Feminist Generation Gap." Open Democracy, March 14. Accessed November 9, 2023 (https://www.opendemocracy.net/en/5050/us-presidential-race-and-feminist-generation-gap/).

Rosenberg, Martha. 2016. "How an Army of Pharma Lobbyists in Washington Have Locked in One of the Biggest Corporate Ripoff Schemes in America." AlterNet, February 22. Accessed August 15, 2017 (alternet.org).

Rosenblatt, Lauren. 2017. "Bias Creeps Up on Older Workers." *Baltimore Sun*, July 2, 19.

Rosenfeld, Steven. 2014. "The Supreme Court's Radical Right Wing Majority: Waging War on Women and Boosting Corporate Power." AlterNet, June 30. Accessed July 26, 2014 (www.alternet.org).

Rosenthal, Bill. 2016. "The Ashley Madison Hack—One Year Later." Logical Operations, July 20. Accessed April 13, 2017 (logicaloperations.com).

Rosenthal, Elizabeth. 2014. "Medicine's Top Earners Are Not the M.D.s." *New York Times*, May 18, SR4.

Rosenthal, Robert, and Lenore Jacobsen. 1968. *Pygmalion in the Classroom: Teacher Expectations and Pupils' Intellectual Development*. New York: Holt, Rinehart, and Winston.

Rosin, Hanna. 2012. "Boys on the Side." *The Atlantic*, September. Accessed January 7, 2013 (www.theatlantic.com).

Ross, Brian, Rhonda Schwartz, and Megan Christie. 2016. "Recovered Madoff Money Now over $11 Billion: More Hidden Away?" ABC News, February 1. Accessed March 1, 2017 (abcnews.go.com).

Ross, Elizabeth N. 2023. "Turning Around Teacher Turnover." Harvard Graduate School of Education, January 19. Accessed June 14, 2023 (https://www.gse.harvard.edu/news/uk/23/01/turning-around-teacher-turnover).

Ross, Martha, Nicole P. Svajlenka, and Jane Williams. 2014. "Part of the Solution: Pre-Baccalaureate Healthcare Workers in a Time of Health System Change." Brookings, July. Accessed December 5, 2014 (www.brookings.edu).

Ross, Terris, Grace Kena, Amy Rathbun, Angelina KewalRamani, Jijun Zhang, Paul Kristapovich, and Eileen Manning. 2012. "Higher Education: Gaps in Access and Persistence Study." National Center for Education Statistics, August. Accessed June 21, 2014 (www.nces.ed.gov).

Rossano, Matt J. 2012. "The Essential Role of Ritual in the Transmission and Reinforcement of Social Norms." *Psychological Bulletin* 138 (May): 529–549.

Rossi, Max, and Lisa Jucca. 2014. "Villages Face 'Slow Death' as Italy's Population Ages." *Baltimore Sun*, March 2, 14.

Rossin-Slater, Maya. 2017. "Maternity and Family Leave Policy." National Bureau of Economic Research, January. Accessed July 19, 2017 (nber.org).

Roth, Steve. 2017. "New Data Reveal the Depressing Truth About How Wealth Is Amassed in America." AlterNet, January 6. Accessed June 4, 2017 (alternet.org).

Rothman, Sheila M. 1978. *Women's Proper Place: A History of Changing Ideals and Practices, 1870 to the Present*. New York: Basic Books.

Rowe-Finkbeiner, Kristin. 2012. "It's Not a 'Mommy War,' It's a War on Moms." Moms Rising, April 14. Accessed April 12, 2014 (www.momsrising.org).

Rubin, Kenneth, William Bukowski, and Jeffrey G. Parker. 1998. "Peer Interactions, Relationships, and Groups." Pp. 619–700 in *Handbook of Child Psychology: Vol. 3. Social, Emotional, and Personality Development*, edited by William Damon and Nancy Eisenberg. New York: Wiley.

Ruetschlin, Catherine, and Dedrick Asante-Muhammad. 2013. "The Challenge of Credit Card Debt for the African American Middle Class." Demos & NAACP, December 4. Accessed March 22, 2014 (www.demos.org).

Rule, James B. 1988. *Theories of Civil Violence*. Berkeley: University of California Press.

Rushe, Dominic, Oliver Milman, Molly Redden, Jamiles Lartey, David Smith, and Oliver Laughland. 2017. "6 Ways Trump Is Dismantling the Country Without Passing a Single Piece of Legislation." AlterNet, July 19. Accessed July 25, 2017 (alternet.org).

Ruth, Jennifer. 2014. "Non-Critical Thinking in China." *Chronicle Review*, February 28, B20.

Rutter, Virginia, and Pepper Schwartz. 2012. *The Gender of Sexuality: Exploring Sexual Possibilities*, 2nd edition. Lanham, MD: Rowman & Littlefield.

Ryan, Camille L., and Kurt Bauman. 2016. "Educational Attainment in the United States: 2015." U.S. Census Bureau, Current Population Reports, March. Accessed June 20, 2017 (census.gov).

Ryan, Camille. 2013. "Language Use In the United States: 2011." American Community Survey Reports, August. Accessed November 20, 2023 (https://www2.census.gov/library/publications/2013/acs/acs-22/acs-22.pdf).

Ryzhkov, Vladimir. 2015. "Russia's New Totalitarianism Depends on Silence." *Moscow Times*, April 23. Accessed June 23, 2017 (themoscowtimes.com).

S

Saad, Lydia. 2001. "Majority Considers Sex Before Marriage Morally Okay." *Gallup Poll Monthly*, No. 428 (May): 46–48.

Saad, Lydia. 2013. "Half in U.S. Support Publicly Financed Federal Campaigns." Gallup, June 24. (Accessed April 16, 2014 (www.gallup.com).

Saad, Lydia. 2013. "U.S. Support for Euthanasia Hinges on How It's Described." Gallup, May 29. Accessed May 16, 2014 (www.gallup.com).

Saad, Lydia. 2016. "Trump and Clinton Finish with Historically Poor Images." Gallup, November 8. Accessed June 27, 2017 (gallup.com).

Saad, Lydia. 2017. "A Third of Investors Help Parent or Grown Child Financially." Gallup, May 31. Accessed July 20, 2017 (gallup.com).

Saad, Lydia. 2017. "Record Few Americans Believe Bible Is Literal Word of God." Gallup, May 15. Accessed July 30, 2017 (gallup.com).

Saad, Lydia. 2019. "What Percentage of Americans Celebrate Christmas?" Gallup, December 23. Accessed June 14, 2023 (https://news.gallup.com/poll/272357/percentage-americans-celebrate-christmas.aspx).

Saad, Lydia. 2020. "Gallup Vault: Black Americans' Preferred Racial Label." Gallup, July 13. Accessed September 26, 2023. https://news.gallup.com/vault/315566/gallup-vault-black-americans-preferred-racial-label.aspx).

Saad, Lydia. 2021. "Americans' Emphasis on Environmental Protection Shrinks." Gallup, April 8. Accessed June 28, 2023 (https://news.gallup.com/poll/344252/americans-emphasis-environmental-protection-shrinks.aspx).

Saad, Lydia. 2022. "Military Brass, Judges Among Professions at New Image Lows." Gallup, January 12. Accessed June 14, 2023 (https://news.gallup.com/poll/388649/military-brass-judges-among-professions-new-image-lows.aspx).

Saad, Lydia. 2022. "'Pro-Choice' Identification Rises to Near Record High in U.S." Gallup, June 2. Accessed April 22, 2023 (https://news.gallup.com/poll/393104/pro-choice-identification-rises-near-record-high.aspx).

Saad, Lydia. 2022. "Public Interest in Having a Third Major Party Dips to 56%." Gallup, October 6. Accessed May 12, 2023 (https://news.gallup.com/poll/402515/public-interest-having-third-major-party-dips.aspx).

Saad, Lydia. 2023. "Economic Pessimism Persists, With Inflation Still Key Concern." Gallup, March 31. Accessed May 12, 2023 (https://news.gallup.com/poll/473132/economic-pessimism-persists-inflation-key-concern.aspx).

Sachs, Jeffrey S. 2005. "Confusion over Population: Growth or Dearth?" *Pop!ulation Press* 11 (Winter/Spring): 17.

Sacks, Vanessa, David Murphey, and Kristin Moore. 2014. "Adverse Childhood Experiences: National and State-Level Prevalence." Child Trends Research Brief, July. Accessed November 25, 2014 (www.childtrends.org).

Sagarin, Edward. 1975. *Deviants and Deviance*. New York: Praeger.

Sagon, Candy. 2017. "Medicine's Gender Issues." *AARP Bulletin*, January–February, 24–27.

Sahgal, Neha. 2014. "Coke, 'America the Beautiful,' and the Language of Diversity." Pew Research Center, February 3. Accessed November 25, 2014 (www.pewresearch.org).

Sahgal, Neha, and Greg Smith. 2009. "A Religious Portrait of African-Americans." Pew Research Center, Pew Forum on Religion & Public Life Project, January 30. Accessed June 21, 2014 (www.pewforum.org).

Saghal, Neha, Johnathan Evans, Ariana Monique Salazar, Kelsey Jo Starr, and Manolo Corichi. 2021. "Religion in India: Tolerance and Segregation." Pew Research Center, June 29. Accessed April 13, 2023 (https://www.pewresearch.org/religion/wp-content/uploads/sites/7/2021/06/PF_06.29.21_India.full_.report.pdf).

Salamone, Frank A. 2005. "Jazz and Its Impact on European Classical Music." *The Journal of Popular Culture* 38 (May): 732–743.

Sales, Nancy Jo. 2016. "How Social Media Is Disrupting the Lives of American Girls." *Time*, February 22–29, 26–27.

Salkin, Allen. 2017. "Take a Load Off. The Robots that Fold Laundry Are Coming." *New York Times*, May 25, B5.

Samuelson, Kate. 2016. "7 Ideas from Other Countries That Could Improve U.S. Elections." *Time*, November 14, 11.

Sanburn, Josh. 2015. "The Joy of Less." *Time*, March 23, 44–50.

Sanburn, Josh. 2016. "The Faults of Oklahoma." *Time*, March 21, 36–41.

Sánchez-Páramo, Ruth Hill, Daniel Gerszon Mahler, Ambar Naray, and Nishant Yonzan. 2021. "COVID-19 Leaves a Legacy of Rising Poverty and Widening Inequality." World Bank Blogs, Let's Talk Development, October 7. Accessed April 13, 2023 (https://blogs.worldbank.org/developmenttalk/covid-19-leaves-legacy-rising-poverty-and-widening-inequality).

Sánchez, Erika L. 2013. "The Challenge of Defining Muslim Feminism." *Huffington Post*, February 10. Accessed October 12, 2013 (www.huffingtonpost.com).

Sandbu, Martin. 2013. "Talkin' 'Bout a Revolution." *Financial Times*, April 19. Accessed October 8, 2014 (www.ft.com).

Sandstrom, Aleksandra. 2015. "Nearly All States Allow Religious Exemptions for Vaccinations." Pew Research Center, July 16. Accessed July 31, 2017 (pewresearch.org).

Sandstrom, Aleksandra. 2016. "Most States Allow Religious Exemptions from Child Abuse and Neglect Laws." Pew Research Center, August 12. Accessed July 31, 2017 (pewresearch.org).

Sandstrom, Aleksandra. 2017. "God or the Divine Is Referenced in Every State Constitution." Pew Research Center, August 17. Accessed August 21, 2017 (pewresearch.org).

Sandstrom, Aleksandra, and Angelina E. Theodorou. 2016. "Many Countries Allow Child Marriage." Pew Research Center, September 12. Accessed July 20, 2017 (pewresearch.org).

Sandstrom, Kent L., Daniel D. Martin, and Gary Alan Fine. 2006. *Symbols, Selves, and Social Reality: A Symbolic Interactionist Approach to Social Psychology and Sociology*, 2nd edition. Los Angeles: Roxbury.

Sang-Hun, Choe. 2016. "American Runs Afoul of Nation's Devotion to Slogans." *New York Times*, March 18, A4.

Sapir, Edward. 1929. "The Status of Linguistics as a Science." *Language* 5 (4): 207–214.

Sarkar, Donna. 2022. "5 of The Biggest Hacks in Cybersecurity History." Discover Magazine, October 18. Accessed March 25, 2023 (https://www.discovermagazine.com/technology/5-of-the-biggest-hacks-in-cybersecurity-history).

Sassler, Sharon, Fenaba R. Addo, and Daniel T. Lichter. 2012. "The Tempo of Sexual Activity and Later Relationship Quality." *Journal of Marriage and Family* 74 (August): 708–725.

Satter, Raphael. 2020. "U.S. Court: Mass Surveillance Program Exposed by Snowden Was Illegal." Reuters, September 2. Accessed February 12, 2023 (https://www.reuters.com/article/us-usa-nsa-spying/u-s-court-mass-surveillance-program-exposed-by-snowden-was-illegal-idUSKBN25T3CK).

Sauer, Meghan. 2022. "270 Health Experts to Spotify: Joe Rogan's Covid Misinformation is 'a Sociological Issue of Devastating Proportions'." *CNBC*, January 14. Accessed January 29, 2023 (https://www.cnbc.com/2022/01/14/experts-confront-spotify-over-misinformation-in-joe-rogan-podcast.html).

Saulny, Susan. 2011. "Counting by Race Can Throw Off Some Numbers." *New York Times,* February 10, A1.

Sawchuck, Stephen. 2012. "Teacher Quality, Status Entwined Among Top-Performing Nations." *Education Week* 31 (16): 12–16.

Sawhill, Isabel V., and Joanna Venator. 2014. "Families Adrift: Is Unwed Childbearing the New Norm?" Brookings, October 13. Accessed October 14, 2014 (www.brookings.edu).

Sayles, Justin. 2021. "The Bloody Bubble." The Ringer, July 9. Accessed September 16, 2023 (https://www.theringer.com/tv/2021/7/9/22567381/true-crime-documentaries-boom-bubble-netflix-hbo).

Scarlett, W. George, Sophie Naudeau, Dorothy Salonius-Pasternak, and Iris Ponte. 2005. *Children's Play.* Thousand Oaks, CA: Sage.

Schachter, Jason P. 2004. "Geographical Mobility: 2002 to 2003." U.S. Census Bureau, Current Population Reports. Accessed May 15, 2007 (www.census.gov).

Schaeffer, Katherine. 2023. "U.S. Congress Continues to Grow in Racial, Ethnic Diversity." Pew Research Center, January 9. Accessed April 28, 2023 (https://www.pewresearch.org/short-reads/2023/01/09/u-s-congress-continues-to-grow-in-racial-ethnic-diversity/#:~:text=When%20it%20comes%20to%20the,and%20one%20is%20American%20Indian).

Schaeffer, Katherine. 2023. "Nearly All Members of the 118th Congress Have a Bachelor's Degree – And Most Have a Graduate Degree, Too." Pew Research Center, February 2. Accessed June 13, 2023 (https://www.pewresearch.org/short-reads/2023/02/02/nearly-all-members-of-the-118th-congress-have-a-bachelors-degree-and-most-have-a-graduate-degree-too/).

Schaeffer, Katherine. 2023. "Key Facts about Americans and Guns." Pew Research Center, September 13. Accessed October 12, 2023 (https://www.pewresearch.org/short-reads/2023/09/13/key-facts-about-americans-and-guns/).

Schaeffer, Robert K. 2003. *Understanding Globalization: The Social Consequences of Political, Economic, and Environmental Change,* 2nd edition. Lanham, MD: Rowman & Littlefield.

Schanzenbach, Diane Whitmore, Lauren Bauer, Megan Mumford, and Ryan Nunn. 2016. "Money Lightens the Load." Brookings, December 12. Accessed December 23, 2016 (brookings.edu).

Schehr, Robert C. 1997. *Dynamic Utopia: Establishing Intentional Communities as a New Social Movement.* Westport, CT: Bergin & Garvey.

Schieman, Scott. 2010. "Socioeconomic Status and Beliefs About God's Influence in Everyday Life." *Sociology of Religion* 71 (Spring): 25–51.

Schiesel, Seth. 2011. "Supreme Court Has Ruled; Now Games Have a Duty." *New York Times,* July 29, C1.

Schiffrin, Holly H., Miriam Liss, Haley Miles-McLean, Katherine A. Geary, Mindy J. Erchull, and Taryn Tashner. 2014. "Helping or Hovering? The Effects of Helicopter Parenting on College Students' Well Being." *Journal of Child and Family Studies* 23 (April): 548–557.

Schlesinger, Izchak M. 1991. "The Wax and Wane of Whorfian Views." Pp. 7–44 in *Influence of Language on Culture & Thought,* edited by Robert Cooper and Bernard Spolsky. New York: Mounton de Gruyter.

Schlesinger, Robert. 2011. "Two Takes: Collective Bargaining Rights for Public Sector Unions?" *U.S. News Weekly,* February 25, 15–16.

Schmall, Emily. 2007. "The Cult of Chick-fil-A." *Forbes,* July 23, 80, 83.

Schmidt, Peter. 2008. "2 Studies Raise Questions about Research Based on Student Surveys." *Chronicle of Higher Education,* November 6. Accessed November 9, 2008 (www.chronicle.com).

Schmidt, Peter. 2014. "Supreme Court Exposes Affirmative Action at Colleges to Continued Political Assault." *Chronicle of Higher Education,* May 2, A3–A4.

Schneider, Eric C., Arnav Shah, Michelle M. Doty, Roosa Tikkanen, Katharine Fields, and Reginald D. Williams II. 2021. "Mirror, Mirror 2021: Reflecting Poorly." The Commonwealth Fund, August 4. Accessed June 10, 2023 (https://www.commonwealthfund.org/publications/fund-reports/2021/aug/mirror-mirror-2021-reflecting-poorly).

Schneider, Mark. 2013. "Higher Education Pays: But a Lot More for Some Graduates Than for Others." College Measures. Accessed June 21, 2014 (www.collegemeasures.org).

Schneider, Monica C., and Angela L. Bos. 2014. "Measuring Stereotypes of Female Politicians." *Political Psychology* 25 (April): 245–266.

Schnittker, Jason. 2009. "Mirage of Health in the Era of Biomedicalization: Evaluating Change in the Threshold of Illness, 1972–1996." *Social Forces* 87 (June): 2155–2182.

Schoch, Marta, Samuel Kofi Tetteh Baah, Christoph Lakner, and Jed Friedman. 2022. "Half of the Global Population Lives on Less Than US$6.85 Per Person Per Day." World Bank Blogs, December 8. Accessed April 13, 2023 (https://blogs.worldbank.org/developmenttalk/half-global-population-lives-less-us685-person-day).

"Schools in Finland 2016." *The Economist,* May 14, 46.

Schramm, J. B., Chad Aldeman, Andrew Rotherham, Rachael Brown, and Jordan Cross. 2013. "Smart Shoppers: The End of the 'College for All' Debate." College Summit. Accessed June 21, 2014 (www.collegesummit.org).

Schrobsdorff, Susanna. 2016. "A Distressing Summer of Workplace Sexism Reminds Us How Far We Have to Go." *Time,* September 5, 55.

Schultz, Ellen E. 2011. *Retirement Heist: How Companies Plunder and Profit from the Nest Eggs of American Workers.* New York: Penguin.

Schur, Edwin M. 1968. *Law and Society: A Sociological View.* New York: Random House.

Schurman-Kauflin, Deborah. 2000. *The New Predator: Women Who Kill.* New York: Algora.

Schutz, Alfred. 1967. *The Phenomenology of the Social World.* Evanston, IL: Northwestern University Press.

Schwartz, Larry. 2014. "8 Mistakes We're Making About Ebola That We Also Made When AIDS Appeared." AlterNet, October 17. Accessed October 18, 2014 (www.alternet.org).

Schwartz, Larry. 2017. "The Depressing Trendlines That Are Driving America's Decline in Life Expectancy." AlterNet, March 9. Accessed March 20, 2017 (alternet.org).

Schwartz, Nelson D. 2017. "Workers Needed, but Drug Testing Takes a Toll." *New York Times,* July 25, A1, A15.

Schwarz, Alan. 2014. "Thousands of Toddlers Are Medicated for A.D.H.D., Report Finds, Raising Worries." *New York Times,* May 17, A11.

Schweizer, Peter. 2013. *Extortion: How Politicians Extract Your Money, Buy Votes, and Line Their Own Pockets.* New York: Houghton Mifflin Harcourt.

Schwyzer, Hugo. 2011. "How Our Sick Culture Makes Girls Think They Have to Be Gorgeous to Be Loved." AlterNet, April 12. Accessed October 29, 2012 (www.alternet.org).

Scleza, Brooke A. et al. 2020." Patterns of Paternal Investment Predict Cross-Cultural Variation in Jealous Response." *Nature Human Behavior,* January (4). Accessed March 25, 2023 (https://www.nature.com/articles/s41562-019-0654-y.epdf?author_access_token=rH76kJTFiJ-HguLimJS89nNRgN0jAjWel9jnR3ZoT-v0MsSSmqyNvbAEWRHfnXIMhn8P1jiFpxDYl4qsnjI45Zd3lSF71fE28Ny8aTmg-tIq9s6PAB0qE78ATYNVKDLUnh3DYY-qPNoLPzExgbcqlYA0qQ%3D%3D).

Scommegna, Paola. 2013. "Aging U.S. Baby Boomers Face More Disability." Population Reference Bureau, March. Accessed May 16, 2014 (www.prb.org).

Scott, Robert E. 2012. "The China Toll." EPI Briefing Paper 345, Economic Policy

Institute, August 23. Accessed April 12, 2014 (www.epi.org).

"Select-a-Faith." 2014. *The Economist*, May 17, 30.

Segal, Nancy. 2021. "Speaking of Psychology: What studying twins can teach us about ourselves, with Nancy Segal, PhD." American Psychological Association. Accessed August 30, 2023 (https://www.apa.org/news/podcasts/speaking-of-psychology/studying-twins).

Semega, Jessica, and Melissa Kollar. 2022. "Income in the United States: 2021." U.S. Census Bureau, September 13. Accessed May 3, 2023 (https://www.census.gov/library/publications/2022/demo/p60-276.html).

Semega, Jessica, Kayla R. Fontenot, and Melissa A. Kollar. 2017. "Income and Poverty In the United States: 2016." Current Population Reports, September. Accessed November 22, 2023 (https://www.census.gov/content/dam/Census/library/publications/2017/demo/P60-259.pdf).

Semuels, Alana. 2010. "Can a Prison Save a Town?" *Los Angeles Times*, May 3. Accessed May 4, 2010 (www.latimes.com).

Semuels. Alana. 2023. "Airlines Are Terrible. Small Cities Are Still Paying Them Millions of Dollars to Stick Around." Time, January 17. Accessed May 12, 2023 (https://time.com/6247052/airlines-deregulation-american-inequality/).

Sengupta, Somini. 2013. "What You Didn't Post, Facebook May Still Know." *New York Times*, March 26, B1.

Sengupta, Somini. 2022. "What Is Greenwashing and How Do You Avoid It?" New York Times, August 23. Accessed July 23, 2023 (https://www.nytimes.com/2022/08/23/climate/climate-greenwashing.html).

The Sentencing Project. 2017. "Trends in U.S. Corrections." March 7. Accessed April 20, 2017 (sentencingproject.org).

Senter, Mary S., and Roberta Spalter-Roth. 2016. "Individual Salary Is Not Enough: Measuring the Well-Being of Recent College Graduates in Sociology." American Sociological Association, Bachelors & Beyond Series, August. Accessed December 23, 2016 (asanet.org).

Senter, Mary S., Nicole Van Vooren, and Roberta Spalter-Roth. 2014. "Sociology, Criminology Concentrations, and Criminal Justice: Differences in Reasons for Majoring, Skills, Activities, and Early Outcomes?" American Sociological Association, April. Accessed October 25, 2014 (www.asanet.org).

Seward, Zachary M. 2012. "58 Countries with Better Voter Turnout Than the United States." Quartz, November 6. Accessed April 20, 2014 (www.qz.com).

"Sex Selection." 2017. *The Economist*, January 21, 49–50.

Shannon-Missal, Larry. 2016. "Tattoo Takeover: Three in Ten Americans Have Tattoos, and Most Don't Stop at Just One." Harris Poll, February 10. Accessed April 14, 2017 (theharrispoll.com).

Shapira, Ian. 2014. "In Arizona, a Navajo High School Emerges as a Defender of the Washington Redskins." *Washington Post*, October 26. Accessed October 27, 2014 (www.washingtonpost.com).

Shapiro, Jenessa R., and Amy M. Williams. 2012. "The Role of Stereotype Threats in Undermining Girls' and Women's Performance and Interest in STEM Fields." *Sex Roles* 66 (February): 175–183.

Shariff, Azim F., and Lara B. Aknin. 2014. "The Emotional Toll of Hell: Cross-National and Experimental Evidence for the Negative Well-Being Effects of Hell Beliefs." *PloS One* 9 (January): e85251.

Sharifzadeh, Virginia-Shirin. 1997. "Families with Middle Eastern Roots." Pp. 441–482 in *Developing Cross-Cultural Competence: A Guide for Working with Children and Families,* edited by Eleanor W. Lynch and Marci J. Hanson. Baltimore, MD: Paul H. Brookes.

Shauk, Zain, and Bradley Olson. 2014. "Where Frackers Are Friendly Neighbors." *Bloomberg Businessweek*, September 8–14, 37–38.

Shea, Christopher. 2014. "The Liar's 'Tell.'" *The Chronicle Review*, October 17, B6–B9.

SHEEO. 2017. "SHEF: FY 2016 State Higher Education Finance." State Higher Education Executive Officers Association. Accessed July 31, 2017 (sheeo.org).

Shellenbarger, Sue. 2013. "Finding the Just-Right Level of Self-Esteem for a Child." *Wall Street Journal*, February 27, D1, D3.

Shepherd, Julianne Escobedo. 2011. "Hard-Partying Rich Boy Kills Two in Hit-and-Run, Buys Himself Out of a Prison Sentence." AlterNet, June 6. Accessed June 10, 2011 (www.alternet.org).

Sheppard, Kate. 2011. "The Hackers and the Hockey Stick." *Mother Jones*, May/June, 33–45.

Sheppard, Kate. 2013. "Scientific Misconceptions: The Junk Science Behind Anti-Abortion Advocates' Wildest Claims." *Mother Jones*, January/February, 14.

Sherman, Erik. 2015. "Report: Disturbing Drop in Women in Computing Field." *Fortune*, March 26. Accessed July 31, 2017 (fortune.com).

Sherwood, Jessica Holden. 2010. *Wealth, Whiteness, and the Matrix of Privilege: The View from the Country Club.* Lanham, MD: Lexington Books.

Shibutani, Tamotsu. 1986. *Social Process: An Introduction to Sociology.* Berkeley: University of California Press.

Shilo, Guy, and Riki Savaya. 2011. "Effects of Family and Friend Support on LGB Youths' Mental Health and Sexual Orientation Milestones." *Family Relations* 60 (July): 318–330.

Shor, Eran, Arnout van de Rijt, and Babak Fotouhic. 2019. "A Large-Scale Test of Gender Bias in the Media." *Sociological Science*. 6 (September): 526-550. Accessed January 29, 2023 (https://sociologicalscience.com/download/vol-6/september/SocSci_v6_526to550.pdf).

Shorto, Russell. 2008. "No Babies?" *New York Times*, June 29, 34.

Shriver, Maria. 2014. *The Shriver Report: A Woman's Nation Pushes Back from the Brink.* New York: Rosetta Books.

Shteir, Rachel. 2013. "Feminism Fizzles." *The Chronicle Review*, February 1, B6–B9.

Siegman, Aron W., and Stanley Feldstein, eds. 1987. *Nonverbal Behavior and Communication.* Hillsdale, NJ: Lawrence Erlbaum Associates.

Silva, Jason R. 2021. "Global Mass Shootings: Comparing the United States Against Developed and Developing Countries." *International Journal of Comparative and Applied Criminal Justice*, March 21 (https://doi.org/10.1080/01924036.2022.2052126).

Silverman, Rachel E. 2003. "Provisions Boost Rights of Couples Living Together." *Wall Street Journal*, March 5, D1.

Silverman, Rachel E. 2013. "Tracking Sensors Invade the Workplace." *Wall Street Journal*, March 7, B1, B2.

Silverstein, Jason. 2015. "Dylann Roof Obsessed over Trayvon Martin and 'White Race'." *New York Daily News*, June 20. Accessed April 16, 2017 (nydailynews.com).

Simmel, Georg. 1902. "The Number of Members as Determining the Sociological Form of the Group." *American Journal of Sociology* 8 (July): 1–46.

Simon, Paula. 2013. "AP Is Not for Everyone." *Baltimore Sun*, August 24, 17.

Singal, Pooja, Adi Rattner, and Meghana Desale. 2014. "Court Ruling Limits a Doctor's Options." *Baltimore Sun*, July 17, 17.

Singer, Natasha. 2013. "Group Criticizes Learning Apps for Babies." *New York Times*, August 8, B8.

Singer, Natasha. 2013. "They Loved Your G.P.A. Then They Saw Your Tweets." *New York Times*, November 10, BU3.

Singh, Tushar, et al. 2016. "Vital Signs: Exposure to Electronic Cigarette Advertising Among Middle School and High School Students—United States, 2014." *MMWR* 64 (52): 1403–1408.

Sjoberg, Gideon. 1960. *The Preindustrial City: Past and Present.* Glencoe, IL: Free Press.

Skinner, Victor. 2013. "What's the Practice Called 'Passing the Trash' That

Allows Teachers to Continue Sexually Targeting Kids?" The Blaze, March 27. Accessed April 3, 2017 (theblaze.com).

Skoning, Gerald D. 2013. "How Congress Puts Itself Above the Law." *Wall Street Journal*, April 16, A15.

Skoumal, Robert J., Michael R. Brudzinski, and Brian S. Currie. 2015. "Earthquakes Induced by Hydraulic Fracturing in Poland Township, Ohio." *Bulletin of the Seismological Society of America* 105 (February): 189–197.

Slater, Dan. 2013. "Darwin Was Wrong About Dating." *New York Times*, January 13, SR1.

Smedley, Audrey. 2007. *Race in North America: Origin and Evolution of a Worldview*, 3rd edition. Boulder, CO: Westview Press.

Smeeding, Tim. 2017. "Cash Matters and Place Matters: A Child Poverty Plan That Capitalizes on New Evidence." *Pathways*, Spring, 14–20.

Smelser, Neil J. 1962. *Theory of Collective Behavior*. New York: Free Press.

Smelser, Neil J. 1988. "Social Structure." Pp. 103–129 in *Handbook of Sociology*, edited by Neil J. Smelser. Newbury Park, CA: Sage.

Smith, Aaron. 2016. "Gig Work, Online Selling and Home Sharing." Pew Research Center, November 17. Accessed June 16, 2017 (pewresearch.org).

Smith, Aaron. 2017. "Record Shares of Americans Now Own Smartphones, Have Home Broadband." Pew Research Center, January 12. Accessed March 14, 2017 (pewresearch.org).

Smith, Aaron, and Monica Anderson. 2015. "5 Facts About Online Dating." Pew Research Center, April 20. Accessed August 14, 2015 (pewresearch.org).

Smith, Aaron, and Monica Anderson. 2016. "5 Facts About Online Dating." Pew Research Center, February 29. Accessed May 19, 2017 (pewresearch.org).

Smith, Aaron, Janna Anderson, and Lee Rainie. 2014. "AI, Robotics, and the Future of Jobs." Pew Research Center, August 6. Accessed October 8, 2014 (www.pewresearch.org).

Smith, Adam. 1776/1937. *An Inquiry into the Nature and Causes of the Wealth of Nations*. New York: Modern Library.

Smith, Dorothy E. 1987. *The Everyday World as Problematic: A Feminist Sociology*. Toronto: University of Toronto Press.

Smith, Erica L. 2022. "Female Murder Victims and Victim-Offender Relationship, 2021." Bureau of Justice Statistics, December. Accessed May 22, 2023. (https://bjs.ojp.gov/female-murder-victims-and-victim-offender-relationship-2021#:~:text=Of%20the%20estimated%204%2C970%20female,victims%20of%20intimate%20partner%20homicide).

Smith, Jane I. 1994. "Women in Islam." Pp. 303–325 in *Today's Woman in World Religions*, edited by Arvind Sharma. Albany: State University of New York Press.

Smith, Gregory A. 2021. "About Three-in-Ten U.S. Adults Are Now Religiously Unaffiliated." Pew Research Center, January 14. Accessed July 30, 2023 (https://www.pewresearch.org/religion/2021/01/14/measuring-religion-in-pew-research-centers-american-trends-panel/).

Smith, Kate. 2020. "Mississippi to Remove State Flag, the Last to Display the Confederate Battle Emblem." *CBS News*, June 29. Accessed February 11, 2023 (https://www.cbsnews.com/news/mississippi-state-flag-remove-confederate-battle-emblem/).

Smith, Morgan. 2021. "These Are the 11 Most In-Demand Health-Care Jobs of the Next Decade." CNBC, October 15. Accessed June 11, 2023 (https://www.cnbc.com/2021/10/15/the-11-fastest-growing-jobs-in-health-care-over-the-next-decade-.html).

Smith, N., and A. Leiserowitz. 2013. "American Evangelicals and Global Warming." *Global Environmental Change* 23 (October): 1009–1017.

Smith, Oliver. 2017. "Mapped: The World's Most Polluted Countries." *Telegraph*, January 2. Accessed August 20, 2017 (telegraph.co.uk).

Smith, Samantha. 2017. "Why People Are Rich and Poor: Republicans and Democrats Have Very Different Views." Pew Research Center, May 2. Accessed July 4, 2017 (pewresearch.org).

Smith, Stacy L., Katherine M. Pieper, Amy Granados, and Marc Choueiti. 2010. "Assessing Gender-Related Portrayals in Top-Grossing G-Rated Films." *Sex Roles* 62 (June): 774–786.

Smith, Stacy L., Marc Choueiti, Katherine Pieper, Traci Gillig, Carmen Lee, and Dylan DeLuca. 2015. "Inequality in 700 Popular Films: Examining Portrayals of Gender, Race, & LGBT Status from 2007 to 2014." USC Annenberg School for Communication and Journalism. Accessed September 29, 2015 (loomberg.usc.edu).

Smith, Tom W., Peter V. Marsden, and Michael Hout. 2011. *General Social Survey, 1972–2010, Cumulative File Codebook*. Ann Arbor, MI: Inter-University Consortium for Political and Social Research. Accessed February 13, 2014 (www.icpsr.umich.edu).

Snell, Kelsey. 2020. "What's Inside the Senate's $2 Trillion Coronavirus Aid Package." NPR, March 26. Accessed May 12, 2023 (https://www.npr.org/2020/03/26/821457551/whats-inside-the-senate-s-2-trillion-coronavirus-aid-package).

Snider, Laureen. 1993. "Regulating Corporate Behavior." Pp. 177–210 in *Understanding Corporate Criminality*, edited by Michael B. Blankenship. London: Garland Press.

Snipp, C. Matthew. 1996. "A Demographic Comeback for American Indians." *Population Today* 24 (November): 4–5.

Soble, Jonathan. 2015. "A Sprawl of Ghost Homes in Aging Tokyo Suburbs." *New York Times*, August 24, A1, A3

Soldatov, Andrei. 2016. "Putin Has Finally Reincarnated the KGB." Foreign Policy, September 21. Accessed June 21, 2017 (foreignpolicy.com).

"Social Media Fact Sheet." 2017. Pew Research Center, January 12. Accessed March 14, 2017 (pewresearch.org).

Sonfield, Adam, and Kathryn Kost. 2013. "Public Costs from Unintended Pregnancies and the Role of Public Insurance Programs in Paying for Pregnancy Infant Care: Estimates for 2008." Guttmacher Institute, October. Accessed May 16, 2014 (www.guttmacher.org).

Southern Poverty Law Center. 2023. "Hate Map, 2021." Accessed February 12, 2023 (https://www.splcenter.org/hate-map).

Spalter-Roth, Roberta, Nicole Van Vooren, and Mary S. Senter. 2013. "Social Capital for Sociology Majors: Applied Activities and Peer Networks." American Sociological Association. Accessed October 12, 2013 (www.asanet.org).

Sparks, Sarah D. 2016. "Summing Up Results from TIMSS, PISA." *Education Week*, December 13. Accessed July 31, 2017 (edweek.org).

Spencer, Herbert. 1862/1901. *First Principles*. New York: P. F. Collier & Son.

Spiker, Ted. 2015. "The Key to Making Your Kid a Star Athlete: Back Off." *Time Magazine*, February 15. Accessed February 22, 2023 (https://time.com/3707641/child-star-athletes-sports-parenting/).

Spinks, Rosie. 2018. "A Global Guide to Queuing Philosophies, from Wimbledon to São Paulo." Quartz, July 7. Accessed September 7, 2023 (https://qz.com/quartzy/1322666/a-global-guide-to-queuing-philosophies-from-wimbledon-to-sao-paolo).

Spitznagel, Eric. 2013. "Men on the Run." *Bloomberg Businessweek*, March 25–31, 82.

Spivak, Howard R., E. Lynn Jenkins, Kristi VanAudenhove, Debbie Lee, Mim Kelly, and John Iskander. 2014. "CDC Grand Rounds: A Public Health Approach to Prevention of Intimate Partner Violence." *MMWR* 63 (January 17): 38–41.

Spotts, Peter N. 2009. "New Climate Change Signal: Oceans Turning Acidic." *Christian Science Monitor*, December 9. Accessed December 12, 2010 (www.csmonitor.com).

Spradley, J. P., and M. Phillips. 1972. "Culture and Stress: A Quantitative Analysis." *American Anthropologist* 74 (3): 518–529.

Sprigg, Peter. 2011. "Marriage's Public Purpose." *Baltimore Sun*, February 2, 27.

Srivastava, Mehul. 2013. "In Bangladesh, Outside Inspectors Are Still MIA." *Bloomberg Businessweek*, November 4–10, 28–30.

The Staff at The Henry Ford. 2022. "Children's Car Seats: The Henry Ford." Accessed March 25, 2023 (https://www.thehenryford.org/collections-and-research/digital-collections/expert-sets/101263/).

Stager, Curt. 2017. "Sowing Climate Doubt Among Schoolteachers." *New York Times*, April 27, A27.

Stanley, Megan, Ife Floyd, and Misha Hill. 2016. "TANF Cash Benefits Have Fallen by More Than 20 Percent in Most States and Continue to Erode." Center on Budget and Policy Priorities, October 17. Accessed May 5, 2017 (cbpp.org).

Starr, Christine R., and Gail M. Ferguson. 2012. "Sexy Dolls, Sexy Grade-Schoolers? Media and Maternal Influences on Young Girls' Self-Sexualization." *Sex Roles* 67 (October): 463–476.

Startz, Dick. 2016. "Teacher Pay Around the World." Brookings, July 11. Accessed July 20, 2017 (brookings.edu).

"The State of American Jobs." 2016. Pew Research Center, October 6. Accessed June 20, 2017 (pewresearch.org).

"The State of Arab Men." 2017. *The Economist*, May 6, 46.

State Higher Education Finance. 2022. "Report – SHEF." Accessed June 13, 2023 (https://shef.sheeo.org/report/#report-highlights).

Statista. 2016. "Daily Media Use Per Capita in the United States in Spring 2016." Accessed March 13, 2017 (statista.com).

Statista. 2017. "Number of Retail Clinics in the United States from 2008 to 2018." The Statistics Portal. Accessed April 3, 2017 (statista.com).

Stebbins, Samuel. 2020. "College Coaches Dominate List of Highest-Paid Public Employees." *USA Today*: Money, September 23. Accessed June 13, 2023 (https://www.usatoday.com/story/money/2020/09/23/these-are-the-highest-paid-public-employees-in-every-state/114091534/).

Steckenrider, Janie. 2023. "Sexual Activity of Older Adults: Let's Talk About It." *The Lancet: Health Longevity* I4 (3). Accessed April 21, 2023 (https://www.thelancet.com/journals/lanhl/article/PIIS2666-7568(23)00003-X/fulltext).

Steel, Emily. 2015. "For Cable Network, Crime Certainly Pays." *New York Times*, January 5, B1.

Steele, Eurídice Martínez, Larissa Galastri Baraldi, Maria Laura da Costa Louzada, Jean-Claude Moubarac, Dariush Mozaffarian, and Carlos Augusto Monteiro. 2016. "Ultra-Processed Foods and Added Sugars in the US Diet: Evidence from a Nationally Representative Cross-Sectional Study." *BMJ Open* 6 (3): 1–9.

Steffensmeier, Darrell J., Jennifer Schwartz, and Michael Roche. 2013. "Gender and Twenty-First-Century Corporate Crime: Female Involvement and the Gender Gap in Enron-Era Corporate Frauds." *American Sociological Review* 78 (June): 448–476.

Stein, Joel. 2015. "Baby, You Can Drive My Car." *Time*, February 9, 32–40.

Stein, Rob. 2017. "Scientists Precisely Edit DNA in Human Embryos to Fix a Disease Gene." NPR, August 2. Accessed September 2, 2017 (npr.org).

Steinberg, Julia R., and Lawrence B. Finer. 2011. "Examining the Association of Abortion History and Current Mental Health: A Reanalysis of the National Comorbidity Survey Using a Common-Risk-Factors Model." *Social Science & Medicine* 72 (1): 72–82.

Steinkopf-Frank, Hannah. 2021. "Women Imams Around the World Challenge Male-Dominated Islam." Worldcrunch. Accessed June 13, 2023 (https://worldcrunch.com/culture-society/women-imams-around-the-world-challenge-male-dominated-islam).

Steinmetz, Katy. 2014. "Disrupted: The Tech Wealth Transforming San Francisco Has Unemployment Down, Evictions Up and Tensions Flaring." *Time*, February 10, 32–35.

Steinmetz, Katy. 2015. "What the Toy Aisle Can Teach Us About Gender Parity." *Time*, August 24, 25–26.

Stepan-Norris, Judith. 2015. "Social Justice & the Next Upward Surge for Unions." *Contexts* 14 (Spring): 46–51.

Stepler, Renee. 2017. "Led by Baby Boomers, Divorce Rates Climb for America's 50+ Population." Pew Research Center, March 9. Accessed July 20, 2017 (pewresearch.org).

Stepler, Renee. 2017. "Number of U.S. Adults Cohabiting with a Partner Continues to Rise, Especially Among Those 50 and Older." Pew Research Center, April 6. Accessed July 20, 2017 (pewresearch.org).

Stepler, Renee, and Mark Hugo Lopez. 2016. "U.S. Latino Population Growth and Dispersion Has Slowed Since Onset of the Great Recession." Pew Research Center, October 26. Accessed September 8, 2017 (pewresearch.org).

Sternbergh, Adam. 2008. "Why White People Like 'Stuff White People Like.'" *New Republic*, March 17. Accessed November 6, 2008 (www.tnr.com).

Steuerle, C. Eugene, and Caleb Quakenbush. 2013. "Social Security and Medicare Taxes and Benefits over a Lifetime: 2013 Update." Urban Institute, November. Accessed July 20, 2017 (urban.org).

Steverman, Ben. 2021. "Divorces and Marriages Tumbled in U.S. During Covid, Study Shows." *Bloomberg*, January 5. Accessed May 10, 2023 (https://www.bloomberg.com/news/articles/2021-01-05/divorces-and-marriages-tumbled-in-u-s-during-covid-study-shows).

Stewart, Madeline. 2021. "Confederate Flag Removal Prompts Dueling Rallies at SC State House." *CBS News*, July 10. Accessed February 11, 2023 (https://www.cbsnews.com/news/mississippi-state-flag-remove-confederate-battle-emblem/).

Stewart, Susan D. 2022. *On the Rocks: Straight Talk About Women and Drinking*. Lanhma, Maryland: The Rowman & Littlefield Publishing Group, Inc.

Stewart, Timothy. 2022. "Overview of Motor Vehicle Crashes in 2020 (Report No. DOT HS 813266). National Highway Traffic Safety Administration. Accessed January 29, 2022 (https://crashstats.nhtsa.dot.gov/Api/Public/ViewPublication/813266).

"Steve Jobs: Adopted Child Who Never Met His Biological Father." 2011. *Telegraph*, October 6. Accessed March 22, 2014 (telegraph.co.uk).

Stillars, Alan L. 1991. "Behavioral Observation." Pp. 197–218 in *Studying Interpersonal Interaction,* edited by B. M. Montgomery and S. Duck. New York: Guilford Press.

Stobbe, Mike. 2022. "'Not Been a Great Time for Romance,' with the Fewest Marriages in the United States Since 1963." *Chicago Sun Times, May 28*. Accessed May 19, 2023 (https://chicago.suntimes.com/2022/5/18/23126072/marriage-down-cdc-covid-pandemic).

Stokes, Bruce. 2017. "What It Takes to Truly Be 'One of Us'." Pew Research Center, February 1. Accessed February 5, 2017 (pewresearch.org).

Stone, Geoffrey R. 2016. "Free Expression in Peril." *Chronicle of Higher Education*, September 16, B9–B11.

Strasburger, Victor C., and Marjorie J. Hogan. 2013. "Policy Statement: Children, Adolescents, and the Media." *Pediatrics* 132 (5): 958–961.

Straus, Murray A. 2011. "Gender Symmetry and Mutuality in Perpetration of Clinical-Level Partner Violence: Empirical Evidence and Implications for Prevention and Treatment." *Aggression and Violent Behavior* 16 (July–August): 279–288.

Straus, Murray A. 2014. "Addressing Violence by Female Partners Is Vital to Prevent or Stop Violence Against Women:

Evidence from the Multisite Batterer Intervention Evaluation." *Violence Against Women* 20 (7): 889–899.

"Streaking." 2005. Wikipedia Encyclopedia. Accessed September 8, 2005 (www.wikipedia.org).

Strean, William B. 2009. "Remembering Instructors: Play, Pain, and Pedagogy." *Qualitative Research in Sport and Exercise* 1 (November): 210–220.

Streib, Jessi. 2011. "Class Reproduction by Four Year Olds." *Qualitative Sociology* 34 (June): 337–352.

Stroebe, Margaret, Maarten van Son, Wolfgang Stroebe, Rolf Kleber, Henk Schut, and Jan van den Bout. 2000. "On the Classification and Diagnosis of Pathological Grief." *Clinical Psychology Review* 20 (January): 57–75.

Strom, Stephanie. 2014. "Coca-Cola to Remove an Ingredient Questioned by Consumers." *New York Times*, May 6, B6.

Strom, Stephanie. 2015. "The Fried Chicken Is O.K., but the Cups Are Delicious." *New York Times*, February 26, B3.

Strom, Stephanie. 2016. "At Chobani, It's Not Just the Yogurt That's Rich." *New York Times*, April 27, B1, B5.

"Study Shows Positive Results from Early Head Start Program." 2002. U.S. Department of Health and Human Services, June 3. Accessed July 10, 2002 (www.hhs.gov).

Sturm, Roland, and Ruopeng An. 2014. "Obesity and Economic Environments." *CA: A Cancer Journal for Clinicians* 65 (5): 337–350.

Substance Abuse and Mental Health Services Administration. 2022. "Table 1.1A – Types of Illicit Drug Use in Lifetime, Past Year, and Past Month: Among People Aged 12 or Older; Numbers in Thousands, 2021." Accessed June 11, 2023 (https://www.samhsa.gov/data/sites/default/files/reports/rpt39441/NSDUHDetailedTabs2021/NSDUH-DetailedTabs2021/NSDUHDetTabsSect1pe2021.htm).

Sudarkasa, Niara 2007. "African American Female-Headed Households: Some Neglected Dimensions." Pp. 172–182 in *Black Families*, 4th edition, edited by Harriette Pipes McAdoo. Thousand Oaks, CA: Sage.

Suddath, Claire. 2013. "So Hard to Say Goodbye." *Bloomberg Businessweek*, November 4–10, 79–81.

Suliman, Adela. 2017. "Going to School? Unlikely If You're a Girl in South Sudan, Report Says." Thomas Reuters Foundation, October 9. Accessed February 12, 2023 (https://www.reuters.com/article/us-women-education/going-to-school-unlikely-if-youre-a-girl-in-south-sudan-report-says-idUSKBN1CF01S).

Sullivan, Andrew. 2011. "Why Gay Marriage Is Good for Straight America." *Newsweek*, July 25, 12–14.

Sullivan, Deirdre. 2017. "How Much Does a Tiny House Really Cost?" The Spruce, June 1. Accessed August 20, 2017 (thespruce.com).

Summers, Nick. 2010. "Do Fines Ever Make Corporations Change?" Newsweek, November 13, 56.

Sumner, William G. 1906. *Folkways*. New York: Ginn.

Sunstein, Cass R. 2009. *On Rumors: How Falsehoods Spread, Why We Believe Them, What Can Be Done*. New York: Farrar, Straus and Giroux.

"Super Bull Sunday." 2015. Snopes.com, January 30. Accessed July 22, 2015 (snopes.com).

Superville, Darlene. 2017. "Trump Lets States Block Some Planned Parenthood Money." *Baltimore Sun*, April 14, 4.

Superville, Denisa R. 2021. "Female Principals Are Paid Less Than Men. That's a Big Concern." Edweek, April 21. Accessed April 16, 2023 (https://www.edweek.org/leadership/female-principals-are-paid-less-than-men-thats-a-big-concern/2021/04).

Sustainable Packaging Coalition. 2016. "2015–16 Centralized Study on Availability of Plastic Recycling." Accessed August 20, 2017 (sustainablepackaging.org).

Sutherland, Edwin H. 1949. *White Collar Crime*. New York: Holt, Rinehart, and Winston.

Sutherland, Edwin H., and D. R. Cressey. 1970. *Criminology*, 8th edition. Philadelphia: Lippincott.

Sutton, Philip W. 2000. *Explaining Environmentalism: In Search of a New Social Movement*. Burlington, VT: Ashgate.

Swift, Art. 2017. "Americans See US World Standing as Worst in a Decade." Gallup, February 10. Accessed March 1, 2017 (gallup.com).

Swift, Art. 2017. "More Americans Say Immigrants Help Rather Than Hurt Economy." Gallup, June 29. Accessed June 27, 2017 (gallup.com).

Sykes, Selina. 2015. "Indian Man Has 39 Wives, 94 Children and 33 Grandchildren All Living Under the SAME Roof." Express, November 9. Accessed August 15, 2017 (express.co.uk).

Szarota, Piotr. 2010. "The Mystery of the European Smile: A Comparison Based on Individual Photographs Provided by Internet Users." *Journal of Nonverbal Behavior* 34 (December): 249–256.

T

Tabachnick, Rachel. 2011. "The 'Christian' Dogma Pushed by Religious Schools That Are Supported by Your Tax Dollars." AlterNet, May 23. Accessed July 12, 2011 (www.alternet.org).

Tabuchi, Hiroko. 2015. "A Tiara? No Thanks." *New York Times*, October 29, B1, B8.

Tabuchi, Hiroko, and Neal E. Boudette. 2017. "Four Carmakers Knew of Hazard, Suit Says." *New York Times*, February 28, B1–B2.

"Tackling Poverty." 2015. *The Economist*, September 5, 31–32.

Taibbi, Matt. 2014. *The Divide: American Injustice in the Age of the Wealth Gap*. New York: Spiegel & Grau.

Tajfel, Henri. 1982. "Social Psychology of Intergroup Relations." *Annual Review of Psychology* 33 (February): 1–39.

Tannen, Deborah. 1990. *You Just Don't Understand: Women and Men in Conversation*. New York: Ballantine.

Tannenbaum, Frank. 1938. *Crime and the Community*. New York: Columbia University Press.

Tanur, Judith M. 1994. "The Trustworthiness of Survey Research." *Chronicle of Higher Education*, May 25, B1-B3.

Tanzi, Alexandre. 2023. "Even on $100,000-Plus, More Americans Are Living Paycheck to Paycheck." *Bloomberg*, January 23. Accessed May 11, 2023 (Bloomberg.com).

"Tattoos in the Workplace: Ink Blots." 2014. *The Economist*, August 2, 22.

Taub, Amanda, and Brendan Nyhan. 2017. "Why Objectively False Things Continue to Be Believed." *New York Times*, March 22, A18.

Tavris, Carol. 2013. "How Psychiatry Went Crazy." *Wall Street Journal*, May 18, C5.

Tavris, Carol. 2013. "The Experiments That Still Shock." *Wall Street Journal*, September 6, C1, C7.

Taylor, Bruce G., and Elizabeth A. Mumford. 2016. "A National Descriptive Portrait of Adolescent Relationship Abuse: Results from the National Survey on Teen Relationships and Intimate Violence." *Journal of Interpersonal Violence* 3 (6): 963–988.

Taylor, Carl S., and Pamela R. Smith. 2013. "The Attraction of Gangs: How Can We Reduce It?" Pp. 19–29 in *Changing Course: Preventing Gang Membership*, edited by Thomas R. Simon, Nancy M. Witter, and Reshma R. Mahendra. National Institute of Justice and Centers for Disease Control and Prevention. Accessed December 18, 2013 (www.nij.gov).

Taylor, Danielle. 2020. "Same-Sex Couples Are More Likely to Adopt or Foster Children." U.S. Census Bureau, September 17. Accessed May 22, 2023 (https://www.census.gov/library/stories/2020/09/fifteen-percent-of-same-sex-couples-have-children-in-their-household.html).

Taylor, Frederick W. 1911/1967. *The Principles of Scientific Management*. New York: W. W. Norton & Company.

Taylor, Jay. 1993. *The Rise and Fall of Totalitarianism in the Twentieth Century*. New York: Paragon House.

Taylor, Paul, and Mark H. Lopez. 2013. "Skepticism About the Census Voter Turnout Finding." Pew Research Center, May 15. Accessed August 8, 2014 (www.pewresearch.org).

Taylor, Paul, Carroll Doherty, Rich Morin, and Kim Parker. 2014. "Millennials in Adulthood: Detached from Institutions, Networked with Friends." Pew Research Center, March 7. Accessed June 21, 2014 (www.pewresearch.org).

Taylor, Paul, et al. 2012. "The Lost Decade of the Middle Class." Pew Research Center, August 22. Accessed February 14, 2014 (www.pewsocialtrends.org).

Taylor, Paul, et al. 2013. "The Rise of Asian Americans." Pew Research Center, April 4. Accessed May 16, 2014 (www.pewsocialtrends.org).

Taylor, Paul, Rich Morin, Kim Parker, and D'Vera Cohn. 2009. "Growing Old in America: Expectations vs. Reality." Pew Research Center, June 29. Accessed April 30, 2010 (www.pewsocialtrends.org).

Taylor, Susan C. 2003. *Brown Skin: Dr. Susan Taylor's Prescription for Flawless Skin, Hair, and Nails.* New York: HarperCollins.

Tefferi, Andrew. 2015. "In Support of a Patient-Driven Initiative and Petition to Lower the High Price of Cancer Drugs." *Mayo Clinic Proceedings* 90 (8): 996–1000.

Tenbrunsel, Ann, and Jordan Thomas. 2015. "The Street, the Bull and the Crisis: A Survey of the US & UK Financial Services Industry." University of Notre Dame and Lebaton Sucharow, May. Accessed September 2, 2015 (www.secwhistlebloweradvocate.com).

Teo, Alan. R., Heather. E. Marsh, Sarah S. Ono, Christina Nicolaidis, Somnath Saha, and Steven K. Dobscha. 2020. "The Importance of 'Being There': A Qualitative Study of What Veterans with Depression Want in Social Support. *Journal of General Internal Medicine*, 1-9. doi: 10.1007/s11606-020-05692-7.

Terando, Adam J., Jennifer Costanza, Curtis Belyea, Robert R. Dunn, Alexa McKerrow, and Jaime A. Collazo. 2014. "The Southern Megalopolis: Using the Past to Predict the Future of Urban Sprawl in the Southeast U.S." *PloS One* 9 (July): e102261.

Tétreault, Mary Ann. 2001. "A State of Two Minds: State Cultures, Women, and Politics in Kuwait." *International Journal of Middle East Studies* 33 (May): 203–220.

Ticeahkie, Sarae. 2022. "Comanche Nation works with United Nations to keep language alive." ABCNews, April 13. Accessed May 3, 2023 (https://www.kswo.com/2022/04/13/comanche-nation-works-with-united-nations-keep-language-alive/).

Thai, Helen, Christopher G. Davis, Wardah Mahboob, Sabrina Perry, Alex Adams, and Gary S. Goldfield. 2023. "Brief Report: Reducing Social Media Use Improves Appearance and Weight Esteem in Youth with Emotional Distress." American Psychological Association, Psychology of Popular Media. (https://doi.org/10.1037/ppm0000460).

Tharoor, Ishaan. 2015. "A Politician in Finland Declared War on Multiculturalism. This Is How His Country Responded." *Washington Post*, July 31. Accessed July 31, 2015 (washingtonpost.com).

Theodorou, Angelina E. 2014. "64 Countries Have Religious Symbols on Their National Flags." Pew Research Center, November 25. Accessed February 5, 2017 (pewresearch.org).

Theodorou, Angelina, and Peter Henne. 2014. "Religious Police Found in Nearly One-in-Ten Countries Worldwide." Pew Research Center, March 19. Accessed June 21, 2014 (www.pewresearch.org).

Thibaut, John W., and Harold H. Kelley. 1959. *The Social Psychology of Groups.* New York: Wiley.

Thies, Breccan F. 2023. "Mounting Evidence Shows Living Together Before Marriage Increases Likelihood of Divorce." *Washington Examiner*, May 1. Accessed May 19, 2023 (https://www.washingtonexaminer.com/restoring-america/community-family/living-together-before-marriage-increases-risk-divorce).

Thio, Alex D., Jim D. Taylor, and Martin D. Schwartz. 2012. *Deviant Behavior*, 11th edition. Upper Saddle River, NJ: Pearson.

Thomas, Anita Jones, Jason Daniel Hacker, and Denada Hoxha. 2011. "Gendered Racial Identity of Black Young Women." *Sex Roles* 64 (April): 530–542.

Thomas, W. I., and Dorothy Swaine Thomas. 1928. *The Child in America.* New York: Alfred A. Knopf.

Thompson, Alexandra, and Tapp. 2022. "Criminal Victimization, 2021." U.S. Department of Justice, September. Accessed March 23, 2023 (https://bjs.ojp.gov/content/pub/pdf/cv21.pdf).

Thompson, Arienne. 2010. "16, Pregnant . . . and Famous: Teen Moms Are Newest Stars." *USA Today*, November 23. Accessed November 24, 2011 (www.usatoday.com).

Thorpe-Moscon, Jennifer, and Alixandra Pollock. 2014. "Feeling Different: Being the 'Other' in US Workplaces." Catalyst. Accessed March 22, 2014 (www.catalyst.org).

Threewitt, Cherise, and Warren Clarke. 2023. "10 Vehicles That Are Almost Self-Driving in 2023." *U.S. News & World Report*, April 20. Accessed June 26, 2023 (https://cars.usnews.com/cars-trucks/advice/cars-that-are-almost-self-driving).

Tilly, Charles. 1978. *From Mobilization to Revolution.* Reading, MA: Addison-Wesley.

Timberg, Craig, and Ellen Nakashima. 2013. "Amid NSA Spying Revelations, Tech Leaders Call for New Restraints on Agency." *Washington Post*, November 1. Accessed November 10, 2013 (www.washingtonpost.com).

Timberlake, Cotten. 2013. "Don't Even Think About Returning This Dress." *Bloomberg Businessweek*, September 30–October 6, 29–31.

Timmermans, Stefan, and Hyeyoung Oh. 2010. "The Continued Social Transformation of the Medical Profession." *Journal of Health and Social Behavior* 51 (March): S94–S109.

Toastmasters International. 2023. "Youth Leadership Program." Accessed February 17, 2023 (https://www.toastmasters.org/education/youth-leadership-program).

Tobar, Hector. 2009. "Language as a Bridge and an Identity." *Los Angeles Times*, September 22. Accessed September 24, 2009 (www.latimes.com).

"Top Spenders." 2017. Center for Responsive Politics, Open Secrets. Accessed June 26, 2017 (opensecrets.org).

Tormey, Simon. 1995. *Making Sense of Tyranny: Interpretations of Totalitarianism.* New York: Manchester University Press.

Torpey, Elka, and Andrew Hogan. 2016. "Working in a Gig Economy." U.S. Bureau of Labor Statistics, May. Accessed June 5, 2017 (bls.gov).

Toth, Emily. 2011. "No Girls Aloud." *Chronicle of Higher Education*, April 1, A26.

Touraine, Alain. 2002. "The Importance of Social Movements." *Social Movement Studies* 1 (April): 89–96.

Tozzi, John. 2014. "Who Should Pay the Bill for Wonder Drugs?" *Bloomberg Businessweek*, July 10–16, 27–28.

Tozzi, John. 2017. "Health Care Has a Goldilocks Problem." *Bloomberg Businessweek*, July 31, 38–39.

Tracer, Zachary, and Hannah Recht. 2017. "Obamacare's Problems Still Need Solving." *Bloomberg Businessweek*, July 24, 37–38.

Trading Economics. 2023. "United States Corporate Profits." Accessed May 12, 2023 (https://tradingeconomics.com/united-states/corporate-profits#:~:text=Fall%20in%20Q4-,Corporate%20profits%20in%20the%20United%20States%20fell%202.7%20percent%20to,gain%20in%20the%20previous%20period).

Traister, Rebecca. 2016. *All the Single Ladies: Unmarried Women and the Rise of an Independent Nation.* New York: Simon & Schuster.

Tran, Van C. 2016. "Social Mobility Among Second-Generation Latinos." *Contexts* 15 (Spring): 28–33.

"Transgender Rights." 2015. *The Economist*, June 13, 31.

Travers, Mark. 2022. "A Psychologist Calls Out the Many Dangers of Helicopter Parenting." *Forbes*, November 30. Accessed May 18, 2023 (https://www.forbes.com/sites/traversmark/2022/11/30/a-psychol)ogist-calls-out-the-many-dangers-of-helicopter-parenting/?sh=6078d8e020d0).

Troitino, Christina. 2017. "Most Of The World Eats Bugs; The Nordic Food Lab Thinks You Should, Too." Forbes, August 30. Accessed February 11, 2023 (http://www.forbes.com/sites/christinatroitino/2017/08/30/unearthing-the-buzz-behind-eating-bugs-with-the-nordic-food-lab/?sh=4d6c937a6364).

Trottman, Melanie. 2013. "Religious Discrimination Claims on the Rise." *Wall Street Journal*, October 28, B1, B4.

Truman, Jennifer L., and Erica L. Smith. 2012. "Prevalence of Violent Crime Among Households with Children, 1993–2010." Bureau of Justice Statistics, September. Accessed May 16, 2014 (www.bjs.gov).

Truman, Jennifer L., and Lynn Langton. 2014. "Criminal Victimization, 2013." Bureau of Justice Statistics, September. Accessed November 6, 2014 (www.bjs.gov).

Truman, Jennifer L., and Lynn Langton. 2015. "Criminal Victimization, 2014." Bureau of Justice Statistics, August. Accessed September 2, 2015 (bjs.gov).

Trumbull, Mark. 2015. "Life in the Squeezed Middle." *Christian Science Monitor Weekly*, November 2, 26–32.

Tsai, Tyjen. 2012. "China Has Too Many Bachelors." Population Reference Bureau, January. Accessed September 19, 2014 (www.prb.org).

Tsukayama, Hayley. 2014. "High-Tech Upgrades May Let Aging Boomers Live Independently in Their Own Homes Longer." *Washington Post*, January 20. Accessed October 8, 2014 (www.washingtonpost.com).

Tucker-Drob, Elliot M., and Timothy C. Bates. 2016. "Large Cross-National Differences in Gene x Socioeconomic Status Interaction on Intelligence." *Psychological Science* 27 (2): 138–149.

Tumin, Melvin M. 1953. "Some Principles of Stratification: A Critical Analysis." *American Sociological Review* 18 (August): 387–393.

Turk, Austin T. 1969. *Criminality and the Legal Order.* Chicago: Rand-McNally.

Turk, Austin T. 1976. "Law as a Weapon in Social Conflict." *Social Problems* 23 (February): 276–291.

Turkle, Sherry. 2015. *Reclaiming Conversation: The Power of Talk in a Digital Age.* New York: Penguin Press.

Turley, Jonathan. 2022. "Cancel Culture Came for Clarence Thomas at George Washington Law. Now, He's Stepped Aside." *USA Today*, July 27. Accessed February 12, 2023 (https://www.usatoday.com/story/opinion/columnist/2022/07/27/clarence-thomas-george-washington-university/10167664002/?gnt-cfr=1).

Turner, Cory. 2022. "6 Things We've Learned about How the Pandemic Disrupted Learning." NPR, June 22. Accessed June 13, 2023 (https://www.npr.org/2022/06/22/1105970186/pandemic-learning-loss-findings).

Turner, Cory. 2023 "10 Things to Know about How Social Media Affects Teens' Brains." NPR, February 16. Accessed September 7, 2023 (https://www.npr.org/2023/02/16/1157180971/10-things-to-know-about-how-social-media-affects-teens-brains).

Turner, Margery A., Rob Santos, Diane K. Levy, Doug Wissoker, Claudia Aranda, and Rob Pitingolo. 2013. "Housing Discrimination Against Racial and Ethnic Minorities 2012." The Urban Institute, June. Accessed March 22, 2014 (www.huduser.org).

Turner, Ralph H., and Lewis M. Killian. 1987. *Collective Behavior*, 3rd edition. Englewood Cliffs, NJ: Prentice Hall.

Twenge, Jean M., Ryne A. Sherman, and Brooke E. Wells. 2017. "Declines in Sexual Frequency Among American Adults, 1989–2014." *Archives of Sexual Behavior* 46 (January): 1–13.

Tynan, Michael A., Jonathan R. Polansky, Kori Titus, Renata Atayeva, and Stanton A. Glantz. 2017. "Tobacco Use in Top-Grossing Movies— United States, 2010–2016." *MMWR* 66 (26): 681–686.

U

Umberson, Debra, Julie S. Olson, Robert Crosnoe, Hui Liu, Tetyana Pudrovska, and Rachel Donnelly. 2017. "Death of Family Members as an Overlooked Source of Racial Disadvantage in the United States." *Proceedings of the National Academy of Sciences* 114 (5): 915–920.

UN Women. 2014. "Facts and Figures: Ending Violence Against Women." Accessed March 2, 2014 (www.unwomen.org).

"Uncle Sam Overpays." 2017. *The Economist*, June 22. Accessed August 15, 2017 (economist.com).

Underhill, Wendy. 2017. "Voter Identification Requirements/ Voter ID Laws." National Conference of State Legislatures, June 5. Accessed June 26, 2017 (ncsl.org).

UNICEF. 2023. "Female Genital Mutilation." Accessed April 21, 2023 (https://www.unicef.org/protection/female-genital-mutilation).

UNICEF. 2023. "Water Scarcity." Accessed June 28, 2023 (https://www.unicef.org/wash/water-scarcity).

United Nations. 2014. *World Urbanization Prospects: The 2014 Revision.* Department of Economics and Social Affairs, Population Division. Accessed September 19, 2014 (www.esa.un.org).

United Nations, 2015. "Redistribute Unpaid Work." Accessed November 22, 2023 (https://www.unwomen.org/en/news/in-focus/csw61/redistribute-unpaid-work).

United Nations. 2022. "Goal 5: Achieve Gender Equality and Empower All Women and Girls." Sustainable Development Goals. Accessed April 29, 2023 (https://www.un.org/sustainabledevelopment/gender-equality/).

United Nations. 2022. "World Population Prospects:' Summary of Results." Accessed June 27, 2023 (https://www.un.org/development/desa/pd/sites/www.un.org.development.desa.pd/files/wpp2022_summary_of_results.pdf).

United Nations. 2020 "The World's Women 2020: Trends and Statistics." United Nations, Department of Economic and Social Affairs, October 20. Accessed April 23, 2023 https://www.un.org/en/desa/world%E2%80%99s-women-2020).

United Nations. 2023. "Goal 2: Zero Hunger." Sustainable Development Goals. Accessed April 13, 2023 (https://www.un.org/sustainabledevelopment/hunger/#:~:text=In%202020%2C%20between%20720%20million,regular%20access%20to%20adequate%20food).

United Nations Committee on the Rights of the Child. 2014. "Concluding Observations on the Second Periodic Report of the Holy See." February 25. Accessed June 21, 2014 (www.ohchr.org).

United Nations Department of Economic and Social Affairs. 2012. "World Urbanization Prospects: The 2011 Revision, Highlights." March. Accessed June 23, 2012 (www.esa.un.org).

United Nations Development Program. 2013. "Humanity Divided: Confronting Inequality in Developing Countries." November. Accessed February 6, 2014 (www.undp.org).

United Nations High Commissioner for Refugees. 2017. "Global Trends: Forced Displacement in 2016." Accessed August 20, 2017 (unhcr.org).

United Nations High Commissioner for Refugees. 2021. "Situation Reports 2018 – 2021: Global Focus" Accessed June 28, 2023 (https://reporting.unhcr.org/situation-reports-2018-2021).

United Nations Office on Drugs and Crime. 2017. "Statistics." Accessed April 23, 2017 (unodc.org).

United Nations Population Division. 2017. "World Population Prospects: The 2017 Revision." Department of Economic and Social Affairs. Accessed August 20, 2017 (esa.un.org).

U.S. Bureau of Labor Statistics. 2021. "Paid Sick Leave Was Available to 79 Percent of Civilian Workers in March 2021." TED: The Economics Daily, October 12. Accessed May 12, 2023 (https://www.bls.gov/opub/ted/2021/paid-sick-leave-was-available-to-79-percent-of-civilian-workers-in-march-2021.htm).

U.S. Bureau of Labor Statistics. 2022. "Occupational Employment and Wages, May 2022, 15-1251, Computer Programmers." Accessed May 10, 2023 (https://www.bls.gov/oes/current/oes151251.htm).

U.S. Census Bureau. 2017. "Voting and Registration in the Election of November 2016." May 12. Accessed June 5, 2017 (census.gov).

U.S. Census Bureau. 2008. "An Older and More Diverse Nation by Midcentury." August 14. Accessed August 18, 2008 (www.census.gov).

U.S. Census Bureau. 2012. *Statistical Abstract of the United States: 2012,* 131st edition. Washington, DC: Government Printing Office.

U.S. Census Bureau. 2021. "2020 Presidential Election Voting and Registration Tables Now Available." April 29. Accessed May 12, 2023 (https://www.census.gov/newsroom/press-releases/2021/2020-presidential-election-voting-and-registration-tables-now-available.html#:~:text=As%20with%20past%20elections%2C%20a,18%20to%2024%20at%2051.4%25).

U.S. Census Bureau, 2021. "Census Bureau Releases New Estimates on America's Families and Living Arrangements." November 30. Accessed April 16, 2023 (https://www.census.gov/newsroom/press-releases/2021/families-and-living-arrangements.html).

U.S. Census Bureau. 2021. "DP02: Selected Social Characteristics in the United States." American Community Survey. Accessed February 22, 2023 (https://data.census.gov/table?tid=ACSDP5Y2021.DP02&hidePreview=true).

U.S. Census Bureau. 2021. "Census Bureau Releases Report on Same-Sex Couple Households." February 24. Accessed May 19, 2023 (https://www.census.gov/newsroom/press-releases/2021/same-sex-couple-households.html).

U.S. Census Bureau. 2021. "Living Arrangements Over the Decades." November 20. Accessed May 18, 2023 (https://www.census.gov/library/visualizations/2021/comm/living-arrangements-over-the-decades.html).

U.S. Census Bureau. 2021. "America Recycles Day: November 15, 2021." November 15. Accessed June 28, 2023 (https://www.census.gov/newsroom/stories/america-recycles-day.html).

U.S. Census Bureau. 2021. "Census: Racial and Ethnic Diversity Index by State." August 12. Accessed September 26, 2023 (https://www.census.gov/library/visualizations/2021/dec/racial-and-ethnic-diversity-index.html).

U.S. Census Bureau. 2022. "Census Bureau Releases New Estimates on America's Families and Living Arrangements." November 17. Accessed May 19, 2023 (https://www.census.gov/newsroom/press-releases/2022/americas-families-and-living-arrangements.html#:~:text=The%20estimated%20median%20age%20to,20.5%2C%20respectively%2C%20in%201947).

U.S. Census Bureau. 2022. "Educational Attainment in the United States: 2021." February 24. Accessed May 3, 2023 (https://www.census.gov/data/tables/2021/demo/educational-attainment/cps-detailed-tables.html).

U.S. Census Bureau. 2022. "Median Household Income." September 13. Accessed June 10, 2023 (https://www.census.gov/library/visualizations/2022/comm/median-household-income.html).

U.S. Census Bureau. 2023. "International Database (IDB)." Accessed June 15, 2023. (https://www.census.gov/data-tools/demo/idb/#/country?COUNTRY_YEAR=2023&COUNTRY_YR_ANIM=2023).

U.S. Census Bureau. 2023. "U.S. Census Bureau Current Population." Accessed June 6. 2023 (https://www.census.gov/popclock/print.php?component=counter).

U.S. Census Bureau, Current Population Survey. 2013. "Annual Social and Economic Supplements." November. Accessed October 20, 2014 (www.census.gov).

U.S. Census Bureau News. 2015. "Census Bureau Reports at Least 350 Languages Spoken in U.S. Homes." November 3. Accessed June 20, 2017 (census.gov).

U.S. Census Bureau Newsroom. 2017. "The South Is Home to 10 of the 15 Fastest-Growing Large Cities." May 25. Accessed August 20, 2017 (census.gov).

U.S. Centers for Disease Control and Prevention. 2022. "Fast Facts: Preventing Child Abuse and Neglect." April 6. Accessed February 18, 2023 (.https://www.cdc.gov/violenceprevention/childabuseandneglect/fastfact.html).

U.S Centers for Disease Control. "Provisional number of marriages and marriage rate: United States, 2000-2021." Accessed May 19, 2023. "Provisional number of marriages and marriage rate: United States, 2000-2021 (https://www.cdc.gov/nchs/data/dvs/marriage-divorce/national-marriage-divorce-rates-00-21.pdf).

U.S. Commission on International Religious Freedom. 2023. "Did You Know... Muslim Constitutions." Accessed June 14, 2023 (https://www.uscirf.gov/publications/did-you-knowmuslim-constitutions).

U.S. Department of Agriculture. 2023. "My Plate." Accessed June 10, 2023 (https://www.myplate.gov/).

U.S. Department of Agriculture, 2020. "Dietary Guidelines for Americans, 2020-2025." Accessed November 17, 2023 (https://www.dietaryguidelines.gov/sites/default/files/2020-12/Dietary_Guidelines_for_Americans_2020-2025.pdf). .

U.S. Department of Education. 2012. "Degrees and Other Formal Awards Conferred." Higher Education General Information Survey (HEGIS). Accessed July 22, 2017 (nces.ed.gov).

U.S. Department of Education. 2016. "2013–2014 Civil Rights Data Collection: A First Look." Office for Civil Rights, October 28. Accessed July 30, 2017 (ocrdata.ed.gov).

U.S. Department of Health and Human Services. 2013. *Child Maltreatment 2012*. Administration for Children and Families, Children's Bureau. Accessed May 16, 2014 (www.acf.hhs.gov).

U.S. Department of Health and Human Services. 2014. "The Health Consequences of Smoking—50 Years of Progress. A Report of the Surgeon General." Office of the Surgeon General. Accessed July 30, 2014 (www.cdc.gov).

U.S. Department of Health and Human Services. 2016. "E-Cigarette Use Among Youth and Young Adults. Office of the Surgeon General. Accessed August 15, 2017 (surgeongeneral.gov).

U.S. Department of Health and Human Services. 2016. *Child Maltreatment 2014*. Administration for Children and Families, Children's Bureau. Accessed March 2, 2017 (acf.hhs.gov).

U.S. Department of Health and Human Services. 2018. Physical Activity Guidelines for Americans 2nd edition. Accessed June 11, 2023 (https://health.gov/sites/default/files/2019-9/Physical_Activity_Guidelines_2nd_edition.pdf).

U.S. Department of Health and Human Services. 2021. *Child Maltreatment 2021*. Accessed May 22, 2023 (https://www.acf.hhs.gov/sites/default/files/documents/cb/cm2021.pdf).

U.S. Department of Health and Human Services. 2022. "New HHS Report Shows National Uninsured Rate Reached All-Time Low in 2022." Accessed June 11, 2023 (https://www.hhs.gov/about/news/2022/08/02/new-hhs-report-shows-national-uninsured-rate-reached-all-time-low-in-2022.html).

U.S. Department of Housing and Urban Development. 2022. "HUD Releases 2021 Annual Homeless Assessment Report Part 1." February 4. Accessed September 17, 2023 (https://www.hud.gov/press/press_releases_media_advisories/hud_no_22_022#:~:text=WASHINGTON%20%2D%20The%20U.S.%20Department%20of,of%20eight%20percent%2C%20from%202020.).

U.S. Department of Justice. 1996. "Policing Drug Hot Spots." National Institute of Justice, January. Accessed May 3, 2005 (www.ncjrs.org).

U.S. Department of Justice. 2011. "U.S. Attorney Announces Drug Endangered Children Task Force." Press Notice, May 31. Accessed July 25, 2011 (www.usdoj.gov).

U.S. Department of Justice. 2021. "Poverty Status of Children by Family Structure." Office of Juvenile Justice and Delinquency Prevention. Accessed May 22, 2023 (https://ojjdp.ojp.gov/statistical-briefing-book/population/faqs/qa01203#:~:text=In%202021%2C%209.5%25%20of%20children,17.4%25).

U.S. Department of Justice. 2021. "Elder Abuse Statistics." Accessed May 22, 2023 (https://www.justice.gov/file/1098056/download).

U.S. Department of Justice. 2023. "2020 Hate Crimes Statistics." February 21. Accessed March 12, 2023 (https://www.justice.gov/crs/highlights/2020-hate-crimes-statistics).

U.S. Department of Veteran Affairs. 2017. "Profile of Post–9/11 Veterans: 2015." National Center for Veterans Analysis and Statistics, March. Accessed August 15, 2017 (va.gov/vetdata).

U.S. Equal Opportunity Employment Commission, 2022. "Sexual Harassment in Our Nation's Workplaces" April. Accessed April 16, 2023 (https://www.eeoc.gov/data/sexual-harassment-our-nations-workplaces#:~:text=EEOC%20Charge%20Data%20(FY%202018,27%2C291%20charges%20alleging%20sexual%20harassment).

U.S. Environmental Protection Agency. 2019. "Our Nation' Air: Air Quality Improves as America Grows." Accessed June 28, 2023 (https://gispub.epa.gov/air/trendsreport/2019/documentation/AirTrends_Flyer.pdf).

U.S. Environmental Protection Agency. 2022. "National Overview: Facts and Figures on Materials, Wastes and Recycling." Accessed June 28, 2023 (https://www.epa.gov/facts-and-figures-about-materials-waste-and-recycling/national-overview-facts-and-figures-materials#:~:text=National%20Overview%3A%20Facts%20and%20Figures%20on%20Materials%2C%20Wastes,Today%20...%208%20Generation%20Trends%20...%20More%20items).

U.S. Environmental Protection Agency. 2022. "Impacts of Climate Change." Accessed June 28, 2023 (https://www.epa.gov/climatechange-science/impacts-climate-change).

U.S. Government Publishing Office. 2020 "Stopping Wasteful Spending." Budget of the U.S. Government, Budget of the United States Government, Fiscal Year 2021, February 10. Accessed March 11, 2023 (https://www.govinfo.gov/content/pkg/BUDGET-2021-BUD/pdf/BUDGET-2021-BUD-4.pdf).

U.S. Holocaust Memorial Museum. 2020. "Documenting Numbers of Victims of the Holocaust and Nazi Persecution." Accessed April 28, 2023 (https://encyclopedia.ushmm.org/content/en/article/documenting-numbers-of-victims-of-the-holocaust-and-nazi-persecution).

U.S. Office of Personnel Management. 2014. "2014 Federal Employee Viewpoint Survey Results." Accessed November 2, 2016 (opm.gov/fevs).

United States Courts. 2023 "Judicial Compensation." Accessed May 12, 2023 (https://www.uscourts.gov/judges-judgeships/judicial-compensation).

U.S. Senate Special Committee on Aging, American Association of Retired Persons, Federal Council on the Aging, and U.S. Administration on Aging. 1991. *Aging America: Trends and Projections, 1991.* Washington, DC: U.S. Department of Health and Human Services.

U.S. Senate. 2004. *Report on the U.S. Intelligence Community's Prewar Intelligence Assessments on Iraq.* Accessed September 13, 2006 (www.gpoaccess.gov).

U.S. Travel Association. 2014. "Working for Free: U.S. Workforce Forfeits $52.4 Billion in Time Off Benefits Annually." October 21. Accessed November 23, 2014 (www.ustravel.org).

USA Life Expectancy. 2016. "Life Expectancy All Races." Accessed July 14, 2017 (worldlifeexpectancy.com).

"Universities." 2015. *The Economist*, November 28, 25–26.

"Unmarried and Single Americans Week: Sept. 17–23, 2017." 2017. U.S. Census Bureau, August 14. Accessed August 20, 2017 (census.gov).

Upton, John. 2014. "Chevron and BP Are Pulling Out of Wind and Solar." Grist, May 30. Accessed August 20, 2017 (grist.org).

The Urban Institute. 2023. "Higher Education Expenditures." States and Local Backgrounders, February. Accessed July 29, 2023 (https://www.urban.org/policy-centers/cross-center-initiatives/state-and-local-finance-initiative/state-and-local-backgrounders/higher-education-expenditures#:~:text=In%202020%2C%20state%20and%20local,and%20local%20direct%20general%20spending.&text=As%20a%20share%20of%20state,spending%20on%20health%20and%20hospitals).

The Urban Institute. 2023. "Criminal Justice Expenditures: Police, Corrections, and Courts." February 27. Accessed March 25, 2023 (https://www.urban.org/policy-centers/cross-center-initiatives/state-and-local-finance-initiative/state-and-local-backgrounders/criminal-justice-police-corrections-courts-expenditures).

Urbina, Ian. 2006. "In Online Mourning, Don't Speak Ill of the Dead." *New York Times*, November 5, 1.

V

"Vacation Habits Around the World." 2013. Expedia 2013 Vacation Deprivation Study, November 18. Accessed April 12, 2014 (www.viewfinder.expedia.com).

Vallance. Chris. 2022. "3D Printed Guns: Warnings Over Growing Threat of 3D Firearms." *BBC News*, November 9. Accessed June 27, 2023 (https://www.bbc.com/news/technology-63495123).

Valerio. Tayleor, Rose M. Kreider, and Wan He. 2021. "Childless Older Adults More Educated, More Likely to Live Alone Than Older Parents." December 14. Accessed May 22, 2023 (https://www.census.gov/library/stories/2021/12/no-kids-no-care-childlessness-among-older-americans.html#:~:text=Declines%20in%20marriage%20and%20fertility,as%20having%20no%20biological%20children).

Vallotton, Claire, and Catherine Ayoub. 2011. "Use Your Words: The Role of Language in the Development of Toddlers' Self-Regulation." *Early Childhood Research Quarterly* 26 (2): 169–181.

Van Agtmael, Peter. 2013. "Laugh After Death." *Time,* November 18, 44–49.

Van Green, Ted. 2022. "Americans Overwhelmingly Say Marijuana Should Be Legal for Medical or Recreational Use." Pew Research Center. November 22. Accessed March 25, 2023 (https://www.pewresearch.org/fact-tank/2022/11/22/americans-overwhelmingly-say-marijuana-should-be-legal-for-medical-or-recreational-use/).

Van Kessel, Patrick, and Laura Silver. 2021. "Where Americans Find Meaning in Life Has Changed Over the Past Four Years." Pew Research Center, November 18. Accessed February 11, 2023 (https://www.pewresearch.org/fact-tank/2021/11/18/where-americans-find-meaning-in-life-has-changed-over-the-past-four-years/).

Vanorman, Alicia G., and Paola Scommegna. 2016. "Understanding the Dynamics of Family Change in the United States."

Population Reference Bureau, July. Accessed July 1, 2017 (prb.org).

Vargas, Edward D. and Gabriel R. Sanchez. 2020. "American Individualism is an Obstacle to Wider Mask Wearing in the U.S." Brookings Institution, August 31. Accessed February 11, 2023 (https://www.brookings.edu/blog/up-front/2020/08/31/american-individualism-is-an-obstacle-to-wider-mask-wearing-in-the-us/).

Van Vooren, Nicole, and Roberta Spalter-Roth. 2010. "Tracking Master's Students Through Programs and into Careers." *Footnotes* (September/October): 10–11.

Varinsky, Dana, and Uwa Ede-Osifo. 2023. "Norovirus Appears to Keep Spreading as Rate of Positive Tests Exceeds Last Year's Peak." *NBC News*, February 9. Accessed June 10, 2023 (nbcnews.com).

Varnum, Michael E.W. 2023." What Personal Space Can Tell Us. Psychology Today, February 18. (https://www.psychologytoday.com/us/blog/unserious-psychology/202301/what-personal-space-can-tell-us).

Vassallo, Trae, et al. 2015. "Elephant in the Valley." Accessed January 22, 2016 (elephantinthevalley.com).

Veblen, Thorstein. 1899/1953. *The Theory of the Leisure Class*. New York: New American Library.

Velasco, Schuyler. 2014. "Food Additives: More Fuss Than Harm?" *Christian Science Monitor*, May 26, 43.

Venkatesh, Sudhir. 2008. *Gang Leader for a Day: A Rogue Sociologist Takes to the Streets*. New York: Penguin Press.

Ventura, Luca. 2022. "COVID-19 Spread Around the World and Made Existing Inequalities Worse." *Global Finance Magazine*, World Inequality Ranking by Country 2022, February 17. Accessed April 13, 2023 (https://www.gfmag.com/global-data/economic-data/world-inequality-ranking).

Ventura, Stephanie J., and Christine A. Bachrach. 2000. "Nonmarital Childbearing in the United States, 1940–99." National Center for Vital Statistics. October 18. Accessed May 22, 2023 (https://www.cdc.gov/nchs/data/nvsr/nvsr48/nvs48_16.pdf).

Vespa, Jonathan, Jamie M. Lewis, and Rose M. Kerner. 2013. "America's Families and Living Arrangements: 2012." U.S. Census Bureau, August. Accessed May 16, 2014 (www.census.gov).

Vespa, Jonathan, Lauren Medina, and David M. Armstrong. 2020. "Demographic Turning Points for the United States: Population Projections for 2020 to 2060." U.S. Census Bureau Current Population Reports, February. Accessed January 29, 2023 (https://www.census.gov/content/dam/Census/library/publications/2020/demo/p25-1144.pdf).

Vicens, A. J. 2015. "Missing in Action." *Mother Jones*, July/August, 13.

The Violence Project. 2021. "Why Are Almost All Mass Shooters Men?" March 29. Accessed February 18, 2023 (https://www.theviolenceproject.org/media/why-are-almost-all-mass-shooters-men/).

Vlasic, Bill. 2015. "G.M.'s Ignition Switch Death Toll Hits 100." *New York Times*, May 12, B3.

Vogels, Emily A. 2023. "A Majority of Americans Have Heard of ChatGPT, but Few Have Tried It Themselves." Pew Research Center, May 24. Accessed June 26, 2023 (https://www.pewresearch.org/short-reads/2023/05/24/a-majority-of-americans-have-heard-of-chatgpt-but-few-have-tried-it-themselves/).

Vogelstein Rachel B., and Erik Fliegauf. 2021. "Women's Power Index: Find Out Where Women Lead—and Why It Matters." Council on Foreign Relations, April 14. Accessed April 21, 2023 (https://www.cfr.org/blog/womens-power-index-find-out-where-women-lead-and-why-it-matters).

Vold, George B. 1958. *Theoretical Criminology*. New York: Oxford University Press.

Volz, Matt. 2012. "$3.4 Billion Indian Land Royalty Settlement Upheld." Salon, May 22. Accessed March 22, 2014 (www.salon.com).

Von Drehle, David. 2014. "Breaking the Marriage Bans." *Time*, March 3, 12.

Vongkiatkajorn, Kanyakrit. 2016. "Prison Labor Is Unseen and 'Utterly Exploitative'." *Mother Jones*, October 6. Accessed March 4, 2017 (motherjones.com).

Vukelich, Lara. 2023. "Pros and cons of crowdfunding for business." Bankrate, May 22. Accessed June 27, 2023 (https://www.bankrate.com/loans/small-business/crowdfunding-pros-cons/).

W

Walker, Laquitta, and Danielle Taylor. 2021. "Same-Sex Couple Households: 2019." American Community Survey, February. Accessed May 19, 2023 (https://www.census.gov/content/dam/Census/library/publications/2021/acs/acsbr-005.pdf).

Wallace, Bruce. 2006. "Japanese Schools to Teach Patriotism." *Los Angeles Times*, December 16. Accessed December 18, 2006 (www.latimes.com).

Wallis, Cara. 2011. "Performing Gender: A Content Analysis of Gender Display in Music Videos." *Sex Roles* 64 (February): 160–172.

Walls, Mark. 2011. "High Court Strikes Down Calif. Law on Violent Video Games." *Education Week*, June 27. Accessed June 28, 2011 (www.edweek.org).

Walmsley, Roy. 2016. "World Prison Population List (Eleventh Edition)." International Center for Prison Studies. Accessed April 15, 2017 (prisonstudies.org).

Walsh, Ben. 2017. "Donald Trump Is Breaking His Promise to Be Tough on Wall Street." *Huffington Post*, January 31. Accessed February 3, 2017 (huffingtonpost.com).

Walsh, Bryan. 2015. "It May Be Too Late to Reverse the Damage of China's One-Child Policy." *Time*, November 16, 23–24.

Wan, William. 2011. "Chinese Dog Eaters and Dog Lovers Spar over Animal Rights." *Washington Post*, May 28, A14.

Wang, Hansi Lo. 2023. "New 'Latino' and 'Middle Eastern or North African' Checkboxes Proposed for U.S. Forms." National Public Radio, January 26. Accessed January 29, 2023 (https://www.npr.org/2023/01/26/1151608403/mena-race-categories-us-census-middle-eastern-latino-hispanic).

Wang, Wendy, and Kim Parker. 2014. "Record Share of Americans Have Never Married: As Values, Economic and Gender Patterns Change." Pew Research Center, September 24. Accessed October 26, 2014 (www.pewresearch.org).

Wang, Wendy. 2012. "The Rise of Intermarriage." Pew Research Center, Social & Demographic Trends, February 16. Accessed March 22, 2014 (www.pewsocialtrends.org).

Wang, Wendy. 2015. "The Link Between a College Education and a Lasting Marriage." Pew Research Center, December 4. Accessed July 20, 2017 (pewresearch.org).

Wang, Wendy. 2022. "The Decline in Church Attendance in COVID America." Institute for Family Studies. Accessed June 14, 2023 (https://ifstudies.org/blog/the-decline-in-church-attendance-in-covid-america).

Wang, Wendy, and Kim Parker. 2020. "The Share of Never-Married Americans Has Reached a New High." Institute for Family Studies, September 9. Accessed May 19, 2023 (https://ifstudies.org/blog/the-share-of-never-married-americans-has-reached-a-new-high).

Wang, Y. Claire. 2015. "The Dangerous Silence of Academic Researchers." *Chronicle of Higher Education*. February 27, A48.

Ward, Nicole, and Jeanne Batalova. 2023. "Frequently Requested Statistics on Immigrants and Immigration in the United States." Migration Policy Institute, March 14. Accessed September 26, 2023 (https://www.migrationpolicy.org/article/frequently-requested-statistics-immigrants-and-immigration-united-states).

Warren, Kenneth R. 2012. "NIH Statement on International FASD Awareness Day." National Institutes of Health, September 5. Accessed November 20, 2013 (www.nih.gov).

Wartik, Nancy. 2005. "The Perils of Playing House." *Psychology Today* (July/August): 42–52.

Wasley, Paula. 2008. "The Syllabus Becomes a Repository of Legalese." *Chronicle of Higher Education* 54, March 14, A1, A8–A11.

Waters, Mary C., and Marisa Gerstein Pineau, eds. 2015. *The Integration of Immigrants into American Society.* Washington, DC: The National Academies Press.

Wealth-X. 2017. "Billionaire Fact File from the Billionaire Census 2017." March 21. Accessed May 13, 2017 (wealthx.com).

Wealth-X and UBS. 2014. "Billionaire Census 2014." Accessed November 10, 2014 (www.billionairecensus.com).

Weathers, Cliff. 2014. "Research Behind Dr. Oz-Endorsed Diet Supplement Was Bogus, Authors Admit." AlterNet, October 22. Accessed December 5, 2014 (www.alternet.org).

Weathers, Cliff. 2015. "It's 2015 and Congress Is Almost Entirely Made Up of White Christian Men." AlterNet, January 6. Accessed September 29, 2015 (alternet.org).

Weaver, Rosanna Landis. 2016. "The 100 Most Overpaid CEOs: Are Fund Managers Asleep at the Wheel?" As You Sow. Accessed June 22, 2017 (asyousow.org).

Weber, Jennifer Beggs. 2020. "Being There (Or Not): Teen Dads, Gendered Age, and Negotiating the Absent-father Discourse." *Men and Masculinities 23 (April): 42-64.* Accessed May 22, 2023 (https://journals.sagepub.com/doi/epub/10.1177/1097184X17747082).

Weber, Lynn, Tina Hancock, and Elizabeth Higginbotham. 1997. "Women, Power, and Mental Health." Pp. 380–396 in *Women's Health: Complexities and Differences,* edited by Sheryl B. Ruzek, Virginia L. Olesen, and Adele E. Clarke. Columbus: Ohio State University Press.

Weber, Max. 1920/1958. *The Protestant Ethic and the Spirit of Capitalism,* translated by Talcott Parsons. New York: Charles Scribner's Sons.

Weber, Max. 1925/1947. *The Theory of Social and Economic Organization.* New York: Free Press.

Weber, Max. 1925/1978. *Economy and Society,* edited by Guenther Roth and Claus Wittich. Berkeley: University of California Press.

Weber, Max. 1946. *From Max Weber: Essays in Sociology,* translated and edited by H. H. Gerth and C. Wright Mills. Berkeley: University of California Press.

Weinhold, Bob. 2012. "A Steep Learning Curve: Decoding Epigenetic Influences on Behavior and Mental Health." *Environmental Health Perspectives* 120 (October): A396–A401.

Weintraub, Arlene. 2010. "Break That Hovering Habit Early." *U.S. News & World Report,* September, 42–43.

Weise, Elizabeth, and Tom Vanden Brook. 2015. "More Ashley Madison Files Published." *USA Today,* August 20. Accessed August 21, 2015 (usatoday.com).

Weisman, Carrie. 2015. "Can a Woman Find Sexual Desire in a Pill?" AlterNet, June 15. Accessed October 15, 2015 (alternet.org).

Weisman, Carrie. 2016. "Avoiding This Behavior Could Help Save Your Relationship." AlterNet, March 4. Accessed December 29, 2016 (alternet.org).

Weiss, Carol H. 1998. *Evaluation: Methods for Studying Programs and Policies,* 2nd edition. Upper Saddle River, NJ: Prentice Hall.

Welding. Lynn, 2022. "Student Loan Debt by Year." Best Colleges, June 3. Accessed June 13, 2023 (https://www.bestcolleges.com/research/student-loan-debt-by-year/).

Weller, Chris. 2015. "Black Market Sells Human Body Parts for Hundreds of Thousands: What Do You Cost?" *Medical Daily,* April 20. Accessed March 1, 2017 (medicaldaily.com).

Wenneras, Christine, and Agnes Wold. 1997. "Nepotism and Sexism in Peer Review." *Nature* 387 (May 22): 341–343.

West, Anne and Jane Lewis 2018. *Helicopter Parenting and Boomerang Children How Parents Support and Relate to Their Student and Co-Resident Graduate Children.* Routledge: New York, NY.

West, Candace, and Don H. Zimmerman. 2009. "Accounting for Doing Gender." *Gender & Society* 23 (February): 112–122.

West, Charlotte. 2021. "An Unnoticed Result of the Decline of Men in College: It's Harder for Women to Get In." The Hechinger Report, October 27. Accessed June 13, 2023 (https://hechingerreport.org/an-unnoticed-result-of-the-decline-of-men-in-college-its-harder-for-women-to-get-in/).

Westcott, Lucy. 2014. "More Americans Moving to Cities, Reversing the Suburban Exodus." The Wire, March 27. Accessed September 19, 2014 (www.thewire.com).

"When Workers Are Owners." 2015. *The Economist,* August 22, 56.

White, James W. 2005. *Advancing Family Theories.* Thousand Oaks, CA: Sage.

White, Nicole, and Janet L. Lauritsen. 2012. "Violent Crime Against Youth: 1994–2010." Bureau of Justice Statistics, December. Accessed May 4, 2013 (www.bjs.gov).

Whiteaker, Chloe, Laurie Meisler, Yvette Romero, and Karen Weise. 2014. "Voter ID States That Make Voting Super Simple—or Really Hard." *Bloomberg Businessweek,* October 27–November 2, 36.

Whitehead, Deborah. 2022. "Women Lead Religious Groups in Many Ways—Besides the Growing Number Who Have Been Ordained." *Colorado Arts and Sciences Magazine,* January 19. Accessed June 14, 2023 (https://www.colorado.edu/asmagazine/2022/01/19/women-lead-religious-groups-many-ways-besides-growing-number-who-have-been-ordained).

Whitehead, Jaye C. 2011. "The Wrong Reasons for Same-Sex Marriage." *New York Times,* May 16, A21.

White House Council of Economic Advisers. 2016. "The Long-Term Decline in Prime-Age Male Labor Force Participation." June. Accessed May 15, 2017 (obamawhitehouse.archives.gov).

The White House. 2022. "FACT SHEET: How The American Rescue Plan Is Keeping America's Schools Open Safely, Combating Learning Loss, And Addressing Student Mental Health." Accessed June 13, 2023 (https://www.whitehouse.gov/briefing-room/statements-releases/2022/03/11/fact-sheet-how-the-american-rescue-plan-is-keeping-americas-schools-open-safely-combating-learning-loss-and-addressing-student-mental-health/).

The White House. 2023. "Fact Sheet: President Biden's Budget Strengthens Border Security, Enhances Legal Pathways, and Provides Resources to Enforce Our Immigration Laws." March 9. Accessed April 26, 2023 (https://www.whitehouse.gov/briefing-room/statements-releases/2023/03/09/fact-sheet-president-bidens-budget-strengthens-border-security-enhances-legal-pathways-and-provides-resources-to-enforce-our-immigration-laws/).

Whitehurst, Grover J. R. 2016. "Hard Thinking on Soft Skills." Brookings, March 24. Accessed March 1, 2017 (brookings.edu).

Whitelaw, Kevin. 2000. "But What to Call It?" *U.S. News & World Report,* October 16, 42.

Whiteman, Shawn D., Alexander C. Jensen, and Susan M. McHale. 2017. "Sibling Influences on Risky Behaviors from Adolescence to Early Adulthood: Vertical Socialization or Bidirectional Effects?" *Directions for Child and Adolescent Development,* 156, 67-85. doi:10.1002/cad.20197.

Whitesides, John. 2014. "Anti-Union Group Pledges to Remain on the Offensive." *Baltimore Sun,* February 18, 2.

Whiting, Kate. 2022. "The Motherhood Penalty: How Childcare and Paternity Leave Can Reduce the Gender Pay Gap." World Economic Forum, May 19. Accessed May 12, 2023 (https://www.weforum.org/agenda/2022/05/reduce-motherhood-penalty-gender-pay-gap/).

Whitlock, Craig, and Bob Woodward. 2016. "Pentagon Buries Evidence of $125 Billion in Bureaucratic Waste." *Washington Post*, December 5. Accessed April 10, 2017 (washingtonpost.com).

WHO. 2016. "Ambient (Outdoor) Air Quality and Health." Fact sheet, September. Accessed August 20, 2017 (who.int).

WHO. 2016. "Children: Reducing Mortality." Fact Sheet, September. Accessed May 13, 2017 (who.int).

"Who Pays the Bill?" 2013. *The Economist*, July 27, 24–26.

"Who Votes, Who Doesn't, and Why: Regular Voters, Intermittent Voters, and Those Who Don't." 2006. Pew Research Center for the People & the Press, October 18. Accessed May 3, 2007 (www.peoplepress.org).

Whorf, Benjamin Lee. 1956. *Language, Thought, and Reality.* Cambridge, MA: MIT Press.

Whoriskey, Peter, and Dan Keating. 2014. "Dying and Profits: The Evolution of Hospice." *Washington Post*, December 26. Accessed October 16, 2015 (washingtonpost.com).

Wicker, Christine. 2009. "How Spiritual Are We?" *Parade,* October 4, 4–5.

Wiesen, Kasee. 2023. "How Many Nurse Practitioners are There in the U.S. – State Wise Data." Nursing Process. Accessed June 11, 2023 (https://www.nursingprocess.org/how-many-nurse-practitioners-are-there-in-the-us.html).

Wiik, Kenneth A., Renske Keizer, and Trude Lappegård. 2012. "Relationship Quality in Marital and Cohabiting Unions Across Europe." *Journal of Marriage and Family* 74 (June): 389–398.

Wigert, Ben. 2022. "The Top 6 Things Employees Want in Their Next Job." Gallup, February 21. Accessed March 11, 2023 (https://www.gallup.com/workplace/389807/top-things-employees-next-job.aspx).

Wilbur, Tabitha G., and Vincent J. Roscigno. 2016. "First-generation Disadvantage and College Enrollment/Completion." *Socius: Sociological Research for a Dynamic World* 2: 1–11.

Wilcox, W. Bradford. 2015. "Knot Now: The Benefits of Marrying in Your Mid-to-Late 20s (Including More Sex!)." *Washington Post*, February 20. Accessed September 30, 2015 (washingtonpost.com).

Wilke, Joy, and Frank Newport. 2014. "Americans' Satisfaction with Economy Sours Most Since 2001." Gallup, January 16. Accessed April 12, 2014 (www.gallup.com).

Wilkinson, Charles. 2006. *Blood Struggle: The Rise of Modern Indian Nations.* New York: W. W. Norton & Company.

Wilkinson, Lindsey, and Jennifer Pearson. 2013. "High School Religious Context and Reports of Same-Sex Attraction and Sexual Identity in Young Adulthood." *Social Psychology Quarterly* 76 (June): 180–202.

Willett, Megan. 2015. "Living in Tiny Homes Was Much Harder Than These People Realized." Business Insider, July 27. Accessed August 20, 2017 (businessinsider.com).

Williams, Frank P., III, and Marilyn D. McShane. 2004. *Criminological Theory,* 4th edition. Upper Saddle River, NJ: Prentice Hall.

Williams, Lauren. 2014. "21 Things You Can't Do While Black." AlterNet, February 12. Accessed March 22, 2014 (www.alternet.org).

Williams, Robin M., Jr. 1970. *American Society: A Sociological Interpretation,* 3rd edition. New York: Knopf.

Williams, Timothy. 2013. "Blighted Cities Prefer Razing to Rebuilding." *New York Times,* November 13, A15.

Williams, Thomas Chatterton. 2022. "How the French Do Christmas." *The Atlantic,* December 24. Accessed February 11, 2023 (https://www.theatlantic.com/ideas/archive/2022/12/french-christmas-paris-traditions-children-adults/672529/).

Wilson, Jason. 2020. "White Nationalist Hate Groups Have Grown 55% in Trump Era, Report Finds." *The Guardian,* March 18. Accessed February 12, 2023 (https://www.theguardian.com/world/2020/mar/18/white-nationalist-hate-groups-southern-poverty-law-center).

Wilson, Thomas C. 1993. "Urbanism and Kinship Bonds: A Test of Four Generalizations." *Social Forces* 71 (March): 703–712.

Wilson, Valerie, and William M. Rodgers III. 2016. "Black-White Wage Gaps Expand with Rising Wage Inequality." Economic Policy Institute, September 19. Accessed June 20, 2017 (epi.org).

Wilson, William Julius. 1996. *When Work Disappears: The World of the New Urban Poor.* New York: Knopf.

Wiltz, Teresa. 2016. "Why More Grandparents Are Raising Children." Pew Charitable Trusts, November 2. Accessed July 14, 2017 (pewtrusts.org).

Winerip, Michael. 2013. "Pushed Out of a Job Early." *New York Times*, December 6. Accessed May 16, 2014 (www.nytimes.com).

WIN-Gallup International. 2013. "Global Index of Religiosity and Atheism, 2012." Accessed June 21, 2014 (www.gallup-international.com).

Wingfield, Adia Harvey. 2010. "Are Some Emotions Marked 'Whites Only'? Racialized Feeling Rules in Professional Workplaces." *Social Problems* 57 (May): 258–268.

Winkler, Adam. 2011. *Gunfight: The Battle over the Right to Bear Arms in America.* New York: W. W. Norton & Co.

Winship, Michael. 2015. "Inside the Cushy, Fabulously Compensated Life of an Ex-Congressman." Alternet, April 8. Accessed October 7, 2015 (alternet.org).

Wintersmith. Saraya. 2022. "Arab Americans Say the Census and Other Forms Don't Consider Their Roots." NPR, March 9. Accessed May 3, 2023 (https://www.npr.org/2022/03/09/1085355634/arab-americans-say-the-census-and-other-forms-dont-consider-their-roots).

Wire, Sarah. 2016. "The government spends $1.7 billion a year on 770,000 empty buildings, and one Central Valley congressman is fed up." *The Los Angeles Times,* March 3. Accessed November 17, 2023 (https://www.latimes.com/politics/la-pol-ca-jeff-denham-government-property-bill-20160303-story.html).

Wirth, Louis. 1938. "Urbanism as a Way of Life." *American Journal of Sociology* 44 (July): 1–24.

Witt-Swanson, Lindsey, Jennifer Benz, and Daniel A. Cox. 2023 "Faith After the Pandemic: How COVID-19 Changed American Religion." The Survey Center on American Life, January 5. Accessed June 13, 2023 (https://www.americansurveycenter.org/research/faith-after-the-pandemic-how-covid-19-changed-american-religion/).

"Wives Who Earn More Than Their Husbands, 1978–2012." 2014. U.S. Bureau of Labor Statistics, Labor Force Statistics from the Current Population Survey, March 24. Accessed April 9, 2014 (www.bls.gov/cps).

Wolf, Martin. J., John W. Emerson, Daniel C. Esty, Alex de Sherbinin, and Zachary A. Wendling. 2022. "Environmental Performance Index." New Haven, CT: Yale Center for Environmental Law & Policy. Accessed June 28, 2023 (https://epi.yale.edu/downloads/epi2022report06062022.pdf).

Wolff, Edward N. 2016. "Household Wealth Trends in the United States, 1962 to 2013: What Happened over the Great Recession?" *The Russell Sage Foundation Journal of the Social Sciences* 2 (October): 24–43.

Wolfson, Mark. 2001. *The Fight Against Big Tobacco: The Movement, the State, and the Public's Health.* New York: Aldine de Gruyter.

Wolven, Jacqueline. 2014. "The 3 Things Middle-Aged Women Need to Stop

Doing When Speaking in Public." *Huffington Post*, September 18. Accessed September 18, 2014 (huffingtonpost.com).

Wolverton, Brad. 2016. "The New Cheating Economy." *Chronicle of Higher Education*, September 2, A40–A45.

"Women at Work." 2017. U.S. Bureau of Labor Statistics, March. Accessed May 10, 2017 (bls.gov).

Wondergem, Taylor R., and Mihaela Friedlmeier. 2012. "Gender and Ethnic Differences in Smiling: A Yearbook Photographs Analysis from Kindergarten Through 12th Grade." *Sex Roles* 67 (October): 403–411.

Wong, Brittany. 2014. "The Top 10 Reasons People Stay in Unhappy Marriages." *Huffington Post*, December 15. Accessed January 1, 2015 (huffingtonpost.com).

Wong, Curtis M. 2013. "Hallmark's 'Holiday Sweater' Keepsake Ornament Omits 'Gay' from 'Deck the Halls' Lyrics." *Huffington Post*, October 28. Accessed November 10, 2013 (www.huffingtonpost.com).

Wood, Julia T. 2015. *Gendered Lives: Communication, Gender, & Culture*, 11th edition. Stamford, CT: Cengage.

Woolf, Steven H., and Laudon Aron. 2013. *U.S. Health in International Perspective: Shorter Lives, Poorer Health*. National Research Council and Institute of Medicine of the National Academies. Accessed July 26, 2014 (www.nationalacademies.org).

World Bank. 2015. "Country and Lending Groups." Accessed October 14, 2015 (data.worldbank.org).

The World Bank. 2021. "Fertility Rate, Total (Births Per Woman) – China." Accessed June 27, 2023 (https://data.worldbank.org/indicator/SP.DYN.TFRT.IN?locations=CN).

The World Bank. 2023. "Current Health Expenditure per Capita (Current U.S Dollars)." Accessed June 10, 2023 (https://data.worldbank.org/indicator/SH.XPD.CHEX.PC.CD?most_recent_value_desc=false).

The World Bank. 2023. "Life Expectancy at Birth." Accessed June 10, 2023 (https://data.worldbank.org/indicator/SP.DYN.LE00.IN).

The World Bank. 2023. "Mortality Rate, Infant (per 1,000 live births)." Accessed June 10, 2023 (https://data.worldbank.org/indicator/SP.DYN.IMRT.IN).

The World Bank. 2023. "Urban Development." Accessed June 10, 2023 (https://www.worldbank.org/en/topic/urbandevelopment/overview).

The World Bank, 2017. "Annual Report 2017." (https://thedocs.worldbank.org/en/doc/908481507403754670-0330212017/original/AnnualReport-2017WBG.pdf).

World Bank Group. 2021. "Women, Business, and the Law 2021." Accessed April 29, 2023 (https://openknowledge.worldbank.org/server/api/core/bitstreams/b2274591-b8fd-5636-ac36-d3ba2ad0100a/content).

World Bank and Collins. 2013. *Atlas of Global Development: A Visual Guide to the World's Greatest Challenges*, 4th edition. Washington, DC, and Glasgow: World Bank and Collins.

World Economic Forum. 2013. "Outlook on the Global Agenda 2014." Accessed February 6, 2014 (www.weforum.org).

World Economic Forum. 2022. "Six Surprising Facts About the Global Gender Pay Gap." March 8. Accessed April 13, 2023 (https://www.weforum.org/agenda/2022/03/6-surprising-facts-gender-pay-gap/#:~:text=Women%20globally%20earn%20around%2037,gender%20gap%20is%20painfully%20slow).

World Economic Forum. 2022. "Our World in Data." Accessed June 30, 2023 (https://www.weforum.org/agenda/2019/07/populations-around-world-changed-over-the-years/).

World Economic Forum, 2016. "Five Surprising Facts About the Global Gender Pay Gap." October 28. Accessed November 22, 2023 (https://www.weforum.org/agenda/2016/10/5-surprising-facts-from-this-years-gender-gap-report/).

World Food Programme. 2020. "5 Facts About Food Waste and Hunger." June 2. Accessed June 27, 2023 (https://www.wfp.org/stories/5-facts-about-food-waste-and-hunger).

World Health Organization. 2021. "Global Database on the Prevalence of Violence Against Women." Accessed February 17, 2023 (https://srhr.org/vaw-data/map).

World Health Organization. 2022. "Child Mortality and Causes of Death." Accessed April 13, 2023 (https://www.who.int/data/gho/data/themes/topics/topic-details/GHO/child-mortality-and-causes-of-death).

World Health Organization and UNICEF. 2014. "Progress on Drinking Water and Sanitation." May. Accessed September 19, 2014 (www.unicef.org).

World Health Organization. 2013. "Global and Regional Estimates of Violence Against Women: Prevalence and Health Effects of Intimate Partner Violence and Non-Partner Sexual Violence." Accessed August 6, 2015 (who.int).

World Health Organization. 2016. "Violence Against Women." November. Accesses May 25, 2017 (who.int).

World Health Organization. 2021. "Violence Against Women." March 9. Accessed July 23, 2023 (https://www.who.int/news-room/fact-sheets/detail/violence-against-women).

World Health Organization. 2022. "Cancer," February 3. Accessed June 10, 2023 (https://www.who.int/news-room/fact-sheets/detail/cancer).

World Health Organization. 2022. "Sanitation." March 21. Accessed June 10, 2023 (https://www.who.int/news-room/fact-sheets/detail/sanitation).

World Health Organization. 2022. "Malaria." March 29. Accessed June 10, 2023 (https://www.who.int/news-room/fact-sheets/detail/malaria).

World Health Organization. 2022. "Violence Against Women." Accessed April 21, 2023 (https://www.who.int/news-room/fact-sheets/detail/violence-against-women).

World Health Organization. 2022. "UN Report: Global Hunger Numbers Rose to as Many as 828 million in 2021." July 6. Accessed June 27, 2023 (https://www.who.int/news/item/06-07-2022-un-report--global-hunger-numbers-rose-to-as-many-as-828-million-in-2021).

World Health Organization 2022 "Drinking-Water." March 21. Accessed June 28, 2023 (https://www.who.int/news-room/fact-sheets/detail/drinking-water).

World Meteorological Organization. 2014. "Atlas of Mortality and Economic Losses from Weather, Climate and Water Extremes (1970–2012)." Accessed September 19, 2014 (wmo.int).

World Meteorological Organization. 2023. "Past Eight Years Confirmed to be the Eight Warmest on Record." January 12. Accessed June 28, 2023 (https://public.wmo.int/en/media/press-release/past-eight-years-confirmed-be-eight-warmest-record#:~:text=GENEVA%2C%2012%20January%202023%20%2D%20The,by%20the%20World%20Meteorological%20Organization).

Women's Refugee Commission. 2019. ""We Have a Broken Heart": Sexual Violence against Refugees in Nairobi and Mombasa, Kenya." October. April 23, 2023 (https://www.womensrefugeecommission.org/wp-content/uploads/2020/04/SVMB-Kenya-Report-10-2019-FINAL.pdf).

World Water Assessment Program. 2009. *The United Nations World Water Development Report 3: Water in a Changing World*. Paris: UNESCO and London: Earthscan.

Worley, Heidi. 2014. "Top 10 Countries Closing Gender Gap." Population Reference Bureau, February. Accessed March 2, 2014 (prb.org).

Wrede, Insa and Kristie Pladson. 2021. "Germany Still Lags Behind on Number of Female Executives." *Deustche Welle*, October 27. Accessed April 21, 2023 (https://www.dw.com/en/more-women-have-joinedgermanys-executive-boards-but-not-many/a-59628774).

Wretman, Christopher J. "School Sports Participation and Academic Achievement in Middle and High School" *Journal of Social Work and Research 8*, 3(Fall): 399-420. Accessed February 18, 2013 (https://www.journals.uchicago.edu/doi/10.1086/693117).

Wright, Wynne, and Elizabeth Ransom. 2005. "Stratification on the Menu: Using Restaurant Menus to Examine Social Class." *Teaching Sociology* 33 (July): 310–316.

WUNRN. 2017. "Women, Slums, & Urbanization." Women's UN Report Network, March 17. Accessed August 20, 2017 (wunrn.com).

Wysong, Earl, Robert Perrucci, and David Wright. 2014. *The New Class Society: Goodbye American Dream?* 4th edition. Lanham, MD: Rowman & Littlefield.

X

Xie, Yu, James Raymo, Kimberly Goyette, and Arland Thornton. 2003. "Economic Potential and Entry into Marriage and Cohabitation." *Demography* 40 (May): 351–367.

Y

Yan, Holly. 2015. "717 People Dead: What Caused the Haji Stampede?" CNN, September 26. Accessed October 25, 2015 (cnn.com).

Yeager, David S., Valerie Purdie-Vaughns, Sophia Yang Hooper, and Geoffrey L. Cohen. 2017. "Loss of Institutional Trust Among Racial and Ethnic Minority Adolescents: A Consequence of Procedural Injustice and a Cause of Life-Span Outcomes." *Child Development* 88 (March/April): 658–676.

Yee, Amy. 2021. "Research on Asian Americans and Pacific Islanders Is Being Stifled." *Scientific American*, July 8. Accessed January 29, 2023 (https://www.scientificamerican.com/article/research-on-asian-americans-and-pacific-islanders-is-being-stifled/).

Yee, Bruce Y. 2018. "Study Shows How Bad People Are at Washing Their Hands." *Forbes,* June 30. Accessed January 30, 2023 (https://www.forbes.com/sites/brucelee/2018/06/30/study-shows-how-bad-people-are-at-washing-their-hands/?sh=a6f01bb24810).

Yeung, Ken. 2016. "Yahoo Reveals New Hack: 'Unauthorized Third Party' Stole Data from More than 1 Billion Accounts." Venture Beat, December 14. Accessed September 2, 2017 (venturebeat.com).

Yeung, 2023. https://www.cnn.com/2023/03/23/china/beijing-population-drop-intl-hnk/index.html

Yglesias, Matthew. 2013. "The End of Cheap Airfare." Slate, February 19. Accessed April 12, 2014 (www.slate.com).

Yokota, Fumise, and Kimberly M. Thompson. 2000. "Violence in G-rated Animated Films." *JAMA* 283 (May 24/31): 2716–2720.

Yongoro, Ester. 2022. "What is Affecting the Girl Child in South Sudan." UNICEF, September 30. Accessed February 12, 2023 (https://www.unicef.org/southsudan/stories/what-affecting-girl-child-south-sudan).

Young, Danielle. 2016 "On Reclaiming 'Boy' and Giving Young Black Men Something to Celebrate." The Root, September 1. Accessed February 11, 2023 (https://www.theroot.com/on-reclaiming-boy-and-giving-young-black-men-somethin-1790856576).

Young, Elise. 2015. "A Rough Ride for Renewable Energy." *Bloomberg Businessweek*, May 25-May 31, 28–29.

Young, Natalie A.E. 2021. "Childhood Disability in the United States: 2019 American Community Survey Briefs." American Community Survey Briefs, March 2021. Accessed June 10, 2023 (https://www.census.gov/content/dam/Census/library/publications/2021/acs/acsbr-006.pdf).

Z

Zagorsky, Jay L., and Patricia K. Smith. 2017. "The Association Between Socioeconomic Status and Adult Fast-Food Consumption in the U.S." *Economics & Human Biology* 27 (November): 12–25.

Zahedi, Ashraf. 2008. "Concealing and Revealing Female Hair: Veiling Dynamics in Contemporary Iran." Pp. 250–265 in *The Veil: Women Writers on Its History, Lore, and Politics,* edited by Jennifer Heath. Berkeley: University of California Press.

Zarya, Valentina. 2017. "Facebook: Iterating Diversity?" *Fortune*, February 1, 55.

Zhang, Qiang, et al. 2017. "Transboundary Health Impacts of Transported Global Air Pollution and International Trade." *Nature* 543 (March): 705–709.

Zhang, Xue, Marcela Gonzalez Rivas, M. Grant, and Mildred E. Warner. 2022. "Water Pricing and Affordability in the US: Public Vs. Private Ownership." *Water Policy* 24 (3): 500–516. Accessed June 15, 2023 (https://doi.org/10.2166/wp.2022.283).

Zhang, Yuanyuan, Travis L. Dixon, and Kate Conrad. 2010. "Female Body Image as a Function of Themes in Rap Music Videos: A Content Analysis." *Sex Roles* 62 (June): 787–797.

Ziccarelli, Gabriella E. 2016. "The Price of Pokemon Go." *Baltimore Sun,* July 15, 15.

Ziegler, Mary. 2023. "Opinion: Idaho's New Abortion Law is First to Challenge the Right to Travel. CNN, April 7. Accessed April 22, 2023 (https://www.cnn.com/2023/04/07/opinions/idaho-abortion-law-right-to-travel-ziegler/index.html).

Zimbardo, Philip G., Christina Maslach, and Craig Haney. 2000. "Reflections on the Stanford Prison Experiment: Genesis, Transformations, Consequences." Pp. 193–237 in *Obedience to Authority: Current Perspectives on the Milgram Paradigm,* edited by Thomas Blass. Mahwah, NJ: Lawrence Erlbaum Associates.

Zimbardo, Philip. G. 1975. "Transforming Experimental Research into Advocacy for Social Change." Pp. 33–66 in *Applying Social Psychology: Implications for Research, Practice, and Training,* edited by Morton Deutsch and Harvey A. Hornstein. Hillsdale, NJ: Erlbaum.

Zimmerman, Jonathan. 2014. "Are E-Cigarettes Ensnaring a New Generation of Smokers?" *Christian Science Monitor Weekly*, March 31, 35.

Zoll, Rachel, and Emily Swanson. 2015. "AP-NORC Poll: Christian-Muslim Split on Religious Freedom." AP-NORC, December 30. Accessed February 17, 2017 (apnorc.org).

Zong, Jie, and Jeanne Batalova. 2017. "Frequently Requested Statistics on Immigrants and Immigration in the United States." Migration Policy Institute, March 8. Accessed June 6, 2017 (migrationpolicy.org).

Zuo, Mandy. 2021. "China's Aging Crisis: New Plan to Encourage Children to Live with Their Parents Met with Resistance." *South China Morning Post,* November 26. Accessed February 11, 2023 (https://www.scmp.com/news/people-culture/social-welfare/article/3157372/chinas-ageing-crisis-new-plan-encourage-children).

Zweigenhaft, Richard L., and G. William Domhoff. 2006. *Diversity in the Power Elite: How It Happened, Why It Matters.* Lanham, MD: Rowman & Littlefield.

Name Index

A

B

C

D

E

F

G

H

L

N

O

P

Q

R

S

T

U

V

W

X

Y

Z

Subject Index

A

B

C

D

E

F

G

H

I

J

K

L

M

N

O

P

Q

R

S

T

U

V

W

Y

Z